Clare Benson.

International Handbook of Research and Development in Technology Education

AF860244

INTERNATIONAL TECHNOLOGY EDUCATION STUDIES

Series Editors
Rod Custer, *Illinois State University, Normal, USA*
Marc J. de Vries, *Eindhoven University of Technology, The Netherlands*

Editorial Board
Piet Ankiewicz, *University of Johannesburg, South Africa*
Dov Kipperman, *ORT Israel, Israel*
Steven Lee, *Taiwan National Normal University, Taipei, Taiwan*
Gene Martin, *Technical Foundation of America, USA*
Howard Middleton, *Griffith University, Brisbane, Australia*
Chitra Natarajan, *Homi Babha Centre for Science Education, Mumbai, India*
John R. Dakers, *University of Glasgow, UK*

Scope
Technology Education has gone through a lot of changes in the past decades. It has developed from a craft oriented school subject to a learning area in which the meaning of technology as an important part of our contemporary culture is explored, both by the learning of theoretical concepts and through practical activities. This development has been accompanied by educational research. The output of research studies is published mostly as articles in scholarly Technology Education and Science Education journals. There is a need, however, for more than that. The field still lacks an international book series that is entirely dedicated to Technology Education. The International Technology Education Studies aim at providing the opportunity to publish more extensive texts than in journal articles, or to publish coherent collections of articles/chapters that focus on a certain theme. In this book series monographs and edited volumes will be published. The books will be peer reviewed in order to assure the quality of the texts.

International Handbook of Research and Development in Technology Education

Alister T. Jones
University of Waikato, New Zealand

and

Marc J. de Vries
Delft University of Technology, The Netherlands

Associate Editor

Cathy M. Buntting
University of Waikato, New Zealand

Section Editors
International Developments: Alister Jones
The Nature of Technology: Marc de Vries and John Dakers
Perceptions of Technology: John Dakers and Marc de Vries
Technology and STSE: Derek Hodson
Learning and Teaching: Robert McCormick
Assessment: Judy Moreland
Teacher Education: P. John Williams
Theoretical and Practical Approaches to Research: Alister Jones

SENSE PUBLISHERS
ROTTERDAM/BOSTON/TAIPEI

A C.I.P. record for this book is available from the Library of Congress.

ISBN 978-90-8790-877-5 (paperback)
ISBN 978-90-8790-878-2 (hardback)
ISBN 978-90-8790-879-9 (e-book)

Published by: Sense Publishers,
P.O. Box 21858, 3001 AW
Rotterdam, The Netherlands
http://www.sensepublishers.com

Printed on acid-free paper

All Rights Reserved © 2009 Sense Publishers

No part of this work may be reproduced, stored in a retrieval system, or transmitted in any form or by any means, electronic, mechanical, photocopying, microfilming, recording or otherwise, without written permission from the Publisher, with the exception of any material supplied specifically for the purpose of being entered and executed on a computer system, for exclusive use by the purchaser of the work.

TABLE OF CONTENTS

Section 2: The Nature of Technology

PREFACE

The *International Handbook of Research and Development in Technology Education* was borne out of discussions over a long period of time over a number of international conferences. It fact, it probably has its roots at the second Jerusalem International Science and Technology Education Conference (JISTEC) in 1996. This conference brought a number of us together to focus not just on research but also to consider implementation in a much wider historical, cultural and political context.

Ten to twelve years later the time appeared right for the writing of an international handbook. International handbooks represent significant markers in their respective fields and this handbook is no different. The field of technology education has reached a particular stage where it is possible to write such a book. In developing this book we have been able to reflect on the key characteristics of the field that is highlighted in the different named sections and also in the nature of the chapters themselves. We have been very fortunate with the enthusiasm of the section editors and the authors of the chapters themselves that they have not only identified the need for a handbook but they have whole heartily committed themselves to it.

This book constitutes a significant collection of work from numerous countries and authors actively engaged in technology education research and development. This book is also unique in that it includes country overviews from developed and developing countries, philosophical discussions of the discipline of technology, as well as considerations of teaching, learning, assessment, teacher education and research. It is impossible to separate the development of a field from its historical, philosophical and social setting. Technology education is no different and this is reflected throughout the chapters.

We have considered it a privilege to work such people of influence in and commitment to the developing field of technology education. We trust that this handbook will contribute to the on-going development of the field and may be of benefit to those who read it.

Alister Jones and Marc de Vries
Editors

MARC DE VRIES

1. THE DEVELOPING FIELD OF TECHNOLOGY EDUCATION: AN INTRODUCTION

INTRODUCTION

The fact that this *International Handbook of Research and Development in Technology Education* is now available indicates that technology education has grown out into a well-established field. Although it is still very much in development, there is a certain maturity that justifies presenting a survey of what has been accomplished in the past. That is why the editors and the publisher considered it to be timely to bring out this *Handbook*.

In this *Handbook*, all the main topics one would expect for an educational field are present: the underlying philosophy, curriculum development, relations with other subject areas, teacher education, assessment, and educational research. The (relative) maturity of technology education is illustrated by the fact that for each of these topics a sufficient number of high quality contributions could be commissioned. This introductory chapter offers an overall survey of the topics and the way they have developed in the (relatively short) history of technology education. Let us first consider what is meant by 'technology education' and related terms.

DEFINING THE FIELD

In this *Handbook*, we use the term 'technology education' to indicate the learning area that deals with the ways in which human beings change their environments to be better suited to their needs and wants, thereby using various types of knowledge. This description fits well with the current state of the philosophy of technology (see Section 2). Technologies surround us, and it makes sense to educate future citizens about this important aspect of our lives. It is another matter, though, to consider how this should be done. Does it mean a separate school subject is necessary? Or can one deal with technology entirely in the context of another school subject, be it science or something else? Does it require interdisciplinary projects at intervals in the normal curriculum? Different answers to these questions are provided in different countries, and for different school levels. For this reason, technology education is still a fairly fuzzy concept in spite of its current state of development. It is important to discuss a number of possible confusions.

The editor-in-chief of the *International Journal of Technology and Design Education* receives a substantial number of submitted articles that are rejected even before they are reviewed because they are not within the scope of the journal. These include articles that describe studies into the ways technology can be used to support the teaching and learning of various school subjects, such as science, mathematics, and languages. Apparently the term 'technology education' in the title of the journal has triggered scholars to submit articles in which technology is not the content of what is taught and learnt, but a means for teaching and learning about other issues. For that, however, the term 'educational technology' is more appropriate and more commonly used. This is a first confusion regarding the nature of technology education. Of course technology education and

A.T. Jones and M.J. de Vries (eds.), International Handbook of Research and Development in Technology Education, 1–9.
© 2009 Sense Publishers. All rights reserved.

educational technology are not entirely unrelated. For instance, educational technologies can be applied in technology education no less than in other educational fields. Perhaps one can even consider technology education to be one of the most obvious learning areas where new technologies can enhance teaching and learning. Secondly, a good understanding of the possible and the appropriate role of technologies in the teaching of science, mathematics, languages, or any other subject area, assumes a good understanding of technology itself – and this is exactly what technology education aims at. Still, one can distinguish clearly between questions that are solely concerned with the effects of new technologies (new media in particular) on teaching and learning, whereby the technology itself is taken as a black box, and questions that concern the nature of the technology, whereby the black box is opened. The former is 'educational technology'; the latter is 'technology education'.

Another term related to technology education is 'technological literacy'. This term is used to indicate an important aim of technology education, namely to provide students with a type of literacy that is important particularly for the technological society in which we live. Not only do people need to be able to read and write (the original meaning of the term 'literacy') and use numbers (numerical literacy), they also need to be able to use and control technological devices and systems. The term 'technological literacy' is not identical to 'technology education' because the aim of helping students to acquire technological literacy can partially be done in the teaching and learning of other domains in which technology is not the main focus. The term 'technology education' is applied to teaching and learning in which technology is the main focus.

A third related term is 'engineering education'. This term, too, must be distinguished from the term 'technology education'. Engineering is the domain of professions concerned with the development and maintenance of technological devices and systems. Although engineers use technology, like citizens also do, this is not their primary role. The work of engineers is to design new technological solutions to practical problems, or to improve existing systems. In order to do this, engineers use knowledge about the physical and chemical phenomena that underlie the functioning of artefacts and systems. They also use knowledge of mathematics for modelling and making calculations. Technology education is broader than engineering education because it also considers what users need to know and be able to do. Another difference that is sometimes articulated is that technology education covers the human and social aspects of technology, whereas engineering education only deals with the 'hard' aspects of technology. This, however, is in conflict with many developments in engineering education today; it has become clear that engineers cannot do a good job without taking into account the human and social aspects. Every technological problem is a socio-technological problem, and therefore needs to be approached as such. This has caused important changes in the education of engineers. It has certainly brought engineering education closer to technology education, as engineers are now often taught to imagine themselves in the position of users. Nevertheless, there is still a difference. In engineering education, students are still primarily prepared to become engineers; in technology education, students are prepared to become citizens, irrespective of their future career.

The term 'engineering education' is closely related to another term than can easily cause confusion about the nature of technology education, namely 'vocational education'. Like engineering, vocational education prepares for a specific profession. Technology education, in contrast, offers an introduction to technology as a component of both our professional life and our life as a consumer and citizen. Consequently, vocational education – like engineering education – deals with technology in a much more detailed and in-depth way. This is only possible due to the limited scope of vocational education. Put another way, technology education is a form of generic education, whereas vocational education is a form of specialised education. Sometimes vocational education is seen as part of technology education. In this *Handbook*, however, the term 'technology education' is used to refer to the type of education that primarily contributes to the general education of all future citizens.

THE DEVELOPMENT OF TECHNOLOGY EDUCATION

Although now sufficiently mature to warrant having its own handbook, technology education is still a latecomer in education. It was only in the past two or three decades that it has been developed on a worldwide scale (see De Vries 2000; 2006). In this respect it is relatively new. But 'new' is not to be taken in an absolute sense. In many cases technology education emerged from an already existing subject. In most countries that subject was craft. The history of technology education can to a large extent be characterised as the effort to move away from this past, both in content and in reputation. However, technology education continues to struggle with inappropriate popular images of what it is. Policy makers, school boards, parents, and industry still tend to think of technology education as a school subject in which pupils learn to make simple devices by processing materials with a variety of hand tools and simple machines. These prejudices are extremely hard to correct.

The development of technology education is also hampered by competition with science education. In principle, science education and technology education need not be competitors, but in practice this is often the case. The simple explanation is again related to terminology: it is commonly said and heard that televisions, mobile phones, computers, and other such devices are the outcome of science, not of technology. In newspapers, one reads that "scientists have invented a new type of device for …", while actually invention is not the work of scientists, but of technologists. This has to do with an old and deeply embedded preference for valuing the cognitive over the practical. We admire the new high-tech devices that we see around us, and because of their sophistication we think that they must be the work of scientists. Technology, in contrast, is associated with technicians and what are considered to be simple tasks, like repairing the toilet or screwing a plug onto a wire. It will take a long time for such prejudices to be outgrown, and in the meantime technology education will continue to struggle against the stronger position of science education in the school curriculum.

Over the course of time, technology educators have adopted different approaches to overcome this prejudicial barrier. One obvious way has been 'if you can't beat them, join them'. This, however, still assumes that science and technology are competitors – not a very fruitful way of thinking. There are good reasons related to the nature of both science and technology that these two learning areas can be related. For instance, science *does* play an important role in technology because it offers insights into the sort of variables designers can consider for optimising their designs. Vice versa, technology offers instruments to scientists for experimenting, measuring, and processing data. If one then also realises that this interaction between science and technology does not take place in a vacuum, but is embedded in a social context, one ends up with a type of education that is often called 'Science, Technology and Society', or STS. In theory, this could be an appropriate way to teach about both science and technology (see Section 4). In practice, however, the central T is crushed between the surrounding S's and the practice has often been to use technology merely as a motivating context for teaching science. It is the social relevance of technology that features in such STS teaching, not the nature of technology itself. STS has thus tended to problematise the development of technology education, rather than acting as a stimulus.

An alternative route for addressing the problem of competition with science education is to develop technology education into a separate subject in which the specific characteristics of technology are strongly emphasised and used to differentiate it from science. There are certainly many opportunities for doing this, and in fact a lot of recent insights in the philosophy of technology deal with such specifics. One such distinguishing element is the normative dimension in technological knowledge, which is absent in scientific knowledge; another relates to the 'designerly' ways of thinking identified by design methodologists ('means-ends' reasoning as opposed to 'cause-effect' reasoning). In general, design as a distinguishing activity in technology has been put forward to promote technology education as a distinct area in the curriculum (as opposed to 'investigating' in science). The manipulation of tools and materials, inherited from past emphasis on crafts, also often features as a distinct characteristic of technology education.

The outcome is that in many countries technology education has become a separate school subject (see Section 1). In some, it has survived serious threats of being removed from the curriculum (e.g., in South Africa and Sweden). When this text was written (November 2008), technology education in Australia faced a similar threat. In many countries it still struggles to maintain its position in the curriculum. Even in the United Kingdom, where 'Design and Technology' (D&T) had a longstanding status as a compulsory subject for pupils of ages 6–18, it is elective for upper secondary education (ages 16–18) and its status is probably already under negotiations when this *Handbook* is published.

Other countries have integrated an existing subject 'technology' (or whatever the name may be) with science education. In these cases, there is potential to create a positive combination in which the nature of both is done justice, balanced with the risk that technology will simply be absorbed by science (this scenario threatens technology education in, for example, the Netherlands). In contrast, technology education is making significant progress in countries like New Zealand; in others it maintains its craft emphasis (e.g., Switzerland, Denmark). In a large country such as the United States of America, we find different practices in different states, and the main activity of the International Technology Education Association (ITEA) in that country has been to create some unification in this palette to strengthen the overall position of technology education in the curriculum across the country. The *Standards for Technological Literacy* (ITEA, 2000/2002/2007) has been a major accomplishment in this context.

MOTIVES AND STAKEHOLDERS IN TECHNOLOGY EDUCATION

The development of technology education in the school curriculum has been driven by several motives. The most obvious, of course, is the fact that technology plays such an important role in society, and that citizens must be technologically literate in order to be able to live in and contribute meaningfully to such a society. A similar, but probably more vulnerable, motive is that technology education is the subject 'par excellence' to stimulate general abilities such as creativity, problem solving, and communication skills. This ambition, however, is not unproblematic. Research studies have shown that these skills are much more context-dependent than was assumed in the past. One can therefore question if learning how to solve a technological problem will 'automatically' also improve pupils' abilities in solving other problems. It seems that one can indeed transfer from one context to another, but still not have achieved a 'general' level of skill.

A third motive for technology education is that pupils need to learn about the nature of technology as a possible field of studies and future career pathway. In many countries, shortage in the number of engineers is a continuous concern and technology education is seen as a means for promoting post-school technological studies and professions. The risk of drawing too heavily on this motive is that it will move technology education in the direction of vocational education, with negative effects on its status and maybe also negative (narrowing) effects on its scope and content.

Ultimately, the primary motivation depends, of course, on the stakeholder that promotes it. Although no data are available on which stakeholders prefer which motive, the following analysis seems plausible: Technology educators themselves tend to put most emphasis on the cultural and social importance of learning about technology. Governments and industries often have primary interest in assuring a sufficient future workforce. Parents and school boards tend to focus on the general educative merits of technology education, such as its contribution to stimulating creativity and problem solving skills. For pupils, a new motive appears in cases where they have encountered effective technology education: It is fun! In such cases, pupils may experience technology education as a welcome change in a curriculum that is otherwise filled with more abstract subjects.

We have already seen that developments in technology education have been hampered from negative images of the subject. A further problem arises from ways in which stakeholders assess technology education against the purposes to which they subscribe. All stakeholders tend to be impatient when it comes to the realisation of the motives for technology education. Although no school subject has ever been removed from the curriculum because it did not result in measurable social effects within a time span as short as one or two decades, or even only one government period, technology education has been expected to meet and fulfil such unrealistic expectations. When, for example, enrolment numbers for engineering studies do not increase within a couple of years, it is assumed that technology education has failed to achieve its goals.

THE LONG-TERM AIM OF TECHNOLOGY EDUCATION AND WHAT IT REQUIRES

The naïve thinking sketched above does not do justice to the fact that the most important aims of technology can only be realised in the long run. For the realisation of all the motives mentioned in the previous section, it is necessary for technology education to result in helping students to develop an appropriate concept of technology, an understanding of the basic concepts in technology, and a positive-critical attitude towards technology (see Section 4). Many technology educators are of the opinion that in order to reach these goals, it is best to give students experiences in undertaking the process of technology. Going through a process of designing is probably the best way to learn that technology is a human endeavour in which decision making, based on normative judgments, and the use of a variety of knowledge domains plays a vital role. This approach has the additional advantage that it connects the various types of literacies that are aimed at in general education; designing entails the use of language, symbols, and numbers. In this way, doing design connects technological literacy with linguistic and numerical literacy. Sometimes this multi-literacy is used as an additional motive for having technology education in the school curriculum. A factor that can further enhance this multi-literacy is the increased role of new media and other technologies in technology education.

This multi-literacy learning can only take place when a number of conditions are fulfilled. In the first place, teachers are needed that have an appropriate attitude and the necessary knowledge and skills to be able to design and implement teaching and learning situations and environments conducive to such learning (see Section 5). Teacher education programmes thus need to be targeted towards educating such teachers (see Section 7). Secondly, teachers, teacher educators, and curriculum developers need to develop curricula and teaching materials that such teachers can use. Thirdly, the assessment of technology education, both formative and summative, must be rich and multifaceted in order to assure that multi-literacy teaching and learning takes place (see Section 6).

Fourthly, industry support is needed to ensure the 'empirical validity' of technology education. Sometimes it is feared that industry involvement in technology education will narrow its scope, but this overlooks the fact that industries in the past decades have learnt a great deal about the importance of human and social aspects of technology. In some countries, like Germany, it is even industry in particular that challenges technology education to emphasise this multi-literacy (for which the term *schlüsselqualificationen* or 'key competencies' is used).

A fifth crucial factor in the development of such learning is that the development of technology is supported by sound educational research. A true 'R&D culture' is thus needed for technology education to continue to grow and develop. It is area in particular that still leaves much to be desired, and this Handbook includes a special focus on research for technology education (Section 8). It is here that it becomes perhaps most clear that the *Handbook* should be seen as a torch rather than as a milestone. Much of what is written needs further development in order to reach its full value, and the *Handbook* is thus forward-looking rather than backward-looking. Alister Jones resumes this point in his concluding section.

A final condition for reaching the long-term aims of technology education is the development of a continuous path of learning through the various levels of education. Many countries are still far removed from this situation. Ideally, pupils would get a first orientation and awareness in preschool and primary education, a learning of basic concepts and processes in lower secondary, a more precise and differentiated image of (the various types of) technology in upper secondary, and specialisation in tertiary education (higher vocational or academic). This condition appears more and more to also affect for the survival of technology education in the school curriculum. For example, in the countries where technology education is or has been under threat, the fact that technology education was only present in some levels of education contributed significantly to its vulnerability. Subjects that are only taught in lower secondary education and are not part of the exam syllabus are almost inevitably doomed to get a low status. This unfortunately holds for technology education in many situations. In addition, having technology education as part of lower secondary education but not in primary education (which is the case in many countries) means that the position of technology education is open to attack. Apart from the educational value, a continuous learning pathway is therefore a prerequisite for the success of technology education.

INTERNATIONAL DEVELOPMENTS

One of the driving forces behind the development of technology education, apart from the ones that have already been mentioned, is the increasing level of international cooperation. One of the most obvious effects of this is the blending of elements from different national traditions in technology education curricula. As Section 1 shows, technology education emerged from a form of craft education in the majority of countries, although the direction in which it emerged differed. In some countries, design was brought in as a new element (as was the case in the United Kingdom). In other countries, such as the countries that were part of the former East Block in Europe, industrial production became the new focus. In yet others, learning concepts (that is, an enhancement of the cognitive element) became the new trend whilst others brought in 'high tech' elements, such as robotics. Finally, some countries emphasised the social aspect as a new ingredient. Thus, a spectrum of approaches in technology education has developed worldwide. Due to international conferences where people representing different approaches have met and exchanged ideas, elements from these approaches were transferred and curricula with richer combinations of elements developed.

One of the first organisations that initiated international exchange of information was the American Industrial Arts Association that changed its name into International Technology Education Association (ITEA) to indicate firstly a shift away from craft-based education and secondly an interest in internationalisation. At around the same period, the series of international PATT conferences started. PATT is the acronym for Pupils' Attitudes Towards Technology. This name referred to attitude research among pupils, which was the original interest of the PATT conferences. Later, the scope of the conferences widened and specific themes were defined for each conference. Cooperation with ITEA was established and since then international sessions are a recurring phenomenon at the ITEA conferences, held annually in the USA. Another organisation that initiated international cooperation and meetings was the World Council of Associations for Technology Education (WOCATE). The early death of its founder, Dietrich Blandow, however, meant a substantial reduction in activities and today its existence is not much beyond a formality only. In the UK, the International Design And Technology Education Research (IDATER) conferences for some years functioned as a platform for exchange of research output, but they no longer exist. To some extent the conferences organised by the UK's teachers association DATA (Design And Technology Association) have taken over this role.

Apart from these conferences and their organising bodies, the international journals act as a vehicle for the internationalisation of technology education. Of these, the *International Journal for Technology and Design Education* is the only one published by an academic (commercial) publisher, namely Springer (formerly Kluwer Academic Publishers). The *Journal of Technology Education* is published under the auspices of ITEA. Then there is the *Journal of Technology Studies*, published by Epsilon Pi Taw, a USA-based fraternity for technology education. The *Journal for Design and Technology Education*, published by DATA, increasingly has an international character, visible in both its editorial board and its contributing authors. Apart from the academic journals, several book series exist for technology education. Sense Publishers, an academic publisher founded fairly recently, publishes the *International Technology Education Series*, of which this Handbook is one volume. Springer publishes the *International Science and Technology Library*, which includes some titles with a strong technology focus.

The result of these efforts to bring together knowledge and experience from different countries has not only been the merging of elements from different approaches to technology education internationally, but also the emergence of international research projects. The DEPTH project, for instance, was initiated at a PATT conference and the results were later published in the *International Journal of Technology and Design Education.* Effort has also been placed on producing an international agenda for technology education research, but this has to date not led to a substantial effort to coordinate international research for technology education. The idea of an international technology education curriculum has also been raised at several conferences, but that has not resulted in any practical outcomes. The differences in (national) contexts appear to be a very real barrier for more drastic internationalisation of technology education. The more modest efforts, though, clearly have had an impact on technology education practice. International contacts have also been instrumental in supporting technology educators in, for example South-Africa and Sweden, in their efforts to save the position of technology education in their national curricula.

TECHNOLOGY EDUCATION AND GENDER

Glancing through the Table of Contents, a reader may wonder why there is no separate section, or even chapter, about the technology education and gender. Most introductory technology education books have such a section, with good reason. This issue has been an important topic for research and development work, attitude studies showing time and again that females hold different concepts of technology – and different attitudes towards it – when compared with males. Usually this impacts negatively on technological study and career choices. However, despite all the efforts made to date, the problem has not been adequately resolved. Gender differences continue to be deeply embedded in our society, and no doubt embedded in our being human. But this does not have to mean that one gender is more suited to technology than the other; some comparative studies of male and female performance in technology education suggest that girls outperform boys in terms of design creativity and sensitivity to usability issues. However, this does not – in general – help when it comes to the social acceptance of women in engineering. The issue is particularly difficult to tackle because on the surface it seems that nothing is wrong; there are no formal restrictions that keep women from being involved in technology. However, references to a 'glass ceiling' in career pathways are common and although this ceiling cannot be seen, women can hit their heads hard against it.

We chose not to dedicate a specific chapter or section to this issue in the *Handbook* in order to acknowledge that it is not – and cannot be seen as – an isolated issue. Neither can it be solved by treating it as an isolated problem; introducing specific 'female friendly' topics in technology education often has no measurable effect. A more fruitful approach could be to develop good technology education that does justice to what technology really is, and then to expect that technology as a human activity – in which both men and women have been involved from the beginning – will enable pupils of both genders to live out their capacities and interests. Good design projects will necessarily draw from creativity and involve an interest for humans, and if this appeals to girls more than boys, so be it. Equally, when this is not true and creativity is evenly distributed between girls and boys, the project will still appeal to girls.

The *Handbook* includes several chapters in which the gender issue is raised, but always in relation to the topic of the particular chapter. Thus we avoid the impression that the gender issue is something that can be dealt with in isolation from other issues.

READERS AND AUTHORS

The purpose of publishing this *Handbook* is to provide a general introduction to the field of technology education those on the relative outskirts of technology education, and as a source of information for those involved in it. We believe it can also serve as a reference text for teacher education programs to introduce future technology teachers to issues relevant to their (future) field of teaching.

In the course of this introduction chapter, the eight sections have been introduced. The editor of each section has written an introduction that provides more detailed information about the chapters in the section, as well as a more in-depth introduction to the topic with which the section is concerned. Finally, Alister Jones bring together the various topics in his closing chapter, presenting a final consideration of the future of technology education that focuses particularly on the quest for the sustainability of technology education as an often threatened learning area.

Readers will notice that the various chapters in this *Handbook* differ in nature. These differences are often related to the various backgrounds from which the authors have contributed their texts. Several of the authors in Section 2 (in particular the philosophers of technology) are not directly involved in technology education but have developed insights that can serve as a resource for those that are. Other authors are involved in technology education, but only as researchers and not in technology education practice. This may sometimes give a certain colour of distance to their texts, although mostly it is not difficult to imagine consequences for technology education classroom practice. Other authors, mainly those in Section 4) write from a science or STS education background. This is at times reflected in their use of the word 'science' to indicate both science and technology. Finally, there are authors that are directly involved in technology education practice, either as teacher educators or as technology teachers. Their chapters, of course, reveal a great degree of practical experience. Many of the teacher educators combine educational research and teacher training. In this context, a direct translation of research output into educational practice is not only possible, but often realised. Perhaps the gap between research outputs and the use of those outputs for improving and enhancing practice is one of the most difficult to bridge. Hopefully this *Handbook*, which brings together issues in technology education research and in technology education practice, can help.

REFERENCES

de Vries, M. J. (2000). Technology education: Towards a new school subject. In B. Moon, S. Brown, & M. Ben-Peretz (Eds.), *The Routledge international companion to education* (pp. 910–920). London: Routledge.

de Vries, M. J. (2006). Two decades of technology education in retrospect. In M. J. de Vries & I. Mottier (Eds.), *International handbook of technology education. Reviewing the past twenty years* (pp. 3–10). Rotterdam, The Netherlands and Taipei, Taiwan: Sense.

International Technology Education Association [ITEA]. (2000/2002/2007). *Standards for technological literacy: Content for the study of technology*. Reston, VA: Author.

Marc de Vries
Faculty of Applied Sciences
Delft University of Technology
Delft, The Netherlands

SECTION 1

INTERNATIONAL DEVELOPMENTS

ALISTER JONES

2. THE DEVELOPMENT OF TECHNOLOGY EDUCATION INTERNATIONALLY

The last 15-20 years have seen the emergence of technology as an area of study in its own right and the development of school curricula in Australia, the United Kingdom, USA, Canada, Europe, South Africa, and New Zealand that emphasise the importance of students developing technological literacy.

The development of technology education within countries and regions is set within the historical, cultural, and political environment. Curriculum, teacher education, and in fact educational research do not sit in isolation from these. Each of the chapters in this section sets out the context for technology education in its respective country and provides a historical and political analysis of the development of technology education as a field of development. The history of technology education is a long one if we consider its development back to the days of craft, and in this section many of the chapters trace the journey from craft through to much broader notions of technology and technological literacy.

In Chapter 3, Clare Benson explores the development of design and technology in the curriculum of England. She traces the development of over 50 years, through the different political and educational agendas from the 1950s until the mid 2000s. The subject of craft, design and technology (CDT) was developed in the 1980s to expand the notion of craft, with its gendered and set pieces to include planning and thinking. However, national assessment and/or monitoring can have a large impact on the possible development and implementation of a new subject and Clare highlights the influence of the Assessment Performance Unit on the development of design and technology curricula. In addition, external professional bodies such as Design Councils and engineering groups can act as advocates for a curriculum, but also limit its development.

Literacy and numeracy initiatives as well as over-crowded curriculum can lead to a reduction in curriculum time in 'new' subject areas such as technology, particularly in primary schools. Different subject subcultures also influence the interpretation of a curriculum area like technology, which can be perceived as an applied science or craft depending on the background of the teacher (Jones, 1997). Clare highlights the different influences on the 1990s curriculum revisions in England and the fact that there was little professional development available for teachers. She concludes that whilst good practice in design and technology is developing in England, teacher knowledge remains a significant barrier to effective national practice.

Jacques Ginestié in Chapter 4 presents the ways in which technology education is represented in France, particularly with respect to technology education research. Jacques also highlights the way that more dominant subject subcultures can influence how technology education is perceived and operationalised in curricula and research. The identification and articulation of knowledge suitable for technology education is an open-ended problem and highlights the fragility of technology education. However, although Jacques indicates that the emerging area of technology may be somewhat fragile in nature, there is a world wide movement to develop a more robust framework.

A.T. Jones and M.J. de Vries (eds.), International Handbook of Research and Development in Technology Education, 13–16.
© 2009 Sense Publishers. All rights reserved.

Tapani Kananoja introduces an historical perspective to technology education in Finland in Chapter 5 and notes that the teaching of practical skills and technology in general education was evident in Europe as the early as the seventeenth century as 'education for work'. Handicrafts were included in the Finnish curriculum in 1866, reflecting the main aim of the curriculum as training the future work force. Industrial arts were included in the 1920s as the technical skills required for work changed.

Building on the notion of industrial arts, William (Bill) Dugger analyses the development of technology education in the United States of America in Chapter 6, noting that in industrial arts evolved from manual arts in the late 1800s and early 1900s. 'Technology' was used in the 1970s and 1980s by particular curriculum development projects. Although curriculum can be agreed at the national level in the USA, the way it is implemented – or even included – is defined at the State and even the school district level. The development of the national curriculum – *Standards for Technological Literacy: Content for the Study of Technology* [STS] (International Technology Education Association [ITEA], 2000/2002/2007) – is discussed in detail. Very rarely do we get such an insight into the development of a national curriculum from an insider's perspective, including the ways in which curriculum and stakeholders interact.

The STS makes a strong call to educators, engineers, scientists, mathematicians, and parents for the notion of 'technological literacy for all'. However, this is not federally mandated in the US, nor is it an actual curriculum for direct implementation in schools but rather a framework for developing a curriculum. The ITEA released *Advancing Excellence in Technological Literacy* in 2003. This document outlines the standards and principles in relation to assessment, professional development, and programme implementation. The integration of these is one of the strengths of this approach to curriculum. Too often in national curriculum development, their isolation becomes a barrier to successful implementation. Bill concludes his chapter with the sobering note that technology education and technological literacy will become a reality for our students of tomorrow only when the different sectors in the country work together to this end.

The implementation of the technology as a curriculum area is always diverse when a country's education system is run at the national and state/provincial/local level. This is true for Canada as well as the US, and in Chapter 7 Anne Marie Hill provides a thorough analysis of the recent research and development in the study of technology in Canada, with its 10 provinces and 3 territories. One of the interesting pieces of analysis to come through in commentaries and research evidence is that students enjoy technology as a subject in schools. Anne Marie provides an historical account of Canada to set the context for curriculum and curriculum development, illustrating the multifaceted nature of education in Canada and the political nature of curriculum development. Although most places internationally have had a history of technical education in secondary schooling and this has moved to broaden into technology education, Anne Marie highlights that if we are to have effective technology education in secondary schools we must start with effective technology education programmes at the elementary/primary end of the school system.

Australia is in some ways similar to Canada in that there is some federal influence on curriculum, although it is mainly a state responsibility. Howard Middleton in Chapter 8 outlines the development of technology as a key learning area at the national level, highlighting the importance of co-operation between researchers, administrators, schools, and teachers. He goes on to report on ways in which researchers are exploring issues in the teaching and learning of technology. The disconnection between having a curriculum but being slow to implement professional development opportunities for teachers is raised. As has been found in other countries, primary teachers are enthusiastic about technology whereas at the secondary school level there has been a much more mixed response. For effective implementation of technology, then, there not only needs to be effective teacher professional development, but effective pre-service teacher education programmes also need to be developed, including the appropriation of government scholarships to encourage people in to technology teaching. Although pre-service and inservice training are important aspects of curriculum implementation, Howard also highlights the productiveness of working collaboratively at the national, state, and local level to implement technology education. In the development and implementation of a new curriculum area there are always tensions and contradictions, particularly around general technology education and vocational education, shortening teacher education programmes, and more instrumental approaches to research.

Although New Zealand and Australia collaborate in areas of education, the national systems are different. Whereas Australia has a federal and state approach to education, New Zealand has a national approach due mainly to its small size. Like other countries, New Zealand has had a history of technical education and in fact has had a national curriculum since the late 1870s. Chapter 9 provides background on the education system both from a historical and more recent perspective. As in other countries, technology emerged as an area in its own right in the 1990s with the development of the national curriculum (which was reviewed from 2001), national research projects, and national assessment and monitoring. In 2006-2008 changes were made to the curriculum to reinforce the notion of technological knowledge and practice, as well as students developing an understanding of the nature of technology.

Associated with the development of the technology curriculum in NZ has been a significant increase in the technology education research over the last 20 years. From 1990 onwards there has been a developing postgraduate research culture in technology education, including a focus on classroom studies. In the early days of the implementation of the curriculum there was a strong national focus on professional development. However, this was only for a limited time. The area of senior secondary technology became a concern and significant resources were obtained to enhance this area. These resources have expanded beyond senior secondary school and have the potential to significantly enhance technology education in the lower secondary level. However, concern must be expressed about the pressures of major curriculum revision by primary teachers, particularly with emphases on literacy and numeracy. Like Australia, the government is responding to the technology teacher shortage by introducing scholarships for pre-service teacher qualifications. Specific attempts have been made to link research, teacher education, professional development, and the curriculum framework together. Only time will tell if this has been an effective way forward.

In Chapter 10, Chitra Natarajan and Sugra Chunawala outline the development of technology and vocational education in India, showing strong links between the two. Similar to other chapters, they provide an historical perspective of the education system through to its current structure. The link between the structure of schooling and the structure of work is an important context in this chapter. In terms of curriculum change, Chitra and Sugra highlight that the current education system does not necessarily encourage collaboration and that there is a tendency to shy away from radical changes. The possible breadth of technology education also causes some difficulties in implementation, as do the inclusion of creativity and choice. This chapter clearly articulates how historical, political, social, and environmental perspective can both facilitate and constrain curriculum change.

Similarly, in Chapter 11 Bangping Ding provides an historical and social analysis of technology education in China, where labour-technical education has been officially recognised as a school subject in primary and secondary schools since the 1980s. However, Bangping highlights that China is no different from other countries in experiencing difficulties implementing a new curriculum. In particular, the influence of exit examinations, the shortage of qualified teachers, lack of resources, the influence of parents, and assessment methods constrain implementation. Given the size of China, Bangping uses the examples of Shanghai and Beijing to describe the emergence of technology education in secondary schools and the growing influence to move away from technical skills to general technological literacy and replacing labour-technical education with technology education.

In Chapter 12 Andrew Stevens again highlights the ways historical and social issues influence curriculum development. South Africa has undergone radical political and social reform in the last two decades, with educational reform being part of this. Teacher education has also been restructured, impacting on the implementation of technology education. The country, as Andrew notes, is still in a period of transition.

Research has mainly focused on the development and implementation of the curriculum and this needs to expand to highlight a way forward for technology education in South Africa.

All curriculum development is set within the social and political fabric of a nation and or state, in fact all curricula are essentially political. The opportunities for research and development are also influenced by the political frame. This section provides a sound basis from which to interpret the development of technology education research, development, teacher education, and curriculum.

REFERENCES

Jones, A. (1997). Recent research in student learning of technological concepts and processes. *International Journal of Technology and Design Education, 7*(1–2), 83–96.

International Technology Education Association [ITEA]. (2000/2002/2007). *Standards for technological literacy: Content for the study of technology*. Reston, VA: Author.

ITEA. (2003). *Advancing excellence in technological literacy: Student assessment, professional development, and programme standards*. Reston, VA: Author.

Alister Jones
School of Education
University of Waikato
Hamilton, New Zealand

CLARE BENSON

3. DESIGN AND TECHNOLOGY: A 'NEW' SUBJECT FOR THE ENGLISH NATIONAL CURRICULUM

INTRODUCTION

This chapter will outline the introduction and development of a 'new' curriculum area into the English National Curriculum in 1990. The background to this major educational change will be given; a rationale for the introduction of a National Curriculum in design and technology will be suggested; the developments from 1989 until 2008 will be discussed; and future pathways will be explored.

BACKGROUND TO THE INTRODUCTION OF A NEW SUBJECT

Not since the 1944 Education Act had such a major reform of educational policy taken place in England, together with Wales, as that which took place following the Education Reform Act 1988. Whilst there may always be a tension between two major aims of education – to provide education as a right to the individual, and to promote education as a means of providing economic success – the 1944 Education Act attempted to cover both purposes (Encarta, n.d.). In theory, schools had freedom to create their own syllabi and identify their own priorities, linked to the needs of individuals, until pupils reached 14 years and national examinations placed certain restrictions on the curricula that were offered. Pupils in primary schools (aged 5-11 years) had opportunities to do craft, woodwork, sewing and cooking. In science they mainly studied the natural world relating to flora and fauna, and areas such as mechanisms, forces, structures, and energy were largely ignored. In secondary schools (pupils aged 11-14 years) studies included woodwork, needlework, cooking, building work, and metal work. Whilst there were no regulations prohibiting girls and boys from taking particular specialisms, often girls were not given the opportunity to take, for example, wood work and metal work and had to opt for needlework and cooking, and vice versa. These handicraft and domestic science subjects were accorded a higher status with the creation of national exams (General Certificate in Education, GCE) but it also led to an almost total lack of creativity as pupils had to produce set pieces and learn different procedures by rote.

During the 1950s and particularly the 1960s there was a continuing movement towards focusing on the provision of education for individuals even if it produced little, if any, return (Central Advisory Council for Education [CACE], 1959). The establishment of the Schools Council in 1964 focused on the needs of pupils who were not following traditional academic courses and the Plowden Committee (CACE, 1967) discussed providing primary children with more informal experiences in the classroom. Teachers were encouraged to follow their pupils' individual interests through topic work to help engage all children in their learning.

A.T. Jones and M.J. de Vries (eds.), International Handbook of Research and Development in Technology Education, 17–30.
© 2009 Sense Publishers. All rights reserved.

With knowledge and understanding constantly evolving to take account of the rapid changes in the designed and made world, recognition was given to the need for the curriculum to prepare pupils to be adaptable, to participate actively in acquiring new knowledge and skills, to be able to solve problems, and to make decisions (Eggleston, 1996). Even in 1955, Friedman wrote of the need to prepare pupils for job changes throughout their careers as a single occupation for life could not be thought of as the norm. Design and technology was well placed even then to provide pupils with the kinds of experiences that would help them throughout their lives.

At the start of the 1970s, economic recession affected the amount of money available for the education budget and tighter controls on spending were introduced. This was coupled with a backlash by some (Cox & Boyson, 1977; Cox & Dyson, 1969) against what they saw as a decline in standards through the laissez-faire attitudes towards educational structures. In 1976, James Callaghan gave his now famous speech at Ruskin College in which he declared that too many pupils, from 16-18 years, were studying arts and humanities and not science and technology, which he felt would not lead directly to increasing economic and commercial wellbeing. However, it was during the 1980s, particularly in secondary education, that a number of initiatives developed curriculum ideas that fundamentally influenced the development of design and technology to what it is today.

One initiative that attempted to redress the balance in favour of science and technology for those pupils who wanted to follow a more vocational path was established following the Manpower Services Commission's brief in 1982. Called the Technical and Vocational Education Initiative (TVEI), it had a clear aim and purpose and was a step towards a more government-controlled education policy where economic purposes were given a higher priority than those of education for the individual. Courses included those in computer studies and information technology. Initially schools were given generous funding, and were then well placed to teach the new subject 'design and technology' as it came on stream.

Other, more powerful initiatives included two projects – Project Technology and the Design and Craft Education Project – that led to a reforming movement in the secondary school curriculum to introduce two new subjects: Technology and Design (Eggleston, 1996). Practical support for teachers came from both projects in the form of published resources, providing much needed ideas and strategies for implementation. By the early 1980s the reforming movement led to the creation of the subject Craft, Design and Technology (CDT). Its acceptance as a subject is outlined in an article published in *The Sunday Times* newspaper (5 March 1989), highlighted the fact that CDT was more than just a practical subject and that it involves a great deal of thought and planning. The subject brought together manual, intellectual, and organisational skills – all of which are vital in preparing pupils for living and working in a technological world (Eggleston, 1996). This notion has continued, and is embedded in the 'Importance' statement in the National Curriculum revision (Department for Education and Science [DES]/Qualifications and Curriculum Authority [QCA], 1999). The insightful article is unusual as even in 2008 many educational press reports and articles fail to draw out the nature of CDT as a subject, focusing instead on the practical, making element of design and technology.

Perhaps the most powerful initiative in the development of secondary design and technology in the late 1970s and 1980s came through the Assessment of Performance Unit project (APU). Whilst this project started to look at assessment in English, mathematics, and science, both in primary and secondary schools, a strong lobby resulted finally in a project that focused on the teaching and assessing of the subject design and technology at the secondary school level. A pilot study was conducted by Nottingham Trent University before a major project was established at Goldsmiths College, London University in 1986. This project was ongoing throughout the development of the

National Curriculum, with findings starting to be published in the early 1990s (Schools Examination and Assessment Council, 1991).

Throughout this period, design and technology did not grow significantly in primary schools, where handicrafts were still an important and enjoyable element of the curriculum for most. The Design Council had promoted the teaching of Design, and where there were individual enthusiasts pupils were given exciting experiences in this area. It was not until the National Curriculum (1990) that most primary schools realised that they had to plan for this curriculum area, despite a majority of teachers having little understanding of the nature and value of the subject.

THE NEW NATIONAL CURRICULUM FOR DESIGN AND TECHNOLOGY

Interestingly, even in the late 1970s, Her Majesty's Inspectorate was still promoting experiences for the individual, and suggested organising the curriculum in areas of experience, including technological experience, rather than traditional subjects (DES, 1977). This was not taken up, and against a climate of change towards more control of the curriculum, the creation of a Consultation group and publication of its findings in a paper (DES/Welsh Office [WO], 1987) marked the end of an era that had highlighted the importance of meeting the needs of all pupils. In 1988, the Education Reform Act led the way for a National Curriculum to be written and published initially for English, mathematics, and science in 1989, followed in March 1990 by Design and Technology and other foundation subjects. Implementation of Design and Technology was to be started from September 1990 in all state schools in England and Wales for children aged 5-16 years. Pupils aged 16-18 years would follow vocational courses or exam syllabi set by the exam boards. Wales continued to be joined to England, but separated after the 1995 National Curriculum. Scotland and Northern Ireland had different curricula and have continued their own development of the subject. Scotland has National Guidance rather than a National Curriculum, and both countries incorporate Technology within broader areas of experience.

The content of each subject area was developed by working parties, mainly working in isolation from each other. This resulted in an overloaded curriculum and lack of coherence, continuity, and progression, especially at primary level. The chairperson of the Design and Technology Working Group, Lady Parkes, was lobbied by different pressure groups as the work progressed – the design lobby led to the inclusion of the word 'design' in the title; the science lobby pushed for a focus that would lead the subject down the 'appliance of science path'; the home economists were afraid of losing food and textiles to design and technology and wanted them to be separate; and the business and vocational lobbies wanted to keep their places in the curriculum, rather than have them integrated into Design and Technology. The working group was composed mainly from business and industry and those with secondary or higher education backgrounds. Primary views were certainly not strongly represented. Indeed, the science working party had much to do with the initial development of primary Design and Technology, creating a different philosophy to that of secondary Design and Technology. However, the group soon handed the subject over to the newly created Design and Technology Working Group and a more coherent approach to both phases of schooling was developed. Damage was done, however, as open discussions from the group resulted in early misconceptions of the nature of the subject in primary schools. The 'appliance of science' view was really only dispelled during the late 1990s.

Finally, on publication the subject was joined with Information Technology into one document entitled *Technology in the National Curriculum* (DES/WO, 1990); it was in two parts: design and technology capability, and information technology capability. If there had been confusion and misconceptions before, this only served to add to the problem. Primary teachers in particular, who mostly had little knowledge and expertise in the area, immediately thought that the subject focused on computers. Unpublished surveys undertaken by the author for the five years after 1990 found that this was a common perception held even by those leading the subject in their primary schools. This was less of a problem for secondary teachers as they only had one subject to focus on – Design and Technology – and they had more time to disentangle the mixed messages contained in the documentation.

The content of the document was organised under four Attainment Targets (DES, 1998b), reduced from five in the interim report (DES, 1998a):

- AT1. Identifying needs and opportunities;
- AT2. Generating a design;
- AT3. Planning and Making; and
- AT4. Evaluating.

Each Attainment Target had a different weighting when assessing pupil progress and the whole process was unwieldy, unworkable, and complex. The layout of the Programmes of Study was confusing and the examples given were all taken from different contexts so that it was impossible to follow through one context and see how an holistic experience could be achieved. Pupils were to be given experiences related to artefacts, systems, and environments – words that meant little to many primary teachers.

As implementation got underway, perhaps the most disappointing development was the way in which the subject was delivered. Teachers both in primary and secondary schools saw the four Attainment Targets as a linear process to be worked through, rather than areas that could be covered in innovative and flexible ways. This was hardly surprising. With little previous knowledge, and little continuing professional development (CPD), teachers looked for strategies that would help them; working through four Attainment Targets gave them a structure, albeit a rigid one.

When the 'new' subject Design and technology was finally acknowledged and created for primary and secondary pupils, the challenge that this laid down for both primary and secondary schools was almost overwhelming in different ways. Against a background of primary teacher subject knowledge and skills primarily in art and craft, nature studies, cooking and needlework, and many secondary teachers' expertise in, for example, domestic science, woodwork, and metalwork, teachers were expected to understand, plan, and deliver a 'new' subject that was different in nature to any existing curricula. In addition, primary teachers had to implement the National Curriculum in eight other subjects, including the core subjects of English, mathematics, and science. Little, if any, support for implementation was organised except for the core subjects; few resources were available, partly due to the short time scale between publication of the National Curriculum and its implementation; CPD was almost non-existent in design and technology, partly due to the very small number of people who could provide relevant courses; and, perhaps most importantly, there was confusion as to the nature of the subject. It would be fair to say that many primary teachers concentrated on implementing the core subjects as the pressure of testing and target-setting grew, and despite more work in the 1980s on the development of the subject design and technology, many secondary teachers hoped that by slightly adjusting their approach they could continue with their 'own' subjects and ignore design and technology.

'GROWING' THE SUBJECT 1990 – 2007

1990–1995

It might be imagined that after such a confusing and pressured start, teachers would be given time to gain an understanding of the new Order, to plan appropriate schemes of work, find suitable resources, trial activities, and evaluate and change practice as they grew in confidence. This was not to be. Almost immediately, concerns were raised about the way in which the subject was developing and the document *Technology in the National Curriculum – Getting it Right* (Smithers & Robinson, 1992), prepared for the Engineering Council, was released into the public domain. They talked about 'Mickey Mouse' activities with egg boxes and suggested that the subject had become generalised problem solving without a specific knowledge base. There was much debate as to the specific knowledge base that was part of the subject, some arguing that it needed to be very limited, whilst others promoted the notion that it should draw on a range of subjects, including science, art, mathematics, IT, home economics, and business studies.

Primary teachers started to create some exciting projects, but as more subjects came on stream in 1991, pressures of time and lack of understanding hindered these developments. An inquiry group was set up under the new Inspection service – Office for Standards in Education [OFSTED] – and their findings became the basis for the new draft Order for Technology (National Curriculum Council [NCC], 1992). Again, progress was slow as a new Secretary for State, John Patten, announced a review of the entire National Curriculum. During 1994 a new draft Order was published for design and technology – not technology – (School Curriculum and Assessment Authority [SCAA], 1994) before the final Order was published (Department for Education [DFE]/WO, 1995). Thus, it seems hardly sur-prising that progress was slow during 1990-1995, and that teachers began to lose sight of the nature and importance of the subject. However, the Schools' Inspectors report 'Subject and Standards 1994-5' indicated that pupils were almost always enthusiastic about the subject and found the work enjoyable and interesting. Primary standards and standards in teaching (satisfactory and above) in English, mathematics, science, and design and technology were all graded at 80% with the exception of Key Stage 2 (7-11 years) teaching; this was graded at 75%. The format of reports changed after this, thus making comparisons in future years impossible.

1995–2000

The new National Curriculum in 1995 clarified the nature of the subject through the statement "pupils should be taught to develop their design and technology capability through combining designing and making with knowledge and understanding in order to design and make products" (p. 2). This was similar to the statement in the 1992 new draft Order (NCC, 1992) but lacked its important emphasis on "good quality products fit for their intended purpose" (p. 5). The content of the 1995 document was slimmed down; the language of the document was simpler and much more easily accessed, especially for primary teachers; the attainment targets were reduced to two: designing and making; and the holistic nature of design and technology was emphasised through the level descriptions.

One of the major positive changes, particularly for primary teachers, was the introduction of the section relating to opportunities through which pupils could develop their capability. Three opportunities were identified: assignments in which pupils design make products (DMA); focused practical tasks in which they develop and practise particular skills and knowledge (FPT); and activities in which they investigate, disassemble, and evaluate simple products (IDEA). This offered a clear structure for teaching, and whilst it did lead to linear teaching programmes, it promoted the idea that teaching pupils skills and knowledge was essential for good quality design and technology. Ive (1997) HMI identified that this structure had been responsible for rising standards, certainly in primary design and technology.

Whilst secondary schools were now left to develop appropriate schemes of work, primary teachers were faced with major changes in 1998. The government, faced with poor standards in literacy and numeracy, introduced two new initiatives – one for literacy (Department for Education and Employment [DfEE], 1998) and one for numeracy (DfEE, 1999) – and whilst they were not mandatory, head teachers felt pressured into implementing them. If schools did not have these in place and standards were shown to be less than satisfactory, then school leaders felt they would be heavily criticised by inspection teams. There were time implications now as schools tried to cover the new, suggested content, and foundation subjects such as design and technology were once again under threat of being marginalised. The government, through the document "Maintaining breadth and balance" (QCA 1998a), did suggest that schools should keep a balance in their curricula, but the threat of poor inspection results did much to negate the content of this document.

The publication of the national 'Schemes of Work' (QCA, 1998b) for Key Stage 1 (5-7 years), Key Stage 2 (7-11 years), and Key Stage 3 (11-14 years) did go some way, particularly at primary level, to offset the damage to the development of design and technology. The schemes of work were created to give schools possible ways of delivering the curriculum; they were only guidance; and they were for schools to use as a tool against which they could evaluate, monitor, and plan new units of work. QCA surveys found that at primary level the scheme for design and technology was the most used scheme, but not at secondary level. It can be argued that for the first time since the National Curriculum was introduced, schools had an overview of a possible design and technology scheme of work for 5-11 years that incorporated flexibility, and alternatives, whilst covering the content of the National Curriculum. As the majority of primary teachers had little knowledge and understanding of the subject (OFSTED, 1998), this resource proved invaluable. However, problems arose when teachers kept rigidly to the suggested activities and did not develop the units to meet the needs of their own pupils. Initially the Schemes therefore did much to help raise standards and, if used appropriately, continue to do so even in 2008.

2000–2008

The five year moratorium on National Curriculum change drew to an end and during 1998-9 fresh debates took place with regard to proposals for the new 2000 Curriculum (DfEE/QCA, 1999b). Perhaps the most significant change in design and technology was the inclusion of an 'importance' statement at the beginning of the programmes of study. There was still some confusion as to the nature and importance of the subject and it was felt to be crucial to have a clear statement that set out the rationale for the subject, together with key elements relating to the nature of the subject. This was not an easy task and as soon as it was published, discussions were underway as to how it could be improved. However, the user, purpose, and function of a product is emphasised and this is something that needs to be focused on throughout all phases of education, but particularly in the primary school. The content of the new document was slimmed down again but there were few fundamental changes. The heading 'Structures' was taken out as it was felt that structures could be included in all projects; there was more emphasis on the use of ICT, including Computer Aided Design and Computer Aided Manufacture (CAD/CAM) as a tool; and the two Attainment Targets became one to emphasise the holistic nature of design and technology and the importance of assessing design and technology as a whole, not as a series of 'can do' statements.

At the same time, a new National Curriculum for children in the Foundation Stage (3-5 years) was published (DfEE/QCA, 1999a). This curriculum was divided up into six areas of experience, and whilst design and technology capability can be developed through all the areas, the main focus for the capability is within Knowledge and Understanding of the World. Through a Department for Education and Skills (DfES) funded projected, *Developing Designerly Thinking in the Foundation Stage*, Benson (2003) identified the lack of opportunities that 3-5 year olds are offered in developing their design and technology capability through the activities they experience. There is till much emphasis on the natural world, and little on the designed and made world. Children are given many activities in relation to making, but few that will develop design skills. If pupils are to develop their capability, then it is vital that they have appropriate experiences from the start of their education.

Since 2000, primary OFSTED inspection reports (e.g., OFSTED, 2001, 2002) indicate that whilst there is much good practice in design and technology, much remains to be done at all levels, particularly in the areas of designing and developing teacher knowledge and understanding. However, schools were not given much time to implement the new curriculum: Primary schools had yet another initiative to contend with in 2003 when the government started a review of the whole curriculum, starting with the publication of 'Excellence and enjoyment – a primary strategy' (DfES, 2003a). Through this strategy, it was hoped that schools might place less emphasis on numeracy and literacy and give children a broader, more balanced curriculum. However, with the subsequent creation of the Primary Strategy (DfES, 2003b), many hopes were dashed. Literacy and numeracy were still emphasised, but the inclusion of the importance of the development of thinking skills, questioning, and problem solving did link with design and technology and are integral to good practice. In addition, misconceptions about the content of the document arose. Schools started to plan using a topic, or thematic, approach based on the perception that the Strategy promoted cross curricular links. This led to a growth in making, or craft, activities rather than design and technology. Topics linked to history often provide the worst examples, with children making Viking boats from card templates and Tudor houses using straw and card boxes. The user, purpose, and functionality of the product are ignored as design is not included in the activities that are undertaken. The latest OFSTED report (OFSTED, 2007) covers the three years 2003-6, provides a clear overview of the state of design and technology, and suggests key areas that need to be addressed. These do not surprise those working to develop primary design and technology and include teachers' limited expertise and confidence; schools' reluctance to allow teachers to attend staff development courses due mainly to funding; and the need to improve assessment, recording, and reporting of pupils' progress.

Secondary schools faced another upheaval in 2006 with the development of a new Key Stage 3 (11–14 years) programme of study (QCA, 2007). The 'importance' statement has been updated to include references to creativity and innovation. The programmes of study have been reorganised into four areas – key concepts, key processes, range and content, and curriculum opportunities – with the intention to promote creativity and innovation. The assessment levels from 4-8 have been rewritten to reflect the changes to the programmes of study.

FUTURE PATHWAYS

It is almost impossible to identify long term developments in the subject, if the past is to be repeated with its constant push for change. In July 2004, the DfES published its 'Five Year Strategy for Children and Learners'; in December 2007, the Department for Children, Schools and Families [DCSF] published its ten year plan, 'The Children's Plan: Building Brighter Futures'. Whilst there are some similar messages, reference to the Five Year Plan was not built into the new Ten Year Plan. Whilst these are not specific curriculum plans, this exemplifies the way in which educational change does not necessarily build on what has gone before.

It can be argued that it is a great tribute to teachers that design and technology has become an established and well-liked subject within primary and secondary school curricula. Without their resilience, hard work, and commitment to the subject, and their belief in its value to the children they teach, the subject could have disappeared after one of the numerous, major changes over the years. However, there are already some indications of changes that will affect the subject. There is a primary review being undertaken of the whole curriculum, led by Sir Jim Rose. A draft of the proposals will be available in November 2008, the final document will be available in the Spring of 2009, and implementation will be in September 2011. One of the main areas for debate is how to organise the curriculum; 'subjects' may disappear and 'areas of experience' may take their place. This links back to the HMI suggestion in the late 1970s and to the way in which the Foundation Stage curriculum is organised. It could be that there is an area – Technological Experience – or that design and technology becomes art, design and technology; or even back to science and technology.

The Science, Technology, Engineering, and Mathematics (STEM) agenda is one that is being promoted strongly by the DCSF. Whilst this could result in closer links being made across the subjects, exciting, relevant projects being created for pupils, and curricula that will prepare young people for the rapidly changing world and technologies that they will experience, it could also lead to a struggle between subjects to maintain their integrity and identity in the curriculum. STEM in primary schools at the present time is not an issue that many working in schools would identify as important. However, increasingly engineering challenges are being presented to primary schools that can lead to confusion as to what is design and technology, and what is engineering? How do they fit together, and where do they fit in the current National Curriculum?

Two other issues will undoubtedly have high priority in the foreseeable future. First, with the high profile given to global warming and issues relating to sustainability, design and technology should offer opportunities for pupils to work through projects that offer possible solutions to enable us to make appropriate changes in our lifestyles. Second, whilst food technology has always been included in the primary curriculum, there have been periods of time when it has been optional in secondary schools, including the present new Orders (2007). However, within six months of their publication the government decided to change this and make food a compulsory element for Key Stage 3 (11–14 year olds) within design and technology from September 2011. The issue of growing obesity levels, an emphasis on healthy eating and lifestyles, and other food initiatives for schools appear to have been the government's motive for this.

From lessons that have been learnt, whatever changes take place, it is vital that the nature of the subject is always clear for those who have to implement the curriculum. Changes should be well supported through the provision of quality CPD and relevant resources. One thing is certain: It will be crucial for design and technology to keep abreast of technologies in the rapidly changing world, enabling the pupils who experience the subject to become "responsible citizens who make a positive contribution to society" (QCA, 2007, p. 51).

INITIAL TEACHER TRAINING

Not surprisingly, there have been many changes in Initial Teacher Training (ITT) since the subject design and technology became part of the English National Curriculum; even the title 'Initial Teacher Training' has been contentious, despite being chosen to reflect the notion that students were involved in a reflective, evaluative course rather than one that trained them to perform in the classroom. It is now ITT but hopefully students are not 'trained', but engage in debate and reflective practice throughout their courses. Updates are constant, so it is important to use the Teacher Development Agency (TDA) website (http://www.tda.gov.uk/) to keep abreast with the constant revisions.

Secondary courses used to focus on Craft, Design and Technology (CDT) as well as individual subjects such as wood work, metal work, and home economics, whilst primary courses included art and craft. With the introduction of the National Curriculum in 1990, ITT had to change the nature of its courses to reflect the required practice in schools. Courses have to be validated before they can be taught, and whilst small changes could be made without the usual lengthy process of revalidation, it took time for ITT to be able to offer courses that reflected these significant changes. Staff also needed to quickly understand the fundamental changes in approach to this new subject. There was little professional development available to them to support the change, and whilst some staff embraced the changes, understanding the value that the new subject had to offer young people, others felt threatened by the perceived loss of their own specialism.

In primary ITT, it was a case of starting a new subject area, not adapting established ones. Again, this caused problems as there were few existing staff who could lead the subject, and no resources were provided to train new staff. Initially, many of the science lecturers were asked to take on the subject, which led to some confusion. Design and technology is a subject in its own right and not to be viewed as the 'appliance of science'. Thus, some students were making lighthouses with electric circuits, buggies with motors, and rubber band powered vehicles that were not designed – the students just followed instructions – without giving consideration to the purpose or user. The situation did change as new staff, often recruited from Education Support Grant (ESG) teams and Local Education Authorities (LEA) and who had devised development programmes for existing teachers, began to create new programmes of work for ITT. During the period 1995-2000, ITT was able to produce some very well qualified teachers of design and technology who had enthusiasm, knowledge and understanding and skills, and who understood the value of the subject for young people. However, research by Davies (2000) highlighted the lack of opportunities for teaching design and technology by trainee teachers during their school experience.

More recently, there have been problems in retaining courses, staff, and specialisms for a variety of reasons. Design and technology requires specialist training resources that are expensive and need updating regularly, and this has been one of stumbling blocks in keeping departments. Falling numbers of applicants, fewer staff with the right qualifications, and the cutting of specialisms in primary ITT have all contributed to a downturn in numbers. Numbers seem more stable now in secondary training; indeed there has been a substantial increase in those wanting to do a food specialism in the last three years. In primary, although there are only a very few courses that have retained specialisms, these are well subscribed and students are in demand on leaving ITT.

At present there are a number of different routes through which ITT in secondary design and technology can be accessed. These routes all lead to qualified teacher status (QTS). They combine theoretical learning with at least eighteen weeks of school placement training. The routes include:

- School-based training schemes, such as School Centered Initial Teacher Training (SCITT), the Graduate Teacher Programme (GTP), Overseas Trained Teacher Programme (OTTP), or the Registered Teacher Training Programme (RTP). These courses are taught in schools with the support of experienced mentors. Lectures or discussions cover the theory related to teaching. Trainees on these routes receive a salary from their school. This is at least equal to the minimum point on the unqualified teacher pay scale.
- Universities and colleges offer both undergraduate and postgraduate courses. They combine theoretical learning with school placements. All undergraduate and postgraduate students are eligible for a means-tested grant from their local authority (LA). Eligible design and technology postgraduate trainee teachers are also entitled to a tax-free bursary. In August 2008 this was £9,000 (£225 per week).
- Online options allow for distance learning courses. All trainees teach and train in at least two schools. When in school, they are supported by trained teachers who provide training and feedback.

Typical ITT courses include taught modules using a combination of workshops, lectures, seminar groups and individual study to support work in school. Education, pedagogy, and professional studies during the training allow trainees to explore a range of educational issues based on school experience. During training it is important to develop professional practise in subject teaching to enable trainees to apply knowledge and understanding gained to teaching design and technology.

Subject pedagogy looks more specifically at design and technology as a specialist subject, addressing the "hows and whys" of teaching the subject including issues such as health and safety, managing workshops, promoting creativity, and developing computer aided manufacture and design skills. Guidance from the 'Minimum Competencies for Training to Teach Design and Technology in Secondary Schools' (Design and Technology Teachers Association [DATA], 2003) divides subject-specific knowledge and skills into core and specialist "materials". All trainees, regardless of the specialist materials that they decide to study, cover the core competencies integrated into their subject specific specialist studies. The core competencies are based on the process and procedure of "designing". The specialist materials are food technology, materials technology, textiles technology, electronics, and communication technology. Different ITT courses offer trainees the opportunity to specialise in one, two, or all of the specialist materials.

Assessment of design and technology ITT is made on the strength of a trainee's ability to meet the requirements of the TDA standards for the award of QTS. This can include written work, school experience, online skills tests, and evidence portfolios. Employment prospects are excellent, as design and technology has been identified as a "shortage subject". Secondary design and technology trainees who successfully complete an ITT course and accept a newly qualified teacher (NQT) position are also eligible for a one-off taxable 'golden hello payment' of £2,500.

For primary teachers, the training and financial support is different. Undergraduate students are awarded the same qualification whether their course is three or four years in length. Most ITT courses now are three years and students study English, mathematics, science, ICT, religious education, and education studies. They can be given a choice between art or design and technology, and history or geography. Few institutions maintain specialist routes, including design and technology, although those that do recruit well. The courses all feature a balance between theory and practice to help develop appropriate personal knowledge, understanding, and skills. Thus, the majority of primary NQTs do not feel confident in teaching design and technology as a subject and cannot help colleagues in schools who look for support. There is also no additional financial support for those students opting for design and technology as a specialism.

For postgraduate students who undertake a one year Post Graduate Certificate in Education (PGCE) to enable them to teach, there are almost no opportunities to specialise in their own subject during ITT. They are expected to make the connections themselves, although a degree, for example in engineering, may not have equipped them with appropriate skills and knowledge for teaching primary children. Some will undertake between 2-10 hours of lectures relating to a subject; others will have none. As the number of lecturers in primary design and technology fall, ITTs are employing teachers from the classroom for a few hours to deliver the subject. Whilst many of these will have excellent classroom practice, it is very difficult for a primary teacher, teaching nine subjects, to keep abreast of new developments and resources in their subject area. In 2011, the new primary National Curriculum will be a legal requirement for all primary schools in England. Consideration needs to be given now as to how staff in ITT will be updated and receive professional development to ensure that students have appropriate courses during their training.

RESEARCH

Disciplines that have a long history, such as science, have an established background of research, supported by funding from a range of sources. It is therefore impossible for a relatively new subject to compete in research terms with those subjects that have been within the curriculum for decades, and even centuries. Nevertheless, research frameworks for design and technology have been quickly established; an annual International Design and Technology Educational Research conference was held at Loughborough University 1998-2001, and journals were published including *Studies in Design, Education Craft and Technology*, *Design in Technology Teaching*, and the *Journal of Design and Technology Education*. This provided subject-specific outlets for burgeoning research, in addition to general research conferences and publications. In 2000, DATA (now the Design and Technology Association) celebrated the millennium with its first research conference and it has continued to hold an annual conference. In 2001, Loughborough University's annual face-to-face conference came to an end but various other options exist on their website. In 1997, as primary design and technology became more established, the Centre for Research in Primary Technology (CRIPT) was created at University of Central England (now Birmingham City University). Its first international biennial conference was held in the same year, and still continues. Conference proceedings produced ahead of all the conferences have provided an excellent resource for research, covering current key issues.

Large-scale funding for research projects has been difficult to acquire for a variety of reasons: the subject is in its infancy; some funders have been unsure as to the nature and value of the subject; other subjects such as science have overshadowed design and technology and its importance; and there appears to be less money available for such research projects. It could be argued that the first major project to be funded was based at Goldsmiths College, London (DES, 1987; 1988). The Assessment of Performance Unit (APU), the then major research arm of the DES awarded Goldsmiths the project to identify national design and technological capability (Kimbell, Stables, Wheeler, Wosniak, & Kelly, 1991). The research findings have, and still do, make a significant contribution to the teaching and learning of design and technology, not only in England but worldwide. Other projects undertaken by the team at the Technology Education Research Unit (TERU) can be viewed on their website (www.goldsmiths.ac.uk/teru) but the latest project reflects very recent changes in teaching and learning – the use of ICT. 'Project e-scape' came out of a project assessing design innovation for QCA. The principal outcome of the project was a developed portfolio assessment system, and the data and findings are in the project report.

Other research projects in technology education include 'Young Foresight' and 'Understanding Creativity'. Young Foresight (2004) was funded by DfES and the Department for Trade and Industry. This involved young people in Year 9 (13–14 years) and focused on designing products and services for the future but not making them. More recently, a large scale project started in January 2005 involved pupils 11–16 years and was entitled 'Understanding Creativity for Creative Understanding'. Funded by Gatsby and based at the University of Cambridge, the research aims were twofold: to develop a model of learning and teaching in design and technology that explicitly develops creativity; and to support heads of departments to become more effective in their role as leaders of learning in the subject, creating and managing the learning environment. The findings and implications have been published in numerous papers (Nicholl & McLellan, 2007; 2008).

The only large scale project for primary work has been funded by the DfES and was undertaken by Centre for Research in Primary Technology (CRIPT) from 2003-2006 (Benson 2003; 2005; 2008). The project – Designerly thinking in the Foundation Stage – supported the development of designerly thinking through the engagement of young children with the designed and made world around them, including designed and made products. Findings included the lack of such practice in the Foundation stage, the active participation of children who had shown little interest in other types of activity, and the development of both speaking and listening skills, together with an increase in and use of technical vocabulary.

With regard to setting a research agenda, Baynes and Johnsey (1997) set out a platform on which to build research in primary education. This was followed in 2003 by a report from Harris and Wilson (2003), funded by the DfES to review, comment on, and offer recommendations for areas that should be funded and systematically researched. Their report provides an excellent, if not complete, picture of research relating to design and technology. They concluded, as many suspected, that there were few large scale projects; most were small scale and case studies that often made generalisation difficult, if not impossible. Whilst the report was favourably received in the main, it has not resulted in more funded projects.

The future of research into design and technology education has difficulties to overcome. It can be argued that the Research Assessment Exercise (RAE), that has been taking place since 1992 and is due to report again in December 2008, has done little to help. Many researchers are based in Faculties of Education within Universities, some of which do not see research as a major activity. A survey of the results of the RAE since it began shows that Faculties of Education do not fare well and are often towards the lower end of the ratings within a University. There are also fewer staff in many Universities who are research active or have research as a major component of their work. Funding is not forthcoming and the research that is often undertaken is small scale and based around a researcher's everyday activity, rather than on areas that have been identified in the Harris and Wilson (2003) report. However, a positive development in recent years has been the growth of action research undertaken by secondary and primary teachers involved in MA courses (e.g., Benson, Lawson, & Till, 2005; Benson, Lawson, Lunt, & Till, 2007) and funded through the TDA. Whilst the research is small scale, it is classroom-based and offers other educators the opportunity to review findings and undertake similar activity in their own schools. This could be a very productive way forward in promoting research and providing findings that can be used to improve practice in schools in the next few years.

REFERENCES

Baynes, K., & Johnsey, R. (1997). Research in preschool and primary technology. In R. Ager & C. Benson (Eds.), *Proceedings of the first international primary design and technology conference* (pp. 43–46). Birmingham, West Midlands: CRIPT.

Benson, C. (2003). Developing designerly thinking in the Foundation Stage. In C. Benson, M. Martin, & W. Till (Eds.), *Fourth international primary design and technology conference proceedings* (pp. 5–8). Birmingham, West Midlands: CRIPT.

Benson, C. (2005a). Developing designerly thinking in the Foundation Stage: Perceived impact on teachers' practice and children's learning. In C. Benson, S. Lawson, & W. Till (Eds.), *Fifth international primary design and technology conference proceedings* (pp. 15–19). Birmingham, West Midlands: CRIPT.

Benson, C., Lawson, S., & Till, W. (Eds.). (2005). *Proceedings of the fifth international primary technology conference.* Birmingham, West Midlands: CRIPT.

Benson, C., Lawson, S., Lunt, J., & Till, W. (Eds.). (2007). *Proceedings of the sixth international primary technology conference.* Birmingham, West Midlands: CRIPT.

Benson, C. (2008). Young children as discriminating users of products: Essential experiences for language development. In H. Middleton & M. Pavlova (Eds.), *Proceedings of the fifth biennial International Conference on Technology Education Research. Exploring technology education: Solutions in a globalised world* (Vol. 1, pp. 19–26). Surfer's Paradise: Griffith University.

Callagan, J. (1976) *The Ruskin Speech 20th anniversary lecture given by Rt Hon Lord Callaghan of Cardiff KG on October 15 1996.* Retrieved from http://guardian.co.uk

Central Advisory Council for Education. (1959). *15–18. The Crowther report.* London: Her Majesty's Stationery Office.

Central Advisory Council for Education. (1967). *The Plowden report.* London: Her Majesty's Stationery Office.

Cox, C. B., & Boyson, R. (Eds.). (1977). *Black paper 1977.* London: Temple Smith.

Cox, C. B., & Dyson, A. E. (Eds.). (1969). *Fight for education: A black paper.* Manchester, Lancashire: Critical Quarterly Association.

Davies, D. (2000). Ten years of universal primary technology education in England and Wales: What have we learnt? *The Journal of Design and Technology Education, 5*(1), 26–35.

Department for Education and Science [DES]. (1977). *Curriculum 11–16.* London: Her Majesty's Stationery Office.

DES. (1977). *Report by Her Majesty's Inspectorate.* London: Her Majesty's Stationery Office.

DES/WO. (1987). *The National Curriculum 5–16: A consultation document.* London: Her Majesty's Stationery Office.

DES. (1988). *National Curriculum.* London: Her Majesty's Stationery Office.

DES. (1988b). *Interim report on the National Curriculum.* London: Her Majesty's Stationery Office.

DES/WO. (1990). *Technology in the National Curriculum.* London: Her Majesty's Stationery Office.

Department for Education and Employment [DfEE]. (1998). *The national literacy strategy.* London: Her Majesty's Stationery Office.

DfEE. (1999). *The national numeracy strategy.* London: Her Majesty's Stationery Office.

DfEE/Qualifications and Curriculum Authority [QCA]. (1999a). *Early learning goals.* London: Author.

DfEE/QCA. (1999b). *The National Curriculum.* London: Author.

Department for Education and Skills [DfES]. (2003a). *Excellence and enjoyment: A strategy for primary schools.* London: Author.

DfES. (2003b). *The primary strategy.* London: Her Majesty's Stationery Office.

DfES. (2004). *Five year strategy for children and learners.* London: Her Majesty's Stationery Office.

Department for Children, Schools and Families. (2007). *The children's plan: Building brighter futures.* London: Her Majesty's Stationery Office.

Department for Education [DFE]/WO. (1995). *Design and technology in the National Curriculum.* London: Her Majesty's Stationery Office.

Design and Technology Association [DTA]. (2003). *Minimum competencies for training to teach design and technology in secondary schools.* Wellsbourne, UK: Author.

Eggleston, J. (1996). *Teaching design and technology.* Buckingham: Open University Press.

Encarta. (n.d.). *Education Act 1944.* Retrieved from http://uk.encarta.msn.com

Friedman, G. (1955). *Industrial society.* New York: Free Press.

Harris, M., & Wilson, V. (2003). *Designs on the curriculum? A review of the literature on the impact of design and technology in schools in England.* London: Department for Education and Skills.

Ive, M. (1997). Primary design and technology in England – Inspection evidence and 'good practice'. In R. Ager & C. Benson (Eds.), *First International Primary design and technology conference proceedings* (pp. 1–6). Birmingham: CRIPT.

Kimbell, R., Stables, K., Wheeler, T., Wosniak, A., & Kelly, V. (1991). *Assessment of performance in design and technology: Final report.* London: Her Majesty's Stationery Office.

National Curriculum Council [NCC]. (1992). *Technology for ages 5 to 16.* London: Department for Education/Welsh Office.

Nicholl, B., & McLellan, R. (2007). The contribution of product analysis to fixation in students' design and technology work. In E. Norman & D. Spendlove (Eds.), *The Design and Technology Education Association and international research conference proceedings* (pp. 71–76). Wellesbourne, UK: The Design and Technology Association.

Nicholl, B., & McLellan, R (2008). 'We're all in this game whether we like it or not to get a number of As to Cs'. Design and technology teachers' struggles to implement creativity and performativity policies. *British Educational Research Journal, 34*(5), 585–600.

Office For Standards in Education [OFSTED]. (2001). *Report* 1999/00. London: Author.

OFSTED. (1998). *OFSTED annual report: Design and technology.* London: Her Majesty's Stationary Office.

OFSTED. (2002). *Report 2000/01.* London: Author.

OFSTED. (2007). *Education for a technologically advanced nation: Design and technology in schools 2003–6.* London: Author.

Qualifications and Curriculum Authority [QCA]. (1998a). *Maintaining breadth and balance.* London: Author.

QCA. (1998b). *A scheme of work for KS 1, 2, 3: Design and technology.* London: Author.

QCA. (2007). *The national curriculum KS 3. Design and technology.* London: Author.

School Curriculum and Assessment Authority [SCCA]. (1994). *Design and technology in the national curriculum.* Draft proposals. London: Author.

Secondary Schools Examination Council. (1991). *The assessment of performance.* London: Author.
Smithers, A., & Robinson, P. (1992). *Technology in the national curriculum: Getting it right.* London: Engineering Council.
Young Foresight. (2004). Retrieved from http://www.youngforesight.org/

Clare Benson
Faculty of Education
Birmingham City University
Birmingham, England

JACQUES GINESTIÉ

4. THINKING ABOUT TECHNOLOGY EDUCATION IN FRANCE

A Brief Overview and some Aspects of Investigations

INTRODUCTION

Like all educational research, that concerned with technological education belongs to a category of research that is socially oriented, that is to say, research findings impact on school organisation. In addition, academic evolution alters and adds depth to the elements studied. Such research can be organised into three categories: research dealing with curriculum developments and the nature of the subject research concerned with the process of rendering these knowledge and research investigating teaching and learning in the classroom. Each of these will be examined in this chapter, in the context the way they have impacted on technology education in Western Europe, particularly France.

The nature of the knowledge used in technology education has been the subject of ongoing debate since the 1960s (see Section 2 of the *Handbook*). The goal is to examine what it is that a technologically literate citizen living in the 21st needs to know in order to understand the environment in which he finds himself. This task is based on the philosophy and history of techniques whilst also attempting to create a specific epistemology of technological knowledge. By constructing a scholarly epistemology in accordance with set objectives in a specific context, the reference knowledge can be transformed into an educational context. An important aspect within this is the organisation of conceptual progression based on the transmission and acquisition of the relevant knowledge, and impacted by the level of difficulty, depth, complexity, and portability. The definition of a subject area and the teaching curricula associated with it comes from this process of transposing reference knowledge into a scholarly epistemological framework that is socially, culturally, ideologically, and philosophically influenced.

Contrary to widespread belief, deciding what has to be taught, why, and to whom has nothing to do with how it is taught or how effectively it is learned. Rather, the teaching and learning process places the epistemological framework in the context of a classroom, where a teacher gives tasks to pupils who then organise how they will act in order to complete each task. Studying this process thus allows us to consider multiple dimensions: curriculum organisation, teaching methods used, and their effectiveness in enabling and enhancing student learning.

This article presents a few of the elements currently influencing research in technology education in Europe, with a particular focus on France. As such, the chapter focuses more on reviewing the types of questions dominating current research than on presenting the findings. This review of questions is underpinned by a reference list selected to reflect the richness and diversity of recent and current work. The list is, however, far from exhaustive. Rather, it provides a starting point from which the reader can widen his or her field of references.

A.T. Jones and M.J. de Vries (eds.), International Handbook of Research and Development in Technology Education, 31–40.
© 2009 Sense Publishers. All rights reserved.

THE NATURE OF TECHNOLOGICAL KNOWLEDGE FOR TECHNOLOGY EDUCATION AT SCHOOL

The question of what to include in technology education gave rise, and continues to underpin, a number of works – sometimes linked, sometimes specific – which all attempt to organise knowledge to be taught in schools in such a way that all children are enabled to participate effectively as part of a technological culture and understand and interact with their environment.

A longstanding Anglo-Saxon tradition of technological philosophy includes studies based on an epistemology of technological knowledge as specific and separate from other domains, most notably science. In this respect, technology education is required for all children because understanding the technological environment depends on the establishment of a specific area of knowledge – technological knowledge – which is separate from specific fields traditionally defined in the sciences, but which relies instead on both the material and the human sciences (Dakers, 2006). De Vries notes that this knowledge has a specificity to it that springs from its use for producing identifiable knowledge artefacts in traditional academic subject areas (de Vries, 2005a, 2005b, 2006a); that said, the study of these academic disciplines, in translating them into academic curricula, is of no help in facilitating comprehension (de Vries, 2006b). Hence, studying the principle laws of electricity in science does not particularly aid understanding of how electrical equipment operates any more than it assists the person to understand why the equipment in question has the characteristics it does.

Somewhat surprisingly, this Anglo-Saxon line of research into the philosophy of technology does not seem to draw upon (and vice versa) equally vast investigations in the same field in Europe. Marc de Vries is one of the few authors to establish relationships and to show closely linked points, as well as differences in approach. In France, works on the mode of existence for technical objects (Simondon, 1989) serve as a reference for thinking about the place and role of technology education in schools. Through a series of works, Deforge goes on to develop this multi-reference approach for technological knowledge. Firstly, with regard to the viewpoint we can have on technical objects (Deforge, 1970, 1985, 1988; Deforge & Chancerel, 1985), then in an attempt to form a basis for introducing specific concepts for work and products (Deforge, 1990, 1993, 1995, 1997). For this author, out of all the points of view we can adopt in order to use technological aids, there are four which are particularly interesting and that are given priority: objects as products of a production system, objects in a system of consumption, machines in a system of usage, and objects as entities in themselves within a system of objects.

If we develop this line of thought, we note that the existence of objects does not take into account how they exist (Simondon, 1989). The mere use or manipulation of an object says nothing about how or why it exists. Hence, there are limits to forming an approach based upon a history of technology stemming from a history of objects (Perrin, 1991a). Again, we see that the result says nothing (or little) about the production process or the way of organising and doing things (Perrin, 1988). In contrast, Deforge (1995) demonstrates the importance of the processes for the production of objects. Others also underline this importance, be it to discuss the history of techniques (Daumas, 1991; Daumas, Guéron, Herléa, & Moïse, 1978; Daumas, Perrot, Ache, & Audin, 1979; Gille, 1992; Jacomy, 1997), technological evolution (Lattès, 1990; Perrin, 1991b), relationships between product and producer from an ethnological point of view (Leroi-Gourhan, 1973; Sigault, 1994), or sociological perspectives (Baudrillard, 1990, 1994; Veltz, 2000).

Using artefact production processes as a starting point in this way hints at a given series of facts or phenomena, in accordance with a certain setup and leading to a clear result. References for technology education become clear with regard to the action based on three factors: conscience of minimal satisfaction, awareness of the possibility for greater satisfaction, and hopes for improvement based on the possible effectiveness of a particular action. Bringing together the means to achieve such an end hinges on this description of a process, in which the actors' practices are set and where these practices are always social.

In this broadening of a social definition for technology education references, one should take note of works published on curricula, including those developed by the mixed research unit affiliated to l'École Normale Supérieure de Cachan (Lebeaume, 2000; Martinand, 1995), which have clearly influenced the definition of technology for the French national curriculum. Their main area of interest concerns the content of teaching syllabuses, and hence technological references. The terms 'social reference practices' (Martinand, 1986) and 'socio-technical reference practices' (Lebeaume, 2000) were introduced by this team to link technology education to human activities and hence to the social setups which organise such activities. This referential approach is based on working objectives, material and intellectual instruments, problems, knowledge, attitudes, and social roles. For example, "small scale building of something in class will allow one to relate to a similar situation in a business context and thus to consider the cultural vocation of such school activities" (Martinand, 1986, p. 138).

More recently, these authors defined some of the attributes of the concept of socio-technical reference practices (Lebeaume & Martinand, 1998), allowing one to question academic practices and pick out the different social practices that give meaning to these scholarly activities. Specifically, they define a relationship of authenticity (separate from identity) and coherence. This provides a way of taking stock of problem areas in technological disciplines. In spite of all this, such a proposal causes problems. Reference to practices evokes the question of how they are modelled in order to implement them in schooling; these are not methods which serve as a reference point for teaching in themselves, but constructions based on such practices. In her thesis from the ENS in Cachan, Paindorge (2005) uses as a starting point the link between task and target, an activity suggested by Lebeaume (2000) to help the model to evolve. This reorientation of referring to practices shows that these are not methods which serve as teaching references, but as knowledge of such practices: "the schooling setup is linked to a specific way of teaching, and hence to a transposition of knowledge, knowledge relating to practical methods" (Johsua, 1996, p. 67). Along with innate knowledge (Chevallard & Johsua, 1991), there is also said to be expertise. This distinction allows one to more clearly define the places where knowledge is institutionalised. Indeed, innate or academic knowledge is produced by identified (or identifiable) scientific communities, whereas expert knowledge aims to take into account other sources of knowledge production that do not collate with these institutional production criteria. From this perspective, knowledge (of an academic or expert nature) is put into textual format using knowledge production sources (Ginestié, 2005c).

The identification of knowledge suitable for forming the basis of technological education remains an open-ended problem. Can we find an epistemology of scholarly knowledge for technology education that is not based on clearly identified epistemologies? The ongoing debates about relationships with sciences – which tend to lead technology back to applied sciences, or even the application of sciences – or its position as a form of initiation and job-finding – which is often insisted on for pupils encountering difficulties at school – or as a group of activities to heighten awareness – which must allow one to associate thought and action – demonstrate the complexity and fragility of the position of such an academic subject area.

ACADEMIC KNOWLEDGE TAUGHT IN TECHNOLOGY EDUCATION

Technology education curricula are the result of consensual constructions that, in the absence of clearly defined epistemological references, reflect the uncertainty linked to their creation. As a result, the teaching of technology differs vastly from one country to the next. This diversity can be defined according to four different approaches: the production of artefacts, the study of existing technological artefacts, the study of the job market and world of work, and the study of how and why technological artefacts are developed and used.

When discussing the production of artefacts, a distinction can be made between objects used domestically or decoratively as part of a tradition of working at home or manually on the one hand, and more industrialised work based on the planning and putting into practice of industrial processes on the other (Crindal, 2001; Manneux, 2004). Pupils' activities are mainly centred on building things, with the views of the end-user often not considered; obtaining an acceptable product in accordance with aesthetic or even artistic criteria constitutes the main aim of this teaching.

The study of existing technological artefacts includes the use of familiar tools right through to using 'high-tech' objects, for example, micro-technologies or information technologies. Here, the aim is to familiarise students with their everyday environment and enhance their understandings of the world of technical appliances (Leroux, 1995; Levin & Mioduser, 1996). Such a setup often hinges on an ambiguity between science and technology activities, with technology often viewed as an applied science, or even the application of sciences (Fourez, 1994; Talis & Ginestié, 2003).

Studying the world of work forms part of a process designed to allow pupils to get to grips with the working world and their own possible career paths. The study of the evolution of the job market is generally based on the development of general skills such as cooperation, teamwork, flexibility, adaptability, and innovative thinking. Activities offered in the classroom thus tend to emphasise the development of behaviour linked to these general skills and often simulate socio-professional organisations (Crindal & Larcher, 2007; Gonnin-Bolo, 2006; Vérillon & Ouvrier-Bonnaz, 2007).

Studying how and why technological artefacts are developed and used includes an analysis of social interactions and is based on processes related to engineering, the life of products, as well as social sciences. Classroom activities such as analysing processes for the design and development of an object and using semantic tools (taken from the systemic approach, functional analysis, an example might be the economic organisation of businesses) aim to place technological artefacts within a social context that needs to be studied (Andreucci, 2006; Brandt-Pomares, 2008; Chatoney, 2008; Cheneval-Armand & Ginestié, 2008).

The diversity of aims within each of the four approaches reflects the difficulty of designing and organising a stable and socially acceptable technology education framework. It also impacts on the social position of a school subject area and the amount of legitimacy it is accorded. In some ways, each approach (and all those concerned with the subject but which are not cited here) attempts to confirm the social necessity of technology education for all, even when schools do not have the means to implement it in ways that are intended (Ginestié, 2006a). Judging from the facts, education technology is not clearly defined, its references are fluctuating, its aims are not stable and, at the end, its social status is problematic.

The development over the last twenty years of technology education for all reveals another dimension, that is, the link between knowledge as defined by the curriculum, and ways in which such knowledge is transmitted. Again, little research has been done in this area. For example, if one adepts a view that technology teaching has something to do with an understanding of industrial production processes (the subject to be taught), one needs to consider this teaching as something which incorporates industrial production in the way the subject is organised and taught. In the absence of a well-established body of knowledge, the difficulty in defining clear areas of reference and the instability of curricula in which the academic knowledge is organised, technology education research is given particular depth. All that seems to be self-evident in any other academic discipline needs to be questioned and there is still much to be studied. There are also multiple, even different, ways of carrying out the research (Graube, Dyerenfurth, & Theuerkauf, 2003).

Because of these broader issues, technology education has developed and will continue to develop based on a different model than that of a traditional dogmatic pedagogy – the teacher teaches and the students memorise – or a pedagogy used in science teaching – the observation of a phenomenon, offering a hypothesis, and the putting in place an experimental protocol and treatment of data to either confirm or question the hypothesis. Rather, foundations will hinge instead on the link between mastering a particular technological skill and constructing meanings for the said skill: action for contemplation and contemplation for action. Technology teaching will find its niche in teaching skills, emphasised in the production of artefacts (Tauke, Story, & Ostroff, 2004). From this point of view, the link between what the pupil conceptualises and what he/she builds plays an important role. This, for example, is what we notice in France with the structuring role played by the industrial projects method (Ginestié, 2002) in which all the process of putting on the market a new product is formalised and decomposed in successive and definite procedures. Similar examples exist in Germany, Holland, Belgium, Italy, and Spain.

Discussions about technology education thus need to involve the link between an approach taken from scientific education centred around the structuring of observable facts on the one hand, and an approach focused on practical education involving mastery of professional skills and the construction of meaning for these skills on the other. Is it a kind of teaching based on lessons about things, where one studies a technological artefact in order to extract the main scientific principles? Is it a matter of looking at how an object exists and the process which allows this object to exist? Is it about studying the social context in which the artefact exists and came into existence? Numerous pieces of research exist that defend one position or another, often with more ideological than scientific reasoning.

TASKS AND ACTIVITIES FOR TECHNOLOGY EDUCATION

Of course, if we consider that schools serve to help students construct and learn knowledge and values, the matter of the pupils acquiring this knowledge becomes vital: It is not enough to simply put social structures in place for the transmission of knowledge; it must also be assured that pupils learn them properly. Such a process brings with it pressure that is greater than the social demands of sharing knowledge: of education for all, and equal opportunities (extending fields of knowledge, academic broadening, availability to other social groups and cultures) (Bautier & Derycke, 2005; Felouzis, 1990; Ginestié, 2006b; Schwartz, Labbé, & Association nationale des directeurs de missions locales, 2007). Putting teaching structures into place is a necessary but insufficient condition: "They [the teachers] teach but what do they [the pupils] learn?" (Altet, 1997, p. 11). The process which links teaching and learning is a specific one, academically oriented, and drawing on the wider process of knowledge transmission/acquisition. Research into the study of this process is out of the ordinary and little is known about it (Ginestié, 2006c).

The knowledge acquired by an individual is mediated by the objects the person creates in order to establish social relationships, and hence to be able to act in and with the material world. This mediation flows directly between individuals and social groups and the objects used are merely tools to aid mediation (Rabardel, 1995). From this perspective, classification of knowledge according to its kind (knowing how to do or be something) is inoperative; knowledge is first and foremost a means of acting upon and with the environment. In the same way, creating a hierarchy between behaviour, capacities, and skills is of little importance given that it is founded on analysing the result of the action rather than the processes by which the result is produced. In these terms, a learning situation is an interaction between the task to be accomplished, knowledge available in order to do so, and the activity employed by the learner. Development of new knowledge rises from a conflicting situation in which a person cannot complete the set task by simply calling upon the knowledge he/she already has. Not managing to find a socially satisfactory solution creates an imbalance, and the individual seeks for ways to find a solution in his/her social environment. Thus, the individual calls upon new things learnt, or else develops new ways of organising things at his/her disposal in order to find a newly adapted answer which in turn constitutes new knowledge. Stabilising this new knowledge depends upon producing sense and meaning of objects, and of symbols of these objects (Vérillon, 2000).

Studying the teaching-learning process leans towards viewing and analysing classroom situations as an interactive network between pupil, teacher, and knowledge. The situation allows the pupil to face the obstacles that he/she will have to overcome in order to learn new knowledge. Pupil activity is directed through the task set by the teacher, organised to make the obstacles salient. The teacher then helps the pupils in their activity, using teaching materials to aid them in overcoming difficulties they face. In terms of research, it is a question of describing and understanding the conditions which enhance (or hinder) the teaching-learning process. This understanding is based upon a methodology which links task analysis and activity analysis (Ginestié, 2007). This analytical framework allows one to qualify the activity actually undertaken by pupils, characterise the difference between what is expected of pupils and what they actually do, and find out the difficulties they face or the obstacles they overcome.

Definition of the task is an essential element which conditions pupil activity and whether or not anything is learned. The task is the clear expression of teaching logic. It shows at the same time what is involved, the context in which the teaching is placed, what is expected of the pupil, and what he/she has to do in order to accomplish the set task. In this way, it unifies a whole range of values, models, theoretical elements, and knowledge that forms the basis of the taught subject and identifies the teacher within the teaching fraternity. Task construction therefore signifies the putting into place of a teaching curriculum in the specific context of a particular class – and this is what is analysed.

Numerous works analysing gaps between the prescribed and the real show that an analysis of the set task can allow one to characterise the fundamental components, whether epistemological, curricular, didactic, or pedagogical (Ginestié, 2005b). However, understanding the process does not stop at this task analysis, which needs to be supplemented by an analysis of pupil activity. For example, students may first read the instructions associated with a particular task. This analysis will in turn result in action being taken. In a formal way, the initial description of the task allows the pupil to analyse the goal to be reached, the way of reaching it, and the tools at his/her disposal for doing so. From this basic plan, he/she organises his/her actions by planning them in an active strategy and by giving him/herself the means of checking that the action to be taken is taken at the appropriate time, that it allows him/her to move towards achieving the goal, and that the partial result of this action contributes to progress (Andreucci, 2008; Chatoney, 2006). Of course, such an action process which is supposedly expertly mastered by the person defines an expert strategy, that is to say, of one who knows. In many cases, the pupil efficiently puts into place a clear strategy in order to successfully carry out a given task. Like any expert, he/she proceeds in a precise way in accordance with a series of routine procedures which he/she masters with perfection (Mioduser & Kipperman, 2002). In other words, he/she does the job of a pupil by accomplishing ordinary scholarly tasks by using a routine strategy for doing so.

A description such as this does not take into account the acquisition of new knowledge that the pupil has not yet mastered. Bringing new knowledge into play leads to the creation of a new task, one that the pupil cannot solve by using the usual process. In this sense, he/she will have difficulty in using actions which will no longer help him/her to accomplish the set task. Effective teaching materials have to help him/her to identify these problems, get the new resources needed in order to progress, and integrate them into a new problem-solving strategy. Research in this case moves towards recording the information relevant to the pupil's activity through his/her way of describing the task (what he/she does or does not find out, and what he/she knows about the teaching environment and hence the resources available), and the action that is subsequently taken in carrying out the task. By observing the pupil's activity, the researcher is able to describe the pupil's task, the strategy used and how it is planned, the pupil's method of planning what he/she is going to do, the difficulties faced, and how these are dealt with (Ginestié, 2005a).

The crossover between task and activity allows one to highlight pupil activity in relation to the characteristics of the task. Hence, one can develop a better idea of the difficulties faced by the pupil and identify those arising from difficulties inherent to the context (task structure, organisation of study conditions, etc.) and those resulting from obstacles to learning. In this way, analysing study conditions in technology education allows us to identify and characterise differences between what is ordained and what is real, whether at a micro or macro level. It is a question of fitting out the conditions of study in technology education. This arrangement bases on the analysis of the impact of the school organizations on the processes of teaching and learning of this knowledge.

SOME PARTIAL CONCLUSIONS ...

How is research in technology education different from other educational research, whether scientific, mathematical, literary, historical, artistic, physical, or sports- based? The first response is to note that it is different because the reference knowledge is different. In this sense, it is the epistemological boundaries which organise academic subject areas, including technology education. However, the kind of reference knowledge, the different ways of separating these references, and the points of view adopted in order to separate them, constitute many specific factors that are not found in other epistemological fields serving as references for other academic disciplines. Thus, from a scholarly point of view, it is interesting to take note of just how fragile and unstable the existence of 'technology education for all' is. At the same time, we cannot ignore just how regularly in everyday life we are confronted with objects, systems, and setups that require our understanding. Technology education research into references available for such an education sheds light upon this question through, for example, philosophy of techniques, of their history, and the sociology of the setups for whom and due to whom these techniques exist.

The process of academically passing on reference knowledge into taught knowledge is specific to technology education. An important area of research has thus been to try to characterise this uniqueness in relation to other school subject areas. Whether it is a matter of defining curricular organisation or analysing the mechanics of transforming reference knowledge to construct academic knowledge to be taught in the classroom, all the results published to date point to the specificity of technology education and technology education research, as well as ways in which they can enrich other subject areas.

A third area of research focuses on studying classroom situations in order to gain a greater understanding of the teaching-learning process in technology education. Focused on questions regarding the effectiveness of tools used in the classroom, this research tends to adopt research methodologies used in work psychology and leans on the theory of activity. This area of research highlights the advantages to be gained from linking the psychological process of knowledge acquisition to the social process of passing on this knowledge. However, these works – as promising as they are – are still under-developed and very few teams in Europe, and around the world, are heading in this direction.

REFERENCES

Altet, M. (1997). *Les pédagogies de l'apprentissage*. Paris: Presses universitaires de France.

Andreucci, C. (2006). La fabrication d'artefacts comme moyen didactique de conceptualisation de la réalité technique. *Aster, 41*, 153–184.

Andreucci, C. (2008). The structuring role of artefacts in thought development. In J. Ginestié (Ed.), *The cultural transmission of artefacts, skills and knowledge: Eleven studies in technology education* (pp. 21–41). Rotterdam, The Netherlands: Sense.

Baudrillard, J. (1990). *Seduction*. London: Macmillan.

Baudrillard, J. (1994). *Le miroir de la production*. Paris: Le livre de poche.

Bautier, E., & Derycke, M. (2005). *Culture(s) et réflexivité*. Saint-Etienne: Presses Universitaires de Saint-Etienne.

Brandt-Pomares, P. (2008). Searching for information on the internet about the link between task and activity. In J. Ginestié (Ed.), *The cultural transmission of artefacts, skills and knowledge: Eleven studies in technology education* (pp. 173–192). Rotterdam, The Netherlands: Sense.

Chatoney, M. (2006). The evolution of knowledge objects at the primary school: A study of the material concept as it is taught in France. *International Journal of Technology and Design Education, 16*(2), 143.

Chatoney, M. (2008). Elementary school and technological education for all in France—Study on technology in the French educational system and teacher training. In C. Benson (Ed.), *International handbook of primary technology, 10 years on...* . Rotterdam, The Netherlands: Sense.

Cheneval-Armand, H., & Ginestié, J. (2008). Teaching the prevention of risks in professional contexts: Procedure & knowledge. In J. Ginestié (Ed.), *The cultural transmission of artefacts, skills and knowledge: Eleven studies in technology education* (pp. 61–79). Rotterdam, The Netherlands: Sense.

Chevallard, Y., & Johsua, M. A. (1991). *La transcription didactique: Du savoir savant au savoir enseigné* (Rééd. rev. et augm. ed.). Grenoble: La Pensée Sauvage.

Crindal, A. (2001). *Enquête sur les figures de la démarche de projet en technologie*. Cachan: STEF.

Crindal, A., & Larcher, C. (2007). Nouveaux dispositifs d'enseignement au lycée. Nouveaux processus d'apprentissage de l'élève; Nouvelles pratiques de l'enseignant. *Didaskalia, 30*, 39–65.

Dakers, J. (2006). *Defining technological literacy. Towards an epistemological framework*. New York: Palgrave MacMillan.

Daumas, M. (1991). *Le cheval de césar ou le mythe des révolutions techniques*. Paris: Edition des archives contemporaines.

Daumas, M., Guéron, J., Herléa, A., & Moïse, R. (1978). *Histoire générale des techniques* (Vol. 4 – Les techniques de la civilisation industrielle, énergie et matériaux). Paris: Presses universitaires de France.

Daumas, M., Perrot, M., Ache, J.-B., & Audin, M. (1979). *Histoire générale des techniques* (Vol. 5 – Les techniques de la civilisation industrielle, transformation, communication, facteur humain). Paris: Presses universitaires de France.

de Vries, M. (2005a). *80 years of research at the Philips Natuurkundig Laboratorium, 1914–1994*. Amsterdam: Pallas Publications (Amsterdam University Press).

de Vries, M. (2005b). The nature of technological knowledge: Philosophical reflections and educational consequences. *International Journal of Technology and Design Education, 15*(2), 149–154.

de Vries, M. (2006a). Technological knowledge and artifacts: An analytical view. In J. Dakers (Ed.), *Defining technological literacy. Towards an epistemological framework* (pp. 17–30). New York: Palgrave MacMillan.

de Vries, M. (2006b). Two decades of technology education in retrospect. In M. de Vries & E. Mottier (Eds.), *International handbook of technology education. Reviewing the past twenty years* (pp. 3–10). Rotterdam, The Netherlands and Taipei, Taiwan: Sense.

Deforge, Y. (1970). *L'éducation technologique*. Paris: Casterman.

Deforge, Y. (1985). *Technologie et génétique de l'objet industriel*. Paris: Maloine.

Deforge, Y. (1988, 2–7 juillet). *Objets techniques et systèmes de production*. Paper presented at the Université d'été: Technologie et Didactique, Marseille.

Deforge, Y. (1990). *L'œuvre et le produit*. Seyssel: Champ Vallon.

Deforge, Y. (1993). *De l'éducation technologique à la culture technique pour une maitrise sociale de la technique*. Paris: ESF éditeur.

Deforge, Y. (1995). La production d'objets originaux. *Skholê, 3*, 7–12.

Deforge, Y. (1997). *Des techniques à la technologie: Recueil d'articles*. Paris: Ministère de l'éducation nationale.

Deforge, Y., & Chancerel, J.-L. (1985). *Système de production et système d'acquisition*. Lausanne: Éditions du Sentier.

Felouzis, G. (1990). *Filles et garçons au collège. Comportements, dispositions et réussite scolaire en 6ème et 5ème*. Thèse de doctorat, Aix-en-Provence: Université de Provence.

Fourez, G. (1994). *Alphabétisation scientifique et technique. Essai sur les finalités de l'enseignement des sciences*. Bruxelles: de Boeck.

Gille, B. (1992). *Les ingénieurs de la renaissance*. Paris: Ed. du Seuil.

Ginestié, J. (2002). The industrial project method in French industry and in French school. *International Journal of Technology and Design Education, 12*(2), 99–122.

Ginestié, J. (2005a). Filles ou garçons, seuls ou à deux, quelle influence sur les activités de production en éducation technologique. *Aster, 41*, 153–179.

Ginestié, J. (2005b). La production attendue n'est peut-être pas celle escomptée. In P. Vérillon, J. Ginestié, B. Hostein, J. Lebeaume & P. Leroux (Eds.), *Produire en technologie à l'école et au collège* (pp. 333–346). Lyon: INRP.

Ginestié, J. (2005c). Résolutions de problèmes en éducation technologique. *Éducation technologique, 28*, 23–34.

Ginestié, J. (2006a). Adéquation emploi, qualification, formation, éducation: Une approche du développement durable. In J. S. Bekale Nze, J. Ginestié, B. Hostein, & C. Mouity (Eds.), *Colloque international: Éducation technologique, formation professionnelle et développement durable* (pp. 193–201). Libreville: Éditions de l'ENSET.

Ginestié, J. (2006b). De l'exigence de qualification à l'exigence d'éducation. In J. S. Bekale Nze, J. Ginestié, B. Hostein, & C. Mouity (Eds.), *Colloque international: Éducation technologique, formation professionnelle et développement durable* (pp. 14–17). Libreville: Éditions de l'ENSET.

Ginestié, J. (2006c). La formation des enseignants de l'enseignement technologique en regard des organisations curriculaires: Aspects problématiques et questions de recherche. In J. Ginestié (Ed.), *Formation professionnelle des enseignants, au-delà des apparences, quelles différences? Une étude internationale sur la formation des enseignants d'éducation technologique* (pp. 9–20). Santiago, Chile: Éditions Los Salesianos.

Ginestié, J. (2007). Konzepte für eine technologische bildung in frankreich (C. Vitale, Trans. from French). In W. Theuerkauf & E. Hartmann (Eds.), *Allgemeine technologie und technische bildung* (pp. 27–55). Braunschweig: TUB Editions.

Gonnin-Bolo, A. (2006). *Entreprises et écoles. Une rencontre de professionnels*. Lyon: INRP.

Graube, G., Dyrenfurth, M., & Theuerkauf, W. (2003). *Technology education international concepts and perspectives*. Frankfurt am Main: Peter Lang.

Jacomy, B. (1997). *Diffusion des sciences et des techniques; recherches en muséologie technique: Conceptions et réalisations muséales dans le domaine des techniques*. Note de synthèse pour l'habilitation à diriger des recherches. Paris: Université Paris Sud.

Johsua, S. (1996). Le concept de transposition didactique n'est-il propre qu'aux mathématiques? In C. Raisky & M. Caillot (Eds.), *Au-delà des didactiques, le didactique; Débats autour de concepts fédérateur* (pp. 57–82). Bruxelles: De Boeck.

Lattès, R. (1990). *Le risque et la fortune: L'aventure de l'innovation*. Paris: J.-C. Lattès.

Lebeaume, J. (2000). *L'éducation technologique. Histoire et méthodes*. Paris: ESF.

Leroi-Gourhan, A. (1973). *Évolution et techniques: Milieu et techniques*. Paris: Albin Michel.

Leroux, P. (1995). *Conception et réalisation d'un système coopératif d'apprentissage. Etude d'une double coopération maitre/ordinateur et ordinateur/groupe d'apprenants*. Thèse de doctorat. Paris: Université Pierre et Marie Curie.

Levin, I., & Mioduser, D. (1996). A multiple constructs framework for teaching control concepts. *IEEE Transactions of Education, 39*(4), 488–496.

Manneux, G. (2004). *Caractérisation des situations de production en technologie au collège*. Cachan: ENS.

Martinand, J.-L. (1986). *Connaître et transformer la matière: Des objectifs pour l'initiation aux sciences et techniques*. Berne: Peter Lang.

Martinand, J.-L. (1995). Rudiments d'épistémologie appliquée pour une discipline nouvelle: La technologie. In M. Develay (Ed.), *Les savoirs scolaires et didactique des disciplines*. Paris: éditions ESF.

Mioduser, D., & Kipperman, D. (2002). Evaluation/modification cycles in junior high students' technological problem solving. *International Journal of Technology and Design Education, 12*(2), 123–138.

Perrin, J. (1988). *Comment naissent les techniques: La production sociale des techniques*. Paris: Publisud.

Perrin, J. (1991a). *Construire une science des techniques*. Limonest: l'Interdisciplinaire.

Perrin, J. (1991b). *Histoire des techniques et compréhension de l'innovation – Séminaire de recherche, Paris les 16 mars, 18 mai, 15 juin, 18 octobre, 14 décembre 1989 et 8 février 1990* (Vol. 1). Paris: Institut national de la recherche agronomique.

Rabardel, P. (1995). *Les hommes et les technologies; Approche cognitive des instruments contemporains*. Paris: Armand Colin Éditeurs.

Schwartz, B., Labbé, P., & Association nationale des directeurs de missions locales. (2007). *Rapport sur l'insertion professionnelle et sociale des jeunes [rapport au premier ministre, septembre 1981]*. Rennes: Éd. Apogée.

Sigault, F. (1994). La technologie, une science humaine. In R. Scheps (Ed.), *L'empire des techniques* (pp. 51–62). Paris: éditions du Seuil.

Simondon, G. (1989). *Du mode d'existence des objets techniques* (Réédition ed.). Paris: Aubier.

Talis, V., & Ginestié, J. (2003). Éducation technologique et systèmes automatisés à partir d'une expérience israélienne. *Éducation technologique, 20*, 18–24.

Tauke, B., Story, M. F., & Ostroff, E. (2004). Good, better, best: Using universal design and digital technology to develop great products. *Technology Teacher, 63*(8), 13–31.

Veltz, P. (2000). *Le nouveau monde industriel*. Paris: Gallimard.

Vérillon, P. (2000). Instruments and cognition: Piaget and Vigotsky revisited in search of a learning model for technology education. *The Journal of Technology Studies, 26*(1), 3–10.

Vérillon, P., & Ouvrier-Bonnaz, R. (2007). Le document d'information sur les métiers: Un instrument au service de l'activité en situation d'orientation? *Éducation Permanente, 171*, 73–88.

Jacques Ginestié
Gestepro – UMR ADEF
Aix-Marseille Université
IUFM Aix-Marseille – Université de Provence
Marseille, France

TAPANI KANANOJA

5. TECHNOLOGY EDUCATION IN GENERAL EDUCATION IN FINLAND

INTRODUCTION

Teaching practical skills and technology in general education occurred in central Europe as early as the seventeenth century as 'education for work', providing skills necessary to society. These techniques were first introduced in Finnish schools at the end of the nineteenth century as 'handicraft education', which involved using materials and skills to produce objects and artefacts. At the same time, the student learned how to 'work according to the rules' and developed various skills needed for working life. Such 'handicraft education' combines carefulness and perseverance, with consequential development of the whole personality.

Traditionally, handicrafts have existed in the curriculum in Finnish schools from its inception in 1866. The subject usually concentrates on copying traditional handicraft objects, albeit with some scope for incorporating pupils' own designs. At present, some schools offer effective and up-to-date technology education in the more modern sense of the word.

Changes in schools are usually quite slow; indeed, without specific incentives to accelerate development or reform of the school curriculum, proposed changes may take decades to implement. There are, however, exceptions. For example, the 'civic school' was created in the 1950s when Finland had to rapidly develop and diversify its industry. In addition to general education, schools needed to provide vocational training and plenty of practical education in the curriculum. At that time, the network of vocational institutions was far from comprehensive and there was a need for entrepreneurial skills. The civic school was able to develop these quickly and effectively.

Motivating pupils to keep their options open with regard to pursuing technology in further studies is mostly the task of the comprehensive and upper secondary schools. These schools need to pay particular attention to developing practical-technological education. In addition, education needs to be sufficiently versatile and flexible in order to cope with the changing industrial structure of the country. Modern technology is needed in more and more vocational branches in order to ensure the technological growth of the country. It is also needed at home. If a nation wishes to grow then it must invest in education, plan ahead, and support new technology in the general school curriculum.

While most Finnish homes and citizens use mobile phones and have access to the Internet, few teachers use these in education and it has been estimated that about half of the unemployed are 'computer-illiterate', lacking even basic computing skills. It has been said that we are governed by 'techno bureaucracy', which is trying to generalise the use of new communication gadgets only. This criticism also applies to general education. Medical science is developing machines that can crawl through the veins. Yet after four decades of human space travel, the practical-technological education at schools is often still limited to crocheting and woodwork.

A.T. Jones and M.J. de Vries (eds.), International Handbook of Research and Development in Technology Education, 41–50.
© 2009 Sense Publishers. All rights reserved.

The effort to help individuals understand new technologies and thereby exercise greater control over their environments is one of the aims of technology education. Another important aim, of course, which has recently received a great deal of publicity, is the realisation of equal opportunities.

CHANGING SKILL DEMANDS

The school as an educational institution has always responded to the challenge of society to teach the appreciation of technology. Originally introduced as handicraft education, technology education has been developed and enhanced to take into account societal needs. These needs change rapidly, however, in step with progress in technology. In Finnish schools, the different categories of practical technological education – handicraft, 'sloyd', and 'technical work' – have traditionally represented the disciplines responsible for education for work, techniques, and technology.

Technological education began as handicraft education around the world, with Finland being the first country in which it was accepted as a compulsory school subject (Cygnaeus, 1866). Accordingly, the school should above all lead through work to work. Handicraft at that stage was connected with self-supporting agriculture. Such do-it-yourself skills continue to be part of the Finnish style of life. The school programme offered starting points at many levels, for example, even if the projects were quite vocational and connected with working life, handicrafts should serve in character building and development of the whole personality (e.g., accuracy, carefulness, and persistence in work). Such handicraft education, as outlined by Cygnaeus, formed an early model for many countries (e.g., the Nordic Countries, England, and the United States, the latter through Otto Salomon, the Swede).

In the international educational literature, handicrafts refer to reproduction and imitative making. With the development of technology, handicrafts no longer provided sufficient guidelines for practical technological education. When machining and other such techniques became more generally used in production, the importance of handicraft skills diminished. Production processes were rationalised and standards regulating the production were made. Industrial production and patent systems were generalised. Knowledge about how to produce was separated from production skills; the former developed in the form of engineering skills, the latter as manual making. The general interpretation of handicrafts as 'non-academic' probably emerged from this background, the main function of such works being to satisfy industrial needs. On the other hand, development of technology led to the need to apply science with handicraft skills and education systems began to include 'industrial arts' – officially incorporated in Finland in the 1920s–1970s in the civic school, the task of which was prevocational and vocational and which created the basis for national industrial development and entrepreneurship.

'Technology education', in the latest global evolution of practical technological education, often refers to education emphasising new technology. However, the new title of the subject was first launched in the world, as far as we know, by Uno Cygnaeus when planning Finnish teacher training programmes (Cygnaeus, 1861).

The formerly-taught 'work skills' are insufficient in modern working life. However, when we do not know exactly what kind of technical demands future industry will place on technology, the challenge must be met by emphasising the important mental processes involved in the development of technology – work tasks will be more and more based on knowledge and theoretical governance of work. The status of creativity will be increased and broad abilities become more important. The new technological world demands more effective, but less academic skills, than school currently focus on. Technological changes also mean that a greater focus is needed on general education than on vocational education for specific roles. It must lead to improved skills in logical, analytical, and abstract thinking, and an ability to use these in practical situations.

Thus, technological development across society demands change within the education system, perhaps especially in 'education for work'. This impacts on both general and vocational education, as well as teachers' pre- and in-service training. The increased mobility in working life also demands workers to be more flexible and to have new attitudes.

Humankind has moved at once closer to a greater abundance of technological artefacts, and to problems caused by technology. The ongoing development of production technology requires professionals, technologists, and engineers, with better and better training in mathematics and science. Education has become a more important factor in this, and also a target for investments. The need for technological education is to *understand the processes*; when you learn to work on materials by hand and machine, you will also better understand the basics of controlling automated machines.

A high standard in examinations is no longer a necessary presupposition for success in the technological world (one needs only to think of Bill Gates or Linus Torvalds). Rather, a key premise in general education might be to give clues about new technology.

Changes in work have brought social changes in the whole of society. Such a technology-centered society or administration can be called 'technocratic'. There can be a danger that some irreplaceable values are sacrificed for technology, for example the intrinsic value of nature, and that technology will become over valued. Some fear that technological determinism might govern individual and societal actions. Decision making about technological options is, however, always based on values. These values depend on knowledge and attitudes, both of which should be fostered through education. Supporting talented pupils in studies equal to their abilities also guarantees a high standard of future engineers and technologists.

Even if the ultimate demands for technology education come mainly from industry, there are other stakeholders. For example, few would argue the importance of an individual living in a technological society and in a home with a range of electrical and electronic gadgets to know how to interact with technology and how to use it effectively, safely, and properly. Thus, the needs for modern technology education are both social and individual – as they were 142 years ago.

Urbanised societies no longer have the needs to justify retaining 'handicrafts' as part of general education. Artefacts are cheap when compared to living standards and teaching how to produce them does not provide meaningful learning; handicraft skills are not needed for survival in the same ways as before. Paradoxically, handicrafts education can even support elitism; artefacts have a demand mainly from the wealthy upper class. Elitism, however, was never the aim when handicrafts education was introduced to schools.

The following section explores other names for technology education, depending on the contexts. For example, 'technological education' is used to define technology more broadly, and tends not to be limited to a new school subject, but rather can be thought to make better connections between different subjects. The roles of terms such as 'practical' and 'technical' are historic compromises; in Finland ongoing development is rather slow.

TECHNOLOGY EDUCATION

The early American philosophy of developing technological education provided a suitable challenge for Finland. Olson (1963) highlighted that handicrafts education, wood as material and as methods the skills based on muscular power, belonged to the 'handicrafts era'. If technology education is supposed to reflect the actual technology of the society, then school subjects must be continuously developed.

Development

After the first initiative by Cygnaeus (1861), technology education has in Finland been proposed many times from 1973 (Kananoja, 1989). This has been motivated on the one hand by the need to internationalise the concepts and on the other hand with the needs to develop education. So far the proposals have not been fully realised.

Cygnaeus (1866) founded 'folk schools', primary schools with 'boys' handicrafts' and 'girls' handicrafts'. The aim was general education; Cygnaeus wrote quite often about his fear that vocational education would take over.

Comprehensive school came into being in Finland in 1970. Civic school and academic junior secondary school were amalgamated as higher comprehensive schools. Allocation for 'education for work' at the primary level was diminished considerably compared to the civic school. The secondary school handicrafts curriculum was the model for the comprehensive school. It already included design, and was no longer based on materials like the civic school. The technical work curriculum in comprehensive schools aimed to realise the ideas of the folk school, civic school, and secondary school. Creativity was emphasised at the lower comprehensive level. Higher comprehensive schooling was, however, still clearly pre-vocational. Diminishing timetable allocations also meant a reduced need for teachers. A new aspect was, however, to satisfy the needs of the whole age group, to create basis for both academic and vocational technological studies. As part of the comprehensive school reform, 'boys' crafts' was renamed 'technical (handi)crafts', the task of which was to emphasise more general education and give 'technological basic skills' (literacy).

In 1975 students' options at higher comprehensive school for technical work were reduced by the Ministry of Education. The subject was renamed 'technical work'; a new curriculum had to be developed and new guidelines for teachers were published. Aspects of work and technology were emphasised a little bit more than before. A guide for electronics was published in 1979. Computers were introduced in practical subjects from 1981. National subject guides, published between 1977 and 1988, supported subject development.

In 1991 the education administration system was reformed, the (old) National Board of General Education becoming a (new) National Board of Education. Earlier that year an expert group for Technology Education had been founded by the (old) National Board. Its task was to develop technology education in light of international guidelines (see Figure 1). The group was founded because it seemed that in the National Board there would soon be some changes in the personnel. However, the 'Committee for Period Allocation' (new National Board, 1992) stated, with total misunderstanding, that changing the subject title 'handicrafts' to 'technology' would result in an approach considered to be too cognitive, and the new title was not adopted! In addition, whether the main ideological background should be underpinned by science or a developed handicrafts curriculum was hotly contested. The proposal naturally was a basic initiative, and continued development has been proposed several times but so far with no full result.

Starting points for Figure 1 were global models. The ellipses describe deeper collaboration between some subjects. In 1991, textile work representatives had especially strong interest in working together with the Arts. However, in the international models textile work was generally more closely aligned with Home Economics. The basic message about the connections and the capacity of technology education as a central subject is still valid in 2008.

The (new) National Board published the *Basics for comprehensive school curriculum* in 1994 (POPS-94). It represented unifying general guidance. However, its three-page guide for handicrafts education was unfortunately written by a group with no expertise in technical work education or research. The guide emphasised equal opportunities; the common, gender-free aspects of the subject; and the possibility to emphasise teaching in different sub areas. The guide could be said to represent 'developed handicrafts education' – with certain reservations. 'Hand skills' were mentioned twice, as was 'modern technology'. Without elaborating on teaching content, the guide is incomplete and also allows for very old-fashioned interpretations. In addition, the long term 'new' aims from 1970, which had not yet been properly realised at the comprehensive school level – pupil-centered teaching, creativity, adopting problem solving approaches, diminishing the influence of the gender, and integration into broader units – were not handled at all. The *Basics for comprehensive school curriculum* returned to the old subject term 'handicrafts education' instead of technical work and textile work. In practice, this could mean returning to the historic basic philosophy of the subject for taking care mostly of personality development and rejecting the development made in technical work during the previous 25 years.

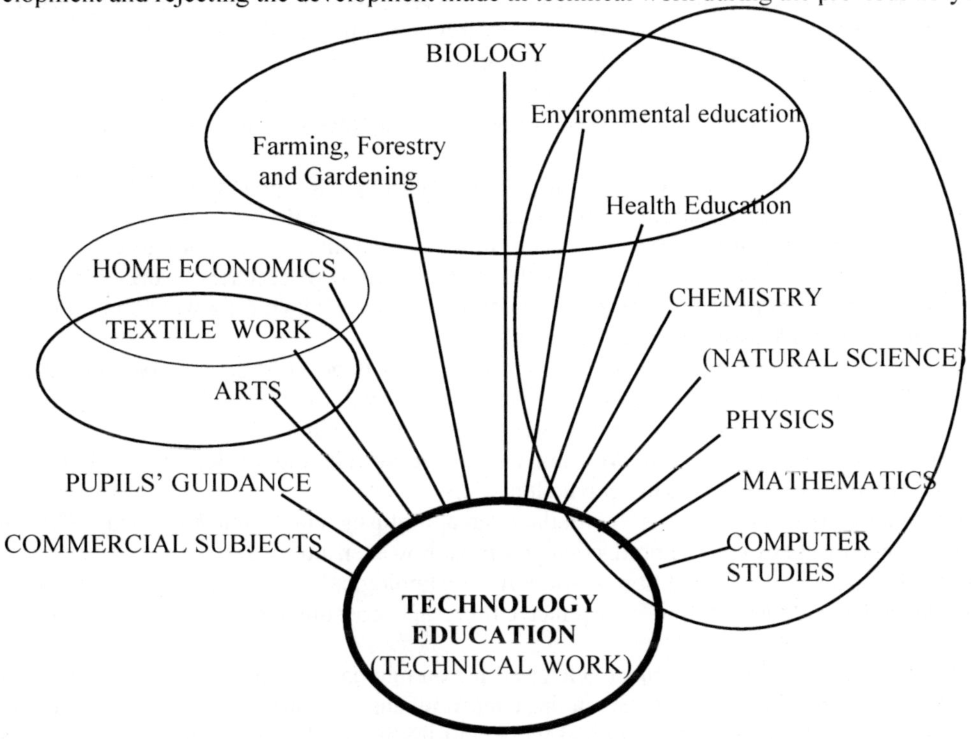

Figure 1. Relationships between technology education and other subjects (Muistio, 1991).

In the *Basics for comprehensive school curriculum*, technology as a concept is represented as one of the general starting points and aims of education as well as an application in many subjects. This is particularly relevant for laboratory courses in science, the teachers of which do not necessarily have the interest or skills to advance practically-oriented technology. Even if the general part of the text is very positive toward technology as a basis for all education, technology is not so much mentioned in handicraft education. This is reflected in the problematic contradiction in understanding the term 'technology' in the memo of the Period Allocation

Committee (Muistio, 1992) and in the *Basics for comprehensive school curriculum* (POPS-94, 1994). There are also conceptual contradictions between the representatives of handicrafts education. For example, textile teachers tend to interpret technology as tools, methods of work, techniques, or machine technology.

Thus, the official terminology continued to support old fashioned approaches. Many individual teachers and schools have, however, been following the international development of the subject area and eagerly adopted the contents and aims of a more-broadly defined technology education. Teacher training in technical work is also no longer limited only to handicraft training as a result of increased research and international influences.

The Present Situation

The writer has recently had the opportunity to discuss the subject development with teachers and teacher trainers, and to check the curricula in Helsinki schools. Both at the lower and higher comprehensive schools, the standard of teaching varies considerably. In both, very good and thoroughly thought about study units, projects, and topics can be found. There is, however, no real coordination of education because of the distributed nature of the education administration, and 'stop-gap' solutions are common, for example, reproduction tasks, particularly at the lower comprehensive school. There is also a lack of appropriate textbooks and near total absence of in-service training.

Materials, wood, metal, textile, machines, and electricity still generally form the basis of handicraft curricula. Some schools have begun to systematically use the revitalised term 'handicrafts', even if it is still divided according to traditional materials. Added to this is the way curriculum content is often described in the form of students' or teachers' work, for example, 'familiarising with, constructing, teaching' in contrast to criteria concerning what should be learnt (which would help guide assessment and evaluation).

The content of technology offered as an optional subject is in many schools based on students' choices – what they want to produce for themselves. This naturally secures motivation and increases the number of students selecting the option, which also will guarantees a safe number of teaching periods for the teacher. This is, however, not the procedure that should define teaching content, which should rather be attached to the curriculum aims.

'Teaching how to produce artefacts' is still the main aim of handicrafts education. This refers either to handicraft-like or industrial-type production. There is, however, no longer such a strong need for this type of learning in Finland. Internationally, technological education has moved towards emphasising greater understanding of artefact culture and consumer needs, rather than teaching production skills as such.

Whilst integration within the subject is aspired for on one hand, connections with other subjects are also sought. Internal integration, which in this context means structuring technical work education in collaboration or closeness with technical work and textile work, has begun in schools. There have been many approaches to foster collaboration between technical work and textile work, for example, the arts and crafts spring exhibitions. According to *Basics for comprehensive school curriculum* (POPS-94, 1994) and the development of technology education internationally, it seems that ongoing integration should be pursued. This will, however, require some issues to be addressed. For example, the definition of the contents of textile work is in most curricula more detailed than of technical work. Textile work education has only one material area to take care of; the traditional scope of technical work is about five times greater. Thus, the aims and practice of teaching/learning will be implemented quite differently if the same number of periods is allocated to each. The need to further define the curricular needs and sub areas in the technology education curriculum seems obvious.

In some schools, a collaboration between technical work and home economics has been implemented. Teaching electronics and computers as part of technical work has been in place for a long time – electronics for 40 years, computers about 25 years. Computers, especially in lower comprehensive technical work, are, however, quite rare.

Entrepreneurship education has recently received much publicity in Finland. In technical work guides it was mentioned as far back as 1977. Surprisingly, it is not so much as mentioned in technical work curricula today. In lower comprehensive schools it should also be interpreted more as 'enterprising activity' or 'primary acquaintances of schools and enterprises'. At the higher comprehensive level, some productive projects could be possible.

After completing higher comprehensive schooling, students select whether to attend vocational institutions or upper secondary (academic) schools. Motivating both groups should be considered in comprehensive school curricula and organisation. However, the vocational aspect has been diminished because of the lack of timetabled periods. Information about continuing studies in metals, wood, machines, electricity, and electronics should of course still be given.

DEVELOPMENT NEEDS

Technology should be seen as having a cultural dimension. In addition to writing, reading, and mathematical skills, skills such as computer skills have come to be regarded more and more as part of our culture. Technology should also be seen as influencing the whole of human culture. Everybody should have the opportunity to undertake a 'basic course in technology' at school. Its content should be to understand – in general – methods and products of technology as well as its impacts and potential impacts. Teaching should motivate students and meet their needs. Technology education should especially be organised for those who will not go on to pursue a career in technology, but who will nonetheless go on in life to make decisions connected with technological choices.

Children's images about technology are based on artefacts which they come across in their everyday lives, including during play. Teaching should thus focus on and analyse everyday artefacts, including toys. Important, too, is that teaching does not only aim at reproduction: "We can reflect the spirit of technology, the attitude and application of search and research, the inventiveness, the creativity, the experimenting and developing, the problem solving, all of which are a part of, and fundamental to, an advancing technology" (Olson, 1963, p. 89). For teaching new technology it is no longer sufficient for the teacher to have 'master' skills such as in the handicrafts era. At that time, teacher training aimed at mechanical activity, which only has limited educational value.

Development of technology is of course continuing, and technological education should be accessible to everybody. The increased and increasing penetration of technology in more and more areas of human life means increased needs for abstract thinking and rationality. Essential in education in the future is, for example, connections between practical applications and using a problem solving approach.

'Technological literacy' is needed in order to survive in the modern society, to understand and use technology. It can be called 'a survival skill'. Valuing computers more than technology in general, and 'computer literacy' beyond technological literacy has been criticised (Dyrenfurth, 1984; Otala, 1986). Rather, technological literacy includes computer skills. If technology means 'making', technological literacy is the skill to use technology appropriately at work and at home.

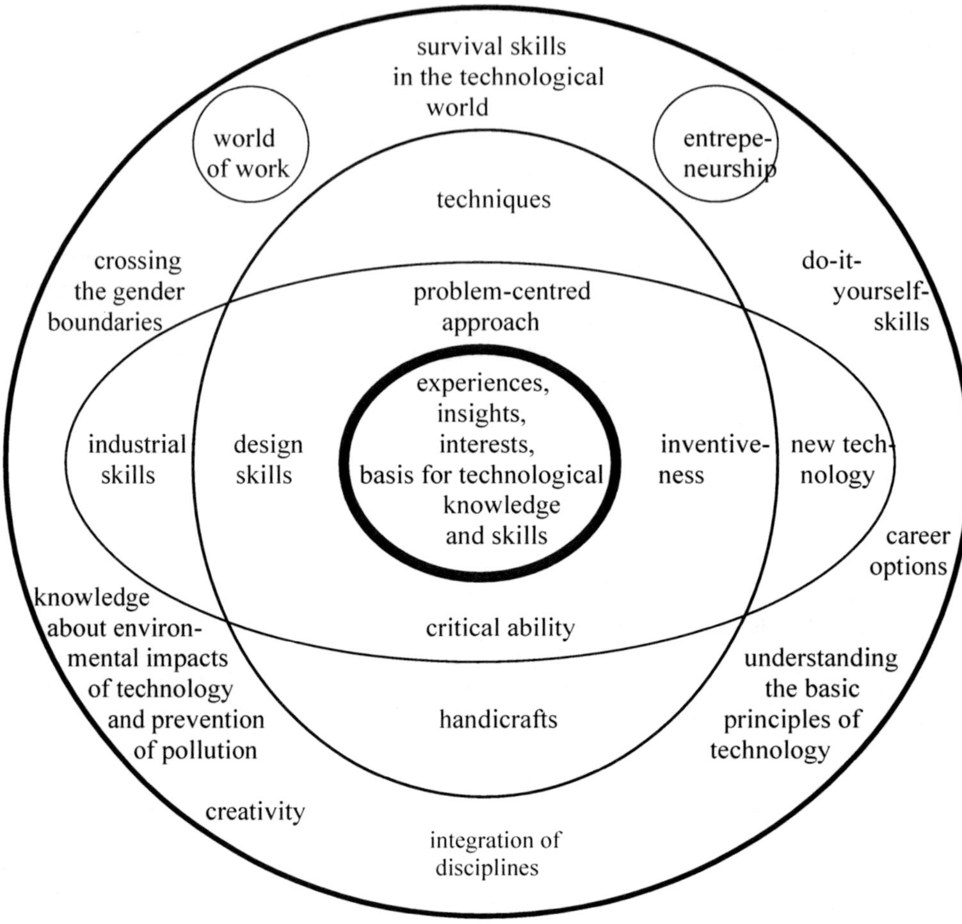

Figure 2. Content of technology education.

Structural changes in industry have mobilised the need to advance technological education internationally. Pioneering work has been done in France, the US, and the UK from the 1960s onwards. From the first time technology was mentioned in an educational context (Cygnaeus, 1861), it has been involved with structural changes in industry. The new title, technology education, has spread quickly through the 1980s and 1990s. It has been adopted in countries such as England, Italy, The Netherlands, Belgium, Australia, and New Zealand. In Sweden the subject is called 'Teknik', in Germany 'Technik' as a part of 'Arbeitslehre', education for work. The common underpinning philosophy has been preparation for survival and thriving in the changing environment, which has become more complex because of technology. It has brought to schools:

- A move from material or skill-based education to a more general education in order to guide students at their own level to understand technology;
- new technologies added to old material-based approaches (e.g., wood, metal, machines, electricity, textiles); and
- the placing of design in students hands instead of and added to the ready made drawings.

Figure 2 summarises the content of technology education for technological literacy. The core of the figure describes the teaching / learning activity, the elements of which should be materialised in every lesson. Most important is to give positive experiences in technology. These bring motivation and enthusiasm for learning, and good results. The balance between knowledge and skills will naturally become more central with increasing years of studies. The inner middle circle describes teaching/learning aims such as they are usually written in the curriculum language. The outer middle circle describes the different development stages of technical work education – from handicrafts, industrial skills, and techniques, to new technologies – which all affect education at present. The emphasis moves from the beginning of the list towards the end. The outer circle points out general background factors for technological education, which influence teaching but are not operational during every lesson. When writing the curriculum and planning education, one has to consider how these factors will impact on learning. Naturally, the basic principles of technological education also include, for example, basic concepts, processes, and systems of technology.

SUMMARY

At the beginning of the new Millennium there are plenty of challenges for the development of education. One is the modernisation of practical technological education in general education in schools. In Finland, this is called handicraft education.

At present the whole spectrum from the 143 year history of handicrafts education is reflected in Finnish schools. Whilst the best schools and teachers have developed teaching in line with current international trends emphasising technological literacy, the majority still teach mostly according to old models representing old technology.

Handicrafts education, which has the main responsibility for technological education, has no national systematic guidance or authorised curriculum. The result is that the standard is varied and uneven. In higher comprehensive schools the situation tends to be better because of the specialised subject teacher system.

At every school stage the subject contents should move from specific reproduction techniques to understanding new technologies and technological literacy. Teaching can be different at different school stages. 'Handicraft technology' can still be defended when it is appropriately assessed. Technology education should also be extended to upper secondary schools.

An indisputable fact remains that the main task of practical technological education in general education continues to be education for work. Also indisputable is that content of all school subjects must be updated when the respective disciplines develop and change.

Education must evolve to reflect the changing needs of society. Every production culture – whether handicraft, industrial, or automatic – seems to prioritise education for the world of work. School education must, however, emphasise content relevant to responsible citizenship as well as for direct employment needs. Internationalisation increases the need for analysing global innovations in education in order to develop a common terminology, which in turn forms the basis for communication.

Whereas handicrafts were once the foundation of production and artefact cultures, it has been replaced by electronics, computer technology, control technology, communication technology, etc. The term 'handicrafts' has during the last 30 years also proved to be powerless at breaking gender boundaries. In addition, innovations for 'technical work' implemented in the 1980s and 1990s are tumbling down under new education emphases. In Finland, any update in the contents of practical technological education needs to coincide with a renaming of the subject as 'technology education'.

ACKNOWLEDGEMENTS

This chapter is drawn from earlier work: Kananoja, T. Alamäki, A., Kantola, J., Lehtonen, M., Parikka, M., & Tiusanen, T. (2000). *Technology education in general education in Finland.* Paper presented at the Finnish Association for research in Technology Education, Helsinki. Available from www.teknologiakasvatus.fi

REFERENCES

Cygnaeus, U. (1861). *A letter.*

Cygnaeus, U. (1866). *The Finnish schools legislation.*

Dyrenfurth, M. J. (1984). *Literacy for a technological world.* Columbus, OH: The National Centre for Research in Vocational Education, Ohio State University.

Helsingin kaupungin peruskoulujen opetussuunnitelmat. (1999). (The curricula of the Helsinki comprehensive schools.) Available as a CD-Rom.

Kananoja, T. (1989). *Työ, taito ja teknologia: Yleissivistävän koulun toiminnallisuuteen ja työhön kasvattamisesta. Turun yliopiston julkaisuja C72. Väitöskirja. (Ks. myös tämän lähteet.)* (Work, skill and technology: About activity education and education for work in general education.) Unpublished doctoral disssertation. Turku University

Muistio. (1991). *Teknologianopetuksen suunnitteluryhmän. Kouluhallitus.* (Memo on Technology Education.)

Muistio. (1992). Tuntijakotyöryhmän. 9. Opetusministeriö. (Memo on Period Allocation.)

Olson, D. W. (1963). *Industrial arts and technology.* Englewood Cliffs, NJ: Prentice Hall.

Otala, M. (1986). Tietotekniikka muuttaa maailmaamme. Teoksessa: Pöyry, S. (Toim.) *Parempi tulevaisuus—haaste tekniikalle* (pp. 52–58). STS 90 vuotta. Jyväskylä: Suomen Teknillinen Seura.

POPS-opas 13 a, 13b, 13d. Tekninen käsityö. 1977, 1979, 1988. Kouluhallitus. (Guides for technical work. Comprehensive school.)

POPS-94. (1994). *Peruskoulun opetussuunnitelman perusteet.* Opetushallitus. (Basics for comprehensive school curriculum.)

Tapani Kananoja
Retired

WILLIAM E. DUGGER, JR.

6. DEVELOPMENTS IN TECHNOLOGY EDUCATION IN THE UNITED STATES OF AMERICA

INTRODUCTION

Technology education has been a school subject in the United States since the 1980s. Prior to that, it was taught as industrial arts education. Curriculum development has been accomplished primarily at the local and state levels since the Constitution of the Unites States of America (U.S.) mandates that education shall be the responsibility of the state and locality. However, more recently there have been some efforts to research and develop curriculum at the national level.

This chapter will briefly discuss how the U.S. transitioned from industrial arts to technology education before focusing research and development that has taken place in technology education in the United States over the past two and one-half decades. Work accomplished by the International Technology Education Association (ITEA) and its Technology for All Americans Project (TfAAP) to develop content standards for the study of technology in the U.S. (a basic ingredient of curriculum) will be presented, as will recent research findings on the status of technology education. A major effort currently underway to develop curriculum at the state and national levels will also be discussed.

TRANSITIONING FROM INDUSTRIAL ARTS TO TECHNOLOGY EDUCATION

In the United States, industrial arts education evolved from manual arts education in the late 1800s and early 1900s. Charles Richards was credited with coining the term 'industrial arts' in 1904 and the name served to represent the discipline until the latter part of the twentieth century.

The term 'technology' was first used in literature in the U.S. when the American Industrial Arts Association (AIAA) titled the theme of its first conference in 1947, *A Curriculum to Reflect Technology*. The term was subsequently referred to numerous times in the 1970s and 1980s by curriculum development projects funded by the federal government as part of the Elementary and Secondary Education Act. In 1985, the AIAA changed its name to the International Technology Education Association (ITEA). Subsequently, most states changed the name of their state educational associations from industrial arts to technology education in the late 1980s and early 1990s.

WHAT IS CURRICULUM?

Curriculum is viewed in the U.S. as a written plan that defines how content is arranged, ordered, and emphasised (ITEA, 2005b). It provides the means by which the teacher and students interact. A curriculum is the plan for delivery of content day-by-day in the laboratory-classroom, which engages students in learning. The curriculum allows for flexibility and freedom so that individual

A.T. Jones and M.J. de Vries (eds.), International Handbook of Research and Development in Technology Education, 51–64.
© 2009 Sense Publishers. All rights reserved.

teachers can adapt to student needs, and ensures that the content is based on the appropriate standards.

In the U.S. the content for the study of technology is defined as what every student should know and be able to do in order to be technologically literate. This is provided in *Standards for Technological Literacy: Content for the Study of Technology* (*STL*) (ITEA, 2000/2002/2007), which will be discussed in detail later in this chapter.

Many educators view curriculum as the 'glue' that bonds together content (what is to be taught), instruction and methodology (how the content is to be taught), and student assessment (how well the content is being learned). In other words, the recipe for teaching includes content as the essential ingredient, instruction and methodology as the process of mixing and combining the content ingredients, and student assessment to allow for adjustments during the learning process or in preparation for the next 'batch'.

DEVELOPMENT OF *STANDARDS FOR TECHNOLOGICAL LITERACY (STL)* AND *ADVANCING EXCELLENCE IN TECHNOLOGICAL LITERACY (AETL)*

In 1994, ITEA created the Technology for All Americans Project (TfAAP). The overarching goal was to increase the technological literacy of all students in compulsory education. Funding was provided by the National Science Foundation (NSF) and the National Aeronautics and Space Administration (NASA). Funding for TfAAP ended in October 2005. The two major publications produced by the project were:

- *Standards for Technological Literacy: Content for the Study of Technology (STL)* (ITEA, 2000/2002/2007); and
- *Advancing Excellence in Technological Literacy: Student Assessment, Professional Development, and Programme Standards (AETL)* (ITEA, 2003).

Other publications were also created by TfAAP to support STL and AETL and are discussed later in this chapter).

Creating a vision for what Standards for Technological Literacy *should be*

In the first funded phase of TfAAP, from 1994 to 1996, ITEA wanted to develop a document that would discuss the power and promise of technology in everyday lives. In addition, it wanted to establish a universal need of technological literacy for all people. This was very important since the study of technology in schools in the U.S. was a relatively new educational effort. ITEA did this in the first phase of the project through the publication of *Technology for All Americans: A Rationale and Structure for the Study of Technology* (*R&S*) (ITEA, 1996). This document provided a structure for what the content in the study of technology could be in the future.

In retrospect, the development of *R&S* was a very valuable tool in grounding the profession in what every student should know and be able to do in order to be technologically literate. The *R&S* document was prepared through assistance from project staff and a group of writing consultants made up from a 25-member National Commission for Technology Education. In developing the various drafts of the document, hundreds of practitioners from technology, engineering, science, mathematics, and other areas served as reviewers of the material. The document went through an extensive revision in 2006 and was retitled *Technological Literacy for All: A Rationale and Structure for the Study of Technology* (ITEA, 2006).

Standards for Technological Literacy: Content for the Study of Technology (STL)

From 1996 to 2000 (Phase 2 of the project), *Standards for Technological Literacy: Content for the Study of Technology (STL)* was developed, reviewed, published, and disseminated. *STL* set forth the vision that all students can and should become technologically literate. Four groups advised and provided input to TfAAP during the development of *STL*—(1) an Advisory Group, (2) a Standards Team, (3) a committee of the National Research Council of the National Academy of Sciences, and (4) a focus group from the National Academy of Engineering. The Advisory Group advised ITEA's TfAAP on best practice for standards development and determined ways for the study of technology to be integrated within the total school curriculum. It was formed by key representatives of the National Council of Teachers of Mathematics (NCTM), the National Science Teachers Association (NSTA), the American Association for the Advancement of Science (AAAS) Project 2061, the National Research Council (NRC), the National Academy of Engineering (NAE), ITEA, and the Foundation for Technology Education. They met semi-annually to provide specific advice on the development of the standards and how technology education could be integrated with other fields of study, especially science and mathematics.

The Standards Team proposed, evaluated, and recommended the content of the standards. The 27-member team consisted of three subteams with nine people each (one team for Grades K–5, one team for Grades 6–8, and one team for Grades 9–12) to provide input to TfAAP staff responsible for the writing, generating, and consensus-building process of the standards. The Standards Team was made up of classroom teachers, supervisors, and teacher educators from technology education, as well as elementary administrators and representatives from science, mathematics, and engineering.

In 1999 and 2000, the NRC and a special focus group of engineers from the NAE carried out a formal review of *STL*. In mid December 1999, the NRC committee issued a final report stating that ITEA/TfAAP had "successfully completed the review process established by the NRC". In early 2000, the NAE committee issued a public statement in support of *STL*. Additionally, the project received funding from the Technical Foundation of America in the development of three standards-related implementation publications for elementary, middle, and high schools.

Overview of STL

The *STL* document begins with a preface that sets the stage for the publication. Chapter 1 provides a broad perspective on preparing students for a technological world. Chapter 2 contains an overview of the features of *STL* as well a guide to its format. Chapter 2 also provides a section explaining the primary users of the standards and recommendations for using the standards for curriculum development. Lastly, Chapter 2 lists administrator guidelines for resources based on *STL*. Chapters 3 through 7 elaborate on each of the five major categories under which the standards were developed (see below). Chapter 8 is a call to action regarding how ITEA can acquire help implementing *STL* from others both within and outside the profession. The document also has an appendix, which includes the history of the project, a compendium that provides a quick overview of the standards and related benchmarks, and an articulated curriculum example for Grades K–12; references; acknowledgements; a glossary; and an index.

Features of STL

STL represents the collective view of hundreds of people regarding what should be the content for the study of technology in Grades K–12. In order to be as broadly valuable as possible, *STL* was created with the following basic features:

- It offers a common set of expectations for what students in technology should learn;
- it offers specific details about what every student should learn about technology.
- it is developmentally appropriate for students;
- it provides a basis for developing meaningful, relevant, and articulated curricula at the local and state/provincial levels; and
- it promotes content connections with other fields of study in Grades K–12.

STL is *not* a curriculum. A curriculum provides the specific details of how the content (*STL*) is to be delivered, including organisation, balance, and the various ways of presenting the content in the classroom, whereas standards describe what the content should be. Curriculum developers, teachers, and others should use *STL* as a guide for developing appropriate curricula, but the standards do not specify what should go on in the classroom.

In laying out the essentials for the study of technology, *STL* represents a recommendation from educators, engineers, scientists, mathematicians, and parents about specific skills and knowledge needed in order to become technologically literate. It is not, however, a federal policy or mandate. Nor does *STL* prescribe an assessment process for determining how well students are meeting the standards, although it does provide criteria for this assessment.

Format of STL

The individual standards presented in *STL* are organised into five major categories:

- The Nature of Technology (Chapter 3);
- Technology and Society (Chapter 4);
- Design (Chapter 5);
- Abilities for a Technological World (Chapter 6); and
- The Designed World (Chapter 7).

Standards in STL

STL includes twenty standards are distributed across the five categories (see Table 1). These are written statements about what is valued in the study of technology that can be used for judging quality. In this way, the document specifies what every student should know and be able to do in order to be technologically literate, and offers criteria to judge progress toward a vision of technological literacy for all students, from kindergarten through Grade 12.

Table 1. The 20 standards for technological literacy (ITEA, 2000/2002/2007).

The nature of technology	
Standard 1.	Students will develop an understanding of the characteristics and scope of technology.
Standard 2.	Students will develop an understanding of the core concepts of technology.
Standard 3.	Students will develop an understanding of the relationships among technologies and the connections between technology and other fields of study.

Technology and society	
Standard 4.	Students will develop an understanding of the cultural, social, economic, and political effects of technology.
Standard 5.	Students will develop an understanding of the effects of technology on the environment.
Standard 6.	Students will develop an understanding of the role of society in the development and use of technology.
Standard 7.	Students will develop an understanding of the influence of technology on history.
Design	
Standard 8.	Students will develop an understanding of the attributes of design.
Standard 9.	Students will develop an understanding of engineering design.
Standard 10.	Students will develop an understanding of the role of troubleshooting, research and development, invention and innovation, and experimentation in problem solving.
Abilities for a technological world	
Standard 11.	Students will develop the abilities to apply the design process.
Standard 12.	Students will develop the abilities to use and maintain technological products and systems.
Standard 13.	Students will develop the abilities to assess the impact of products and systems.
The designed world	
Standard 14.	Students will develop an understanding of and be able to select and use medical technologies.
Standard 15.	Students will develop an understanding of and be able to select and use agricultural and related biotechnologies.
Standard 16.	Students will develop an understanding of and be able to select and use energy and power technologies.
Standard 17.	Students will develop an understanding of and be able to select and use information and communication technologies.
Standard 18.	Students will develop an understanding of and be able to select and use transportation technologies.
Standard 19.	Students will develop an understanding of and be able to select and use manufacturing technologies.
Standard 20.	Students will develop an understanding of and be able to select and use construction technologies.

Benchmarks in STL

Benchmarks play a vital role in *STL* by providing the necessary elaboration of the broadly-stated standards. Benchmarks, which are statements that enable students to meet a given standard, are provided for each of the 20 standards at the K–2, 3–5, 6–8, and 9–12 grade levels. The benchmarks are followed by supporting sentences that provide further detail, clarity, and examples. (See Table 2 for an example.) Each student is required to meet all standards and their associated benchmarks, and teachers should feel free to add to the benchmarks to further enhance the ability of the student to meet a given standard.

Table 2. A representative standard and two benchmarks (shown in bold type in C and D).

Standard 8—Students will develop an understanding of the attributes of design.
In order to realize the attributes of design
C. **The design process is a purposeful method of planning practical solutions to problems.** The design process helps convert ideas into products and systems. The process is intuitive and includes such things as creating ideas, putting the ideas on paper, using words and sketches, building models of the design, testing out the design, and evaluating the solution.
D. **Requirements for a design include such factors as the desired elements and features of a product or system or the limits that are placed on the design.** Technological designs typically have to meet requirements to be successful. These requirements usually relate to the purpose or function of the product or system. Other requirements, such as size and cost, describe the limits of a design.

Research in education has demonstrated that if previously learned knowledge is tapped and built upon, it is likely that students will acquire a more coherent and thorough understanding of the processes than if they are taught them as isolated abstractions (NRC, 1999). With this in mind, the benchmarks are articulated, or 'ramped', from Grades K–12 to progress from very basic ideas at the early elementary school level to more complex and comprehensive ideas at the high school level. Certain content 'concepts' – such as systems, resources, requirements, optimisation, trade-offs, processes, and controls – extend across various levels to ensure continual learning of an important topic related to a standard.

Advancing Excellence in Technological Literacy: Student Assessment, Professional Development, and Programme Standards (AETL)

In March 2003, ITEA released *Advancing Excellence in Technological Literacy (AETL)* at its 65th annual conference in Nashville, Tennessee. *AETL* is based on *STL* and is designed as a companion to *STL*. *AETL* was developed by TfAAP from 2000 to 2003.

AETL provides standards and guidelines that address student assessment, professional development, and programme enhancement. The primary goal of all the standards is to help students achieve technological literacy. The 11-person TfAAP Advisory Group provided valuable counsel in best practice of standards development to the project. They met annually in Washington, DC.

The TfAAP Standards Writing Team was made up of 27 people (three teams of nine). They provided detailed input in fashioning the initial draft of *AETL*, and their continued review and input added strength and quality to the final document. The development and refinement of *AETL* took place over three years and involved hundreds of educators and experts in the fields of technology, mathematics, science, engineering, and other disciplines. Their input was attained through various methods, including hearings, web-based electronic document review, and individual reviews through the mail and in person. Three formal drafts of *AETL* were developed and reviewed before the final draft was prepared in autumn 2002.

Overview of AETL

Chapters 1 and 2 of *AETL* provide valuable introductory material. Chapter 1 is an overview that presents the rationale of need and conceptually introduces Chapters 3, 4, and 5. Chapter 2 discusses relevant principles and definitions. Subsequent chapters focus on the three separate, but interrelated, sets of standards that make up *AETL*:

- Student Assessment Standards (Chapter 3);
- Professional Development Standards (Chapter 4); and
- Program Standards (Chapter 5).

Finally, Chapter 6 invites users to participate in the visionary basis of *STL* and *AETL*.

The standards in *AETL* are based upon *STL*. To fully and effectively implement the content standards in *STL*, all of the *AETL* standards presented in Chapters 3, 4, and 5 must be met through the guidelines. While *AETL* is designed to leave specific curricular decisions to educators, teachers, professional development providers, and administrators, *STL* and *AETL* should be used as guides for advancing technological literacy for all students.

Student assessment standards (chapter 3). The definition for student assessment presented in *AETL* is "the systematic, multi-step process of collecting evidence on student learning, understanding, and abilities and using that information to inform instruction and provide feedback to the learner, thereby enhancing student learning" (ITEA, 2003, p.18). The primary audience for the student assessment standards is teachers. It is important to note that the standards are applicable to those who educate students on any aspect of technology.

The five organisational topics for the student assessment standards are:

- Consistency with *STL;*
- intended purpose;
- research-based assessment principles;
- practical contexts; and
- data collection.

While the student assessment standards (see Table 3) define how assessment of technological literacy should be designed and implemented, Chapter 3 does not provide a test, quiz, or other handy instrument to be photocopied and used in the laboratory-classroom. This task is left – as it should be – to individual teachers and others.

Users of the student assessment standards should recognise that student assessment should be *formative* (ongoing) as well as *summative* (occurring at the end). Further, users should recognise that the assessment process should be informative, that is, it should inform students and teachers about progress toward technological literacy and provide data on the effectiveness of instruction and the teaching and learning programme. Teachers should use student assessment data to improve classroom practices, plan curricula, develop self-directed learners, report student progress, and research teaching practices.

Table 3. Student assessment standards (ITEA, 2003).

A–1.	Assessment of student learning will be consistent with *Standards for Technological Literacy: Content for the Study of Technology* (*STL*).
A–2.	Assessment of student learning will be explicitly matched to the intended purpose.
A–3.	Assessment of student learning will be systematic and derived from research-based assessment principles.
A–4.	Assessment of student learning will reflect practical contexts consistent with the nature of technology.
A–5.	Assessment of student learning will incorporate data collection for accountability, professional development, and programme enhancement.

Professional development standards (chapter 4). Chapter 4 presents criteria for professional development providers (including teacher educators, supervisors, and administrators) to use in planning professional development. Professional development includes a continuous process of lifelong learning and growth that begins early in life, continues through the undergraduate, pre-service experience, and extends through the in-service years.

The standards are applicable to those who prepare teachers on any aspect of technology, including teachers whose primary focus may be another subject area.

The seven organisational topics for the professional development standards are:

- Consistency with *STL*;
- students as learners;
- curricula and programmes;
- instructional strategies;
- learning environments;
- continued professional growth; and
- pre-service and in-service (see Table 4).

Users of this document should focus on preparing teachers to continue to pursue professional development to keep up to date with changing technologies and research on how students learn. The necessity to address issues of technological literacy is pertinent to all programmes that prepare teachers of every grade level, including K–5 elementary teachers and teachers of science, mathematics, social studies, language arts, and other content areas. Therefore, faculty members in every teacher preparation programme can use *STL* and *AETL* to determine how the technological literacy of teacher candidates can be enhanced.

Table 4. Professional development standards (ITEA, 2003).

PD–1.	Professional development will provide teachers with knowledge, abilities, and understanding consistent with *Standards for Technological Literacy: Content for the Study of Technology* (*STL*).
PD–2.	Professional development will provide teachers with educational perspectives on students as learners of technology.
PD–3.	Professional development will prepare teachers to design and evaluate technology curricula and programs.
PD–4.	Professional development will prepare teachers to use instructional strategies and enhance technology teaching, student learning, and student assessment.
PD–5.	Professional development will prepare teachers to design and manage learning environments that promote technological literacy.
PD–6.	Professional development will prepare teachers to be responsible for their own continued professional growth.
PD–7.	Professional development providers will plan, implement, and evaluate the pre-service and in-service education of teachers.

Program standards (chapter 5). *AETL* defines the programme refers to everything that affects student learning, including content, professional development, curricula, instruction, student assessment, and the learning environment implemented across grade levels. The system-wide technology programme manages the study of technology in technology laboratory-classrooms as well as in other content area classrooms. The primary audience for the programme standards are twofold: (1) teachers and (2) administrators (including supervisors). As a result of this, the guidelines are divided for addressing these two audiences.

Chapter 5 in *AETL* presents criteria for teachers and administrators (including supervisors) responsible for the technology programme and system-wide technology programme. The standards are applicable to those who organise the learning of students on any aspect of technology. The five organisational topics for the programme standards are:

- Consistency with *STL*;
- implementation;
- evaluation;
- learning environments; and
- management.

Users of the programme standards should recognise that thoughtful design and implementation of technology programmes at school levels and of system-wide technology programmes at district levels are necessary to provide comprehensive and coordinated experiences for all students across grade levels and disciplines, including science, mathematics, social studies, language arts, and other content areas. The programme standards (see Table 5) call for extending technology programmes beyond the domain of the school. Technology programmes should, for example, involve parents, the community, business and industry, school-to-work programmes, and higher education as well as professionals in engineering and other careers related to technology. And finally, it is essential that adequate support for professional development be provided by administrators to ensure that teachers remain current with the evolving fields of technology and educational research.

Table 5. Programme standards (ITEA, 2003).

P–1.	Technology programme development will be consistent with *Standards for Technological Literacy: Content for the Study of Technology* (*STL*).
P–2.	Technology programme implementation will facilitate technological literacy for all students.
P–3.	Technology programme evaluation will ensure and facilitate technological literacy for all students.
P–4.	Technology programme learning environments will facilitate technological literacy for all students.
P–5.	Technology programme management will be provided by designated personnel at the school, school district, and state/provincial/regional levels.

Guidelines as enablers to meet the standards. Guidelines play a vital role in *AETL*. They are presented under each standard, and must be addressed to enable the user to meet a given standard. ITEA does not recommend that users eliminate any of the guidelines, although users may add to the guidelines if there is a need to accommodate local differences. A sample standard (A–4) with related guidelines is presented in Table 6.

TECHNOLOGY STANDARDS-BASED ADDENDA

The ITEA Addenda series (to *STL* and *AETL*) is part of the standards package for technological literacy. They were produced by the TfAAP with special assistance from ITEA's Centre to Advance the Teaching of Technology and Science (CATTS). The addenda are based on the standards but include concrete processes or suggestions for incorporating national, state, and/or local technological literacy standards into the programmes of all students throughout Grades K–12. Additionally, all of the documents contain worksheets for educators to use to make changes specific to their locality and situation. The addenda series marks another pioneering effort in educational reform as it provides a supplement to educational standards that focuses on the entire picture of programme reformation rather than concentrating solely on curricula. The addenda are:

- *Measuring Progress: Assessing Students for Technological Literacy* (ITEA, 2004) (student assessment);
- *Realizing Excellence: Structuring Technology Programmes* (ITEA, 2005c) (programmes);
- *Planning Learning: Developing Technology Curricula* (ITEA, 2005b) (curricula); and
- *Developing Professionals: Preparing Technology Teachers* (ITEA, 2005a) (professional development).

Table 6. Sample standard with related guidelines (ITEA, 2003, p.30–35).

Standard A—4: Assessment of student learning will reflect practical contexts consistent with the nature of technology.	
Guidelines for meeting Standard A–4 require that teachers consistently:	
A.	Incorporate technological problem solving.
B.	Include variety in technological content and performance-based methods.
C.	Facilitate critical thinking and decision making.
D.	Accommodate for modification to student assessment.
E.	Utilize authentic assessment.

RESEARCH BY ITEA ON THE STATUS OF TECHNOLOGY EDUCATION IN THE UNITED STATES

ITEA conducted research on the status of technology education in the United States in 2006-07 (ITEA, 2007). This was the third study conducted by ITEA on the condition of the study of technology in all 50 states. The previous studies were completed by ITEA's TfAAP in 2000-01 and 2003-04. The reports of the previous two studies were published in *The Technology Teacher* (Dugger, 2007; Meade & Dugger, 2004).

Questionnaires were sent electronically via e-mail in October 2006 to all 50 state technology education supervisors. In cases where no supervisor was available, alternate contacts in the state education departments were used. Two additional follow-up surveys were e-mailed in January and March 2007 to those states that had not returned their responses. Finally, telephone follow-up calls were conducted in April and May 2007 to attempt to gather unreported data from the states that had not responded and to clarify responses as necessary. There were 46 states (out of 50, or 92%) that responded to the 2007 survey.

The survey consisted of 10 questions. Questions 1, 2, and 4 were duplicated from the Newberry (2001) study and questions 5 and 6 were added in the 2004 survey. Questions 3 and 7 through 10 were added to the 2006-07 instrument.[1] The specific questions were:

1. Is technology education in your state framework? (Yes or No)
2. Is technology education required in your state? (Yes or No)
3. If you answered Yes to question #2, is it:
   __ Under local control
   __ An elective
   __ A requirement that is pending/proposed
   __ At what grade level? ______________________
4. How many technology education teachers are in your state? __________

5. Have you used *Standards for Technological Literacy: Content for the Study of Technology (STL)* in any of the following ways? (Select all that apply)
   __ Not used at all
   __ Placed in your state standards
   __ Adopted "as is" for your state standards
   __ Used in your curriculum guides
   __ Conducted workshops using the standards
   __ Other, please specify ________________
6. Have you used *Advancing Excellence in Technological Literacy: Student Assessment, Professional Development, and Programme Standards (AETL)* in any of the following ways? (Select all that apply)
   __ Not used at all
   __ Placed in your state standards
   __ Adopted "as is" for your state standards
   __ Used in your curriculum guides
   __ Conducted workshops using the standards
   __ Other, please specify ________________
7. Are you doing *Standards for Technological Literacy* assessments in your state at this time? (Yes or No) (If Yes, please share how used). __________
8. What course title(s) best describe the secondary school level technology education curriculum being taught in your state? __________________
9. Do you have a technology education state curriculum guide(s)? (Yes or No)
10. What best describes where technology education programme funding comes from in your state (i.e., relationships to local, state, national programmes)?

In response to Question 5, which asked whether and how *STL* was used, 42 states (91.3% of those reporting) in 2006-07 reported using *STL* at either the state or local school district level (see Table 7). Two states (4.3%) stated that they did not use *STL*; two states reported they were not sure whether they used it or not; and four states did not report. In 2004, 41 states (78.8%) reported using *STL*, with two states reporting "unknown." This compares very favourably to the Ndahi and Ritz (2003) findings that 43 states (83%) were using *STL*. Both the 2004 survey and the Ndahi and Ritz survey showed that seven states (13.5%) were not using *STL*. Averaging these data indicates that *STL* is used by over four out of every five states across the nation. Table 8 summarises how states reported using *STL*. Only one state (2%) reported that *STL* was not used at all. There were 14 states (30%) that said that *STL* was placed in their state standards. When asked if *STL* was adopted "as is" for their state standards, 11 states (24%) reported that it was. There were 22 states (48%) that reported that *STL* was used in their state curriculum guides. When asked if they had conducted workshops using *STL*, 18 states (39%) answered in the positive. State supervisors were also asked to describe other ways that *STL* was used in their states. Of the 13 responses (28%), *STL* was used primarily as a resource or reference and as a guideline for technology and engineering education.

Table 7. Programme summary of the 2007 and 2004 ITEA studies and the 2003 Ndahi and Ritz report on the usage of national technological literacy standards in the United States.

	STL Used? 2004	*STL* Used? 2003*	*STL* Used? 2007	*AETL* Used? 2004	*AETL* Used? 2007
Yes	41	43	42	22	29
No	7	7	2	23	13
Unknown	2	0	2	5	2
No response	1	0	4	1	4
No answer	1	0	0	1	2

Ndahi and Ritz (2003)

Table 8. Responses from state supervisors on how STL *is used.*

Questionnaire options	# respondents	% respondents
Not used at all	1	2%
Placed in state standards	14	30%
Adopted "as is" for state standards	11	24%
Used in curriculum guides	22	48%
Conducted workshops using the standards	18	39%
Other, please specify	13	28%

In response to Question 6, AETL was reported as being used in 29 of the states (63% of those reporting; see Table 7). Only 13 states (28.3%) of those reporting have not yet used *AETL*. The difference between *STL* and *AETL* usage is not unexpected considering that *AETL* had been published only four years prior to the time that the survey was conducted. Table 9 indicates some of the ways that *AETL* may be used in states. Eleven states (25% of those reporting) said that they did not use *AETL* at all. Five states (11%) reported using *AETL* in their state standards, with three states (7%) stated that *AETL* was adopted "as is" in their state standards. Eight states (18%) reported that *AETL* was used in their state curriculum guides, while nine states (20%) said that they had conducted workshops for teachers on *AETL*. When asked what other ways *AETL* was being used, 15 (34%) of state supervisors stated that it was used as a reference or resource and as a document to provide guidance to local school districts.

Questions 8 and 9 specifically address technology education curriculum in various states. A brief summary of the major findings for these two questions are as follows:

– When asked in Question 8 – "What course titles best describe the secondary school technology education curriculum taught in your state?" – state supervisors provided a wide variety of answers. Many of the supervisors stated that the local school districts have the responsibility to provide course titles. The response that was provided the most frequent for course title(s) was "Technology Education."
– Question 9 was "Do you have a technology education state curriculum guide(s)?" Twenty-seven states (59% of those reporting) answered that they had technology education curriculum guides. There were 19 other states (41%) that reported they did not have any curriculum guides for technology education.

Table 9. Responses from state supervisors on how AETL *is used.*

Questionnaire options	# respondents	% respondents
Not used at all	11	25%
Placed in your state standards	5	11%
Adopted "as is" for your state standards	3	7%
Used in your curriculum guides	8	18%
Conducted workshops using the standards	9	20%
Other, please specify	15	34%

ENGINEERING BY DESIGN™ (EBD): A STANDARDS-BASED MODEL PROGRAMME/CURRICULUM

ITEA's Centre to Advance the Teaching of Technology and Science (CATTS) has developed the only standards-based national model curriculum for Grades K–12 that delivers technological literacy. The model, Engineering by Design™ (EbD), is built on *Standards for Technological Literacy* (ITEA, 2000/2002/2007); *Principles and Standards for School Mathematics* (NCTM, 2000); and *Project 2061, Benchmarks for Science Literacy* (AAAS, 1993). EbD publications can be viewed at www.iteaconnect.org.

CHALLENGES FOR THE FUTURE

ITEA's TfAAP and its work to generate *STL*, *AETL*, and other related materials do not represent an end, but rather a beginning. In other fields of study, developing content standards and the resulting curriculum has often proven to be the easiest step in a long, arduous process. Therefore, getting these technology standards and curriculum accepted and implemented in Grades K–12 in every school will be far more difficult and daunting than developing them has been. Only through the combined efforts of educational decision makers throughout the U.S. will we be able to ensure that all students develop higher levels of technological literacy.

NOTES

[1] Complete data tables with comments are viewable online at www.iteawww.org.

REFERENCES

American Association for the Advancement of Science, Project 2061. (1993). *Benchmarks for science literacy*. New York: Oxford University.

Dugger, W. E. (2007). The status of technology education in the United States: A triennial report of the findings from the states. *The Technology Teacher*, *67*(1), 14–21.

International Technology Education Association [ITEA]. (1996). *Technology for all Americans: A rationale and structure for the study of technology*. Reston, VA: Author.

ITEA. (2000/2002/2007). *Standards for technological literacy: Content for the study of technology*. Reston, VA: Author.

ITEA. (2003). *Advancing excellence in technological literacy: Student assessment, professional development, and programme standards*. Reston, VA: Author.

ITEA. (2004). *Measuring progress: Assessing students for technological literacy*. Reston, VA: Author.

ITEA. (2005a). *Developing professionals: Preparing technology teachers*. Reston, VA: Author.

ITEA. (2005b). *Planning learning: Developing technology curricula*. Reston, VA: Author.

ITEA. (2005c). *Realizing excellence: Structuring technology programmes*. Reston, VA: Author.

ITEA. (2006). *Technological literacy for all: A rationale and structure for the study of technology*. Reston, VA: Author (CD format only).

ITEA. (1985). *Technology education: A perspective on implementation*. Reston, VA: Author.

Meade, S., & Dugger, W. E. (2004). Reporting on the status of technology education in the U.S. *The Technology Teacher*, *64*(2), 29–35.

National Council of Teachers of Mathematics. (2000). *Principles and standards for school mathematics*. Reston, VA: Author.

Newberry, P. B. (2001). Technology education in the U.S.: A status report. *The Technology Teacher*, *61*(1), 1–16.

William E. Dugger, Jr.
International Technology Education Association
USA

ANN MARIE HILL

7. THE STUDY OF TECHNOLOGY IN CANADA

Landscapes, 'Lifescapes', and Curricula

ABSTRACT

This chapter examines recent research and developments in the study of technology in Canada and provides an analysis of these developments. Canada is comprised of 10 provinces and 3 territories, each with different landscapes, 'lifescapes', and politics. Education is under the jurisdiction of each province/territory. As such, each province/territory is responsible for (1) envisioning and implementing its own elementary and secondary school curricula, (2) establishing policy on what are core/compulsory and complementary/elective courses, (3) setting secondary school graduation requirements, and (4) envisioning how the study of technology factors into school curricula and graduation requirements.

Within this political framework, each province/territory approaches the study of technology in different ways and independently names its programme of study for technology. Elementary school curricula across regions is not consistently inclusive of the study of technology, and one finds the study of technology at the secondary school level known by many different names, for example, technological education, technological studies, technical studies, technical education, technology education, practical arts, vocational education, and career studies. The different names represent different grade levels, philosophical underpinnings, and purposes for this curricular area. This chapter uses the term 'study of technology' as an overarching term when referring to general discussion and specific names when representing a specific province/territory.

A HISTORICAL OVERVIEW

Canada is comprised of 10 provinces and 3 territories. They are, from west coast to east coast: British Columbia, Yukon, Alberta, Northwest Territories, Nunavut, Saskatchewan, Manitoba, Ontario, Québec, New Brunswick, Nova Scotia, Prince Edward Island, and Newfoundland and Labrador. Education in Canada is under the jurisdiction of each province/territory, not the federal government. This policy was established by Canada's Constitution Act of 1867 and is still in place today. This said, the federal government has special programmes in place that influence provincial/territorial policy for education, for example, the Prime Minister's Award for Teaching Excellence in Science, Technology and Mathematics. In addition, intergovernmental bodies exist such as the Council of Ministers of Education, Canada (CMEC), founded in 1967 by provincial/territorial ministers of education. This system of responsibility for education in Canada is seen as effective in meeting provincial/territorial educational needs:

> While there are a great many similarities in the provincial and territorial education systems across Canada, there are important differences that reflect the geography, history, culture, and corresponding specialized needs of the populations served. The comprehensive,

A.T. Jones and M.J. de Vries (eds.), International Handbook of Research and Development in Technology Education, 65–84.
© 2009 Sense Publishers. All rights reserved.

> diversified, and widely accessible nature of the education systems in Canada reflect the societal belief in the importance of education. (CMEC, 2005, p. 1)

One similarity in education systems across Canada is the structure of the school systems, that is, the systems of governance and funding. Another is that students progress through school grade-by-grade (i.e., Kindergarten, Grade 1, Grade 2, etc.) and must successfully complete a grade to move on to the next grade. At the end of secondary school, students graduate and are awarded a secondary school diploma. From secondary school, students can enter the workforce or enter post-secondary education at the college or university level.

To understand the idea of differing educational needs in provincial/territorial education, it is important to understand Canada's early European settlement. Eastern Canada was settled first and then settlement moved west. The teaching of technology was important for both individual survival and economic development. It corresponded to the geographic area of settlement and local needs, for example, boat building was taught on the east coast (and later on the west coast); cabinet making, masonry, shoe and dress making were taught in urban areas; and agriculture was taught in rural areas. Most early Canadian settlers from Europe received no more than an elementary school education. Gardner and Hill (1999a, 1999b) provide a historical perspective on the development of the study of technology in Ontario and discuss the beginnings of the study of technology in Canada:

> In one sense, technological education in the North American continent began about 15,000 years ago, when the first human beings from Asia crossed the ice bridge covering the Bering Strait, and entered what today is called Alaska. These people taught their children how to build shelters, hunt animals, trap fish and make cloths. Remnants of this early form of technical education can still be observed today: the Athabaskan mother teaching her daughter to skin, stretch and tan a moose hide to prepare it for making moccasins and mukluks is continuing an educational process that began in pre-historic times.
>
> In its more modern sense, of an enterprise located in schools and taught by people other than members of one's family and tribe, technological education in Canada originated in various forms of technical education in the early days of French settlement. Young (1992) notes that Jean Talon brought skilled artisans and teachers of rug-making from France to Quebec in the mid 17th century, and Bishop Laval opened trade schools to teach cabinet-making, masonry, shoe-making, sculpting and painting. Agricultural instruction was given on a model farm, and various schools taught girls domestic skills and boys the use of tools. By the end of the century, Jesuit colleges were teaching mathematics, surveying and hydrography to train ships' pilots and captains, explorers and surveyors.
>
> Formal vocational education institutions were set up in the 19th century, with the establishment of the Artisans' Institute in Montreal in 1828. In Ontario, the Society of Artists' School, founded in 1876, taught trade skills as well as art education. These efforts received some financial support from provincial governments. (1999a, pp. 105–106)

Since the 19th century, the study of technology has been offered in varied forms across Canada. Today, provincial/territorial needs still influence local curricula, and these needs are based on local landscape, 'lifescapes', and politics.

CONTEMPORARY PROVINCIAL/TERRITORIAL CURRICULA FOR THE STUDY OF TECHNOLOGY IN SECONDARY SCHOOLS

Various authors have reported on the complex, multifaceted nature of the study of technology in Canada (Chinean, Oaks, & Boutin, 1995; Gradwell, 2003; Hill, 2003; Kozolanka & Medway, 1996). Many of the provinces/territories have undergone or are presently undergoing changes to the study of technology since I last reported on this topic (Hill, 2003a). During the writing of this chapter, frequent updates were made because provinces/territories released new policy and curricula documents. In this chapter, for general discussion or discussion at a national or Canadian level, a general term 'the study of technology' is used. Reporting for each province/territory uses terminology specific to each province/territory. This method serves two purposes: (1) to clearly portray a snapshot of the complex, multifaceted nature of the study of technology in Canada, and (2) to document the landscape, lifescapes, and political nature of the curricula in each province/territory. Reporting is from the west to the east coast of Canada.

British Columbia

> British Columbia offers the study of technology from Grade 8 to 12 using a variety of terms. Opportunities are available also for secondary school students to complete their secondary school studies through in-school and/or out-of-school programmes. Out-of-school programmes include Accelerated Credit Enrollment in Industry Training (ACE IT), Secondary School Apprenticeship (SAA), and Work Experience Courses offered as independent directed studies (IDS). Work Experience Courses can be taken as part of an ACE IT or a SSA, or as a single course. The term 'Technical Training' is used for these programmes. ACE IT, an industry training programme for high school students, offers courses that give both high school graduation credits and a head start towards completion of an apprenticeship programme. . . . ACE IT programmes are developed and offered as partnerships between school districts and post-secondary institutions, and classes are often taught at colleges. (Industry Training Authority [ITA], 2008)

A Secondary School Apprenticeship (SSA) is:

> ... an educational program jointly supported by the Ministry of Education and the Industry Training Authority (ITA). Through participation in the SSA program, students earn credit toward secondary school graduation and begin earning work-based training hours toward provincially and nationally recognized industry training program credentials. Individuals will complete an industry training program and become a certified tradesperson by combining both work-based training and technical training in an apprenticeship program. Typically industry training programs consist of 80% work-based training under the guidance of a certified tradesperson or equivalent and 20% technical training. A typical apprenticeship program takes four years to complete. (British Columbia Ministry of Education, 2007a, p. 1)

In British Columbia, all students must take a course called *Planning 10*, a course in life skills, to graduate from secondary school. The four curricular organisers of this course are: Graduation Program; Education and Careers; Health; and Finances (British Columbia Ministry of Education, 2007b, p. 4). The Graduation Program organiser is comprised of a number of components, one of which is called Focus Areas. It "provides a means for students to consider a broad range of education and career options, and to see how their secondary school course selections can lead to specific post-secondary education and training programs and career options" (p. 53). There are eight Focus Areas: (1) Business and Applied Business; (2) Fine Arts, Design, and Media; (3) Fitness and Recreation; (4) Health and Human Services; (5) Liberal Arts/Humanities; (6) Science and Applied Science; (7) Tourism, Hospitality, and Foods; and (8) Trades and Technology. ACE IT and SSA programmes can be used to complete the Focus Areas requirement.

In addition to ACE IT and SSA programmes, British Columbia schools offer students in-school courses for the study of technology under the curriculum subject area called *Applied Skills*. Programmes for *Technology Education* are found under this subject area. In Grades 8 to 10, learning outcomes for Technology Education are grouped under five curriculum organisers: Self and Society, Communications, Production, Control, and Energy and Power (British Columbia Ministry of Education, 1995). In this course, all organisers receive equal attention. In 2004, a supplemental document to the 1995 document was released to describe different curriculum organisers for the Grade 10 Technology Education course. These organisers are: Drafting and Design, Electronics, Mechanics, Metalwork, and Woodwork (British Columbia Ministry of Education, 2004). In Grades 11 and 12, the subject called *Applied Skills* consists of three programme groupings: Business Education (and Economics), Home Economics, and Technology Education. Thus, the term *Technology Education* continues through Grades 11 and 12. There are six Technology Education programmes: (1) Automotive Technology, (2) Carpentry and Joinery, (3) Drafting and Design, (4) Electronics, (5) Industrial Design, and (6) Metal Fabrication and Machining. Each programme has its own integrated resource package, or curriculum document (British Columbia Ministry of Education, n.d.a). In addition to the Applied Skills curriculum subject area, where Technology Education is found, there is also a curriculum subject area called Information and Communications Technology offered in Grades 8 through 12 (British Columbia Ministry of Education, n.d.b).

Yukon

The Yukon, a Canadian territory, uses the British Columbia curricula for *Applied Skills*. The study of technology is located under this umbrella as a subject area called *Technology Education.* The Yukon is a partner in the Western and Northern Canadian Protocol (WNCP), which supports the development of common curriculum frameworks for Western and Northern Canada: "Within these frameworks, the British Columbia program of studies forms the basis of the Yukon curriculum. This curriculum is frequently adapted to reflect local needs and conditions" (Department of Education, Government of Yukon, 2008a).

The Yukon also offers a Yukon Secondary School Apprenticeship Training programme where secondary school students can explore a skilled trade and complete high school at the same time. This programme recognises secondary school students' part-time work in trades designated for apprentice training while they attend school or during the summer break period:

> Those students working in a trade can be registered as apprentices with their employer and receive credit for their trade experience, even if the work is on a part-time or short term basis. Once they complete high school (defined as not returning to high school for the following semester), their apprenticeship agreements convert to a regular apprentice training schedule. They then continue their on-the-job training and access in-school technical training as regular apprentices. (Department of Education, Government of Yukon, 2007, p. 1)

There are 48 occupations under the Apprentice Training Act, for which apprenticeship training and certification are available. They are: Automotive Painter, Automotive Service Technician, Baker, Barber, Bricklayer, Cabinetmaker, Carpenter, Communication Electrician (Construction),

Communication Electrician (Switching), Communication Electrician (Toll), Construction Electrician, Cook, Community Antenna Television Technician, Electric Motor System Technician, Electronic Technician (Consumer Products), Floorcovering Installer, Gasfitter (1st and 2nd Classes), Glazier, Hairstylist, Heavy Duty Equipment Technician (Off Road), Heavy Equipment & Truck & Transport Technician, Industrial Electrician, Industrial Instrument Mechanic, Industrial Mechanic (Millwright), Industrial Warehouse Worker, Insulator (Heat & Frost), Lather (Interior Systems Mechanic), Machinist, Motor Vehicle Body Prepper, Motor Vehicle Body Repairer, Motor Vehicle Body Repairer (Metal & Paint), Oil Burner Mechanic, Outdoor Power Equipment Technician, Painter and Decorator, Partsperson, Plumber, Power System Electrician, Powerline Technician, Recreation Vehicle Service Technician, Refrigeration and AC Mechanic, Roofer, Sawfitter/Filer, Sheet Metal Worker, Sprinkler System Installer, Steamfitter/Pipefitter, Transport Trailer Technician, Truck and Transport Technician, and Welder (Department of Education, Government of Yukon, 2008b).

Alberta

In April 2008, Alberta released, in draft form, new educational policy and curricula documents; curricula for the study of technology were included in the revisions. Alberta categorises all courses in secondary school as either core curriculum or complementary curriculum. Courses for the study of technology are categorised as complementary curriculum and are offered under the umbrella programme of Career and Technology Studies (CTS). The 2008 draft of the new CTS programme organises CTS into five clusters: (1) Media, Design & Communication Arts – MDC, (2) Business, Administration, Finance and Information Technology – BIT, (3) Natural Resources – NAT, (4) Trades, Manufacturing & Transportation – TMT, and (5) Health, Recreation & Human Services – HRH. To date, the curriculum for two of these clusters, BIT and TMT, have been developed in detail and implementation is planned for September 2009. Implementation for MDC, NAT, and HRH are scheduled for September 2010. The curricula documents for the two clusters that have been developed to date (BIT and TMT) display a vast array of course listings and make suggestions for sequencing courses into pathways. The pathways:

> are connected to competencies specific to an occupational designation or a desired career path. There are two kinds of pathways:
>
> – Specialized skill pathways can provide students with the knowledge, skills and attitude for employment or further education. These pathways can be created or customized to meet student, school, or community program needs. An example would be a framer pathway in the TMT cluster. Courses can be substituted in this or other pathways to prepare a student for specific community or job-site skills.
> – Credentialed pathways can provide students with post-secondary and/or business and industry credentials or articulation. For students to obtain the desired credential or articulation, all specified course outcomes within this pathway must be met. An example would be the Carpenter Apprenticeship pathway that would qualify a student to complete first period apprenticeship training. (Alberta Education, 2008a, p. 4)

Here, a 'cluster' is "a group of pathways and courses related to the nine categories of the National Occupational Classification (NOC)" (Alberta Education, 2008a, p. 1). There are more than 700 courses in the five clusters of CTS. The pathways and courses for each of the five clusters are further organised into three levels of achievement: introductory level courses for life skills and an introduction to a cluster; intermediate level courses that build on introductory courses and offer courses in cluster related careers; and advanced level courses that offer more in-depth expertise leading to cluster-related workplace and post-secondary education.

To illustrate an example, the 2008 draft document (Alberta Education , 2008b) for the TMT cluster lists 41 courses at the introductory level, 76 courses at the intermediate level, and 106 courses at the advanced level – at total of 223 courses in the TMT cluster alone. The document then portrays examples of how these courses are organised into the nine categories of the NOC related to this cluster, and suggests pathways for each of the nine NOC categories for the different achievement levels.

Northwest Territories

The Northwest Territories (NT) is a Canadian Territory. In the NT, the study of technology in Grades K to 6 is called *Science and Technology Education.* The curriculum is found in the curriculum document, *K-6 Science and Technology Curriculum* (Government of the Northwest Territories, Department of Education, Culture and Employment, 2004). This document is influenced by science education development carried out by the Council of Ministers of Education, Canada (1997) and uses the 1998 foundation for science and technology education in Ontario elementary schools (Ontario Ministry of Education, 1998). The curricular strands are: life systems, matter and materials, energy and control, structures and mechanisms, and earth and space systems. In 2007, Ontario produced a new curriculum document for science and technology. This latter document is presently not used in the NT.

In Grades 7 to 12, the Alberta curriculum for *Career and Technology Studies* (CTS) is used in the NT. The Government of the Northwest Territories, Department of Education, Culture and Employment's website (http://www.ece.gov.nt.ca/Divisions/kindergarten_g12/indexK12.htm) indicates for Career and Technology Studies, "Please refer to the Alberta learning website link above for courses". The link takes the reader to the Alberta CTS curriculum that existed prior to 2008, where CTS was organised into 22 strands: Agriculture, Career Transitions, Communication Technology, Community Health, Construction Technologies, Cosmetology Studies, Design Studies, Electro-Technologies, Energy and Mines, Enterprise and Innovation, Fabrication Studies, Fashion Studies, Financial Management, Foods, Forestry, Information Processing, Legal Studies, Logistics, Management and Marketing, Mechanics, Tourism Studies, and Wildlife. Apprenticeship and occupation training programmes also are offered, but as post-secondary study in the college system.

Nunavut

Nunavut, another of Canada's Territories, presently uses Alberta's curriculum for the study of technology. As such, the term *Career and Technology Studies* (CTS) is used. Some files developed by the Northwest Territories (NT) are also used. The Department of Education's website indicates:

> The files on this site are the programs of study and curricula approved for use in Nunavut schools from grades 7 to 12. These programs and curricula are the ones that were approved for use in Nunavut schools at the time of division from the Northwest Territories. New programs and curricula are currently being developed for our schools, and this information will be updated on an annual basis.
>
> The files from Alberta were downloaded in PDF format from the Alberta Learning web site (www.learning.gov.ab.ca). The NT files for the most part were scanned and converted to PDF. (Government of Nunavat, Department of Education, n.d.)

The CTS programme documents listed on this website indicate that Nunavut is using Alberta curricula available prior to the revised 2008 curricula. Vocational-related education appears to be at the post-secondary, college level.

Saskatchewan

Saskatchewan uses the term *Practical and Applied Arts* (PAA) as an umbrella term for the study of technology (Government of Saskatchewan, Saskatchewan Learning, 2006). The PAA curriculum is composed of modules configured into suggested courses. PAA courses are mapped to six occupational clusters as follows:

- Agriculture (Value-Added)
  - Agriculture Studies; Cow/Calf Production; Dairy Production; Feedlot Production; Field Crop Production; Food Studies; Horticulture; Machining; Pork Production; and Sheep Production.
- Care and Hospitality
  - Accounting; Commercial Cooking; Cosmetology; Life Transitions; Tourism, Hospitality; and Entrepreneurship.
- Communication
  - Communication Production Technology; Electrical and Electronics; Information Processing; Photographics, Photography and Graphic; and Theatre Arts.
- Design (Construct/Fabricate)
  - Clothing, Textiles, and Fashion; Construction and Carpentry; Design Studies; Drafting and Computer-Aided Design; Housing; Interior Design; Upholstery; and Welding.
- Resources (Natural)
  - Energy and Mines; Forestry Studies; and Wildlife Management.
- Transportation
  - Autobody; and Mechanical and Automotive.

In addition to the in-school PAA curriculum, Saskatchewan also offers transition-to-work, work study, and apprenticeship opportunities, all through the flexible modular system of PAA (Government of Saskatchewan, Saskatchewan Learning, 2006).

Manitoba

Manitoba uses the umbrella term *Technology Education* for the study of technology. Under this umbrella term, clusters are found for apprenticeship options, business education, marketing education, home economics, vocational industrial education (light industrial cluster, heavy industrial cluster, and human services cluster), and industrial arts. The three designations for the vocational industrial education cluster and the industrial arts cluster are directly related to what other provinces/territories would consider the study of technology. Each cluster has a number of programmes associated with it, as listed below (Manitoba Education, Citizenship and Youth, 2007, pp. ix–x):

- Cluster: Vocational Industrial (Light Industrial-LI) has 14 programmes:
  - Programmes: Advertising Art, Audio Recording Production and Performance, Drafting, Electronic Communications, Electronics, Fashion Technology and Design, Graphic Arts, Industrial Electronics, Jewelry Arts, Microcomputer Systems, Photograph, Production Art, Radio and Television Broadcasting, and Technical Drafting.

- Cluster: Vocational Industrial (Human Services-HS) has 13 programmes:
  - Programmes: Child Care, Cosmetology (Hairstyling – Nails), Dental Assisting, Dental Technology, Family Community Services and Health Programs, Food Services, Health Care Assistant, Health Services Admissions, Hotel Hospitality, Human Services, and Pharmacy Technician.
- Cluster: Vocational Industrial (Heavy Industrial-HI) has 20 programmes:
  - Programmes: Agriculture, Auto Body Repair, Building Construction, Building Custodial Services, Building Operations and Power Engineering, Electricity, Heavy Duty Mechanics, Horticulture and Landscaping, Industrial Controls and Automation, Industrial Electricity, Industrial Maintenance, Machine Shop, Metals and Aerospace, Mineral Science, Plastics Technology, Power Mechanics, Refrigeration and Air Conditioning, Resources and Environmental Management, Sheet Metal Works, and Welding.
- Cluster: Industrial Arts in secondary school is comprised of four courses:
  - Programmes: Comprehensive-Technology Education, Construction, Drafting, and Electricity/Electronics.

The Manitoba Education, Citizenship and Youth website (http://www.edu.gov.mb.ca/k12/docs/policy/sthte/tecguide07.pdf) then provides a listing of specific courses for each programme found within a cluster. For example, the Advertising Art programme in the Light Vocation Industrial cluster has 20 courses associated with it:

- Cluster: Vocational Industrial (Light Industrial-LI):
  - Programme: Advertising Art
    - Courses: 8882 Introduction To L.I. Tech., 8883 Introduction To L.I. Tech.5m, 8379 Game Design & Development, 8380 Intro To Advertising Art, 8381 Intro To Advertising Art, 8382 Drawing, 8383 Design, 8384 Production Art, 8385 Illustration, 8386 Advertising Layout, 8387 Type Design, 8388 Advanced Production Art, 8389 Typography, 8390 Graphic Design, 8391 Advanced Applications, 8392 Portfolio Management, 8393 Applied Advertising Art, 8394 L.I. Tech Initiative, 8395 Animation, 8396 Digital Imaging.

The study of technology can also be found in the Middle Years, or Grades 7 to 9. Here it is called *Industrial Arts* and is comprised of four courses: Graphic Communications, Power/Energy, Manufacturing, and Construction (Manitoba Education, Citizenship and Youth, 2003). There is no study of technology before Grade 7.

Ontario

Ontario has undergone numerous significant changes over the past 25 years (Hill, 2006). Prior to 1998, the study of technology in elementary schools in Ontario was known as *Industrial Arts*, then *Design and Technology*, then *Technology Education*. As of 1998, the study of technology as a discrete subject area in the elementary school was eliminated and changed to *Science and Technology* (Ontario Ministry of Education, 1998). A new *Science and Technology* curriculum document for elementary schools was released in 2007. This new document organises the subject content for Grades 1 to 8 into seven fundamental concepts: matter; energy; systems and interactions; structure and function; sustainability and stewardship; and change and continuity (Ontario Ministry of Education, 2007a~~b~~). The focus in this new document is more on science from a Science, Technology, Society (STS) perspective.

Prior to the late 1980s, Ontario had two routes for the study of technology in secondary schools. First, *Industrial Arts*, which was later renamed *Design and Technology,* and then *Technology Education* – but with little curricula changes. Second, *Technological Studies*, where approximately 77 discrete technical courses were offered.

In the late 1980s the curricula for technological studies changed. The term *Technological Studies* changed to *Technological Education*. The change represented a philosophical shift in the understanding of the study of technology. Discrete technical courses were disbanded and replaced in Grades 10 to 12 by seven *Broad-Based Technology* courses: Communications Technology, Construction Technology, Health and Personal Services, Hospitality and Tourism, Manufacturing Technology, Technological Design, and Transportation Technology. In Grade 9, an introductory course named Integrated Technologies was also offered.

Changes also occurred with two other related curricula subjects: Technology Education (formerly Industrial Arts, then Design and Technology), which continued from elementary school study into secondary school study, and Computer Studies. For the first time in Ontario, both were housed under the *Technological Education* umbrella. The former became one of the Broad-based Technologies – Technological Design – while the latter remained a separate content area under the umbrella. The seven Broad-Based Technology courses and the Computer Studies courses within the *Technological Education* framework continued until 2008.

As of November 2008, Ontario has new courses in Technological Education at the secondary school level, with a September 2009 full implementation date (Ontario Ministry of Education, 2008). New curricular documents are expected to be released over the winter of 2009. The new curriculum represents a revised vision of the framework with regard to: (1) the expansion of the number of Broad-Based Technology clusters, (2) the return to a number of courses classified under each Broad-Based Technology cluster, and (3) the placement of some of the study of computers back into Technological Education from Computer Studies.

In the new 2008 framework for Technological Education, there are 10 Broad-Based Technology clusters, increased from seven. They are: Communications Technology, Computer Technology, Construction Technology, Green Industries, Hairstyling and Aesthetics, Health Care, Hospitality and Tourism, Manufacturing Technology, Technological Design, and Transportation Technology. There will be exploratory courses for each cluster in Grade 9: Exploring Communications Technology, Exploring Computer Technology, Exploring Construction Technology, Exploring Green Industries, Exploring Hairstyling and Aesthetics, Exploring Health Care, Exploring Hospitality and Tourism, Exploring Manufacturing Technology, Exploring Technological Design, and Exploring Transportation Technology. In Grade 10, introductory courses to the 10 new Broad-Based Technology clusters will be offered and will use the name of the Broad-Based Technology clusters. It is in Grades 11 and 12 that more detailed courses are offered, with each of the 10 Broad-Based Technology clusters expanded into courses (Ontario Ministry of Education, 2008):

- Cluster: Communications Technology
  - Grade 11 Courses: Communications Technology University/College Preparation; and Communications Technology Broadcast and Print Production.
  - Grade 12 Courses: Communications Technology University/College Preparation; and Communications Technology Digital Imagery and Web Design.
- Cluster: Computer Technology
  - Grade 11 Courses: Computer Technology Workplace Preparation; and Computer Engineering Technology University/College.
  - Grade 12 Courses: Computer Technology Workplace Preparation; and Computer Engineering Technology University/College.
- Cluster: Construction Technology
  - Grade 11 Courses: Construction Engineering Technology College Preparation; Construction Technology Workplace Preparation; and Custom Woodworking Workplace Preparation.
  - Grade 12 Courses: Construction Engineering College Preparation; Construction Technology Workplace Preparation; and Custom Woodworking Workplace Preparation.

- Cluster: Green Industries
  - Grade 11 Courses: Green Industries Workplace Preparation; and Green Industries University/College Preparation. Farm Production; Floristry; Forest Industry Worker; Horticulture Technician; Landscape Construction and Maintenance.
  - Grade 12 Courses: Green Industries Workplace Preparation; and Green Industries University/College Preparation.
- Cluster: Hairstyling and Aesthetics
  - Grade 11 Course: Hairstyling and Aesthetics Workplace.
  - Grade 12 Course: Hairstyling and Aesthetics Workplace.
- Cluster: Health Care
  - Grade 11 Courses: Health Care College Preparation; and Health Care University/College Preparation.
  - Grade 12 Courses: Health Care College Preparation; Health Care University/College Preparation; Child Development and Gerontology College Preparation; and Health Care Support Services Workplace Preparation.
- Cluster: Hospitality and Tourism
  - Grade 11 Courses: Hospitality and Tourism Workplace Preparation; and Hospitality and Tourism College Preparation.
  - Grade 12 Courses: Hospitality and Tourism Workplace Preparation; and Hospitality and Tourism College Preparation.
- Cluster: Manufacturing Technology
  - Grade 11 Courses: Manufacturing Technology Workplace Preparation; Manufacturing Technology College Preparation; and Manufacturing Engineering Technology University/ College Preparation.
  - Grade 12 Courses: Manufacturing Technology Workplace Preparation; Manufacturing Technology College Preparation; and Manufacturing Engineering Technology University/ College Preparation.
- Cluster: Technological Design
  - Grade 11 Courses: Technological Design and the Environment; Technological Design University/College.
  - Grade 12 Courses: Technological Design in the Twenty-first Century; and Technological Design University/College.
- Cluster: Transportation Technology
  - Grade 11 Courses: Transportation Technology Vehicle Ownership; and Transportation Technology College Preparation.
  - Grade 12 Courses: Transportation Technology Vehicle Maintenance Workplace Preparation; and Transportation Technology College Preparation.

It is interesting to note that the 2007 draft document outlining curricular changes in Technological Education (Ontario Ministry of Education, 2007b) provided a number of emphasis options, or discrete technical courses, under each cluster. These discrete technical courses disappeared in the official 2008 release outlined above. Clearly, the move back to discrete technical courses did not successfully clear all levels of government.

Québec

At the elementary school level in Québec, the study of technology is found in the subject area of *Mathematics, Science and Technology*. This subject area's content is broken down into (a) mathematics, and (b) science and technology. The science and technology content is organised around essential knowledges that students must acquire. Essential knowledges in science and technology are divided into three main categories: the material world, Earth and space, and living things (Gouvernement du Québec, Ministère de l'Éducation, 2001, pp. 173–177).

Secondary school in Québec is comprised of Cycle One [Grade 7 (Secondary One), Grade 8 (Secondary Two), and Grade 9 (Secondary Three)], and Cycle Two [Grade 10 (Secondary Four) and Grade 11(Secondary Five)].

> Secondary school offers five years of general education, divided into two cycles. Cycle One, which lasts three years, enables students to consolidate the learning acquired in elementary school and to begin to think about their career options. From the third year on, optional subjects are added to the general curriculum, giving students the opportunity to explore various subject areas (sciences, arts, etc.). At the end of the fifth year of secondary education, students are awarded a Secondary School Diploma (SSD) that provides access to college, but does not lead directly to university. . . .
>
> Various vocational training programs leading to a trade are offered in Cycle Two. Some of these programs start in the third year of secondary school. Over 170 vocational training programs in 21 sectors allow young people and adults to earn a Diploma of Vocational Studies (DVS) and then, if they so choose, an Attestation of Vocational Specialization (AVS). After Secondary Three [Grade 9], students may also enroll in a program leading to an Attestation of Vocational Education (AVE) in order to prepare for a semiskilled occupation. (Gouvernement du Québec, Ministère de l'Éducation, du Loisir et du Sport, 2006, p. 5)

In the five years of secondary school general education, Québec uses the term, *Mathematics, Science and Technology*, to develop subject-specific competencies related to this subject area. The subject area content, as at the elementary school level, is broken down into (a) mathematics, and (b) science and technology. There is no stand-alone subject area for technology education.

> The content of the Science and Technology program for Secondary Cycle One [Grades 7 to 9] is essential to helping students develop and use the competencies. ... It is based on the elementary-level Science and Technology program, and aims to develop scientific and technological literacy. In Cycle One, there are four major areas for the science and technology curriculum, each with distinct general concepts. (Gouvernement du Québec, Ministère de l'Éducation, 2004, pp. 240–241)

The four major areas are:

- The Material World
  - Properties
  - Changes
  - Organization
- The Earth and Space
  - General characteristics of the Earth
  - Geological and geophysical phenomena
  - Astronomical phenomena
- The Living World
  - Diversity of life forms
  - Survival of species
  - Life-sustaining processes
- The Technological World
  - Engineering

- Technological systems
- Forces and motion

> Secondary Cycle Two [Grades 10 and 11], poses the same educational challenges as the preceding cycles, with the addition of certain issues that arise from this cycle's role to serve as a transition from a renewed basic education to vocational training and college education. (Gouvernement du Québec, Ministère de l'Éducation, du Loisir et du Sport, 2008a, p. 3)

This is accomplished through the diversification of educational choices and the exploration of occupational fields by offering a common *and* a diversified curriculum through three paths:
- a General Education Path;
- an Applied General Education Path; and
- a Work-Oriented Training Path consisting of two learning options
  - prework training
  - training for a semiskilled trade.

> The Work-Oriented Training Path enables students to join the work force fairly quickly, but it also allows them to receive training in a semiskilled trade, go into vocational training or continue their studies in general education. The General Education Path and the Applied General Education Path can both lead to further studies in vocational training, college-level education or university (Gouvernement du Québec, Ministère de l'Éducation, du Loisir et du Sport, 2008a, p. 23)

In the two General Education Paths for Secondary Cycle Two, the Mathematics, Science and Technology subject area consists of the same two content areas as in Secondary Cycle One: (a) mathematics, and (b) science and technology. In the General Education Path, the same four major areas continue from Cycle One, but the compulsory concepts for each change (Gouvernement du Québec, Ministère de l'Éducation, du Loisir et du Sport, 2008b, see p. 67). In the Applied General Education Path, Year One (Grade 10) has only three major areas of study: The Material World, The Living World, and The Technological World. Year Two (Grade 11) has all four major areas of study (Gouvernement du Québec, Ministère de l'Éducation, du Loisir et du Sport, 2008c, see pp. 44-45 and 60-61 respectively).

The third path, the Work-Oriented Training Path, is offered in work-study programmes and can lead to work, vocational training programmes, or continued studies in general education. Students have elective courses to broaden their educational choices and to take advantage of the various bridges between the paths. The Work-Oriented Training Path offers 170 vocational training programmes in 21 sectors. The 21 sectors are: Administration, Commerce and Computer Technology; Agriculture and Fisheries; Food Services and Tourism; Arts; Woodworking and Furniture Making; Chemistry and Biology, Buildings and Public Works; Land Use Planning and Environment; Electrotechnology, Motorized Equipment Maintenance; Mechanical Manufacturing; Forestry and Pulp and Paper; Communications and Documentation; Maintenance Mechanic; Mining and Site Operations; Metallurgical Technology; Transportation; Fashion, Leather and Textiles; Health; Social, Educational and Legal Services; and Beauty Care (see http://www2.inforoutefpt.org/guide/anglo/programmes_sec.asp).

The Atlantic Provinces

Canada's Atlantic provinces (New Brunswick, Newfoundland and Labrador, Nova Scotia, and Prince Edward Islands) established The Atlantic Provinces Education Foundation (APEF) in

September 1995 with the mandate to establish common educational foundations for Grades 1 to 12 in Atlantic Canada. The common foundations include a foundational curriculum and Essential Graduation Learnings (EGL's) required for students to graduate from secondary school. In addition, curricular groups within the Atlantic Provinces Education Foundation developed overarching guideline for each subject area. The document titled *Foundation for the Atlantic Canada Technology Education Curriculum* (APEF, 2001) sets foundational policies for the study of technology in all Atlantic Provinces. The study of technology in each Atlantic province draws on the Foundation document and the needs of the province.

New Brunswick. In the recent past, New Brunswick used the term *Broad-Based Technology* to encompass all programs and courses for the study of technology. More recently, the term *Technology/Vocational Education* is used. Under this umbrella, the term *Technology Education* is used for Grades K-8 courses. In Kindergarten to Grade 5, the study of technology is software applications. In middle school, Grades 6 to 8, Technology Education is a blend of hands-on modules and software applications, with a focus on wood, manufacturing, culinary, and information technologies. The goal is that students be able to use the different technologies and understand their relationships. In Grades 9 and 10, a course titled *Introduction to Applied Technology* (with some schools still using the prior term *Broad-Based Technology*) provides a modular, hands-on, computer-directed approach. The purpose of Technology Education in the middle school years is career exploration (New Brunswick Ministry of Education, 2000). In Grades 11 and 12, Technology Education focuses on specific knowledge, skills, and safety. A more advanced level to the *Introduction to Applied Technology* course is offered, again using a modular approach. Some schools offer, in combination with this course, concentrated technical courses, for example, concentrated courses in automotive, welding, construction, and metal (New Brunswick Ministry of Education, n.d.).

There are also *Vocational Education* courses available in some schools. Examples of these courses are Residential Finish, Framing and Sheathing, Mill and Cabinet Work, Internal Combustion Engines, Automotive Electrical Systems, Robotics and Automated Technology, Micro Electronics, and Introductory Electronics. The concentrated skill courses in various technical areas and the Vocational Education course offerings vary widely across the province, and decisions about course offerings appear to be at the level of individual secondary schools. One can grasp the decentralised nature of course offerings by visiting the websites of the secondary schools in this province. Apprenticeship programmes are also available outside of school time through the New Brunswick Youth Career Connections Program (NBYCC).

Newfoundland and Labrador. Newfoundland and Labrador uses the term *Technology Education.* Five general curriculum outcomes in the document *Foundation for the Atlantic Canada Technology Education Curriculum* (APEF, 2001) define the intent of the programme: technological problem solving, technological systems, history and evolution of technology, technology and careers, and technological responsibility. At the intermediate level (Grades 7 to 9 the Technology Education curriculum consists of five technology modules: A Communications Technology Module in Grade 7; two modules (1) Production, and (2) Control Technology in Grade 8: and two modules (1) Energy and Power, and (2) Biotechnology in Grade 9 (Government of Newfoundland and Labrador, Department of Education, n.d.a).

At the senior level (Grades 10 to 12), a programme called *Technology Education and Industrial Arts* is offered, which includes six courses: design technology, communications technology, computer technology, integrated systems, woodworking, power mechanics, and home maintenance (Government of Newfoundland and Labrador, Department of Education, n.d.b).

Nova Scotia. Nova Scotia uses the term *Technology Education.* As with other Atlantic Provinces, the course content is based on the five general curriculum outcomes defined in the document *Foundation for the Atlantic Canada Technology Education Curriculum* (APEF, 2001). Technology Education begins in Grade 7 where the focus is on Communication Technology. In Grade 8, Production Technology, and Energy, Power and Transportation are studied. In Grade 9, students learn Control Technology and Biotechnology. In addition, Information Technologies is offered as a content area from Grades 6 to 7 (Nova Scotia Department of Education, 2004). At the secondary school level, courses include Communications Technology; Construction Technology; Computer related studies; Data Processing; Design; Electrotechnologies; Energy, Power and Transportation; Film and Video Production; Exploring Technology; Production Technology; Word/Information Processing; and Word/Information Technology.

There is also a programme called *Options and Opportunities* (O2) where students in Grades 10 to 12 can link school learning to the workplace (Nova Scotia Department of Education, n.d.a). O2 programs are based on the concept of Career Academies in the high school, or 'schools-within-schools'. Some of the trades and technologies offered through O2 are: Business Education, Health and Human Services; Hospitality and Tourism; IT; and Arts, Culture, and Recreation. Presently, *Expanded Course Options* are: LifeWork Transitions; Tourism; Design; Electrotechnologies; Agriculture/Agrifood; Business Management; Film and Video Production; Food Science; Health/Human Services; Multimedia; Geomatics; Food Services; Textile Technology; and Food Studies/Hospitality (Nova Scotia Department of Education, n.d.b).

Prince Edward Island. In elementary school, the study of technology in the province of Prince Edward Island (PEI) is related to the integration of information and communications technologies across all subject areas and students' learning (Prince Edward Island Department of Education, 2008a). The study of computer technology continues through the intermediate years – Grades 7 to 9 – in a subject area called *Communication and Information Technology* (Prince Edward Island Department of Education, 2007). In Grades 10 to 12, under the umbrella of English Programs five courses in this subject area are offered: Information Technology Communication; Creative Multimedia; Introductory Computer Studies; Computer Studies; and Computer Literacy (Prince Edward Island Department of Education, 2008b). In Grades 7 to 9, an additional subject area called *Intermediate Industrial Technology* is added to the curriculum:

> Intermediate Industrial Technology Education provides modular curriculum components designed to achieve the general curriculum outcomes (GCOs) for technology education.
> A. Technological Responsibility
> B. Technological Systems
> C. History and Evolution of Technology
> D. Technology and Careers
> E. Technological Problem Solving. (Prince Edward Island Department of Education, 2007, p. 18)

These GCOs are from the document *Foundation for the Atlantic Canada Technology Education Curriculum* (APEF, 2001). Unlike Nova Scotia, PEI implements Intermediate Industrial Technology courses in a very generic manner. The curriculum document suggests list of materials and tools but does not specify any grade-by-grade delineation of the curriculum, nor does it suggest any specific technical area through which the GCOs could be learned.

At the secondary school level, prior to 2008, two kinds of courses were offered for the study of technology. *Pre-vocational Technical Courses* offered introductory courses in Woodworking Fundamentals, Welding Fundamentals, Drafting Fundamentals, Power Mechanics, and Electricity and Electronics. *Career/Technical Courses* offered courses in Carpentry, Hospitality and Tourism, Cosmetology, Drafting Technician, Electrical Wiring, Motor Vehicle Repair, Welding, and Industrial Education.

In April 2008, the Prince Edward Island Department of Education released a discussion paper that effectively re-envisions the study of technology in PEIs high schools: "The Grade 10 curriculum would place an emphasis on personal introspection and career exploration while Grade 11 and Grade 12 career-related curricula would focus more on community experiences and transition planning" (Prince Edward Island Department of Education, 2008c, p. 18). The term *Technology Education* is mentioned within the Grades 11 and 12 context, as are terms such as community-based learning experiences and course/credit, career academies (schools within schools), apprenticeship learning opportunities and apprenticeship courses/credits, and real world use of technology. The consultation process ended in June 2008 and will no doubt lead to a very different programme for the study of technology in PEI's senior high schools.

COMPARISONS AND CONTRASTS

An analysis of the above reporting reveals that regional differences influence the interpretation of the study of technology and the courses offered within this subject area. For example, in Alberta and Saskatchewan, agricultural and natural resource courses are important components of the study of technology. Some provinces/territories offer both academic and vocationally-oriented routes to study technology, while others offer only academic routes, offering vocationally-oriented programmes only as postsecondary education, for example, the Northwest Territories and Nunavut. Developments since I last reported on the study of technology in Canada (Hill, 2003) have been significant, but not consistent. Some provinces appear to be moving towards a more vocational interpretation of the study of technology, while others have moved away from this interpretation. More than half of the provinces/territories have undergone significant review of the curriculum for the study of technology, and some are still in the process of revision. One important feature of new curricula is the focus on the environment, sustainability, and environmental stewardship.

All provinces/territories have required courses and optional courses for graduation from secondary school. Science is always listed as a required course for graduation. One province, Alberta, and one territory, the Northwest Territories, require the study of technology, in some form, for graduation. The other provinces/territories do not require the study of technology for graduation. Rather, it is linked to a selection *between* courses, as with British Columbia, Yukon, Nunavut, Saskatchewan, Ontario, Québec, New Brunswick, Newfoundland and Labrador, and Nova Scotia, which means a student could leave secondary school without a course in the study of technology. In Prince Edward Island and Québec there is an alternative vocation graduation route where vocational education credits are required. At the other extreme, Manitoba does not require the study of technology, and the academic route in PEI does not require the study of technology. This said, all provinces include the study of technology in their listings of optional courses for graduation. Table 1 displays required graduation courses as they pertain to the study of technology.

*Table 1. Requirement of the study of technology for graduation.**

Province/Territory	Technology as a graduation requirement
British Columbia	Fine Arts **or** Applied Skills
Yukon	Applied Skills
Alberta	Knowledge and Employability Occupational course, **or** Career and Technology Studies (CTS), **or** Locally developed course with an occupational focus **AND** Knowledge and Employability Workplace Practicum course, **or** Work Experience course, **or** Green Certificate course, **or** Registered Apprenticeship Program (RAP) course
Northwest Territories	Career and Technology Studies
Nunavut	Uses Alberta Curriculum
Saskatchewan	Arts Education **or** Practical and Applied Arts
Manitoba	None
Ontario	One Science (Grade 11 or 12) **or** Technological Education (Grades 9-12). Two other credits of Science are already required. Additional courses in Technological Education can be obtained through Optional Credits
Québec	Science and Technology
New Brunswick	Science to be selected from **one of** Physics, Biology, Chemistry, Environmental Science, Robotics and Automated Technology, Automotive Electrical Systems, Micro Electronics, Introductory Electronics, or Geography
Newfoundland & Labrador	In addition to existing Mathematics and Science requirements, additional courses in Mathematics **or** Science **or** Technology Education
Nova Scotia	In addition to existing Mathematics and Science requirements, additional courses in Mathematics **or** Science **or** Technology
Prince Edward Island	1. Regular high school graduation – None 2. High school graduation with a focus on vocational education – 8 vocational credits required

* All provinces/territories have the study of technology as elective courses required for graduation.

One of the most important developments in Canada over the past two decades is the focus of the study of technology at the elementary school level, whether it is as science and technology, the use of computers, or a discrete subject area. More provinces/territories have introduced the study of technology as a new content area in the elementary schools, while others have re-envisioned it. This focus, in the long term, can only benefit the study of technology at the secondary school level as graduates from this system become more aware of technology, its importance to the human-built world, and the need to think of human impacts on the natural world. The historical pattern in Canada of landscape, lifescapes, and politics influencing the provincial/territorial curricula for the study of technology will no doubt remain, as the Canadian government is committed to meeting local educational needs.

CRITIQUE AND CONCLUSION

As is the case internationally, the study of technology in Canada is both vocational and academic in nature. The place of the study of technology in the curriculum was the theme of the 2003 PATT 13 conference in Glasgow, Scotland. In response to what seemed an artificial need for determining if the study of technology was vocational or academic, Hill (2003b) wrote: ". . . in a larger picture, technology is many different things. History documents this, and the world outside of school shows us this" (p. 91). In conclusion, she stated:

> If the study of technology offers an array of possibilities and students learn with both head and hand as part of the full understanding of any one technology, then the question, Is it vocational (typically associated with hand) or academic (typically associated with head)? becomes a mute point. What the student chooses to do with their study of technology becomes the actual deciding factor as to whether their study of technology is vocational, craft, or academic. (pp. 91–92)

The point here is that schools offer the study of technology and students take technology courses for a variety of reasons – and in more modern times, the study of technology embraces both head and hand, regardless of whether it is set up as vocational or academic. This is the case in Canada.

Regardless of orientation, over the past 25 years we have moved forward from an industrial revolution model for the study of technology. Higher-order thinking skills are critical to think about technology and its place in the human-built world, and to consider the impact of artefacts or systems on humans and the environment. The study of technology is no longer about producing widgets thoughtlessly, as in an industrial revolution model. Instead, it concerns itself with thinking about technology and acquiring skills and knowledge through critical lenses. For example, we now know about the importance of conceptual and procedural knowledge (McCormick, 1997). In Canada, an important impetus for this lens is the infusion of sustainability and the environment into school curriculum. Hill (1998) writes about using courses in the study of technology to connect the real-world to the school-world of our youth as a way to address such issues.

In examining curricular documents from the provinces/territories of Canada, it becomes clear that there is variation in modern interpretations of the study of technology. Some provinces offer two distinct pathways – one academic and one vocational, some only one, and others offer programmes that are a 'melting pot' of both orientations. In the case of schools with vocational orientations, programmes are offered in special high schools or in regular schools as 'schools-within-schools'. This latter model seems to be growing in popularity in several provinces. On a very pragmatic note, in Canada, where the study of technology offers various orientations, more students complete secondary school; more students' needs are met, leading to motivation and self-confidence; schools address a wider variety of student intelligences and learning styles; and in situations where a variety of technological processes are used to think about technology, creativity is enhanced. Schooling thus takes on characteristics of authentic learning environments (Hill & Smith, 2005). In being more inclusive of varied student needs, schools become more democratic places for our youth (Dewey, 1916).

In provinces/territories where an academic orientation to the study of technology is favoured, heed must be given to Arendt's (1958) words where she expresses the interconnections of thought and action: "the thought process by itself no more produces and fabricates tangible things, such as books, paintings, sculptures and compositions, than usage by itself produces and fabricates houses and furniture" (p. 169).

In provinces/territories where the study of technology is interpreted as a fairly rigid design process, heed must be given to other countries who have adopted this route and who have focused on design paperwork; where students who take technology courses see technology as 'more paperwork'. Students typically take courses in technology for reasons other than paperwork alone; they get lots of this in other courses. Rather, the experiences students get in technology courses must be ones that connect to students' lives and is important to them as developing adults (Dewey, 1977). Further, narrow experiences do not lead to as deep an understanding of the study of technology as when both conceptual and procedural knowledge are used: "Looking at technology as formalised practice links technology directly to culture, because culture is a set of socially accepted practices and values" (Franklin, 1993, p. 15). Wiener (1993) states that invention is complete only after ideas are refined, as they are worked out in materials and processes. Technology is, after all, the human-built world to meet human needs and should be examined in a context of human and environmental considerations through open-ended problem posing (Lewis, Petrina, & Hill, 1998) and problem solving (Hill, 1998). In this way, as Wiener suggests, ideas are refined through materials and processes.

Recent trends in some provinces/territories have been to move a few steps back towards specialisation. Implementation of such changes needs to be interpreted in modern understandings of and research in the study of technology to ensure that we are not moving backwards, but instead are learning from the past and moving forward. In doing so, perhaps we have made a next step in our understanding of this curricular area – understanding that the study of technology does not have to be either vocational or academic. Perhaps a modern understanding embraces both, as suggested by Arendt (1958), Franklin (1993), and Wiener (1993).

REFERENCES

Arendt, H. (1958). *The human condition*. Chicago: University of Illinois Press.

Alberta Education. (2008a). *Senior high school programs publishers' draft: Career and technology studies. Program rational and philosophy*. Edmonton, AB: Author.

Alberta Education. (2008b). *Senior high school programs publishers' draft: Career and technology studies. Trades, manufacturing & transportation (TMT). Scope and sequence*. Edmonton, AB: Author.

Atlantic Provinces Education Foundation [APEF]. (2001). *Foundation for the Atlantic Canada technology education curriculum*. Halifax, NS: Author.

British Columbia Ministry of Education. (1995). *Technology education 8 to 10. Integrated resource package 1995. IRP 014*. Victoria, BC: Author.

British Columbia Ministry of Education. (2004). *Supplement to the technology education 8 to 10 IRP (1995). Required program model content for technology education 10: Drafting and design, electronics, mechanics, metalwork, woodwork 10*. Victoria, BC: Author.

British Columbia Ministry of Education. (2007a). *Program guide for secondary school apprenticeship*. Victoria, BC: Author.

British Columbia Ministry of Education. (2007b). *Planning 10. Integrated resource package 2007. IRP 152*. Victoria, BC: Author.

British Columbia Ministry of Education. (n.d.a). *Applied skills integrated resource packages for Grade 11 and 12 Technology Education*. Retrieved February 1, 2009, from http://www.bced.gov.bc.ca/irp/irp_appskills.htm

British Columbia Ministry of Education. (n.d.b). *Information and Communication Technology integrated resource packages for Grades 8 to 10, and 11 and 12*. Retrieved February 1, 2009, from http://www.bced.gov.bc.ca/irp/irp_infotech.htm

Chinean, C. A., Oaks, M. M., & Boutin, F. (1995). A national consensus on technology education in Canada. *Journal of Industrial Teacher Education, 32*(2), 76–92.

Council of Ministers of Education, Canada [CMEC]. (1997). *K to 12. Common framework of science learning outcomes*. Toronto, ON: Author.

CMEC. (2005). *Education in Canada*. Toronto, ON: Author.

Department of Education, Government of Yukon. (2007). *Yukon secondary school apprenticeship training*. Retrieved April 1, 2008, from http://www.education.gov.yk.ca/advanceded/apprenticeship/yssap.html

Department of Education, Government of Yukon. (2008a). *Curriculum*. Retrieved April 1, 2008, from http://www.education.gov.yk.ca/psb/curriculum.html

Department of Education, Government of Yukon. (2008b). *Yukon designated occupations for certification*. Retrieved April 1, 2008, from http://www.education.gov.yk.ca/advanceded/apprenticeship/certification.html

Dewey, J. (1916). *Democracy and education*. New York: MacMillan.

Dewey, J. (1977). *Experience and Education* (20th printing). New York: MacMillan Collier.

Franklin, U. (1990). *The real world of technology*. Toronto, ON: Anansi Press.

Gardner, P., & Hill, A. M. (1999a). Technology education in Ontario: Evolution, achievements, critique and challenges. Part 1. *International Journal of Technology and Design Education, 9*(2), 103–136.

Gardner, P., & Hill, A. M. (1999b). Technology education in Ontario: Evolution, achievements, critique and challenges. Part 2. *International Journal of Technology and Design Education, 9*(3), 201–239.

Gouvernement du Québec, Ministère de l'Éducation. (2001). *Québec education program (approved version). Preschool education, primary education.* Québec, QU: Bibliothèque nationale du Québec.

Gouvernement du Québec, Ministère de l'Éducation. (2004). *Québec education program. Secondary school education. Cycle one.* Québec, QU: Bibliothèque nationale du Québec.

Gouvernement du Québec, Ministère de l'Éducation, du Loisir et du Sport. (2006). *Education in Québec: An overview*. Québec, QU: Bibliothèque nationale du Québec.

Gouvernement du Québec, Ministère de l'Éducation du Loisir et du Sport. (2008a). *Québec education program. Secondary school education, Cycle two. A curriculum for the 21st century.* Québec, QU: Bibliothèque nationale du Québec. Retrieved July 28, 2008, from http://www.mels.gouv.qc.ca/sections/programmeformation/secondaire2/ medias/en/1_QEP_Chap01.pdf

Gouvernement du Québec, Ministère de l'Éducation du Loisir et du Sport. (2008b). *Québec education program. Secondary school education, Cycle two. Science and technology.* Québec, QU: Bibliothèque nationale du Québec. Retrieved July 28, 2008, from http://www.mels.gouv.qc.ca/sections/programmeformation/secondaire2/medias/en/6c_QEP_ ScienceTechno. pdf

Gouvernement du Québec, Ministère de l'Éducation du Loisir et du Sport. (2008c). *Québec education program. Secondary school education, Cycle two. Applied science and technology.* Québec, QU: Bibliothèque nationale du Québec. Retrieved July 28, 2008 from http://www.mels.gouv.qc.ca/ sections/programmeformation/secondaire2/medias/en/6d_ QEP_ApplicTech.pdf

Government of Newfoundland and Labrador, Department of Education. (n.d.a). *Program of studies 2007–2008: Intermediate program.* Retrieved February 1, 2008, from http://www.ed.gov.nl.ca/ edu/sp/pos/2007-08/7-intermediate.pdf

Government of Newfoundland and Labrador, Department of Education. (n.d.b). *Program of studies 2007–2008: Senior high program*. Retrieved February 1, 2008, from http://www.ed.gov.nl.ca/edu/sp/pos/ 2007-08/8-senior-high.pdf

Government of the Northwest Territories, Department of Education, Culture and Employment. (2004). *K-6 Science and technology curriculum.* Yellowknife, NMT: Author.

Government of Nunavut, Department of Education. (n.d.). *Program of Studies Grades 7–12*. Retrieved March 4, 2008, from http://www.gov.nu.ca/education/eng/css/progstudies7_12.htm

Government of Saskatchewan, Saskatchewan Learning. (2006). *Practical and applied arts handbook.* Retrieved April 1, 2008, from http://www.sasklearning.gov.sk.ca/docs/paa/PAAHandbook/index.html

Gradwell, J. B. (2003). Technology education in Canadian secondary schools. *Canadian Journal of Science, Mathematics and Technology Education, 3*(1), 17–35.

Hill, A. M. (2003a). Technology and its study in Canada: A mosaic. *Canadian Journal of Science, Mathematics and Technology Education, 3*(1), 5–16.

Hill, A. M. (2003b). An analysis of the debate: Has the study of technology a vocational or academic purpose? In J. R. Drakers & M. J. de Vries (Eds.), *Pupils attitudes toward technology: Thirteenth international conference on design and technology* (pp. 87–92). Glasgow, UK: University of Glasgow, Faculty of Education.

Hill, A. M. (1998). Problem solving in real-life contexts: Alternatives for design in technology education. *International Journal of Technology and Design Education, 8*(3), 203–220.

Hill, A. M. (2006). Reflections on twenty years of wandering through the pathways and forest of technological education in Ontario, Canada. In M. de Vries & I. Mottier (Eds.), *International handbook of technology education: Reviewing the past twenty years* (pp. 133–170). Rotterdam, The Netherlands: Sense.

Hill, A. M., & Smith, H. A. (2005). Research in purpose and value for the study of technology in secondary schools: A theory of authentic learning. *International Journal of Technology and Design Education, 15*(1), 19–32.

Industry Training Authority. (2008). *Youth ACE IT program.* Retrieved April 1, 2008, from http://www.itabc.ca/Youth-ACEIT.php

Kozolanka, K., & Medway, P. (1996). *Technical and technological education in Canadian schools: An interview survey in nine provinces.* Unpublished manuscript.

Lewis, T., Petrina, S., & Hill, A. M. (1998). Problem posing – Adding a creative increment to technological problem solving. *Journal of Industrial Teacher Education, 36*(1), 6–36.

Manitoba Education, Citizenship and Youth. (2007). *Subject table handbook. Technology education student records system and professional school personnel system. September 2007–June 2008.* Winnipeg, MB: The Crown in Right of Manitoba.

Manitoba Education, Citizenship and Youth. (2003). *Middle years home economics/industrial arts: Linking learning to living.* Winnipeg, MB: The Crown in Right of Manitoba.

Manitoba Education, Citizenship and Youth. (n.d.). *Subject table guide. Technology education. Effective date – September 2007-June 2008.* Retrieved March 4, 2008, from http://www.edu.gov.mb.ca/k12/docs/policy/sthte/ tecguide07.pdf

McCormick, R. (1997). Conceptual and procedural knowledge. *International Journal of Technology and Design Education, 7*(1–2), 141–159.

New Brunswick Ministry Department of Education. (2000). *Grades 9/10 companion document.* Fredericton, NB: Curriculum Development Branch.

New Brunswick Ministry of Education. (n.d.). *Vocational/Technology education.* Retrieved June 1, 2008, from http://www.gnb.ca/cnb/televox/video/techvoc_e.asx.

Nova Scotia Department of Education. (2004). *Learning outcomes framework: Grades 7 to 9.* Retrieved April 1, 2008, from http://www.ednet.ns.ca/pdfdocs/outcomes/by_grade/grades_7-9.pdf

Nova Scotia Department of Education. (n.d.a). *Options & opportunities. O2 description.* Retrieved April 1, 2008, from http://www.ednet.ns.ca/O2/e/index.shtml

Nova Scotia Department of Education. (n.d.b). *Options & opportunities. How does it work? Expanded course options.* Retrieved April 1, 2008, from http://www.ednet.ns.ca/pdfdocs/outcomes/by_grade/grades_7-9.pdf

Ontario Ministry of Education. (1998). *The Ontario curriculum grades 1–8: Science and technology.* Toronto, ON: Queen's Printer for Ontario.

Ontario Ministry of Education. (2007a). *The Ontario curriculum grades 1–8: Science and technology.* Toronto, ON: Queen's Printer for Ontario.

Ontario Ministry of Education. (2007b). *Consultation documents for technological education.* Toronto, ON: Author.

Ontario Ministry of Education. (2008). *Memorandum: Revised course descriptions for grades 10–12 computer science, grade 9–12 science, and grades 9-12 technological education.* Toronto, ON: Author.

Prince Edward Island Department of Education. (2007). *Intermediate program of studies and authorized materials. 2007–2008.* Retrieved August 1, 2008, from http://www.gov.pe.ca/photos/ original/ed_ips_0708.pdf

Prince Edward Island Department of Education. (2008a). *Elementary program of studies and authorized materials. 2008–2009.* Retrieved February 16, 2009, from http://www.gov.pe.ca/photos/ original/ed_elemps_0809.pdf

Prince Edward Island Department of Education. (2008b). *Senior high program of studies and list of authorized materials. 2008–2009.* Retrieved February 16, 2009, from http://www.gov.pe.ca/ photos/original/ed_sps_0809.pdf

Prince Edward Island Department of Education. (2008c). *Senior high school model. Discussion paper.* Retrieved June 1, 2008, from http://www.gov.pe.ca/photos/original/ed_shsm_08.pdf

Wiener, N. (1993). *Invention: The care and feeding of ideas.* Cambridge, MA: MIT Press.

Ann Marie Hill
Faculty of Education
Queen's University
Kingston, Ontario
Canada

HOWARD MIDDLETON

8. TECHNOLOGY EDUCATION AND RESEARCH IN AUSTRALIA: A COMING OF AGE IN THE 21ST CENTURY

INTRODUCTION

This chapter provides an analysis of the development of the field of technology education in Australia over the last decade. Three lines of argument are advanced. Firstly, while there have been significant advances since the introduction of the Technology Key Learning Area (KLA), there are important challenges to its further development. Secondly, there is a greater level of cooperation between technology education researchers, administrators, schools and teachers, than previously, aimed at improving the quality of technology education in schools. Finally, technology education researchers are discovering, through necessity, new ways of researching teaching and learning in technology education, which are contributing to educational research methods generally. However, there remain within this picture tensions and contradictions that need to be resolved.

A little over ten years ago I presented an analysis of technology education in Australia (Middleton, 1997). At that stage the eight key learning areas were still relatively new, with the National curriculum documents having been produced three years before (Curriculum Corporation, 1994a; 1994b). Agreement existed across the states and territories that the Technology KLA would become the basis for individual states' curriculum and syllabus documents, at least up to the end of junior high school (Year 10, or approximately 16 years of age). The picture in terms of research at that time was less encouraging, with few research projects and fewer technology education academics possessing doctoral degrees or the research experience necessary to be successful with research grant applications.

The new Technology KLA was defined as including industrial technology, information and communications technology, home economics, business, agriculture, and media. A professional association, the Technology Education Federation of Australia (TEFA), was established with a board made up of representatives from existing subject associations and a number of cross-curricular associations such as the Design in Education Council of Australia (DECA). TEFA was established to provide support for the new curriculum directions, and was a response to the then-federal government desire to channel funding for such activities as professional development through professional associations rather than state departments of education. As a single national body, TEFA acted as the broker between the federal government and the various professional associations, which, being largely state-based, consisted of over thirty groups.

Progress in implementing the Technology KLA has been relatively slow for a variety of reasons. In primary schooling, the introduction of the eight KLA's determined that change would be required in all areas, not just in technology. States adopted a staged approach, which determined that the introduction of technology was governed by its place in the schedule for implementation within each state. Other issues such as a lack of sufficient professional development for teachers who had not previously taught technology education, and who were not

A.T. Jones and M.J. de Vries (eds.), International Handbook of Research and Development in Technology Education, 85–92.
© 2009 Sense Publishers. All rights reserved.

trained to do so, also tended to act as a brake on progress. Despite this, there has been a lot of enthusiasm from primary teachers (Williams, 2001, 2006), who see the value of technological activities in the learning experiences of primary-aged students and the potential of the Technology KLA as an integrator with other areas.

At the secondary level the picture has been mixed, with schools responding in various ways. Some regarded the Technology KLA, with its emphasis on a design process, as a natural extension and endorsement of the existing approach already adopted. Others saw the KLA as a threat to what they regarded as a quasi-vocational area of the curriculum and a new direction they were ill-equipped to address. Added to this was the issue in some schools of working out how to create a single Technology KLA in a situation where a school might have five existing subject areas that could make a claim to the title.

TECHNOLOGY EDUCATION IN THE 21ST CENTURY

Moving forward to 2008, the picture has predictably changed again in all states. The most positive development over the last decade has been the more general introduction of technology education into primary schooling and primary teacher education programmes. However, the most pervasive current issue is the vocational education and training (VET) agenda, introduced by the previous conservative government and re-emphasised by the current labour government. Added to this is the critical shortage of technology teachers and the difficulty of attracting students into teacher education programmes. This shortage is matched by a shortage of technology education academics.

Australia has a governance and education system/s similar to the USA in that the federal government provides some of the funding for education but the organisation and delivery of education is undertaken by state governments. Given this, it would be impractical to deal with the minutia of differences in policy and practice across the states. I have taken the approach of outlining changes that are occurring across all states. Although there may be variation in how they are occurring, they are still there. For example, there is a strong VET agenda in all states, although there is some variation across states in the details of how it is being implemented. The balance of this paper will draw examples from Queensland to support or illustrate my arguments.

Primary Technology Education

One of the most exciting developments over the last decade has been the implementation of technology education in primary schooling. The introduction has not been widespread and some of the practices may be questionable, such as instances where technology education is equated with the more narrowly defined information and communication technology (ICT). However, where it has been introduced in the spirit and intent of the Technology KLA, the results in terms of student learning have been impressive. For example, interviews with the principal and some of the teachers at a particular primary school, and observations of a year 4/5 composite class presenting their technology projects (Williams, 2001), suggested that students had a sophisticated understanding of the nature of technology and of how they might use that knowledge in living their lives and making decisions. The projects being undertaken in the school were such that they integrated other areas of the curriculum so there appeared to be no sense in which teachers thought that technology education was something extra to fit into their teaching programmes. It needs to be said, however, that the school in question had sought and received significant professional development support.

Primary Teacher Education

A related development over the last ten years has been the introduction of technology education into primary teacher education programmes (Chester & Barnes, 2002). This has been significant, given that Watts (1997) found that many primary teachers felt they were not sufficiently qualified to teach either science or technology and as a consequence tended to avoid or minimise these areas of the curriculum. Watts (ibid) further noted that the background of many primary teachers contributed to this feeling of a lack of capacity, given that many come to teaching with a social science/humanities background and inclination.

Chester reported in 2002 that of the 33 out of 39 universities in Australia offering primary teacher education programmes, most included a technology education component and about a third stipulated technology education as a compulsory course. Taking the Bachelor of Education (Primary) at Griffith University as an example, the response to these courses appears to have been positive, despite the course being only 10 credit points over one semester. Specifically, students in the Bed (Primary) have requested further courses, and a number take courses in the secondary Bachelor of Technology Education programmeme (BTechEd) as electives.

The Vocational Education and Training (VET) Agenda

Over the last decade, an increasing shortage of skilled workers has occurred in Australia in a range of industries, including building and construction, mining, and trade areas generally. The main reason for this is a strengthening economy flowing from increasing raw materials exports, principally to China and India. This has led to strong demand for skilled labour in all areas of the minerals, energy, and other raw materials and associated industries. The strong demand has led to attractive salaries for work in these areas. However, many of these jobs are in regional and remote areas of Australia and have attracted skilled workers from major population centres. This has led to calls for increased temporary migration visas, an increase in vocational education and training (VET) provision, and initiatives to make VET qualifications more attractive (Curriculum Corporation, 2005; Smith, 2004). This has been achieved by a combination of strong marketing and mechanisms such as the establishment of pathways between VET qualifications and university programmes (Pavlova, in press). For example, the previous conservative government responded to the shortage by introducing a number of initiatives designed to introduce VET courses into the upper years of schooling to encourage more students to enter trades.

With the election of the labour government in November 2007, the VET in schools initiative gained greater impetus, with the government committed to having a new VET centre in all high schools. Communications with the new Minister for Education (Williams, P.J., personal communication, 2008) suggest that there is little understanding of the concept of technology education as a component of general education. In the response to a letter to the Minister seeking clarification of the role the new government sees for technology education, the Minister appeared to see technology education as either computer education or VET. Whilst it is still early days for the government, this view represents a serious threat to the work of establishing technology education as a key learning area.

Teacher Education Programmes

An issue that has emerged over the last three years, related to the skills shortage noted above, is the difficulty of attracting sufficient numbers of suitably qualified students into technology education programmes. Most technology teacher education programmes have intakes that consist of varying proportions of school leavers and mature entry students (Chester & Barnes, 2002). Many of this latter group have a trade or industry background, and this usually gains them credit and reduces the length of the programme for them. However, with the increasing skill shortages and increasing wages on offer to tradespeople, the mature entry cohort has virtually disappeared in many universities. In addition, the possibility of obtaining generous wages for even semi-skilled or unskilled occupations has had the effect of reducing interest in university study overall, including technology education programmes, by qualified school leavers.

Governments and universities have responded to the shortage of technology teachers in three principal ways. Firstly, they are retraining teachers from areas of the curriculum not in such high demand, such as the social sciences or business, to enable them to teach in areas of high demand. Currently, these include mathematics, science, and technology education. Starting in 2008, Education Queensland has also provided 200 scholarships for teacher retraining, with 30 allocated specifically to retraining teachers to become technology teachers. Secondly, universities are developing one year, graduate diploma courses to attract graduates from areas, such as engineering, architecture, industrial design, and drafting, into technology teaching. Graduate diplomas in other areas have proven to be popular with students. Thirdly, existing programmes are being reviewed and revised with the major purpose of shortening them or making parts available by external mode to reduce the cost to students and to produce graduates in a reduced time frame.

Technology Education Academics

A looming issue in Australia that appears to be mirrored in most western countries is the growing shortage of technology education academics (Williams, 2002). The shortage is occurring in teacher education generally and appears to be related to salaries and career advancement. For example, the salary of a teacher with a basic bachelor's degree can quickly rise to a level that is higher than a mid-range academic, who will need to have three degrees to progress to a level where their salary is equal to a high school head of department or higher. Further, the requirement of doctoral qualifications precludes the previous practice of recruiting experienced and innovative teachers as lecturers. As a consequence, it is often the case that when lecturing positions are advertised, there are no suitable applicants and the positions go unfilled. Although some technology education teachers are currently undertaking doctoral studies, these are almost always part-time so there is a long lead time between commencement of studies and completion, typically around seven years.

NEW LEVEL OF COOPERATION BETWEEN RESEARCHERS, SCHOOLS, AND EDUCATION AUTHORITIES

One factor that has contributed to an increase in technology education research is the connections being developed between researchers, schools, teachers, teacher professional organisations, and education departments. One example to illustrate this is the *Working Together to Improve Practice* project (Middleton & Gooch, 2002). This was an action research project to examine the introduction of new teaching initiatives in six selected schools throughout Queensland. The project was funded by the Industrial Technology and Design Association of Queensland (INTAD). INTAD organised the call for expressions of interest to participate in the project and funded the six school-based projects as well as the parallel action research. The collaboration was important in providing some interesting and unexpected findings. For example, because the project involved schools in both metropolitan and rural locations, a monthly teleconference was built into the project. This was seen initially as a convenient way for teachers to provide feedback and was conceptualised as a form of focus group. However, an interesting phenomenon occurred over the course of the project: Although teachers were initially reluctant to share anything but good news about the project, by about the third teleconference this changed and teachers were engaging with each other, seeking help and ideas, and providing feedback to the research team. The teleconferences thus developed a life of their own that was nevertheless consistent with an action research methodology.

The national research project chaired by Williams (2001) is another example of collaboration where researchers had high levels of cooperation from schools in conducting a national study of the status of technology education in primary and secondary schools across Australia. Although such access may not seem unusual to researchers outside Australia, state governments within Australia have generally been reluctant to become involved in comparative studies that may show one state in an unfavourable light with respect to another.

Evidence of collaboration has extended more recently to state departments of education, with technology education researchers invited and funded to undertake research by these departments. A recent example is the *Researching School Change in Technology Education* project (Middleton, Chester, Finger, Jamieson-Proctor, & Pavlova, 2004). This research project examined teacher change and adjustment of practice in response to the new Technology syllabus. The project brief was developed collaboratively by Griffith academic staff and technology education consultants from Education Queensland (EQ), and EQ officers facilitated access to schools for data collection. This was crucial in terms of rural schools where no established relationship existed or would have been impractical without EQ assistance.

Another example can be found in the decision by Education Queensland to provide full funding for 20 exemplary technology teachers (a mix of primary and secondary teachers) to attend the International Technology Education Research Conference (TERC08) hosted by Griffith University and held every two years on the Gold Coast. This is a considerable commitment to ensuring that the connection between technology education researchers and technology teachers is established and maintained.

A final example to illustrate the levels of cooperation is the access to schools being provided to university students, including honours, masters and doctoral research students. This is particularly encouraging given that it is occurring at a time of increasing constraints in terms of issues such as ethics requirements, where schools might well have responded that it was too difficult to have researchers working with teachers, and particularly students. If anything, the opposite is happening, with some schools initiating contact with researchers. A recent project by Walmsley (2008) in schools that was reasonably invasive involved a complex data collection system including simultaneous videorecording of student classroom activity with one camera and teacher activity with another, with a system for recording both activities as parallel data streams. This was followed by stimulated recall sessions with the teachers. Walmsley was able to achieve high levels of cooperation from teachers and school administrators, who were prepared to go to some lengths to accommodate his research and were interested in exploring the implications of his findings for classroom practice.

NEW WAYS OF RESEARCHING TECHNOLOGY EDUCATION

Research in technology education in Australia is a relatively recent phenomena related to changes in the higher education sector. Prior to the early 1990s, higher education in Australia was comprised of universities and colleges of advanced education (CAE). Most research was conducted in universities and bodies like the Australian Research Council did not provide research funding for the CAE sector. In addition, CAEs were restricted to providing undergraduate and masters level programmes, so the kinds of research associated with a doctoral programme and the academic staff associated with it did not generally occur within CAEs.

The significance of this for education generally, and technology education specifically, was that most education faculties and all technology education programmes were within the Colleges of Advanced Education (CAE) sector. This changed as a result of the recommendations contained in the Dawkins (1987) paper. Dawkins was the federal Minister for Education and the paper recommended that there be one higher education sector consisting of universities; CAEs were to either amalgamate with existing universities (as happened with the Hunter Institute for Higher Education and Newcastle University) or become universities themselves (as happened with the Canberra CAE, now the University of Canberra). This change determined that research funding and the imperative to undertake research became part of the role for technology education academic staff.

As a result of the change, technology education academic staff began to engage in research, initially to gain doctoral qualifications. However, given the limited number of PhD-qualified academic staff in technology education, and thus few doctoral supervisors within technology education, the research undertaken in these early days was rooted in the traditions and methods of the humanities and social sciences and paid little heed to the particular features of learning and teaching in technology. As an example, much of the early research made the assumption, common in social science research, that human thinking is represented via speech or verbal renditions of thought. This assumption was made despite the fact that much thinking in technology education is multi-modal. That is, thinking is represented by both abstract codes, analogous to text, and by visual mental imagery.

By 2008 this has changed. For example, in a book on research methods and techniques (Middleton, 2008a) six Australian authors (as well as authors from a number of other countries) present chapters on research projects. However, the focus of the book was not on research per se, but on new research methods or techniques (Middleton, 2008b), with 'new' being defined as new to technology education research, or new in the sense that the way the method was being used was new. The chapters included descriptions and analyses of methods including: knowledge elicitation techniques for exploring expertise in Computer Aided Design (CAD) usage (Chester, 2008); cultural-historical approaches to understanding primary students' concept formation in technology education (Fleer, 2008); a new approach to comparative research methods (Pavlova, 2008); a case study approach to uncovering contextualised knowledge (Stevenson, 2008); an information technology enhanced stimulated recall technique for understanding what 'presses' technology education students into engaging in higher-order thinking (Walmsley, 2008); and a method for capturing and analysing concurrent verbal and visual data or technological problem-solving activity (Middleton, 2008c).

TENSIONS AND CONTRADICTIONS

Despite significant developments in technology education over the last decade, a number of tensions and contradictions exist in the current state of technology education in Australia, including:

- Whilst stronger, more coherent bachelor degrees of sufficient length to provide an appropriate balance of breadth and depth of technological knowledge for students have been developed over the last ten to fifteen years, these are being threatened by the strong demand for shorter programmes, such as graduate diplomas, which have difficulty in covering the breadth of content required by a technology education teacher;

- Whilst the introduction of Technology education into primary teacher education programmes has been a recent success, the reduction in four-year primary teacher education courses and the introduction of one-year graduate diplomas reduces the exposure to technology education by students to a portion of one integrated course;
- As a result of research (Pavlova, 2003) and practice (Martin & Middleton, 2003) over the last decade, technology education academics have a clearer concept of what a good technology education programme contains, but against this is the strong vocational direction of both state and federal governments;
- Improved research connections with schools have the potential to be threatened or seen as irrelevant by the more instrumental approach embodied in the competency-based approach to vocational education; and
- Whilst there are now demonstrably better and more research being undertaken into technology education, the current government's emphasis on vocational education means that the possibility of changed priorities for research funding cannot be ruled out.

CONCLUSIONS

This chapter explores the issues and developments in technology education research and practice in Australia in the ten years to 2008. The argument is advanced that important developments in terms of curricula, research, and collaboration between technology education researchers have occurred. However, the chapter also argues that significant challenges exist in terms of shortages of technology teachers and teacher educators, but that important work has been and is being undertaken in terms of research and dissemination of that research within schools. Finally, the chapter identifies a number of tensions and contradictions that confront technology educators as we move forward.

REFERENCES

Chester, I. (2008). Researching expertise development in complex computer applications. In H. E. Middleton (Ed.), *Researching technology education: Methods and techniques* (pp. 70–88). In *International Technology Education Series*. Rotterdam, The Netherlands and Taipei, Taiwan: Sense.

Chester, I., & Barnes, R. (2002). The transformation of Queensland junior secondary school manual arts into technology education: A qualitative analysis of the teacher's role. In Middleton, H. E., Pavlova, M., & Roebuck, R. (Eds.), *Learning in technology education: Challenges for the 21st Century. Proceedings of the 2nd Biennial Technology Education Research Conference*. (Vol. 1, pp. 17–24). Surfer's Paradise: Centre for Technology Education Research, Griffith University.

Curriculum Corporation. (1994a). *Technology statement*. Melbourne, Victoria: Curriculum Corporation.

Curriculum Corporation. (1994b). *Technology statement and profiles*. Melbourne, Victoria: Curriculum Corporation.

Curriculum Corporation. (2005). New framework for vocational education in schools: A comprehensive guide about pathways for young Australians in transition. In *Policy Directions*. Melbourne, Victoria: Curriculum Corporation & Ministerial Council on Employment, Education and Youth Affairs.

Dawkins, J. (1987). *Higher education: A policy discussion paper*. Canberra, ACT: Australian Government.

Fleer, M. (2008). A cultural-historical perspective on research in design and technology education. In H. E. Middleton (Ed.), *Researching technology education: Methods and techniques* (pp. 89–98). In *International Technology Education Series*. Rotterdam, The Netherlands and Taipei, Taiwan: Sense.

Martin, E., & Middleton, H. E. (Eds.). (2003). *Initiatives in technology education: Comparative perspectives. Proceedings of the American-Australian Technology Education Forum*. Brisbane, Queensland: Technical Foundation of America & Griffith University.

Middleton, H. E. (1997). *Technology education in Australia: Moving towards a new basic*. Keynote address at the International Conference on Technology Education in the Asia/Pacific Region (ICTE 97). National Taiwan Normal University, Taipei, Taiwan.

Middleton, H. E. (Ed.). (2008a). *Researching technology education: Methods and techniques.* International Technology Education Series. Rotterdam, The Netherlands and Taipei, Taiwan: Sense.

Middleton, H. E. (2008b). Researching technology education: New ways of understanding teaching, learning, and knowledge. In H. E. Middleton (Ed.), *Researching technology education: Methods and techniques* (pp. 1–5). International Technology Education Series. Rotterdam, The Netherlands and Taipei, Taiwan: Sense.

Middleton, H. E. (2008c). Examining design thinking: Visual and verbal protocol analysis. In H. E. Middleton (Ed.), *Researching technology education: Methods and techniques* (pp. 193–212). International Technology Education Series. Rotterdam, The Netherlands and Taipei, Taiwan: Sense.

Middleton, H. E., Chester, I., Finger, G., Jamieson-Proctor, R., & Pavlova, M. (2004). *Researching school change in technology education. Final report.* Brisbane, Queensland: Griffith University.

Middleton, H. E., & Gooch, D. (2002). Working together to improve practice: The INTAD project. In H. E. Middleton, M. Pavlova, & R. Roebuck (Eds.), *Learning in technology education: Challenges for the 21st Century. Proceedings of the 2nd Biennial Technology Education Research Conference.* (Vol. 2, pp. 52–59). Surfer's Paradise: Centre for Technology Education Research, Griffith University.

Pavlova, M. (2003). Questioning the nature of technology education in the context of social change. In, E. Martin & H. E. Middleton, (Eds.), *Initiatives in technology education: Comparative perspectives* (pp. 46–60). Brisbane, Queensland: Technical Foundation of America & Griffith University.

Pavlova, M. (2008). Comparative analysis as a research method in technology education. In H. E. Middleton (Ed.), *Researching technology education: Methods and techniques* (pp. 116–154). International Technology Education Series. Rotterdam, The Netherlands and Taipei, Taiwan: Sense.

Pavlova, M. (in press). University–TAFE collaboration: Responses to the labour market. In D. Johnson & R. Maclean (Eds.), *Vocational and Higher Education,* Springer.

Smith, E. (2004). Vocational education and training in schools in Australia: What are the consequences of moving from the margins to mainstream? *Journal of Vocational Education and Training, 56*(4), 559–582.

Stevenson, J. C. (2008). Capturing knowledge and activity. In H. E. Middleton (Ed.), *Researching technology education: Methods and techniques* (pp. 155–171). International Technology Education Series. Rotterdam, The Netherlands and Taipei, Taiwan: Sense.

Walmsley, B. (2008). Using stimulated recall techniques in technology education classes. In H. E. Middleton (Ed.), *Researching technology education: Methods and techniques* (pp. 172–192). International Technology Education Series. Rotterdam, The Netherlands and Taipei, Taiwan: Sense.

Watts, D. (1997). *A report on the readiness of Australian schools to meet the demands of teaching the curriculum areas of science and technology in the compulsory years of schooling.* Perth, WA: University of Notre Dame.

Williams, P. J. (2001). *The teaching and learning of technology in Australian primary and secondary schools. Working report to the Department of Education, Science and Training (DEST).* Canberra, ACT: AGPS.

Williams, P. J. (2002). Crisis in technology education in Australia. In H. E. Middleton, M. Pavlova, & R. Roebuck. (Eds.), *Learning in technology education: Challenges for the 21st Century. Proceedings from the 2nd Biennial Technology Education Research Conference* (Vol. 2, pp. 272–281). Surfer's Paradise: Centre for Technology Education Research, Griffith University.

Williams, P. J. (2006). Technology teacher education in Australia. In P. J. Williams (Ed.), *International technology teacher education: 55th Yearbook* (pp. 1–22). New York: Glencoe.

Howard Middleton
School of Education and Professional Studies
Griffith University
Queensland, Australia

ALISTER JONES AND VICKI COMPTON

9. REVIEWING THE FIELD OF TECHNOLOGY EDUCATION IN NEW ZEALAND

INTRODUCTION

In this chapter we discuss the development of technology as a field of study within compulsory education in New Zealand. We argue that technology education has found a place in the national curriculum, research, and teacher education, resulting in technology classroom practice to some degree in all New Zealand schools.

We report first on curriculum developments in terms of the shift from technical subjects to the initial technology curriculum (Ministry of Education, 1995), and on to the development of the second technology curriculum released in 2007 (Ministry of Education, 2007). An overview is also presented of the research undertaken to inform and evaluate curriculum development and implementation, followed by a discussion of research in technology education undertaken to inform teaching and learning generally. The chapter moves on to discuss the history of technology teacher education through professional development support. Implementation of the 1995 technology curriculum was initially supported with significant in-service teacher professional development, and pre-service technology teacher education programmes were in place by 1998. However, little ongoing support was provided for this group. We suggest that the resultant teacher education has been varied and variable, although a number of initiatives are seeking to address this. Finally, we discuss issues and challenges for technology education in the future, and briefly outline planned developments that may begin to address these.

FROM TECHNICAL TO TECHNOLOGICAL

New Zealand has a long history of technical education in the senior primary and secondary school (Burns, 1992). A national school system was established in New Zealand in 1877, and in 1890 technical education was introduced. Technical education was strongly skills-based and reflected the social perspective of the time, particularly regarding what was gender appropriate: metal and woodwork for boys; cooking, needlework, and/or laundry for girls. In 1900 the new Manual and Technical Instruction Act was passed into law (New Zealand Statutes, 1900) and secondary schools were required to extend their traditional academic curriculum to incorporate technical subjects. However, parents were generally reluctant to enrol their children in these subjects due to their being "stamped with the label of social inferiority" (McKenzie, Lee, & Lee, 1990, p. 4) and, as such, limiting their child to later employment as low-paid, unskilled, or manual workers. Soon after this, technical high schools were developed. Despite Government intentions, these schools focused on vocationally-oriented, skill-based programmes that served to channel working class children into manual and trade employment and reinforce the divide between 'white' and 'blue' collar workers in New Zealand.

A.T. Jones and M.J. de Vries (eds.), International Handbook of Research and Development in Technology Education, 93–104.
© 2009 Sense Publishers. All rights reserved.

By the mid 1920s, education advocates were calling for the leaving age of secondary school students to be raised from 12 to 15 years and for the abolishment of technical schools. In their place, comprehensive post-primary schools were promoted that offered a 'multi-lateral' curriculum (Compton, 2001). These schools were to provide *all* students with a 60% common content-based programme, with the remaining time being divided into two areas of semi-specialisation; either 'academic' or 'non-academic'. Technical education was sited in specialist wood and metal workshops and homecraft (food and textiles) rooms and the classroom curricula placed importance on the creation of products with high "take home value'. The pedagogical practices of technical teachers centred on the 'lock-stepped' construction of products in accordance with teacher predetermined plans in woodwork and metalwork, patterns in textiles, and recipes in cooking. This approach was embedded in behaviourist theories of learning in order to support students to attain the 'desired competencies' (Harwood & Compton, 2007).

Manual-technical training was criticised at the time as: "… the most glaring instances of rigid formalism . . . that quite overlooked the creative impulses of children … it certainly implanted sterile conceptions of art and handwork (Campbell, 1941, pp. 94–95). Despite these criticisms, teachers of woodwork and metalwork were not encouraged to change their behaviourist pedagogical practices until 1975, when Workshop Craft was introduced and mandated as the curriculum for technical education for hard materials (woodwork and metalwork) for Year 7–10 students. The homecraft curricula for home economics and clothing and textiles did not follow the hard material lead, however, and continued to endorse behaviourist practices (Harwood & Compton, 2007).

During the 1980s there were further moves to disrupt the purely skills-focused programmes and behaviourist approaches of technical education through the promotion of a stronger design focus and the use of a wider range of materials. This saw the emergence of such subjects as Workshop Technology and Graphics and Design. During this time there were also attempts to break down the gender-stereotyped nature of technical education by having girls and boys taking all technical/technology subjects in senior primary and junior high school. However, at senior high school these subjects still tended to be very gender specific. During this time there was also an increasing emphasis on 'technology' in existing school subjects. For example, science teachers were including a 'technology as applied science' aspect, social studies teachers were beginning to focus on technology as a 'determining' social force, and information technology teachers were focusing on understanding technology as hardware – primarily computers.

All of the above signalled a shift from technical education to a more technological education. However, the remnants of past technical curricula initiatives and teaching approaches resulted in a limited range of skills, processes and knowledge from a narrow and often gendered perspective. As a consequence, students right through to the early 1990s had little opportunity in New Zealand for broad and coherent experiences in technology that were considered conducive to developing citizens who could contribute successfully in a contemporary society.

TECHNOLOGY – A LEARNING AREA IN ITS OWN RIGHT

Technology as a separate subject area emerged as a key outcome from the 1990 curriculum reforms, which in turn were influenced by earlier curriculum reviews undertaken in the 1980s, including a 1984 review of the curriculum for schools (Department of Education, 1987). The public discussion and consultation was substantial – 21,500 submissions initially, with 10,000 submissions to the draft report from a total population of approximately 3.4 million. The resulting recommendation was that there be a national common curriculum for all schools, from new entrants to Form 5 (Years 1–10), and that it include national curriculum principles as well as knowledge, skills, attitudes, and values. In 1990, a new government embarked on a project to revise the curriculum in primary and secondary schools under the banner of The Achievement Initiative (Ministry of Education, 1991). Many of these ideas were influenced by the curriculum reforms taking place in England and Wales at the time. The policies emphasised raising standards, developing levels of attainment, the notion of progression linked to accountability, and the contracting out of the curriculum development process (Bell, Jones, & Carr, 1995). As part of an educational review process, a Ministerial Task Group was set up jointly by the Minister of Education and the Minister of Research, Science and Technology in June 1991 to review science and technology education. Some of the recommendations from the task group concerning technology education were that: a technology curriculum be developed as an area in its own right; there be adequate teacher training and resourcing for technology education; technology curricula should not be imported from overseas; and the inclusiveness of technology education be emphasised, including Maori[1] input and the use of Maori language (Ministry of Research, Science and Technology, 1992). This report was endorsed by members of the business community and by two Ministers of the Crown.

In 1992, the Ministry of Education contracted the Centre for Science and Mathematics Education Research (CSMER) at the University of Waikato to develop a policy framework for technology education in New Zealand. The contract required that: there be wide consultation; best practice be taken into account, nationally and internationally; it be consistent with other government policies in education; it take account of resources, teacher change, teacher development, qualifications frameworks, etc.; and it where possible provides a range of options. The development of the policy thus had to fit within the structure of the New Zealand Curriculum Framework (Ministry of Education, 1993) in terms of levels, strands, achievement aims, and objectives.

The policy decided the general aims of the technology curriculum, the technological areas, classroom and implementation directions, as well as approaches to teaching, including initial assessment ideas. This policy work drew extensively from the Learning in Technology Education (LITE) research project (Jones, Mather, & Carr, 1995), in which ideas within the initial policy paper were explored with teachers and students in the primary and secondary sector prior to the development of the draft technology curriculum. The LITE research also analysed teacher interview data captured during technology unit planning and delivery, providing insights into teacher understandings of the philosophical and theoretical base of technology education. In addition, extensive student data was gathered before, during, and after technological experiences and analysed in terms of technological practice as well as broader concepts of technology and technology education (Mather, 1995).

The CSMER was then contracted to develop the first draft technology curriculum (Ministry of Education, 1993). The draft included three general aims, six strands with related achievement objectives, seven technological areas, and six contexts for learning. It also included a variety of suggested classroom activities under each strand. The draft curriculum statement was published for consultation in 1993 and trialled in schools during 1994. This provided teachers, tertiary educators, professional technologists, and members of the general public with an opportunity to respond to the ideas contained in the draft. In general, teachers and professional technologists were supportive of the underpinning structure and philosophy of the document. However, a reduction in the number of objectives and strands was recommended given the number of other curriculum documents teachers were working with. There was general consensus that the strands should reflect directly the three original aims of the technology curriculum: technological knowledge and understanding, technological capability, and technology and society. The reduction in strands from six to three did not change the underlying philosophy, but in fact strengthened the notion of technological practice as the conceptual base of the curriculum. These three strands would work together to support the new holistic overall single aim of technological literacy. Other contexts were added and the number of learning experiences was reduced but an attempt was made to make these more comprehensive and strand inter-related. The final *Technology in the New Zealand Curriculum* (Ministry of Education, 1995) thus included a single aim of developing student technological literacy through three learning strands of:

- Technological Knowledge and Understanding;
- Technological Capability; and
- Technology and Society.

These three interrelated strands provided a framework for teachers to develop coherent programmes of technology as part of a balanced curriculum.

The seven technological areas were: materials technology; information and communication technology; electronics and control technology; biotechnology; structures and mechanisms; process and production technology; and food technology. There were also nine contexts: personal life, the home, the school, recreation, the community, the environment, energy production and supply, business, and industry. Eight achievement objectives were developed for each of the three strands, and an attempt was made to write these across eight levels. Units of work were to be developed based on technological learning outcomes and reflective of achievement objectives from all three strands as well as one or more context and technological area (Ministry of Education, 1995). The dominant learning theories underpinning the 1995 technology curriculum were socio-cultural in nature with pedagogical approaches seeking to embed student learning in authentic and empowering contexts. While published in 1995, the curriculum was not mandated for full implementation until 1999. Implementation was supported in various and variable ways, discussed later in the chapter.

TIME TO TAKE STOCK

In 2001 a national curriculum stocktake was undertaken in New Zealand that included reviews of learning area curricula, international evaluations, and analysis of teachers' experiences of the curricula in practice. This stocktake was accomplished through the *National School Sampling Study* (NSSS). Along with other learning areas, the NSSS provided an opportunity for teachers who had been involved in implementing technology in the New Zealand curriculum to share their experiences (Jones, Harlow, & Cowie, 2004). Key aspects that were investigated included: background and experience of teachers; general issues related to implementation; practice; professional support; the curriculum documents; and impact and compliance issues. The NSSS sought teacher feedback through national focus groups, questionnaires and case studies.

Findings from the study were reasonably positive in relation to teachers' experiences in the implementation of the technology curriculum in New Zealand schools from Years 1–13. However, there were variations between teachers in different kinds of schools and within school types, especially between primary and secondary school teachers. There was a general degree of satisfaction with the curriculum statement, although one third of teachers (33%) wanted to make changes. The most common change was to 'make it [the curriculum] more simple to understand' and to include 'more and better developed learning and assessment examples'. The 1995 technology curriculum statement was reported as having been most helpful for planning, gaining an overview of the progression of key technological ideas, achieving consistent understanding of the curriculum levels, and assessing student achievement. However, half of all teachers reported that they only 'sometimes' found the curriculum statement helpful for gaining an overview of the progression of key ideas.

Although 70% of teachers reported in the questionnaires that the curriculum statement was 'always' or 'sometimes' helpful in assessing achievement, many also reported difficulties with assessment in technology. The most popular approach to assessing student learning in technology was the use of 'practical tasks'. Large classes, the 'time' factor, and establishing level accuracy were also issues, but these were only noted by less than 20%. Secondary school teachers were more concerned with the amount of paperwork required than teachers in other sectors. They were also influenced by the perception of qualification requirements. At the primary level, teachers were concerned about finding appropriate forms of assessment for the junior years and felt that they needed more guidance, both in planning and in assessment.

Nearly three-quarters of the teachers (73%) reported receiving some form of professional development in technology. For almost 50% of respondents, 'other teachers in the school' had been the most useful source of knowledge. For 75%, professional development had been helpful, and 28% found it had given them sufficient depth of knowledge to plan and implement the technology curriculum as well as helping them to gain confidence in teaching technology. However, across all school types, two thirds of the teachers expressed a medium level of confidence in teaching technology and only one-fifth a high level of confidence. Teachers were most interested in receiving future professional development in the specific technological areas. Information on planning and teaching skills was requested and respondents were also interested in knowing more about progression, assessment, and reporting of student achievement.

According to 88% of teachers, all students up to and including Year 10 studied technology in their respective schools. The majority of teachers considered that technology should be compulsory for all students to the end of Year 10 as it provided students with important skills. Over 60% of schools were integrating technology with other learning areas. This was particularly evident in primary schools (71%). Most of these teachers tended to integrate technology into languages and/or science. Secondary school technology teachers taught technology in blocks or modules or as a new subject with its own timetable slot.

Teachers detailed a wide range of approaches that had been successful in their teaching of technology. These included: choosing topics of relevance to students; practical, hands-on learning activities; a 'problem-solving approach'; and group or co-operative learning approaches. Teachers tended to favour a student-centred approach to teaching technology. Fifty-three percent of secondary school (Year 9–12) teachers placed greater emphasis on technological capability than the other strands. Concerns about the level of student knowledge and skills needed to cope with curriculum requirements were expressed by 31% of secondary school teachers.

Materials technology and food technology were reported to be the most widely-taught technological areas in New Zealand schools. However, all technological areas were reported as being taught in primary schools to some degree, with biotechnology the least common. Fewer teachers in intermediate/middle schools taught food technology or electronics and control technology – perhaps because specialist technology teachers often covered these areas. Overall it appeared that apart from the more traditional technical-related technological areas of food and materials technology, secondary school schools had not developed courses in other technological areas to the same extent as in primary schools. In keeping with the findings from the primary sector, biotechnology was the least common area taught.

In response to a question regarding the major challenges faced in implementing the technology curriculum, the prime concern of teachers was the difficulty of resourcing the necessary equipment (50% overall; 60% of primary teachers). Although many secondary school teachers had new facilities in which to work, they still reported difficulties establishing technology in their schools because of timetable constraints, management decisions, or lack of enthusiasm on the part of former home economics and wood/metal teachers. Secondary school teachers were also grappling with the introduction of new national qualification requirements, although in many cases the technology achievement standards provided additional guidance and incentive for implementing technology in junior secondary programmes. What was termed a 'crowded curriculum' was found to be a major challenge for 32% of all respondents, particularly in the primary sector. As teachers who teach all areas of the national curriculum, primary teachers reported overcoming this to a certain extent by integrating technology with other subject areas. Understanding the 1995 technology curriculum statement was listed as one of the major challenges by almost a quarter of the respondents.

TECHNOLOGY REFINED AND REDEVELOPED

Results in technology achievement standards and findings from national research projects, including the National Education Monitoring Project (NEMP), indicated that school-based technological practice was leading to: high levels of student engagement; increasing levels of ownership of learning; increasing levels of empowerment resulting in an enhanced ability to make decisions; and effective collaboration with others to make a difference to their own lives and developments in their immediate community (e.g., Compton & Harwood 2003; Jones & Moreland, 2003). However, teachers reported and technology achievement standard results showed that student technological literacy was often limited in breadth and depth of knowledge, and lacking in the level of critical analysis required for more informed decision making.

Further policy work was therefore undertaken in 2004 as part of a review of the national curriculum, known as the *New Zealand Curriculum Marautanga Project* (NZCMP). This work sought to provide a stronger focus on the philosophical basis of technology and technological knowledge (Compton, 2004; Compton & Jones, 2004). Based on research and new policy thinking, it was decided to restructure technology around three new strands – technological practice, technological knowledge, and the nature of technology. These strands were still seen as working together to realise the aim of developing student technological literacy, but the nature of this literacy was reconceptualised to be broader, deeper and more critical (Compton & France, 2007a).

The concept of technological practice was considered to be well defined from understandings gained from classroom-based research carried out in New Zealand over the last ten years. Key constructs such as the 'components of [technological] practice', and supporting 'indicators of progression' within them, had also been developed, trialled, and validated within New Zealand classrooms (Compton & Harwood, 2004, 2005) and were available for development into levelled achievement objectives.

The *Technological Knowledge and Nature of Technology* (TKNoT) research project sought to establish the key components of the remaining two strands – technological knowledge and the nature of technology. A central part of the technological knowledge focus was to work with key technology networks representing all significant technological sectors within New Zealand, including leading academic and/or industry biotechnologists, engineers, food technologists, control technologists, information and communication technologists, architects, product developers, and designers (Compton & France, 2007b). While these participants also discussed aspects of the nature of technology, international experts involved in the philosophy of technology provided considerable input into the philosophical underpinnings of the nature of technology focus (Compton & France, 2007b). This access to international experts allowed discussion of developing ideas outside of the New Zealand context. Notions of technological knowledge and the nature of technology were also explored with teachers and students (Compton & France, 2007c). As a result of the TKNoT research, two components were established within the nature of technology strand and three components within the technological knowledge strand (see Table 1).

This framework provided a more robust philosophical and theoretical base for ongoing curriculum development work, and addressed all concepts discussed by Mitcham (1994) as being focused on in technology education in varying ways. A draft of the introductory statement and a series of levelled achievement objectives for each of the identified components was prepared for consultation in 2006 and revised for release the following year in the New Zealand Curriculum (Ministry of Education, 2006, 2007). Socio-cultural and constructivist theories of learning provided an underpinning basis for learning, with pedagogical approaches seeking to embed student learning in authentic and empowering contexts and provide students with established understandings and practices from which to undertake more informed practice and make more critically reflective decisions.

Table 1. Redeveloped technology strands (Compton & France, 2007a, p. 268).

Technological practice	Technological knowledge	Nature of technology
- Brief development - Planning for practice - Outcome development and evaluation	- Technological modelling - Technological products - Technological systems	- Characteristics of technology - Characteristics of technological outcomes

RESEARCH INTO TEACHING AND LEARNING

Technology education research in New Zealand has increased significantly over the last two decades. Early work in the 1980s, tended to focus on using technology to enhance the learning of science, such as using technological applications to teach science (e.g., Jones & Kirk, 1990). Policy moves to include a broad notion of technology in the curriculum in the late 1980s and early 1990s were associated with research commissioned by the Ministry of Education including research on student attitudes to technology based on the PATT questionnaires (Burns, 1992) and the Learning in Technology Education (LITE) project mentioned earlier.

Early LITE research focused on teachers' perceptions of technology and explored the influence of teachers' existing subcultures on the ways in which teacher might be prepared to implement the technology curriculum (Jones & Carr, 1992). Classroom investigations explored ways in which students carried out technological activities; these were strongly influenced by classroom culture and student expectations (Jones, Mather, & Carr, 1995; Mather, 1995). The final phase of the project explored effective characteristics of technology professional development for teachers (Compton & Jones, 1998; Jones & Compton, 1998).

From 1990 onwards there was an emergence of Masters and PhD research in the area of technology education as the area gained prominence in both schools and tertiary education, and a strong focus on teacher development was seen to coincide with full implementation of the 1995 curriculum. The targeting of resources for the professional development meant that there was no government-funded research during this time. Two national professional development programmes did, however, build on teacher development research (see below). They also included a research/evaluative component.

In 1998 there the focus of research shifted to assessment, with a strong emphasis on formative interactions to support 'assessment for learning'. Two projects were funded by the Ministry of Education in 1999 and 2000 to undertake research in this area: the *Learning in Technology Education – Assessment* (LITE-Assessment) project (Jones & Moreland, 2003; Moreland, Jones, & Northover, 2001) and the *Technology Education Assessment in Lower Secondary* (TEALS) project, which was extended to the primary and senior secondary sectors in its second year (Compton & Harwood, 2003, 2005). These projects shifted research back into New Zealand classrooms and resulted in moving understandings of teaching and learning forward.

Classroom-based research from 1992-1997 (as reported by Jones, 1999) had shown that strategies developed by the teachers in their classrooms when implementing technological activities were often positioned within that particular teacher's teaching and subject sub-culture. For a new curriculum area such as technology, this presented particular challenges for teachers as they searched to construct a coherent content base in addition to pedagogical understandings, including appropriate assessment practices. More recent research in New Zealand has continued to focus on student and/or teacher concepts as evidenced by the increase in externally-contracted research and graduate research projects.

Technology is an activity that involves not just the social context, but also the physical context, with thinking being associated with and structured by the objects and tools of action. This means research in student learning has to consider not just conceptual and procedural understandings in isolation, but also the way technological tools and objects influence and interact with student thinking and doing. More research is required in this area to understand and support teacher pedagogical content knowledge for teaching in the technology classroom, particularly in keeping with the 2007 curriculum changes. To this end, the *Technological Knowledge and Nature of Technology: Implications for Teaching and Learning* (TKNoT:Imps) project, currently underway, seeks to explore the components of the two new strands of the reviewed technology curriculum in order to validate and further describe the nature of progression. A second project, *Interactions in Science and Technology Education* (InSiTE), is a longitudinal study focused on formative learning interactions (Cowie, Moreland, Jones, & Otrel-Cass, 2008).

THE VARIABLE STORIES OF TEACHER EDUCATION

With the release of the 1993 draft technology curriculum, teacher education institutions began to grapple with the introduction of technology education and the implications for pre-service teacher programmes. Little coordinated professional development support was provided for personnel within these institutions, although pre-service teacher educators were often invited, along with in-service teachers educators, to Ministry of Education national meetings during 1995 and 1996. These meetings were, however, primarily to *inform* this group of current developments and Ministry of Education expectations. Technology Education New Zealand (TENZ) was established in 1995 in an effort to fill this gap and the biennial conference run since 1997 have provided a site for researchers, teacher educators, and teachers to meet and debate technology education issues.

During 1995–1997, the Ministry of Education funded a range of regional and national professional development programmes for in-service teachers in an attempt to support the implementation of technology. The regionally-based programmes were selected from proposals submitted in response to a Ministry of Education request for trial or pilot programmes. This resulted in a variety of programmes being accepted, the diversity of which is illustrated in the following examples of programme focus: biotechnology, technology links to English literacy, electronics and robotics, information and communication technology, and distance learning in technology. The majority of these programmes only ran during 1995 although in a few cases funding was extended until 1997. As a result, teachers in different regions experienced quite different professional development in technology and in some cases links to the 1995 technology curriculum were not a strong feature of the programme.

The two national programmes included the *National Facilitator Training Programme* and the *Technology Teacher Development Resource Package* (Compton & Jones, 1998; Jones & Compton, 1998). Both took into account past national and international research in teacher development as well as recent technology education baseline research carried out in New Zealand schools. They also included an evaluative research component. The National Facilitator Training

Programme worked with a total of 30 participants from around New Zealand during 1995 and 1996. Participants included primary, intermediate and secondary teachers and teacher educators. Of the teacher educators, the majority were involved in in-service education although some also had pre-service responsibilities. The aim of the programme was to ensure every region in New Zealand had at least one teacher education expert in technology to lead and inform subsequent regional and school-based professional development programmes. The Technology Teacher Development Resource Package was developed to provide teacher educators and all schools access to professional development support materials for use in school-based teacher development aligned to the 1995 technology curriculum. This package included industry/community technological practice videos, classroom practice videos, and written support material.

Both these programmes sought to ensure a consistency across the country. Unfortunately, the potential impact was never to be fully realised as the Ministry of Education decided in 1997 to cut further funding for technology professional development. Although the Technology Teacher Development Resource Package was sent to all schools, without facilitation support, it tended to sit pristine in its box in the school resource room. The trained facilitators returned to their previous positions or moved on in their careers. Many of them have since moved into in-service or pre-service teacher education positions within tertiary institutions. By 1998 technology education formed a part of pre-service education for all students enrolled in primary teacher education programmes and was an option for suitably-qualified students in secondary teacher education programmes. However, differences in emphasis existed depending on the structure of the institution and the background of those employed to staff these programmes. In an attempt to address issues of inconsistency across regions, the *Technology Education Assessment National Professional Development* (TEANPD) project was funded by the Ministry of Education to ensure research findings from the TEALS and LITE-Assessment project were disseminated. This project involved teacher educators (both in-service and pre-service) as well as lead teachers in schools. In 2001, the Ministry of Education also funded limited resource development for Year 1–10 teachers through the Technology National Exemplar Project. An Exemplar Matrix for Technology Education developed as part of this project was controversial, however, as it neither reflected the 1995 technology curriculum, nor was in keeping with the LITE-Assessment or TEALS research findings. The matrix was abandoned in 2003 although a number of the exemplars were produced in an attempt to support more consistent understanding between teachers.

In 2004, a significant amount of funding was accessed from the Ministry of Economic Development to support technology education. This funding was managed by the Ministry of Education and was used to set up the *Growth and Innovation Framework-Technology Initiative* (GIF-Technology) (Dinning, 2007). Relatively stable funding with a ten-year lifespan (until 2013) has allowed the Initiative to be strategic on a level unprecedented in previous attempts to provide teacher professional development support for technology education in New Zealand. While funds were appropriated for the support of senior secondary students (Years 11 – 13) work with students from Years 7 – 13 was subsequently included on the basis that coherent programmes developed across this range would make gains in the senior secondary sector more likely. Major professional teacher education projects established to date include: resource and materials development and the development of Techlink, an online portal for technology educators and students; the Technology Beacon Practice Project to prepare case studies of exemplary classroom practice; a Technology Leader Support Programme including regional and national meetings for those responsible for translating the 2007 curriculum into appropriate school technology programmes; the 2007 Curriculum Support Package which provides an explanation of and guidance for the 2007 technology curriculum; and the establishment of a National Technology Professional Development Manager (NTPDM).[2] A joint learning community of teacher educators, as well as separate in-service and pre-service teacher educator learning communities, established by the NTPDM, has resulted in a growing level of shared understandings and a collaborative and supportive network between regions and institutions.

The result of this more coherent support is beginning to result in strong regional programmes. In addition, a new purpose-developed graduate programme has been supported by the Ministry of Education at the University of Waikato. This programme recognises that the majority of teachers currently teaching technology trained without any formal technology education courses and seeks to redress this. Other graduate courses in technology are being developed and are aligned to the expertise of the provider. However, with greater alignment between different pre-service teacher educators, it is expected that these programmes will all serve to provide more consistent secondary technology teacher graduates. The Ministry of Education has also recently awarded 100 scholarships to support technology pre-service teacher education in the secondary sector. Of these, 59 are 'career changer' scholarships, meaning a significant influx of technologists entering the technology teaching fraternity.

INTO THE FUTURE

Technology education in New Zealand has never been stronger. After more than twenty years of research, trialling, and development we have established a robust curriculum framework, a growing body of researchers, and an increasingly collaborative network of technology teacher educators. Classroom practice in technology exhibited in many schools is of a high level. We have a qualification system with pathways into University, and technology scholarship results that reflect innovative and critical thinking and quality outcomes.

However, significant issues remain. In the secondary sector, teacher understandings of the 2007 technology curriculum and teacher supply remain two major challenges. Another significant issue is senior secondary qualifications. A new phase of development of Technology Achievement Standards is about to begin with the aim of ensuring that assessment tools align with the 2007 technology curriculum. Past experience has provided strong lessons as to the impact high-stakes assessment can have on programme design throughout the secondary sector. It is therefore critical that gains made in national curriculum development are not undermined by inadequate and/or inappropriate assessment tools and guidance within senior secondary programmes.

In the primary sector, the impact of an over-emphasis on numeracy and literacy, and more recently 'key competencies', threatens to undermine past gains through squeezing technology (and other marginalised curriculum areas, such as science and the arts) out of primary school programmes. Teacher understandings of the revised 2007 technology curriculum are also a challenge in the primary sector. As pre-service institutions re-allocate more and more time from specialist learning areas such as technology to numeracy and literacy, primary teachers are graduating with very little understanding of technology and technology education and minimal technology pedagogical content knowledge.

While there are many examples of students from all sectors being involved in innovative and exciting classroom programmes and achieving high levels of technological practice, this is yet to be the 'norm' in all schools across the country. The type of student achievement and teacher practice illustrated by the Beacon Practice case studies, for example, is yet to be exhibited by the 'critical mass' of students and teachers needed to ensure sustainable future development.

New developments and an extension of the sorts of projects established by the GIF-Technology Initiative are planned to address these issues. However, due to the limitations of the GIF-Technology fund focusing on senior secondary technology, concerns in the primary sector are more difficult to address. The biggest challenge in this regard is to secure further funding to focus on the primary sector. In keeping with this, plans have been made to attempt to increase the general public's understanding of technology, as well as increase technology-related industries' understanding of, and commitment to, the aims of technology education. Such plans seek to increase the pool of support for technology at all levels of the national curriculum, and in so doing ensure a level of political 'will' to allow the potential of the 2007 technology curriculum to be realised. As many other countries can testify, such political will is key to the health of a new curriculum such as technology and is thus a fundamental issue to address.

In conclusion, we argue that technology has established a space within New Zealand's national curriculum, education research community, and to some extent at least, the lived world of students in compulsory levels of schooling. We look to the future in which this space is not eroded, but rather supported and allowed to evolve further.

NOTES

1 Maori are the indigenous people of New Zealand and under the Treaty of Waitangi (1840) were guaranteed exclusive and undisturbed rights in terms of preservation of land, fisheries, forest and language. Maori has been an official language in New Zealand (along with English) since 1987.

2 Details of these and other projects funded by the GIF-Technology Initiative can be found at www.techlink.co.nz.

REFERENCES

Bell, B., Jones, A., & Carr, M. (1995). The development of the recent national New Zealand Science Curriculum. *Studies in Science Education, 26*, 73–105.

Burns, J. (1992). Technology—What is it, and what do our students think of it? *The NZ Principal, 6*(3), 22–25.

Compton, V. J. (2001). *Developments in technology education in New Zealand 1993–1995: An analysis of the reflections of key participants*. Unpublished doctoral dissertation. University of Waikato, New Zealand.

Compton, V. J. (2004). *Technological knowledge: A developing framework for technology education in New Zealand*. Briefing paper prepared for the New Zealand Ministry of Education Curriculum Project.

Compton, V. J., & France, B. (2007a). Redefining technological literacy in New Zealand: From concepts to curriculum constructs. In J. Dakers, W. Dow, & M. J. de Vries (Eds.), *PATT-18, pupils' attitudes towards technology. International conference on design and technology education research book. Teaching and learning technological literacy in the classroom* (pp. 260–272). Glasgow, Scotland: Glasgow University.

Compton, V. J., & France, B. (2007b). Towards a new technological literacy: Curriculum development with a difference. *Curriculum Matters, 3*, 158–175.

Compton, V. J., & France, B. (2007c). Exploring the nature of technology: Students' intuitive ideas as a starting point. In J. Dakers, W. Dow, & M. J. de Vries (Eds.), *PATT-18, pupils' attitudes towards technology. International conference on design and technology education research book. Teaching and learning technological literacy in the classroom* (pp. 250–259). Glasgow, Scotland: Glasgow University.

Compton, V., & Harwood, C. (2003). Enhancing technological practice: An assessment framework for technology education in New Zealand. *International Journal of Design and Technology Education, 13*(1), 1–26.

Compton, V. J., & Harwood, C. D. (2004). Moving from the one-off: Supporting progression in technology. *SET Research Information for Teachers, 1*, 23–30.

Compton, V. J., & Harwood, C. D. (2005). Progression in technology education in New Zealand: Components of practice as a way forward. *International Journal of Design and Technology Education, 15*(3), 253–287.

Compton, V., & Jones, A. (1998). Reflecting on teacher development in technology education: Implications for future programmes. *International Journal of Technology and Design Education, 8*(2), 151–166.

Compton, V. J., & Jones, A. (2004). *The nature of technology*. Briefing paper prepared for the New Zealand Ministry of Education Curriculum Project.

Cowie, B., Moreland, J., Jones, A., & Otrel-Cass, K. (2008). *The classroom InSiTE project: Understanding classroom interaction to enhance teaching and learning in science and technology in Years 1–8*. Teaching and Learning Research Initiative. Wellington, NZ: New Zealand Council for Educational Research.

Department of Education. (1987). *The curriculum review. Report of the committee to review the curriculum for schools*. Wellington, NZ: Government Printer.

Dinning, N. (2007). Enabling technological literacy in New Zealand: The GIF—Technology Education Initiative. *Proceedings of the pupils' attitudes towards technology (PATT) 18 conference: Teaching and learning technological literacy in the classroom* (pp. 273–281). Glasgow, Scotland: University of Glasgow.

Harwood, C. D., & Compton, V. J. (2007). Moving from technical to technology education: Why it's so hard! *Proceedings from technology education New Zealand 6th biennial conference* (pp. 105–115). Auckland, NZ: Technology Education New Zealand.

Jones, A. (1999). The influence of teachers' subcultures on curriculum innovation. In J. Loughran (Ed.), *Researching teaching* (pp. 155–171). London: Falmer.

Jones, A., & Carr, M. (1992). Teachers' perceptions of technology education—implications for curriculum innovation. *Research in Science Education, 22*, 230–239.

Jones, A., & Compton, V. (1998). Towards a model for teacher development in technology education: From research to practice. *International Journal of Technology and Design Education, 8*(1), 51–65.

Jones, A., Harlow, A., & Cowie, B. (2004). New Zealand teachers' experiences in implementing the technology curriculum. *International Journal of Technology and Design Education, 14*(2), 101–119.

Jones, A., & Kirk, C. (1990). Introducing technological applications into the physics classroom: Help or hindrance to learning? *International Journal of Science Education, 12*(5), 481–490.

Jones, A., & Moreland, J. (2003). Developing classroom-focused research in technology education. *Canadian Journal of Science, Mathematics and Technology Education, 3*(1), 51–66.

Jones, A., Mather, V., & Carr, M. (1995). *Issues in the practice of technology education* (Vol. 3). Hamilton, NZ: Centre for Science and Mathematics Education Research, University of Waikato.

Mather, V. J. (1995). *Students' concepts of technology and technology education: Implications for practice*. Unpublished masters dissertation, University of Waikato, New Zealand.

McKenzie, D., Lee, G., & Lee, H. (1990). The transformation of the New Zealand technical high school. *Delta: Research Monograph No.10*. Palmerston North, NZ: Massey University.

Mitcham, C. (1994). *Thinking through technology: The path between engineering and philosophy*. Chicago and London: University of Chicago Press.

Ministry of Education. (1991). The achievement initiative. *Education Gazette, 70*(7), 1–2.

Ministry of Education. (1993a). *The New Zealand curriculum framework*. Wellington, NZ: Learning Media.

Ministry of Education. (1993b). *Technology in the New Zealand curriculum. Draft*. Wellington, NZ: Learning Media.

Ministry of Education. (1995). *Technology in the New Zealand curriculum*. Wellington, NZ: Learning Media.

Ministry of Education. (2006). *Draft technology curriculum materials for consultation*. Wellington, NZ: Learning Media.

Ministry of Education. (2007). *The New Zealand curriculum*. Wellington, NZ: Learning Media.

Ministry of Research, Science and Technology. (1992). *Charting the course: The report of the ministerial task group into science and technology education*. Wellington, NZ: Government Printer.

Moreland, J., Jones, A., & Northover, A. (2001). Enhancing teachers' technological and assessment practices to enhance student learning in technology: A two year classroom study. *Research in Science Education, 31*(1), 155–176.

New Zealand Statutes. (1900). *Manual and technical instruction act* (Vol. 39). Government of New Zealand.

Alister Jones
School of Education
University of Waikato
Hamilton, New Zealand

Vicki Compton
Uniservices
University of Auckland
Auckland, New Zealand

CHITRA NATARAJAN AND SUGRA CHUNAWALA

10. TECHNOLOGY AND VOCATIONAL EDUCATION IN INDIA

INTRODUCTION

> Our education has got to be revolutionised. The brain must be educated through the hand. If I were a poet, I could write poetry on the possibilities of the five fingers. Why should you think that the mind is everything and the hands and feet nothing? Those who do not train their hands, who go through the ordinary rut of education, lack 'music' in their life. All their faculties are not trained. (Gandhi, 1994)

Imagine the scenario: An electrician and a plumber walk into an affordable house-under-construction, the owners articulate their needs, and service providers end up giving them switches most convenient to use and easily maintainable plumbing that gets water-on-tap. Surely the owners and the service providers must have all had an education in design. Imagine again that you are visiting an Indian metropolis and find it well laid-out. Everything around you seems to have been *designed* to work: bus routes and stops, trains, stations, airports, all artefacts and organisations, schools, and curricula. You talk to a spectrum of people from different walks of life and learn that the planning has been participatory and carried out with insight. The general education of all people in the metropolis must have included courses on design and technology, which was then extended within vocational education. This situation in India, unfortunately, is a long way coming.

The history of education in India can be traced to the 3rd century BC, when sages and scholars imparted education orally to a select few. Later, palm leaves and bark of trees were used to spread the written word. Temples and community centres served as schools accessible to a limited population. Buddhism opened up education to all people and led to the rise of educational institutions of global fame: Nalanda, Vikramshila, and Takshashila. (Nalanda University is said to have had around 10,000 resident students and teachers on its roll at one time.) Around the 11th century the Muslim rulers established elementary and secondary schools in India and universities were also established at Delhi, Lucknow, and Allahabad. During the Medieval period the Indian and Islamic traditions interacted in several knowledge areas including theology, philosophy, the fine arts, mathematics, medicine, and astronomy.

The craft tradition has stayed outside schools. In the Hindu Varna system of society, occupations were handed down through family lines and community groups. Broadly, the *Brahmins* were keepers and disseminators of knowledge, *Vaisyas* were traders, *Kshatriyas* were warriors, and the *Sudras* were assigned menial labour. There were further divisions in terms of goldsmiths, blacksmiths, carpenters, and so on. Artisans were trained through apprenticeship in the family, while master artisans were recognised by their rigorous training and work.

The arrival of the British in India made education based on existing Western models available in the country. However, these colonial education models introduced the idea of certified and valid knowledge alienated from productive work that satisfied existing social needs. To this day, those who produce wealth by working with their hands often lack access to formal education, while formal education denigrates productive manual work and fails to provide the necessary skills for

A.T. Jones and M.J. de Vries (eds.), International Handbook of Research and Development in Technology Education, 105–116.
© 2009 Sense Publishers. All rights reserved.

the same. In this context, it is useful to recall excerpts from Mahatma Gandhi's address to teachers at the National Education Conference held under his leadership, at Wardha on 22 October 1937:

> What I am going to place before you today is not about a vocation that is going to be imparted alongside education. Now, I wish to say that whatever is taught to children, all of it should be taught necessarily through the medium of a trade or a handicraft. ... Look at takli [spindle] itself, for instance. The lesson of this takli will be the first lesson of our students through which they would be able to learn a substantial part of the history of cotton, Lancashire and the British empire. ... How does this takli work? What is its utility? And what are the strengths that lie within it? Thus the child learns all this in the midst of play. Through this he also acquires some knowledge of mathematics. ... And the beauty is that none of this becomes even a slight burden on his mind. (Gandhi, 1994)

Dilution of such ideas was evident in the conception of the education system even before independence. The following section discusses the form of technical and technology education in school and tertiary levels in post-independent India. It is a commentary on whether and how Mahatma Gandhi's ideas influenced technical education.

OVERVIEW OF TECHNICAL/VOCATIONAL EDUCATION IN POST-INDEPENDENT INDIA

After India's independence in 1947, it became a national priority to generate human resources through education to fulfil the future needs of the country, which was envisioned as an industrialising one. This led the Kothari Commission, set up in 1966, to link education to productivity and to propose that science education, work-experience, and vocational education be an integral part of school education. It is within these three subjects that education in and about technology resides rather nebulously in the context of Indian schooling, even today. A brief overview of processes of curricular change involving commissions and their recommendations, and recent national exercises of drawing up curricular frameworks is presented below.

Structure of Education

Following the recommendations of the Kothari Commission in 1968, a common structure for education was adopted in the whole country (Figure 1). Typically, schooling is for a span of 12 years: 10 years of primary and secondary schooling, and 2 years of senior/higher secondary (referred to as junior college, intermediate, or +2). College graduation is normally after 3 years of study. This is popularly denoted as the 10+2+3 pattern. The National Council of Educational Research and Training (NCERT) is the apex body that formulates the curricular framework and materials for Central schools. State Councils of Educational Research and Training (SCERTs) closely follow NCERT's framework in formulating syllabi and curricular materials for State Board affiliated schools.[1] At the end of secondary (Grade X) and higher secondary stages (Grade XII), students are assessed through public examinations conducted either by the Central or State Boards.

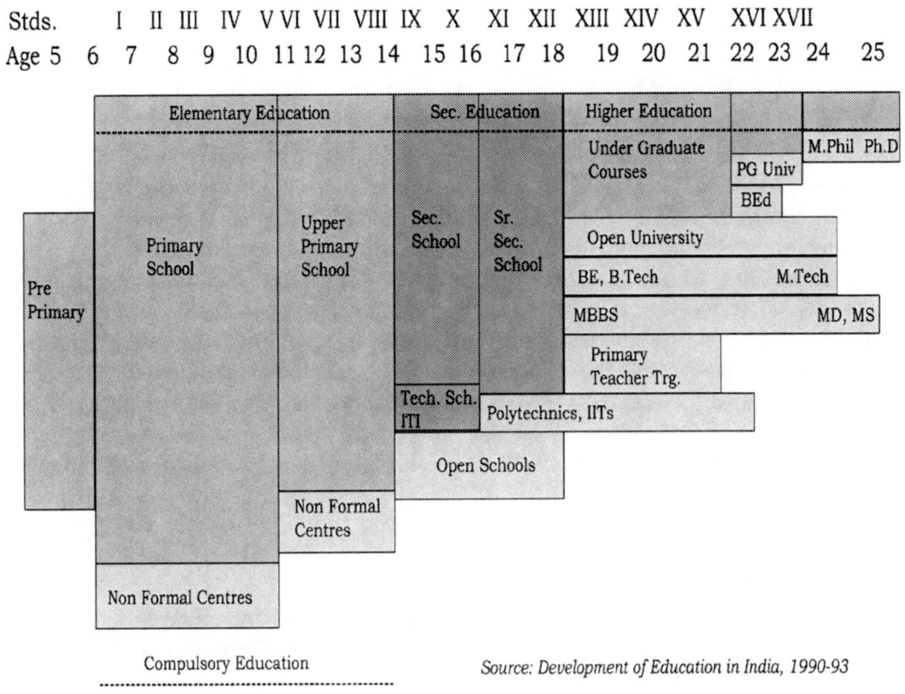

Figure 1. Structure of education in India.

Science Education

Education frameworks prior to 2000 had explicitly recognised scientific and technological literacy as a goal of science education. The National Curricular Framework (NCF) was the first document to explicitly recognise science and technology (S&T) as organically linked and linked to society (NCF, 2000). The textbooks based on the framework were titled *Science and Technology* and included applications of science scattered among concepts, processes, procedures, and activities. Technology was narrowly defined as the 'application of science to meet human needs'. Through all these changes in frameworks and textbooks, it was teaching as usual in the science classes, as neither the laboratories nor the examination systems had undergone any radical change to include testing on technological literacy.

NCF 2005 for the first time referred to design and technology as part of teaching and learning science at the school level in its Position Paper of the National Focus Group on Teaching of Science (NFG-ToS, 2006). For the first time in the history of curricular frameworks there was a recognition that

> Technology as a discipline has its own autonomy and should not be regarded as a mere extension of science. ... Technological solutions are guided as much by design, aesthetic, economic and other practical considerations as by scientific principles. (p. 2)

In the Position Paper of the National Focus Group on Work and Education (NFG-W&E, 2007), the role of productive work and design opportunities in education were linked: "A systematic study of design and technology can provide opportunities for learning a broad spectrum of generic skills and competences" (p. 30).

Work Education

Work experience, as conceived in the Kothari Commission Report (GOI, 1968), was defined as "participation in productive work in the school, in the home, in a workshop, on a farm, in a factory or in any other productive situation" (p. 13). The National Policy on Education (NPE) 1986 extended work education to secondary school level. It sought to reorient Gandhi's *Basic Education* programme to activities that would cater to the needs of an S&T based society. The introduction of work experience was intended to reduce the growing distance between the 'working' classes and the 'educated' elite, and between rural and urban students. However, the existing social stratification only served to reinforce the alienation. The prevailing curricula in various school systems did not reflect an interweaving of work into education and the radical vision of Gandhi was trivialised in practice. In academic discourse, too, work experience is often confused with vocational education, something which Gandhi, as quoted above, warned us about in his Wardha Conference address.

Vocational Education

Despite vocational education being considered by all stakeholders of education as an inferior form of education at the school level, the Education Commission proposed that secondary education be vocationalised (GOI, 1968). The purpose was to enhance individual employability, reduce the mismatch between demand and supply of skilled human resource, and provide an alternative for those pursuing higher education. It was also perceived as an important step for enriching education and making it more meaningful and attuned to the realisation of national goals. The National Working Group on Vocationalisation of Education (MHRD, 1985) reviewed the programme and recommended the development of the Centrally Sponsored Scheme (CSS) on 'Vocationalisation of Secondary Education' (VSE), which began to be implemented from 1988.

Figure 2 depicts the entry at various levels of education into the world of work and into vocational courses by NPE 1986. Of those who enrolled at Grade I, 60% were expected to complete Grade X. Of these, 25 to 30% were expected to take up vocational courses of 1 to 3 years. However, a mere 5% enrolment has been achieved in vocational courses.

The Vocational Education Programme (VEP) at different stages of education has been known by a variety of names. At the primary and middle school levels (Grades I-VIII) it was called 'work experience' or 'socially useful and productive work' and 'pre-vocational education' in Grades IX and X. Vocational education appeared either as a distinct stream or as the generic vocational programme of courses within the academic stream of education at Grades XI and XII. The latter has been a non-starter for several reasons. For one, only a few schools implemented this course. Besides, those who did adopt it allocated far less time than the 12.5 to 20% of the school time recommended. Other difficulties included challenges of assessing millions of students, insufficient funds, and non-availability of raw materials, equipment, etc.

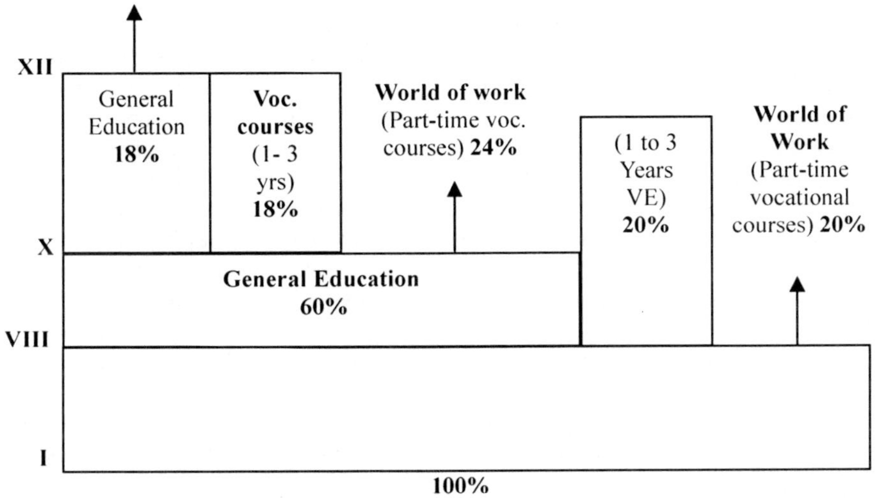

Figure 2. Entry (envisaged) into the world of work at different stages of education (NFG-W&E, 2005, p. 98).

There is also a dearth of adequately trained teachers. In most States, students are streamed into vocational education on the basis of state-level standardised examinations in Grade X. The premium placed on general secondary and higher education by students and their parents, leads one to believe that students joining the vocational system are possibly most likely to be those who perform poorly in the Grade X examination.

As a compelling need for living in the computerised world, over the last decade information and communication technology (ICT) has superseded in importance and urgency the earlier three facets of technology education at school: science, work education, and vocational education. In India, besides being a step towards development, ICT was also considered essential for employability. Subsequently, education about technology is often viewed as learning with and about digital technologies. Under the Private-Public-Partnership scheme, this has been translated in the classroom as basic awareness about computers, while textbooks describe proprietary software for word processing and using spread sheets. The authors think that examples from open source software would be both ethically correct and even empowering for first-time learners of computers.

CONTROLS, STANDARDS, AND THE ECONOMICS OF TECHNOLOGY EDUCATION

The major initiatives towards technical education in the country began well before independence, with the establishment of Indian Institute of Science (IISc) in Bangalore (1909); Institute for Sugar (1936), Textile (1914), and Leather Technology in Kanpur; National Council of Education in Bengal (1905); and industrial schools in several provinces. The Technical Education Committee of the Central Advisory Board of Education was constituted in 1943 and the All India Council for Technical Education (AICTE) was formed as an advisory body in 1945. Following NPE 1986, and by the AICTE Act No. 52 of 1987, AICTE was vested with statutory authority for planning, formulation, and maintenance of norms and standards, accreditation, funding, monitoring, and evaluation of technical education in the country.

Education is a concurrent subject in the Constitution. Education is split between ministries: Primary, secondary, higher, and professional education. Vocational education falls under the purview of the Ministry of Human Resources Development (MHRD). The overall implementation of the vocational programme rests with the State governments, which in most cases work through the State Departments of Education. Some States have separate Directorates of Vocational Education, while others implement the programme through the Directorate of Industrial Training. Since 1994, AICTE has had several offices around the country to help oversee technical and vocational education at the national level in different regions. Facilities for the introduction of vocational courses have been created in the six major areas of agriculture, business and commerce, engineering and technology, health and paramedical, home science and humanities, and science and education. The Pandit Sunderlal Sharma Central Institute for Vocational Education, a constituent but autonomous unit under the NCERT, is responsible for developing the courses and course materials.

It took ten years after the recommendation in 1968 to introduce vocational education courses (Table 1, first row) and another ten years to have a Centrally Sponsored Scheme (CSS) in 1988 to provide financial assistance of about Rs.7 billion to the States and Union Territories for implementing vocational courses in the 6,800 schools. In addition, for the one-year apprenticeship course a stipend is paid to students, which is shared equally by the Central Government and the collaborating institutions. The financial assistance began to be discontinued in 1997-99, resulting in a setback to vocational education programme in several States (NFG-W&E, 2007). The recent report by the National Knowledge Commission (NKC, 2007) has recommended launching a National Skill Development Mission with an outlay of Rs.312 billion to increase capacity from 2.5 million to 10 million students per annum.

Table 1. Institutions, enrolment capacities, and student strengths in vocational education. (Compiled from MHRD, 2003)

Programme	Required education	No. of courses	Enrolment capacity	Course duration	No. of institutions
(A) Formal Programme					
Vocational Education (Vocational Schools)	X	160	972,000 (enrol= 450,000)	2 yrs	6800
Vocational Training (ITIs)	VIII, X	67	672,000	1 to 2 yrs	4591
Technical Education (Polytechnics)	X	40	188,000	3 yrs	1224
Undergraduate level	XII	42	NA	3 yrs	1850
(B) Non-Formal Programme					
Jan Shiksha Sansthans	No bar	NA	NA	Short courses	108
Community Polytechnic	No bar	NA	450,000 per year	3-9 months	675

Unlike vocational education, vocational training programmes in India – with student intake capacity of 740,000 – fall outside the formal schooling cycle. Vocational training is institution-based, with varying entry requirements as well as course durations. The proportion of practical to theoretical instruction in vocational training programmes is also higher than in vocational education. Open to students who have completed anywhere from Grades VIII to XII, the Craftsmen

Training Scheme operates through over 5000 Industrial Training Institutes (ITIs) financed and managed by the State Labour Ministries, and Industrial Training Centres (ITCs) managed by private organisations or NGOs.

Design is a relatively recent area of specialisation available only at the tertiary level and offered by Schools of Art and Architecture. The Indian Institutes of Technology (IITs), IISc (Bangalore), and the National Institute of Design (Ahmedabad) offer masters and doctorate degrees in design. Many of these institutions conduct considerable research in the area of design and design education, although all are focused at the tertiary level. Several institutions offer certificate and diploma courses in a variety of design areas like fashion (clothes), jewellery, and footwear, among a host of others.

PERCEPTIONS ABOUT ENGINEERING AND TECHNOLOGY

The various stakeholders of engineering and technology in India have differing perceptions. As mentioned earlier, the general social view in India is that vocational education is an inferior form of education. The divide between the mind and the hand is acute and the mental overrides the manual. While such attitudes are perhaps found across the world, the Indian sociocultural background privileges the intellect to a larger extent as reflected in the caste system. Thus educational policy makers are often accused of vocationalising general education and making the vocational education too generic and irrelevant for employment.

Employers have a stake in skilled labour, a large number of whom graduate from the ITIs and ITCs. Most employers in three States surveyed reported that ITI graduates did not perform well enough in the use of computers, practical use of machines, communications, and team work practices (World Bank, 2008). A Karnataka State study found that employers felt that ITIs produce personnel who lack basic scientific/technical understanding of their trades (World Bank, 2002). Many occupations still taught are obsolete, and engineering trades – in high demand among students – have outdated syllabi and trainers who are out of touch with current technology and work organisation needs.

In addition, the quality of ITIs across the different States varies significantly. One of the reasons for the mismatch between skills of personnel and industrial needs is the lack of involvement of the industry in the management of ITIs. Whilst ITIs and ITCs rely on formal certification by an independent authority, this certification is perceived to be flawed. There has also been a tendency to create new institutions, while existing ones remain under-funded. Besides, the government has prime responsibility for formal sector training, and there is no pressure and few incentives for employers to be involved in training.

Given that professional courses in engineering and technology attract students while technical courses are looked down upon, it is useful to study school students' ideas about technology (Bhattacharya, 2004; Rajput, Pant, & Subramaniam, 1990). However, there have been far too few studies of this type. Surveys conducted at Homi Bhabha Centre for Science Education (HBCSE) probed the ideas that students hold about technology in a variety of ways. As technology is not an independent school subject, students' ideas may be influenced by a complex set of in-school and out-of-school factors. In one study, about 60 students from Grades V to IX, from 30 different schools in and around Mumbai city, participated in a poster making contest on 'images of science' and 'images of technology' (Mehrotra, Khunyakari, Chunawala, & Natarajan, 2003). The analysis showed that students perceived S&T more in terms of the objects used or created than as processes, and S&T was largely viewed as beneficial to individuals and society.

Another survey, of 654 middle school students of Grade VIII from eleven English and Marathi medium schools in and around Mumbai (Khunyakari, 2008), also indicated a positive view of technology. Students were proud of their country possessing nuclear technology, yet worried about its consequences for people and the environment. Pictorial questionnaires that presented technology through objects and activities were used among Grade VI students (Mehrotra, Khunyakari, Chunawala, & Natarajan, 2007). Findings indicate that students considered objects and activities related to communication and transport, especially modern ones, as technological. Utilitarian and human-made aspects were emphasised by students as reasons for associating objects and activities with technology.

Few studies exist that examine engineering and technology students' attitudes towards their own profession and the state of technology education.

PROFESSIONAL DEVELOPMENT OF TECHNICAL/VOCATIONAL EDUCATION TEACHERS

During 1965–67, four National Institutes of Technical Teachers' Training and Research were established in four metropolises around the country to develop, organise, and evaluate quality training programmes, research studies, and learning packages for technical and vocational education. Since they have all been accorded the Deemed University status, they function in collaboration with the AICTE to offer professional development opportunities – from certificate courses to doctorate level – to teachers of technical and vocational education. Qualifications of teachers in the generic vocational stream (Grades XI and XII) are similar to those of general secondary school teachers. Full-time teachers need to have a masters-level qualification, and teachers of vocational subjects are often the same as those who teach general subjects. In addition, part-time teachers are also hired by institutions to teach specific courses.

State directorates of school education largely work in isolation from other relevant educational departments and institutions of the State. As the State governments were cash-strapped after the discontinuation of the funding from the Centre, they resorted to appointing of part-time teachers – usually unemployed graduates without either industry background or experience in vocational/ technical education. In addition, they were not given in-service training (NFG-W&E, 2007). Thus, there exists in technical and vocational education a bleak picture of staff inadequacy both in the number staff and their professional preparation. Recent reports have held the AICTE and State monitoring bodies accountable for this situation (NKC, 2007). There have been repeated attempts at improving the situation through increased funding for conducting teacher preparation courses, summer camps, etc. However, the fragmentation of higher education and a lack of coordination between the generic vocational education at school levels and vocational/technical education at higher secondary level and beyond have often negatively impacted on implementation.

LOCAL EFFORTS IN CONTEXTUALISING TECHNOLOGY/ TECHNICAL EDUCATION

A number of NGOs have been providing education through skills and training to target groups that vary across out-of-school children, unemployed adults, women, and tribal and other marginalised groups. They often run short courses on technology, home science, commerce, etc. Some NGOs receive funds from the MHRD for conducting programmes in vocational education. Ministries of Social Welfare and Justice, Rural Development, Labour, as well as agencies like the Council for Advancement of People's Action and Rural Technology, also provide financial assistance to NGOs to run non-formal vocational education and training programmes (NFG-W&E, 2007). Other institutions running vocational education and training programmes in the non-formal sector include 'Krishi Vigyan Kendras' (KVKs), Khadi and Village Industrial Commission (KVIC), Jan

Shikshan Sansthan, and Community Polytechnics. Some vocational courses are also offered by National Open School through a number of institutions accredited by it. *Vigyan Ashram* at Pabal in Maharashtra State created a curriculum that allows rural students to become entrepreneurs through skill training and education using appropriate and income-generation technologies (Natarajan, Khunyakari, & Chunawala, 2004).

Some of the issues faced by the non-formal sector include lack of certification and accreditation of institutions and courses; problems of equivalence with formal education and training programmes; and lack of infrastructure, both human and material, for proper running of the programmes.

CONCEPTUALISING A SCHOOL-LEVEL DESIGN AND TECHNOLOGY EDUCATION

About 20% of the country's population of over a billion is in the school-going age group of 5 to 18. A very large number of these either have never enrolled themselves in schools or have dropped out at various stages of education: 52% of those who enrol in Grade I drop out by Grade VIII, and more than 70% drop out by Grade X. Only 20% reach Grade XII, while a mere 7% go for higher education (NFG-W&E, 2007; NKC, 2007). Only a small fraction of these enter technology or design courses. Yet those who drop out of the education system join (or aspire for) the world of work, most without acquiring employable skills.

What does school education provide in terms of employable skills such as knowledge and process, procedural, and team work skills? With an emphasis on studying theoretical principles and observing experiments conducted primarily by the teacher, school science is seldom experienced as an activity at all, let alone a participatory one. As such, students do not view science as a collaborative activity (Chunawala & Ladage, 1998). Work experience, or any of its school-based variations, involves making socially-useful objects using recipes suitable for production in large numbers. It has little scope for design or examining contexts of use.

Technology education needs to involve multidisciplinary perspectives and multiple skills. As such, school education that merely emphasises knowledge about technology can stifle innovation. In contrast, enriching school curricula with explicit opportunities for authentic problem solving and multiple expression modes valid in a variety of classroom contexts could help develop future technology innovators. Technology education thus has a scope wider than either vocational education or work experience, and it transcends science for its disciplinary grounding. As a subject in its own right, it makes links between craft, empirical indigenous knowledge, the natural sciences, and the social sciences. Design, which is the core of technological activity, is a way of seeing the world that is quite distinct from the way the sciences or the humanities see the world. Hence, technology with an emphasis on design deserves subject status in Indian general education (Cross, 2002; Roberts, 1999).

Technology education research at HBCSE has been guided by theoretical perspectives on collaborative learning; cognition and action; concerns of sociocultural and gender appropriateness; and the development of language, quantitative, and problem solving skills. This approach shifts the emphasis from the dominant views of technology education as applied science or ICT to collaborative participation of students in design and evaluation of artefacts and systems (Kimbell, Stables, & Green, 1996). Design, craft, and technology as part of general education (at the school level) will help attract a greater number and diversity (rural, girls, etc.) of students to careers in technology, as well as equip them with the trainable skills of perceiving and defining needs, and *designing* to satisfy them.

Figure 3. A model of D&T education centred on collaboration and communication.

Ideally, the core of technology education at school level in the Indian context will be design and technology (D&T) activities that use a variety of skills and draw upon the knowledge of key concepts traditionally taught within other disciplines. The activities would integrate aspects of affect (wants, desires, and aesthetics) and judgements (making strategic alliances, choosing materials, and evaluating products). They will go beyond episteme (knowledge) and techné (skill), and include phronesis (practical wisdom) (Dunne, 1993). The hallmarks of technological empowerment will include the ability to design, innovate, make, evaluate, and communicate the technological goals, processes, and products through collaboration within a complex sociocultural environment as indicated in the model in Figure 3 (Choksi, Chunawala, & Natarajan, 2006). A stage-wise curricular framework for a collaborative and communication-centred D&T education has also been developed at HBCSE.

THE CURRENT SCENARIO AND POSSIBLE FUTURE DIRECTIONS

India's vocational education and training programmes have never adequately met the needs for which they were created. The reasons are complex and arise from academic, organisational, economic, and sociocultural contexts. Following a boost in financial outlay of the Eleventh Plan (2007-2012) for higher education, three more autonomous IITs (professional technology institutes) have been established. There has also been a greater allocation of funds for revamping the ITIs (after Grade X and XII). Recommendations over the last decade have included introduction of design in ITIs, Polytechnics, and other technical education institutes; strengthening the innovation component in technical education; and encouraging and training faculty to conduct research through seminars and workshops.

The National Knowledge Commission (NKC), constituted by the Government of India in 2005 to study the various aspects of education, at different levels, has recommended more flexibility in vocational education and training (VET). It also pointed to the need for quantifying and monitoring the impact of vocational education, and suggested re-branding to increase its perceived value and ability to command higher incomes. It further proposed a complete overhaul of the regulatory structure for higher education, with the creation of the Independent Regulatory Authority for Higher Education and abolishment of the AICTE. For technical and vocational education to gain status, it will need teachers with better professional preparation, an emphasis on creativity and entrepreneurial skills in the technical and vocational courses, and an attractive remuneration package for staff. This implies teaching elements of design and critical thinking about technology and society not only to those in the vocational education stream, but also to all students in the generic vocational courses.

Making technology education inclusive assumes significance in the complex sociocultural context of India, where there is:

- Immense innovation potential across the country among the schooled and the unschooled, the formal and non-formal sectors;
- continuing traditions of indigenous local technologies;
- wide cultural and resource differences between different parts of the country including the rural-urban divide; and
- classrooms with students from multi-lingual backgrounds.

Consider that over 80% of the students enrolling in Grade I enter the world of work with either generic or part-time vocational courses, and a considerable proportion have dropped out even before Grade VIII. This would imply that a productive and creative work force is possible only by addressing the need for innovation and creativity at as young an age as possible, preferably from the primary years. It would also be more equitable to introduce D&T education across the country. At the same time, the nature of activities within the curriculum and its contextualisation needs to be addressed in order to retain uniform standards of process and cognitive skill development.

The present education system, which does not encourage collaboration and which constrains modes of expression, can be alienating to a large proportion of learners (The probe team, 1999). The D&T model developed at HBCSE proposes stage-wise increases in the complexity of activities, which are designed integrate skills and processes across knowledge domains. A well-planned D&T curriculum, especially in mixed-ability and multicultural classrooms, should be an inclusive rather than an exclusive endeavour – for the children of the rich as well as the dispossessed, for those in the indigenous or the modern mould, for girls and boys.

The educational community has a tendency to avoid radical changes, even if its present state is a scary one. Creativity and critical thinking are difficult to manage in classrooms, difficult to assess, and can threaten political systems. Parents and social systems are also often uncomfortable with it. Technology education poses an additional challenge by its sheer breadth. Engineering is neat and definable; the domain can be subdivided into mechanical, civil, etc. In addition, its acceptability is founded on its links with science and mathematics, the disciplines of prestige and power in all modern cultures. In contrast, technology as a notion is fluid and somewhat ambiguous. It is changeable with time and context – geographical, cultural, and socio-historical. Hence, it has myriad teaching possibilities that imbue the subject with possibilities of inclusivity and equity, creativity and critical thinking, ecology of thought and action. This is also what makes it an empowering school subject. The authors thus call for inclusion of design and technology in Indian school curricula, and hope that all its empowering possibilities will one day be realised.

NOTES

[1] A majority of schools in India are either affiliated to the Central Board of Secondary Education (Central schools) or to respective State Boards of Education.

REFERENCES

Bhattacharya, P. K. (2004). *Technology and children*. New Delhi: NCERT.

Choksi, B., Chunawala, S., & Natarajan, C. (2006). Technology as a school subject in the Indian context. In K. Volk (Ed.), *Articulating technology education in a global community. Conference proceedings of the international conference on technology education in the Asia Pacific Region* (pp. 280–290). Hong Kong: The Hong Kong Polytechnic University.

Chunawala, S., & Ladage, S. (1998). *Students' ideas about science and scientists. Technical Report No. 38*. Mumbai: Homi Bhabha Centre For Science Education.

Cross, N. (2002). The nature and nurture of design ability. In G. Owen-Jackson (Ed.), *Teaching design and technology in secondary schools* (pp. 99–108). London: Routledge Falmer.

Dunne, J. (1993). *Back to the rough ground: 'Phronesis' and 'techne' in modern philosophy and in Aristotle*. South Bend, IN: University of Notre Dame Press.

Gandhi, M. (1994). *The collected works of Mahatma Gandhi*. New Delhi: Publications Division, Ministry of Information and Broadcasting, Govt. of India.

GOI. (1968). *National policy on education–1968*. New Delhi: Ministry of Education.

Khunyakari, R. (2008). *Investigating middle school students' perceptions of technology and developing design and technology education units to study students' design productions*. Unpublished Ph.D. thesis, Tata Institute of Fundamental Research, Deemed University, Mumbai.

Kimbell, R., Stables, K., & Green, R. (Eds.). (1996). *Understanding practice in design and technology*. Buckingham: Open University Press.

Mehrotra, S., Khunyakari, R., Chunawala, S., & Natarajan, C. (2003). *Using posters to understand students' ideas about science and technology. Technical Report No. 1 (02-03)*. Mumbai: Homi Bhabha Centre for Science Education.

Mehrotra, S., Khunyakari, R., Chunawala, S., & Natarajan, C. (2007). Using pictures and interviews to elicit Indian students' understanding of technology. In J. R. Dakers, W. Dow, & M. de Vries (Eds.), *PATT-18, International conference on design and technology education research* (pp. 152–161). Glasgow: Faculty of Education, University of Glasgow.

MHRD. (1985). *V.C. Kulandaiswamy committee report*. New Delhi: MHRD, GOI.

MHRD. (2003). *Draft project implementation plan of national programme in TVET 2003*. New Delhi: MHRD, GOI.

Natarajan, C., Khunyakari, R., & Chunawala, C. (2004). *Vidnyan Ashram, its work and HBCSE: Comments in Dr. Shrinath Kalbag memorial seminar report*. Mumbai: Marathi Vidnyan Parishad.

NCF. (2000). *National curriculum framework for school education*. New Delhi: NCERT.

NFG-ToS. (2006). *Position paper of the national focus group on the teaching of science*. New Delhi: NCERT.

NFG-W&E. (2007). *Position paper of the national focus group on work and education*. New Delhi: NCERT.

NKC. (2007). *National knowledge commission report to the nation 2007*. New Delhi: GOI.

Rajput, J. S., Pant, S. C., & Subramaniam, K. B. (1990). Pupils' attitudes towards technology in rural schools in India. In R. Stryjski & H. Szydkauski (Eds.), *Technology and schools. Report of the PATT conference in Poland* (Vol. 2, pp. 38–54). Eindhoven, The Netherlands: Eindhoven University of Technology.

Roberts, P. (1999). From making things to making sense: The educative roles of practical making activity. In *Learning and making: Transcripts of two linked one-day conferences examining the value of creative practical education in Britain*. Great Britain: Crafts Council.

The Probe Team. (1999). *The public report on basic education in India*. New Delhi: Oxford University Press.

World Bank. (2002). *Karnataka: Secondary education and the new agenda for economic growth. World Bank Report No. 24208-IN*.

World Bank. (2008). *Skill development in India: The vocational education and training system. World Bank Report No. 42315*.

Chitra Natarajan
Sugra Chunawala
Homi Bhabha Centre for Science Education
Tata Institute of Fundamental Research
Mumbai, India

BANGPING DING

11. THE DEVELOPMENT OF TECHNOLOGY EDUCATION IN MAINLAND CHINA

INTRODUCTION

School technology education in Mainland China was embedded as part of a curriculum area titled 'Integrated Curriculum of Practical Activity (ICPA)' during the curriculum reform of basic education (Grades 1-12), which commenced around the turn of the millennium (Ministry of Education [MoE], 2001). ICPA, dictated as a required curriculum area for both primary and secondary schools, was initially composed of 4 curriculum sub-areas: (1) information technology, (2) research-based learning (*yanjiuxing xuexi*), (3) n community service and social practice, and (4) labour-technical education (MoE, 2001, p. 6). As such, school technology education includes (1) and (4) – information technology and labour-technical education – within a somewhat incoherent framework; however, in practice, information technology has come to develop into a separate school subject of its own and been given high priority by schools, governments, and society at large because of its perceived importance in contemporary education as well as in national and international economical competitiveness. Research-based learning (*yanjiuxing xuexi*) is claimed to be a learning style as well as a curriculum sub-area and concerns itself mainly with thematic learning in relation to students' life based on their interests (Zhang, 2001). Community service and social practice is activity-oriented and is carried out outside of the classroom with the aim of developing in students an enthusiasm for serving society or for public affairs.

Labour-technical education (LTE) has been officially recognised as a school subject in primary and secondary schools since 1981, although it had been available in a variety of forms and other titles (Chen, 1989) since the founding of the People's Republic of China (PRC) or New China (as opposed to Old China, which existed before the founding of the PRC on October 1, 1949). As an independent curriculum sub-area in both primary and secondary schools, LTE is much more visible than other curriculum sub-areas of ICPA except for information technology despite the rhetoric of recent curriculum reforms.

The focus of this chapter is on LTE and its assessment as practiced in primary and secondary schools, as one of the two mainstays of technology education (the other being information technology) nationwide. Some comments will also be made on what is called general technology education (*tongyong jishu jiaoyu*), which is still experimental in nature and being trialed in senior high schools in some of the provinces of China.

Because of historical, ideological, political, and economic reasons, technology education in China's schools in the form of LTE is, so to speak, sort of *sui generis* in terms of its comparison with not only other countries, but also areas within Greater China, like Hong Kong (Volk, 2003), where technology education has been influenced by the British model, and Taiwan (Yu & Hsu, 2003), where technology education has been patterned after the U.S. model. Historically, Old China had been characterised by a separation of productive labour from schooling, and hence the separation of working classes from the few educated elite who became rulers or officials in the

A.T. Jones and M.J. de Vries (eds.), International Handbook of Research and Development in Technology Education, 117–130.
© 2009 Sense Publishers. All rights reserved.

Confucian tradition. Ideologically, schooling in New China has adopted the educational theory of Marxism-Leninism and Maoism, which takes as its guiding rationale the combining of schooling and productive labour, a principle embedded in the state educational guideline.

Early in the 1950s, technology education in primary and secondary schools in New China was for a time under the influence of the former Soviet model of polytechnic education (Jiang, 1996). Politically, schooling is part of the superstructure of the state and also an integrative part of China's socialist cause, and therefore it must serve the politics of the proletarian class. Economically, it was – especially from the 1950s through to the 1970s – an imperative for students and teachers to be involved in productive labour so that economic gains could satisfy part of their basic needs for lack of funding from the government (Fouts & Chan, 1997).

Given this historical background, it is understandable that the school subject called LTE emerged early in the 1980s, and has been available in the school timetable in many parts of China, including Shanghai, the biggest metropolitan city where school reforms have been under way since the mid-1980s independently of the national educational reform agenda.

TRADITIONS OF CHINESE CULTURE AND EDUCATION IN RELATION TO TECHNOLOGY AND TECHNOLOGY EDUCATION

In order to understand the uniqueness of technology education in New China's schools, it is necessary to include a brief introduction to traditions within Chinese culture and education. These take several forms, reflecting the historical times in which they originated.

First, the Confucian tradition goes back more than 2500 years. As a great educator, Confucius (551-479 BCE) opposed the idea that his students take part in agricultural labour (and techniques associated with it) when he was asked by one of his students, Fan Chi, how to grow crops. Confucius said, honestly but with irritation, that he was not good at growing crops, and later, when the student was not in his presence, he commented that Fan Chi was 'a small man', whereas in the great master's mind every one of his students should have become a superior man, or a would-be ruler leading the country (Confucius, 1994). Although there has been a dispute as to whether Confucius despised husbandry (and hence agricultural techniques), it seemed certain that nearly all Confucian followers were not involved with manual labour and technology. Thus, the social order in ancient China placed intellectual-officials in the highest position, followed by peasant farmers, with artisans and businessmen – ancient Chinese technologists – in the third position. While traditional technology in ancient China is considered as having reached a high level of development as a result of basic needs, this was not attributable to Confucian scholars but rather through the efforts of artisans who developed their skills by way of apprenticeship. Unlike in the West, however, where masters and apprentices worked in the organisations called guilds since the Middle Ages, in ancient China the relationships between master and apprentice were mainly established within family-clans (*jiazu*) as fathers-sons, since ancient Chinese society was based on family-clans rather than other social organisations.

A second cultural and education tradition is one that developed in the 20th century as an antithesis to the Confucian tradition distaining manual labour and hands-on activities, a tradition advocated by Mao Zedong and his communist regimes and which was partly predicated on Marxism-Leninism and partly grew out of the agricultural culture that had existed tens of thousands of years in China. Influenced by Marxism-Leninism, (manual) work/labour was associated with working classes and therefore both working class people (peasants, artisans, and factory workers) and their vocations were elevated to a much higher position in social status. To counteract the idea of distaining manual work/labour, Mao and his communist regimes encouraged students to do manual labour together with peasants while they were in school early in the 1930s and 1940s in Yan'an times (Cleverley, 1991) – giving rise to the concept of 'manual labour education'. During the 1950s, when New China was developing brotherly relationships with the Soviet Union, this new tradition of emphasising manual labour education in school was considerably strengthened for a period of time to such an extreme degree that considerably more time was devoted to it rather than to classroom teaching as China followed a similar path to the Soviet Union in emphasising labour in schooling. It was at this time that Mao put forth an important instruction: "Schooling must serve the proletarian politics, and it must also be combined with productive labour" (Deng & Fu, 1991).

In addition, 'polytechnic education', a concept born out of the educational theory of Marxism-Leninism and implemented in Soviet schools also had an impact on China. However, as China was much more backward in productivity at that time, the idea of polytechnic education could not be realised in schooling and China's educators proposed the idea of 'education in basic productive skills' (*jiben shengchan jishu jiaoyu*) in its place (Chen, 1989; Jiang, 1996) to suit the conditions of the then agricultural China, where most students had to be participating in agricultural labour or factory work. Later, many schools ran agricultural farms or factories of their own on their school campuses, resulting in the work-study programs and school enterprises (Fouts & Chan, 1997; Cleverley, 1991) that appeared on the Chinese educational landscape for decades.

A third tradition started in the early 1980s when China broke with the Leftist line of politics under Maoist rhetoric of 'class struggle' and turned towards economic construction and modernisation. This transformation in the political climate in China, following the Communist Party's open and reform policy that had begun in 1978 when Deng Xiaoping resumed his position as a top leader, indicates that school curriculum in general and school technology education in particular were to change direction in order to serve the modernisation agenda. In 1981, a circular of the Ministry of Education stipulated that a new school subject called 'Labour-Technical Education' (LTE) was to be introduced into the school timetable. LTE as a separate school subject is different from former concepts – such as manual labour education, polytechnic education, and education in basic productive skills – in that it is less ideological in nature and tends to hold technology in much more prominent regard. Nevertheless, one can still detect a trace of the ideological flavour from the construction of the new term, and this influence from the past revolutionary tradition, together with the narrow conception of technology embedded within it, foreshadows ways in which the structure and practice of this newly-established school subject would be shaped as demonstrated by some of the case studies presented below.

These traditions show that manual labour education and/or labour-technical education as a school subject has been highly influenced by ideological theory because of its historical, ideological, political, and economic underpinnings, which,, to a great degree, shape the policy, methods, and content of technology education at the school level. In addition, the lack of knowledge base within this school subject also restricts teaching, learning, and assessment approaches as outlined in the next section.

LABOUR-TECHNICAL EDUCATION AND GENERAL TECHNOLOGY EDUCATION IN SCHOOLS: RHETORIC AND REALITY

Implementing Labour-Technical Education in Primary and Secondary Schools

As a new school subject, labour-technical education is distinct and different from other subjects such as the Chinese language, mathematics, physics, or even physical education, in that it has less clear content to teach. Broadly speaking, labour-technical education is composed of labour education and technical education, but here labour seems just to mean manual or physical labour, which is meant to represent: (1) productive labour in the factory or on the farm, (2) labour for social and public good, and (3) labour for self-service. All of this is intended to be for the purpose of political-ideological-moral education, that is, to cultivate in students a love for labour and the people associated with it and a habit of work and an appropriate attitude toward labour. Technical education, on the other hand, is usually taken to mean some sort of education in various techniques or technologies associated with agricultural or industrial production. In the vast rural areas of China, technical education, if it is given weight, is usually connected with the techniques and/or technologies prevalent and necessary in the local area where the school is situated. Mostly the content of the subject is focused on the technologies of husbandry, processing of products, and repairs of home electrical and electronic appliances. In city schools, labour education and technical education are usually combined in the centres for labour-technical education that cater to students studying technologies. Although it is a required school subject according to the government policy, in many schools the time allocated to it is often taken up by other supposedly more important subjects like mathematics or science, particularly in senior high schools.

As a means of political-ideological-moral education, labour-technical education has experienced fluctuations over the past decades. For example, at one stage it was added to the list of moral education (including political-ideological education), intellectual education, physical education, and esthetical education, which together make up the so-called all-round education for Chinese students.

In spite of this rhetoric, labour-technical education is difficult to implement in all schools for several reasons. First, as labour-technical education has nothing to do with the entrance examinations for admission to senior high schools or colleges and universities, few educational administrators, school principals, teachers, students, and parents treat it as seriously as they do other subjects. What is considered more important is how many students can move up to higher levels of the educational ladder through examinations.

Second, there has been a shortage of qualified labour-technical teachers, a lasting issue that has yet to be solved. According to a survey conducted by Song (1995), out of 4,249 labour-technical teachers in the city of Tonghua, Jilin Province in the Northeast of China, only 380 were teachers of labour-technical education, with the rest being moonlight teachers (*jianzhi jiaoshi*), that is, teachers that teach other subjects as well as labour-technical lessons. Out of the 380 teachers considered to be specialist teachers of labour-technical education, 330 of them were aged 50 or older, and most were assigned to teach the subject not because they were qualified to do so, but because they could not teach other subjects well, either. As yet, no teacher education programme is provided for labour-technical education by higher education institutions throughout China. A third issue that has prevented labour-technical education from making much progress over the past two decades is concerned with curriculum. Although a national syllabus was developed in 1994, it was expected that the subject matter and/or textbooks would be left to local authorities so that the actual needs and possibilities of the local technological practice could be reflected. As there is not much expertise in relation to technology and technology education at the local level, it is not surprising that few textbooks of labour-technical education exist. Thus, labour-technical education holds no attraction to students, their schools, or their parents (Bao, 1997).

A fourth issue, connected with the curriculum issue, is assessment. As the curriculum content varies from province to province, county to county within provinces, and even from school to school within a county or district, testing or examinations are at best casual. For example, in the city of Tonghua, Jilin Province, where labour-technical education was regarded as exemplary, it was Song (1995) found in his survey that 65% of schools offered mid-term and end-of-the-term testing for labour-technical education, but student performance of the subject was not added to the their performances of other subjects for the means and ratios of pass rates; no testing was administered for labour-technical education in the remaining 35% of schools.

Centres of Labour-Technical Education

According to Xu (2002, 2004), labour-technical education has undergone considerable change in terms of content and school practice over the past two decades. The general trend has been to move increasingly away from the political-ideological underpinnings and focus more and more on the technical and/or technological dimensions. This is particularly true of Shanghai, the largest city of China, where educational reforms have been undertaken independently of the national reform agenda, and where technical-labour education has been allowed to develop on its own.

As a new development of labour-technical education, the late 1980s and early 1990s saw the establishment of centres of labour-technical education, schools that specialise in labour-technical education and that have been set up in many cities across the country. In Shanghai, for example, each district in the city has a centre of LTE that attracts students from schools within its catchment area. In Huangpu District Center of Labour-Technical Education, one of the earliest and most distinguished centres of its kind in Shanghai, teaching, learning, management, teacher training, and competition of labour skills in the field of LTE are all concentrated for schools within the whole district. It adopts as its curricular goals for students to "be good at hands-on activities, able to design, and loving of labour" (Shanghai Municipal Education Commission, 2004), and at the same time it also emphasises moral education. Students from many schools of this district come to learn technical/technological skills such as designing and making everyday products of various kinds (e.g., multi-purpose removable desks, handicrafts made of cloth with colour and design, containers for holding writing tools, etc.). In Changning District Center of LTE, another well-run base for LTE in Shanghai, labour-technical education is combined with a project on environment, population and sustainable development (EPD) in an effort to enhance in students coming from the various schools of its catchment area their ideological-moral quality, scientific-cultural quality, physical and mind quality, esthetical quality, and labour-technical quality (Shanghai Municipal Education Commission, 2004).

The practice of technology in the world outside the classroom covers a diverse range of activities from agriculture through to the production of synthetic materials and electronics. Technology education in schools must reflect this diverse practice and not limit itself to designing and making with a limited range of materials. The development of technology education must reflect relevant technological activities (Jones, 2003). However, one of the features – and also, in my view, shortcomings – in the current LTE in China's schools is that it limits itself to hands-on activities that lead to the production of craft products that the students are allowed to take home.

Emergence of General Technology Education in Senior High Schools

In senior high schools (Grades 10–12), the practice of technology education becomes more complex. Before 2001, senior high schools were required to provide labour-technical education; however, the practice of information technology developed so quickly and pervasively that it soon appeared in the timetable of nearly all senior high schools. Information technology education has since been a required curriculum subject in senior high schools. According to a survey (Liu, Gu, & Yu, 2005), by the end of 2001 12,000 senior high schools across the country had offered courses on information technology, amounting to 92% of all general senior high schools.

The reform of the new curriculum of basic education was also announced in 2001. Labour-technical education and information technology education became two parts of the so-called ICPA, as mentioned at the beginning of this chapter. Meanwhile, the *Framework for the Curriculum Reform of Basic Education* (MoE, 2001) stipulates that senior high schools should provide courses on general technology education. As a result *Standards of Technology Education Curriculum for Senior High Schools* (MoE, 2003) was developed. Since 2004, general technology education has been implemented on an experimental basis in senior high schools of four chosen provinces, and now several more provinces have joined in the efforts.

According to Gu (2005), a professor of technology education of Nanjing Normal University, Nanjing, Jiangsu Province, and also a panel leader for the development of the *Standards of Technology Education Curriculum for Senior High Schools*, general technology education in senior high schools is now connected with labour-technical education in primary schools (Grades 1–6) and junior high schools (Grades 7–9) in terms of curricular goals. For senior high schools, the general goal is expressed as follows:

> By learning this curriculum, the students will (1) broaden their perspectives on technology, learn and/or master some basic knowledge and skills in general technology, and master some ideas and methods of technology and design; (2) develop and maintain lifelong interest in and learning desire for technology, formulate correct views of technology and a strong consciousness of innovation; (3) cultivate a habit of using technology in a positive, responsible, and safe way, and develop preliminary technological capacity and some ability in planning their careers, so that they will lay a foundation for the preparation of meeting future challenges, improving their life quality and realising their lifelong development (MoE, 2003, p. 34).

Connected with this general goal, the standards also propose five sub-goals: The curriculum will develop in students:

> (1) the capacity in technological understanding, application, improvement and decision-making; (2) the capacity of the expression of ideas and the capacity of translating ideas into operational schemes; (3) the capacity of integrating, applying, and materializing knowledge; (4) the ability of creative imagination, critical thinking and problem-solving; and (5) the ability of understanding, evaluating, and selecting of technological culture. (MoE, 2003, pp. 34–35)

Obviously, this general goal and its sub-goals are in striking contrast to the goals of labour-technical education.

Based on the *Standards of Technology Education Curriculum for Senior High Schools*, the content of general technology education consists of nine modules:

- technology and design I;
- technology and design II;
- electronic control technology;
- the making of simple robots;
- auto driving and maintenance;
- clothing and its design;
- home economics and living technology;
- architecture and design; and
- modern agricultural technology ([MoE], 2003).

Of the nine modules, only the first two are required course content, with the rest being selective curriculum content. Each module takes 36 credit hours, serving as two credits.

As a new school subject, the question has arisen as to whether general technology should be one of the subjects on which all students should be examined for admission to college matriculations. If it is not involved in the examination system, then it is likely that nobody – students, parents, teachers, administrators and others – will treat it seriously.

Debates over Labour-Technical Education and New Trends Toward General Technology Education

As the current curriculum reform moves on, the Chinese educational community seems to increasingly be of the opinion that labour-technical education is out of date in terms of its conception, content, and pedagogy. A debate has arisen that suggests that many researchers, administrators, and teachers concerned with technology education perceive something confusing with ICPA or labour-technical education. For example, some (e.g., Xu, 2004) feel that the current conception of ICPA narrows the space allocated to labour-technical education by making it just part of the ICPA curriculum area. Xu (2002, 2004) admits that the meaning of LTE has changed since the mid-1980s, moving away from the political-ideological-moral dimension of schooling and increasing its elements of technology. Yu (2001) also argues that labour-technical education is asynchronous with the times in terms of curriculum goals, content, implementation, evaluation, teacher professionalism, and facilities and infrastructure. It is time, she categorically asserts, that the subject is reformed – but this is still a call for reform under the conventional framework.

Other scholars argue for a more radical reform oriented more towards the international trend of restructuring the subject (Lewis, 1995), that is, to establish a new subject of technology education predicated on technological literacy. Zhou, Yang, and Ni (2004), for example, published an article titled 'The importance of technology education', arguing that technology education should be added to the curriculum of China's basic schooling in parallel with science education. Writing with a more international perspective, Ma (2005) offers the following critique:

> What is lacking in the practical exploration of labour-technical education over the past two decades is nothing less than the awareness of technology. The function of technology education was not played out, and the emphasis was on labour education rather than on technology education, to such an extreme degree that labour education was in complete placc of technology education. The main reason behind it is the unclearness in meaning of the concept 'labour-technical education', which makes no distinction between labour education and technology education, while the more in-depth reason for the confusion is that we are still not out of the linguistic context in which was born the concept of labour-technical education, which makes labour education overshadow technology education, hence exposing the perceived limitations and problems that are obvious to all. (pp. 46–47)

Having clearly analysed the problem inherent in the terminology of labour-technical education, Ma goes on to argue unequivocally for technology education in place of LTE, a reformulation of the objectives of technology education, and a reconstruction of the technology curriculum.

On the side of educational administration, Meng, a vice-head of department of education of Heilongjiang Province in the Northeast of China, has launched a set of reform efforts replacing labour-technical education with technology education in junior high schools (Grades 7–9) of that province since 2004. Meng (2005) asserts that the developments of our times and our nation, the solution of the so-called 'three *nongs* issues' (this Chinese term refers to the developmental issues of agriculture, the countryside, and the peasant farmers), and the improvement of the quality of citizenship all call for a need of technology education at the level of basic education. In an interview with *China's Educational Daily*, he is quoted as saying:

> We must admit that, compared with developed countries, the most lacking part in the quality of our citizens is technology literacy. This could be proved by the simple fact, for instance, that advanced productive lines imported from other countries cannot, as expected, produce the most advanced products at home.

The introduction of technology education into the curriculum of junior high schools in Heilongjiang Province is affected under the *Framework for the Curriculum Reform of Basic Education* (MoE, 2001) that stipulates that local district curriculum (*defang kecheng*) is part of the curricula of basic education, along with the national curriculum and school-based curriculum. So far, Heilongjiang Province is the only province in China that has replaced labour-technical education with technology education at the level of junior high schools (Grades 7–9). Perhaps, together with the increasing adoption of technology education in senior high schools, this signals a trend for technology education to replace of the old-style labour-technical education all over China?

ASSESSMENT PRACTICE IN CENTRES FOR LABOUR-TECHNICAL EDUCATION: THE CASES OF BEIJING AND SHANGHAI

Although in many schools (especially rural schools in the countryside) LTE is marginalised for various reasons, there are some exceptions. By way of example, the following two case studies illustrate assessment practice of LTE in Beijing, the capital of China, and in Shanghai, the largest and probably one of the most modernised cities in China.

Assessment of Labour-Technical Education: The Case of Beijing

Not all schools in Beijing place much importance to LTE – in fact, many schools have only nominal LTE lessons in their school timetable. One of the reasons for inattention to LTE is the high competitive pressure on schools in preparing their students to move to higher levels of education by teaching to the test, whereas LTE is one of the auxiliary subjects not involved in testing of any kind. Like other auxiliary subjects, such as music and the fine arts, LTE is marginalised in the school curriculum, especially in junior high schools. This, in turn, leads to the fact that LTE has been considered subordinate to more academic subjects in the school curriculum since it has been the Chinese tradition that "examinations sanctioned knowledge and created the Chinese nobility" (Llasera, 1987, p. 24).

In spite of this, LTE is relatively valued in some schools and they have been selected for elaboration here as models. Typically, LTE is not undertaken in individual schools in Beijing, but rather concentrated in Labour-Technical Education Centres in each district – self-contained, single-purpose schools founded in the 1990s in Beijing to cater for primary and secondary school student education in labour-technical skills.

The first centre of LTE that was observed for the purpose of this report is Haidian Central School District Centre of LTE, which serves the 12 surrounding primary schools and sometimes other schools outside the district. Located at the hub of Haidian Central School District, the Centre owns a 4-story teaching building with 12 classrooms. There are some 20 LTE teachers in the Centre, three quarters of whom are full-time classroom teachers, with the rest employed on a temporary basis. Most of the teachers are graduates from the arts and crafts programmes of teacher education institutions and are adept at designing and making. (There are practically no technology education programmes currently available in any of China's universities or colleges.)

Every week day, about 300 students of Grades 4 and 5 (lower primary and Grade 6 students don't take LTE lessons) come to this centre for learning LTE. The Centre offers 15 technical/ technological programmes, encompassing ceramics, woodwork, paperwork, knitting, leatherwork, electronics, environmental studies, and making models of vehicles. These programmes feature practical, hands-on activities and national culture. The students usually study for a whole day and they are allowed to take home the products they create. Unlike in their own schools where the size of class is usually large (40 or more students in each classroom), at Haidian Central District Centre of LTE the number of students is limited to about 20 in each classroom and students sit around worktables in groups for the convenience of cooperative learning. This makes it much easier for students to learn and be assessed in class. LTE teachers also teach the subject in a different way: they don't talk as much; rather learning/teaching is task-driven. Although it was observed to still be teacher-centred, it was to a much lesser degree.

Another centre of LTE that was observed is Xicheng District Centre of LTE, Beijing, a small school with only 15 full-time LTE teachers and adjoining an ordinary high school. Here, students in Grades 7 and 8 and Grades 10 and 11 from some 20 secondary schools in this district come to learn technical/technological practice. (Students of Grades 9 and 12 are in graduating classes in junior and senior schools respectively, and for this reason they do not take LTE courses.)

Founded in 1990, Xicheng District Centre of LTE receives some 7,000 students every academic year, with each studying for an intensive consecutive four-day period each year. The centre offers traditional technology programmes as well as modern ones, such as woodwork, clothing, knitting, electrics and electronics, and design by computers. In interviews with the head of the centre and a director responsible for teaching, it was revealed that the purpose of these programmes was not to prepare students for vocational training but rather to "instill in them work/labour awareness".

Assessment of LTE in Haidian Central School District Centre of LTE was observed to take place in a range of ways. For example, formative assessment in LTE classes took the form of allowing individual students to speak out the intensions/aims of their design, self-assess their products in terms of successes and weaknesses, think of ways to improve their products, and assess how they cooperated in groups. Peer assessment was also encouraged. In the process of teaching/learning, the role that teachers played in the formative assessment was particularly noted while students were engaged in a task, although teachers were barely conscious that this was a kind of assessment (they were seldom introduced to such an idea of formative assessment, or assessment for learning, in their school-based in-built professional development sessions). For instance, they moved from group to group and made observations, identified good ideas and/ or designs, and provided encouragement, comments, and/or guidance so that students were better informed to finish their tasks. When individual problems were identified, guidance rendered by the teacher was usually given to individual students, but when common problems arose in class the teacher would stop students to provide whole-class explanations or demonstrations. In Xicheng District Centre of LTE, students were sometimes allowed to discuss among themselves the causes of identified problems and develop collective solutions.

As practical activity was prioritised, summative assessment, it was disclosed in the interviews with the director responsible for teaching of Xicheng District Centre of LTE, was conducted in terms of learning attitudes shown in class, activeness in thinking, learning and answering questions in class, and the quality of products finished in class. Students were given a grade of excellence, good, satisfaction, or failure in the summative report. In general, the final judgment of an individual student's achievement in LTE courses was made not only according to the quality of his/her finished products, but also in relation to his/her attitude and effort during class time. Encouragement given by teachers was prioritised over negative comments, and partly because most students were engaged and interested in learning in LTE lessons and partly because achievement in

LTE has nothing to do with students' admission into a high school or college/university, about half received a final grade of excellence.

Another form of summative assessment prevalent outside of Centres of LTE in Beijing (and in other cities as well) is for a selected group of students to take part in competitive contests of LTE products, administered annually either citywide or nationally, in order to motivate and encourage more students to engage in innovations in technology. The problem with this kind of competitive contests is, in my view, that they cater only to a few interested students rather than engaging all students in technology learning. In many cases they also represent a showcase of the participating school, or for the teachers who coach the few selected students taking part, as both schools and teachers/students of the winning products benefit in terms of honour.

Assessment of Labour-Technical Education: The Case of Shanghai

As mentioned earlier, educational reform in Shanghai has come to pass independently of the whole country since the mid-1980s, partly because it has a unique place, that is, a municipality directly under the Central Government enjoying a parallel status to provinces, and partly because it has a much more advanced educational research expertise of its own. Like other cities and provinces, school technology education in Shanghai from the primary to the senior high school still keeps the old title of labour-technical education, as evidenced from the recently published standards for the subject (Shanghai Municipal Education Commission [SMEC], 2004). In addition, LTE is mostly implemented at the Centres of LTE in each of they city's district rather than in each school classroom. However, the curriculum objectives, contents, and practices of teaching, learning, and assessment are fairly unique within China as a whole. These, to a certain degree, reflect the technological practice prevalent in Shanghai, a much more technological community than others in China.

According to *The standards for labour-technical curriculum for Shanghai schools* (SMEC, 2004), labour-technical education is provided to students in three consecutive stages: Grades 4–5 in the primary school stage, in which students experience 68 periods of classes of LTE, each of which lasts 35 minutes; Grades 6-9 in the middle school stage, in which students take 170 periods of LTE, each of 40 minutes; and Grades 10-12 in the senior high school stage, in which students are offered 102 periods of classes of LTE, each of 40 minutes The *Standards* in Shanghai define LTE as follows:

> In the stage of basic education, labour-technical curriculum is a basic course that aims at letting students learn technological knowledge, master technological operations, increase technological awareness, and improve technological literacy through a series of work experiences and practical inquiry-oriented technological activities while they are independently or cooperatively doing activities such as designing, applying, making and repairing (SMEC, 2004, p. 25).

In line with this curriculum definition, LTE in Shanghai emphasises three characteristics of the subject: practical, comprehensive, and innovative. The nature of the subject, together with its characteristics, makes the assessment of LTE distinct from that of other subjects in that it emphasises more of the process of learning in technology. As the content of technology learning cover three fields: technological knowledge and skills, technological process and methods, and emotions and values in technology, so does the assessment.

Compared with the assessment of LTE in Beijing, the assessment of LTE in Shanghai is much more systematic, with documents defining assessment that should be applied in all schools throughout the metropolitan city. In the *Standards for Labour-technical Curriculum for Shanghai Schools* (SMEC, 2004), for example, it specifies that:

- A well-established system of assessment for LTE should be in place, so that assessment of LTE becomes part of the whole assessment system in basic education. In terms of content, assessment of LTE should cover such areas as knowledge of labour-technology, learning and applications in technological operations; awareness, attitude, and habit formation in labour-technology; and the quality of products resulting from the learning in labour-technology.
- Developmental assessment should be the main approach to assessment in labour-technological learning. It should stick to the principle that the subjects, approaches, processes, and means should be favourable to the development of students. In other words, assessment should be attentive to the change and development in students that arise in the process of technology learning, as well as the products resulting from such technological practice. Positive comments should be made to all students who are involved in the labour-technological learning and practice and who finish the basic learning tasks set to them. To those students whose products are outstanding and excellent in design and making, or those who have made innovations in technological practice, a special system of prizes should be established in order to encourage and honour them.
- Methods of assessment should be flexible and various: apart from the end-of-term examination, other forms of assessment are to be used as well, such as display of products, writing of learning experiences, thematic activities, mutual communications among students, appraisal of products, self-appraisal, and everyday observation, etc.
- The process of assessment in labour-technological learning is also a process in which students receive education once more, and a process in which they are actively engaged in learning. The subjects of assessment should be students themselves, but also encompass teachers, parents and other people concerned as well. The assessment conducted by teachers should take into account the results of self-assessment and peer-assessment of students themselves in a timely way; meanwhile teachers also need to combine these assessments with those made by parents and others concerned. In this way, they can properly adapt their teaching methods and strategies in order to assure the realization of the teaching/learning objectives.
- The assessment/examinations in LTE can be carried out ongoing in the process of learning/teaching, at the end of a unit and/or at the end of the course. The content of the assessment/examination covers technological knowledge, technological operations/skills, and technological awareness, in combined forms of written tests and practical operation (performance). Everyday assessment can focus on certain aspects of technology learning, end-of-term or end-of-academic-year tests should reflect the achievements of an individual student's whole term or year comprehensive achievements. Note that assessment comments and grades should be recorded and filed for each student for the subject, together with ongoing everyday assessment materials and products collected.
- Based on students' habits and attitudes towards labour-technical/technological learning, with a focus on technological operation and design, a system of testing and certification is in place for LTE, combined with the career certificates promulgated by the Municipal Bureau of Labour. For primary students, their achievements in LTE through assessment and testing will be carried in the handbook of assessment and record files. For junior and senior high school students, they must sit in on examinations/testing for LTE, and the qualified, graded certificates that they have obtained will serve as one of the justifications for their graduation from high schools, for their admission to higher level of schooling or for their employment. (pp. 70–71)

CONCLUSION

It seems that implementing technology education in Mainland Chinese schools is replete with many issues and problems. One of the root causes is that technology education in the form of LTE, although with a history of at least two decades, has not been recognised by educational authorities or society at large as worthy of the same value as core school subjects such as the Chinese language and mathematics. The weakness and lack of clarity in the knowledge base of technology may contribute to the marginalisation of the subject in the school timetable, and a lack of teacher preparation for the subject in teacher education institutions of higher learning means that the subject is extremely short of qualified teachers. In addition, Chinese cultural and education traditions may represent ideological and epistemological obstacles to the formulation of technology as a new school subject, as technology is generally not viewed in China as a sanctified knowledge but rather as a mere technique associated with manual labour.

Outside of Mainland China, the past two decades have seen technology education form as a revamped school subject in many other countries. Although some Chinese comparative educators, such as Professor Mingxuan Zhang (1991), introduced Chinese policy makers and educators to developments of technology education internationally, school technology education in Mainland China has not kept abreast of the initiatives in countries such as England, the U.S.A., Canada, Australia, New Zealand, and the Netherlands.

The debate over the reform of LTE moving away from its ideological underpinnings in labour education and more towards technology education in the internationally agreed sense of the word in the 2000s has raised for Chinese educational policy makers the importance of technology education in high schools and a new project of general technology education as a high school course has been introduced on a trial basis over the past several years (Gu, 2005). However, this is not enough. The past three decades of openness and reform have resulted in rapid development of technology in all spheres of society in Mainland China, and technology innovations and applications have permeated every walk of life, including schooling. Technology education in primary and secondary schools as a foundation subject also seems to be necessary if China is to maintain and increase its economic competitiveness on the world stage. Therefore, it seems immensely important for Chinese educational policy makers to acknowledge that:

- Technology education should be another basic school subject, as important as mathematics, the Chinese language, and science;
- technology education should be taught in both primary and secondary schools and in colleges and universities;
- the technology curriculum should be revised so that the aim of the subject is to prepare citizens with technology literacy;
- school technology education should form part of general or basic schooling rather than, as before, merely part of vocational education;
- new teachers of technology education should be prepared, as teachers of other school subjects, in teacher education programmes in colleges and universities;
- in-service teachers of technology education must be provided with chances to learn more about technology and technology education;
- given that little research in technology education has ever been done in China, it is urgent that planning of research in technology education be considered so that current and prospective curriculum reforms and innovations in teaching, learning, assessment, and teacher education and teacher development in technology are research-based; and

– as assessment (both formative and summative) is a leverage over the success of reform and innovation in any school subject, high priority should be given to the research and practice in technology education, so that it would be accorded a proper status in the school curriculum.

REFERENCES

Bao, L. (1997). An inquiry into the direction of development of labour-technical subject. *Curriculum, Teaching Material and Method, 17*(5), 51–52 [in Chinese].

Chen, G. S. (1989). On the evolution of labour-technical education in primary and secondary schools in China. *Curriculum, Teaching Material and Method, 9*(10), 13–16 [in Chinese].

Cleverley, J. (1991). *The schooling of China.* Sydney, New S Wales: Allen & Unwin.

Confucius. (1994). The Confucian Analects. In *The Four Books* (J. Legge, Trans.). Changsa: Hunan Publishing House.

Deng, Z. J., & Fu, G. L. (Eds.). (1991). *A study of labour-technical education.* Tianjing: Tianjing Science and Technology Translation Publishing Company [in Chinese].

Fouts, J. T., & Chan, J. C. K. (1995). Confucius, Mao and modernization: Social studies education in the People's Republic of China. *Journal of Curriculum Studies, 27*(5), 523–543.

Fouts, J. T., & Chan, J. C. K. (1997). The development of work-study and school enterprises in China's schools. *Journal of Curriculum Studies, 29*(1), 31–46.

Gu, J. J. (2005). Curricular goals of general technology education in senior high schools. *People's Education, 19*, 33–35. [in Chinese].

Hoiman, C. (1992). Modernity and revolution in Chinese education: Towards an analytical agenda of the Great Leap Forward and the Cultural Revolution. In R. Hayhoe (Ed.), *Education and modernization: The Chinese experience* (pp. 73–102). Oxford, UK: Pergamon.

Jiang, S. Y. (1996). Technology education in primary and secondary schools. *Educational Research, 17*(4), 43–46 [in Chinese].

Jones, A. (1996). *Technology education in the New Zealand curriculum: From policy to practice.* Hamilton, NZ: University of Waikato.

Jones, A. (2003). The development of a national curriculum in technology for New Zealand. *International Journal of Technology and Design Education, 13*(1), 83–99.

Lewis, T. (1995). From manual training to technology education: The continuing struggle to establish a school subject in the USA. *Journal of Curriculum Studies, 27*(6), 621–645.

Llasera, I. (1987). Confucian education through European eyes. In R. Hayhoe & M. Bastid (Eds.), *China's education and the industrialized world* (pp. 21–31). New York: Sharpe.

Liu, M., Gu, J. J., & Yu, H. Y. (2005). A survey and analysis of the status quo of technology curriculum implementation in China's senior high schools. *Journal of Puyang Vocational and Technical College, 18*(2), 12–15 [in Chinese].

Ma, K. J. (2005). A reflection on and reconstruction of labour-technical education. *Journal of the Chinese Society of Education, 26*(9), 45–48 [in Chinese].

McCormick, R. (1992a). The coming of technology education in England and Wales. In F. Banks (Ed.), *Teaching technology* (pp. 42–55). London: Routledge.

McCormick, R. (1992b). Technology education in the United Kingdom. In R. McCormick, M. Murphy, & M. Harrison (Eds.), *Teaching and learning in technology* (pp. 15–27). England: Addison-Wesley.

Meng, F. J. (2005). The epoch, nation, and citizenship, and development: Thoughts on strengthening technology education in the field of basic education. *People's Education, 17*, 28–30. Retrieved from http://edu.people.com.cn

Ministry of Education. (2001). Framework for the curriculum reform of basic education. In Q. Q. Zhong, Y. H. Cui, & H. Zhang (Eds.), *Interpretations of the framework for the curriculum reform of basic education.* Shanghai: East China Normal University [in Chinese].

Ministry of Education. (2003). *Standards of technology education curriculum for senior high schools.* Beijing: People's Education [in Chinese].

Reid, M. S. (2000). Towards effective technology education in New Zealand. *Journal of Technology Education, 11*(2), 33–47.

Shanghai Municipal Education Commission. (2004). *The standards for labour-technical curriculum for Shanghai schools.* Shanghai: Shanghai Century Publishing Group and Shanghai Education [in Chinese].

Song, J. W. (1995). The dilemma and the way forward of school labour-technical education. *Teaching and Management, 1*, 11–13 [in Chinese].

Volk, K. (2003). Technology Education in Hong Kong. *The Technology Teacher, 63*(3), 21–22.

Xie, Y. G., & Ma, Y. P. (2008). Problems and solutions of curriculum implementation of technology education in senior high schools. *Educational Research, 29*(2), 66–74 [in Chinese].

Xu, C. F. (2002). The meaning and approach to the development of labour-technical education. *Educational Research, 23*(12), 75–79 [in Chinese].

Xu, C. F. (2004). Development of labour and technical education in China, *Educational Research, 25*(12), 11–16 [in Chinese].

Yu, H. Y. (2001). Thoughts on the deepening the curriculum reform of labour-technical education. *Educational Research, 22*(12), 47–51 [in Chinese].

Yu, K. C., & Hsu, H. G. (2003). Higher vocational education in Taiwan. *International Journal of Vocational Education and Training, 11*(2), 61–76.

Zhang, H. (2001). On the nature of integrated curriculum of practical activity (ICPA). *Global Education, 30*(8), 10–18.

Zhang, M. (1990). Technology: A new foundation subject in England's school curriculum. *Comparative Education Review, 11*(1), 40–43 [in Chinese].

Zhou, Q., Yang, H. X., & Ni, J. C. (2004). On the importance of technology education. *Curriculum, Teaching Material and Method, 24*(9), 64–68 [in Chinese].

Bangping Ding
School of Education
Capital Normal University
Beijing, China

ANDREW STEVENS

12. THE INTRODUCTION AND DEVELOPMENT OF TECHNOLOGY EDUCATION IN SOUTH AFRICA

INTRODUCTION

A superficial reading of the relatively thin literature on technology education reveals that it is a new phenomenon in educational curricula worldwide. Although there is growing consensus on the nature of the learning area, there remains room for debate about many of the specific issues and even among experts there are often sharp differences about fundamental questions. The following often-cited typology gives a sense of the range of 'flavours' in which technology education programmes are available throughout the world:

- a craft-oriented approach;
- a production-oriented approach;
- an applied science approach;
- a high-tech approach;
- a technological concepts approach;
- a design approach;
- a key competency approach;
- a science, technology, society (STS) approach. (de Vries, 1994, pp. 153–154)

The South African General Education and Training (GET) curriculum for learners from Grade 1 to 9 – in spite of drawing fairly extensively on the pioneering Design and Technology curriculum of the United Kingdom (England and Wales) and those of other Commonwealth countries such as Australia and New Zealand – would be difficult to categorise into any one of de Vries' categories. The intended curriculum may be that of the 'design approach', but the implemented one is probably more a 'craft-orientated' approach with aspects of the 'STS' approach grafted on. The reasons for this are complex and have as much to do with the history of South Africa's education policy formulation and implementation as with curriculum theory. It is perhaps worthwhile to spend some time reviewing the history of technology subjects in the SA curriculum in order to gain a deeper understanding of current policy and practices.

A BRIEF HISTORY

Prior to 1994, the year of the first democratic election in the country, education in South Africa was organised on racial lines with separate schools, universities, teacher colleges, and administration systems for each of the four main race groups as defined by the apartheid state, namely black, white, coloured, and Indian. To complicate matters further, there were four so-called 'independent homelands' within the borders of the country for four of the main black population groups, each having their own educational ministry and administration. Although the curricula in each of these systems were similar – and theoretically equal – the huge differential in state funding for the different systems made a mockery of the apartheid state's claim of 'separate

A.T. Jones and M.J. de Vries (eds.), International Handbook of Research and Development in Technology Education, 131–137.
© 2009 Sense Publishers. All rights reserved.

but equal' treatment for all races. The legacy of these discriminatory practices continues to plague South African education to the present.

Given the glaring disparities between schools for the different race groups, it was not surprising that it was an education issue (the compulsory use of the Afrikaans language in black schools) which sparked the famous 'Soweto Uprising' in June 1976 which signalled the beginning of the end of apartheid. One of the responses of the state was a comprehensive investigation into education by the Human Sciences Research Council (HSRC). Its comprehensive findings, published as 'Provision of Education in the RSA' (HSRC, 1981), attempted to shift the focus in education policy-making from formal education based on the traditional 'academic' arts and sciences curriculum towards a more 'appropriate' skills-based vocational curriculum (Kraak, 2002). This applied particularly to the education of black South Africans, which the commission found to be particularly lacking in these respects. A recurrent criticism of black education was that it was too theoretical, that it lacked a practical dimension, and that it appeared to be largely irrelevant to the lives of young South Africans.

Throughout the 1980s, a period characterised by widespread unrest in black education, the vocationally-oriented ideas of the HSRC report began to take root in much of the official discourse of the time. The Education Renewal Strategy (ERS) of 1991, a product of the pre-democratic government, recommended the introduction of a number of new subjects into the general formative curriculum (South Africa. Department of National Education, 1991). Amongst these were three innovations to the South African curriculum: Economics, Technology and Arts Education, the rationale being that these three subjects would provide education relevant to the needs of learners and society as well as contributing to the person-power requirements of the country.

Although the ERS was in favour of giving a vocational slant to the curriculum, it was also aware of research pointing towards the dangers of vocationalising education too early in a learner's school career, and it was careful not to propose too clear a differentiation between academic and vocational pathways in the compulsory phases of education (i.e. Grades 1–9). However, in the proposed post-compulsory phase (Grades 10–12) it is clear that vocational education was to assume far greater significance. Although never implemented, some of these proposals by the apartheid government were to find strong echoes in the policies and legislation of the new democratic order.

When the African National Congress (ANC) swept to power in the first democratic elections in 1994, much was expected of the new government, particularly in the long neglected area of educational transformation. At the level of policy, Kraak (2002) identified the following three pillars underpinning the new dispensation:

- An integrated education and training system – the new government committed itself to eradicating the difference in status and privilege that a differentiated 'academic' vs. 'technical/ vocational' system promoted;
- a single qualifications structure – a new statutory body, the South African Qualifications Authority (SAQA), was established in 1995 to co-ordinate and manage the new National Qualifications Framework (NQF); and
- a new curriculum framework.

The new curriculum, named Curriculum 2005 (C2005) for the year in which implementation was to be accomplished, is the first single curriculum for all South Africans. Education, for the first time in South Africa's history, was to be compulsory for all learners for nine years, the newly named General Education and Training Band (GET). Thereafter would follow three years of Further Education and Training (FET) that would provide for more differentiated general, vocational, and work-based education and training.

THE NEW CURRICULUM FOR SCHOOLS

A feature of the new C2005 was the introduction of eight new compulsory learning areas (replacing the label 'subjects' with the new term 'learning areas' was an attempt to encourage the integration of disparate 'disciplines'). These new learning areas were: Language, Literacy and Communication; Mathematics; Human and Social Sciences; Natural Sciences; Technology; Arts and Culture; Economics and Management Sciences; and Life Orientation. For the first time, Technology, in a form corresponding largely to the British Design and Technology model, was to be part of every learner's education up the end of Grade 9 (age 15).

The new curriculum represented a radical break with the past. In addition to embodying the broad political aims of access, equality, and redress, the new curriculum was designed around the 'new' philosophy of constructivism and 'outcomes-based education' (OBE). Among the many new features to the South African school curriculum education were the notions of integration of discrete 'subjects' into broader 'learning areas', the idea of learner-centredness and the teacher as facilitator (rather than sole authority), and the policy of continuous assessment rather than the reliance on final examinations to determine progression through the system (South Africa, Department of Education, 2002a).

Underlying the whole educational system are twelve 'critical outcomes' which all learning programmes are presumed to encompass. These include problem solving, working co-operatively, time management, communication in various modes, using science and technology effectively, etc. The fact that the new technology learning area, itself a product of recent educational thinking, clearly incorporated most of these critical outcomes was certainly one of the factors that prevented it from being removed from the curriculum when the curriculum was reviewed in the year 2000 (see South Africa. Department of Education, 2000).

Another important feature of the new system, with a bearing on technology education, was the introduction – after the *formative, general* education of the GET Band – of *specialised* education at the FET levels. The new FET Curriculum (South Africa. Department of Education, 2002b) introduced a number of radical changes, including:

- reducing the number of subjects from 124 to 35 (24 subjects plus 11 languages) and getting rid of differentiated curricula within these subjects;
- formulating the subject curricula in terms of outcomes;
- introducing compulsory mathematics (in the form of 'pure' Mathematics or the new subject 'Mathematical Literacy') and Life Orientation as two of part of every learners' seven-subject curriculum;
- attempting to increase mobility of learners across pathways and institutions in the FET Band; and
- attempting to decrease the generality and perceived lack of relevance of ordinary 'academic' high school, and also to widen the offerings of the traditionally narrow vocational colleges.

The new specialised offerings at the FET level include the subjects Design, Computer Applications Technology, Mechanical Technology, Electrical Technology, Civil Technology, Engineering Graphics and Design, and Consumer Studies. Noticeably absent is a general technology subject suitable for ordinary 'academic' schools. This absence was deemed sufficiently serious by the Technology Association as to provoke an unsuccessful petition to the Minister of Education (see Stevens, 2004). One consequence of the absence of such a subject is that relatively few learners in the final three years of high school have any exposure to 'design and make' or practical forms of education, since many of the specialised subjects mentioned above are only offered in the relatively few schools that are equipped to provide for them.

Although it is still too early to judge the impact of the FET innovations (the first cohort of Grade 12 learners will write their final external examinations at the end of 2008), it is fair to question the level of implementation of the new technology learning area insofar as the GET Band is concerned. This is by no means an easy question to address, since there is very little published research into this issue. One of the few studies conducted on a national level (Mouton, Tapp, Luthuli, & Rogan, 1999) made some illuminating findings which point to the potential of technology as a transformational pedagogical tool:

> Most teachers believe in the learning area as being beneficial not only to their learners, but to themselves. Teachers' enthusiasm for and dedication to Technology is one of the most consistent and impressive findings from this evaluation. The positive attitude of teachers was fed, in part, by the enthusiasm of their learners. Most teachers indicated that they would like to continue teaching Technology. More than that, many seemed pleased to be able to break out of the old modes of teaching and reconceptualise their notions of what it means to be a teacher/facilitator. (pp. 157–158)

Anecdotal evidence suggests that it is not possible to generalise about the progress of the learning area: there is some evidence that, contrary to expectation, the progress in some formerly 'advantaged' schools is modest, while some relatively poorly resourced schools have embraced the subject and made relatively good progress. This suggests that resources are not necessarily the limiting factor to progress that they are usually made out to be. Nevertheless, from the reports of teachers, it appears that a large stumbling block to progress is the shortage of resources and basic equipment for hands-on activities. There is clearly still a big disparity between schools in this regard, with certain traditional independent schools able to boast 'state of the art' facilities comparable with the best in the world, while many rural schools still lack the bare necessities such as electricity and running water. Perhaps the greatest impediment to progress, however, remains the limited number of teachers with an adequate grounding in technology education. The next section attempts to analyse this by placing technology teacher education in context.

AN OUTLINE OF TECHNOLOGY TEACHER EDUCATION

Teacher education in South Africa, like the whole of the education system, is still in transition. After 1994, the new government focused on the need to develop a post-apartheid school curriculum. Once a range of enabling legislation had been put in place (South African Qualifications Act of 1995, National Education Policy Act of 1996, South African Schools Act of 1996, amongst others), the new curriculum, C2005, was introduced at the beginning of 1998.

Once the new GET school curriculum had been launched, attention could be turned to other areas of the system such as the Further Education and Training Band (Grades 10–12) and Higher Education and Training, under which teacher education resided. In teacher education particularly, the new government inherited a complex, even chaotic situation. Under apartheid, teacher education took place largely in colleges of education, which were conceived of as institutions located somewhere between secondary and tertiary institutions and were administered by a range of national and provincial departments, depending on the race of the group for which they catered. Although there was tight ideological control over these institutions, there was little or no emphasis on quality assurance or even on the numbers of teachers trained. A consequence was that many of these institutions, particularly in rural areas, acted as a type of 'finishing school' where school leavers went for some form of government-sponsored further education that did not necessarily have as its main purpose the professional development of teachers.

A national audit conducted in 1995 (Hofmeyr & Hall, 1995) revealed that there were some 150 publicly-funded institutions providing teacher education to over 200,000 students, including 70,000 initial teacher trainees at the 93 'contact' colleges alone. Considering that the South African system employs around 355,000 teachers, it is clear that too many teachers were being trained during the eighties and early nineties. A major restructuring of the sector reduced the number of contact colleges to 25 and the number of distance colleges to two. All these institutions have now been incorporated into universities and are seen as an integral part of the higher education sector. This process, only recently completed, has coincided with recent legislation requiring all new teachers to be in possession of a degree before being employable in the state system. Prior to this, a three-year college diploma had been sufficient to qualify as a professionally trained teacher and this remains the most common qualification of South African teachers.

When technology was introduced into schools as part of C2005 in 1998, there were very few formally trained technology teachers in the system. Those who had received training had received it either at the hands of a nationally appointed task team (T2005) or at the hands of the few non-governmental organisations (NGOs) that had been involved in pioneering the introduction of technology into SA before it became part of the new curriculum. Although a few of these NGOs have managed to survive in the difficult funding environment of the new democratic South Africa, most of the formal technology teacher education now takes place under the auspices of the higher education institutions. Here, the major emphases since 1994 have been on upgrading the qualifications of teachers who had been disadvantaged under apartheid education and providing in-service programmes in learning areas for which there are shortages, notably technology, mathematics, and science. Such teachers may enrol for the two-year, part-time Advanced Certificate in Education (Technology Education) or, more recently, a three-year part-time Bachelor of Education degree course. Most higher education institutions also offer the BEd degree as a four-year, full-time initial teacher qualification. Within these programmes students have the option to study technology as one of their teaching method subjects. Another option for students intending to become secondary or further education teachers is to enrol in a three-year degree programme (B.A., B.Sc., B.Com., etc.) and follow this with a 'capping' post-graduate certificate in education (PGCE). During this single-year, full-time course, technology may be offered as a teaching method but since very few South African universities offer suitable undergraduate courses in 'technology' or some suitable cognate subject, this option has not produced many technology teachers.

Although technology is now firmly established in the compulsory GET phases, the recent restructuring of the FET (Grades 10–12) sector has resulted in the abandonment of the generalist and integrative approach of the GET Band and given the last three years of schooling a markedly specialised and vocational slant. From 2006, learners interested in technology-orientated fields will, at the relatively early age of 15–16 years, be required to choose to pursue studies in one or more of the following five subjects: Electrical Technology, Mechanical Technology, Civil Technology, Engineering Graphics and Design, and Computer Applications Technology. Those who prefer a 'softer' approach to technology may choose the new subject 'Design', which has emerged from the Arts and Culture traditions and most closely resembles the technology learning area of the GET Band. The five specialised technology subjects appear to be reformulations of the technical subjects which were offered in the vocationally-orientated technical schools and colleges of the past; it is doubtful whether many ordinary secondary schools, which form the vast majority of schools in the country, will have the capacity, either in terms of physical or human resources, to be able to offer these subjects as part of their programmes. The teacher education challenges posed by these changes have yet to be faced since many of the colleges which provided this specialised training no longer exist or have been incorporated into higher education institutions which do not have a tradition of skills-based education and training. Again, there is a shortage of information on how this aspect of the FET will be implemented, but it is difficult to imagine an ordinary high school having the capacity to offer any of these rather specialised technology subjects.

Because technology was introduced in the late nineties at the same time as the traditional teacher colleges were being closed, incorporated, or merged with other institutions, the normal supply of teachers was severely disrupted, while the supply of newly qualified technology teachers was choked off. Although many of the teacher education institutions (universities and universities of technology) now have some capacity in technology education, it is fair to say that the area is severely understaffed in relation to the needs of the system. Indeed, the whole structure of teacher education was the subject of a two-year study by a specially appointed Ministerial Committee on Teacher Education whose findings were published in 2005 but have not yet been implemented. It is safe to say that the demand for continuing professional development of teachers and initial teacher education in technology far outstrips the supply capacity of the current providers.

TECHNOLOGY EDUCATION RESEARCH AND DEVELOPMENT

As has been suggested above, the major focus of providers of technology education (both in the government and NGO sectors) has been on ensuring sufficient initial and continuing professional education for teachers. There has been little time in the decade since its introduction for sustained research on the scale undertaken, for example, by the Assessment of Performance Unit in the UK (Kimbell, Stables, Wheeler, Wozniak, & Kelly, 1991). The only research-based attempt to gain a bird's eye view of the status of the emerging learning area was the evaluation study of the Technology 2005 project (Mouton et al., 1999). The major positive finding of this study was that, in spite of the host of problems confronting its introduction into schools, technology enjoyed high levels of support from principals, teachers, and learners alike. It was this positive impact on schools that perhaps more than other factors saved technology from being excised from the curriculum as was recommended by the committee set up to review the curriculum in the year 2000 (South Africa. Department of Education, 2000). However, this committee did make a range of proposals that were adopted and that led to a streamlining and simplification of the curriculum structure, including that of the technology curriculum. The number of learning outcomes was reduced to only three and the assessment standards were also significantly reduced. In addition, the overall tone and language of the curriculum was simplified by removing some of the OBE jargon. However, it is interesting to note that although this work was undertaken by an expert team, this team was not able to draw on much empirical research to support its work. This remains a shortcoming of much curriculum development in South African education generally, and particularly in technology education.

In comparison to established learning areas such as mathematics and science education, technology education has as yet attracted very few postgraduate students, although some higher education institutions are developing their capacity in this respect. A gauge of this is the volume of papers presented at the annual conference of the Southern African Association for Research in Mathematics, Science and Technology Education (SAARMSTE), the only regional organisation devoted exclusively to research in the three fields. In the 15 years of its existence, the number of contributions from mathematics or science typically outnumbers those of technology by a factor of twenty or more.

Also, in terms of journal publications, the research output in technology remains low. An internet search of the contents of six local journals devoted to educational research going back to the year 2000 revealed only eight published articles with a bearing on technology education. A comparable scan of three international journals of technology education reveals a similar picture, with under a dozen articles published over the past ten years by South African authors. Although there is one local journal on science, mathematics and technology education research, there is no journal exclusively focused on technology education. This might be a necessary developmental step in encouraging and creating a research and development culture in the field.

Technology education researchers in South Africa are presently focused on issues of curriculum development and implementation. This is understandable given the newness of the learning area and its somewhat precarious status. A number of new part-time researchers (who are mainly teachers and teacher educators) are engaged in studies which attempt to flesh out areas of the curriculum that differ from international curricula. An example is the area of indigenous knowledge systems and technologies, which has attracted considerable governmental funding support. Although this area is an explicit part of the curriculum, the available learning support material is thin and merits attention from writers and researchers. In many respects, South Africa's curriculum reflects Western, even 'globalised' worldviews, and this curriculum innovation is an attempt to introduce an alternative and specifically African counterpoint to this dominance. However, this is only one of several challenges for the small but growing technology education community. There remains a great need for both large-scale systemic research and for smaller, fine-grained studies to illuminate our way forward.

REFERENCES

de Vries, M. J. (1994). Teacher education for technology education. In M. Galton & B. Moon (Eds.), *Handbook of teacher training in Europe* (pp. 153–164). London: David Fulton.

Hofmeyr, J., & Hall, G. (1995). *The national teacher education audit. Synthesis report*. Johannesburg: CEPD.

Human Sciences Research Council. (1981). *Provision of education in the RSA. Report of the main committee of the investigation into education*. Pretoria: Author.

Kimbell, R., Stables, K., Wheeler, T., Wozniak, A., & Kelly, V. (1991). *The assessment of performance in design and technology – The final report of the APU design and technology project*. London: SEAC.

Kraak, A. (2002). Discursive shifts and structural continuities in South African vocational education and training: 1981–1999. In P. Kallaway (Ed.), *The history of education under apartheid, 1948–1994* (pp. 74–93). New York: Peter Lang.

Mouton, J., Tapp, J., Luthuli, D., & Rogan, J. (1999). *Technology 2005: A national implementation evaluation study*. Stellenbosch: CENIS.

South Africa. Department of Education and Culture (DEC). (1990). *The evaluation and promotion of career education in South Africa. Main report of the committee chaired by Dr S.W. Walters*. Pretoria: Government Printer.

South Africa. Department of Education. (2000). *A South African curriculum for the twenty-first century: Report of the review committee on Curriculum 2005*. Pretoria: DoE.

South Africa. Department of Education. (2002a). *Revised national curriculum statement grades R-9 (schools)*. Pretoria: DoE.

South Africa. Department of Education. (2002b). *National curriculum statement grades 10-12 (schools). Overview (Draft)*. Pretoria: DoE.

Stevens, A. (2004). Getting technology into the FET. In A. Buffler & R. Laugksch (Eds.), *Proceedings of the 12th annual meeting of the Southern African Association for Research in Mathematics, Science and Technology Education*. Cape Town: University of Cape Town. [CD-Rom]

Andrew Stevens
Department of Education
Rhodes University, Grahamstown
South Africa

SECTION 2

THE NATURE OF TECHNOLOGY

MARC J. DE VRIES AND JOHN R. DAKERS

13. PHILOSOPHY OF TECHNOLOGY FOR RESEARCH IN TECHNOLOGY EDUCATION

INTRODUCTION: WHY A SECTION OF PHILOSOPHY OF TECHNOLOGY IN THE HANDBOOK?

Two sections in this *Handbook* deal with perceptions of technology. In terms of academic disciplines, the first section (introduced here) is based primarily on insights from the philosophy of technology; the second section (introduced in chapter 18) primarily draws from psychology. The first section focuses on how technology is perceived and conceptualised in the philosophy of technology. One may be tempted to call this the 'correct' concept of technology. This, however, would be a simplification. There are substantial differences between the images that different philosophical schools use to present their view on technology. What they all have in common, however, is that they represent the views of scholars on technology. This differs from the second section on philosophy in the *Handbook*, where the focus is on the concepts of technology held by pupils and teachers in school-based education. Knowledge of these concepts is relevant as we may wish not only to adapt our teaching according to the preliminary views that pupils hold, but also to assess whether any changes in their concepts have occurred during the time spent in technology education classes.

There is a clear relationship between the two sections. The views that philosophers present can serve as references for identifying what kind of views are held by teachers and pupils. They represent the spectrum of different perspectives on technology that are possible. By studying the way philosophers write about technology we can identify the sorts of issues that determine how people think about technology. These issues also feature in the ways that teachers and pupils think about technology.

Having a section on the philosophy of technology in this *Handbook* also serves another purpose, apart from its use in investigating teachers' and pupils' perceptions of technology. It also provides information about the way curricula for technology education can support and stimulate certain perspectives on technology. The philosophy of technology can therefore be a useful resource for research into technology education curriculum development. Systematic reflection on the structure and content of technology education curricula and comparison with what philosophers write about technology can provide indications for the sorts of perceptions technology pupils will (hopefully) acquire as a result of working through a particular curriculum. In a similar way, the philosophy of technology can serve as a reference for developing teacher education programmes. When the ultimate aim of educating teachers is the acquisition of a valid image of technology, philosophy of technology can help us recognise what is essential and what is not. Finally, insights from the philosophy of technology can provide clues for what is relevant to include in the assessment of curricula, teachers, and pupils.

In the early days of technology education, there was little interest in philosophical issues. This was because of the practical nature of the subject. Later, technology educators began to realise that

A.T. Jones and M.J. de Vries (eds.), International Handbook of Research and Development in Technology Education, 141–149.
© 2009 Sense Publishers. All rights reserved.

perceptions of technology do not emerge 'automatically' as a result of the practical activities engaged in by pupils. Nowadays we also value theoretical and cognitive elements in technology education, and this is a further stimulus to take into account the views of technology philosophers. This is why we find frequent references to philosophical publications in current articles in scholarly journals devoted to technology education. It will thus not come as a surprise to regular readers of such journals to find that there is a section dedicated to the philosophy of technology in this *Handbook*.

MAIN STREAMS IN CONTEMPORARY PHILOSOPHY OF TECHNOLOGY

The philosophy of technology is a relative newcomer to philosophy. The philosophy of science, for instance, is about half a century older than the philosophy of technology. This is perhaps surprising because if philosophy is a systematic reflection on the reality in which we live, it would seem more likely that technology should feature in that reflection than science. Probably the most important reason for philosophers' original lack of interest in technology is that science is related to human thinking while technology is related to doing and since the ancient Greeks, the first has had a higher social status than the second. In addition, the idea that technology is merely the application of the natural sciences developed and, if so, what new ideas could technology as a human effort provide that would make its philosophical study worthwhile?

Sadly enough, an interest in technology began to develop among philosophers only when the role of technology in culture and society became problematic. The way in which WWII had been ended had removed all doubts about the enormous potential impact of technological artefacts on the course of human history. Although that role had always existed, our history books had systematically ignored it. History was traditionally described in terms of the actions of kings, governments, merchants, and the like. The only exception is the Industrial Revolution, in which engineers had evidently played a major role. The atomic bomb, however, had a much more immediately dramatic impact and made historians and philosophers acutely aware of the ways in which technology can have a decisive effect on the course of human history. As a result, the philosophy of technology as a distinct domain of systematic reflection emerged in the 1950s. Since then it has developed rapidly and is now a fairly mature scholarly domain, although many fundamental discussions remain.

The relative maturity of the philosophy of technology can be demonstrated by the fact that (1) all major philosophical streams now have a counterpart in the philosophy of technology, and (2) all philosophical sub-disciplines have a counterpart in the philosophy of technology. Let us first focus on the philosophical streams (see also de Vries, 2005).

Phenomenology

One of the first philosophical streams to begin constructing a view on technology is phenomenology, and probably the best known example of an early philosophy of technology is Heidegger's (1977a) view on technology and the way it influences our perception of reality. For Heidegger, technology played a negative role in that perception. The overall presence of technology in our society made us believe that all we see around us is merely a resource that we need to continue to work on and improve. Rather than appreciating reality as it is, we look for opportunities to change it for what we think is the better. We do not admire the tree we see because of its intrinsic beauty; instead we start counting how many planks we can saw out of it. This way of looking at reality has become so strong that we can no longer hope to be rid of it. Heidegger (1977b) stated that 'only a god can save us' from this view, and as he did not believe in God or a god, it is clear that he meant to say that we will never be able to free ourselves from this distorted view of reality.

A second observation that Heidegger made concerning our perception of reality is that technical artefacts tend to disappear from our conscious perception when we become experienced users. His famous example is that of the hammer. A carpenter sees the hammer on the table when he looks for his tool to insert a nail into a piece of wood, but as soon as he has picked it up and starts hammering his focus is entirely on the nail and the wood. The hammer disappears from consciousness. It becomes, as it were, part of his hand. One could call this embodiment. Only when the head of the hammer flies off, for example, does the hammer return to his conscious awareness and only then will he realise again the distinctness of both the hammer and his hand.

This observation was later picked up by Don Ihde, a more contemporary philosopher of technology, who positions himself in phenomenology and draws extensively from Heidegger (see chapter 14). He extended the idea of embodiment into four different ways in which technological artefacts can influence our perception of our lifeworld. When we wear a pair of glasses, we do not perceive them as separate entities; they have become like parts of our body. Only when a drop of water hits one of the lenses do we become aware of their presence and separateness from our body. This 'embodiment relation' is the first of four ways in which technological artefacts can serve as intermediaries between us and our perceived world.

The second relation Ihde identifies is the 'hermeneutic relation'. An example of this is a control panel in a power station. The operator observes the panel and, as it were, sees through it and imagines all that is behind the panel. In this case the technological artefact has not become one with the body, but with the lifeworld that is observed. The artefact, though, makes a certain 'translation' of the lifeworld to focus our attention on certain issues. We have to be aware of that in order not to be misled by this technology-mediated representation of the lifeworld. Ihde uses the example of medical imaging to illustrate this. An MRI scan of a human body is not a 'normal' picture of the body, but it presents computerised data about the body in such a way that we recognise it as a body. There are, however, quite different ways of selecting and presenting data, depending on the medical specialist for whom the image is produced. Each of these specialists knows that (s)he only gets to see a selection of what is in the body, presented in a certain way. A third example is a picture of stars: if a layman sees the picture, (s)he may think that there are red, green, and blue stars, whereas the colours actually represent temperatures. As long as we understand that we have a hermeneutic relation, the picture can enhance our perception of reality because we can see things that our own eyes do not perceive. But if we do not realise this, we get a wrong idea of our lifeworld. Here we can see that Ihde is more nuanced in his appreciation of technology than Heidegger was. For Ihde, the intermediary role of technology can enhance our understanding of reality as long we understand the relation between us, the technological artefact, and the lifeworld.

The third relation Ihde identifies is the 'alterity relation', in which the technology makes changes in the lifeworld, so that we get to see a lifeworld that does not actually exist. Examples are a science fiction movie or a computer game. It needs no explanation that here, too, it is important that we know that this relation exists.

Finally, there is the 'background relation', in which the technological artefact produces observable phenomena of which we are unaware, but which nevertheless influence our perception of reality. An example is the optical pollution which hampers our view of the stars in the sky. Even the darkest evening is spoiled by the dispersed light from street lights and other light sources, but being aware of the relation that occurs helps us gain a better image of our lifeworld.

Another contemporary pupil of Heidegger who worked out a phenomenological philosophy of technology is Albert Borgmann (1984). He takes up the rather dark image of technology that Heidegger himself had sketched. He focuses on the way technological devices are used to provide us with certain commodities, thereby reducing our engagement with reality. Instead of having a rich experience with our lifeworld by going into the forest, chopping the wood for our stove, and lugging it to our home, we now use only two fingers to slightly adjust the knob on our thermostat and – without knowing how it works – the temperature is adjusted to our desires. The only way to escape from this 'device paradigm' is to regain engagement by taking part in 'focal activities'. Rather than buying a meal that we only need to warm up in the microwave, we have to buy basic ingredients and cook our own meal. This enriches our experience of reality because it gives us back direct engagement with reality and also undoes the standardisation that resulted from the device paradigm (all microwave meals taste the same, whereas in cooking our own, we can generate all sorts of variations). Borgmann realises that our current economy does not allow for an approach that is fully based on focal activities, but he makes a plea for at least reserving niches in which focal activities have a place so that we remain aware of the disengagement that takes place the rest of the time.

Another philosopher of technology that picked up this theme is Hubert Dreyfus (2001). In his book, *On the internet*, he emphasises that having only a virtual experience with reality does not do justice to the fact that humans are both body and psyche. In education, for instance, the bodily presence of teacher and pupil in one room can never be fully replaced by virtual learning environments only. Here, too, the intermediary role of technology results in a reduction of our lifeworld experience. Both Borgmann and Dreyfus see this as a negative impact of technology on human life.

Marxism and Critical Theory

Contrary to most other philosophical streams in the 1950s (including phenomenology), Marxism was very optimistic about the role of technology in society. According to Marx, technology was the ultimate means by which the desired (and according to him, unavoidable) destruction of capitalism would take place. As time passed, however, it became evident that his Hegelian view on the necessity of this transformation was not materialising. The neomarxists (also referred to as adherents of 'Critical Theory') therefore continued his view on the desirability of this transformation, but noted that it would not happen automatically. Rather, it would require a certain consciously-orchestrated use of technology in society. Technology could be used both to consolidate and enforce the primacy of capitalism, and to bring about its fall.

Langdon Winner (1986) has provided a number of case studies in which he shows that technology can indeed enhance capitalism and other forms of social discrimination. The viaducts on Long Island, New York, were designed to be low enough to prevent the buses, used by the black people, from using the roads to the beaches. As a result the beaches were kept 'free' from black people. Although there are doubts about this example, it illustrates well the point Winner wishes to make.

Another contemporary philosopher of technology who was inspired by Critical Theory is Andrew Feenberg (see chapter 15). He has elaborated a two-step process through which technology can be developed in such a way that it serves as a catalyst for desirable social changes. In the phase of 'primary instrumentalisation', a social need is translated into a technical problem and solved by engineers. In the phase of 'secondary instrumentalisation', this solution is embedded in society, whereby important changes can still be made by the users. Feenberg illustrates the feasibility of this latter phase by the way the French Minitel system was transformed by its users from a one-way information channel operated by the French government into a two-way communication tool for citizens. Thus he demonstrates that technology can play the role ascribed to it by Marx, but only when the phase of second instrumentalisation becomes effective.

Pragmatism and Postmodernity

The philosophical stream of pragmatism developed particularly in the USA, with John Dewey as one of its better known exponents. Philosopher of technology Larry Hickman (2001) followed in Dewey's footsteps, elaborating a pragmatist view on technology. In this view, technology is the activity 'par excellence' that shows the value of experimenting with reality and thus letting truth and usefulness emerge 'naturally', rather than starting from certain predefined truths and values. In this respect, he sees technology as a model for social action in general. As such, he holds that governments were to determine their policies based not on given ideologies and values but on collaborative experimentation with engineers to establish what works and what does not, a lot of problems would be prevented.

Another philosophical stream that refuses to work with given truths and values is postmodernism. Jean-Francois Lyotard applied postmodernism to epistemology to indicate that we all construct our own truths and that we may even hold conflicting truths as this fits our needs in our compartmentalised lives. In addition, Jean Baudrillard has shown how technology in particular plays a role in developing symbolic, virtual images of reality that we now tend to accept as our truths. Journalists can produce any reality they want, and there is no way we can find out which is the 'true' one. The Internet, moreover, is a sophisticated tool for developing such symbolic versions of reality, whereby it is impossible for us to retrace 'the one and only truth' among all these different virtual worlds. This of course makes the whole notion of truth questionable.

The Analytical Approach

All the previously described streams belong to what is called 'Continental' philosophy. However, this term has become very problematic as many of the contemporary philosophers in this realm do not live or work in the European Continent (as was the case in the past with such names as Heidegger, Sartre, Marx, etc.). In general, the focus of Continental philosophy has been on what one can call the 'big questions' of life. In the Continentally-oriented philosophy of technology the focus is on the role of technology in society and culture.

Even more confusing is the fact that the alternative approach is called the 'analytical' approach. No geographical location is mentioned here. 'Analytical' instead indicates the function of philosophy in this approach: to analyse, define, argue, and conceptualise. The aim of analytical philosophy is to create a consistent conceptual framework for describing reality. Such a framework can be useful for discussion about the role of technology in culture and society.

The debates about technology as developed in Continentally-oriented philosophy of technology often suffered from conceptual inconsistencies. Often, for instance, technology was seen as one homogeneous whole. Empirical studies have shown, however, that developments in technology can differ substantially between different types of technology (for instance, architectural engineering works quite differently from mechanical engineering, both in terms of processes and knowledge content). Continental philosophers usually had no engineering background whatsoever. In analytical philosophy of technology, however, we find philosophers who also have an engineering background. This enables them to understand and make effective use of empirical case studies on specific technological developments. In order to describe their contribution to the philosophy of technology, it makes sense to first outline the main sub-domains in the philosophy of technology.

Sub-domains in the Philosophy of Technology

In his classic book, *Thinking through technology*, Carl Mitcham (1994) presented a history of the philosophy of technology, highlighting what were identified as the four main areas in the philosophy of technology as they emerged: artefacts, knowledge, activities, and technology as part of our being humans ('volition'). Although he does not mention this explicitly, these four areas at first sight seem connected (loosely) with four main domains in philosophy. For example, much of what has been written about technical artefacts deals with the ontology of artefacts; what has been written about technological knowledge naturally connects with epistemology; writings about activities refer to methodology as a sub-domain in philosophy; and the 'volition' area draws mostly from metaphysics, ethics, and aesthetics. The fact that all the main sub-domains of philosophy are thus represented in this survey of the philosophy of technology indicates the relative maturity of the philosophy of technology, as mentioned previously.

The Continentally-oriented philosophy of technology has dealt mainly with the 'volition' area. To appreciate the contribution of the analytical approach, it is necessary to shift our attention to the areas of (the nature of) technical artefacts, (the nature of) technological knowledge, and (the nature of) processes.

When it comes to the nature of technical artefacts, one of the major research programmes in the philosophy of technology in the Netherlands has resulted in conceptualising artefacts in terms of a 'dual nature' (Kroes & Meijers, 2006). On the one hand one can give a description of an artefact in terms of its physical nature: its structure, materials, size, shape, weight, colour, etc. This, however, cannot be a complete description of the artefact as it leaves open various possibilities for its function. On the other hand, a description of the same artefact in functional terms only is equally incomplete as it leaves open various physical realisations. Only a description of the artefact that contains both its functional and its physical nature can be complete. Related to this 'dual nature' concept, several reflections have focused on the concept of functions. Here, a difference between proper and accidental functions emerges that seems to be absent in a biological conceptualisation of functions (the function of the heart, for instance, whereby it is hard to see how one can consider the organism to be the 'user' of its own heart).

In reflections on technological knowledge, the notion of 'normativity' seems to make an original contribution to the philosophy of technology (with perhaps even consequences for more general debates about the nature of knowledge in epistemology). Normativity is essential to certain types of technological knowledge. Although this is evident for knowledge of technical standards and requirements in design processes, normativity also features in knowledge of functions – which does not refer to what the artefact actually does, but to what it *ought* to do. For example, even though the car is broken, we know that it is a car because of what it ought to do. Similarly, an engineer can claim to 'know' that a particular drilling machine is good, whereas no physicist can claim to 'know' a good electron. Knowledge about electrons refers only to what electrons actually do, not to what they are supposed to do (if they do not behave in that manner, we do not believe them to be electrons, not 'bad' electrons). The normativity in the knowledge that engineers have makes engineering science a peculiar type of science that often falls in between traditional categories of sciences. For instance, it does seek only for general laws but also describes the specific (it is neither strictly nomothetic nor ideographic). There remains a lot of work to be done in the philosophy of technology to achieve a more complete picture of the peculiarity of engineering sciences.

Reflections on technological activities have mainly focused on the design process, and here we can see connections with the academic discipline of design methodology as it emerged from practitioners reflecting on their own design practice. We have learnt that design processes are context-dependent and much more complex than the design schemes that the early days of design methodology suggested. Design is an activity that deals with a variety of conditions – scientific, technological, psychological, social, and more – and for each specific context a design solution needs to be found that matches with the whole set of conditions. The role of scientific knowledge in the design process varies significantly between different types of products and technologies, and various taxonomies for technologies have been developed on the basis of these differences. This means that although design processes have commonalities, they also have a context-bound character. The same holds for methods that are used in certain phases of the design process. All of these methods are based on assumptions about what we know or can do. If we try to apply a method in a situation in which these assumptions are not fulfilled, the method does not work properly and needs to be adapted.

Reformational Philosophy of Technology

In addition to the Continentally-oriented and the analytical approach in the philosophy of technology, there is a philosophical 'school' that integrates both approaches. Developing from a protestant base, reformational philosophy has developed a view of reality that brings together many of the insights that have been developed in the approaches described earlier. In particular, the combination of the complexity of reality and the regularities in reality contribute significantly to our understanding of technology. As such, an artefact is not only a physical thing, but also functions in many other aspects of reality (such as the spatial aspect, the linguistic aspect, the economic aspect, the social aspect, the juridical aspect, etc.) and in each of these aspects there are certain 'laws' that govern the functioning of the artefact. This is what makes technological design so complex.

At the same time, the variety of aspects and their regularities or laws create a wealth of possibilities to play with in order to develop a design solution that fits with all, or at least sufficient, conditions. Part and parcel of reformational philosophy is also its interest in normative issues. Technological developments are always driven by motives, and we need to question these motives and critically assess whether they do justice to reality and certain values we want to hold. This normativity is in the laws that we find associated with the various aspects of reality. Many of these laws are not based purely on cause-effect relations but are related to human intentions. This, too, is typical of technology: not only are we dealing with laws that are cause-effect based (such as the laws of physics), but also with 'laws' that are based on human intentionality (such as psychological or social theories).

The Philosophy of Mind

One philosophical domain that is particularly related to technology is the philosophy of mind. This domain deals with questions related to what the mind is and what thinking is. The emergence of Artificial Intelligence was a particular stimulus for this domain as it raised lots of new questions. Can computers think? Are humans fundamentally different from computers or are they two species of the same kind? How is the interaction between humans and computers?

The last question is also related to the phenomenon of the cyborg, an entity that is human but with lots of technology built in. Is it still a human? How many artificial parts are needed to turn it into a machine? Is there a boundary, anyway? One cyborg-related publication that became famous is Donna Haraway's *Cyborg manifesto*. In this manifesto she applauds the idea that it is possible to create a being that is really half-human and half-technology. What she appreciates particularly is the absence of the concept of human beings using technology to dominate nature. Domination, she argues, is what made men rule over women, and as a feminist writer she sees the cyborg as a sign of hope that the domination ideal can be abandoned. This is one of the few occasions in the philosophy of technology where we the issue of gender is made explicit. Of course, questions can be raised as to whether or not a cyborg really is a sort of 'in between' being in which there is no domination by humans, but in general Haraway's attempt to raise the gender issue in the philosophy of technology has been appreciated by many.

THE CHAPTERS IN THIS SECTION

The chapters in this section span the variety of approaches sketched above. The phenomenological approach is represented in Chapter 14 by Don Ihde, one of the most prominent contemporary philosophers of technology in this philosophical stream. The text has been published before in *Techné, the Society for Philosophy of Technology* (SPT) scholarly journal (Ihde, 1997). This journal is a very useful resource for *Handbook* readers seeking additional information on developments in the philosophy of technology. Chapter 15 by Andrew Feenberg represents the stream related to the Critical Theory philosophers. Feenberg, too, is one of the leading philosophers of technology in our time. His text was published previously in Dakers (2006). Dennis Cheek's offers a survey of the sort of studies analytical philosophers of technology use for conceptualising technology and related issues in Chapter 15. He describes how academic disciplines such as history and sociology have produced empirical studies that provide insights into how technology developed in practice. Such studies serve as resources for analytical philosophers of technology to ensure that the concepts they develop are not only internally consistent, but also empirically valid. Finally, Chapter 17 by Pierre Verillon focuses on one of the main fields of interest developed in the analytical philosophy of technology, namely the nature of technological knowledge. Verillon shows how both historical and psychological considerations can help philosophers to shape their ideas about this nature. Readers wanting to learn more about the reformational approach are referred to de Vries (2005), and Strijbos and Basden (2006).

REFERENCES

Borgmann, A. (1984). *Technology and the character of contemporary life: A philosophical inquiry*. Chicago: University of Chicago Press.

Dakers, J. (Ed.). (2006). *Defining technological literacy. Towards an epistemological framework*. New York: Palgrave MacMillan.

de Vries, M. J. (2005). *Teaching about technology. An introduction to the philosophy of technology for non-philosophers.* Dordrecht, The Netherlands: Springer.

Dreyfus, H. (2001). *On the internet*. New York: Routledge.

Feenberg, A. (2006). What is philosophy of technology? In J. R. Dakers (Ed.), *Defining technological literacy: Towards an epistemological framework* (pp. 5–16). New York: Palgrave Macmillan.

Heidegger, M. (1977a). *The question concerning technology* (W. Lovitt, Trans.). New York: Harper and Row.

Heidegger, M. (1977b). Only a god can save us now (D. Schendler, Trans.). *Graduate Faculty Philosophy Journal, 6*(1), 5–27.

Hickman, L. (2001). *Philosophical tools for technological culture*. Bloomington, IA: Indiana University Press.

Ihde, D. (1997). Whole earth measurements. *Techné: Research in Philosophy and Technology, 2*(2), 61–74.

Kroes, P. A., & Meijers, A. W. M. (2006). Introduction. The dual nature of technical artefacts. *Studies in the History and Philosophy of Science, 37*, 1–4.

Mitcham, C. (1994). *Thinking through technology. The path between engineering and philosophy.* Chicago: University of Chicago Press.

Strijbos, S., & Basden, A. (Eds.). (2006). *In search for an integrative vision for technology*. Dordrecht, The Netherlands: Springer.

Winner, L. (1986). *The whale and the reactor: A search for limits in an age of high technology*. London: University of Chicago Press.

Marc de Vries
Faculty of Applied Sciences
Delft University of Technology
Delft, Netherlands

John Dakers
Department of Educational Studies
University of Glasgow
Scotland

DON IHDE

14. WHOLE EARTH MEASUREMENTS[1]

HOW MANY PHENOMENOLOGISTS DOES IT TAKE TO DETECT A "GREENHOUSE EFFECT"?

Let us take a very commonplace, often discussed and critical topic within our conversations regarding critical environmental issues: Are we detecting a Greenhouse Effect, and related to this, is it exacerbated by 'homogenic factors', i.e., human actions? At this occasion I suspect that most would be inclined to give a positive answer to both of these questions. But, if pushed philosophically, what would be the evidence, and how well grounded would it be for such affirmations?

Within scientific communities and associated scientifically informed circles, the answers have to be somewhat more ambiguous, particularly when rigorous questions concerning evidence are raised. Were scientific truth to be a matter of consensus and some contemporary philosophers of science argue that scientific truth often turns out to be just that then it is clear that there is beginning to be a kind of majoritarian consensus among many earth science practitioners that the temperature of the Earth, particularly of the oceans, is indeed rising and that this is a crucial indicator for a possible Greenhouse Effect.

Most of these scientists admit that the mean oceanic temperature has risen globally in the last several decades. But this generalisation depends upon (a) how accurate measurements may be, not just for samples, but for the whole Earth. Hot spots, for example the now four year old hot spot near New Guinea which is part of the El Nino cycle, does not count by itself because it might be balanced by cold spots elsewhere. And the fact of the matter is that 'whole earth measurements' are still rare and primitive in the simple sense that we simply do not have enough thermometers out. (b) Secondly, even if we have enough thermometers, a simply synchronic whole earth measurement over three decades is but a blip in the diachronic history of Ice Age cycles over the last tens of thousands of years. (c) Thirdly, even if we know that the Earth is now heating up, has an ever increasing Ozone Hole, and from this computer modelled strange weather effects can be predicted, how much of this is due to homogenic factors, such as CFCs, CO_2 increases, hydrocarbon burning, and the like? Is it the case, as *Science* magazine claimed in 1990, that "24% of greenhouse encouraging gases are of homogenic origin"?

As I have described the current debate it does not sound on the surface to be very philosophical; instead it seems empirical and thus within the domain of scientific discourse. However, in the way in which the story was cast, there lurk some deep epistemological issues which relate to my ironic first subtitle, "How many phenomenologists does it take to detect a greenhouse effect?"

What I wish to do in this paper is to look both at 'classical' phenomenology (Husserl in particular) and at Heideggerian hermeneutics regarding the theme, 'environmental phenomenology', and show that both approaches are to be found wanting with respect to the Greenhouse Effect phenomenon. Then I wish to show how a rather radically modified hermeneutic and phenomenological epistemology can address this problem. This modification entails two positive concepts, both of which, I will argue, are necessary for a phenomenological approach to environmental issues: namely, Technoscience as a thoroughly technologically embodied science as a necessary concept

A.T. Jones and M.J. de Vries (eds.), International Handbook of Research and Development in Technology Education, 151–158.
© 2009 Sense Publishers. All rights reserved.

to get at presumed sub-perceptual entities, and Earth-as-Planet as a necessary presupposition for dealing with whole earth measurements.

Classical phenomenology, I argue, lacks the first of these concepts, and Heideggerian romanticism rejects the latter. Let us begin with a very simple phenomenological question: from what standpoint or perspective can the issue of whole earth measurements be made? If, at base, our very knowledge is constituted by way of our bodies and through perception as both Husserl and Merleau-Ponty seem to contend, then what is claimed about whole earth measurements becomes problematical in two senses.

First, how do we get a sufficiently encompassing perspective to talk about a whole earth? And, second, given that the greenhouse producing gases are, in fact, not perceived at least in Husserl's primordial dator sense or at the level of primary lifeworld perception, how do we constitute these?

One approach might be to recognise that the scientific discourse concerning greenhouse gases seems itself to be constructed in terms consistent with "modern" as opposed to phenomenological epistemology. Indeed, all the entities to be measured, ozone, CFCs, CO_2, are sub-perceptual in any direct perceptual sense. One cannot directly sense them. In early modern (Cartesian) terms these entities are "inferred." And, indeed, these constructs, even in Husserl's version of Galilean science, are thought to be distant from the primary perceptions of the lifeworld and are, instead, results of a mathematisation process.

I cannot take the time here to walk through the whole of Husserl's version of mathematisation and Galilean science, but the elements which are of importance for this argument revolve around the virtually exclusive interpretation of science through mathematisation. (a) At the heart of Husserl's version is a strong distinction between primary perceived plena which are directly sensorily perceived, and mathematised objects (the pure shapes of geometrisation) which are both abstract or idealised, and only indirectly available.

> [Science in] ... all this pure mathematics has to do with bodies and the bodily world only through an abstraction, i.e., it has to do only with abstract shapes within space-time, and with these, furthermore, as purely 'ideal' limit-shapes. Concretely, however, the actual and possible empirical shapes are given, at first, in empirical sense-intuition, merely as 'forms' of a 'matter' of a sensible plenum; thus they are given together with what shows itself, with its own gradations, in the so-called 'specific' sense qualities: color, sound, smell, and the like. (Husserl, 1970, pp. 29–30)

Husserl's mathematised science thus presupposes this strong distinction between the fully perceived plena and the merely idealised abstractions of pure shapes. (b) Thus, to incorporate the perceived world into and under the scientific world, there must be an indirect process.

> Now with regard to the 'indirect' mathematization of that aspect of the world which in itself has no mathematizable world-form: such mathematization is thinkable only in the sense that the specifically sensible qualities (plena) that can be experienced in the intuited bodies are closely related in a quite peculiar and regulated way with the shapes that belong essentially to them. (Husserl, 1970, p. 35)

In short, there is needed some form of association between the idealised abstracta and the plena, a relation which Husserl terms indirect.

If we now return to our environmental problem, the measured changes occurring with greenhouse gases, were Husserl right then the gases are themselves constituted through mathematisation with its abstractive process and indirect or inferential geometrisation. But there is something radically wrong with this approach because it implicitly leaves out the very materiality of the gases themselves and with it any possible realism which drives the earth sciences as well as any other late modern science.

The flaw, however, lies not with the description or the presumed indirect constitution of the gases; it rather lies with Husserl's flawed concept of science. And while Husserl is not alone he in fact remains close in his interpetation to many other 'Cartesian' or theory-weighted interpreters of science, including the positivists of his time and Descartes himself it is his concept of science which has undergone the abstraction. Of course science requires measurement, quantification, and the processes of analysis which occur in mathematisation but it equally requires a material relation with the 'things themselves' and this occurs in actually embodied science. That embodiment is the technological extension of primary perception through instrumentation. Husserl's Galileo is a Galileo without the telescope. And the relation between Galileo's satellites of Jupiter and rings of Saturn and early modern science is not mathematical or indirect; it is rather mediated and instrumentally real through this technological extension of bodily perception. This technologically mediated perceivablity reduces and transforms the presumed strong distinction between Husserl's plenary perception and the equally presumed indirect and abstract reduction to merely pure shapes. CFCs, CO_2, and ozone are not pure shapes, but are instrumentally presentable, material entities through science's increasingly sophisticated technological embodiments. They, too, are plena, albeit micro-plena instrumentally presentable. Simply put, science's mathematisation does not fit the Husserlian version of it, and, moreover, a purely or only mathematised science is itself a disembodied science. Contrarily, a technologically embodied science is a science which may operate through an instrumental realism which allows a mediated bodily perception of micro-entities rather than of abstract, pure shapes. I am here arguing that only such a technoscience can allow for the realism needed to take a greenhouse effect seriously.

ONLY BY BEING PROFOUND CAN HEIDEGGER BE PROFOUNDLYWRONG:

Husserl has not often been invoked for discussion of environmental issues. But Heidegger has; indeed, he is often termed 'green' and he has been invoked particularly by 'deep ecologists'. There are two surface reasons why this is the case. (a) In the example here, a Greenhouse Effect, if it is the case that there are homogenically originated causes arising from escaped CFCs, hydrocarbon burning processes, and increased CO2 accumulations, it is largely because of industrial processes or Technology with the capital T Heidegger gives to it. Technology is thus culpable in the environmental crisis. (b) The second reason for the invocation of Heidegger is a 'romantic' one and lies in his honorific use of the term, 'Earth'. (Has anyone noted that the Heideggerian concept of Earth as groundedness actually fits a still geocentric universe?) (c) I find, however, that there is a deeper reason why Heidegger is thought to be relevant to the environmental crisis. It lies in his much more material realism concerning the artefactual, which I contended was lacking in Husserl.

This realism was very early applied to the role of artefacts in his concept of the Zuhanden. Tools or instruments were ways in which human actions were mediated into the world. The hammer 'withdraws' into the sense of bodily action which produces the project in its materiality. Much later, in the context of an emergent philosophy of technology, Heidegger included science itself as a kind of institutionalised Zuhanden process. One must also remember that the tools which Heidegger analyses in Being and Time play a special role where they are the indicators of the "Worldhood of the World". The covered railway platform "takes account of the weather", and Nature itself is shown via these artefacts. But, later, technology, now elevated to a metaphysical mode of seeing, makes science itself "technological":

> It is said that modern technology is something incomparably different from all earlier technologies because it is based on modern physics. ... Meanwhile we have come to understand more clearly that the reverse holds true as well: modern physics, as experimental, is dependent upon technical apparatus and upon progress in the building of apparatus. (Heidegger, 1977, p. 30)

Science is technologically embodied. This is the origin of its instrumental realism and although Heidegger did not follow through on this concept, he may be said to have thereby corrected the Husserlian abstract view of science.

In Heidegger's view, however, science is doubly instrumental. If, in its actions, it is through its instruments or artefacts that it can act upon nature, it itself is an instrument of a certain metaphysical way of seeing Technology namely, with the capital. It is the means by which Nature is challenged and taken as "standing reserve" or a "resource well" [Bestand]. In short, it is through the artefactual or material instrumentality of science that this can occur. By extension, then, there can be a homogenically related Greenhouse Effect, through the very challenging process.

Normatively, this is part of the danger found in Technology: by being materially efficient it can pose a threat to the preservation of Earth, another of Heidegger's central concepts. Technology, now become materially efficient through the artefactual, reveals nature as a resource well which can be acted upon. Or, to put it another way, nature in the modern sense becomes what it is through the possibilities of material instrumental realism.

Now I want to turn to another, subterraneanly-close look at Heidegger's materially efficient or realist uses of the artefactual. Tools are not the only artefacts Greek temples are also artefactual. And, like earlier Zuhanden uses of tools, they reveal a world.

> Standing there, the building rests on the rocky ground. This resting of the work draws up out of the rock the mystery of that rock's clumsy yet spontaneous support. Standing there, the building holds its ground against the storm raging above it and so first makes the storm manifest in its violence. (Heidegger, 1977, p. 42)

This artefact carries lots of freight it reveals, or even "makes" the storm reveal itself as storm, the very sun its grace, the light its light, the invisible space of the air ... tree, grass, eagle, bull, snake, cricket "come to appear as what they are" (p. 42). The temple-work (artefact) "opens up a world and at the same time sets this world back again on earth, which itself only thus emerges as native ground" (ibid.).

But while this Wagnerian, Nietzschean romantic tone rings much louder than the accounting for the weather of Being and Time's Zuhanden, it plays a similar role. It is through the artefact (work) that the world is revealed and now, settled, upon the Earth. The temple is a kind of 'nice' technology. "To be a work means to set up a world. ... [but this ultimately also] lets the earth be an earth" (Heidegger, 1977, pp. 42, 46). Heidegger then goes on to create a dialectic between 'world' and 'earth'. Other than to reiterate that the dialectic itself is revealed through the work or artefact, I shall now leave the commentary and turn to what I think to be profoundly wrong with this view.

The first error lies in the romantic amplification of the deeper insight from the Zuhanden period. If it is through the artefactual in its praxical context that a "world" is revealed, then the later Heideggerian selective romanticisation of this in his ontological aesthetic is one which fails to fully reveal the world which gives rise to the artwork, but which instead in its romantic selectivity covers over precisely the full set of involvements which the artwork potentially reveals.

It was the historian J. Donald Hughes (1975) who first drew my attention to this problem. Although I doubt he knew of Heidegger's work, he opens his *Ecology in Ancient Civilizations* with the following remark:

> Those who look at the Parthenon, that incomparable symbol of the achievements of an ancient civilization, often do not see its wider setting. Behind the Acropolis, the bare dry mountains of Attica show their rocky bones against the blue Mediterranean sky, and the ruin of the finest temple built by the ancient Greeks is surrounded by the far vaster ruins of an environment which they desolated at the same time. (p. 1)

As I read this passage, a new revealing suggested itself instead of storm, snake and bull, I began to see rock surpassing forest, species-counts changing or disappearing, and soil irreversibly eroding into sea around Heidegger's temple. But there is worse: were one to look closely, much more closely, at the temple itself, one would find in its very stone testimony to the Greek destruction of an environment. The first temples were made of wood which became increasingly scarce as deforestation continued, not under chainsaws but under mere bronze axes, probably well crafted in a techne way and the temple on the hill evidences this past:

> The Greeks persisted in using the architectural forms that they were accustomed to from the days of wood. All the details of the entablature, with its cornices, friezes, ... were copies in stone of wooden structural elements. Greek builders even imitated the pegs that held the ancestral wooden structure together by adding little stone knobs called 'drops'. (de Camp, 1963, p. 93)

In short, even the temple itself, read closely, evidences the process of earlier deforestation, and all of this before modern machine technology.

The temple as art work, at least in Heidegger's description of it, does not, will not, perhaps cannot reveal its wider and more negative context. The reason, I contend, is the choice of a romanticised version of such works. But, deeper, the art work is simply the other side, the obverse of a technology or a technological artefact. Both are material; both are focal entities which reveal fields of involvements and references; and both thus "gather the fourfold" to use a Heideggerian overstatement.

There is, in the Heideggerian context, good artefactuality the artwork which reveals the romanticised interplay of the gods and mortals, earth and sky, and bad artefactuality technologies which rush us across rivers (the steel bridge), or dam the same river (hydroelectric dams), or reduce the Holocaust to modern agriculture (both of which treat humans as standing reserve). While I argue that there is something profoundly wrong with such romantic reductionism, I want now to make one more post-Heidegger move before returning to the Greenhouse Effect.

Good artefactuality the temple reveals Earth as ground, that upon which a world can rest. Bad artefactuality in this case I shall introduce the now familiar global imaging made possible by such technologies as earth shots from the Moon makes of the world a 'world as picture': "Hence world picture, when understood essentially, does not mean a picture of the world, but the world conceived and grasped as picture" (Heidegger, 1977, p. 129). This, presumably, is bad. What makes it bad is the reduction of the world to picture, whereas world (itself) cannot be so reduced.

Behind this critique, of course, lies a lot of ancient and still modern garbage, which beginning with Plato's copy-epistemology and its transformation into modern representationalist-epistemology, needs weeding out. Heidegger's insight that world is revealed as picture is better than that, but it still misses the point. I wish to argue, contrarily, that to see by means of imaging technologies, is not to look at a picture, it is rather to look through the image. That, at least, is how imaging including whole earth measurements is used in scientific praxis.

WHOLE EARTH MEASUREMENTS

I now return to the initial problem: How can one tell anything about a Greenhouse Effect, phenomenologically? I opened this venture with the assertion that two concepts are necessary as a frame for detecting a Greenhouse Effect: science as technoscience and a perspective, earth-as-planet, for making whole earth measurements.

The first is necessary if there is to be any referential realism regarding the phenomenon. Measurements must be measurements of something, and what an instrumentally embodied science gives one is precisely the opening to micro and macrophenomena through the mediated bodily perceptions made possible by instrumentation. Instruments give access to the phenomena, and that often in mediated perceptual form. Greenhouse gases are not inferred, they are instrumentally perceived. I argued that the Husserlian science of the Crisis lacked this conception of science, although the praxical realism of Heidegger's earlier work saw its possibilities.

The second conception needed is perspectival in the sense that one must have a sense of a "whole earth" as that field which is measurable, and this is what I am calling earth-as-planet. The idea of earth as planet, as a finite sphere, is itself quite ancient: it is anticipated by Aristarchus, is assumed by Copernicus, and its size has been measured along with these notions, both in Greek and early modern times. But one could argue that it is not seen as planet until it is so embodied in the earth shots with which we are now all familiar. And it is here that I enter the last set of arguments with classical phenomenology, Husserlian and Heideggerian.

Put crassly, what might Husserl and Heidegger see when they look at a moonshot of earth-as-planet? Or, when they look at other imagings of environmental phenomena? If the answers are "pictures" or "images", then the look involved is both naive and cast in terms of modern epistemologies which remain caught with a passive theory of perception which sees only surfaces and representations. And if, in the Heideggerian case, there is more penetration, but a penetration which yields the referent phenomenon as itself a picture, then the seeing remains short of scientific seeing as evident in science praxis. Indeed, such a look would remain short of even the primary insights of a phenomenological epistemology claimed by both but restricted to everyday or pre-scientific lifeworld regions. Heidegger notes, "What we 'first' hear is never noises or complexes of sounds, but the creaking wagon, the motorcycle. We hear the column on the march, the north wind, the woodpecker tapping, the fire crackling" (Heidegger, 1962, p. 207). This auditory plenum precedes its abstract resolution into distinct qualia, the gestalt precedes its separable parts. "It requires a very artificial and complicated frame of mind to 'hear' a 'pure noise'" (ibid.).

Similarly, I am arguing that it likewise takes an artificial and complicated frame of mind to see a 'mere picture' or an 'image' this is an effect of a modern epistemology. Instead, in scientific seeing, a "seeing which understands" to again paraphrase Heidegger, gestalts are "seen through" in the processes of imaging and measuring. And no scientific seeing takes the image for the referent itself, nor confuses the world with its picture. Rather, all pictures and images are themselves instrumental and are seen or read through. I return, now, to the Greenhouse Effect.

SCIENCE PRAXIS AS A PHENOMENOLOGICAL HERMENEUTIC

Contemporary technoscience does its work by means of its technological embodiments, its instrumentation. Its range of instruments is wide and diverse. With respect to some of these referring to whole earth measurements, attempting to discern a possible Greenhouse Effect, there are various sorts of recording thermometers such as buoys in the oceans which record both surface and depth temperatures. Similarly, thermometers are placed in the atmosphere (balloons, aircraft probes, etc.). The recorded results are then fed into computer models, mathematically analysed and reduced to graphs and trajectories to determine overall patterns. This process falls under what I have earlier called 'hermeneutic relations' whereby the referent is mediated by digital, numerical, or graphed results which are non-isomorphic with the items themselves referred to.

Imaging processes, however, are of special interest because imaging mediates perceivable patterns which often do refer isomorphically, or partially isomorphically with the referent intended. Thus satellite photography in geosatellites depicts earth phenomena which in more visually gestalted forms show the state of the whole earth. I choose this special form of instrumentation here because it displays a much more phenomenological morphology in its very closeness to perceivability.

In the cases of straightforward isomorphism, in which colour photography remains representational with true colours, one can detect what is happening on the planet. One can 'see' the deforestation process in South America, and a satellite photo this fall dramatically showed the smoke plume hundreds of miles to sea from the forest fires on my own Long Island. And, with enough satellites, one can scan the whole planet. But, scientifically, such straightforward visual isomorphism is not enough.

What I call 'instrumental phenomenological variations' show more: by turning to the sophisticated varieties of false colour, which varies from human vision, one may learn more. Infra-red imaging shows much more dramatically the changing ratios of organic to inorganic surfaces thereby showing where desertification is taking place. Thermal imaging shows temperature variations; magnetomometry shows even that which is below the surface; and with each variation a richer and more complex set of phenomena emerge. I am suggesting in the strong sense that this deliberate variational praxis is, in science, very much akin to an embodied form of phenomenological variational praxis.

But it is also critically hermeneutic. One must learn to read the interrelations. For example, the current ability to get closer to synchronic whole earth measurements must also be related to diachronic readings of the past. These are made possible by the material 'calendars' left by the earth itself. Tree rings reveal past periods of drought and wet seasons, deep glacial ice shows past temperatures, etc. I am suggesting here a 'hermeneutics of things', not merely of languages and texts. (Indeed, I take the current penchant to make of everything a 'text' to be analogous to the Heideggerian fear of taking the world as 'picture'.)

Finally, scientific praxis even if it still sometimes self-interprets its knowledge in modern epistemological language is no longer modern. It is, rather, reconstructive. Instrumental perceivability is not passive observation, it is actively constructive. To get the truest imaging, the result must be manipulated, constructed. Here two techniques may be noted in passing. First, the use of contrastive techniques is known in all types of imaging processes. False colour or some other contrastive technique is used to sharpen, enhance, and bring forth features which otherwise would lie undetected. Here manipulation away from the passively real is needed to bring out the phenomenon. Second, computerised enhancement also may sharpen and create more true or hyperreal imaging. I contend that this technologically constructed approach is beyond the limits of modern epistemology and is more akin to a phenomenological constitution which in its classical sense did not realise its scientific potential.

Imaging, as one example of the technological embodiment of science, is not some set of pictures; it is the praxical, instrumental Zuhanden of that which reveals a world. Insofar as science is not merely some set of mathematised theory, but is embodied as a technoscience which materially relates to a world, and insofar as the earth is a perceivable planet, and then, insofar as our measurements are critically reconstructed through instrumentally constitutive praxis, then phenomenologists can detect the possible Greenhouse Effect.

NOTES

[1] This article is a reprint of Ihde, D. (1997). Whole earth measurements. *Techné: Research in Philosophy and Technology, 2*(2), 61–74. Reproduced with permission from *Techné: Research in Philosophy and Technology*, published by Digital Library and Archives (DLA) of Virginia Polytechnic Institute and State University, on behalf of the Research in Philosophy & Technology.

REFERENCES

de Camp, L. S. (1963). *The ancient engineers*. New York: Dorset.
Heidegger, M. (1962). *Being and time* (J. Macquarrie & E. Robinson, Trans.). New York: Harper and Row.
Heidegger, M. (1971). *Poetry, language, thought* (A. Hofstadter, Trans.). New York: Harper.
Heidegger, M. (1977). *The question concerning technology* (W. Lovitt, Trans.). New York: Harper.
Hughes, J. D. (1975). *Ecology in ancient civilizations*. Albuquerque, NM: University of New Mexico Press.
Husserl, E. (1970). *The crisis of European sciences and transcendental phenomenology* (D. Carr, Trans.). Evanston, IL: Northwestern University Press.

Don Ihde
Department of Philosophy
State University of New York at Stony Brook
USA

ANDREW FEENBERG

15. WHAT IS PHILOSOPHY OF TECHNOLOGY?[1]

INTRODUCTION

In this chapter I will attempt to answer the question posed in my title from two standpoints, first historically and then in terms of contemporary options in the field, the various different theories that are currently under discussion.[2] But before I begin, I would like to clear up a common misunderstanding: philosophy of technology is *not* closely related to philosophy of science. Science and technology share a similar type of rationality based on empirical observation and knowledge of natural causality, but technology is concerned with usefulness rather than truth. Where science seeks to know, technology seeks to control. However, this is by no means the whole story.

Our image of premodernity is shaped by the struggles between science and religion in the early modern period. From those struggles we derive the notion that traditional societies restrict questioning of their basic customs and myths. In the premodern West, the principle of authority was the basis not just for church doctrine, but for knowledge of the world as well. Modern societies emerge from the release of the power of questioning against such traditional forms of thought. The 18th century Enlightenment demanded that all customs and institutions justify themselves as useful for humanity. Under the impact of this demand, science and technology become the new basis for belief. Eventually, technology becomes omnipresent in everyday life and scientific-technical modes of thought predominate over all others.[3] In a mature modern society, technology is taken for granted much as were the customs and myths of traditional society. Scientific-technical rationality has become a new culture.

This culture is clearly 'useful' in all its details in the sense the Enlightenment demanded, but it is now so all encompassing that larger questions can be asked about its value. We judge our technological civilisation as more or less worthy, more or less ethically justified, more or less fulfilling. Modernity itself authorises, even demands such judgment. We need to understand ourselves today in the midst of technology and neither scientific nor technical knowledge can help us. Insofar as our society is technological at its basis, philosophy of technology is its theoretical self-awareness. Philosophy of technology teaches us to reflect on what we take for granted most of all, that is, the rationality of modernity. The importance of this perspective cannot be overestimated.

GREEK ORIGINS

The question of technology is raised at the very origins of Western philosophy, not as we pose it today of course, but at a metaphysical level. Philosophy begins in ancient Greece by interpreting the world in terms of the fundamental fact that humanity is a laboring animal constantly at work transforming nature. This fundamental fact shapes the basic distinctions that prevail throughout the tradition of Western philosophy.[4]

A.T. Jones and M.J. de Vries (eds.), International Handbook of Research and Development in Technology Education, 159–166.
© 2009 Sense Publishers. All rights reserved.

The first of these is the distinction between what the Greeks called *physis* and *poiésis*. *Physis* is usually translated as nature. For the Greeks, nature creates itself, emerges from out of itself. But there are other things in the world, things which depend on humans to come into being. *Poiésis* is the practical activity of human production. We call the beings so created artefacts and include among them the products of art, craft, and social convention.

The word *techné* in ancient Greek signifies the knowledge or the discipline associated with a form of *poiésis*. For example, sculpture is a *techné* that creates out of stone; carpentry is a *techné* that builds from wood. Each *techné* includes a purpose and a meaning for its artefacts. For the Greeks, *technai* show the 'right way' to do things in a very strong, even an objective sense. Although artefacts depend on human activity, the knowledge contained in the *technai* is no matter of opinion or subjective intention. Even the purposes of artefacts share in this objectivity insofar as they are defined by the *technai*.

The second fundamental distinction is that between existence and essence. Existence answers the question whether something is or is not. Essence answers the question what the thing is. *That* it is and *what* it is appear to be two independent dimensions of being. In the tradition of Western philosophy, existence has been a rather hazy concept. We know the difference between what exists and what does not, for example, as immediate presence or absence, but there is not much more to say. Most of the attention is given to essence and its successor concepts as developed by the sciences because this is the content of knowledge.

These distinctions are self-evident. They form the basis of all philosophical thought in the West. But the relation between them is puzzling. The source of the puzzle is the Greek understanding of *techné*, the ancestor of modern technology. Strange though it seems, they conceived nature on the model of the artefacts produced by their own technical activity.

To show this, I will analyse the relation between the two basic distinctions that I've introduced, *physis* and *poiésis*, and existence and essence. The difference between existence and essence in *poiésis* is real and obvious. The thing is present first as an idea and only later comes into existence through human making. For the Greeks the idea is not arbitrary or subjective but rather belongs to a *techné*. Each *techné* contains the essence of the thing to be made prior to the act of making. The idea, the essence of the thing is a reality independent of the thing and its maker. Although humans make artefacts, they do so according to a plan and for a purpose that is an objective aspect of the world.

But the corresponding distinction between existence and essence is not obvious for natural things. The thing and its essence arise together and exist together. The essence does not have a separate existence. The flower emerges along with what makes it a flower: that it is and what it is 'happen', in a sense, simultaneously. We can define a concept of the flower, but this is our notion, not something essential to the flower as a concept or plan is to artefacts. Indeed, the very idea of an essence of the things of nature is our construction. Unlike the knowledge that is active in *techné*, which participates in bringing into existence the objects it defines, the essences identified by *epistemé*, the science of nature, appear to be purely human doings to which nature itself would be indifferent. Or is it? Here is where the story gets interesting.

Although essences quite obviously relate differently to *physis* than to *poiésis,* since Greek times philosophers have struggled to efface that difference in a unified theory. For Plato, who started the tradition on this path, the concept of the thing, its 'idea', exists in some sense prior to the thing itself and allows it to exist and us to know it. This is exactly the pattern familiar from *techné*, but Plato does not reserve his theory for artefacts; rather, he applies it to all being. He relies on the structure of *techné* to explain not only artefacts, but nature as well.

Plato understands nature as divided into existence and essence just as are artefacts and this becomes the basis for Greek ontology. In this ontology there is no radical discontinuity between technical making and natural self-production because they both share the same structure. *Techné* includes a purpose and a meaning for artefacts. The Greeks import these aspects of *techné* into the realm of nature and view all of nature in teleological terms. The world is thus a place full of meaning and intention. This conception of the world calls for a corresponding understanding of man. We humans are not the masters of nature but realise its potentialities in bringing a meaningful world to fruition. Our knowledge of that world and our action in it is not arbitrary but is in some sense the completion of what lies hidden in nature.

What conclusion follows from these historical considerations on ancient Greek philosophy? I will be provocative and say that the philosophy of technology begins with the Greeks and is in fact the foundation of all Western philosophy. It was the Greeks who first interpreted being as such through the concept of technical making. This is ironic. Technology has a low status in the high culture of modern societies but it was actually there at the origin of that culture and, if we believe the Greeks, contains the key to the understanding of being as such.

MODERN ALTERNATIVES

I will now leave these historical considerations and turn to the status of philosophy of technology in our era. At its inception, Descartes promised that we would become "the masters and possessors of nature" through the cultivation of the sciences, and Bacon famously claimed that "knowledge is power". Here technology no longer fulfils nature's potentialities as it did for the Greeks, but rather it realises human plans. Clearly we are in a different world from the Greeks. And yet we share with them the fundamental distinctions between essence and existence, and between the things that make themselves, nature, and the things that are made, artefacts. But our understanding of these distinctions is different from theirs. This is especially true of the concept of essence. For us essences are conventional rather than real. The meaning and purpose of things is something we create not something we discover. The gap between man and world widens accordingly. We are not at home in the world, we conquer the world. This difference is related to our basic ontology. The question we address to being is not *what* it is but *how* it works. Science answers this question rather than revealing essences in the Greek sense of the term.

Note that technology is still the model of being in this modern conception. This was particularly clear in the Enlightenment when philosophers and scientists challenged the medieval successors to Greek science with the new mechanistic worldview. Eighteenth century physical science identified the workings of the universe with a clockwork mechanism. Thus strange though it may seem, the underlying structure of Greek ontology survived the defeat of its concept of essence.

In the modern context technology does not realise objective essences inscribed in the nature of the universe as does *techné*. It now appears as purely instrumental, as value free. It does not respond to inherent purposes, but is merely a means serving subjective goals. For modern common sense, means and ends are independent of each other: "Guns don't kill people, people kill people". Guns are a means independent of the users' ends, whether it be to rob a bank or to enforce the law. Technology, we say, is neutral, meaning that it has no preference as between the various possible uses to which it can be put. This 'instrumentalist' philosophy of technology is a spontaneous product of our civilisation, assumed unreflectively by most people.

Technology in this scheme of things encounters nature as raw materials, not as a world that emerges out of itself, a *physis*, but rather as passive stuff awaiting transformation into whatever we desire. This world is there to be controlled and used without any inner purpose. The West has made enormous technical advances on this basis. Nothing restrains us in our exploitation of nature. Everything is exposed to an analytic intelligence that decomposes it into usable parts. Under this assumption, our means have become ever more efficient and powerful. In the 19th century it became commonplace to view modernity as an unending progress toward the fulfilment of human needs through technological advance.

But for what ends? The goals of our society can no longer be specified in a knowledge of some sort as they were for the Greeks. The Greeks lived in harmony with the world whereas we are alienated from it by our very freedom to define our purposes as we wish. So long as no great harm could be attributed to technology, this situation did not lead to serious doubts beyond small circles of intellectuals. But as the 20th century proceeds from world wars to concentration camps to environmental catastrophes, it becomes more and more difficult to ignore the strange aimlessness of modernity. This has led to a crisis of civilisation from which there seems no escape: we know how to get there but we do not know why we are going or even where. It is because we are at such a loss that so many 20th century philosophers of technology became critics of modernity.

I want now to present the various alternatives so far discussed and others as well in a chart that puts order in the discussion (see Figure 1). Technology is defined here along two axes reflecting its relation to values and agency. The vertical axis offers two alternatives: either technology is value neutral, as we typically assume in modern times, or it is value-laden as the Greeks believed and, we will see, as some philosophers of technology believe today. The choice between these views is not obvious. From a common sense perspective a technical device is simply a concatenation of causal mechanisms. No amount of scientific study will find anything like a purpose in it. But perhaps common sense misses the point. After all, no scientific study will find money in a $100 billion. Not everything is a physical or chemical property of matter. Perhaps technologies, like bank notes, have a way of containing value in themselves as social entities.

On the horizontal axis technologies are signified as either autonomous or humanly controllable. To say that technology is autonomous is not of course to say that it acts alone. Human beings are involved, but the question is, do they actually have the freedom to decide how technology will be applied and develop? Is the next step in the evolution of the technical system up to human decision-makers or do they act according to a logic inscribed in the very nature of technology? In the latter case technology can rightly be said to be autonomous. On the other hand, technology would be humanly controllable if we could determine the next step in its evolution in accordance with intentions elaborated without reference to the imperatives of technology.

The intersection of these two axes defines four types of theories. Instrumentalism, the occupant of the box in which human control and value neutrality intersect, has been discussed above. This is the 'standard' modern view according to which technology is a tool or instrument of the human species as a whole. As noted in the chart, this view corresponds to the liberal faith in progress which was such a prominent feature of mainstream Western thought until fairly recently.

Technology, on the determinist account, is rooted on the one side in knowledge of nature and on the other in generic features of the human species. This is why it can be described as neutral, as a rationally constructed tool serving universal human needs. Some determinists argue that technologies simply extend human faculties: the automobile extends our feet while computers extend our brains. It is not up to us to adapt technology to our whims but on the contrary, we must adapt to technology as the most significant expression of our humanity. Determinism is a widely held view in social science where it supports the pretensions to universality of both capitalist and communist societies.

Substantivists attribute substantive values to technology in contrast with instrumentalism and determinism which view technology as neutral. The contrast here is actually more complex than it seems at first sight. The neutrality thesis to which instrumentalists and most determinists subscribe does admit that technology embodies a value but it is a merely formal value, efficiency. Using technology for this or that purpose would not be a specific value choice in itself, but just a more efficient way of realising a pre-existing value of some sort. A substantive value on the contrary involves a commitment to a specific conception of the good. According to substantivism, the values embodied by technology are the pursuit of power and domination. These values track technology like a shadow and show up in everything it touches. If technology embodies these values, it is not merely instrumental and cannot be used for the various purposes of individuals or societies with different ideas of the good. According to substantivism, insofar as we use technology we are committed to a technological way of life.

Technology is:	Autonomous	Humanly Controlled
Neutral (complete separation of means and ends)	Determinism (e.g. modernisation theory)	Instrumentalism (liberal faith in progress)
Value-laden (means form a way of life that includes ends)	Substantivism (means and ends linked in systems)	Critical Theory (choice of alternative means-ends systems)

Figure 1. Technology's relation to values and agency. (Drawn from Feenberg, 1999, p. 9)

There are obvious relations between substantive theory of technology and determinism. In fact most substantive theorists are determinists as well. But the position I have characterised as determinism is optimistic and progressive. Both Marx (in the commonplace interpretations) and the modernisation theorists of the post-war era believed that technology would save humanity. Substantive theory is not so optimistic and regards autonomous technology as threatening if not malevolent rather than benign. Once unleashed, technology becomes more and more imperialistic, taking over one domain of social life after another. In the most extreme imagination of substantivism, a *Brave New World* such as Huxley describes in his famous novel converts human beings into mere cogs in the machinery. This is not utopia – the 'no place' of an ideal society, but dystopia – a world in which human individuality has been completely suppressed. Here people become, as Marshall McLuhan once said, the "sex organs of the machine world" (1964, p. 46).

Martin Heidegger was the most famous substantive theorist. He argued that the essence of modernity is the triumph of technology over every other value. He noted that Greek philosophy had already based its understanding of being on technical making and argued that this starting point culminates in modern technology. Where the Greeks took *techné* as the model of being in theory, we have transformed beings technically in practice. Our metaphysics is not in our heads but consists in the real technical conquest of the earth. This conquest transforms everything into raw materials and system components, including human beings themselves (McLuhan, 1964). Not only are we constantly obeying the dictates of the many technical systems in which we are enrolled, we tend to see ourselves more and more as devices regulated by medical, psychological, athletic, and other functional disciplines. Our bookstores are full of 'operating manuals' for every aspect of life: love, sex, divorce, friendship, raising children, eating, exercise, making money, having fun, and so on and so forth. We are our own machines.

But, Heidegger argues, although we may control the world through our technology, we do not control our own obsession with control. Something lies behind technology, a mystery we cannot unravel from our technological standpoint. Where we are headed is a mystery too. The West has reached the end of its rope. His last interview concludes that "Only a God can save us" (Heidegger, 1977b).

Heidegger's views contrast sharply with critical theory of technology. Critical theory agrees with substantivism that technology is not the unmixed blessing welcomed by instrumentalists and determinists. It recognises the catastrophic consequences of technological development but still sees a promise of greater freedom in a possible future. The problem is not with technology as such but with our failure so far to devise appropriate institutions for exercising human control over it. We could tame technology by submitting it to a more democratic process of design and development.

The economy offers an encouraging parallel to this view of technology. A century ago mainstream political and academic thought conceived the economy as an autonomous power operating according to inflexible laws. Today we know the contrary, that we can influence the direction of economic development through democratic institutions. Critical theory of technology argues that the time has come to extend democracy to technology as well. It thus attempts to save the Enlightenment values that have guided progress for the last several hundred years without ignoring the resulting problems.

As can be seen from the chart, critical theory shares traits with both instrumentalism and substantivism. Like instrumentalism critical theory argues that technology is in some sense controllable, but it also agrees with substantivism that technology is value-laden. This seems a contradictory position since, in the substantivist view, precisely what cannot be controlled are the values embodied in technology such as efficiency and domination. If this is true, the choices within our power would be like those we make in the supermarket between different brands of soap, i.e. trivial and delusory. How then can we conceive the value-ladenness of technology such that human control matters?

The Critical Theorist, Herbert Marcuse, sketched an answer I have tried to develop in what I call a critical theory of technology. According to critical theory the values embodied in technology are socially specific and are not adequately represented by such abstractions as efficiency or control. Technology can frame not just one way of life but many different possible ways of life, each of which determines a different choice of designs and a different range of technological mediation. Does this mean that technology is neutral, as instrumentalism believes? Not quite: modern societies must all aim at efficiency in those domains where they apply technology, but to claim that they can realise no other significant values besides efficiency is to overlook the tremendous social impact of differing design choices. What is worse, it obscures the difference between the current miserable state of technological societies and a better condition we can imagine and for which we can struggle. One must look down on mankind from a very great height indeed not to notice the difference between efficient weapons and efficient medicines, efficient propaganda and efficient education, efficient exploitation and efficient research! This difference is socially and ethically significant and so cannot be discounted as thinkers who share Heidegger's Olympian view sometimes claim.

Nevertheless, the substantivist critique of instrumentalism does demonstrate that technologies are not neutral tools. Means and ends are connected. Thus even if some sort of human control of technology is possible, it cannot be understood on the same terms as instrumental control of particular devices. In critical theory technologies are not seen as mere tools but also as frameworks for ways of life. Thus we cannot agree with the instrumentalist that "Guns don't kill people, people kill people". Supplying people with guns creates a social world quite different from the world in which people are disarmed. We can choose which world we wish to inhabit through legislation.

This is not the sort of control over technology instrumentalism generally claims we have. Its model of control is based on a restricted notion of use of individual devices, not choices between whole technological systems with different social consequences. This is a meta-choice, a choice at a higher level determining which values are to be embodied in the technical framework of our lives. Critical theory of technology opens up the possibility of reflecting on such choices and submitting them to more democratic controls. We do not have to wait for a god to save us as Heidegger expostulated but can hope to save ourselves through democratic interventions into technology.

But critical theory is not naïve about the difficulties that stand in the way of democracy. Technology gradually subverts the capacity for democracy even as it destroys the objective world. Thus the practical difference between substantivism and critical theory is not as great as it seems at first.[5] Critical theory is relatively skeptical about the capacity of human beings to get technological civilisation under reasonable control, but at least it does not exclude the possibility in principle as does substantivism. This is why it is necessary to talk not in terms of a utopian democracy of technique but more modestly of democratic interventions into technology.

What is meant by the concept of democratic interventions? Clearly, it would not make sense to hold an election between competing devices or designs. The voting public is not sufficiently concerned, involved, and informed to choose good politicians, much less good technologies. So, in what sense can democracy be extended to technology under current conditions? People affected by technological change ever more frequently protest or innovate. Where it used to be possible to silence all opposition to technical projects by appealing to Progress with a capital 'P', today communities mobilise to make their wishes known, for example, in opposition to nuclear power plants or toxic waste dumps in their neighbourhood. In a different vein the computer has involved us in technology so intimately that our activities have begun to shape its development. Email, the most used function of the Internet, was introduced by skilled users and did not originally figure in the plans of the designers at all. Similar examples can be adduced from medicine, urban affairs, and many other domains where technology shapes human activity.

Critical theory of technology detects a trend toward greater participation in decisions about design and development in examples such as these. The public sphere appears to be opening slowly to encompass technical issues that were formerly viewed as the exclusive preserve of experts. Can this trend continue to the point where citizenship will include the exercise of human control over the technical framework of our lives? We must hope so for the alternative is likely to be the eventual failure of the experiment in industrial society under the pressure of untrammeled competition and national rivalries. If people are able to conceive and pursue their intrinsic interest in peace and fulfilment through the political process, they will inevitably address the question of technology along with many other questions that hang in suspense today. We can only hope this will happen sooner rather than later.

NOTES

1. This article is a reprint of Feenberg, A. (2006). What is philosophy of technology? In John R. Dakers (Ed.), Defining technological literacy: Towards an epistemological framework (pp. 5–16). New York: Palgrave Macmillan. Reproduced with permission from Palgrave Macmillan.
2. This short chapter can give only a hint of the richness of the field. For a thorough account, see Mitcham (1994).

[3] This is a summary of ideas developed in Habermas (1984). While the usual contrast of premodern dogmatism with modern reflexivity is no doubt overdrawn, this is hardly the moment to drop it entirely. For a critique of this position see Latour (1991).

[4] The discussion in this section is loosely derived from Heidegger's (1973) history of being.

[5] For a comparison of Heidegger and Critical Theory, see Feenberg (2005).

REFERENCES

Feenberg, A. (1999). *Questioning technology*. New York: Routledge.

Feenberg, A. (2005). *Heidegger and Marcuse: The catastrophe and redemption of history*. New York: Routledge.

Habermas, J. (1984). *The theory of communicative action*. Boston: Beacon Press.

Heidegger, M. (1973). *The end of philosophy* (J. Stambaugh, Trans.). New York: Harper and Row.

Heidegger, M. (1977a). *The question concerning technology* (W. Lovitt, Trans.). New York: Harper and Row.

Heidegger, M. (1977b). Only a god can save us now (D. Schendler, Trans.). *Graduate Faculty Philosophy Journal*, *6*(1), 5–27.

Latour, B. (1993). *We have never been modern* (C. Porter, Trans.). Cambridge, MA: Harvard University Press.

McLuhan, M. (1964). *Understanding media: The extensions of man*. New York: McGraw Hill.

Mitcham, C. (1994). *Thinking through technology: The path between engineering and philosophy*. Chicago: University of Chicago Press.

Andrew Feenberg
School of Communication
Simon Fraser University
British Columbia, Canada

DENNIS W. CHEEK

16. SOCIAL ASPECTS OF TECHNOLOGY

INTRODUCTION

Technology is most commonly known by its products and their effects on society (Landers, 2003). Many moderns rarely consider the manner in which these products come to be or what they represent beyond their surface features and readily observable effects. Ancient philosophers such as Plato and Aristotle recognised that technological artefacts did not appear by themselves but were the ultimate outcomes of a distinct process whereby human beings conceived, created, refined, and produced objects to achieve human-desired purposes or fulfil human wants and needs (Kaplan, 2004). Even at the level of artefacts produced by a single individual for himself or herself, the stamp of an individual's creativity can be seen, for example, in ancient spears, which reflect a multitude of variations in terms of the manner in which the spearheads were shaped, the distinct wrapping configurations, and the whittled shape, length, width, and weight of the shaft. Such human distinctiveness at the level of the individual craftsman can be seen whether one is inspecting ancient spears, ancient pots, or a variety of other artefacts uncovered by the archaeologist's spade. Similarly, assemblages of stoneware, pottery, writing, brickwork, stonework, waterwheels, or other ancient technologies can be grouped into different 'schools' of creativity and production with clear commonalities between many artefacts within a given locale, region, and/or time period. These commonalities remind us that while ancient technologies could be quite personal they also developed and evolved within a given social context as well as a physical environment that combined to form a series of constraints and possibilities within which human creativity could thrive and manifest itself. The attendant features of the resultant artefacts that survived the ravages of time frequently make it possible for experts to identify distinct ancient cultures on the basis of the physical remains and trace these cultures and their interactions with other ancient cultures over time and space, with or without the additional help of historical records.

ANCIENT AND MEDIEVAL TECHNOLOGIES

Technologies far back in recorded history and prehistory were constrained, focused, shaped, sustained, spread, and evolved within particular social matrices. Clay tablets from ancient Sumer, Babylon, Assyria, and records preserved in the tombs of Egyptian pharaohs as well as within the Hebrew Bible, for example, give witness to the many ways in which ancient rulers and governmental structures shaped the types of technologies that were developed, enhanced, diffused, and imposed. Elaborate systems of barter, monetary exchange, and maintenance of accounts developed using a myriad of technologies working in combination to enable goods to move efficiently throughout the Fertile Crescent along the King's Highway and all the way to the upper cataracts of the Nile River in modern Sudan. Evidence exists for similar but less well-understood trade routes and exchanges with civilisations throughout sub-Saharan Africa, Southeast Asia and China, and the northern Russian steppes. Similarly complex systems evolved independently in other areas of the world. Coterminous

A.T. Jones and M.J. de Vries (eds.), International Handbook of Research and Development in Technology Education, 167–174.
© 2009 Sense Publishers. All rights reserved.

with the formal or informal development of media of exchange, human language (a preeminent technology), record keeping, measurement systems, and other technologies, we also see the development of intricate networks of major and secondary roads that facilitated the movements of rulers and their entourages and armies in annual marches of conquest and tribute collection or punishment for nonpayment of tribute. Of course, these routes also provided a means for merchants to speedily move goods to new and emerging markets. Herodotus writes with admiration of the Persian roads that allowed the Persian kings to move effortlessly between the twin capitals of Susa and Persepolis as well as throughout their kingdom in a manner that reinforced for local residents kingly power, wisdom, and divine presence. The siting of the road and, conversely, those areas that the road did not encompass, exerted a direct influence on the development of particular towns and cities and the consequent diminishment or social/ economic/political marginalisation of other locales. We see here the principle that technologies not only are a product of social creation but they also by their very nature and use favour some users and impose burdens or unintended (or intended) consequences on other users/nonusers (Cheek, 2005).

With the rise of the Roman Empire, systematisation of technologies and technological systems was taken to new heights and evidenced itself throughout the Empire such that many structural remains stand in mute testimony even today to the engineering and technical prowess of multitudes of unknown Roman craftsmen, artisans, and engineers. Roads were carefully and systematically surveyed, prepared, and maintained throughout the Empire with signposts that always indicated one's distance from the Eternal City of the Seven Hills. An advanced postal system, guard towers, inns, markets, and associated secondary roads enabled people, armies, tribute, and commerce to move swiftly throughout the Empire – at least when a major war or uprising was not in effect. Many of these roads were built with such care and attention to engineering matters that modern roads follow the exact same traces and in some instances are even sitting overtop of prepared surfaces first laid by Roman workers centuries earlier. French engineers in the 19th century carefully studied preserved segments of Roman roads as a basis for developing modern highway engineering techniques. These roads, coupled with Roman systematisation of sea routes, linkages to ports, and largely clearing the Mediterranean Sea of pirates, provided the means to keep the masses within Rome itself satiated with foods from around the known world and gladiatorial and other pageants and festivities that seemed to know no bounds. Few today stop to consider the enormous array of technologies both near and far that made such spectacles possible on an unprecedented scale that would not be equalled until well into the Industrial Revolution (Landels, 2000). The various 'needs' or technological problems that the creation and maintenance of such events imposed spawned all kinds of technological solutions, many of which are still utilised today.

Architectural remains and the observable layouts of entire cities within the ancient Roman Empire (e.g., Rome, Ephesus, Corinth, and Caesarea Maritima) witness to the manner in which local customs, mores, and aesthetics shaped the evolution and implementation of various technologies and technological systems as experts can readily distinguish the regional nature of various innovations by the subtle and not so subtle differences in materials, orientation, decoration, etc. evident in these structures. These differences remind us that some technological innovations fail to spread outside of the culture that creates them due not to their technological impotence but because of differences in social values, such as perceptions about the social status or desirability of association with the creators of the technologies in question, aesthetic considerations, compatibility with religious and social mores, and adaptability to differing physical, environmental, and/or socio-political contexts (Cowan, 1997; McClellan & Dorn, 2006).

The decline and 'fall' of the Roman Empire brought social disruptions and caused many of the elaborate technological systems of the Romans to disintegrate, although a large portion of these types of structures and systems persisted in the Eastern Roman Empire for many more centuries with quite a few passing into the hands of the Mohammedan conquerors of Byzantium. Within Western Europe the smaller kingdoms that emerged after the Roman Empire continued to use vestiges of what remained but were unable to effectuate the kinds of monumental efforts that had been common in the past.

The most technologically advanced achievement of the medieval period in Western Europe would be the cathedral. Cathedrals were far more than religious edifices for the collective worship of Christians. They were the means by which large numbers of illiterate people were instructed in the Christian faith, a focal point of community life in an age full of rampant disease and unsettling economic and social life, and a focal point for human technological innovation and creativity (Bandmann, 2005). Largely anonymous architects, stonemasons, carpenters, and craftsmen advanced well beyond their classical counterparts, creating flying buttresses, arches, suspended ceilings, and a myriad of architectural and engineering innovations that continue to influence the way we think about design and inspiring visitors to these mammoth structures down to the present age. A medieval cathedral demonstrates on a massive scale the social nature of technology by the ways in which inanimate materials have been combined with light and informed by religious ideals and social values to instruct and inspire generations both within the Christian communities that sponsored them and to a wider society who lived, and continue to live, around them.

RENAISSANCE AND EARLY MODERN TIMES

The Renaissance witnessed the further evolution of technological developments germinated in the ancient and medieval periods, as well as the creation of entirely new technologies. The careful work of historian Fernand Braudel (1981–1984) of the French Annales School reminds us in singular fashion of the many ways in which cultural values, societal institutions, and individuals *par excellence* shaped various technologies during this period to transform the political, social, and economic worlds of the time. In particular, this period witnessed the importance of sponsorship in the creation and spread of technological innovations, a key facet of technological innovation and diffusion right down to the present era. So many inventions, discoveries, and entire technological systems owed their existence, spread, or startling evolution to the patronage of wealthy families, royal personages, and individuals without whom the creative work of Michelangelo, Raphael, DaVinci, and others would never have surfaced, thrived, and survived.

Basic techniques around mass production of goods also arise in this period, especially in manifestations such as Venetian shipbuilding that pioneered many techniques that would later be seen in industrialised societies. The printing press is another excellent example that builds upon earlier work that dates all the way back to ancient China but does not flower until this period, when movable type and the simplicity of scripts in Western languages makes possible the rapid production and dissemination of knowledge in the form of the printed page and bound printed book rather than the hand-copied letter or manuscript. Historians have noted that without the printing press there likely would have been no Protestant Reformation or, at least, not as we now know it. Books, for the first time, come within the tangible reach of the common person with an attendant interest in acquiring literacy, although it would take several centuries for this evolution to be fully realised. The long gestation period for printing is a reminder that social factors influence the widespread deployment of technologies, right down to the nature of the script that a particular language utilises and the ease with which it can be put to use in large-scale production.

INDUSTRIAL AND POST-INDUSTRIAL SOCIETIES

What are now known as the first and second Industrial Revolutions spur new inventions that harness the power of water in varied forms, coal, oil, and other natural resources and amplify many times over the capabilities of human beings (Landes, 2003). While historians, sociologists, economists, engineers, and others will continue to debate the many specific cases and the convoluted histories of each innovation and its associated sociocultural milieu and attendant environmental factors, there can be no dispute that technological innovation is never fully determined by technological factors alone nor that societal values are the sole shapers of technological patterns, products, and systems. Some innovations, like the steam engine or the factory production system, evolve over many years with a multitude of scientific, technical, social, political, economic, aesthetic, religious, and cultural factors that interweave in a complex manner to influence technological innovation and evolution. No one factor predetermines the final outcome, and it is only in retrospect that any technological artefact or system appears inevitable since the many twists and turns it might have taken largely disappear or are impossible to recreate in an adequate enough manner to understand fully what might have been (Verbeek, 2005).

A former generation of scholars who studied technologies in society frequently advanced an *internalist* view that certain technical features of the technologies themselves largely predetermined technologies. A competing view that followed in due course held that technologies were entirely socially (i.e., *externally*) determined in a manner that reflected standard sociological theories of the time, which focused on power relations, social groups, and peer influence. Careful case studies of technologies, especially from the Industrial and Post-Industrial ages, have led modern scholars to realise that both views are naïve in the face of actual information from genuine cases that have been patiently investigated and documented (Clark, 2005; Ong & Collier, 2005). Robert Pool (1997), in a well-known study of the nuclear power industry in a monograph produced for the Sloan Technology Series, provides ample documentation for how the nuclear energy industry and its attendant technological developments were both shaped by society and shaped society. The various technologies that were developed, the manner in which they were evaluated and deployed, the ways in which the government regulated the industry, and the markedly different evolutionary pathways and varying degrees of public acceptance one finds for these technologies even today in different nations all speak to the powerful social influences on technologies, as well as the ways in which particular technologies create differentiated impacts in turn on human populations and ways of thinking about the technologies in question. Another clear principle from this case study is that often technologically superior solutions do not command wide assent due to nontechnical factors, such as sponsorship, political power, the ability of the inventor to scale the innovation, confidence people have in the inventor versus his/her competitors, and a host of other factors. The steam car, for example, was vastly superior in its technical qualities to the best gasoline engines in the early days of the automobile in the 1930s, yet it lost out to inferior products because of a combination of factors – both avoidable and unavoidable – that combined to doom its chances in the marketplace. Historical and sociological studies that dovetail with principles seen in the story of nuclear energy abound in contemporary scholarship, including those dealing with genetic engineering, the development of pharmaceuticals, electric power generation, highway construction, nuclear weapons, and a myriad of other industries and developments in the modern world (Pursell, 1995, 2005).

CONTEMPORARY UNDERSTANDING OF SOCIAL ASPECTS OF TECHNOLOGY

Modern technologies are complex and exhibit features and developments that invoke the discoveries of science while also being heavily influenced by engineering practices and principles (Vandenburg, 2005). The role of markets, availability of capital to finance ventures, and the specific impact of entrepreneurs who take on business risk in order to launch new enterprises designed to transform markets, has accelerated the pace of technological innovation in the nineteenth century and beyond.

Modern technologies bear only limited relationship to pre-modern ones and exhibit a number of qualities that are unique to the present time. Nearly all modern technologies reside within a complex socio-technical system of production and use that is both necessary to their successful operation and that also creates complex relationships such that ongoing evolution of the technology in question requires cosympatic evolution in one or more associated technological systems within which the artefact is embedded (Nye, 2006). Frequently, contemporary technologies are opaque to the user, exhibiting 'black box' qualities that are impervious to user understanding and intervention and requiring sophisticated analysis, troubleshooting, and repair by an expert. Some are not even designed to be repaired but are merely replaced. This has led some scholars, such as Albert Borgmann (2006), to make the case that these devices by their very nature distance us from reality, that is, the natural world as it truly is, and require us to live in a human-designed world that is largely shaped by ubiquitous technologies that are automated, regulated, articulated, and operated in a manner that requires a high degree of social trust and allows for very little user engagement or alteration in the kinds of experiences the technologies manifest and create.

Despite the opaqueness of these devices, their ready availability at a cost that is highly affordable creates a consumer culture of unprecedented scale, which is a hallmark of the modern world with even people of modest means possessing goods that in prior generations were only available to the rich and famous. This massive consumer culture, in turn, creates the opportunity for more rapid technological innovations and the expansion of economic opportunities that, in turn, fuels yet further conspicuous consumption. More recent recognition of the environmental costs associated with this rapid escalation in global consumption patterns has triggered legislative action as well as the growth of 'green tech', a technological solution to yet another set of problems partially brought about by prior technological development.

All technologies in our present day reflect inherent tradeoffs that confer benefits upon individuals and groups and impose burdens upon other individuals or groups. Their very complexity also results in unintended consequences that cannot be anticipated but that lead to further technological 'solutions', which in turn perpetuate ongoing evolution of the technologies in question (Rosner, 2004). Large-scale technologies require societal involvement to an unprecedented degree as compared to earlier and simpler technologies (although we should not think that ancient systems like those found within the Roman Empire were 'simple' in any straightforward sense of the word). Societal involvement in technological developments are largely expressed through intermediary institutions and individuals, including federal, regional, and local government, international laws and treaties, non-profit agencies that advocate particular platforms regarding technological developments, social critics, and technical associations and individuals with technical expertise in one or more areas implicated in the development of the technology in question (Veak, 2006).

Social Construction and Actor Network Theories

In recent decades, scholars have sought to understand more precisely the nature of the relationship between technology and society. This body of work has been a reaction to a prior orientation towards technological development largely focused on the presumed manner in which technologies inevitably and deterministically influenced society and individuals. Philosophically, we can view current research as moving one step farther back in the trajectory of technological development and studying how human beings and socio-technical systems exert influences on the development, evolution, and spread of new technologies. Two major theories have been formulated and subjected to various types of tests, although admittedly these 'tests' tend to take the form of case studies of particular technologies within particular cultures at specific time periods. The first view places a priority on the human social construction of technology, that is, social forces – which embody particular cultural beliefs and mores, desires, and acceptable approaches – define and largely determine the manner in which new technologies both arise and develop. These social forces are brought to bear on new developments through the individual values, mores, skills, and orientations of the technologists themselves but also by social groups (Norman, 2004).

This 'social construction' approach to technological development was heavily influenced by work in the sociology of science, especially David Bloor's (1991) theory of social causation. Well-known advocates of the social construction view of technological development are sociologists Wiebe Bijker and Trevor Pinch (1987). The theory not only provides a framework for explaining technological development but has also evolved a distinct methodology for analysis, following closely the so-called 'strong programme' in the sociology of science. This requires an analysis that carefully looks at how technical criteria are determined – and by whom – in the course of the development process, as well as a myriad of other steps along the path toward full implementation.

Many scholars these days adhere to a weak rather than a strong version of the sociology of technology, pointing out that these are very complex questions and not all alternative explanations for the varied paths chosen can be adequately understood or weighed. In part, this weak form was a response to the first trenchant critique of this position by Langdon Winner (1993) who found the strong programme too limiting in the scope of its inquiry and too simplistic in its notions of what factors might be relevant and how such relevancy is assessed.

The contrasting viewpoint to the strong or weak sociology of technology, known as actor network theory (ANT), accepts that human beings do indeed exert considerable influence on the development and evolution of technologies but also views nonhuman agencies, such as manufacturing systems and corporations, as possessing and exercising agency that also influences the role, direction, scope, and pace of technological change. It extends the sociology of technology approach by recognising the enshrinement of human values and approaches within legislation, financial markets, and a myriad of other sociocultural institutions and processes. ANT has been argued at greatest length and cogency by Michel Callon (2006; Callon & Raberharisoa, 2004) of the Ecole des Mines de Paris. Historian Thomas Hughes (2004) has also substantially contributed to this approach by his careful historical accounts of the evolution of large socio-technical systems and their impacts on varied actors. The degree to which these various actors are rational or irrational is one subject of current debate as well as how the relative influence of these varied actors can be measured and evaluated in terms of their impact on the technology in question. Johan Schot and colleagues (Geels & Schot, 2007; Kemp, Schot, & Hoogma, 1998; Schot, 2001; Schot & Geels, 2007) recently advanced an evolutionary model for technological development that seeks to investigate all factors within an evolutionary framework and the role that competition, succession, and other evolutionary concepts might contribute to further elucidation of these dilemmas of understanding technological change within society. They have also advanced a preliminary typology of socio-technical transition pathways for critical feedback. Whether actors act in a rational or irrational manner within technological systems design, creation, and implementation is under considerable discussion (e.g., Law & McNeish, 2007), as is ANT's usefulness in addressing contemporary developments in biotechnology and human biotechnological evolution (Baillie & Casey, 2005; Lantos & William, 2006; Smith & Morra, 2006).

CONCLUSION

Anthropologists, sociologists, philosophers, engineers, and social critics have all contributed to our knowledge about the complex manner in which society affects technological development and use, and the ways in which society in turn is affected by the technologies that are created and brought to the marketplace. Important English language journals that routinely contain articles germane to this chapter include *Bulletin of Science, Technology & Society*; *Comparative Technology Transfer and Society*; *Engineering & Technology*; *Science, Technology & Human Values; Techné*; *Technology & Culture*; *Technology in Society*; and *Zygon*, which explores in particular religious and theological considerations concerning modern science and technology. Although the roles of various actors in the technological drama of innovation and change can be described with increasing precision, many of the outcomes of technological change still cannot be predicted with any degree of reliability. Technology has from the beginning – and will continue into the future – to be a fundamental substrate of society.

REFERENCES

Baillie, H. W., & Casey, T. K. (Eds.). (2005). *Is human nature obsolete? Genetics, bioengineering, and the future of the human condition*. Cambridge, MA: MIT Press.

Bandmann, G. (2005). *Early medieval architecture as bearer of meaning*. New York: Columbia University Press.

Bloor, D. (1991). *Knowledge and social imagery* (2nd ed.). Chicago, IL: University of Chicago Press.

Borgmann, A. (2006). *Real American ethics: Taking responsibility for our country*. Chicago, IL: University of Chicago Press.

Braudel, F. (1981–1984). *Civilization & capitalism, 15th-18th Century* (Vol. 3). New York: Harper and Row.

Callon, M. (2006). Can methods for analyzing large numbers organize a productive dialogue with the actors they study? *European Management Review*, *3*, 7–16.

Callon, M., & Rabeharisoa, V. (2004). Gino's lesson on humanity: Genetics, mutual entanglements and the sociologist's role. *Economy and Society*, *33*(1), 1–27.

Cheek, D. W. (2005). Roads and highways. In Mitcham, C. (Ed.), *Encyclopedia of science, technology and ethics* (Vol. 3, pp. 1649–1654). New York: Macmillan Reference USA

Clark, D. (Ed.). (2005). *Theory of technology*. New Brunswick, NJ: Transaction.

Cowan, R. S. (1997). *A social history of American technology*. New York: Oxford University Press.

Geels, F. W., & Schot, J. (2007). Typology of sociotechnical transition pathways. *Research Policy*, *36*, 399–417.

Hughes, T. P. (2004). *Human-built world: How to think about technology and culture*. Chicago, IL: University of Chicago Press.

Kaplan, D. M. (Ed.). (2004). *Readings in the philosophy of technology*. Lanham, MD: Rowman & Littlefield.

Kemp, R., Schot, J., & Hoogma, R. (1998). Regime shift to sustainability through processes of niche formation: The approach of strategic niche management. *Technology Analysis & Strategic Management*, *10*(2), 175–195.

Landels, J. G. (2000). *Engineering in the ancient world* (Rev. ed.). Berkeley, CA: University of California Press.

Landers, J. (2003). *The field and the forge*. New York: Oxford University Press.

Landes, D. S. (2003). *The unbound Prometheus: Technological change and industrial development in Western Europe from 1750 to the present* (2nd ed.). New York: Cambridge University Press.

Lantos, J. D., & William, L. M. (2006). *Neonatal bioethics: The moral challenges of medical innovation*. Baltimore, MD: Johns Hopkins University.

Law, A., & McNeish, W. (2007). Contesting the new irrational actor model: A case study of mobile phone mast protest. *Sociology*, *41*(3), 439–456.

McClellan, J. E. III, & Dorn, H. (Eds.). (2006). *Science and technology in world history: An introduction* (2nd ed.). Baltimore: Johns Hopkins University Press.

Norman, D. A. (2004). *Emotional design: Why we love (or hate) everyday things*. New York: Basic Books.

Nye, D. E. (2006). *Technology matters; Questions to live with.* Cambridge, MA: MIT Press.

Ong, A., & Collier, S. J. (Eds.). (2005). *Global assemblages: Technology, politics, and ethics as anthropological problems.* Malden, MA: Blackwell.

Pinch, T. J., & Bijker, W. E. (1987). The social construction of facts and artifacts: Or how the sociology of science and the sociology of technology might benefit each other. In W. E. Bijker, T. P. Hughes, & T. J. Pinch (Eds.), *The social construction of technological systems: New directions in the sociology and history of technology* (pp. 17–50). Cambridge, MA: The MIT Press.

Pool, R. (1997). *Beyond engineering: How society shapes technology.* New York: Oxford University Press.

Pursell, C. (1995). *The machine in America: A social history of technology.* Baltimore, MD: Johns Hopkins University Press.

Pursell, C. (Ed.). (2005). *A companion to American technology.* Malden, MA: Blackwell.

Rosner, L. (Ed.). (2004). *The technological fix: How people use technology to create and solve problems.* New York: Routledge.

Schot, J. (2001). Towards new forms of participatory technology development. *Technology Analysis & Strategic Management, 13*(1), 39–52.

Schot, J., & Geels, F. W. (2007). Niches in evolutionary theory of technical change: A critical review of the literature. *Journal of Evolutionary Economics, 17*, 605–622.

Smith, M., & Morra, J. (Eds.). (2006). *The prosthetic impulse: From a posthuman present to a biocultural future.* Cambridge, MA: MIT Press.

Vandenburg, W. H. (2005). *Living in the labyrinth of technology.* Toronto, ON: University of Toronto Press.

Veak, T. J. (Ed.). (2006). *Democratizing technology: Andrew Feenberg's critical theory of technology.* Albany, NY: State University of New York Press.

Verbeek, P.-P. (2005). *What things do: Philosophical reflections on technology, agency, and design.* University Park, PA: Pennsylvania State University Press.

Winner, L. (1993). Upon opening the black box and finding it empty: Social constructivism and the philosophy of technology. *Science, Technology & Human Values, 18*(3), 362–378.

Dennis W. Cheek
Ewing Marion Kauffman Foundation
Kansas City, MO
USA

PIERRE VÉRILLON

17. TOOLS AND CONCEPTS IN TECHNOLOGICAL DEVELOPMENT

INTRODUCTION

In this chapter, I attempt to clarify for the reader as well as for myself some ideas concerning the nature, structure, and embodiment of thought and knowledge in technological communities and settings. This question has been a concern for me ever since the early 1980s when, as a psychologist, I joined a team of researchers and teachers involved in what we call in France the didactics of technology and vocational education. The term didactics doesn't have in French the pejorative connotation it has in English. For us, didactics refers to the scientific study of the teaching/learning process inasmuch as it pertains to a particular area of school knowledge. Thus, French didacticians (or educationalists) conduct subject-specific studies on such things as curriculum design, curriculum delivery, student/teacher interactions, pupils' misconceptions, teachers' epistemologies, etc. As a psychologist intent on contributing to the didactics of technology, my focus has mainly been on how people – especially children and adolescents – evolve within technological environments, how they develop technological knowledge and skills, and to what extent the incorporation of technology influences the way they interact with their material and social environment.

In order to develop some understanding of these questions, it is apparent to me that a twofold agenda needs to be addressed. First, the status of technological activity and artefacts within psychological theories and models of human interaction requires clarification, especially in relation to cognition and development. Second, reliable epistemological guidelines concerning the nature of technology as a specific social sphere of human thought, knowledge, and action need to be developed. This chapter reflects my progression in trying to address this agenda, including references to the ideas and scholars that influenced my views in the process. It is therefore necessarily a personal and – not being a philosopher or a technologist – a debatable vision. As such, it should be considered as an unpretentious contribution inviting discussion and incidentally giving a glimpse of some Francophone scholarship on the subject.

SCHOLARLY DISINTEREST FOR TECHNICS

In the early 1980s, it seems that questions relating to the influence of the technological environment and tool use on human learning and development were no longer in the focus of French psychological research. In educational psychology, literature concerning a child's development of technological and knowledge and skills was scarce when compared to the considerable amount of writing then available concerning early cognitive development in such fields as science and mathematics. As Weil-Barrais, a noted psychologist working in the didactics of physics, wrote in 1995:

> ... one can but observe the dearth of research concerning the interaction of children with technical objects... The specific nature of technical objects is generally unrecognised by developmental psychologists although it would justify elaborating a psychogenetic theory of

A.T. Jones and M.J. de Vries (eds.), International Handbook of Research and Development in Technology Education, 175–197.
© 2009 Sense Publishers. All rights reserved.

> the technical object ... This lack of scholarship is attested by the fact that no literature or handbooks of psychology are available on this subject. (p. 59)[1]

Concerning the epistemology of technology, there was also little to go on. In 1978, Gille published an *Essay on technical knowledge* at the end of his monumental *History of techniques*. In the introduction, Gille apologises for referring to this paper as an essay but, as he goes on to explain, the scarcity of learned discourse concerning the topic prevented him from writing anything more ambitious. In 1991, Godelier, a leading anthropologist, remarked that the existence of several "solitary magnum opuses" (p. 7) – he mentions those of Gille (1978), Leroi-Gourhan (1965), and Haudricourt (1987) – could hardly hide the fact that there lacked in the scientific community a collective and structured effort to develop a "science of technique" (p. 7).

This lack of interest in psychology and the social sciences for human technological cognition struck me as puzzling in a society so dependent on technology and in which technology is a major theme of everyday discourse in the media and in conversations. One could also wonder that the works of French authors such as Mauss (1936), Wallon (1941), Rey (1935), and Meyerson (1987), who in the 1930s had researched technical intelligence in apes and humans, seemed almost forgotten.

The Minority Status of Technics

Today, it seems to me that there are many converging reasons that account for this situation. One important reason is linked to the inferior social status conferred in our societies to those social agents most closely connected with technical skills: industrial workers, technicians, craftsmen, and artisans directly involved in the production process. Another reason is linked to the neo-positivist, 'intellectualocentric'[2] bias that considers the norms of scientific reasoning, knowledge and discourse to be the accepted or 'natural' standard for all modes of thought, and that 'deviant' forms of cognition have to be explained in reference to this standard. Significantly, Gille (1978) in his paper mentioned above, begins his reflection on technical knowledge by asking whether it is possible, legitimate, and conceptually correct to conceive of a form of knowledge that is not scientific. In the next section, I look into this problem of the legitimacy of non-scientific ways of thinking and knowing and try to shed some light on the historical conditions that led Gille to ask such a question.

School culture is partly responsible for fostering the idea of scientific thinking as the norm of all thinking. For example, French secondary school folk tradition distinguishes three categories that enable classifying students according to their 'natural' capacities and aspirations. Teachers, parents, and the students themselves seem to consider that one necessarily belongs to one of the three: 'scientific', 'literary', or by default 'technician'. It is tacitly agreed that because of their talent for mathematics, 'scientific' individuals exhibit intellectual superiority and that the 'literati' best represent the valued humanities culture. 'Technicians' on the other hand are generally considered to deal with matters both less demanding intellectually and of minor cultural importance.

Contempt for the Mechanical Arts

This partition of human intellectual achievement instituting an inferior status for technical knowledge and skills is a deeply rooted social and historical construction in the French – and perhaps the Western – worldview. For example, Simondon (1969) argues that culture developed as a defensive system against technics. Several French historians and philosophers of technology (e.g., Sigaut, 1987; Stiegler, 1996; Vérin, 1993; Vernant, 1965) argue that this situation finds its origin in ancient Greece with the emergence of modern lay scholarly discourse when philosophers began to break away from the theological and mythical explanations that then dominated traditional conceptions of the world. In their view, the Platonic and Aristotelian tradition laid down what were to become the basic characteristics of learned inquiry:

- the legitimate problems and thematics: knowledge, beauty, the city, the cosmos, …;
- the adequate methodology and media for addressing them: debate, discourse (the logos);
- the explanatory framework: order, equilibrium, …;
- the institutions of learned production and transmission: schools, academies, …; and
- the economic conditions: freedom from want and work (*skholè*).

Under these conditions *tekhnè* was ruled out both as a possible source of knowledge and as an object of inquiry. This ineligibility was as a result of three primary reasons. Firstly, theory considered knowledge as proceeding from a contemplative and speculative relationship to ideas. In contrast, material things could only be assessed through the senses or through material action which inevitably led to delusion. Secondly, from the point of view of ethics, knowledge was to be open, disinterested, and sought for its own sake. This disqualified *tekhnè*, which was considered to be servile, crafty, and calculating. Finally, at the social level, learned occupations belonged by right solely to free citizens (hence, the 'liberal' arts) whereas mechanical arts remained the vocation of slaves.

This negative perception of technical skills by the learned dominated until the Renaissance period. During the middle ages, scholasticism confirms the pre-eminence of discourse over empirical evidence. Attempts to alter natural phenomena or to experiment with nature were frowned upon by the Church and seen as adulteration or witchcraft (Thuillier, 1992). Similarly, military nobility considered contemporary engineering novelties in warfare to be un-chivalrous: the Old French word 'ingeniour' (engineer) is one of the terms then used to refer to Satan (Vérin, 1993).

With the advent of the Renaissance and the Reformation, things changed. Progress in agriculture allowed an increase in population size and the rise of cities. Improved navigation expanded the known world. Intellectually, a new mind-set developed in line with the ideas of early proponents of empirical inquiry of the medieval period, then often condemned for their works (e.g., Grossetête, Bacon, and Buridan). Galileo confirmed the break away from Aristotelian physics, although his ideas remained expressed in the form of Platonic dialogue. Francis Bacon and Descartes claimed that man must become "nature's lord and master" (Descartes). The ban on tampering with nature was lifted and 'natural philosophy' became a legitimate pursuit which would eventually lead, with Newton, to the development modern physics. Nevertheless, technology remained outside the sphere of scholarly interest. Layton (1991) mentions that for a brief period during the Renaissance – with Da Vinci, for instance – there seemed to be an aspiration towards a unification of science and technology but, as he goes on to say, the ambition of science soon became the production of knowledge *per se* regardless of its practical potentialities. The unification did not take place, and technology and science continued to follow their own separate paths of development. As Layton (ibid.) points out, the industrial revolution that developed in Great Britain a century after the scientific revolution owed, in fact, little to scientific knowledge.

In France, the period of Enlightenment furthered the process of the "secularization of knowledge" (Vernant, 1965, p. 178) begun with the Greek philosophers and revived by the Renaissance. However it often expressed a similar lack of learned interest – if not outright contempt – for technics. Sigaut (1987) mentions Voltaire's wonder that mankind's most intricate and useful artefacts were owed to 'stupid' craftsmen. In contrast, the Encyclopaedist movement sprung by Diderot and D'Alembert's monumental work constitutes a notable – if relatively short-lived – exception stemming from the awareness by a marginal fraction of the intelligentsia of its shameful ignorance pertaining to the world of industry and crafts. It is probably also the result of the State's long unheeded plea to the savant community that they inventory existing technology and contribute to the dissemination of technological skills and knowledge accumulated in the different fields of industry.

During the preceding century, Colbert, Louis the XIV's brilliant minister, had seen the craft corporations' tradition of technical secrecy as a danger to France's economic development and had consequently funded a project for a systematic description of all existing arts and crafts. At the time it hadn't raised much enthusiasm, so that the Encyclopaedia, a century later, can be seen as the first instance in France of a prolonged and thorough scientific inquiry into the technology of the industrious crafts and trades. It met with and overcame several novel theoretical and methodological difficulties: getting the craftsmen to talk about their trade, accessing their technical knowledge, and expressing in written form things that, as D'Alembert points out, are not learned through books. However, Gille (1978) claims that the breakthroughs in the verbal and graphical descriptions of the mechanical arts made by the Encyclopaedists in their attempts to give a comprehensive account of the crafts and technologies of their day, however adroit, nevertheless met their limits. For example, he considers that the Encyclopaedists were unable to accomplish more than a mere – although noteworthy – compilation of existing technology because they lacked a unified system of description and classification based on a theory of artefacts. Sigaut (1991a) shows that isolated attempts were later made in Europe at evolving such a theory of artefacts (for instance, by Beckman, Reuleaux, Babbage, Laffite, and Simondon). But, according to him, such a 'theoretical technology' – what Laffite (1972) terms a "mechanology" and Simondon (1969) a "general technology" – has yet to be invented. For Sigaut, this theory would be quite distinct from engineering theory or applied science in the sense that the latter can only tell us how artefacts work (what phenomena underlie the way they operate) but not what they are or how they come to exist (as phenomena in themselves).

An Epistemological Flaw

The Encyclopaedists' project revealed an ontological and epistemological deficiency in their approach to technical knowledge and artefacts that in a way still fails to attract attention and concern. It is the fact that technology constitutes a world in itself that has to be approached on its own within a specific framework. Auguste Comte's Positivism, that soon triumphed in the wake of the Enlightenment within the developing spheres of science and engineering, was insensible to this weakness because, as Sigaut (1991b) phrases it, for positivists "in a sense, there is no technology, there are only applications which are each but a subordinate sub-section of a given specific scientific field" (p. 76).

For Perrin (1991), "a conceptual shortage for thinking technology" continues to be experienced and, as a consequence, "the model which today dominates and inspires much technological research and policy, sees technology as the application of the natural sciences to the construction of artefacts" (p. 381). This echoes Schön's (1983) question: "How comes it that in the second half of the twentieth century we find in our universities, embedded not only in men's minds but in the institutions themselves, a dominant view of professional knowledge as the application of scientific theory and technique to the instrumental problems of practice?" (p. 30). In education, too, the positivist view blurs for students any vision that is conveyed of the science-technology relationship. Frey's (1991) warning that "this misconception about the nature of science and technology and about the relationship between them can be misleading at best and fatal at worst for technology education" (p. 16) would be quite relevant to the French situation.

My position is that the dominant positivist 'applied science' perception of technology occults the fact that in our western societies technology and science have, as Layton (1991) suggests, indeed followed fairly diverging paths since the Renaissance period. Communities of technicians, engineers, and industrialists have historically developed alongside communities of scientists but have actually remained quite oblivious of each other because ordinarily each community carries out its activities without having to rely directly on the other. As Vincenti (1990) points out, even in the fields of advanced technologies relationships between the two communities are scarce and opportunistic. The reason, for me, lies in the fact that the "overriding aims" (Jarvie, 1966, p. 387) that guide and organise these two communities differ quite fundamentally and impact both the way each community relates to empirical reality and the way it constructs its bodies of knowledge. In the next section, I attempt to analyse and illustrate these differences.

COGNITION IN SCIENCE AND TECHNOLOGY

The Differing Overriding Aims of Science and Technology

Science aims at developing an understanding of the world whereas technology aims at producing material advantages. Skolimowski (1966) sees science as motivated by cognition: it produces knowledge. It focuses on what is: reality. Theoretical development is what inspires scientific production; consequently, scientific progress results from the continuous adjustment of theories for the empirical data they attempt to explain.

In contrast, technology has a practical motive: it focuses on what should be or what could be. It creates reality (artefacts) according to design. What motivates artefact design and fabrication is the pursuit of anticipated interesting outcomes. As a result, technological progress is measured in terms of increased efficiency – the extent to which artefacts are able to produce the expected, advantageous outcomes.

The Differing Relationship of Science and Technology to Existing Reality

Science and technology share a realist vision of the natural world. They both presuppose the existence of an outside reality possessing sufficiently stable characteristics and predictable behaviour to allow for actions to be conducted with reliably anticipated outcomes. It is this shared agreement concerning the existence and the regularity of the world that has led to technology being seen as applied science. However the specific aims of science and technology lead each to develop a distinctive relationship to this world. As Frey (1991) points out, science considers the world as something to be known whereas technology sees it as something to be used. In its quest for unbiased truths, science strives to develop a neutral and disinterested relationship to reality, notably by keeping circumspect of subjective points of view and value judgements. Similarly, no aspect of reality is considered *a priori* as unworthy of scientific enquiry; anything can, potentially, become an object of knowledge.

Technology on the other hand sees reality either as an exploitable resource or, on the contrary, as an obstacle to overcome. As a result, for any given technical action, certain aspects of reality are valued, some repressed, and others ignored. For example, on a vehicle, technology will strive to reduce friction whenever motion is sought but will try to maximise it when braking is needed. Thus, technology's relationship to the environment varies according to the situation and the advantage that is sought. This subjection to situatedness characterises technology and contributes to the development of an opportunistic and cunning frame of mind, a sort of situation awareness that the

Greeks called *métis* and the Romans *ingenium*, and that was considered to be typical of technicians and engineers (Vernant, 1965; Vérin, 1993) yet scorned by academics.

Science and technology both have an interventionist approach to outside reality: both proceed through active and intentional alterations of the material world. However, these alterations are essentially different. Concerning human intervention on the world, Cellérier (1979) distinguishes what he terms 'epistemic transformations' from 'pragmatic transformations'. The former refer to acts aimed at detecting phenomena (probes, soundings, polls, ...) or at questioning reality (through scientific experiments as well as through a child's manipulations of his/her material and social environment). These epistemic actions aim at evidencing some distinguishing aspect of reality: through the transformation effected, an invariant (or a systematically variant) characteristic of reality is revealed. For example, a baby becomes aware through multiple trials that the objects he/she throws tend to have a downward trajectory. Generically speaking, epistemic transformations characterise scientific enquiry. Pragmatic transformations, on the other hand, aim at bringing about conditions or rearrangements of reality that are considered interesting because of the advantage they provide. Cooking food, contriving a tool or a shelter, and synthesising a new substance or organism are examples of such transformations that confer to some given matter certain desired features, not occurring naturally, thus giving it added value. Such transformations are characteristic of technology. Note that these transformations produce artefacts. The pragmatic manipulation and rearrangement of physical, chemical, or biological invariants in nature give rise to artificial beings conferred with novel, intentional, and anticipated characteristics.

The Status and Process of Knowledge in Science and Technology

Epistemic and pragmatic alterations of the material world are not identical when considered in their relation to knowledge. The former are performed in order to reduce uncertainty in relation to some aspect of reality, hence to produce information and explanations (theories). Action in this instance is in the service of knowledge. Pragmatic alterations, on the other hand, are carried out in order to produce an interesting effect, an advantageous condition, or a useful artefact. Here, knowledge serves action. Is pragmatic knowledge – that is, knowledge in the service of action – essentially different from descriptive or explanatory knowledge?

Pragmatic and epistemic alterations of nature are both prompted by conjectures or hypotheses, which are themselves suggested by the awareness of invariant properties acquired through observation and experience. However, in science, this conjecture (hypothesis) is in fact a first step in the process of producing theories, that is, statements relating to some area of reality that are considered to be true by the scientific community. It is a plausible anticipation of some truth suggested by observed invariants, and possibly by existing theory. The next stage in the process – the epistemic transformation embodied in the experimental protocol – is designed to verify or falsify the conjecture. If it is verified then the theorem can be adopted and it contributes to theoretical development (see Figure 1).

In contrast, in the field of technology, conjectures are design hunches or insights concerning a possible novel arrangement of invariant phenomenal properties enabling the production of some interesting effect or device (see Figure 2).

Models of the process or of the device are then built and tested to see to what extent the desired outcome is obtained. Feedback from the tests either invalidates the design hypothesis or contributes to the amelioration of the initial design and finally to the production of a new artefact. During the design process new knowledge is produced but, as Vincenti (1990) and Perrin (1991) point out, it is rarely produced *per se* and when it is, for example when development requires technological research, it is always motivated by – and fed back into – problematic situations connected with artefact design, fabrication or maintenance.

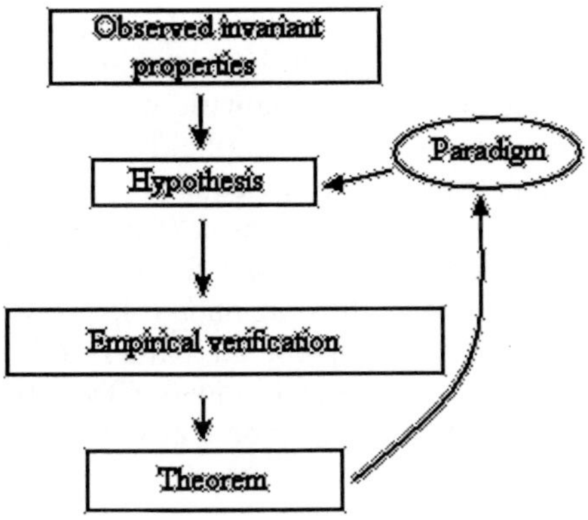

Figure 1. The epistemic process of knowledge production.

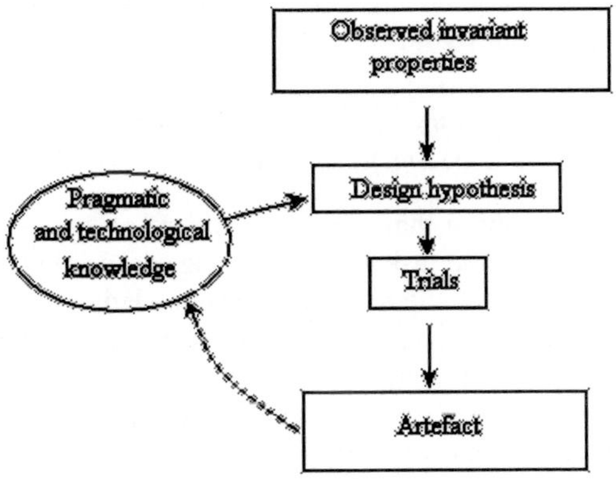

Figure 2. The pragmatic process of knowledge production.

Thus, the use and production of knowledge is theory-driven in science while in technology it is driven by performative action, artefact design, and tool use. Scientific knowledge is connected with the establishment of structured bodies of statements considered to be true and able to be empirically validated. These bodies of knowledge aim at consistency, non-contradiction, exhaustiveness, and generalisation. They generate new questions and conjectures that can in turn be validated or refuted. Scientific knowledge, because it is empirically grounded, can in principle be put to use for practical action and artefact design. But as Staudenmaier (1985) argues, scientific knowledge has to be deeply restructured before it can become appropriate for a technological context. Similarly, referring to the chemical industry, Layton (1991) points out that transferring research findings from the lab to the production plant requires solving considerable practical problems.

The performative or practical knowledge of technicians is, like scientific knowledge, an interpretation of invariants observed in dealing with reality. Just as with scientific knowledge, it is involved in the formation of conjectures. However, what is at stake is not the establishment of 'truth' but the attainment of success and effectiveness: "It works!" is the technician's equivalent of Archimedes' "Eureka!"

According to Bunge (1985), scientific knowledge is expressed in a nomological mode, that is to say, by the means of assertions that can be declared true or false. In contrast, technological knowledge, and performative knowledge in general, is worded in nomopragmatic form, such as in recipes: "In order to obtain X, do Y". Nomopragmatic statements are not judged in terms of veracity but in terms of effectiveness. Also, they implicitly refer to an agent, whether human or artefact, whereas nomological statements are expressed in an impersonal form.

From a true nomological assertion ("water freezes at 0°C") it is possible to derive efficacious nomopragmatic statements ("in order to prevent water from freezing, keep the temperature above 0°C"). The reverse is not true. For example, the pragmatic rule applied by the Montgolfier brothers in 1783 for the invention of their successful hot air balloon ("in order to obtain the upward movement of a balloon, fill it with the smoke from a straw fire") was based on an incorrect conjecture ("smoke as well as other vapours possess a levitation capacity"). However, in contrast with their contemporary, the physicist Charles who gave a correct account of the phenomenon in terms of atmospheric pressure, the Montgolfier's aim was not intelligibility but obtaining a spectacular effect. Historically, such instances of successful artefact design grounded on incorrect theory are numerous, even in breakthrough technologies (Vincenti, 1990). For example, many seamen today, although they know better, navigate with a geocentric conception of the solar system, viewing the sun, stars and moon as revolving over the horizon.

THE INSTRUMENT: A CONCEPT FOR THE PSYCHOLOGICAL ANALYSIS OF TOOL-USE AND TOOL DEVELOPMENT.

In the preceding sections I have tried to show that while the sphere of scientific work has in our societies been institutionally invested with the task of discovering, describing, and accounting for natural and social phenomena, it does not have the monopoly of knowledge production. The design, production, and use of material goods and services, which is the specific object of technology, also produces knowledge – which differs in its characteristics from scientific knowledge. Of course, other historically formed spheres of human activity exist under this conception such as, for example, those concerning artistic or religious activities. Each of these spheres or 'worlds' (Cassirer, 1942/1991) pursue their own 'overriding aim' and harbour their own worldviews, techniques, institutions, and evidently their own forms of knowledge. Each obviously also engages and determines cognition and purposive action in a specific way and the point I want to make now is that this has raised problems for psychological theory.

The Problem with General and Dualist Theories of Cognition

In the 1980s, as a result of a growing awareness of relatively autonomous 'worlds' and consequently regional ontologies and épistemès, some Francophone psychologists – especially educational (Vergnaud, 1987) and occupational psychologists (Leplat & Pailhous, 1975) – expressed dissatisfaction with general and unitary theories of cognition and development. Vergnaud pleaded for the development of subject-specific theories of learning. He contributed within the Piagetian paradigm to such theory for mathematics education. Leplat, on the other hand, a leading figure in work psychology, expressed dissatisfaction with the Piagetian framework itself, then dominating French psychology, arguing that tool-use, for example, could hardly be adequately described in terms of Piaget's assimilation-accommodation cycle.

Pierre Rabardel and I, as psychologists working together at the time in the field of vocational and technology education, were quite sensitive to the dilemmas exposed by Vergnaud and Leplat. Much like Vergnaud we felt we needed a theory of cognition able to take into account the specific characteristics of technological tasks and environments. However, along with Leplat we questioned the relevance of the Piagetian model for analysing processes involving technical artefacts (Vérillon & Rabardel, 1995; Vérillon, 2000).

The main problem posed by Piaget's (1937; 1967) framework for a psychology of technology lies in the impossibility for artefacts to acquire a theoretical status within this model[3]. Artefacts are an irrelevant and meaningless category in Piaget's system. I think there are three main reasons for this. The first is that Piaget's declared aim was to evolve a genetic theory of what he termed "the epistemic subject", meaning the subject as a knowledge builder. To this end, through clinical experimentation, he systematically studied the child's progressive construction of cognitive categories such as space, time, motion, logic, etc. He very convincingly showed the importance of the child's actions in the process of his/her conceptualisation of reality. However, his approach omitted the study of what might, in contrast, be called "the pragmatic subject", that is, the subject whose actions are not aimed at gaining knowledge but at creating rearrangements of reality useful and desirable in themselves (artefacts, in a wide sense).

A second reason for Piaget's omission of artefacts is that he approached the problem of psychogenesis much more as a biologist than as an anthropologist. He saw the development of intelligence as an adaptative process very much homologous to the biological evolution of organisms in changing environments. Artefacts in this framework are naturalised. In Piaget's experiments, identical to liquids, pieces of clay, wooden blocks, or plastic tokens, artefacts constitute an opportunity for the child to actively discover invariant properties both of surrounding reality and of the operations underlying his/her alterations of this reality. There is no social or historical dimension as such in this naturalistic conception of subject-object interaction, hence no specific status for the made world.

The third reason is that Piaget viewed his epistemic subject's relation to the world as a purely dualistic, unmediated, face-to-face interaction. In spite of the fact that, in real life, children interact with their environment through artefacts and people, Piaget, like other psychologists of naturalistic inspiration, studied and theorised cognition through a methodology focusing on the behaviour of single unaided individuals confronted with laboratory tasks that were not very meaningful. Artefact-mediated activity in this framework is unthinkable.

Of course this situation is not typical of the French context. In the United States, Bruner (1990), Norman (1991), Hutchins (1995), and others addressed similar criticism to cognitive theories seeking their inspiration in artificial intelligence. The positivist *epistemè* that we have seen underlying the dominant conception of technology as applied science, it can be argued, also underlies these logicist and empiricist theories that dominate psychology and approach cognition through the traditional dualistic partition of mind and matter.

Artefacts as Organisers of Activity

In view of the contradictions raised by these paradigms rooted in biology or information processing, I have tried to adopt a more anthropological outlook. This outlook is grounded on a phylogenetic consideration. It assumes that, while among the animal species the basic organisers of activity are formed, preserved, and transmitted endosomatically (i.e., through genetics), with the advent of the human species this process becomes exosomatic in the sense that organisers – tools, language, symbols, and also formalised spheres of knowledge, institutions, etc. – are externalised in the material and memorial culture. Thus, tools and symbols are artificial mediators that enable humans to carry out material and mental activities in their social and physical environment. They are also historical and cultural constructs that communities have evolved in order to regulate their members' interactions with one another and with things. As such, they constitute a network of pre-existing normative means of doing and thinking that are available to new members (youngsters, strangers, ...). During the process of integrating into the community, these new members eventually appropriate these organisers and mediators of activity. Lave and Wenger (1991) have shown that this appropriation takes place essentially through a joint novice-expert educative interaction organised by adults and/or peers within more or less formal institutional frameworks. In this sense, the transmission of a community's cultural mediators involves a 'reconstructive'[4] process through which every individual reproduces for him/herself, with the aid of the group and under its supervision, the characteristics and uses of that community's heritage.[5]

Clearly, language – along with other symbolic and iconic sign systems – plays a leading role among these pre-constructed organisers of activity in the sense that carrying out any joint action requires some sort of collective agreement regarding the aims, means, and conduct of that action and therefore some degree of verbal interaction among participants. However, in addition to this primary performative function, language enables secondarily the development of reflexive thought. Words can be used not only to refer to things but also to comment on and think about them. As such, they assume a conceptual function and enable the construction of organised and shared world-views within the community.[6]

Yet if language plays a major role in human activities, tools evidently constitute the most prominent and, so to speak, the paradigmatic class of cultural mediators of human activity. By this generic term we refer to all the material utilitarian artefacts produced by a community in the process of bringing about valued[7] and deliberate transformations of its environment. Artefacts are, as a consequence, both the means and products of these transformations.

The idea that human reflexive and performative activity is organised by historically and socially transmitted material, discursive, and semiotic tools contrasts quite significantly with Piaget's view of cognitive growth as a solipsist and autogenetic process. For example, within this and other naturalistic and cybernetic models, it is difficult to understand how learning and development can lead to anything other than idiosyncratic forms of cognition and behaviour. Whereas in the viewpoint I adopt, learning is on the contrary precisely a process – never fully attained – of drawing away from idiosyncratic modalities of interaction. For example, the baby's initially unorganised and uncoordinated perceptive and motor activity is progressively structured and shaped to social standards by adult-initiated verbal and gestural play. Later, through imitation and symbolic games with peers and adults, as well as through schooling, the child picks up the technical usages and skills of his/her community.

Learning is, then, essentially a dialogic process (Bakhtine, 1984) in the sense that our own gestures, words, and tools are originally those of others. They are the words and tools that we have seen others use in different situated contexts. With these situations in mind, we reconstruct them as vehicles (Bates, 1979) of our own activity. Tutors help us develop and use our imitative abilities (Bates, 1976, 1979; Vygotsky, 1934/1985) to replicate situated verbal utterances and technical manipulations. Word-use and tool-use are always both a replica of – and a response to – others' uses of those words and tools. They always refer to a pre-existing corpus of verbal and technical occurrences.

This model, which is inspired by activity theory (Clot, 2000; Engeström, 1987; Leontiev, 1975; Vygotsky, 1931/1978; 1934/1985), differs quite extensively from the previously discussed dualist and unitary models:

- because culture is a key element of the framework, it allows for the existence of multiple forms of cognition and knowledge;

- it affords artefacts a decisive psychological status as mediators and organisers of material and mental activity (artefacts materialise our thoughts and knowledge but in turn they shape our cognition); and
- it stresses the similarities between material and semiotic artefacts.

In this 'materialist epistemology' (Baird, 2002), for example, concepts are considered, like material tools, as artefacts. They mediate our mental activity. They help us organise and connect our representations of reality. They enable sharing and discussion of these representations with others. Inversely material artefacts are, like concepts, also tools for thought. Thought is also in our material actions. Tool-use can be seen as 'realising' our thought in the material world in a way similar to the way words 'realise' our thought in speech (cf. Vygotsky, 1934/1985).

The idea of mediation is of course what essentially characterises this model and sets it apart from others. Two intertwined forms of mediation are seen as organising our relationship to the world: mediation by others and mediation by artefacts. Figure 3 shows how we share a common world with others by sharing artefacts.

Recognising the importance of mediation discloses new questions for psychological research. One is understanding how these cultural organisers of material, discursive, and semiotic activities are transmitted and appropriated within human groups. In continuation with Vygotsky's (1934/1985) seminal idea of proximal development zones, writers such as Bruner (1983), Wertsch (1985), Griffin & Cole (1984), and Lave and Wenger (1991) have contributed to documenting and conceptualising the role of others in cultural mediation.

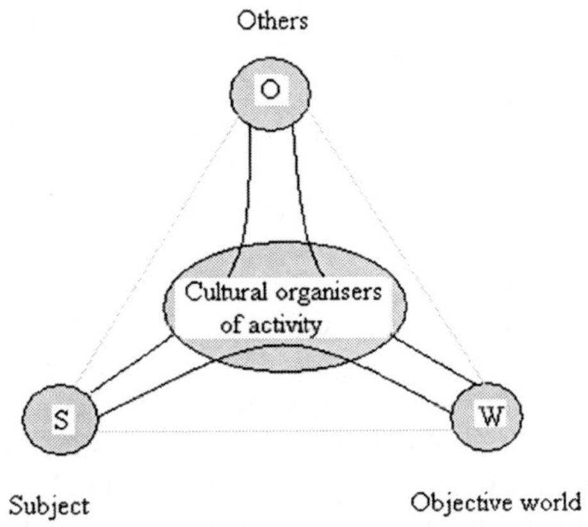

Figure 3. Cultural mediation.

Instrumental Genesis

Just as important as the issue of their transmission, is the issue concerning how at the individual level these external organisers (language, concepts, tools) are actually internalised in order to function as mediators in our dealings with the social and material world. In this section, I will present a psychological theory of instruments that Pierre Rabardel and I have been developing since 1985 and which addresses this issue. The idea we have tried to put across is that the individual (re)construction of artefacts and their uses requires a joint and articulated process of function assignment and semiotic and schematic development. Rabardel (1995, 1999) has termed this process "instrumental genesis".

Perhaps before going into an analysis of individual instrumental genesis, I can first give a hint of what it is by illustrating the effects it has in the long term on the evolution of artefacts and artefact use. To me, the anthropological and historical evidence of technical evolution shows that instrumental genesis is a very basic human process. Technical evolution is not an intrinsic and autonomous tendency of artefacts as Simondon (1969) sometimes seems to argue. Nor can it be seen as the sole result of professional design and invention. When we consider the 'normal' artefact-mediated intercourse of people at work, play or study, we see that neither the individual users of a technology, nor the technical devices themselves, remain unchanged during tool-use. Despite their physical and/or normative and conventional properties, which favour their enduring nature, tools (and concepts in a similar way) continue undergoing changes and alterations when they are put to use – especially in novel or unusual contexts. In order to carry out their activity, individual and collective users are continuously extending, restricting, and otherwise transgressing conventional uses and, when possible, altering artefacts[8]. Technical change (and linguistic change) eventually occurs when some of these innovations become normalised, thus contributing to the renewal of both artefacts and usages.

On the other hand – and conjointly – during technical change, the users themselves evolve as they alter their means of action: they develop skills, modify their world-views, expand and deepen their scope of action, anticipate new possibilities, etc. Thus, at the individual level, the outcome of technical change is a threefold process of development: artefactual development, cognitive development of the individual/collective user, and development of the target activity. This is what we mean by instrumental genesis.

Here is an example from history. Writing, as a technology for recording and communicating discourse, provides many historical instances displaying this mutual link between cognitive change, artefactual innovation,[9] and the development of activity. For example, considering the activity of reading, Illich (1991) claims that up to the 7th century, due to the absence of breaks between written words, silent reading as we do it was unimaginable: "There was almost no other way of reading than rehearsing the sentences aloud and listening to hear whether they made sense... The sense remained buried in the page until it was voiced" (p. 36). Consequently, reading was necessarily a collective activity. Only when breaks were introduced for didactical reasons – they were supposed to help novices acquire Latin vocabulary – did silent reading become, as a side-effect, an unintended possibility. In the *Confessions*, Augustine tells of his surprise as he discovered that he could read noiselessly at night without waking his brethren. As the practice of silent reading and writing spread among learned communities, a personal – that is, no longer public – relationship to texts became possible. It enabled readers to distance themselves 3and develop independent interpretations of written material. In return, these new reading practices led to further artefactual innovation: Paragraphs were introduced, chapter and verse were numbered, etc. In present-day school contexts, similar changes can be observed at the individual level among young children developing reading skills. We have also observed such changes among pupils in technology education (Rabardel & Vérillon, 1985).

Let us now see how we can analyse this process of instrumental genesis at the individual cognitive level. In order to account for the process of cognitive and physical integration of artefacts in tool-use underlying instrumental genesis we have conceived of a psychological entity that we have termed 'instrument' (Rabardel, 1995; Vérillon & Rabardel, 1995). Mounoud (1970) had already used the term to denote "any object that a person links to his/her action in the process of carrying out a task" (p. 16).

In everyday situations, instruments (utensils, tools, machines…) are material objects designed for the purpose of carrying out a particular class of actions. As such, an instrument in Mounoud's sense is what we call a material artefact – that is, literally, a product of human design. And an artefact-mediated action, in the context of a given task, is intrinsically an intentional material transformation aiming to bring about a desired state of ambient reality. It is a physical, chemical, or biological transformation that confers to the object of action some advantageous characteristics (cooking meat, turning metal, transporting goods…). In this sense, an artefact can be defined as "a structured set of invariants (physical, chemical, morphological, relational, conventional…) intentionally organised to bring about an anticipated transformation of the material or social environment, when brought into play by a user" (Vérillon, 1996, p. 62).

A Two-Fold Entity

From a psychological point of view, the fact that an artefact has to be brought into action and operated purposively by a person in order to perform a desired effect is obviously the crucial point and has led us to redefine the notion of instrument. In our view, the fact that tool-use requires some degree of operative participation on the part of the user indicates that the artefact, as a material or semiotic construct, is in itself only a partial component of instrumented action. The user's often complex cognitive and gestural contribution to the process is the other component. Thus, an artefact becomes an instrument – is instituted as an instrument – when it is purposefully appropriated by someone in order to mediate his/her action. In other words, an ancient tool that no one knows the meaning of is no longer an instrument, though we know that it is still an artefact.

An artefact becomes meaningful and instrumental only through the intentions, the representations, and the operations activated by the user in the process of engaging it in some purposeful activity. This is why we consider that an instrument is in fact a two-fold entity (Rabardel, 1995): Both artefactual and psychological. Instrumental genesis, then, is the cognitive process that underlies the conversion of an artefact into an instrument. It is as a two-way developmental process well conveyed by the idea of appropriation. On the one hand, in the direction of the subject, it entails appropriation in the sense of having to integrate a new organiser of activity within one's existing means of action (motor schemas, representations, etc.). On the other hand, in the direction of the artefact, it entails appropriation in the sense of assigning a function to the tool and adjusting it to the ambient characteristics of the task at hand. In other words, instrumental genesis reveals a zone of potential development (Vygotsky, 1934/1885) created by the tension between the constraints imposed by tool-use on the user's manual, perceptive, and cognitive skills and the expanded possibilities of action afforded by the artefact (Vérillon & Rabardel, 1995).

Instrument and Technical Development

Consider middle school pupils familiarising themselves with a hitherto unknown artefact, a lathe (Vérillon, 2000). Their attention and first manipulations initially focus on the transversal and longitudinal handwheels. Manipulation of the wheels causes the slide-rest to move and leads pupils to identify the possibility of directing tool motion. However, coordination of the two wheels in order to obtain control requires a bit of time. In terms of instrumental genesis, this process involves mostly subject-oriented development, that is, integrating a novel organiser of their activity within existing schema and representations.

Pupils then draw their attention to tool-workpiece interaction (instrument-object interaction). A number of pupils initially view this interaction in terms of abrasion (metal cutting seems unimaginable to them). Their procedure consists, with one wheel, in bringing the tool in contact with the rotating workpiece and then, very slowly, with the other wheel, in "wearing away" (their terms) matter along the part. These pupils, when asked how a tapered form might be obtained, suggest progressively longer wearing periods towards the extremity of the piece. Here we see artefact-oriented development: The pupils assign a function to the lathe – shaping matter through abrasion. Note that subject and artefact-oriented development are interdependent: Assigning functions is linked to existing representations (you can't cut metal). At this stage, these pupils have converted an artefact into an instrument. Evidently, through further practice on their own, or more likely with the teacher's help, they can eventually develop it into an instrument more in line with what is considered in the community of technicians as turning on a lathe.

As shown in this example, the construction of an instrument is often a difficult process that can – in areas like work and sports – develop over a long period. What I want to stress now is that, seen as a zone of potential development, it is an important and probably neglected source of knowledge. From what we have presented it is possible to distinguish two different, albeit partly connected, zones of potential development for technological knowledge:

- A zone of operative development, resulting from what Rabardel (1995) terms the "instrumentation" of the artefact. As artefact use develops, new unsuspected effects are obtained that, in order to be reproduced and controlled by the user, require that he/she construct new schemes. Conjointly, new characteristics (new invariants) are revealed that entail a modification or a renewal of existing representations bearing on the artefact, as well as on the reality with which it interacts and on the ongoing transformations. New aims of action can then be construed by the user, possibly powered by new motives; and
- a zone of artefactual development, linked to the fact that artefact use is in general guided by the expectation of a valued effect that by design the artefact is supposed to produce. Thus, during artefact use, the user is led to assign purpose and functions to the artefact. In the course of this process, termed "instrumentalisation" by Rabardel (ibid), the user to some extent constructs causal links articulating the expected effect, certain ambient properties of the situation, and certain morphological and physical properties of the artefact. These causal links may be quite different from the ones the designer actually had in mind, as in catachreses (using a wrench as a hammer, for example).

In general, because contexts change, the functional adjustment of the artefact in use varies from one instance to another and never completely complies with the aims of the designer. Consequently, one might say that artefacts, because of differing contexts, are in a way being continually redesigned by their users. As Rabardel (1995) points out, each singularly situated use of an artefact actually constitutes a continuation of the design process. While not necessarily involving an alteration of the artefact, it nevertheless results in a reassignment by the user of the causal and functional properties initially conceived by the designer. Sometimes this functional reassignment can go quite far – through catachresis, for example (i.e., by deliberately transgressing the conventional or prescribed usage attached to it, as in using a plastic canoe on a ski slope). And, of course, this reassignment can eventually lead to an actual material rearrangement of the artefact by the user (e.g., altering a garment, customising a car). In doing this, users expand the instrumental field – the scope of enabled activities – originally intended by the designers.

Another example of the relationship between instrumental genesis and technological development was highlighted earlier with Illich's (1991) example of the development of medieval reading practices. There is thus a dialectical relationship between instrumental genesis and activity. As Leontiev (1975) points out, "An activity realised is always richer and truer than the consciousness that anticipated it" (p. 143). Unsuspected possibilities are revealed that bring about changes in the user's representations, schemes, aims, and motives. Thus, a cyclical movement of mutual development and renovation is set off in which the development of activity calls for added artefactual and operative progress that in turn unveils novel and unanticipated prospects for future activity.

TECHNOLOGICAL CONCEPTS

The last part of this paper deals with artefacts that are designed to mediate action no longer upon material reality but upon other people. Human activity, because it generally involves peers, also entails communication that can be seen as an intentional transformation of another's mental state or representations. The instrument of this transformation – a part of the body (mouth, hands, …) or an artefact (writing, graphics, signals, …) – fulfils a semiotic function. It generates specific stimuli (gestures, figures, sounds…) that bear a shared, socially-constructed meaning among addressors and addressees. We refer to this kind of artefact as a 'semiotic artefact'. Semiotic artefacts mediate actions of different kinds. Utterances like "go walk the dog", "trim the mainsail", or "scalpel!" are for instance acts aimed at obtaining from someone an action regarding some component of the environment referred to as "dog", "mainsail", or "scalpel". The mediating artefacts in these examples are words. They mediate speech acts (Austin, 1962), which we call orders. They are shared semiotic devices, specific sounds we learn to make and that by convention and communal agreement refer to situations, objects, and actions. Other semiotic devices such as gestures or traffic lights work in the same way. Like all artefacts they are assigned agentive functions: They are structured to enable us to do things. Searle (1995) calls this agentive function a status function in the sense that the purpose of these objects is, by communal agreement, to represent or stand for something else.

Orders are of course not the only acts we can carry out with words and other signs although they are among the first instrumental acts we learn to master in infancy (Bates, 1979). Words can also be used to convey information, discuss, or make an inquiry as in the following utterances: "Dogs are mammals", "Mainsails are not necessarily the largest sails", and "What is a scalpel?" In such instances words are used to mediate categorisation, a complex mental operation that distinguishes between objects according to significant (meaningful) characteristics. It enables generalising the characteristic invariants of a class to a sub-class (for example, mammals to dogs). It implies the ability to extract invariants (the stable class of things called sails), to construct relevant descriptors (size is a meaningful descriptor of sails), to note variations (sails can be big or small), and to construct and compare classes (main and other sails). In all cultures children are trained for these operations at a very early age. Awareness of invariance also enables theories – that is, predictions – and anticipations that in turn enable the organisation and development of efficient material and virtual (in the mind) actions.

Concepts as Organisers of Activity

When words are used to refer to and organise knowledge, they contribute to conceptualisation and can take on the status of concept. At a primary level words (dog, sail, scalpel) refer to objects. At a secondary level, the same words refer to knowledge about objects. Conceptualisation entails the secondarisation of language (Bakhtine, 1984; Bronckart, 2001), that is, the ability to use language as a means for reflection.

According to Vergnaud (1985) concepts articulate three entities: Situations, invariants, and signs. Concepts are related to situations that give them meaning in the sense that the production, transmission, and use of knowledge is necessarily situated. It is notably related to the social institutions and spheres of activity (see above) that can organise and use knowledge in specific ways. For example, school science knowledge is not organised in the same way as scientific knowledge. Similarly, the designer's concept of 'sail' is necessarily different in structure, content, and extension from that of the sailor who handles it, although they may overlap. Consequently, thinking about dogs as mammals has a specific sense in the context of school biology, just as thinking about sails has in a design situation, and about scalpels in a surgery ward.

'Invariants' are the central component of concepts (Vergnaud, 1985). They refer to the stable aspects of reality upon which we can ground operations, rules, and anticipations. For example, Vergnaud (ibid) shows that when adding objects to an already counted collection, children are able at an early age to state the new total without having to recount the whole. He argues that in such a case the children use a 'theorem in action' that says that the cardinal of a collection resulting from the uniting of two subsets is equal to the sum of the cardinals of each subset. A teacher can perceive the child's action as evidencing the concept of addition.

Signs are the semiotic devices that according to Vergnaud (1985) "represent" the concept. In the above example, the additive transformations are carried out through the use of vocalised number words, written digits, diagrams, and of course fingers. Vergnaud (ibid), in the Saussurian or Piagetian manner distinguishes "signifiers" – the signs – and the "signified" – here the invariants.[10]

As mentioned above, all spheres of human activities require knowledge and socialised forms of exchanging knowledge, so that all resort in variable extent to conceptualisation and concepts. However, not all knowledge is easily expressed in words and – especially in practices involving gestural skills – knowledge remains tacit and has to be learned through media other than books or oral teaching, such as apprenticeship. Consequently, specific communities have developed specific relationships to knowledge in relation to their purposes and means. The advent of communities specifically organised for the purpose of developing and organising knowledge *per se* – scientific or scholarly communities – is fairly recent. It is predated by communities of practice involved in material production, which produce knowledge in order to enable efficient actions. It may be interesting to compare these relationships.

Scientific communities produce knowledge in discursive forms which is ultimately made public within the community and beyond through written publications. Conceptualisation as a means of organising knowledge is a dominant activity in this context. It underlies the design of theories and determines the methodological and technological means that enable evidencing new invariants in reality. Scientists rely extensively – like members of other communities – on material technology in the pursuit of their aims. Electron microscopes, space probes, and synchrotrons enable extremely complex material transformations that are not sought for themselves, but for the invariant realities they reveal and for the theoretical and conceptual arrangements of knowledge that they enable either to validate or disconfirm. In the sphere of science, technology as well as the specific forms in which knowledge is expressed – networks of propositions, mathematical equations, models, and diagrams – is determined by the overriding aim of producing and presenting truthful, general, connected, non-contradictory, and sparsely formulated knowledge.

Some scientists and epistemologists seem to believe that scientific knowledge – because it aims at universality – could, should, and eventually will replace other forms of knowledge which in a way they consider to be vestigial. This posture overlooks the fact that the generation and use of knowledge in other spheres of activity are also determined by the tension between that sphere's overriding and generic design and its ambience (Staudenmaïer, 1985). The way knowledge is encapsulated in language and tool-use is dependent on the specific aims, needs, and constraints of the work process pertaining to that sphere.

The sphere of technology, because it shares with science a realist approach to reality, could be imagined as an obvious consumer and user of scientific knowledge and concepts. The history and sociology of technology show that this is not the case. Vérin (1993), in her study of medieval and post medieval engineers, argues that while scientists strive to find unity in the face of diversity, technicians have to deal with particular situations and how they evolve locally. This is reflected by the fact that scientific concepts tend to free themselves from local determinations whereas technological concepts are useless if they can't be particularised. Similarly, Staudenmaier (1985) remarks that "design concepts are radically incomplete when they remain on the abstract level" (p. 104). Consequently, in order for scientific concepts to contribute to technological knowledge, they must be appropriated and restructured according to the specific demands of technological design.

Vincenti (1990) investigates the knowledge of aeronautical engineers in the first half of the 20th century. He shows that even in this advanced field of technology, much of the knowledge needed for designing propellers or airplane wings is generated inside the field and not imported from science. He also shows that compared with scientific knowledge, technological knowledge is generally partial knowledge because once a sufficing solution to a problem exists, nearby areas of ignorance remain unattended. This reveals the distinctly opportunistic nature of technological knowledge. Engineering theory may be what most resembles scientific knowledge in structure and in its forms of expression – propositions, mathematical equations, etc. – however, as Layton (1974) notes, engineering science makes statements about devices, not about nature. The conceptual focus is on artefacts in an *ad hoc* and pragmatic perspective quite at odds with that of science.

Pragmatic Conceptualisation

The form of technological knowledge most distinct from scientific knowledge is represented by the bodies of experientially-learned knowledge and skills of the practitioners that are the most closely engaged in the material transformations mediated by technology. This knowledge is encapsulated into rules of thumb, in maxims describing processes, and in general principles expressed in non-theoretical language. Pastré (1999), a psychologist involved in adult vocational training, has coined the term "pragmatic concept" to refer to this kind of technological knowledge that owes little to codified scientific or technological knowledge. He began using the idea of pragmatic conceptualisation to refer to the shop-floor terminology used by operators of injection presses in the plastic industry. In order to communicate about injection problems and discuss conjectures about the invisible mechanical and physical transformations and phases during the process, they used a connected network of expressions not derived from the vocabulary of engineering or even that of technical training.

According to Vidal-Gomel & Rogalski (2007), pragmatic concepts are "schematic and operative representations constructed for – and through – activity and which are the outcome of a historical and collective process transmitted essentially through experience and among fellow practitioners" (p. 51). Pragmatic concepts have been mostly studied in situations of process management and monitoring such as in the nuclear and chemical industries, piloting aircrafts, or fighting forest fires. In such contexts, workers need to construct a working mental image of what is going on. They also need descriptors of the ongoing process in order to be able to intervene appropriately so that the process can be carried out smoothly. These descriptors concern knowledge of the desirable and undesirable configurations of the process in order keep it from drifting towards an undesirable situation and also in order to recover from such a situation. In addition, a process often undergoes a change of phase during which the phenomenology of the situation is altered, for example, during landing in aeronautics or during casting in metallurgy. These phases have to be identified because phenomenological changes alter the conditions of action and modify the appropriate responses. Pragmatic concepts contribute to forming this working mental image that Pastré (1999) terms the conceptual structure of a situation, an extended interconnected network of concepts, indicators, and operations. In a way, pragmatic concepts resemble what Vygotsky (1934/1985) called "everyday concepts" in the sense that they are experientially rich, refer to much tacit knowledge, and would be useless in a scientific context.

The ancient sphere of sailboat transport, with its specific terminology, provides a good illustration of pragmatic knowledge and concepts. Working a sailing vessel is a collective activity and the different participants need to share a common representation of the ongoing process. Three major interacting entities are concerned: the boat, the sea, and the wind. Each is described with a very elaborate vocabulary. Wind – the driving element – can be a scientific concept for meteorologists, an engineering concept for sail designers, but for sailors it is a pragmatic concept. Two descriptors of the wind are particularly important: One is the 'lay', or direction of the wind, which among other things determines the course of the boat, the desirable configuration of sails, the relationship to 'leeward' obstacles, etc. The other is the 'strength' or 'force' of the wind. Seamen develop a very keen perception of these parameters directly through their own senses, through their effects on the boat ('list'), and through artefacts such as wind vanes and anemometers. The Beaufort scale was a pragmatic attempt in the early 19th century to devise a numerically-graded descriptor of wind force. It is a 12-point scale originally based on the amount of sails a ship could carry. In the 20th century, the scale was correlated to wind velocity. Although the wind is described by the ship's instruments and weather forecast in terms of wind speed – that is, the distance covered by air in one hour – it continues to be conceptualised by sailors in terms of force. Descriptors of the boat itself and of its behavioural characteristics are numerous. The spatial description of the vessel in terms of 'forward', 'aft', 'port', and 'starboard' afford an external reference frame for the crew members thereby eliminating confusions resulting from the use of individually centred references such as 'left' and 'right'.

According to the ship's heading in relation to the lay of the wind, the physical interaction between the boat and the wind and sea changes. It alters the boat's behaviour and way the boat has to be steered and the sails trimmed. Sailors distinguish and can identify several states or phases that determine the crew's actions: 'Reaching' when the wind comes from the side, 'running' when it is astern, and 'tacking' when the boat is sailing windward. Sailing windward requires a lot of attention from the crew because sailing too close to the wind can stall the boat, a situation from which it is sometimes difficult to recover. On the other hand, steering away from the wind in this situation can cause the boat to list excessively, which is also undesired. Under such conditions, the novice's attention is drawn to the physical perception of the wind, the 'feel' of the helm, the list of the boat, and a flutter of the sails, which signals that the boat is being 'caught by the lee' and nearing the stall condition.

From this example, it is possible to understand how pragmatic concepts develop. Expert seamen extend their conceptual field by increasing and enriching their knowledge of relevant phenomena through experience of varied situations and of their own varied involvement. Through a dialogue with renewed situations – but also with other seamen (and through internal reflective dialogue) – they acquire and generate new descriptors and refine old ones. They also develop an increasingly richer network of relationships among concepts and descriptors. Relating and associating observed phenomena and assigning them relevance appears to be a major characteristic of pragmatic knowledge. Causality, on the other hand, seems less developed in pragmatic conceptualisation.

Educationalists and scientists generally tend to think that pragmatic thinking can be enhanced by integrating concepts from engineering and science. Better causal understanding through scientific explanations and models might lead practitioners to more efficiency. In vocational training this is, of course, an important issue. It raises at least two questions: To what extent does causal understanding lead to greater efficiency? And can scientific and everyday concepts come together? (The latter is a question already asked by Vygotsky, 1934/1985.)

Understanding the physics that makes a sailboat work doesn't mean one can steer one. But, on the other hand, many good sailors may have no, or even false, explanations for their efficient action. Pragmatic concepts emerge and develop during the course of action in situated contexts. They acquire functional meaning in the situation. This means that scientific or engineering models could theoretically become meaningful locally if, during action, they could graft on to some existing functional knowledge. In a way, this is what happens when university trained technicians begin to acquire on-the-job experience. In the course of practical action, part of the scientific knowledge learned in class acquires functional meaning becomes 'pragmatised' (Pastré, 2005; Vidal-Gomel & Rogalski, 2007). In sailing schools, notions and diagrams from mechanics are sometimes used to study the functioning of sails. The resultant force produced by the sail is graphically shown to be resolved into two components: one in the direction of the desired course (drive) and one normal to it (leeway force). Depending on the setting of the sail and the care taken in trimming it, the diagram 'shows' that it is possible to maximise drive in force and direction and minimise leeway. My experience is that such a model can be helpful under two conditions. The first is that the diagram (as a semiotic device) is understood and mastered, that is, it becomes a semiotic instrument for the practitioner. This requires quite a bit of abstract school knowledge. The second is that it be articulated with existing experience. In this case, perceptive descriptors already acquired through pragmatic experience of sail behaviour can merge with the new descriptor and take on a new dimension. As Vygotsky (1934/1985) wrote: "An everyday concept must acquire a certain level of spontaneous development before the superiority of the scientific concept can become evident" (p. 286).

CONCLUSION

In this chapter, I have argued that the knowledge developed and shared by communities of technologists differs from that of other communities, and notably from that of scientific communities. Drawing from Staudenmaier (1985) and others, I have suggested that cognition in technology is specifically determined by the tension between technical design and its ambience. This tension can be experienced at any level of technological action. For example, design, as a cognitive activity, cannot be restricted to the emerging process through which artefacts are developed, that mainly concerns designers. On the contrary, with Rabardel (1995), I argue that artefact design is a wider process that continues beyond this emerging phase and concerns users as well as designers. For example, initial design work is prolonged whenever artefacts require maintenance or when users adapt them to their own needs and local ambient constraints. In this sense, the production and activation of pragmatic and ambience-specific knowledge constitutes an essential process of all technological praxis.

This raises questions for philosophers as well as for psychologists and educationalists, especially when dealing with vocational and technological education. Epistemic cognition underlying the processes of knowledge production and transmission in scientific contexts has implicitly become the norm to which all human knowledge is compared and through which it is assessed. In France, technology education is increasingly associated with science education and could well end up being a field of application for scientific knowledge. The claim that technology requires and produces specific and indigenous forms of knowledge not necessarily derived from scientific knowledge is not an accepted idea, even among technology teachers. This makes it necessary to take a closer look at the nature of technological cognition: How it develops, how it works, and how it is transmitted.

Technical intelligence – "artful practice" as Schön (1983) names it – has often been denied or demeaned. From ancient Greece through the medieval period, craftsmen and 'artists' (technicians) were originally considered by the learned either as ignorant users of routine on the one hand or devilish tricksters on the other – characteristics that endure implicitly in some present day ideologies concerning technology. During the Renaissance, however, a shift occurred with the advent of modern philosophy and mechanics that fostered the idea – still dominant among positivists – that technical problems could – and, in principle, should – be solved by science. Within both these world-views there was little allowance for the possible existence of autonomous forms of technological knowledge, which explains the relative paucity of literature on cognition in technological contexts.

In the past decades technology has become more sensitive economically and politically and there has been a renewal of interest in philosophy and the social sciences to explain the ways in which people and technology interact and, in the process, change. Drawing from the literature and from my own work as a psychologist involved in technology and vocational education research, I have attempted to give my personal view on how human thought is simultaneously challenged and shaped by technological problems, mediated and developed through technological artefacts, and generalised and shared through technological and pragmatic concepts.

In all domains of human activity, problems are at the outset of cognitive change. As such, problem-solving in technology has long been seen as a simple process of applying existing scientific knowledge to problematic situations. However, as I have tried to argue, problematic situations are not at all similar in science and in technology because the overriding aims, the relationship to reality, and the function of knowledge in each domain are quite essentially different. The aim of science is epistemic, whereas technology seeks to bring about advantageous transformations.

Artefacts can be seen as materialised solutions to technological problems. As such, artefacts constitute a specific form of technological knowledge. They embody structural and functional answers to problems arising whenever a desired state is not 'naturally' existent or available. They are a bridge between our organism and the world. Artefacts, like concepts, are organisers of activity. They mediate human transformations of the environment. Like concepts, they are pre-existing cultural constructs that must be reconstructed and appropriated by their users. Referring to my work, I have presented this reconstruction as a psychological process of instrumental genesis. I have shown that it involves an intertwined process of subjective and artefactual development.

Concepts are a particular kind of artefact. They are socially–transmitted, semiotic devices that enable human communities to share a common description of their surrounding world. Conjointly, they are also tools that mediate mental operations. Undoubtedly, scientific, research-based knowledge has contributed to the development of technological concepts and theories. Staudenmaier (1985) points out, however, that the purpose of these concepts and theories is to explain the behavioural characteristics of artefacts and materials with a view to put them to advantageous use. The pragmatic aims of technological cognition clearly supersede and replace the initial epistemic perspective.

Besides the technological and engineering concepts taught in schools and universities, communities of practitioners develop their own concepts which Pastré (1999, 2005) has termed "pragmatic" concepts. They share similarities with Vygotsky's (1934/1985) "spontaneous" concepts. They are needed within groups in order to communicate and comment on ongoing activities and processes that have not been addressed in formal training or schooling. In this chapter, I have tried to show how, in order to guide appropriate action in complex situations, they encapsulate an 'operative model' and description of the ongoing process in question.

NOTES

[1] Citations from publications in French are the author's translation.
[2] Bourdieu (1980) cautions structuralists against what he termed the "scholastic bias" in their tendency to forget that their social and linguistic models are constructs and that as such they cannot be attributed to the people or community under study.
[3] Two noteworthy attempts were made by Rey (1935) and Mounoud (1970).
[4] This term is borrowed from Bronckart (1997) to distinguish our position from strictly constructivist views.
[5] There are, of course, moments of solitary acquisition of experience and knowledge but they always occur on the background of previously internalised culture and so in reference to virtually present "others".
[6] Contrasting these two aspects of language use, Bronckart (1997) – after Bakhtine (1984) – distinguishes primary and secondary genres of discourse.
[7] "Value" points to the intrinsic economic dimension of artefacts as potential terms of exchange within and across communities.
[8] This is consistent with the idea that we appropriate the artefacts of others and populate them with our own intention (cf. Bakhtine, 1984).
[9] Alphabetical writing for communication purposes derives according to Havelock (1963) from mnemonic graphics used in Homeric Greece by storytellers.
[10] In my conception of semiotic artefacts as mediators of cognition, signs are more than labels.

REFERENCES

Austin, J. L. (1962). *How to do things with words.* New York: Oxford University Press.

Bakhtine, M. (1984). *Esthétique de la création verbale.* Paris: Gallimard.

Baird, D. (2002). Thing knowledge – Function and truth. *Techné, 6*(2), 13–27.

Bates, E. (1976). *Language and context: The acquisition of pragmatics.* New York: Academic Press.

Bates, E. (1979). *The emergence of symbols: Cognition and communication in infancy.* New York: Academic Press.

Bourdieu, P. (1980). *Le sens pratique.* Paris: Editions de Minuit.

Bunge, M. (1985). *Philosophy of science and technology.* Dordrecht, The Netherlands: D. Reidel.

Bronckart, J.-P. (1997). Action, discours et rationnalisation; L'hypothèse développementale de Vygotski revisitée. In B. Schneuwly, C. Moro, & M. Brossard (Eds.), *Outils et signes; perspectives actuelles de la théorie de Vygotski* (pp. 199–221). Berne, Switzerland: Peter Lang.

Bronckart, J.-P. (1999). La conscience come «analyseur» des épistémologies de Vygotski et Piaget. In Y. Clot (Ed.), *Avec Vygotski* (pp. 17–43). Paris: La Dispute.

Bronckart, J.-P. (2001). La psychologie ne peut être que sociale et la didactique est l'une de ses disciplines majeures. In J.-P. Bernié (Ed.), *Apprentissage, développement et significations.* Pessac, France: Presses Universitaires de Bordeaux.

Bruner, J. (1983). *Savoir faire, savoir dire.* Paris: PUF.

Bruner, J. (1990). *Acts of meaning.* Cambridge, MA: Harvard University Press.

Cassirer, E. (1942/1991). *Logique des sciences de la culture.* Paris: Cerf.

Céllerier, G. (1979). Structures cognitives et schèmes d'action. *Archives de psychologie, 47*(180/181), 87–122.

Clot, Y. (2000). *La fonction psychologique du travail.* Paris: PUF.

Engeström, Y. (1987). *Learning by expanding: An activity-theoretical approach to development research.* Helsinki, Finland: Orienta-Konsultit.

Frey, R. E. (1991). Another look at technology and science. *Journal of Technology Education, 3*(1), 16–29.

Gille, B. (1978). *Histoire des techniques.* Paris: La Pléiade.

Godelier, M. (1991). Préface. In J. Perrin (Ed.), *Construire une science des techniques* (pp. 7–10). Limonest: L'interdisciplinaire.

Griffin, P., & Cole, M. (1984). Current activity for the future: The Zo-ped. In B. Rogoff & J. V. Wertsch (Eds.), *Children's learning in the zone of proximal development: New directions for child development* (pp. 45–63). San Fransisco: Jossey-Bass.

Haudricourt, A. G. (1987). *La technologie science humaine.* Paris: M.S.H.

Havelock, E. (1963). *Preface to plato*. Cambridge, MA: Harvard University Press.

Hutchins, E. (1995). *Cognition in the wild*. Cambridge, MA: MIT Press.

Illich, Y. (1991). A plea for research on lay literacy. In D. Olson & N. Torrance (Eds.), *Literacy and orality* (pp. 28–46). Cambridge, MA: Cambridge University Press.

Jarvie, I. C. (1966). The social character of technological problems. *Technology and Culture*, *7*(3), 384–390.

Lave, J., & Wenger, E. (1991). *Situated learning: Legitimate peripheral participation*. Cambridge, MA: Cambridge University Press.

Layton, D. (1991). Science education and praxis: The relationship of school science to practical action. *Studies in science education*, *19*, 43–79.

Layton, E. (1974). Technology as knowledge. *Technology and culture*, *15*(1), 31–41.

Laffite, J. (1972). *Réflexions sur la science des machines*. Paris: Vrin.

Leontiev, A. (1975). *Activité, conscience, personnalité*. Moscou, URSS: Editions du Progrès.

Leplat, J., & Pailhous, J. (1975). Conditions cognitives de l'exercice et de l'acquisition des habiletés sensori-motrices. *Bulletin de Psychologie*, *321*(XXVII), 205–211.

Leroi-Gourhan, A. (1965). *Le geste et la parole*. Paris: Albin Michel.

Mauss, M. (1936). Les techniques du corps. *Journal de Psychologie*, *3–4*(XXXII), 365–386.

Meyerson, I. (1987). *Ecrits 1920–1983: Pour une psychologie historique*. Paris: PUF.

Mounoud, P. (1970). *Structuration de l'instrument chez l'enfant*. Lausanne, Switzerland: Delachaux & Niestlé.

Norman, D. A. (1991). Cognitive artefacts. In J. Carroll (Ed.), *Designing interaction* (pp. 17–38). New York: Cambridge University Press.

Pastré, P. (1999). La conceptualisation dans l'action, Bilan et perspectives. *Education permanente*, *131*, 13–35.

Pastré, P. (2005). *Apprendre par la simulation. De l'analyse du travail aux apprentissages professionnels*. Toulouse, France: Octarès.

Piaget, J. (1937). *La construction du réel chez l'enfant*. Neuchatel, Switzerland: Delachaux et Niestlé.

Piaget, J. (1967). *Biologie et connaissance*. Paris: Gallimard.

Perrin, J. (1991). Sciences de la nature et sciences de l'artificiel: Deux processus différents de production de connaissances. In J. Perrin (Ed.), *Construire une science des techniques* (pp. 381–397). Limonest, France: L'interdisciplinaire.

Rabardel, P. (1995). *Les hommes et les technologies*. Paris: Armand Colin.

Rabardel, P. (1999). Le langage comme instrument? Éléments pour une théorie instrumentale élargie. In Y. Clot (Ed.), *Avec Vygotski*. Paris: La Dispute.

Rabardel, P., & Vérillon, P. (1985). Relations aux objets et développement cognitif. In A. Giordan & J. L. Martinand (Eds.), *Acte des septièmes journées internationales sur l'éducation scientifique* (pp. 189–196). Paris: LIRESPT, Université Paris VII.

Rey, A. (1935). *L'intelligence pratique chez l'enfant*. Paris: Félix Alcan.

Schön, D. (1983). *The reflective practitioner*. New York: Basic Books.

Searle, J. (1995). *The construction of social reality*. London: Penguin.

Sigaut, F. (1987). Préface. In A. G. Haudricourt (Ed.), *La technologie science humaine* (pp. 9–32). Paris: M.S.H.

Sigaut, F. (1991a). Les points de vue constitutifs d'une science des techniques, essai de tableau comparatif. In J. Perrin (Ed.), *Construire une science des techniques* (pp. 31-45). Limonest, France: L'interdisciplinaire.

Sigaut, F. (1991b). Aperçus sur l'histoire de la technologie en tant que science humaine. In J. P. Bompard & J. Perrin (Eds.), *Histoire des techniques et compréhension de l'innovation* (pp. 67–79). Paris: INRA.

Simondon, G. (1969). *Du mode d'existence des objets techniques*. Paris: Aubier-Montaigne.

Skolimowski, H. (1966). The structure of thinking in technology. *Technology and Culture*, *7*(3), 371–383.

Staudenmaïer, J. (1985). *Technology's storytellers*. Cambridge, MA: MIT Press.

Stiegler, B. (1996). *La faute d'Epiméthée*. Paris: Galilée.

Thuillier, P. (1992). Techniques et sciences: Le message de Hughes de Saint Victor. *La Recherche*, *23*(240), 170–179.

Vérin, H. (1993). *La gloire des ingénieurs*. Paris: Albin Michel.

Vérillon, P. (1996). A problematica do instrument: um cuadro para reflexao do ensino do grafismo/La problématique de l'instrument: un cadre pour penser l'enseignement du graphisme. *Graf & Tec*, 57–78.

Vérillon, P. (2000). Instruments and cognition: Piaget and Vygotsky revisited in search of a learning model for technology education. *The Journal of Technology Studies*, *26*(1), 3–10.

Vérillon, P., & Rabardel, P. (1995). Cognition and artefacts: A contribution to the study of thought in relation to instrumented activity. *European Journal of Psychology of Education*, *X*(1), 77–101.

Vernant, J. P. (1965). *Mythe et pensée chez les Grecs*. Paris: Maspéro

Vergnaud, G. (1985). Concepts et schèmes dans une théorie opératoire de la représentation. *Psychologie française*, *30*(3/4), 245–252.
Vergnaud, G. (1987). Les fonctions de l'action et de la symbolisation dans la formation des connaissances chez l'enfant. In J. Piaget, P. Mounoud, & J. P. Bronckart (Eds.), *Psychologie* (pp. 821–844). Paris: Gallimard.
Vidal-Gomel, C., & Rogalski, J. (2007). La conceptualisation et la place des concepts pragmatiques dans l'activité professionnelle et le développement des compétences. *@ctivités*, *4*(1), 49–84.
Vincenti, W. G. (1990). *What engineers know and how they know it*. Baltimore: John Hopkins University Press.
Vygotsky, L. S. (1931/1978). *Mind and society*. Cambridge, MA: Harvard University Press.
Vygotsky, L. S. (1934/1985). *Pensée et langage*. Paris: Éditions sociales.
Wallon, H. (1941). *L'évolution psychologique de l'enfant*. Paris: Armand Colin.
Weil-Barrais, A. (1995). Genèse des rapports de l'enfant à l'objet technique. *Actes du 68° Congrès de l'AGEEM*, Metz.
Wertsch, J. V. (Ed.). (1985). *Culture, communication and cognition: Vygotskian perspectives*. Cambridge, MA: Cambridge University Press.

Pierre Vérillon
UMR ADEF
Université de Provence / Institut National de Recherche Pédagogique
France

SECTION 3

PERCEPTIONS OF TECHNOLOGY

JOHN R. DAKERS AND MARC J. DE VRIES

18. PERCEPTIONS OF TECHNOLOGY AND TECHNOLOGY EDUCATION

INTRODUCTION

> By these words, "the primacy of perception", we mean that the experience of perception is our presence at the moment when things, truths, values are constituted for us; that perception is a nascent *logos*; that it teaches us, outside all dogmatism, the true conditions of objectivity itself; that it summons us to the tasks of knowledge and action. It is not a question of reducing human knowledge to sensation, but of assisting at the birth of this knowledge, to make it as sensible as the sensible, to recover the consciousness of rationality. This experience of rationality is lost when we take it for granted as self-evident, but is, on the contrary, rediscovered when it is made to appear against the background of non-human nature. (Merleau-Ponty, 1964, p. 25)

In this quote, Maurice Merleau-Ponty offers us two important things to consider when regarding perception. First, perception is an embodied form of experience. In other words, we perceive the world through our bodies and our bodies influence the way we perceive that world. Moran (2000) invites us

> . . . to consider what our notion of the world of external objects would be like if we had eyes on either side of our heads rather than positioned so they both see the same things. Our whole understanding of the world is grounded in our corporeal nature. (p. 419)

Merleau-Ponty then argues, borrowing from Gestalt theory, that the various forms of truth, values, or objects that we perceive are always situated against some background and it is this background that gives rise to meaning.

The four chapters in this section of the *Handbook* consider both teacher and student perceptions of the concept of technology and, thus, their perceptions about technology education. Two chapters discuss empirical studies: one in the primary or elementary sector (Chapter 19) and one in the secondary or middle and upper school sector (Chapter 20). Whilst these two studies take place in English school settings[1], they nevertheless highlight issues that are relevant to an international audience (see, for example, Dakers, 2006; de Vries, 2005; de Vries, Custer, Dakers, & Martin, 2007; Ginestié, 2008). The third study (Chapter 21) takes the form of a discourse analysis of policy documentation from five English speaking G8 countries. Finally, Chapter 22 offers a more epistemologically-orientated account that explores how the conceptualisation of technology education impacts upon student and teacher perception.

Prior to discussing the chapters themselves, however, it is important to broadly define what we mean by perception and how this relates to technology education.

A.T. Jones and M.J. de Vries (eds.), International Handbook of Research and Development in Technology Education, 201–210.
© 2009 Sense Publishers. All rights reserved.

PERCEPTION

The way that human beings perceive the world defines it for them. Their actions upon the world are guided by their perceptions of it. It was not so very long ago, for example, that human beings perceived the world to be flat. This impacted upon their seagoing travel arrangements, as Columbus and his crew would have no doubt testified. It was also not so very long ago that Galileo upset the received wisdom of the Church when he challenged its perception that the Earth stood at the centre of the universe. Church doctrine together with some ominous life-threatening arguments convinced Galileo to recant his position, albeit reluctantly. This had the effect of leaving the Church's perceptions (or perhaps more accurately, doctrine about the nature of the universe) intact – for a time at least.

Current general perceptions relating to religion and gender offer two contemporary illustrations of the way perceptions can vary and define the world for the perceiver. For example, perceptions relating to gender can be seen to have a significant influence on areas relating to technology and science. It was in late sixteenth and early seventeenth century England that Sir Francis Bacon, the man considered to be the architect of modern science, offered us the metaphor of the 'marriage' between knowledge and power. He envisioned this marriage as "a chaste and lawful marriage between mind and nature" (Keller, 1995, p. 31). The aims of science were, for Bacon, the domination and control of nature; and nature, in Bacon's metaphor, was unequivocally feminine (nature as 'she' or 'mother'). Given that Bacon was a student of Oxford and had a strong leaning towards Puritanism (he was tutored by a future Archbishop of Canterbury), it is perhaps no surprise that his views that man (*sic*) should dominate nature follow biblical canon: "Man (*sic*) was not only given dominion over the earth, He (*sic*) was also instructed to 'subdue' it" (Genesis I: 26–28).

It is further interesting to note that at around the time of Bacon, as a new unambiguously 'masculine' scientific rationality was beginning to set about dominating nature, women were considered to be irrational. Indeed, witches – by being seen to invoke the spiritual, sexual, and non-rationale side of nature – continued to be feared by man: "The reality of witchcraft effectively attested to the gravity of the dangers represented by women – dangers against which reason and the new science promised protection" (Keller, 1995, p. 60).

Keller (1995) goes on to argue that during the seventeenth and eighteenth centuries, ideologies about gender shifted dramatically. Definitions of male and female became more and more polarised as evidenced in the woman's role as 'housewife' and the man's role in the workplace. Even as recently as 20 years ago, boys continued to study technical education at school whilst girls studied home economics (and still do, many argue). Moving forward from the Enlightenment towards the industrial revolution, this polarisation of genders resulted in a "severe curtailment of the economic, political, and social options available to women of all classes, and especially to women of the middle and upper classes" (p. 62). By the nineteenth century the wild, rapacious female of the seventeenth century had become completely dominated and domesticated in the service of man.

> In sympathy with, and even in response to, the growing division between male and female, public and private, work and home, modern science opted for an ever greater polarisation of mind and nature, reason and feeling, objective and subjective; in parallel with the gradual desexualisation of women, it offered a deanimated, desanctified, and increasingly mechanized conception of nature. (p. 63)

Thus, not only was nature subject to man's will, so too was the female: "With one reduced to its mechanical substrate, and the other to her asexual virtue, the essence of *Mater* could be both tamed and conquered; male potency was confirmed" (p. 64).

This separation of subject and object follows the Modernist paradigm, which, we would argue, continues to be integral to Western thought to this day. Modernism, in line with Platonism, seeks to categorise and classify the world by separating the "ideal and the real, the original and the copy, the conceptual and the material, and, ultimately, the opposition between man and woman" (Grosz, 1993, p. 169). Woman and nature in this masculine world become something 'other' from 'man'. Paechter (2007), citing Morgan (1972), offers an amusing but significant anecdote as a representation of 'otherness' in terms of the female:

> There is a line in an old folk song that runs: 'I call my donkey a horse gone wonky'. Throughout most of the literature dealing with differences between the sexes there runs a subtle underlying assumption that woman is a man gone wonky; that woman is a distorted version of the original blueprint; that they [men] are the norm, and we [woman] are the deviation. (p. 7)

The 'otherness' of the female in terms of education is realised in multiple forms, one of which is the sexual metaphor. Branches of knowledge like science and technology that are represented as factual and objective are considered to be 'hard' whereas those branches which are more normative in nature, those that are synonymous with hermeneutics and subjectivity, are considered to be 'soft'. The innuendo requires no explanation. "A woman thinking scientifically or objectively is thinking 'like a man'; conversely, a man pursuing a non-rational, non-scientific argument is arguing 'like a woman'" (Keller, 1995, p. 77). These stereotypes have potency, especially for the young. Writing over 35 years ago – and we would argue that the stereotype still remains today – Hudson (1972) argues that:

> The arts are associated with sexual pleasure, the sciences with sexual restraint. The arts man is seen as having a good looking, well dressed wife with whom he enjoys a warm sexual relation; the scientist as having a wife who is dowdy and dull, and in whom he has no physical interest. Yet the scientist is seen as masculine, the arts specialist as slightly feminine. (p. 83)

Substitute the word technologist or engineer for scientist and, we argue, the meaning will be consistent. Masculinity, like science and technology, is considered to be related to notions of independence and objective thought whereas femininity is more related to notions of collegiality and intersubjectivity. These gendered themes that separate objectivity from subjectivity will be seen to arise in the following chapters, albeit implicitly.

Yet another potent and current example concerns perceptions relating to the economy: those who hold right wing political ideologies tend to perceive economies as best led by the free market, whereas those who hold more left wing views perceive that the market should be more regulated and controlled. These opposing perceptions regarding the economy are not a matter of genetics or nature, they are culturally embedded. As such they are value laden and so subject to influence. They are, indeed, normative in nature.

Perceptions, so considered, are clearly subject to cultural interpretation and alteration and so serve to enable potential change in the way we experience the world. This being so, the various perceptions we build up of the world as it appears 'out there', are not fixed and immutable. It was in the Middle Ages, for example, as Ihde (1990) illustrates for us, that mountains were perceived of as being "threatening, ominous and dangerous" (p. 11) and rarely, if ever, thought of as beautiful:

> Until the eighteenth century, for example, the Derbyshire peak region in England was considered wild and unfit for human eyes. In 1681, the poet Charles Cotton described it as a 'country so deformed' that it might be regarded as 'Natures pudenda'. Travellers in those days were advised to keep their coach blinds drawn while traversing the region so as not to be shocked by its ugliness and wildness. (Dubos, 1980, cited by Ihde, 1990, p. 11)

However, this somewhat strange and dystopic Middle Ages regard for mountains was completely reversed by the nineteenth century when mountains were perceived to be "mystically beautiful" as illustrated in the artwork of the time (Ihde, 1990). Indeed, if we consider the way we perceive mountains today, for some they form the basis of our mountaineering leisure activities – nature to be conquered as made possible by evolving technologies. The significant issue here is that our perceptions of the world are subject to change over time, and technology in general, and modern technology in particular, plays a considerable role in shaping our perceptual framework of the world. This being so, the local and global techno-cultural societies of which human beings now form part are continually being (re)defined and (re)constituted as a result of these human-technology relationships (see, for example, Borgmann, 1984; Feenberg, 1999; Haraway, 1991; Ihde, 1990; Latour, 2005; Winner, 1992). Perceptions have, therefore, considerable potency. But how are these perceptions formed?

Sekuler and Blake (2002) inform us that perception is "guided by what [we] see, hear, feel, touch, smell and taste" (p. ix). They claim that perception "comes naturally and effortlessly" (ibid.). Essentially, their entire book is devoted to exploring perception as biological and thus sensorial in nature. For them, when we see a tree, for example, we perceive it sensorially as mediated by our eyes. Given that our brain has some concept of a tree as given by experience, we will then conceive the tree to be a tree. Ihde (1990) suggests, however, that perception involves more than just the biological. He argues, moreover (along with Borgmann, 1984 and Feenberg, 1999), that in order for human beings to engage in a more critical way with the world, perception cannot be considered to be simply 'natural and effortless'. For Ihde, perception involves two contingent dimensions: macro perception and micro perception. Micro perception follows the biological sensory schema as outlined by Sekuler and Blake above, but for Ihde there is simultaneously a form of macro perception that occurs in conjunction with micro perception and is hermeneutic in nature. Whatever it is that we sense, whether the horizon at the edge of the ocean suggesting to us a flat earth, or the sunrise at dawn suggesting to us a static earth, our perception is open to multiple interpretations that will vary according to the context or 'background' that situates us as perceivers in relation to that which we perceive.

As already noted, the received wisdom of several hundred years ago was that the world was flat. The church was blind to the arguments set out by Galileo because his claim challenged the established church doctrine of the day. However, very few human beings, if any, hold these views today. Using the example provided above in relation to Sekular and Blake, moreover, we can see how multiple interpretations can alter how individuals perceive a tree. For example, a child in the forest may see a tree as something to climb; a Druid or a Pagan sees a tree as something mystical; a lumberjack sees a tree as a potential form of income; a human with a wood-burning stove sees a tree as a source of heat; the Clan used a tree as a gallows! This hermeneutic dimension of perception situates that which we perceive within some background that is context dependent. And it is this context that shapes our perception of the world. A white child playing in a forest in the deep south of the USA early last century would have had a completely different perception of tree to that of an African American about to be hanged by the Clan.

Our perceptions of the human and the non-human world help to shape our identity. The human-technology interface also has a profound influence on our development as human beings in the world: try to conceive a world without personal computers!

TECHNOLOGY AND PERCEPTIONS

The human-technology interface has interesting implications for the ways that students who study technology education at school will perceive the technological world that they both occupy and, significantly, have presented to them as prescribed by the school curriculum and the teacher. Outside the school environment, they will have, and will continue to develop, a general perception of the techno-cultural world they occupy. This will come about through interaction with their material and social world. This world will shape and determine their perceptions of it (a flat or round world, for example) over the course of their lives.

At the time of writing this chapter it is likely that school age students in modern industrialised societies will perceive the modern technology that they knowingly interact with in a positive light. They will, for the most part, be competent users of a variety of personal digital technologies such as computers, mobile telephones, and digital games platforms. They were born into a technological era that is replete with these technologies. They welcome new technologies that they consider to be relevant to their lives. Their parents and grandparents, who will have developed different perceptions of the world, are likely to have been less comfortable with the relatively recent dawning of the digital age.

It is when these spontaneous learning experiences (Vygotsky, 1978) encountered outside the classroom, are conflated with technology as prescribed and delivered inside the classroom, that the potential for perceptual disequilibrium may arise. If, like Galileo, the student's perception of technology serves to challenge the technologically-received wisdom of the school curriculum, technological development within the student is likely to be compromised. This disharmony is most likely to arise when the technological issues raised at school are presented by 'experts' as being resolute, immutable, and not open to debate. In this epistemological model, technological knowledge cannot be open to challenge, thus defining it as value free. Perceptions so developed lack the hermeneutic dimension as outlined by Ihde above and thus serve to stifle the development of any critical faculty within the student.

A car, for example, can be presented by the teacher as a technology that has a primary designated function of transporting people from one location to another along a network of roads. The context or background, in this view, is instrumental: the car is simply an assemblage of inert physical components that, when brought together, constitutes what we perceive to be a car. The car, considered only in this instrumental way, is denoted as having a narrow and universal functional dimension: to get from A to B. However, a car such as a Ferrari or an Aston Martin constitutes more than this basic instrumental functionality: these cars have a visceral emotional context inscribed into their design and this value goes well beyond 'getting from A to B'. This metaphysical aspect designates – or as Baudrillard (2005) suggests, signifies by connotation – that the owner has a certain status, wealth, and sexual quality. Whilst a Ferrari will certainly serve the instrumental function of getting its passengers from A to B, they will do so in a qualitatively different way from their fellow humans who travel from A to B in a Ford Ka, for example. The technologies are perceived as having a quality that transcends simple materiality.

Both scenarios allow for multiple interpretations. Feenberg (2006) uses the American Rifle Association's mantra 'guns don't kill people, people kill people' to clearly illustrate this point. The ARA members perceive guns to be purely instrumental; that is, neutral. A gun, for them, has no agency and cannot kill without a human operator, hence, 'people kill people'. However, an alternative perception is offered by Feenberg (and many others) in which he considers this scenario to be set within a different context. A human being with a gun is both quantitatively and qualitatively different from one without a gun: quantitatively because they have a gun; qualitatively because they have a gun for a reason. It is interesting (or perhaps disturbing) to note here that more human beings die every year from gunshot-related incidents in the USA where guns are legal and readily available, than in the whole of Europe, where they are not so readily available and have far greater regulatory control attached to their use. To further illustrate this, consider a potential technology in the form of a body suit that is worn as underwear and offers no physical restrictions to its user. Let us suppose that it is fabricated from a newly developed material that is bullet-proof and will even protect the wearer from any impact pain or bruising caused by a bullet fired from a gun. If every USA citizen got into the habit of wearing this new type of underwear, how keen might they then be to own a gun for protection?

There are thus two distinct positions that emerge. One is that technology is perceived as context free, instrumental, and neutral, "as mere physical objects abstracted from all relations" (Feenberg, 1999, p. 213). But this argument is flawed. Ihde (1990), for example, brings to our attention that fact that "[t]he temptation may be strong here to leap to a contextless conclusion that the 'technology' as such is 'neutral' but takes on its significance dependent upon different 'uses.' But such conclusion remains at most a kind of disembodied abstraction" (p. 128). He goes on to present arguments in support of the second position: that "[t]he technology is only what it is in some use-context. Even the discarded technologies, whether in museums of science and industry, ruins against a landscape, or refitted into a bricolage construction, continue to indicate their perceivable 'usefulness'" (ibid.).

If knowledge development in technology education is perceived by students to be purely neutral and functional, devoid of any ethical, political, or hermeneutic qualities, they are more likely to develop into adults who hold an uncritical and restricted view of the technologically-mediated world they occupy. Feenberg (1999) expresses serious concerns that engineers exhibit this restricted view of technology. He postulates that:

> Once function is selected, the engineer has the last word on its implementation and the humanist interpreter is out of luck. This is the view of most engineers and managers; they are at home with 'function' but have no place for meaning. (p. 84)

If Feenberg's concerns hold any validity, and we suggest they do, this has significant implications relating to epistemological development within a school subject that overtly aligns itself with engineering. These arguments are illustrated and given further elucidation in the chapters that make up this section.

OVERVIEW OF THE FOUR CHAPTERS

When as technology educators we reduce technology education to the purely functional, free from values and social contexts, we encourage learners to perceive technology as such. Therefore, as Julie Lunt suggests in Chapter 19, *Primary school children's perceptions of design and technology*, we need to move away from a pedagogical model that is prescriptive towards one in which greater account is taken of the learners' own perceptions about technology. However, as she indicates, studies relating to children's perceptions about technology are relatively few in number, particularly for younger children. In addition, the empirical studies referred to by Lunt indicate that children do tend to perceive technology education as more functionally orientated. Issues relating to values and social contexts do not appear to be given any place in the primary school learning space.

Lunt argues that research into education has moved over the past 15 years from considering the learner as a passive recipient of knowledge to one in which the learner has a greater degree of agency. She makes a case for the emergence of 'pupil voice' and goes so far as to suggest that children should be afforded some space to "criticise the dominant power structures".

Lunt suggests that children in the elementary sector do appear to enjoy technology education (although this level of enjoyment appears to diminish somewhat by the secondary school stages, according to the findings of Nicholl and McLellan, which are considered later.) Enjoyment notwithstanding, Lunt does, however, express serious concerns that the delivery of technology education in the primary sector appears to emphasise the "practical rather than reflective dimensions of technology education". Lunt's research also reveals a need for more authentic learning experiences to be embedded into the learning space. These experiences, she concludes, need to involve "wider learning objectives, such as developing a critical understanding of the technological world and its impact on how we live our lives". She goes on to advocate a technology curriculum that involves the development of a technological literacy that empowers learners by engaging them in a dialogue about issues relating to ethics and sustainability. These concerns that the technology curricula lack any hermeneutic dimension will be seen, indeed, to emerge throughout every chapter in this section.

In Chapter 20, entitled *This isn't my project (work). It's...just do it...You just do research* by Bill Nicholl and Ros McLellan, we see some of the issues raised by Lunt also emerging in the pedagogical frameworks of the secondary sector. At the very outset, Nicholl and McLellan make a case for student voice in the delivery of technology education. Their research indicates that technology education lessons tend to be very much teacher-led, "where pupils follow a set of procedures to complete tasks which led to all pupils making similar, if not identical outcomes". This follows exactly the same paradigmatic framework that can be encountered in Scottish technology education departments (as witnessed by one of the authors of this chapter) and, we would suggest, many other technology education departments around the world. This narrow focus on procedural knowledge development as revealed by Nicholl and McLellan's research "do[es] not allow students to become 'metacognitively' wise". They go on to suggest, moreover, that this functional approach restricts the development of any creative endeavour because creativity cannot be expressed through the development of procedural knowledge alone.

An interesting issue that Nicholl and McLellan reveal links motivation to perception. For example, they categorise classroom programmes where students are required to follow a set of prescribed routines in order to solve a task being low in ambiguity and low in risk. This pedagogical framework, they claim, serves to de-motivate students. In contrast, however, they found that where tasks were high in ambiguity and high in risk, students were much more engaged and motivated. This category of task requires that the students go beyond knowing 'how' to undertake a task, to developing an understanding of the 'why' and the 'where' aspects. The low ambiguity/low risk framework resonates strongly with Feenberg's concerns (introduced above) about engineers giving primacy to functionality over meaning.

The two empirical research projects discussed thus far suggest that the pedagogy employed in technology education classrooms places strong emphasis on the development of procedural knowledge with little or no emphasis given to the development of conceptual knowledge. Is this the fault of teachers, policy makers, or curriculum guidelines? This is an extremely difficult and complex question to answer and has been subject to much recent debate.

The two chapters discussed above suggest the need for a new form of pedagogy in the delivery of technology education. Leo Elshof, however, places more emphasis on the need for a change in thinking relevant to policy and curriculum development in Chapter 21, entitled *Perceptions and practices: Technology education and the environment*. Elshof undertakes a discourse analysis of curriculum documentation from five G8 countries. His findings are very revealing. Citing another recent study of technology teachers' perceptions about the need to teach issues relating to sustainable development, Elshof demonstrates that whilst teachers strongly supported the need for this to be undertaken, they were, somewhat paradoxically, less enthusiastic about actually teaching these issues. He suggests that this may be due to the fact that "technology teachers had idiosyncratic understandings of sustainable development . . . that few teachers had taken formal courses in environmental science in their professional training".

In terms of the documentation analysed, Elshof found that terms like 'responsibility', 'pollution', and 'toxic' were rarely found, whereas terms like 'management', 'business', and 'skills' were prolific. More disturbingly, perhaps, Elshof reveals that there is a complete disregard of ethical issues relating to the developing world evident in the documentation. In addition, scant interest appears to have been expressed relating to energy efficiency. These are important issues for any citizen living in this time and are, in our view as well as that of Elshof, worthy of greater consideration in any technological curriculum.

Issues relating to pedagogy are also discussed, with Elshof pointing to the lack of dialogue regarding these issues as a barrier to the development of any critical capacity in students of technology education. The documentation, in this respect, appears to promote an 'expert' transmission model of technological *a priori* facts. The concept of revealing multi interpretations of technological systems, as espoused above, appears, according to Elshof, to be sadly lacking. Citing Martin (2002), Elshof points out that technology education is "shaped by the values of those in authority and the existing political culture . . . [when technology education] fails to recognise the importance of values and attitudes [it] is reduced to technical training". This is a view that is gaining much support around the world in terms of technology education research.

This view is indeed further elucidated in Chapter 22, *Conceptual learning in and through technology*, where Eric Parkinson and Gill Hope explore issues relating to what it is that constitutes technological knowledge and how this affects perceptions about technology. They do this by considering the conceptual side of technology and argue, not unlike Ihde, that you cannot develop knowledge relating to technology by practical 'doing' experiences alone. The conceptual or hermeneutical dimension is always present, either implicitly or explicitly. Following Gilbert Ryle's (1949) now famous distinction between the concept of knowing 'how' (procedural knowledge) and knowing 'that' (conceptual knowledge), Parkinson and Hope argue that given the complexity of the human mind, our "interactions with the material world – which underlies technological capability – cannot be reduced to facts, techniques, and routine strategies." They go on to argue that the act of becoming proficient, or even expert, at some given operation in order to change the material world in some way is not enough. Students of technology education need to be afforded the opportunity to "analyse, generalise, and synthesise perceptions, observations, actions, and knowledge". Moreover, and in concert with the conclusions reached by Nicholl and McLellan, Parkinson and Hope advocate for a classroom environment that encourages risk taking, which will serve to encourage innovative design practice. They argue that "[t]he greatest design thinkers are those with the broadest conceptual store, not necessarily the greatest factual knowledge". Many of the arguments presented by Parkinson and Hope resonate very strongly with the findings revealed in the other three chapters.

Given, as the evidence suggests, that the development of procedural knowledge currently holds primacy over that of conceptual knowledge, and given that formal summative assessment procedures tend towards 'measuring' tangible concrete outcomes rather than normative ones that are, by their very nature, value driven, and given further, as Elshof reveals, that policy documentation outlining curriculum content lacks any true commitment to the study of ethical and sustainable issues, it should perhaps come as no surprise that technology teachers are reluctant to engage with the hermeneutic and conceptual dimensions relating to the impact of technology upon the world.

It is apparent from the chapters in this section that technology education is indeed in a condition of stasis, or perhaps stagnation, and that procedural knowledge development continues to be the dominant orthodoxy. The overwhelming single issue to emerge from the chapters is a plea to change the pedagogical framework employed in the delivery of technology education. This change in emphasis requires that student voice be given more agency and the conceptual and hermeneutic aspects of technology and society be afforded more prominence in the learning space. This move from a Weberian behaviourist pedagogy that has its genesis in functional rationalism, to one that is more socially constructed, dialogical, and humanistic in nature, will require significant perceptual reform. The means by which this can be achieved may, however, prove difficult to determine.

It may be that the best solution is for those involved in technology education, whether as academics, teachers or students, to attempt to find ways to reveal our own perceptions, rather than attempting to change the perceptions of others about the technologically mediated world we inhabit. Some of our implicitly held perceptions about technology and technology education, once revealed, may surprise us. This, however, can only ever be achieved through dialogue, and never through instruction. What is required, as made explicit in the chapters in this section, is the development of a critical pedagogy that enables all participants to develop the skills required to become critical rather than passive citizens. Exposure to multiple perspectives about technology will, as Giroux (1992) argues, provide

> ... students with the skills they will need to locate themselves in history, find their own voices, and provide the convictions and compassions necessary for exercising civic courage, taking risks, and furthering the habits, customs, and social relations that are essential to democratic public forms. (p. 74)

NOTES

[1] Design and technology, as it is known in England, is taught over four 'key stages'. Key Stage 1 and Key Stage 2 are taught in the primary or elementary school sector. Key Stage 3 is taught in the equivalent of middle school and Key Stage 4 is taught to upper school. For more information, go to 'design and technology' on the QCA web site: http://www.qca.org.uk/.

REFERENCES

Baudrillard, J. (2005). *A system of objects*. New York: Verso.

Borgmann, A. (1984). *Technology and the character of contemporary life*. Chicago: University of Chicago Press.

Dakers, J. (2006). *Defining technological literacy: Towards an epistemological framework*. New York: Palgrave MacMillan.

de Vries, M. J. (2005). *Teaching about technology: An introduction to the philosophy of technology for non-philosophers*. Dordrecht, The Netherlands: Springer. Germany.

de Vries, M., Custer, R., Dakers, J., & Martin, G. (2007). *Analyzing best practices in technology education*. Rotterdam, The Netherlands: Sense.

Dubos, R. (1980). *The wooing of earth*. New York: Charles Scribner's Sons.

Feenberg, A. (1999). *Questioning technology*. London: Routledge.

Feenberg, A. (2006). What is philosophy of technology? In J. Dakers (Ed.), *Defining technological literacy. Towards an epistemological framework* (pp. 5–16). New York: Palgrave MacMillan.

Ginestié, J. (2008). *The cultural transmission of artefacts, skills and knowledge: Eleven studies of technology education in France*. Rotterdam, The Netherlands and Taipei, Taiwan: Sense.

Giroux, H. (1992). *Border crossings: Cultural workers and the politics of education*. London: Routledge.

Grosz, E. (1993). A thousand tiny sexes: Feminism and rhizomatics. *Topoi*, *12*, 167–179.
Haraway, D. (1991). *Simians, cyborgs, and women*. London: Free Association Books.
Hudson, L. (1972). *The cult of the fact*. New York: Harper and Row.
Ihde, D. (1990). *Technology and the lifeworld: From garden to earth*. Bloomington, IA: Indiana University Press.
Keller, E. F. (1995). *Reflections on gender and science*. New Haven, CT: Yale University Press.
Latour, B. (2005). *Reassembling the social: An introduction to actor-network theory*. Oxford, UK: Oxford University Press.
Martin, M. (2002). Values and attitudes in design and technology. In S. Sayers, J. Morley, & B. Barnes (Eds.), *Issues in design and technology teaching* (pp. 208–223). London: Routledge.
Merleau-Ponty, M. (1964). *The primacy of perception* (J. Edie, Trans.). Evanston, IL: Northwestern University Press.
Moran, D. (2000). *Introduction to phenomenology*. London: Routledge.
Morgan, E. (1972). *The descent of woman*. London: Corgi.
Paetcher, C. (2007). *Being boys being girls: Learning masculinities and femininities*. Berkshire, UK: Open University Press.
Ryle, G. (1949). *The concept of mind*. Harmondsworth: Penguin.
Sekuler, R., & Blake, R. (2002). *Perception*. New York: McGraw-Hill.
Winner, L. (1992). *Autonomous technologies: Technics-out-of-control as a theme in political thought*. Cambridge, MA: MIT Press.
Vygotsky, L. (1978). *Mind in society: The development of higher psychological processes*. Cambridge, MA: Harvard University Press.

John Dakers
Department of Educational Studies
University of Glasgow
Scotland

Marc de Vries
Faculty of Applied Sciences
Delft University of Technology
Delft, The Netherlands

JULIE LUNT

19. PRIMARY SCHOOL CHILDREN'S PERCEPTIONS OF DESIGN AND TECHNOLOGY

INTRODUCTION

During the last fifteen years there has been a significant increase in the amount of educational research that pays attention to children's own accounts of their experiences of school. This trend reflects a change in the way in which children are viewed, moving from a model of children as passive recipients of education to one which positions children as active participants of educational processes and key stakeholders in education (Rudduck & Flutter, 2000). As design and technology educators, it is important that we take account of children's perceptions in order to develop appropriate curricula and classroom experiences. Research into children's perceptions in the field of primary design and technology education is at a very early stage. However, these preliminary studies, together with work from a broader field, offer insights that can illuminate our understanding and provide a foundation for future research.

This chapter reviews the work carried out in the field of design and technology with children of primary age, and draws upon evidence collected in two empirical studies conducted in England with children aged 9–11 (Benson & Lunt, 2007a; Lunt, 2007). While it is acknowledged that children are individuals with unique perceptions of their own particular worlds, it is still possible to provide indications of the value which children place on their experience of learning and participating in design and technology and the chapter concludes by considering what we, as educators and policy-makers, have to learn from children's own accounts of their experience.

RESEARCHING CHILDREN'S PERCEPTIONS

The majority of work on children's perspectives of their experience of education has until recently focused on social processes and issues of social differentiation (Erickson & Shultz, 1992). Curriculum content and learning processes have generally been relatively peripheral issues in studies focusing on, for instance, socialisation and peer culture (e.g., Davies, 1982; Pollard & Filer, 1999). Further, the majority of work has been carried out with secondary-aged pupils. In England, the longitudinal Primary Assessment and Curriculum Experience (PACE) project (1989–1997) was a groundbreaking study in that it was the first major study in the UK that paid serious attention to the perspectives of primary children with regard to their experience of learning in school. Children's views were sought on their experiences of the curriculum, individual curriculum subjects, teacher-pupil relationships, their teachers' teaching, learning, and assessment (Pollard & Triggs, 2000). Consulting pupils as part of school improvement initiatives has also made a major contribution to the literature in this area (e.g., Arnot, McIntyre, Pedder, & Reay, 2004; McIntyre, Pedder, & Rudduck, 2005; Rudduck & Flutter, 2004).

There are a variety of arguments made for researching children's perceptions and these influence the nature of the research and how the findings are used. For example, a romantic

A.T. Jones and M.J. de Vries (eds.), International Handbook of Research and Development in Technology Education, 211–222.
© 2007 Sense Publishers. All rights reserved.

argument is founded on a view of children and childhood in which children are characterised by innocence and naïvety and are positioned as beings in need of adult patronage. From this perspective children's views are collected and sentimentalised, but not used to make a serious contribution to educational debates or to bring about significant change in their circumstances. Despite appearing harmless, this position undermines children's status as people whose views should be taken seriously. In contrast, a critical argument is founded on the notion that children are central actors in the teaching and learning process and as such have very particular insights to offer in helping to develop our understanding as educational practitioners. This argument positions children as experts of their own experience with important and potentially critical contributions to make to current education debates (Pollard, Thiessen, & Filer, 1997).

Although there are many reasoned arguments that support the principle of researching children's perspectives, there are also balancing considerations to be made. Some would argue that children are too immature or naïve to shed light on the complexities of teaching and learning – that they simply do not have the cognitive or communicative competence to make a contribution to education debates. However, an increasing number of research studies involving primary- aged children suggest that young children are capable of providing valuable insights into their experience (e.g., Gipps & Tunstall, 1998; McCallum, Hargreaves, & Gipps, 2000). Furthermore, the kinds of changes to teaching and learning activities which pupils suggest when consulted tend to be very sensible and realistic (Jeffrey, 2003; McIntyre et al., 2005).

Naturally the most articulate pupils are the ones most able to express their views and consequently there is a risk that the overall view of children's perspectives might be biased. Rudduck and Flutter (2000) argue that we should do more to enable children to talk about their learning and about themselves as learners. With the increased interest in researching children's perspectives, more attention has been given to developing appropriate methodologies for researching children's views and consulting with pupils in schools (e.g., Christensen & James, 2000; Greig & Taylor, 1999). However, of all the challenges involved in researching pupils' perspectives of their experience in school, perhaps the greatest is that, given a voice, pupils may well criticise the dominant power structures (Pollard et al., 1997). This then raises the issue of how this is handled and how pupils' perspectives are interpreted and made use of by researchers and teachers. Although pupils' perspectives are not the only ones, effective teaching is underpinned by a recognition of the concerns, interests, and motivations of learners.

RESEARCHING CHILDREN'S PERCEPTIONS OF DESIGN AND TECHNOLOGY

Researching children's perceptions of design and technology (d&t) has as yet received minimal attention within the research community. In three reviews of educational research in design and technology (Eggleston, 2000; Harris & Wilson, 2003; Kimbell, 1996) there are only passing references to eliciting and considering pupils' views and, in the studies where this does occur, it is used as a supplementary method of data collection rather than as a focus of research. The work which exists is small-scale and the majority of studies relate to secondary-aged pupils.

Two recent studies will be discussed in more detail here. The first, a children's survey, was a pilot project conducted in England with 304 children aged 9–11 (Benson & Lunt, 2007a). Both qualitative and quantitative data were collected through a questionnaire using a variety of methods to elicit children's responses. The children were invited to answer questions in their own words (e.g., 'Why do you think you have to learn d&t in school?'); indicate a set response to a question (e.g., 'Do you enjoy d&t?' – always, usually, sometimes, never), and indicate a set response to a statement made by someone else about d&t (e.g., 'It puts a smile on your face when you have made something of your own' – agree, disagree, not sure). The children were also asked to identify their three favourite subjects and the three subjects they thought they were best at. A further qualitative response was required of children through asking them to imagine they were describing design and technology to a new classmate who had never experienced it before. The study is on-going and the data discussed here is from the initial pilot. However, it provides some insights into children's perspectives in this age group in relation to the design and technology curriculum in England and highlights a number of themes and issues worthy of broader consideration.

The second research study is an on-going investigation into writing tasks in design and technology with children aged 9–11. Sixty-five children from three schools in England were interviewed before and after a unit of work in an action-research study to ascertain their perceptions of design and technology in general and specifically in writing tasks in design and technology. Writing tasks were defined as any task in which children are required to write in design and technology. Writing often takes place in conjunction with other modes of representation, such as drawing or discussion, and is usually brief, sometimes using spatial organisers such as tables or mind maps. Typical examples of children's writing include annotated drawings, lists, labels, story-boards, comparison charts, and web diagrams. The interview data are supported by data collected through a variety of methods including observer participation and interviews with teachers.

In the next section, the evidence from these studies will be discussed in the light of other studies in design and technology and those relevant to general education with this age group.

CHILDREN'S VIEWS OF THE NATURE AND PURPOSES OF DESIGN AND TECHNOLOGY

The evidence from both the children's survey and the writing study suggests that children clearly perceive design and technology as a curriculum subject distinct from other subjects. The children's survey showed design and technology to be among the four most liked subjects, together with art, physical education (PE), and information communication technology (ICT). In both these studies the majority of children were extremely positive about design and technology, a finding supported by other small-scale studies which have elicited pupils' views (e.g., Barlex, Welch, & Taylor, 2005; Twyford & Burden, 2000). In the writing study interviews the children were asked what they thought of their design and technology lessons in general. The most frequently articulated word was "fun", with many children using phrases such as "very good", "really exciting", "interesting", "brilliant", and "I look forward to it". Only one child out of the 65 interviewed was anything less than enthusiastic. However, design and technology was perceived by the children of both studies as more than just an opportunity for fun in the classroom; it was also regarded as educationally worthwhile. In the children's survey, the majority of children recognised the educational benefits of design and technology, positioning it as both fun and work. Eighty-one percent thought that they always or usually learnt a lot, 83% that they had to think really carefully in design and technology lessons, while 86% thought they learnt to be creative in design and technology.

Researchers in the PACE project found that many children constructed their meaning of the subject design and technology in opposition to formal school work (Pollard & Triggs, 2000). They observed that in contrast to the core subjects of English, mathematics, and science, in design and technology teachers allowed their children more autonomy, took a more relaxed approach to pupil grouping, encouraged collaborative rather than individual learning, and engaged with 'mess and doing'. It is suggested that this change in the teaching and learning frame might have contributed to the children's perceptions of this subject. Design and technology was frequently linked with the criteria 'fun' and 'interesting', and comments from individual children giving reasons for their subject preferences referred to the opportunity to make things, having fun, playing, and the active nature of design and technology. Although this study was carried out when design and technology was in its infancy as a curriculum subject, it appears that its characteristics as a practical and creative subject were already proving popular with these children.

In the writing study the practical dimension of design and technology was dominant in children's reflections about their experience of the subject. For example, when asked to reflect on how the unit of work they had just completed had helped them to get better at design and technology, more than 50% of the responses in each of the schools referred to developing technical 'know-how' and skills in making. However, to varying degrees, the children were conscious of other elements of their progress in design and technology. Designing featured quite prominently in one of the schools, whereas in another school more general skills such as teamwork, problem-solving, and memory were mentioned. This varied response might have been due to the emphasis placed on these elements by the individual teachers or the specific nature of their most recent design and technology project. Attitudes such as perseverance, overcoming problems, and confidence were also recognised by a small number of children in each school.

In the children's survey, a large majority of children recognised the value of designing as well as making with 86% thinking that designing was really important if you wanted to make a good product. However, there was some inconsistency in the results between schools in response to the statement "I would prefer just to get on and make, not to design". The responses for 'disagree' ranged from between 25% and 77%. There are no data to explain why this might be so, but the results might reflect how designing is taught in these schools, the pressure on time for making, or the differing needs and interests of these children.

Children in both studies were asked why they thought they had to learn design and technology in school. In the children's survey, the responses indicate that the children view design and technology as relevant to their future lives – both in jobs or careers and in other elements of adult life, for example, as parents. This is supported by the findings of the writing study where there was a consistent pattern across schools with the top three categories being "preparing for adult life", "developing their ability to design and make things", and "preparing for a later stage of education". In their responses the children mentioned a wide range of jobs and activities including architect, builder, cook, designer, engineer, mechanic, and making biscuits and food packaging. It is likely that these were influenced not just by their most recent project or the views of their teacher but also by their experiences beyond school. A narrow vocational view of design and technology is one which frequently causes concern in secondary education (e.g., Dow, 2006). However, in the context of these primary schools, the children seemed to be reflecting an interest in themselves and their changing identities as they moved on to secondary school and looked ahead to adult life, rather than a narrow prescription of activity in design and technology lessons. Listening to the children, it appeared that this sense of relevance was a motivating factor which ran alongside the more immediate gratification of their enjoyment in learning and participating in design and technology.

> It's good fun – I like working in groups. When I'm older I might become an engineer or go into making biscuits or food packaging – and also it's enjoyable. (girl, 10)

> It helps you when you go on (to secondary school). It's important for you to be able to do lots of different things not just English and maths – D&T involves lots of things that aren't included in other subjects. (girl, 10)

> To help you when you get older – you can build things and fix things. (boy, 10)

> When we're older, when our kids ask us how to do all that we can tell them. (girl, 11)

However, an analysis of the data also reveals the lack of any mention concerning the development of technological literacy, in which one might see the development of skills necessary for active and critical engagement with an increasingly technological world. This might be a reflection of children's engagement with the practical rather than reflective dimensions of design and technology, although it is more likely to be a reflection of current curriculum practice in English primary schools.

CHILDREN'S VALUES AND CONCERNS IN LEARNING DESIGN AND TECHNOLOGY

The data from these two studies show that these children perceived their experience of design and technology in the classroom as a highly valued activity which they enjoyed and found worthwhile. In the writing study, the children's interviews were analysed and coded to identify reasons which children articulated for enjoying design and technology. The codings were developed from those used by Pollard and Triggs (2000) when analysing primary-aged children's judgements of the curriculum:

- Success/ease;
- interesting;
- activity;
- autonomy;
- educational (mention of learning/challenge/future value); and
- fun.

During the analysis of the interviews in this study, new categories emerged and modifications to the existing set were made. In the children's responses, 'interesting' and 'fun' often occurred together. For the purposes of this analysis these words were grouped together to describe a positive affective response, a personal engagement with what was happening in the moment. In addition, there were three reasons articulated for enjoying design and technology not covered by the original categories: 'novelty/difference', 'goal-oriented', and 'social'. 'Novelty/difference' was created to include responses which suggest that design and technology is valued by children because it is perceived as different from other subjects and can be viewed as a welcome change from some other types of educational experience. 'Goal-oriented' refers to the notion that in a design and technology project, children are working towards designing and making a successful product. In England, the national exemplar schemes of work (Qualifications and Curriculum Authority, 1998) have promoted a particular structure for design and technology projects, in which children prepare for their designing and making assignment with a series of shorter activities to develop relevant understanding and skills. Although this model followed uniformly can be criticised for being repetitive and rigid, it also has several advantages, including an internal coherence where every task is related to the achievement of the main goal – the successful designing and making of a particular product. The final additional category of analysis is 'social'. This category emerged because, for some children, the opportunity to work with, sit next to, or talk to friends during their design and technology activity was a significant reason for enjoying it. Figure 1 shows how frequently these categories of response were mentioned by the children during the course of their interviews. This provides insights into what was at the forefront of these children's minds rather than relying on researcher-prescribed categories. Many of the children gave multiple reasons.

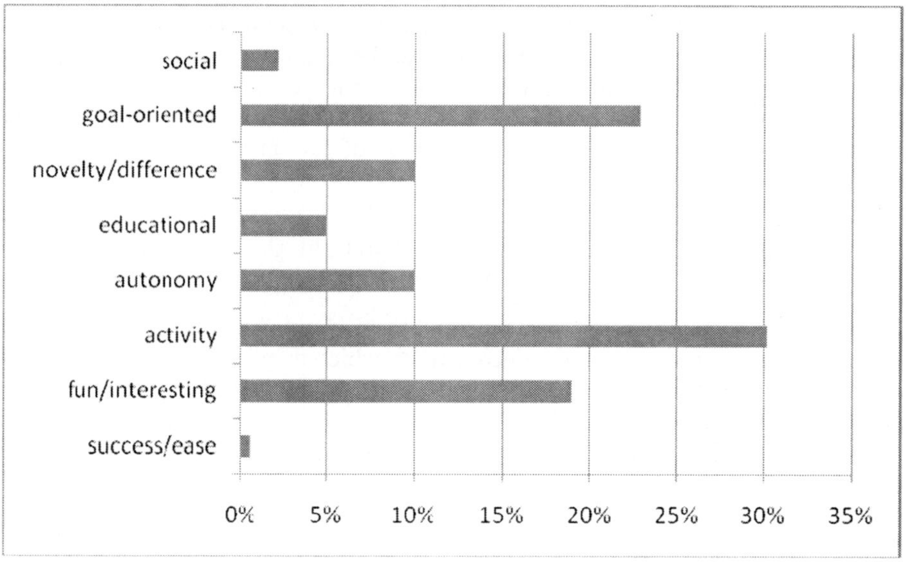

Figure 1. Reasons given by children for enjoying design and technology (n=65).

The active nature of design and technology was the most frequently mentioned reason for enjoying design and technology. These children enjoyed practical activity: they liked making things and using tools, solving practical problems, tasting food, gluing, sewing, painting, building, and constructing. However, it was not just the process of these activities that they enjoyed, it was also the fact that they were making a particular product and all their activities led to this. In each design and technology project, these children had a particular goal they were aiming for and this seemed to give momentum and purpose to their lessons.

The high frequency of 'fun/interesting' might be expected in design and technology. Pollard and Triggs (2000) observed that 'fun' had a range of meanings for children. It was most often linked to 'activity' and 'autonomy' and less frequently with 'interesting'. Liking something because it was 'fun' was also linked with seeing it as play, finding it enjoyable because you could 'have a laugh', and sometimes because it was sociable and you could 'have a chat'. The results here show that these children made stronger links between fun and activity, creating things, autonomy, and novelty/difference than working with friends. This might have been a reflection of the way in which design and technology was organised in these classrooms, although further research as a participant observer with these children and their teachers suggest that this is not the case. It might just have been that this was not the reason at the forefront of their minds.

The representation of 'educational', although small, suggests that some children enjoy design and technology because of what they learn and are able to do as a result of it. The very low representation of 'success/ease' contrasts with the high rating attributed to it by the children in the PACE study. However, in that study the pressure of testing and assessment in the core subjects of English, maths and science was probably an influencing factor on the overall categories.

So what do children not enjoy in design and technology? In the writing study, the least liked aspects of design and technology were defined as 'writing', 'time issues', 'waiting for equipment or the teacher', 'unnecessary activities', and 'problems with making'. These were echoed in the children's survey, although the responses of children from different schools were inconsistent. For example, in one school 74% of children identified writing as the aspect they liked least in design and technology compared to 2% in another. This suggests that there might be a marked difference in practice between the two schools in the presentation of writing tasks. 'Problems with making' was another significant category, but again the response was variable between schools, possibly suggesting a different teaching approach. The highest overall category was 'nothing', which again demonstrates the popularity of this subject for children.

Writing tasks in design and technology are seen by some children as detracting from the enjoyable business of designing and making. However, the more in-depth work carried out in the writing study shows that this is not necessarily the case (Lunt, 2005, 2007). The majority of children in this study came to view writing tasks in design and technology positively. A significant factor in pupils' perceptions was the relevance of the writing task to the successful achievement of their designing and making assignment. McCormick and Davidson (1996) highlight the central place that the creation of products has in the domain of design and technology and the motivational effect that has on pupils. The pupils in the writing study strongly confirm this view. The children's comments about the helpfulness of the writing tasks were directly related to the contribution they made to the specific creative act of this unit of work, rather than to wider educational goals such as learning to design. The writing tasks judged least positively were those where pupils could not see a clear purpose to benefit themselves as designers and makers or where they were required to write in what they considered to be an inappropriate format, for example, paragraphs when bullet points were more appropriate. Quantity of writing did not in itself seem to be a factor as long as there was a strong connection with achieving the designing and making assignment. In fact writing was most positively viewed in the school where the children did the greatest amount of writing but the teacher ensured that it was closely integrated into designing and making. The level of challenge might also have been a factor in that writing which was considered easy by the children was also perceived as less helpful. Many children talked about particular writing tasks making them think and moving their ideas forward. They also spoke of the outcomes of their writing tasks as external objects that could help them, for example, "It tells you what to do" and "If you forget you can just get it out and it helps you to remember".

From the children's survey, particular features of classroom practice emerged as being highly valued by children. In the two schools where design and technology was regarded the most positively, a large majority of the children perceived that in d&t they were able to move around the classroom more than in other lessons (78%) and that they were able to work with other children (81%). In the writing study, the majority of children stated a preference for working with others as opposed to individually, although some articulated difficulties in working with others, such as "falling out" or "taking over".

> I like mixing with other people, being sociable. (girl, 10)
>
> It's easier with someone else – we have half a job each. You get to decide with your friends and you can help each other when you get stuck. (boy, 10)
>
> I prefer it with someone else unless you have arguments. You get to talk and use the other person's skills. (boy, 11)
>
> It depends on how many is in a group. Pairs is OK. I like it on my own but not all the time. We are four friends. (girl, 11)

The majority of children in the children's survey also enjoyed the sense of autonomy that came with being able to develop their own ideas rather than being told what to do by the teacher (80%). This also contributed to a sense of fulfilment gained through creating a product themselves (90%). These perceptions indicate something inherent in the nature of design and technology which children find appealing but also suggest that the teaching and learning approaches used are significant in affecting children's perceptions and levels of engagement with the subject. The lack of physical constraints and opportunities to exercise their own agency in the classroom and work with peers suggests particular teaching and learning approaches and a less pressurised classroom atmosphere than might be found in many of the other lessons experienced by children in the later stages of primary school when there can be intense preparation for end of key stage tests.

In the writing study, three main concerns emerged for children: time, relevance, and control. Time was the most frequently mentioned category of response when children were asked what they would wish to change about their design and technology lessons. These children wanted more of it and in larger blocks to avoid the frustrations of being interrupted in the flow of their ideas or making and the repeated clearing away of equipment. Children from each school referred to having to wait for equipment or while the teacher talked to the whole class. There was a sense of impatience and urgency to get on with their activity. This might have been an immature desire for immediate gratification as the children enjoyed making, but in this study it also related to a perceived scarcity of time and the threat that this posed to the successful completion of their product. Time pressure has been identified by other studies into children's perceptions of their experience in school particularly since the introduction of a standardised curriculum (Pollard & Triggs, 2000; Rudduck & Flutter, 2004). Relevance is a related concept in that the children were less positive about any tasks which detracted from their primary purpose of designing and making a successful product, for example, writing tasks which did not directly contribute to their ability to design and make. It also referred to the level of challenge of particular tasks, with more children commenting on tasks which were too easy for them and therefore perceived as a waste of time than those which were too demanding. In terms of control, negative comments were about tasks which the children found constraining or being interrupted by the teacher; positive comments related to being able to do things for themselves and making their own decisions. A sense of autonomy has been found by many researchers to be closely associated with a greater level of engagement (e.g., Rudduck & Flutter, 2004).

IMPLICATIONS FOR TEACHING AND LEARNING

The evidence from these studies suggests that there might sometimes be a mismatch between teachers' and children's expectations of design and technology. Although many children at the top end of the primary school perceive design and technology as educationally worthwhile, their main focus of attention is located in their immediate experience of designing and making a particular product with a strong internal driver to achieve success. In contrast, teachers' perceptions are likely to cover a much broader range of objectives than the short-term objectives of the children (Cooper & McIntyre, 1996). In the context of design and technology, the goal-focused objective of designing and making a successful product can act as both help and hindrance to the teacher. It is undeniably motivating for the children, but can also be constraining in two main ways. Firstly, the desire for a successful product outcome can sometimes lead teachers to restrict children's options in an attempt to take greater control over their designing and making. This leads to fewer opportunities for authentic designing together with less risk-taking and genuine problem-solving (McCormick & Davidson, 1996). Children are less likely to learn from their mistakes and develop the necessary creative and problem-solving skills which form such an essential part of design and technological capability. Secondly, insufficient attention is often given to wider learning objectives, such as developing a critical understanding of the technological world and its impact on how we live our lives. The challenge for primary teachers is to work with the grain of pupil motivation in order to develop more effective teaching and learning situations. As teachers we need to recognise the potency of the creative act of designing and making a product and weave into this the necessary opportunities for children to experiment, to problem-solve, and to make their own decisions in such a way as to support success. We also need to consider the best ways of incorporating other important dimensions of learning into design and technology, such as technological literacy, whilst maintaining the high levels of engagement that children have with the subject.

Children's enjoyment of opportunities to work with others in design and technology lessons reminds us of the importance of children's social identities in the classroom. For them lessons are not merely about learning objectives and learning activities. How they go about tasks in relation to one another as social beings and the potential for fun can be critical factors for engagement. Add to this the significant role which discussion plays in effective designing (Kimbell, Stables, Wheeler, Wosniak, & Kelly, 1991) and learning (Mercer, 2000), and there emerges a strong argument for teachers to look for more opportunities to enable discussion and collaborative working in design and technology. However, what children also tell us is that as teachers we need to be aware of group dynamics and the difficulties that are sometimes experienced when working with others – particularly the 'falling out' or 'taking over' syndromes.

Several research studies have found that time is an issue for children, experienced by many as a commodity in short supply. There are often feelings of stress and anxiety caused by having to rush and not having adequate time to accomplish work (Doddington, Flutter, & Rudduck, 1999; Pollard & Triggs, 2000). Time was an issue for many of the children in the studies discussed here. The children either wanted more time for design and technology because they enjoyed it or felt anxiety about not being able to complete the task successfully in the time available. Time has also been identified as an essential component of a creative climate for developing creativity in the design and technology classroom (Nicholl, McLellan, & Kotob, 2008).

The findings from the writing study and the children's survey identify several features of effective learning which concur with more general studies of children's perceptions. The literature suggests a high level of consensus among children of various age groups about the characteristics of lessons and learning situations that help or hinder learning. Rudduck and Flutter (2004), in a summary of findings from their own studies in primary and secondary schools over a ten year period, identify four broad categories that describe pupils' views about what makes a good lesson:

- Opportunities for participation and engagement;
- Active lessons with a variety of learning tasks;
- Challenge (that is exciting but not overwhelming); and
- Opportunities to exercise autonomy. (p. 79)

These categories are represented in various ways in the responses of the children in these studies. In particular, the children talked about the active nature of design and technology, challenge, and autonomy. Nicholl et al. (2008) interviewed students aged 11–16 about their experience of design and technology and also found that autonomy and challenge were two key values for children. Where levels of autonomy and challenge were insufficient, students became de-motivated and were more likely to become disengaged with learning processes (see also Chapter 20). This confirms earlier findings from small-scale studies in secondary schools (e.g., Hughes, 2001; Neale, 2003). It could be argued that challenge and autonomy are two inherent characteristics of design and technology and yet what the children remind us is that we do not always get the levels right and that these are factors which have a significant impact on levels of engagement and consequently achievement.

IMPLICATIONS FOR CURRICULUM POLICY

The children in the writing study and the children's survey appeared to find their experience of design and technology relevant to their lives in three important ways: as children in the present moment, as future learners, and as future adults. However, their focus was emphatically on designing and making with minor references made to broader skills such as team-working, and attitudes such as perseverance. Although children showed an awareness of how design and technology might prepare them for future learning, jobs, careers, and parenthood, there was no reference to preparing to be a citizen. Technological literacy and the ethical issues so central to design and technology, such as sustainability, did not feature in the perceptions of these children. This might reflect the emphasis of the current curriculum although there is an increasing requirement for schools to develop children's awareness of these issues and practical resources are beginning to be developed to support this dimension of primary design and technology (e.g., Benson & Lunt, 2007b).

It can be argued that future developments in the curriculum for primary design and technology need to ensure a high degree of authenticity – both personal and cultural. For personal authenticity, learners have to be involved and the learning has to be meaningful for them. Listening to the pupils' voice will be essential to achieve this. For cultural authenticity the teaching and learning activity in school should relate to design and technology in the world outside school (McCormick, 2004). An increased emphasis on current technologies and technological literacy might prove to be a challenge for primary teachers working with children who perceive design and technology in terms of the designing and making model which has become so popular. Changes in the curriculum on a national, local or school basis will need to take account of children's perceptions. Whilst not the only stakeholders, they represent an important voice in the debate. Our understanding of children's perceptions of design and technology is a benchmark against which to judge the activities which we include in the curriculum. If these activities do not match children's current perceptions or values, we will need to consider the most effective ways of expanding children's current construct of design and technology education.

CONCLUSION

Research into primary school children's perceptions of design and technology is still at an early stage but the available studies confirm anecdotal evidence that design and technology is a subject which children find engaging, enjoyable, and educationally worthwhile. The research distinguishes between inherent characteristics of the subject – such as practical activity and its goal-oriented nature – and features of particular teaching approaches – such as the level of challenge, the degree of pupil autonomy, and the amount of collaborative working. More research is needed in order to investigate pupils' perceptions across the full age range. Research into children's perceptions of specific elements of design and technology such as designing or the use of ICT might also provide valuable insights to help us to refine our teaching of the subject.

The research discussed here was conducted in England and therefore only reflects one particular model of design and technology education. Research conducted in different countries would provide a richer picture of children's perceptions related to different curricular experiences and cultural contexts. The research shows that investigating children's perceptions can help to reveal implicit subject constructs, values and concerns, as well as gaps in provision. The construct of design and technology as a goal-oriented activity focused on the creative act of designing and making is a powerful model which needs to be considered in any future development of the curriculum as it is highly valued by children. The more we can learn from children the more we can fine tune the teaching and learning experiences we provide in order to maximise the learning potential of this dynamic and exciting subject.

REFERENCES

Arnot, M., McIntyre, D., Pedder, D., & Reay, D. (2004). *Consultation in the classroom: Developing dialogue about teaching and learning*. Cambridge, UK: Pearson.

Barlex, D., Welch, M., & Taylor, K. (2005). 'Never keep your ideas in your head': Elementary pupils' views of portfolios in technology education. In C. Benson, S. Lawson, & W. Till (Eds.), *Fifth International Primary Design and Technology Conference* (pp. 11–14). Birmingham, West Midlands: CRIPT.

Benson, C., & Lunt, J. (2007a). 'It puts a smile on your face!' What do children actually think of design and technology? Investigating the attitudes and perceptions of children aged 9–11. In J. R. Dakers, W. J. Dow, & M. J. de Vries (Eds.), *PATT 18 Conference* (pp. 297–305). Glasgow, Lanarkshire: University of Glasgow.

Benson, C., & Lunt, J. (Eds.). (2007b). *Developing sustainability through primary design and technology*. Birmingham, West Midlands: CRIPT.

Davies, B. (1982). *Life in the classroom and playground: The accounts of primary school children*. London: Routledge and Kegan Paul.

Christensen, P., & James, A. (Eds.). (2000). *Research with children: Perspectives and practices*. London: Falmer.

Cooper, P., & McIntyre, D. (1996). *Effective teaching and learning: Teachers' and students' perspectives*. Buckingham, West Midlands: Open University Press.

Doddington, C., Flutter, J., & Rudduck, J. (1999). *Improving learning: The pupils' agenda. A report for primary schools*. Cambridge, UK: Homerton College.

Dow, W. (2006). Implicit theories: Their impact on technology education. In J. R. Dakers (Ed.), *Defining technological literacy: Towards an epistemological framework* (pp. 239–250). New York: Macmillan.

Eggleston, J. (Ed.). (2000). *Teaching and learning design and technology: A guide to recent research and its applications*. London: Continuum.

Erickson, F., & Shultz, J. (1992). Students' experience of curriculum. In P. W. Jackson (Ed.), *Handbook of research on curriculum* (pp. 465–485). New York: Macmillan.

Gipps, C., & Tunstall, P. (1998). Effort, ability and the teacher: Young children's explanations for success and failure. *Oxford Review of Education, 24*(2), 149–165.

Greig, A., & Taylor, J. (1999). *Doing research with children*. London: Sage.

Harris, M., & Wilson, V. (2003). *Designs on the curriculum? A review of the literature on the impact of design and technology in schools*. London: Department for Education and Skills.

Hughes, M. (2001). Using students' views of design and technology to inform curriculum review at Key Stage 3. *The Journal of Design and Technology Education, 6*(2), 167–170.

Jeffrey, B. (2003). Countering learner 'instrumentalism' through creative mediation. *British Educational Research Journal, 29*(4), 489–503.

Kimbell, R. (1996). Technology. In P. Gordon (Ed.), *A guide to educational research* (pp. 175–196). London: Woburn.

Kimbell, R., Stables, K., Wheeler, T., Wosniak, A., & Kelly, V. (1991). *The assessment of performance in design and technology*. London: SEAC/HMSO.

Lunt, J. (2005). Pupils' views of writing tasks in design and technology: Purposeful activity or just more paperwork? In C. Benson, S. Lawson, & W. Till (Eds.), *Fifth International Primary Design and Technology Conference* (pp. 65–68). Birmingham, West Midlands: CRIPT.

Lunt, J. (2007). Investigating pupils' perceptions of writing tasks in design and technology. In C. Benson, S. Lawson, J. Lunt, & W. Till (Eds.), *Sixth International Primary Design and Technology Conference* (pp. 77–82). Birmingham, West Midlands: CRIPT.

McCallum, B., Hargreaves, E., & Gipps, C. (2000). Learning: The pupil's voice. *Cambridge Journal of Education, 30*(2), 275–289.

McCormick, R. (2004). Issues of learning and knowledge in technology education. *International Journal of Technology and Design Education, 14*, 21–44.

McCormick, R., & Davidson, M. (1996). Problem solving and the tyranny of product outcomes. *The Journal of Design and Technology Education, 1*(3), 230–241.

McIntyre, D., Pedder, D., & Rudduck, J. (2005). Pupil voice: Comfortable and uncomfortable learnings for teachers. *Research Papers in Education, 20*(2), 149–168.

Mercer, N. (2000). *Words & minds: How we use language to think together.* London: Routledge.
Neale, J. G. (2003). A longitudinal study of pupils' attitudes to design and technology at key stages three and four in the National Curriculum. In J. R. Dakers & M. J. de Vries (Eds.), *Pupil Attitudes Towards Technology: International Conference on Design and Technology Educational Research* (pp. 24–34). Glasgow, Lanarkshire: University of Glasgow.
Nicholl, B., McLellan, R., & Kotob, W. (2008). *Understanding creativity for creative understanding.* Cambridge, UK: University of Cambridge.
Pollard, A., & Filer, A. (1999). *The social world of pupil career: Strategic biographies through primary school.* London: Cassell.
Pollard, A., Thiessen, D., & Filer, A. (Eds.). (1997). *Children and their curriculum: The perspectives of primary and elementary school children.* London: Falmer.
Pollard, A., Triggs, P., Broadfoot, P., McNess, E., & Osborn, M. (2000). *What pupils say: Changing policy and practice in primary education.* London: Continuum.
Qualifications and Curriculum Authority. (1998). *Design and technology: A scheme of work for Key Stages 1 and 2.* London: Author.
Rudduck, J., & Flutter, J. (2000). Pupil participation and pupil perspective: Carving a new order of experience. *Cambridge Journal of Education, 30*(1), 75–89.
Rudduck, J., & Flutter, J. (2004). *How to improve your school: Giving pupils a voice.* London: Continuum.
Twyford, J., & Burden, R. (2000). 'Is it really work?' Primary school pupils' conceptions of design and technology as a National Curriculum subject. *The Journal of Design and Technology Education, 5*(2), 101–105.

Julie Lunt
Centre for Research and Curriculum Development in Primary Design and Technology (CRIPT)
Birmingham City University, UK

BILL NICHOLL AND ROS MCLELLAN

20. 'THIS ISN'T MY PROJECT [WORK]. IT'S ... JUST DO IT ... YOU JUST DO RESEARCH'

What Student Voice Reveals about the Nature of Design and Technology Lessons in English Schools, and the Implications this has on their Motivation and Learning of Complex tasks

INTRODUCTION

The topic of student voice has become increasingly popular with researchers and policy makers in recent years (for an overview see, for example, Rudduck & Flutter, 2004; Rudduck & McIntyre, 2007). Much of the research, however, reminds us that until recently pupils' voices have gone unheard in education and that students, when given the appropriate forum to voice their opinions, offer invaluable insights into their school experiences and in particular to what it is like being a learner, what gets in the way of being a learner, and what helps them learn (Rudduck & Flutter, 2004). The importance of such insights is deemed by many to be a crucial part of education. For example, Cook-Sather (2002) argues that, "It is time that we count students among those with the authority to participate both in the critique and in the reform of education" (p. 3). Rudduck and Flutter assert that pupils are the 'expert witnesses' with respect to providing insights into what helps them learn and what gets in the way of their learning, as well as a means for teachers to find out what interests and motivates them. Thus, "schools need to systematically find out as much as they can about the interests and concerns of their young people and use this information as a starting point to debate and work with young people in designing the arrangements of schooling" (Jamieson & Wikeley, 2000, p. 446).

Surprisingly, there has been little research that has investigated students' attitudes and perceptions of design and technology (D&T). Benson and Lunt's (2007) study surveyed primary students aged 9–11 years and found among other things that students enjoy D&T and enjoy making the most (see also Chapter 19 by Julie Lunt). Grover, Mitchell, and Fuller (2003) also surveyed a cohort of 13–15 year old students in order to elicit their views about D&T as part of a curriculum intervention.

In the empirically-based study presented in this chapter, we focus on 11–16 year old students' perceptions of learning in D&T in England. To help frame this study, we have focussed on four of Rudduck, Chaplain, and Wallace's (1996) six 'conditions of learning'. They include autonomy, intellectual challenge, support, and security. Rather than referring to an absolute state, autonomy is considered "as a right in relation to physical and social maturity" (Rudduck & Flutter, 2004, p. 133). Autonomy is one of the key things that "directly affect[s] the student's sense of self-as-learner" (p. 135). Students want "more opportunity to make decisions about what they do in class" and "want to work things out for themselves" (p. 83). 'Intellectual challenge' is defined as a learning activity that is "dynamic, engaging and empowering" for the student (p. 134). The final two conditions are where students feel supported in "relation to both their academic and emotional concerns" and

A.T. Jones and M.J. de Vries (eds.), International Handbook of Research and Development in Technology Education, 223–232.
© 2009 Sense Publishers. All rights reserved.

secure in that they "feel confident about learning from their mistakes and misunderstandings" (p. 134). These four conditions provide a useful framework in which we will introduce what students have told us about D&T. Later in the chapter, we return to these conditions in section, drawing on the psychological literature on motivation and relating this to Doyle's (1983) influential work on the nature of academic work.

THE APPROACH

The data used for this chapter were collected in schools in England and involved a number of different data sets. It is part of an on-going GATSBY funded research project on creativity in design and technology 11–16 years. Although this research focuses on creativity, the data analysed and discussed in this chapter were collected during the preliminary phase of the research (2005/6), where our aim was to find out what was happening in D&T classrooms generally.

A mixed-methods approach to data collection was taken, including focus group interviews with 126 students across six schools[1]; a survey of 4996 students aged 11–14 across 11 schools; observation of lessons (10 in total); and interviews with 14 teachers across six schools. Interview data were transcribed and coded using Nvivo, and a statistical package (SPSS) was used to help analyse the survey data. Given the discussion on student voice outlined above, the student focus interview data are foregrounded and much of the other data are used to complement points raised in the analysis.

WHAT STUDENTS SAID ABOUT DESIGN AND TECHNOLOGY

Students were asked during interviews to talk about projects they had recently undertaken in D&T. Their responses were then used to elicit further responses about their perceptions of classroom teaching and learning in D&T. It will become clear that the four conditions of learning discussed previously are interrelated. For presentation purposes however, we have decided to make two broader categories: challenge and autonomy, and support and safety.

Challenge and Autonomy

Many students talked about the tasks set by teachers saying, "In most of them [D&T lessons] you're told what to do" (male, 12 years) and, "Yeah, they give you a design and you've gotta do that exact design" (male, 15 years). This approach, where pupils follow a set of procedures to complete a task, resulting in all pupils making similar if not identical outcomes, was not limited to one area of D&T. Students spoke of similar approaches in all areas of D&T[2] and across all age groups, as these excerpts testify:

> Cos you just have to make a box out of wood and follow instructions all the same. (male, 14 years)
>
> Because you are all doing the same thing [in electronics]. Like you have got to make connections, a sequence of lights or something like that, and then you are all doing the same thing. (male, 14 years)
>
> In, like, electronics you get certain things to do and you make like little circuits and each one is basically the same apart from one or two components. (male, 12 years)

> When we do make cakes we're only allowed to put certain things in it...chocolate chips or something...coz in food tech you're not allowed to pick what you wanna make, you get told fruit salad and how...what fruit and all that stuff. (female, 12 years)

Given the procedural nature of the tasks described above, where students were given very little choice or autonomy over their work, it is hardly surprising that many students told us that their design folios and three-dimensional outcomes were identical, as these excerpts reveal:

> The only difference was how you painted it [the colour]. (male, 15 years)

> They all came out exactly the same. Unless you made a mistake! (male, 15 years)

> I think if you look at everyone's folios and all the research is the same, you've got a specification and you've got an analysis and you've probably got a moodboard. Everyone's done that you know and it doesn't take any talent to do that at all. (female, 16)

Design problems that were not relevant in the eyes of students and that allowed little opportunity for them to have a say in their learning were said to be 'boring' as these excerpts reveal:

> Those projects were a bit tedious because there were a number of things that we weren't allowed to do. (male, 14)

> It puts me off D&T and I think that's where people think it is boring. (female, 16)

There were a number of items on the questionnaire that supported students' perceptions of lack of autonomy. For example, only 57% thought they had a choice about the work they did in D&T. Several items in the questionnaire asked about choice in relation to certain aspects of D&T. For example, only 53% of students thought that teachers allowed them any choice with respect to materials/ingredients they could use in their work, and less than half (48%) felt they did not have any choice with respect to the research they did. The lack of autonomy and the transmission model of teaching knowledge as a set of procedures has implications for teaching and learning that are discussed later, but it is summed up here by this advanced skills teacher:[3]

> I think a lot of work that we do and including that here is repetitive, it's irrelevant and it's boring. It doesn't fire the kids up! (D&T teacher)

The lack of relevancy was an interesting observation made by this teacher, which we thought was worthy of further enquiry. Many students thought that the work they did was not very relevant. Students spoke about relevancy in terms of the overall nature of the task or design problem they were given, as this excerpt reveals: "You made a glass holder and nobody's got glasses in your home" (male, 12 years). Other students questioned the authenticity of smaller tasks within a design problem they were asked to solve. For example, one student questioned a task the class had been given:

> Guess what we did last lesson? We copied joints out of a textbook. Then guess what we did next lesson? We made the joints! What's the point in that?" (male, 14).

Furthermore, three female students who had just completed their GCSE Coursework[4] questioned the relevancy of doing tasks such as questionnaires, an end of project evaluation (which was a narrative, rather than an on-going evaluation to help inform design), and a plan for making (which one student reported was not how she made her final design). They admitted to fabricating much of their coursework because they were 'told to do it' by their teacher:

> To be honest, it didn't help me. I thought I had to do it. I made this up [questionnaire results] completely. That was a complete waste of time because he [the teacher] insisted on that as well. I did that the night before it had to be handed in. (female, 16)

> This is my evaluation. That I didn't like doing at all. It took me absolutely ages. But it was telling the story of it. I felt what was the point in me doing all this and explaining what I'm doing [the work]. (female, 16)

> I found that was irrelevant. This was the kind of making which, again, I didn't do it like that at all. (female, 16)

It is clear that these students were aware that such tasks do not lead to authentic learning experiences, as pages were retrospectively inserted into the portfolio even after the final artefact had been made. The students told us that they thought this was 'a waste of time' and they 'knew this at the time':

> I'm not a slack student but when you have to do all this...product analysis, moodboards and research that didn't really help me in the end. Because you have to do textbook things like that. (female, 16)

> Well when I was doing it [research] … this isn't my project [work]. It's...just do it...you just do the research. (female, 16)

The significance of the phrase "this isn't my project" is linked to this student's sense of ownership or autonomy, and the lack of authentic learning she experiences in D&T. Students were able to articulate the difference between tasks for getting grades and passing exams on the one hand, and authentic learning experiences, where they were allowed to think for themselves on the other:

> I enjoyed developing it [my idea] like this. Because you are thinking about it and this is me and this is my work not Mr X's. This isn't textbook where that is fully just copy and paste it off the internet. Look at the difference between it you can see anyone can do it. (female, 16)

> I would have found it better if we'd gone straight into the design ideas and then do the research as you need it. Because I've done more research than is actually relevant to my project. (female, 16)

Responses to related items on the questionnaire suggest students wanted opportunities to make decisions in their own learning. For example, the vast majority of students surveyed (89%) agreed that they felt "happiest about tasks when I can make my own decisions". Furthermore, 76% liked to "play with ideas in D&T and see how far they could go", and 66% agreed with the statement, "I like D&T problems where I can try out my own way of solving them". This was backed up by further interview data where students told us again and again that they wanted a say in their own learning, and this meant having choice and ownership over the learning experiences:

> I like when we get to design our own stuff. We get to make what we want and sort of thing. (male, 16)

> I like the fact we get to choose what we make...like...a whole range of options. (male, 16)

Given the importance stressed by students about having opportunities to think for themselves, it is worrying that as many as 17% of students surveyed disagreed with the statement that their teachers "encourage me to think for myself". When students were given tasks that allowed them to make some decisions in their own learning and therefore to think for themselves, they really enjoyed the experience. Furthermore, students could see the relevancy of the work, that is, they could understand how research could help inform their design thinking:

> But when you're into your own project and you're actually doing it and making it and thinking for yourself, it's then that you can do the research that was relevant to it because you want to do it. (female, 16)

> I like doing this sort of thing. It's quite difficult and I'm willing to put the effort in. But I didn't mind doing this sort of thing because I enjoy it. (female, 16)

This student quite rightly suggests that thinking for yourself can be quite difficult or, in other words, challenging. Students told us that they wanted tasks that were challenging, and that allowed them to think for themselves. For example:

> I think it would be quite nice to do something a bit more adventurous [with food tech] because usually the things we do are a bit basic. (male, 14)

Crucially however, students acknowledged the important role of the teacher in supporting them when they are undertaking such challenging tasks:

> If I was left to do it totally by myself I don't think it would have turned out as good (female, 16).

Support and Safety

Although students wanted tasks that were challenging and that allowed them a degree of autonomy, students also needed to be secure in the knowledge that they could rely on the support of their teacher from the outset:

> Put something from me into it [project] and not be scared that the teacher won't help me if I'm not choosing his idea. (female, 16)

> He sort of said do that one [idea]. I really didn't want to do it at first. But I thought if I didn't do it then he wouldn't give me the help that I needed to develop it [my idea]. (female, 16)

Not getting the teacher's support in this particular instance was a distinct possibility, as during one focus group interview the three girls told us how one of their friends chose to make an idea he had generated, against his teachers advice, and was told, "You're on your own now". Conversely, another student we interviewed told us about a task he had been doing as part of a textiles lesson. He wasn't sure he could realise his design idea using the very unforgiving technique of tie-dying. He told us how his teacher encouraged him to try it to see if it would work:

> When we had to make the pillow cases...I was trying to do the tie-dye and actually write words in it and that really didn't work very well. (male, 14)
> But the teacher let you have a go? (researcher)
> Yes which I thought was good. (male, 14)

The important point here is that this student was supported by his teacher and thus felt secure in pursuing his idea even though he was unsure whether it would work. Furthermore, when it did not work very well, the student clearly thought this was a valuable learning experience, as did his teacher. We shall return to this later when we discuss Doyle's (1983) work on the nature of academic work and students' understanding.

Supportive and secure 'climates' were not typical in some of the lessons we observed where students were expected to generate ideas for design problems they had been asked to solve. Ideas were closed down relatively early by teachers saying to pupils in various classes (11–14 years): "No, you can't do that", "that's too complicated", "do this", and "add a rubber band [to part of a design]". In several classes the teacher took the pencil from the hand of students and proceeded to sketch or adapt the students' design ideas for them without engaging in any dialogue. There may have been many good reasons why teachers thought it necessary to intervene in this way, for example, the culture of performativity which exists in English schools (Nicholl & McLellan, 2008) or teachers' implicit theories (Dow, 2006). The point we are making here, however, is that in none of the cases outlined above (except in the tie-dying example) was the teacher supportive or encouraging of students to try out an idea to see if it would work. This lack of support and encouragement, along with the lack of challenge and autonomy discussed previously, where tasks are reduced to a set of procedures, we argue, has serious implications for student motivation and students' learning in D&T. We expand on these issues in the next section.

IMPLICATIONS FOR TEACHING AND LEARNING IN D&T

There are clear parallels between Rudduck et al.'s (1996) 'conditions of learning' and the wider literature of motivation. Firstly, a perceived lack of autonomy and lack of challenge are linked to the research on motivation, which in turn has been shown to be key to creativity (Amabile, 1983, 1996), the focus of our wider research. It is thus important for teachers of D&T to understand the nature of motivation.

Deci and Ryan's (2002) influential theory of self-determination suggests that motivation can be explained by the fact that (i) we have an innate propensity towards active engagement with our environment (we want to interact with it, explore it) and (ii) there is a need for the existence of three underlying core psychological needs: competence, autonomy, and relateness. These needs must be fulfilled for us to persist and thrive. If the environment allows these three core needs to be satisfied, we experience well-being and psychological growth occurs. If these needs are thwarted, we feel frustration and will disengage. Clearly students in the classroom must have these core needs satisfied. Each need is now discussed in turn.

Competence is the need to feel effective when interacting with the environment, and to experience opportunities to express capabilities. This will lead students to seek challenges that are optimal for their skills and capacities. Autonomy is the need to feel that we are the source of our own behaviour; that we are acting in accordance with our interests and values and are self-determining. The data above suggest that there were many instances when students felt that they did not have autonomy. Students therefore need to feel that they are offered choices in the classroom. Relatedness refers to feeling connected to others – caring for and being cared for by others. Hence students need to have a sense of belonging in the classroom context and feel they will get support when they need it.

When students told us they lacked autonomy, were not appropriately challenged, and were not supported, their needs were not being met. Hence they were likely to be de-motivated. Other research has reported similar findings. For example, in the largest co-ordinated study ever conducted in the UK on teaching and learning, Rudduck and McIntyre (2007) reported four recurring themes students gave when asked what makes a good lesson: "the avoidance of tedium, the pursuit of meaningful learning, the need for togetherness [working and being supportive within a peer group] and the aspiration for autonomy" (p. 59). Furthermore, a number of studies in design and technology education report that students wanted more freedom (Neale, 2003); and freedom and challenge (Doppelt & Barak, 1999).

But classrooms do not have to be de-motivating. Research indicates it is possible to create motivationally supportive classrooms that enable creativity to flourish. Motivation researchers such as Anderman and Young (2002) have focused on a number of dimensions, including authority (who is responsible in the classroom). This is supported by the excerpts above; students enjoy having a sense of control and doing challenging work. At the same time, however, students told us that they wanted to be encouraged and supported by their teachers (and each other) when undertaking such tasks.

Whether students do have appropriate challenges, choices, and are supported by their teachers is not the main point, however. Rather, it is their perceptions of these that are of issue. The analysis of the data presented here suggests that their needs are not being met when undertaking tasks in D&T and it is their perceptions, as opposed to those of their teachers, that affect their motivation when undertaking D&T tasks. It would therefore appear that many students are de-motivated in D&T lessons. Clearly, there is a link between the types of tasks students do and their motivation, that is, their engagement (or otherwise) in doing these tasks. Hence, we now turn to the nature of the tasks that students are undertaking in D&T.

Doyle (1983) classifies academic tasks along two dimensions, namely, ambiguity and risk. He defines ambiguity as the "extent to which a precise answer can be defined in advance or a precise formula for generating an answer is available", whereas risk concerns the "stringency of the evaluation criteria a teacher uses and the likelihood that these criteria can be met on a given occasion" (p. 183). Tasks that have low ambiguity and low risk involve the use of memory, where students recall information or use a simple formula or a pre-defined set of procedures to solve simple problems.

It is clear from the data above that students perceived many of the tasks they were asked to do in D&T as low in ambiguity and low risk, as students followed a 'set of routines' to solve a 'problem', often making identical outcomes. An example from our studies is the task of making a wooden box, where students were allowed to choose the colour they wanted to paint the box, or the pattern with which they wanted to embellish/decorate the lid; this was usually a clichéd logo, something the authors have reported elsewhere (see Nicholl & McLellan, 2007a, 2007b). Galton (2002) argues that such tasks focus on the acquisition of procedural knowledge, and has been reported by others in education generally (e.g., Alexander, Schallert, & Hare, 1991) and in D&T education specifically (e.g., Dow, 2004, 2006; McCormack, 1997; McLellan & Nicholl, in press). Such tasks allow students to know 'how to' follow a set of steps to complete a task, for example, make a box.

Highly ambiguous and high risk tasks, on the other hand, seek to develop understanding by not just teaching 'how to' follow a set of procedures, but by exploring 'why' and 'where' this knowledge might be useful when solving more complex tasks. Since complex tasks are ambiguous and high risk and, as such, are more challenging and allow for many possible answers. This is especially valuable as one of the learning goals of tasks for understanding is learning to evaluate the appropriateness of outcomes (Galton, 2002). This was the case with the example of the textiles student who used tie-dying letters discussed referred to earlier in the chapter. Such tasks may also be viewed as more motivating by students as they potentially meet the need for challenge.

Doyle (1983) argues that procedural and comprehension tasks use different cognitive processes; this is an important point to explore in the context of D&T. Although Doyle's analysis preceded much of the work on metacognition (Flavell, 1987), the parallels are clear. Metacognition can be defined as the ability to think about one's thinking. Alexander et al. (1991) argue that to become an independent thinker, students must have knowledge of their own cognitive processes. It is comprehension-type tasks that allow students to develop understanding over time, so that they become "metacognitively wise" (Galton, 2002, p. 71), that is, are able to think independently from their teachers or peers.

Students have told us that they are doing mainly procedural-type tasks and this has implications for the teaching and learning in D&T. Firstly, procedural tasks do not allow students the opportunity to become 'metacognitively wise'. Secondly, according to Alexander et al. (1991), students need to become metacognitively wise in order to develop subject expertise within a domain, that is, to think like a designer/engineer/technologist. Therefore, tasks need to be carefully planned so that students can experience and develop a wider range of cognitive processes than they report experiencing at present.

It should be noted that Doyle (1983) and the authors are not saying procedural-type tasks are not appropriate. Indeed, there might be occasions where they are appropriate. Furthermore, procedural tasks can inform more ambiguous and high-risk tasks. But a diet of procedural-type tasks only is severely limiting to students developing understanding of complex tasks, as well as to their motivation. Dow (2006) presents a good case for moving away "from the transmission of facts or demonstration of skills towards the development of active, autonomous learners" (p. 309) in the context of D&T education. Deforges (1995) also highlights the relative limitations of direct instruction and procedural knowledge for deeper understanding more generally in education: "Direct instruction is best used for knowledge transmission, for showing, telling and demonstrating. It is never on its own sufficient to ensure deeper understanding, problem solving, creativity or group work capacities" (p. 129).

Deforges' (1995) reference to creativity above is of particular interest as it is the focus of the authors' wider research. Research on creative cognition (see Finke, Ward, & Smith, 1992; Ward, Smith, & Finke, 1999), which examines the fundamental cognitive processes that operate when people think creatively, suggests that such processes can never be deployed in procedural-type tasks as creativity can never be an algorithm (Amabile, 1983, 1996). Furthermore, Boden (2003) highlights the limitations of procedural knowledge in developing creativity, as "the more diverse types of knowledge (concepts, not just facts) a person requires, the richer their mental source for making novel combinations" (p. 96) and that concepts cannot be taught in a "rigid, departmentalised way" (ibid.). To develop creative understanding, then, tasks must go beyond the teaching of procedural knowledge. (For a wider discussion on metacognition and creative thinking, see Moseley, Baumfiled, Elliott, Gregson, Higgins, Miller, et al.'s (2005) work on *Frameworks for Thinking*.

CONCLUSIONS

Students in England have told us that they do not feel challenged, have little autonomy, and do not get support for their ideas when undertaking tasks in D&T. This, they have told us, affects their motivation. Furthermore, procedural-type tasks on their own are limiting in terms of deploying the necessary cognitive processes necessary for creative thinking, becoming 'metacognitively wise', and eventually developing design and technology expertise. In contrast, tasks that provide challenge, some autonomy, and appropriate support, that is, complex tasks, enhance student motivation. In other words, the nature of tasks students do and their motivation to engage with the tasks are interdependent. Teachers must listen to these 'expert witnesses', and plan for and support students' learning by engaging them in tasks that allow them to think creatively and become 'metacognitively wise'.

ACKNOWLEDGEMENTS

We would like to thank Dr Wafa Kotob for reading and commenting on initial drafts of this chapter.

NOTES

1 Student focus group interviews were conducted in single sex, single age groups.

2 D&T in England is divided into 4 broad areas: designing and making with materials (such as wood metal and plastics), electronics, textiles, and food. At the age of 14, students can choose to specialise in one of these areas.

3 Advanced skills teachers are teachers who have been selected for their exemplary classroom teaching practice. Although they are based in one school, their 'expertise' is used to develop classroom practice in other schools, both regionally and nationally.

4 GCSE is the formal examination to be taken at the end of formative schooling in England, aged 16. In Design & Technology, students are required to complete a design and make project which makes up 60% of their final GCSE grade. The other 40% is made up of a more formal written examination.

REFERENCES

Alexander, P., Schallert, D., & Hare, V. (1991). Coming to terms: How researchers in learning and literacy talk about knowledge. *Review of Educational Research, 61*(3), 315–343.

Amabile, T. (1983). *The social psychology of creativity*. New York: Springer-Verlag.

Amabile, T. (1996). *Creativity in context*. Boulder, CO: Westernview.

Anderman, E., & Young, A. J. (1994). Motivation and strategy use in science: Individual differences and classroom effects. *Journal of Research in Science Teaching, 31*(8), 811–831.

Benson, C., & Lunt, J. (2007). 'It puts a smile on your face!' What do children actually think of design and technology? Investigating the attitudes and perceptions of children aged 9–11. In J. Dakers, W. Dow, & M. de Vries (Eds.), *Pupils' attitudes towards technology: International conference in design and technology educational research. Conference proceedings* (pp. 296–305). Glasgow, Lanarkshire: Glasgow University.

Boden, M. A. (2003). Creativity and knowledge. In A. Craft, B. Jeffrey, & M. Liebling (Eds.), *Creativity in education* (pp. 95–102). London: Continuum.

Cook-Sather, A. (2002). Authorizing student perspectives: Towards trust dialogue and change in education. *Educational Researcher, 31*(4), 3–14.

Deci, E., & Ryan, R. (Eds.). (2002). *Handbook of self-determination research*. Rochester, NY: University of Rochester Press.

Deforges, C. (Ed.). (1995). *An introduction to teaching: Psychological perspectives*. Oxford, UK: Blackwell.

Doppelt, Y., & Barak, M. (1999). Pupils perspectives on the most influential characteristics and major outcomes of a rich technological learning environment. In I. Mottier & M. de Vries (Eds.), *Impacts of technology education. Proceedings of the PATT-9 conference* (pp. 36–44). Eindhoven, The Netherlands: PATT-Foundation.

Dow, W. (2004). The role of implicit theories in the development of creative classrooms. In E. W. L. Norman & D. Spendlove (Eds.), *DATA international research conference 2004. Creativity and innovation. Conference proceedings* (pp. 1–6). Wellesbourne, UK: DATA Wellesbourne.

Dow, W. (2006). The need to change pedagogies in science and technology subjects: A European perspective. *International Journal of Technology and Design Education, 16*(3), 307–321.

Doyle, W. (1983). Academic work. *Review of Educational Research, 52*(2), 159–199.

Finke, R. A., Ward, T. B., & Smith, S. M. (1992). *Creative cognition. Theory, research and applications*. Cambridge, MA: MIT Press.

Flavell, J. H. (1987). Speculations about the nature and development of metacognition. In F. Weinert & R. Kluwe (Eds.), *Metacognition, motivation and understanding* (pp. 21–29). Hillsdale, NJ: Erlbaum.

Galton, M. (2002). *Learning and teaching in the primary classroom*. London: Sage.

Grover, P., Mitchell, A., & Fuller, L. (2003). Pupils' values and perceptions of design and technology. In J. Dakers, W. Dow, & M. de Vries (Eds.), *Pupils' attitudes towards technology. International conference in design and technology educational research. Conference Proceedings* (pp. 244–250). Glasgow, Lanarkshire: Glasgow University.

Jamieson, I., & Wikeley, F. (2000). School effectiveness and consistency. *School effectiveness and school improvement, 11*(4), 435–452.

McCormack, R. (1997). Conceptual and procedural knowledge. *International Journal of Technology and Design Education, 7*(1), 141–159.

McLellan, R., & Nicholl, B. (in press). 'If I was going to design a chair, the last thing I would look at is a chair'. Product analysis and the causes of fixation in students' design work 11–16 years. Manuscript submitted to *Asia-Pacific Journal of Teacher Education.*

Moseley, D., Elliott, J., Gregson, M., & Higgins, S. (2005). Thinking skills frameworks for use in education and training. *British Education Research Journal, 31*(3), 367–390.

Neale, J. G. (2003). A longitudinal study of pupils' attitudes to design and technology at key stages three and four in the National Curriculum. In J. Dakers, W. Dow, & M. de Vries (Eds.), *Pupils' attitudes towards technology. International conference in design and technology educational research. Conference Proceedings* (pp. 24–34). Glasgow, Lanarkshire: Glasgow University.

Nicholl, B., & McLellan, R. (2007a). 'Oh yeah, yeah you get a lot of love hearts. The year 9s are notorious for love hearts. Everything is love hearts.' Fixation in pupils' design and technology work (11–16 years). *Design and Technology Education: An International Journal, 12*(1), 34–44.

Nicholl, B., & McLellan, R. (2007b). Fixated on popular culture and other things: What can students tell us about generating ideas in D&T. In J. Dakers, W. Dow, & M. de Vries (Eds.), *Pupils' attitudes towards technology. International conference in design and technology educational research. Conference Proceedings* (pp. 168–175). Glasgow, Lanarkshire: Glasgow University.

Nicholl, B., & McLellan, R. (2008). 'We're all in this game whether we like it or not to get a number of As to Cs'. Design and Technology teachers' beliefs about creativity and working within a climate of performativity. *British Educational Research Journal, Special edition*, 585–600.

Rudduck, J., Chaplain, T., &Wallace, G. (1996). *School improvement: What can pupils tell us?* London: David Fulton.

Rudduck, J., & Flutter, J. (2004). *How to improve your school: Giving pupils a voice*. London: Continuum.

Rudduck, J., & McIntyre, D. (2007). *Improving learning through consulting pupils*. London: Routledge.

Ward, T. B., Smith, S. M., & Finke, R. A. (1999). Creative cognition. In R. J. Sternberg (Ed.), *Handbook of creativity* (pp. 189–212). Cambridge, UK: Cambridge University Press.

Bill Nicholl
Ros McLellan
Faculty of Education
University of Cambridge, UK

LEO ELSHOF

21. TECHNOLOGY EDUCATION AND THE ENVIRONMENT: PERCEPTIONS AND PRACTICES

INTRODUCTION

Transforming our perceptions and understandings of how our technological culture interacts with the natural environment is emerging as arguably one of the most significant problems for human societies. Teachers' perceptions, worldviews, experiences, and sub-cultures all play a significant role in shaping technology education's stance with respect to sustainability and the environment. Historically, concern with the environment and sustainable dimensions of technological practice has not been an area of curricula emphasis, but this is changing.

Technological education, it can be argued, has a greater responsibility in terms of its orientation to sustainability issues and the responsible use of its subject than any other subject in the curriculum. Before beginning an examination of how technology education curricula represent 'the environment' we turn to the role of our worldviews in shaping our perception of the world.

WORLDVIEWS

> School education seems to be mostly stuck in an outdated industrial era worldview, unable to sufficiently address the significance and increasing rapidity of changes to humanity that are upon us. An integrated forward-looking view should, now more than ever, be of central importance in how we educate. (Gidley & Hampson, 2005, p. 255)

Psychologists have argued that parallels exist between the core beliefs held by individuals about their personal worlds and the collective worldviews of larger social groups. Societies, acting through their formal education systems and their informal customs, teach their young to see the world in culturally, historically, and socially constructed ways. This worldview formation process, while seldom explicitly acknowledged as such, is an inescapable part of maturation and cognitive growth. Our worldviews are the ways we interpret reality and are constructed through socialisation, taking shape as a 'cognitive macrostructure' consisting of images and basic assumptions that give coherency to our understanding of the world (Aerts, 2001; Cobern, 1991).

Some schooling practices merely reproduce cultural institutions with their inherent power inequities (Bourdieu, 1977). As Eisner (1992) suggests, schooling is a 'normative enterprise' aimed at promoting particular virtues at the exclusion of others. 'Curriculum ideologies' operate as "belief systems that provide the value premises from which decisions about practical educational matters are made" (Eisner, p. 302). Although our worldviews form the basis of "stable cognitive templates that produce regularities in expectations and interpretations of events" (Eidelson & Eidelson, 2003, p. 182), they can also be habituated to the point where they take the form of deeply entrenched immutable patterns for understanding the world, which in turn influence perceptions emotions and behaviours in problematic and destructive ways. For example, a number of

A.T. Jones and M.J. de Vries (eds.), International Handbook of Research and Development in Technology Education, 233–253.
© 2009 Sense Publishers. All rights reserved.

problematic worldview assumptions and presuppositions underlie many technology curricula. These include, in part:

- The 'product paradigm' provides the only viable orientation to technological education;
- economic growth is an unqualified 'good' and capitalist *laissez faire* economic systems are the only possible economic systems;
- technology is in essence a 'neutral tool';
- technology operates deterministically in society; and
- environmental and social dimensions of technology are of ancillary importance for technological education (Elshof, 2006, 2007).

Pavlova (2001) in an analysis of curricula from Australia, the UK, the USA, and Russia found that all four documents lacked a critical approach to technology insofar as they all reflected a 'strong belief' in technological progress and determinism, while failing to:

- Reinforce a "reasonable doubt" in technological progress and that "technology can escape from peoples control";
- question "technocratic consciousness" and the predominant categories of "purposive rational action and adaptive behavior" as key conceptual components in the framework for technology education;
- question the problematic relationship between power and technological knowledge; and
- acknowledge that uncertainty and precaution are key elements of technological systems and need to be taken more seriously within curricula. (p. 105)

Education for Sustainable Development and Sustainability

The term 'sustainable development' (SD) is not only a normative principle but also one that has emerged as a new normative orientation for Western society (Kemp & Martens, 2007). The interrelationships between economic, technological, social, and environmental issues have been incorporated into various definitions of sustainability and sustainable development. Adding to the complexity is the fact that many competing versions of sustainability exist along the shallower-deeper, weaker-stronger continuum and, as Mulvihill and Milan (2007) point out: "To some it is an urgent and achievable imperative to be taken quite literally, while to others it is a compass pointing to a distant goal to be achieved many generations from now, if ever" (p. 659).

SD is also a concept fraught with internal contradictions and controversy as hundreds of definitions abound, with some of the most popular schematics representing it as three pillars of: economy, environment, and society. Increasingly, the orientation towards sustainable development has shifted the focus from "saving the world" to "remaking the world". Ecological modernisation is one such broad theoretical framework that advances the idea of 'remaking the world' within sustainable development discourses. It attempts to describe emerging new technology-intensive modes of environmental reform that include concepts like industrial ecology, environmentally-conscious manufacturing, and ecological design (Allenby, 2003; Cohen, 2006).

Some have criticised the notion of 'education *for* sustainable development' (Jickling & Wals, 2008), arguing that the phrase has lost its critical social and environmental justice dimensions and has effectively been co-opted by the global business community to become "people, planet, profits" (Kemp & Martens, 2007, p. 6). There are merits to both arguments, but for the purposes of this discussion it is more important to focus on whether the curriculum antecedents and co-requisites of education for sustainable development are present, rather than the phrase itself. Pavlova (2007) argues that an 'effective conceptualisation' of education for SD sits somewhere between the idea that technology will fix the sustainability crisis (the techno-fix) and the wholesale transformation of human attitudes and values towards the natural world.

Analysis of Technology Education Curriculum Documents

Content analysis is any technique for making inferences by objectively and systematically identifying specified characteristics of texts (Bazeley, 2003). As such, content analysis is an analytical tool in which words and phrases, or 'units of analysis', are counted and categorised to determine the "manifest and latent contents" of a text (Petrina, 1998, p. 30). Petrina applied a similar technique to conduct a critical content and discourse analysis of the *Journal of Technology Education, Volumes 1–8.*

We will examine the extent to which the concepts, ideas, and processes that are associated with the SD concept are present in curriculum documents from five developed, wealthy, English-speaking countries. As G8 countries account for about 65% of the world economy, it is appropriate that three of the curricula analysed here are from G8 countries, including Canada, England, and the United States. In addition, curricula from Australia and New Zealand will also be examined. The documents are:

- *Standards for Technological Literacy: Content for the Study of Technology* (International Technology Education Association [ITEA], 2000, 258 pages);
- *Design Studies: Career and Technology Studies/Secondary Guide to Standards and Implementation* (Alberta Education, 1997, 283 pages);
- *The Ontario Curriculum Grades 9 and 10 Technological Education* (Ontario Ministry of Education and Training, 1999) and *The Ontario Curriculum Grades 11 and 12 Technological Education* (Ontario Ministry of Education and Training, 2000; 266 pages combined);
- *Technology and Enterprise Curriculum Framework Curriculum Guide* (Curriculum Council Western Australia, 2005, 66 pages);
- *Technology Curriculum Support* (New Zealand Ministry of Education, 2007, 108 pages); and
- *Design and Technology Key stages 1–4. The National Curriculum for England.* (Department for Education and Skills, & Qualifications and Curriculum Authority, 2004, 50 pages).

It should be emphasised that different curricula use language in unique, socially-situated and culturally-contextualised ways. Factors that will influence the word/phrase count include the fact that the curriculum documents are not all the same length nor are they attempting to accomplish the same things. For example, since education in Canada is a provincial responsibility, each province has its own curriculum; there has not to date been a 'Pan Canadian' technological education document produced similar to the American *Technology for All Americans Project.* Ontario and Alberta are two of Canada's richest provinces, representing the second largest manufacturing region and one of the largest oil and gas production regions in North America respectively. For this analysis, the frequency of selected words and phrases were determined from the most current published documents available through the internet. In the case of the province of Ontario, in order to more accurately represent the contents of the entire secondary school curriculum, two curriculum documents were analysed and reported under the heading 'Ontario'.

TECHNOLOGY EDUCATION AND SUSTAINABLE DEVELOPMENT

There is relatively little research literature to date that examines how technology teachers conceptualise and enact 'sustainable development' or environmental principles in the classroom. When Cross (1998) investigated a mixed subject group of teachers about their understandings of sustainable development, he found that the teachers were:

> … generally unaware of underlying theoretical issues surrounding the 'sustainable development' movement; they took the concept at face value [and they] were inhibited by a lack of knowledge of the complexities of the issues and how their teaching might contribute. (p. 50)

The teachers in Cross's study also identified a concern that they might inadvertently promote "learned helplessness" and a sense of fatalism within their students and thus some avoided the issue altogether. Given the situation where professional development and resources to tackle these issues are in short supply, as in Nova Scotia, it is understandable that some teachers take an ad-hoc approach to sustainable development, introducing concepts where they seem applicable but avoiding engaging students in any systemic exploration of the issue as it relates to technology.

Daugherty's (2005) survey of technology teachers found very strong agreement with the statement: "In the future, technology teacher education programs should prepare individuals to teach 'the effects of technology on the environment'". Daugherty did not articulate how these 'effects' might translate into curricula changes or teacher education programs. Elshof (2005) found that not only were technology teachers not very interested in learning about sustainable development issues as they might arise in the context of their classroom practices, but that they believed their students would not be very interested either. Hill and Elshof (2007) found that technology teachers had idiosyncratic understandings of sustainable development; while it was not a commonplace phrase in their teaching practices, many were aware and committed to some of the principles contained in the very wide umbrella of sustainable development. They also found that few teachers had taken formal courses in environmental science or design in their professional training, and that departments and boards of education offered very little or nothing that could be deemed professional development in 'sustainable practices'. The sections below report on the number of instances where specific concepts are mentioned within the curriculum documents included in the document analysis.

Technological Education to Create Products and Systems

Table 1 summarises the frequency of words and phrases related to the design, development, and manufacture of products. It is evident that all of the documents emphasise the word 'product' and 'design'. Given how much societal emphasis is placed on consumption, products, and the culture of consumerism in all of these capitalist democracies, it is perhaps not surprising to see how infrequently – if at all – these terms are referenced or problematised within the respective curriculum documents. 'Craft' or craft dimensions of technological work are rarely mentioned; this may be related to the fact that in terms of curriculum design, technology education has been attempting to distance itself from aspects of craft.

Shannon (1994) made one of the most compelling arguments for re-conceptualising 'design-based' education to have meaning and utility beyond the mere production of artefacts. Shannon talked about design as 'education for life on earth', basing it on:

> multi-dimensional purposive consciousness – 'designmind' whose overarching theme is the development of a sustainable future [whose design objectives] are:
> – To imbue a curriculum with compelling meaning, and
> – To develop humane, ecologically-grounded designers of a sustainable future. p. 293)

Shannon's scheme would see sustainability as a major sub theme integrated into developmental phases of technological education emphasising:

- Identity – deep internalisation of sustainability values;
- Context – the practice of hands-on, project-based, constructivist pedagogy; and
- Designing – learning that is thoroughly integrated with community life. p. 293)

Table 1. Technological education to create products and systems.

Word/phrase	Alberta	Ontario	United States	Australia	New Zealand	England
Product	64	101	195	20	107	22
Design	1989	416	586	83	167	132
Ingenuity	0	0	2	0	0	0
Creativity/creative	76	11	57	16	12	6
Aesthetic	59	32	5	37	16	5
Craft/crafted	1	0	11	0	2	0
Innovator/ Innovation	18	9	93	10	35	1
Conservation	1	4	19	0	0	0
Consumption	1	4	8	2	0	0
Consumerism	1	0	0	0	1	0
Pollution	0	1	20	2	2	0
Toxic/Toxin	2	1	2	0	0	0
Rebound effect	0	0	0	0	0	0
Climate change/ global warming	0	0	1	0	3	0
Responsible/ responsibility	28	10	10	0	6	2
Social responsibility	2	0	1	0	1	0
Risk (assessment)	2	4	15	19	26	6

None of the curriculum documents studied refers to the need to educate students about 'sustainable consumption' in terms of technological education, while the word 'responsibility' is found infrequently. All but one of the documents ignore the words 'pollution' and 'toxic' despite the fact that these are systemic global side-effects of our production systems. Nor is the 'rebound effect' identified, the rebound effect occurs when the aggregate consumption of a product increases after it has undergone an eco-efficiency improvement (e.g., using less material and/or energy to produce it), leaving a neutral or larger net environmental impact than before the improvement. As Huesemann and Huesemann (2007) point out, technological optimists point to the I=PAT equation believing that dramatic improvements in eco-efficiency will reduce the technology (T) factor, resulting in a decline in environmental impact (I) even as population (P) and per capita affluence (A) continue to grow. What is missing, according to Huesemann and Huesemann, is that technological change enables population and affluence to both grow even more rapidly, so they suggest a more appropriate formulation is I=P(T)*A(T)*T.

High levels of material consumption do not directly correlate with high measures of happiness or well being. In fact, some measures of relative happiness between populations indicate that citizens of rich G8 countries are less happy than countries much less technologically developed (Marks, Abdallah, Simms, & Thompson, 2006). From an ethical and moral standpoint, the world's poor deserve the same aspirations of the right to a long life and the same levels of material prosperity as those enjoyed by the majority of citizens in the world's developed countries. However, realising these aspirations for even a minimum majority of citizens in developing countries will be virtually impossible without creating even more widespread and long-term ecological damage if citizens of developed countries will not dramatically reduce their consumption levels in order to free up more 'environmental space' (Rees, 2002, 2003, 2006; Rees & Westra, 2003).

Despite the fact that a critical part of long-term sustainability involves learning to understand and take responsibility in terms of how we design, use, and dispose of technologies, the word 'responsibility' is under-represented in all of the curriculum documents. There is perhaps no better example of irresponsible technological policy than the continued reluctance of Canada, the United States, and Australia to implement the Kyoto Accord, the only international treaty attempting to set binding limits on the emissions of greenhouse gases. Only two of the curricula even mention global warming or climate change, a problem that the majority of the world's scientific community believe is the most significant environmental crisis facing humanity and the planet (Intergovernmental panel on climate change [IPCC], 2001, 2007). Climate change is directly related to human mis-development in terms of how we have designed our economies, technologies, and lifestyles to consume fossil energy and emit enormous quantities of greenhouse gases into the atmosphere. Human systems of consumption and production are actively changing the planet's natural biogeochemical systems without full appreciation of what the future consequences might be.

Technology Education and Fast Capitalism

Table 2 summarises words and phrases related to student learning concerning the workplace skills and understandings demanded by the business and industrial communities. Lankshear, Gee, Knovel, and Searle (1997) describe the business and management discourse found in popular media and government policy circles as 'fast capitalist texts'. Fast capitalist discourses become the mainstream popular culture stories concerning the 'proper' nature of social, work, and economic relationships for our so-called 'new times'. They trumpet the superiority of individualism, 'just-in-time' employment, and a relentless quest for novelty and branding: "You're CEO of Me Inc., reinventing yourself every few years while balancing a series of provocative, fascinating projects" (Brady, 2007, p. 72).

The importance of fast capitalist texts and the worldview they promote has enormous implications for the future of education, the nature of the workplace, and the environment. Their power lies in the fact that: "How we think and write about the world has a great deal to do with how we act in it and, thus, what it becomes in reality" (Lankshear et al., 1997, p. 25). These discourses become powerful in the hands of managers, education bureaucrats, and politicians, who work to enact fast capitalist ideas through their decisions and their prescriptions for the workplace, schools, and society. Fast capitalist discourses are "profoundly imperialistic insofar as they take over practices and social identities which previously were connected to churches, communities, universities, and governments" (Lankshear et al., 1997, p. 26). These discourses can be found in daily circulation in the editorials of large corporate newspapers, magazines, television news, and the internet. According to Lankshear, Gee, and Hull (1996):

> Fast capitalism is strong on encouraging and enabling 'critical reflection' on the part of employees at the level of understanding relations and processes *internal* to the organization's systems. … Fast capitalists do not, however, want to promote critical reflection in the sense of questioning systems as wholes and in their political relations to other systems. On the contrary, they are keen to pre-empt this … (p. 15).

Table 2. Technology and 'fast capitalism'.

Word/phrase	Alberta	Ontario	United States	Australia	New Zealand	England
Management	75	74	31	37	31	1
Client	10	21	0	3	56	1
Standards	139	149	*	54	3	2
Rules	3	19	6	6	3	1
Customer	10	34	3	0	3	0
Advertising	3	8	12	0	5	0
Solutions	106	124	104	42	5	2
Markets/ marketing	10	11	21	30	7	0
Industry/ industrial	20	273	45	34	7	10
Business	48	25	14	3	3	1
Skills	339	343	55	82	63	44
Quality	48	57	57	18	16	21
Procedures	98	144	37	29	9	2
Subsidies (government)	0	0	0	0	0	0
Taxes	0	0	0	0	0	0
Union(s)	0	0	0	0	0	0

* The American document is concerned with setting 'standards' for the profession.

Table 2 contains the highest frequency of words/phrases of any of the major categories explored in this study. With only a few notable exceptions, most of the documents contain a heavy emphasis on 'management', 'solutions', 'industry/industrial', 'business', 'quality', and 'skills'. The two Canadian documents contain the highest frequency of 'fast capitalism' terms, while those from England and New Zealand reflect these terms the least. These findings are consistent with Pavlova's (2005) assertion that 'late-modern' educational systems are dominated by concerns with human capital, the social capital, and social control discourses, as evidenced by concern with the international economy, knowledge competition, and a displacement of the political by the economic. The Ontario curriculum is unique for the frequency of the fast capitalism jargon of 'standards', 'solutions', 'skills', 'industry/industrial', 'procedures', and 'customers'.

None of the documents mention the words 'taxes' or 'subsidies'. The impact of taxes and government subsidies have an enormous influence on the nature of the mix of technologies that become commonplace on our highways, provide energy to our homes and businesses, and shape our built environment. For example, North Americans have continued to drive large, fuel-inefficient vehicles and live in increasingly larger houses largely because of the way in which a variety of subsidies keep fuel, transportation infrastructure, and energy costs below prices found in other industrialised nations (Lovins, 2005). Some of the most important public sustainability policy debates emerging in the near future will involve the question of how to tax and end public subsidies to inefficient technologies that pollute, and to figure out how to promote a greater number of 'green technologies', such as renewable solar and wind energy technologies, through subsidies and the tax system (REN21 Secretariat & Worldwatch Institute, 2008). This is especially true in developing countries where subsidies for green technologies could help communities bypass, or 'leapfrog', the 'dirty' development phase which characterised industrial growth in developed countries.

Young people need to understand that the so-called 'winning' technological products in the marketplace do not simply reflect the best ideas available or possible, but rather that the contest is often heavily influenced through politics and public policy. This dimension of community 'social innovation' with respect to technology is often neglected within technological education despite its crucial importance to sustainability.

Dimensions of Sustainable Development

Table 3 compiles the frequency of words/phrases that deal with the main facets of sustainable development. What is also noteworthy is how frequently the term 'environmental' is used relative to the more specific and local terms 'ecology' or 'ecosystem'. Although the word 'environmental' is found in all of the documents, very little or nothing at all is said concerning the specifics of how 'environmental' concerns and considerations should be dealt with in terms of technological education. Only two documents specifically mention 'sustainable development' and then only infrequently. None of the documents widely use the general terms 'sustainable' or 'sustainability' in the context of describing curriculum objectives and goals. The Ontario and U.S. documents have the highest frequency of economic references. The New Zealand document is one of the only ones to make use of the terms 'sustainable' and 'sustainability' more than once or twice.

Table 3. Dimensions of sustainable development.

Word/phrase	Alberta	Ontario	United States	Australia	New Zealand	England
Environmental	74	69	37	56	16	4
Environmental concern(s)	0	5	7	0	0	0
Ecology/	0	4	1	0	2	0
Ecological ecosystem	0	0	11	0	0	0
Economy/ economic	10	54	66	27	11	2
Sustainable development	0	0	0	0	2	1
Sustainability (environmental)/	0	1	2	2	11	1
Sustainable	0	0	2	0	5	2
Social/society	27	94	206	85	128	15

Technology as a Sociocultural Process

Table 4 summarises dimensions of technology as a sociocultural process. The Ontario and U.S. documents make frequent reference to local dimensions of the curriculum, while England has very few. The U.S. document also has references to the 'political' dimension of technology while the Ontario and English documents make no reference at all. What is interesting is how infrequently, if at all, these documents refer to the need to connect technological education to 'citizenship' considerations or the 'democratic' processes involved in technological development. Williams (2007) has argued convincingly that the democratisation of technological decision-making is essential if the worst effects of colonisation and globalisation are to be challenged by citizens.

As an evolving concept, sustainable development has emerged from a growing understanding that a better balance is necessary between economics, social progress, and a just and responsible stewardship of the environment. The report, *Learning for work, citizenship and sustainability*

(UNESCO-UNEVOC, 2004), identified the need to incorporate a different notion of citizenship in technical and vocational education:

> In a globalizing world, we must also re-visit our understanding of citizenship. It can no longer be narrowly political or narrowly national in scope but must embrace a wider conception of our rights, freedoms and responsibilities. The challenge of sustainable development gives fresh meaning to the notion of global citizenship. At the same time, it forces us to look at the practical exercise of democratic citizenship differently – not as an occasional visit to the polling station but involving, on a daily basis, choices about what we consume, how we dispose of our waste, where we work, what means of travel and communication we use, and how we interact with our fellow citizens and neighbors. (Matsuura, 2004, p. 20)

Table 4. Technology as a sociocultural process.

Word/phrase	Alberta	Ontario	United States	Australia	New Zealand	England
Local/locally	12	51	46	11	12	3
Community	42	26	49	16	49	7
Political	2	0	33	8	11	0
Democratic	0	0	2	0	0	0
Citizenship	0	2	0	0	1	7

References to the type of sustainability and citizenship consciousness described by Matsuura are under-represented in most of the documents examined here. This despite the fact that all of the documents examined in this chapter were developed after the 1992 Rio Earth Summit and some even after the 2002 Earth Summit +10 global conferences to raise awareness of these issues.

Technology and the Human Condition

Table 5, 'Technology and the human condition', includes the crucial dimensions of sustainability that relate to ethics, values, social justice, and an appreciation of the inequities in our technological world. The New Zealand document clearly makes the most frequent references to the inclusion of values, ethics, and morals in the curriculum, while the Ontario and English documents make the fewest.

Table 5. Technology and the human condition.

Word/Phrase	Alberta	Ontario	United States	Australia	New Zealand	England
Values	3	0	12	50	63	6
Ethics/ethical	22	28	21	25	30	1
Moral	2	0	3	0	6	7
History/historical	26	6	83	11	45	0
Social justice	0	0	0	2	0	0
Fairness	1	0	0	0	0	0
Equity/equitable	1	0	1	3	2	0
Inequality	0	0	0	0	0	0
Gender	0	0	4	7	0	4
Inclusive/inclusiveness	0	0	0	0	0	1
Developing world	0	0	0	0	0	0
Indigenous (technology)	0	0	0	0	2	0
Appropriate technology	0	0	0	1	0	0

As Martin (2007) points out: "Design and technology is an inherently controversial subject as it deals with the development, manufacture, use and effects of technology in human, social and environmental contexts" (p. 165). The report *Education commitment* (UNESCO, 2002) identified ethical values as the most effective agent of transformative change and is worth quoting at length:

> Education for sustainable development cannot be concerned only with disciplines that improve our understanding of nature, despite their undoubted value. Success in the struggle for sustainable development requires an approach to education that strengthens our engagement in support of other values – especially justice and fairness – and the awareness that we share a common destiny with others. Ultimately, sustainability will depend on changes in behavior and lifestyles, changes which will need to be motivated by a shift in values and rooted in the cultural and moral precepts upon which behavior is based. Without change of this kind, even the most enlightened legislation, the cleanest technology, the most sophisticated research will not succeed in steering society towards the long-term goal of sustainability. (p. 11)

What is notable here is that all of the curriculum documents ignore the phrase 'developing world' and nearly completely ignore acknowledgement of indigenous peoples' technologies and the entire 'appropriate technology' movement (Hoepfl, 2001; Wicklein & Kachmar, 2001). Only the Australian document references 'social justice', while 'inequality' and inequity' rarely appear anywhere in relation to individuals, communities, or what (Petrina & O'Riley (2001) call 'the economics of technology'. The phrase 'social justice' only appears in the Australian document while 'fairness' is absent from all of them. The U.S. document ranks highest in terms of acknowledging the historical dimensions of technology, while the English one, the home of the industrial technological revolution, makes no reference to this history at all.

Sustainable Energy Technologies

The emerging climate change crisis is a result of our continued over dependence upon carbon-based energy sources. Table 6 is interesting in terms of how nearly all of the documents ignore nearly all of the technologies that many experts believe we need to rapidly implement if we are to avoid catastrophic climate change. For example, 'energy efficiency' is only mentioned in two

Table 6. Sustainable energy technologies.

Word/Phrase	Alberta	Ontario	United States	Australia	New Zealand	England
Renewable	0	0	10	1	0	0
Energy efficiency	0	0	0	4	2	0
Solar	0	1	19	4	2	1
Photovoltaic (PV)	0	0	1	0	0	0
Wind (energy, turbines)	0	0	4	5	3	0
Biomass	0	0	1	0	0	0
Biofuel	0	0	0	0	0	0
Geothermal	0	0	2	0	0	0
Tidal Energy	0	0	0	0	0	0
Human powered	0	0	0	1	0	0

documents. Perhaps what is what is most significant is how infrequently, if at all, these future-critical technologies are even mentioned. The 'Bonn Declaration' formulated by participants in the UNESCO 'Learning for work, citizenship and sustainability' conference identified the need for modernisation and increased scope in technical and vocational education in "'sustainability industries' such as environmental conservation, cultural heritage site preservation and renewable energy production" as well as "quality skills development that leads to economically viable, environmentally sound and sustainable communities" (UNESCO-UNEVOC, 2004).

Technology and the Biosphere

Table 7, 'Technology and the biosphere', contains some of the key concepts, orientations, and principles related to a broader and more holistic approach to understanding of the relationship between humans, their technological systems, and the environment.

None of the documents make any explicit reference to any natural limits related to how much damage and disruption natural systems can sustain and remain functional despite technological extraction, production, pollution, development, and habitat destruction practices. The technological provision of some of these free 'ecosystem services', if even possible at all, would cost untold trillions of dollars and in nearly all cases provide poorer quality benefits than those provided by the natural systems we neglect and abuse today (Gutman, 2007; Howarth & Farber, 2002; Olgyay & Herdt, 2004).

Table 7. Technology and the biosphere.

Word/Phrase	Alberta	Ontario	United States	Australia	New Zealand	England
Limits (natural or environmental)	0	0	2	0	0	0
Uncertainty	0	0	0	0	0	0
Complexity	7	2	2	2	1	0
Relationship	20	20	4	13	34	0
Precautionary principle	0	0	0	0	0	0
Unintended (consequences)	0	0	4	0	5	0
Stewardship	0	0	0	0	0	0
Holistic	0	0	0	0	6	0
Future	38	7	38	1	122	2
Future generations	0	0	2	0	0	0
Interdependence (environmental)	0	0	0	0	0	0
Worldview	0	0	0	0	0	0

There was no mention of the need to consider the concept of 'interdependence' or 'stewardship' in terms of designing technologies that impact on the natural world. None of the documents mention that although technological design attempts to minimise 'uncertainty' in terms of environmental impact, it is nevertheless an inherent dimension of technological development. As socio-technological systems become more and more complex, the potential for adverse emergent effects and behaviours also rises (Homer Dixon, 2006). Students might consider uncertainty in terms of the adverse impacts of technologies on natural systems that span long periods of time and great distances. Most of the documents did not even mention 'unintended consequences' in relation to technological design and development. The 'precautionary principle' has been developed as a guide to action when the risk associated with unintended consequences of technological development is unacceptable. None of the documents mention the word 'precaution' or the phrase 'precautionary principle'. The precautionary principle is a well established environmental harm reduction strategy (Harremoës, 2001) and, along with the concepts of 'polluter pays' and 'reverse onus', deal directly with the practices of 'risk export' and 'responsibility evasion', which have become a pervasive byproduct of global production-consumption systems (Princen, 2005; Princen, Maniates, & Conca, 2002).

Only the American document mentioned 'future generations', but it is significant that the New Zealand document emphasises consideration of the future far more than any of the others. The concept of sustainable development is rooted in a future-oriented vision in which humankind is living in a different relationship with the natural systems of the planet, a future in which the basic needs of all people are met without destroying the systems that ultimately support all. 'Critical futures' study is concerned with critical and productive discourse about the character, assumptions, worldviews, and likely directions embedded within the dominant culture (Slaughter, 1996). It involves 'deep questioning' that moves away from a reductionist view of the future to a "complex multi-factorial, layered, multi-worldview of the future" (Inayatullah, 2002, p. 297).

'Stewardship' has been described as "the act of entrusting the careful and responsible management of the environment and natural resources for the benefit of the general community" (Bergeron & Archambault, 2005, p. 3). Stewardship, like 'sustainability', is a broader, more encompassing term than 'environment' and includes ethics as well as consideration of future generations:

> Stewardship is the moral obligation to care for the environment and the actions undertaken to provide that care. Stewardship implies the existence of an ethic of personal responsibility, an ethic of behavior based on reverence for the Earth and a sense of obligation to future generations. To effectively care for the environment, individuals must use resources wisely and efficiently, in part by placing self-imposed limits on personal consumption and altering personal expectations, habits and values. Appropriate use of natural resources within the stewardship ethic involves taking actions that respect the integrity of natural systems. (Dixon, quoted in Siemer, 2001, p. 20)

None of the documents mention the word 'stewardship' or what it might mean as an evolving 'ethic of care' that attempts to take a balanced account of societies' interests, future generations, and other species, and "accepts significant answerability to society" (Worrell & Appleby, 2000, p. 263).

Understanding Products within Ecosystems

Table 8 summarises the frequency of words/phrases that indicate an awareness and understanding of the need to think about material, waste, and energy flows involved in the manufacture of products in a cyclical or 'ecosystem thinking' manner. These processes involve 'life cycle' awareness and auditing of these flows in all phases of product development, including extraction, processing, transport, manufacture, use, and reuse of materials. Some of the terms, like 'life cycle analysis' and 'design for the environment', represent widely recognised and emerging eco-design industrial practices. Even the four well established concepts of recycle, reuse, remanufacture, and repair do not appear in all the documents.

Table 8. Understanding cycles, systems, and interdependence.

Word/Phrase	Alberta	Ontario	United States	Australia	New Zealand	England
Systems thinking	0	0	10	0	0	0
Clean(er) production	0	0	0	0	0	0
Life cycle analysis (LCA) (product-environmental)	0	0	2	0	4	0
Life cycle management (LCM)	0	0	0	0	0	0
Embodied energy of materials	0	0	0	0	0	0
Ecological or carbon footprint	0	0	0	0	0	0
Bioregion/ bio-regional	0	0	0	0	0	0
Construction ecology	0	0	0	0	0	0
Design for the environment (DfE)	0	0	0	0	0	0
Eco-efficiency	0	0	0	0	0	0
Closed loop design (materials)	0	0	0	0	0	0
Permaculture	0	0	0	0	0	0
Industrial ecology	0	0	0	0	0	0
Biomimicry/ bionics	0	0	0	0	0	0
Reuse (materials)	0	0	7	3	0	0
Remanufacture	0	0	0	0	0	0
Recycle	0	0	2	4	0	0
Repair	6	41	26	0	5	5
Dematerialisation	0	0	0	0	0	0

The need to reduce materials and energy throughput through our economic systems by recycling materials and remanufacturing products are becoming an essential part of responsible and smart business practice (Hawken, 2007; Pernick & Wilder, 2007).

Critical Thinking and Participatory Approaches

Table 9 lists words and phrases related to pedagogical approaches to teach students to think critically about technology. Dialogic classroom activities can include discussions, debates, and 'town-hall'-type meetings. These in part constitute important components of participatory democratic processes, prerequisites for effective involvement in community technological decision-making.

Table 9. Critical thinking and participatory approaches.

Word/Phrase	Alberta	Ontario	United States	Australia	New Zealand	England
Discuss/ discussion	34	0	33	17	69	1
Debate	0	0	2	0	7	0
Conflict/ conflicting	1	4	3	1	0	0
Philisophical/ philosophy	21	2	5	0	8	0
Town hall	0	0	0	0	0	0
Plan/planning	160	75	24	71	44	12
Forecasting/	0	0	3	0	2	0
backcasting	0	0	0	0	0	0
Scenario	0	0	1	0	0	0
Cooperate/ cooperative	12	15	5	14	0	0
Collaborate/ collaboration	16	6	12	12	13	0
Analyz(s)e/ analysis	8	77	29	18	54	7
Interpret/ interpretation	4	24	13	6	11	2
Strategy/ strategies	26	40	11	72	16	2
Awareness	8	7	2	6	5	0
Research	67	30	113	9	53	4
Judgment	0	0	3	0	3	3
Discern/ discernment	0	0	4	0	0	0
Critical thinking	2	2	6	4	0	0
Critical reflection	0	0	0	0	2	0
Critical analysis	0	0	0	1	10	0
Critical evaluation	0	0	1	0	8	0
Critical understanding	0	0	0	0	5	0
Critical literacy	0	0	0	0	1	0

The New Zealand document clearly emphasises these processes, while the Ontario and English documents are notable for ignoring these terms. The curricula also ignore for the most part any mention of the fact that conflicts and conflicting interests are often an inherent part of public technological decision-making. Open and participatory discussions of conflicts are an essential part of any sustainability discourse, as Lundegard and Wickman (2007) argue: "Value judgments dealing with human conflicts of interest are a foundation upon which the dialogue on sustainable development is constructed, and … they seem to be a prerequisite if the dialogue is not to come to a standstill" (p. 1). This may be a reflection that technology teachers have had little opportunity either through pre-service or in-service professional development to develop more complex understandings of the relationships between technology and sustainable development (Elshof, 2005; Hill & Elshof, 2007).

Developing and analysing scenarios is an important dimension of technological forecasting (Chermack, 2005; Selin, 2006). They have also been applied widely in environmental and sustainability planning. As Raskin et al. (2002) point out:

> Scenarios are stories with a logical plot and narrative about how the future might play out. Scenarios include images of the future – snapshots of the major features of interest at various points in time – and an account of the flow of events leading to such future conditions. (p. 14)

Educational scenario work in collaboration with some of the other critical thinking processes mentioned above constitute an important dimension of technological education, namely one that is future oriented and concerned with the development of confident citizens. None of the documents explicitly refer to scenario work in the context of technological education.

DISCUSSION

Pavlova (2005) describes a broad and creative vision of technological education for social change. It is worth quoting at length and includes:

> … the involvement of students in democratic debates on the future outlines of technological development; development of their social and ecological sensitivities; avoiding orienting their solutions exclusively to the standard of business efficiency and profitability criteria; helping them to distinguish real needs from desires; discussing the role of designed objects in the life of contemporary society; putting more emphasis on other than the aesthetic aspects of life that can provide existential meaning for people; challenging the way people are manipulated through advertising and cultivation of their desires; developing an active/creative attitude towards problems (not reactive); teaching students to formulate problems (not only being involved in problem solving); challenging consumer-oriented design; looking at design as one source of inspiration, not as a source of economic utility; and developing social responsibility. (p. 199)

Other progressive visions of technological education embracing 'multiliteracies' (Williams, 2007) or 'multidisciplines' (Petrina, 1998) are not as commonplace as this analysis indicates. Some curricula remain firmly entrenched within a narrow, instrumental, 'fast capitalist' vision of what technological education should and could be.

In terms of technology education and sustainable development, educators need to engage students beyond the simple techno-trend 'forecasting' activities that dominate the popular media. Many of these are largely concerned with the accuracy and precision of predictions, which are arrived at by looking at current trends and linearly extrapolating into the future. Rather, involving students with scenario planning entails considerations of alternative futures or "contouring the unknown" (Inayatullah, 2002, p. 297) to map out futures preferred as well as ones to be avoided with current and future technologies. These forms of 'backcasting' (Roorda, 2001) exercises begin with the premise that young people need to identify characteristics of preferred futures so that they might begin to consider the technological, social, and economic changes that need to also happen concurrently to realise a preferred future. This type of activity is critical to move beyond the dominant popular fallacy of the quick 'techno-fix' as a viable solution to the sustainability problem.

Breiting, Oslash, and Mogensen (1999) suggest that environmental issues appear at three levels of conflicting interest in a community and that youth need to struggle with all three:

- on the individual level conflict exists between incompatible needs and wishes, often expressed as personal dilemmas;

- on the social level conflicting interests exist between various groups and/or individuals and;
- conflicting interests can be regarded as conflicts at the structural level of society, e.g. conflicts between political decisions and market forces or economical mechanisms. (p. 350)

The technology education curricula examined in this chapter only address the first category of environmental issues identified by Breiting et al., that is, the individual consideration level of conflict – and then only superficially. All the documents fail to explicitly identify the hugely inequitable and unsustainable rates of consumption between developed and developing countries as a barrier to any genuine conception of sustainable development.

CONCLUSIONS

In terms of technology education, there are dangers in using phrases like 'environmentally friendly' or 'sustainable development' without careful consideration of what the terms might mean or entail. These catch phrases may serve as 'curriculum placeholders' to signify and acknowledge to other stakeholders that 'the environment' has been given due consideration in the development of the curriculum. However, the lack of specifics concerning how phrases like SD and 'environmental concerns' translate into emerging and well established sustainable technological practices suggests an underlying cursory or superficial concern with the environmental dimensions of technology.

In order to understand how our technologies are implicated in changing the planet's environment we increasingly depend on crucial 'whole earth' environmental measurements, such as carbon dioxide levels and ozone levels. These are: "sub-perceptual in any direct perceptual sense. One cannot directly sense them" (Idhe, 1997, p. 63). The only way for us to understand these sub-perceptual indicators is through 'embodied science', which Idhe explains as the "technological extension of primary perception through instrumentation" (p. 63). Indeed it is through this form of technologically-mediated instrumental realism, for example, that enables us to understand how minute changes in the order of parts per million (ppm) in atmospheric carbon dioxide concentrations can cause macroscopic changes such as permafrost and pack ice melting. Other examples include mathematical models and the super computers that run them, which enable predictions and scenarios about possible future climates. This is consistent with the idea that technological tools are not only artefacts but also work to "reveal a world" (Idhe, p. 66), and support Idhe's assertion that modern technoscience is fundamentally technologically embodied.

It would be inaccurate and unfair to characterise any or all of these curriculum documents as being 'anti-environment', because clearly they are not. 'Environment' and 'environmental concern' are clearly mentioned in the context of teaching young people about technology. There is unfortunately little or no guidance in most of the curriculum documents that could be used by teachers to better understand how these phrases translate into changing technological thinking or practices in the classroom. Research with technology teachers in Nova Scotia and Ontario indicates that very little in the way of professional development has been made available to teachers who would like to learn more about sustainability and technology (Hill & Elshof, 2007). Teachers clearly need more opportunities to discuss concepts related to the environment and technology within their own professional communities. This is necessary to in turn engage young people in critical discussions about their concerns and hopes for existing and future technologies.

Technological education, in choosing to focus attention on particular aspects of our technoculture while ignoring or downplaying other dimensions, contributes to a way of understanding the world that is largely congruent with dominant cultures, politics, and neoliberal economic and power relationships. There is little evidence that these technology curricula even reflect a commitment to 'ecological modernisation'. They certainly do not reflect an ecocentric worldview, nor do they encourage the development of ecocentric technological processes. Further, there is little to suggest that these curricula encourage any type of critical analysis of technology rooted in 'political ecology' at least as it relates to the environment (Petrina, 2000). As Martin (2002) points out, technology education is "shaped by the values of those in authority and the existing political culture" and that technology education which "fails to recognize the importance of values and attitudes is reduced to technical training" (p. 208).

Breiting et al. (1999) identify two basic approaches to considering environmental education. The first is the traditional behaviour modification approach, which "aims at prescribing certain of the pupils' behavioural patterns which we here and now believe will contribute to solving current environmental problems" (p. 350). The second is the 'action competence' approach: "related to developing a critical, reflective and participatory approach by which the developing adult can cope with future environmental problems" (p. 350). To develop such action competence entails "developing their ability and will to take part in democratic processes concerning man's [sic] exploitation of and dependence on natural resources in a critical way" (p. 350).

The documents examined here suggest that in terms of considering the environment within technology education, most are aligned with the behaviour modification and not the action competence approach. Behaviour modification is concerned with teaching students to use materials efficiently, to design 'greener' products, and perhaps to avoid creating too much waste or treat hazardous products in an inappropriate manner. The dynamics of action competence – which include a critical participatory and democratic exercise in technological decision making and the uncomfortable questions it raises concerning social justice, future generations, inequity and power – remain largely unaddressed.

It is clear that the concepts supporting more sustainable development through sustainable technological practices are not widely infused in the curriculum documents examined. In summary, this glimpse into the content of curricula from five countries reveals some consistent patterns and themes, which include little or no curriculum attention paid to:

- The concepts of sustainable development or 'sustainability';
- problematising consumerism and promoting more sustainable consumption practices;
- pollution and environmental collateral damage as a result existing technological practices;
- a very strong emphasis in some countries on 'management', 'solutions', 'skills,' 'standards', and industrial practices without critical attention paid to the corresponding environmental fallout from these existing practices;
- a citizenship dimension of technological practice or democratic and participatory practices related to technological decision-making;
- appropriate technology, indigenous technologies, and the developing world;
- social justice and inequality as it relates to current patterns of technological design and use;
- renewable energy technologies;
- consideration of future generations and stewardship principles related to the design and use of technology;
- the precautionary principle and a more realistic and holistic portrayal of technology with its uncertain and complex effects on societies and ecosystems; and
- the need to change to a lifecycle and ecological understanding of production-consumption systems.

Zerubaval (2006) suggests that the 'distinctive traditions' held by all professions and disciplines impact and socialise their members in terms of how they work in focusing attention on some parts of 'reality' while simultaneously ignoring others. As he states: "Noticing and ignoring are not just personal acts, since they are always performed by members of particular social communities with particular social conventions of attention and communication" (p. 20).

Given the enormous potential for advancing environmental sustainable 'good work' through technological education, it is incongruous that the formal technology curricula of wealthy developed nations contain so little in the way of specifics concerning sustainable practices. Technological education has struggled to establish a permanent place in the crowded curriculum; this study has revealed that it has much more to do in advancing technological education as a humanistic and sustainable practice. It needs to move toward practices that will help young people transform the built environment; re-imagine and recreate design, production and consumption systems; and most importantly help us live within our environmental means in a socially-just manner. As Fry (2003) states:

> Our failure to confront what we destroy when we create is obviously one reason why we lack the ability to recognise the critical ... we elevate and value futuring (we live in a world of desires and dreams for commodified utopias), yet we are oblivious to what the realisation of our projected futures defuture, that is we are mostly unable to see the critical destruction accompanying the produced ... there can be no real possibility of 'the good' unless one confronts 'the bad' as it is materialised rather than moralised. (p. 2)

If, as Pinar, Reynolds, Slattery, and Taubman (1995) suggest, "curriculum is a[n] extraordinarily complicated conversation" (p. 848), then these findings indicate that in terms of technological education and sustainability, the community conversation needs to become both more complicated and critical.

REFERENCES

Aerts, D. (2001). Transdisciplinary and integrative sciences in sustainable development. In *Our fragile world, a forerunner of the encyclopedia of life support systems*. Oxford, UK: Aldates.

Alberta Education. (1997). *Design studies: Career and technology studies/Secondary guide to standards and implementation*. Edmonton, Canada: Government of Alberta.

Allenby, B. (2003). Industrial ecology. In A. Molella & J. Bedi (Eds.), *Inventing for the environment* (pp. 339–372). Cambridge, MA: MIT Press.

Bazeley, P. (2003). Computerized data analysis for mixed methods research. In A. Tashakkori & C. Teddle (Eds.), *Handbook of mixed methods in social & behavioral research* (pp. 385–422). London: Sage.

Bergeron, S., & Archambault, É. (2005). *Canadian stewardship practices for environmental nanotechnology. A report prepared for Environment Canada*. Montréal, Canada: Science-Metrix.

Bourdieu, P. (1977). *Reproduction in society's education and culture*. London: Sage.

Brady, D. (2007). Creating brand you. *BusinessWeek, 4047*, 72.

Breiting, S., Oslash, R., & Mogensen, F. (1999). Action competence and environmental education. *Cambridge Journal of Education, 29*(3), 349–353.

Chermack, T. J. (2005). Studying scenario planning: Theory, research suggestions, and hypotheses. *Technological Forecasting and Social Change, 72*(1), 59–73.

Cobern, W. W. (1991). *World view-reality as viewed by students: A synopsis of methodology*. Working Paper No. 101 of the Scientific Literacy and Cultural Studies Project (SLCSP). (ERIC document Reproduction Service No. SK# 344 745)

Cohen, M. J. (2006). Ecological modernization and its discontents: The American environmental movement's resistance to an innovation-driven future. *Futures, 38*(5), 528–547.

Cross, R. T. (1998). Teachers' views about what to do about sustainable development. *Environmental Education Research, 4*(1), 41–54.

Curriculum Council Western Australia. (2005). *Technology and enterprise curriculum framework: Curriculum guide*. Osborne Park, Western Australia: Author.

Daugherty, M. K. (2005). A changing role for technology teacher education. *Journal of Industrial Teacher Education, 42*(1), 41–58.

Department for Education and Skills, & Qualifications and Curriculum Authority. (2004). *Design and technology key stages 1–4. The national curriculum for England*. London: The Crown.

Eidelson, R. J., & Eidelson, J. I. (2003). Dangerous ideas: Five beliefs that propel groups toward conflict. *American Psychologist, 58*(3), 182–192.

Eisner, E. W. (1992). Curriculum ideologies. In P. W. Jackson (Ed.), *Handbook of research on curriculum* (pp. 302–326). New York: Macmillan.

Elshof, L. (2005). Teachers' interpretations of sustainable development. *International Journal of Technology and Design Education, 15*(2), 173–186.

Elshof, L. (2006). Productivism and the product paradigm in technological education. *Journal of Technology Education, 17*(2), 17–31.

Elshof, L. (2007, June). *Moving beyond 'artifactual knowing': Emergent ideas for a 21st century technology education.* Paper presented at the Pupils Attitudes Towards Technology PATT 18 International Conference on Design and Technology Educational Research, Glasgow.

Fry, T. (2003). *Redirection and remaking the ground of design.* Retrieved Jan 10, 2008, from www.edf.edu.au

Gidley, J. M., & Hampson, G. P. (2005). The evolution of futures in school education. *Futures, 37*(4), 255–271.

Gutman, P. (2007). Ecosystem services: Foundations for a new rural-urban compact. *Ecological Economics, 62*(3–4), 383–387.

Harremoës, P. (2001). *Late lessons from early warnings: The precautionary principle 1896–2000. Environmental issue report No 22.* Copenhagen: European Environment Agency.

Hawken, P. (2007). *Blessed unrest.* New York: Viking.

Hill, A. M., & Elshof, L. (2007). *Sustainable practices as an aspect of technological literacy: Research findings from secondary school teachers and their classrooms.* Paper presented at the Pupils Attitudes Towards Technology PATT 18 International Conference on Design and Technology Educational Research, Glasgow.

Hoepfl, M. (2001). Design criteria for developing appropriate technology. In R. C. Wicklein (Ed.), *Appropriate technology for sustainable living, 50th yearbook. Council on technology teacher education* (pp. 92–112). New York: Glencoe McGraw-Hill.

Homer Dixon, T. F. (2006). *The upside of down: Catastrophe, creativity and the renewal of civilization* (1st. ed.). Toronto, ON: A.A. Knopf.

Howarth, R. B., & Farber, S. (2002). Accounting for the value of ecosystem services. *Ecological Economics, 41*(3), 421–429.

Huesemann, M., & Huesemann, J. (2007). Will progress in science and technology avert or accelerate global collapse? A critical analysis and policy recommendations. *Environment, Development and Sustainability, 10*(6), 787–825.

Inayatullah, S. (2002). Reductionism or layered complexity? The futures of futures studies. *Futures, 34*(3), 295–302.

Idhe, D. (1997). Whole earth measurements. *Philosophy & Technology, 2*(2), 61–72.

Intergovernmental Panel on Climate Change [IPCC]. (2001). *Climate Change 2001: The Scientific Basis, Summary for Policymakers.* Retrieved from http://www.ipcc.ch

IPCC. (2007). *Climate change 2007: Impacts, adaptation and vulnerability working group II contribution to the intergovernmental panel on climate change. Fourth assessment report. Summary for policymakers.* Geneva: United Nations IPCC Secretariat.

International Technology Education Association [ITEA]. (2000). *Standards for technological literacy: Content for the study of technology.* Reston, VA: Author.

Jickling, B., & Wals, A. E. J. (2008). Globalization and environmental education: Looking beyond sustainable development. *Journal of Curriculum Studies, 40*(1), 1–21.

Kemp, R., & Martens, P. (2007). Sustainable development: How to manage something that is subjective and never can be achieved? *Sustainability: Science, Practice, & Policy, 3*(2), 5–14.

Lankshear, C., Gee, J. P., & Hull, G. (1996). *The new work order behind the language of the new capitalism.* Boulder, CO: Westview.

Lankshear, C., Gee, J. P., Knovel, M., & Searle, C. (1997). *Changing literacies.* Buckingham, UK and Philadelphia: Open University Press.

Lovins, A. B. (2005). *Winning the oil endgame.* Snowmass, CO: Rocky Mountain Institute.

Lundegard, I., & Wickman, P.-O. (2007). Conflicts of interest: An indispensable element of education for sustainable development. *Environmental Education Research, 13*(1), 1–15.

Marks, N., Abdallah, S., Simms, A., & Thompson, S. (2006). *The UnHappy index.* London: New Economics Foundation.

Martin, M. (2002). Values and attitudes in design and technology. In S. Sayers, J. Morley, & B. Barnes (Eds.), *Issues in design and technology teaching* (pp. 208–223). London: Routledge.

Martin, M. (2007, June). *Role of product evaluation in developing technological literacy.* Paper presented at the Pupils Attitudes Towards Technology International Conference on Design and Technology Educational Research, Glasgow.

Matsuura, K. (2004, October). *Learning for work, citizenship and sustainability: UNESCO's role and contribution.* Paper presented at the Learning for Work, Citizenship and Sustainability UNESCO International Experts Meeting Final Report 2004. Retrieved from www.unevoc.unesco.org/sustainable

Mulvihill, P. R., & Milan, M. J. (2007). Subtle world: Beyond sustainability, beyond information. *Futures, 39*(6), 657–668.

New Zealand Ministry of Education. (2007). *Technology curriculum support.* Wellington, NZ: Author.

Olgyay, V., & Herdt, J. (2004). The application of ecosystems services criteria for green building assessment. *Solar Energy. The American Solar Energy Society's Solar 2004 Special Issue, 77*(4), 389–398.

Ontario Ministry of Education and Training. (1999). *The Ontario curriculum grades 9 and 10 technological education.* Toronto, ON: Author.

Ontario Ministry of Education and Training. (2000). *The Ontario curriculum grades 11 and 12 technological education.* Toronto, ON: Author.

Pavlova, M. (2001). Interpretation of technology in curriculum documents—A comparative analysis. *IDATER Archive.* Retrieved from https://dspace.lboro.ac.uk/dspace-jspui/handle/2134/954

Pavlova, M. (2005). Social change: How should technology education respond? *International Journal of Technology and Design Education, 15*(3), 199–215.

Pavlova, M. (2007). *Two pathways, one destination—TVET for a sustainable future. Paper presented at a Virtual conference October 22 – November 10, 2007.* Final report for UNESCO International Centre for Technical and Vocational Education and Training (UNESCO-UNEVOC) in collaboration with Griffith Institute for Educational Research, Australia. Griffith Institute for Educational Research: UNESCO-UNEVOC International Centre for Technical and Vocational Education and Training.

Pernick, R., & Wilder, C. (2007). *The clean tech revolution.* New York: Collins.

Petrina, S. (1998). Men at work: Inspecting the foundations of technology education. *Journal of Industrial Teacher Education, 36*(1), 1–13.

Petrina, S. (2000). The political ecology of design and technology education: An inquiry into methods. *International Journal of Technology and Design Education, 10*(3), 207–237.

Petrina, S., & O'Riley, P. (2001). Economics of appropriate technology. In R. C. Wicklein (Ed.), *Appropriate technology for sustainable living* (pp. 22–55). New York: Council on Technology Teacher Education/Glencoe McGraw Hill.

Pinar, W. F., Reynolds, W. M., Slattery, P., & Taubman, P. M. (1995). *Understanding curriculum: An introduction to the study of historical and contemporary curriculum discourses.* New York: Peter Lang.

Princen, T. (2005). *The logic of sufficiency.* London: MIT Press.

Princen, T., Maniates, M., & Conca, K. (2002). Confronting consumption. In T. Princen, M. Maniates, & K. Conca (Eds.), *Confronting consumption* (pp. 1–20). London: MIT Press.

Raskin, P., Banuri, T., Gallopin, G., Gutman, P., Hammond, A., Kates, R., et al. (2002). *Great transitions: The promise and lure of the times ahead.* Boston: Stockholm Environment Institute.

Rees, W. E. (2002). Globalization and sustainability: Conflict or convergence? *Bulletin of Science, Technology & Society, 22*(4), 249–268.

Rees, W. E. (2003). Economic development and environmental protection: An ecological economics perspective. *Environmental Monitoring and Assessment, 86*(1), 29–45.

Rees, W. E. (2006). Globalization, trade and migration: Undermining sustainability. *Ecological Economics, 59*(2), 220–225.

Rees, W. E., & Westra, L. (2003). When consumption does violence: Can there be sustainability and environmental justice in a resource-limited world? In J. Agyeman, R. D. Bullard & B. Evans (Eds.), *Just sustainabilities. Development in an unequal world* (pp. 99–124). Cambridge, MA: MIT Press.

REN21 Secretariat, & Worldwatch Institute. (2008). *Renewables 2007 Global Status Report.* Paris and Washington, DC: Author.

Roorda, N. (2001). Backcasting the future. *International Journal of Sustainability in Higher Education, 2*(1), 63–69.

Selin, C. (2006). Trust and the illusive force of scenarios. *Futures, 38*(1), 1–14.

Shannon, M. J. (1994). Designing education for life on earth. *International Journal of Technology and Design Education, 4*(3), 289–296.

Siemer, W. F. (2001). Best practices for curriculum, teaching, and evaluation components of aquatic stewardship education. In A. J. Fedler (Ed.), *Defining best practices in boating, fishing, and stewardship education* (pp. 18–36). Alexandria, VA: Recreational Boating & Fishing Foundation.

Slaughter, R. (1996). Critical futures study as an educational strategy. In R. Slaughter (Ed.), *New thinking for a new millennium* (pp. 137–154). London: Routledge.

UNESCO. (2002, June). *Education for sustainability from Rio to Johannesburg: Lessons learnt from a decade of commitment. World summit on sustainable development.* Paris: UNESCO.

UNESCO-UNEVOC. (2004). *Learning for work, citizenship and sustainability: UNESCO international experts meeting. Final report, October 2004.* Bonn: Author.

Wicklein, R. C., & Kachmar, C. J. (2001). Philosophical rationale for appropriate technology. In R. C. Wicklein (Ed.), *Appropriate technology for sustainable living* (pp. 3–21). New York: Council on Technology Teacher Education/Glencoe McGraw Hill.

Williams, J. P. (2007). Technological literacy: A multiliteracies approach for democracy. *International Journal of Technology and Design Education.* Advance online publication. doi: 10.1007/s10798-007-9046-0

Worrell, R., & Appleby, M. (2000). Stewardship of natural resources: Definition, ethical and practical aspects. *Journal of Agricultural and Environmental Ethics, 12*, 263–277.

Zerubavel, E. (2006). *The elephant in the room: Silence and denial in everyday life*. New York: Oxford University Press.

Leo Elshof
Acadia University
Wolfville, Nova Scotia
Canada

ERIC PARKINSON AND GILL HOPE

22. CONCEPTUAL LEARNING IN AND THROUGH TECHNOLOGY

INTRODUCTION

Technology broadly concerns thoughts and actions enabling the construction and manipulation of concrete and virtual worlds. These acts do not occur in isolation. They rely and are built upon various forms of socially constructed knowledge, which lie at the heart of what it is to be human.

The following chapter explores some of the issues surrounding the idea of technological knowledge itself before moving on to consider what might be termed the 'conceptual' side of technology in relation to learning. This conceptual face of technology is seen to be based upon abstract ideas and contained within the mind. It will be shown that the sensory-informed or 'perceptual' side of experience, practical 'doing' experiences, and the conceptual dimension of experience are at times difficult to disentangle and indeed contribute to the special qualities of technology education itself.

KNOWING AND DOING

Traditionally, the priorities of philosophers have been to recognise and establish ways of thinking that are seen to be 'higher' and sufficiently different to the ordinary workaday world of craftsmen or the domestic realm of the production of food, clothing, and shelter. We see this in contemporary Western culture in terms of the separation of 'academic' activities from 'practical' studies in education. However, despite this artificial separation of the 'thinking mind' from the 'doing mind', the problems that technology addresses are frequently complex and non-trivial; are created, reformulated, and re-interpreted in contexts of the actor's own making; and necessitate both original and routine reflection (Schlanger, 1997; de Vries & Tamir, 1997).

Part of the debate about conceptual knowledge within technology education concerns the wider issue of the nature of knowledge within technological thought and action. Whatever kind of field of knowledge technology may be, it may be said to be essentially a *doing* field (Keirl, 2000). But what kind of 'doing', and is this 'doing' a form of knowledge? The 'doing' of technology is most obvious in practical situations, where technology can be seen in action, when materials are cut, shaped, joined, and caused to interact with one another. In the modern industrial world, technological knowledge has become increasingly dependent on electronic implements used to manipulate virtual realities captured through the interaction of software and hardware, which then are applied to production and communication processes. These perhaps are electronic replacements for some of the physical manipulation-based craft skills of the past.

A.T. Jones and M.J. de Vries (eds.), International Handbook of Research and Development in Technology Education, 255–264.
© 2009 Sense Publishers. All rights reserved.

TECHNOLOGY AND SCIENCE

Western science has had a deep impact on the knowledge base and working methods of industrial production around the world. However, it is also the case that much technological activity, historically and globally, has developed (and continues to develop) with very little reference to the kinds of scientific knowledge or processes that have evolved in Europe and America in the past 250 years.

The harnessing of scientific knowledge has, unfortunately, led to technology being often perceived as a derivative activity. This view not only implies the down-grading of the creative and innovative impulses that underpin practical activity, but equates 'technology' with modern industrial engineering, and 'technology education' to being tied strongly to vocational education for the industrial workplace, as Ginestié (2005) demonstrates. This 'science-technology tag' is applied to the most advanced forms of industrial innovation, especially computing, to the detriment of other forms of technological activity (Jones, 1997; de Vries & Tamir, 1997).

One of the crucial markers for technological activity is that it is eclectic, using anything that comes to mind or hand in order to solve a current problem or take advantage of a newly perceived opportunity. Thus, technological activity is opportunistic in its mode of operation at both cognitive and practical levels. It is a supremely ingenious form of activity, in comparison to which the routine and predictable construction actions of other animals who manage their environment (such as beavers, bees, coral polyps, and ants) are massively outclassed. Humans are creative in a way that other creatures are not (excepting perhaps to some extent our closely-related primate cousins).

FORMS OF KNOWLEDGE

In the Western philosophical tradition, epistemological debate has mostly centred upon the assertion that rationality, as expressed in symbolic language, is the highest form of knowledge. Ryle (1949), however, offered a classification of knowledge into knowing how (procedural knowledge) and knowing that (conceptual knowledge) that is beguiling to technologists since, within a technical context, 'know how' relates to the learning of skill-based procedures for the manipulation of tools and materials, and – crucially – may also account for aspects of strategy choice within problem solving.

Further to this, within design repertoires, this procedural knowledge is intimately linked with conceptual knowledge due to cognitive capacities (such as extrapolation, insight, and analogy) that arise through the interrelationships between thinking and making, artefact creation, materials handling, and acts of reflection that occur within any design, construction, and evaluative cycle.

Among the critiques of Ryle's scheme, Hamlyn (1970) asserts that elements of knowledge require both formulation and application, such that procedural 'know how' might precede the conceptual 'know that' component and that some forms of knowledge are both theoretical and practical at the same time. It may even be that attempts to increase students' factual knowledge may impede the development of intellectual skills (Balfanz, 1991). Hope (2000) added 'know relevance' as a third category to Ryle's scheme, which enabled 'know how' and 'know that' to contribute to the strategy knowledge required to conduct technological activity. In her scheme, analogy, metaphor, and extrapolation are also seen as members of the 'know that' family of knowledge, which 'know how' can use to solve technological problems. Such complexity within human cognition and in our interactions with the material world – which underlies technological capability – cannot be reduced to facts, techniques, and routine strategies.

Evidence from Cognitive Archaeology

This claim is illustrated most cogently by considering our closest, but extinct, species cousins, *Homo neanderthalensis*, who accumulated a great deal of factual and technical knowledge that was passed down from generation to generation. These creatures knew about the movement of animals and the growth of plants; they had knowledge of where to find particular materials such as flint, chert, or obsidian for tool-making; and possibly even made observations about the moon and stars. Their knowledge of where to locate the best materials led to its transportation over long distances from the source. Further, they possessed such a high degree of know-how about working in stone that archaeologists find it difficult to distinguish between sharp-faceted artefacts produced by *Homo neanderthalensis* and those created by early *Homo sapiens*.

In relation to Palaeolithic technology (*pre-Homo sapiens*), the French cognitive archaeologist Pelegrin (1991) distinguished and attributed both *connaissances* (conceptual knowledge), consisting of mental representations of forms that are considered ideal, and *savoir-faire* (procedural knowledge) of the properties and techniques of working the materials involved, which Karlin and Julien (1994) further define as "… aris[ing] from ideation, based upon either memory and intelligence or physical movement and bodily skill" (p. 152).

This begs the question: If pre-sapiens hominid species had both *connaissances* (know that) and *savoir-faire* (know how), then wherein resides the difference between Them and Us? If we teach only facts (know that) and technique (know how), will we be relying on learners themselves to supply some magic ingredient that will serve to unite and integrate knowledge? If we wish to educate young *Homo sapiens*, what else is it that we need to teach them that pre-sapiens did not teach their offspring?

For some answers, we can turn again to the past. We now consider the builders of ancient stone monuments, whose activities leave enigmatic traces across the world. One of the most impressive monuments to the Neolithic Age in Britain is Stonehenge, a vast complex of standing stones, which has parallels to similar monuments in other parts of north-west Europe, chiefly Ireland, Scotland, Wales, and Brittany in northern France.

Pitts (2000) claims the builders of Stonehenge combined traditional stone-working techniques with the established traditions of construction of timber-based circular monuments known as 'henges'. The below-ground shapes of the standing stones (megaliths) of Stonehenge bear testimony to the way in which the builders conceived the shaping of the massive, upstanding stones as if they were giant hand tools (lithics), flaking off large shards as only Stone Age expert stone-workers would. Pitts suggests that the architectural genius of Stonhenge is that by

> … taking a small everyday object, using everyday technology but in ways previously unimagined, people created something on an unprecedented scale … [Stonehenge] is also a play on materials. Timber turns to stone, and well-known local rock joined by alien geology [referring to the combination of local sarsen and Welsh bluestone]. (pp. 216–217)

Even if he is wrong, we can see the plausibility of his argument. We can make a leap of the imagination and assume that our Neolithic forebears could too. It is the ability to make this conceptual leap that places us and the Neolithic folk of 4,000 years ago together but separate from the non-sapiens of 400,000 years ago. This is technology, not just technique.

There was, obviously, no single drivers for the cascade into the more powerful modes of thought that typify our species and separated 'Us' from the hominid 'Rest' on the savannas and in volcanic valleys of Africa some 40,000 or more years ago, but self-reflection appears to be close to the nub of the issue. Such self-reflection implies self-awareness, consciousness, and meta-cognition. As Ihde (1979) suggests: "Technologies are *relativistically* transformational and whatever knowledge we gain through them *reflexively* transforms the world which we discover through them and embodied beings which we are in using them" (p. 74).

Reflexivity thus implies the ability to observe, evaluate, modify, and evaluate the modifications of one's own thoughts, perceptions, and concepts; to take hold of one's thoughts and turn them around and over; to chip away at them; to consciously change them as if they were a piece of flint or bone in the hand. It implies the ability to use metaphorical language such as that in the last sentence.

Evidence from Neuropsychology

The centrality of reflexivity to technological capability is supported by recent research in neuropsychology. Broks (2003) suggests that we navigate our way through our world with and through reference to our individual construction of 'self'. This self-awareness is not only central to our social negotiations, but also informs our interactions with the physical world. Broks summarises:

> An emerging theme in neuropsychology is that, just as it has functional systems devoted to perception of, and interaction with, the physical environment, so the brain has evolved systems dedicated to social cognition and action. ... As part of this process it assembles a 'self', which can be thought of as the device we humans employ as a means of negotiating the social environment. (p. 51)

This viewpoint implies the profundity of the relationships that may exist between representation and reflection. This view offers an image of cognitive systems, informed by sensory interaction with the environment, which enable *self-representation*. This self-representation might be seen as an inter-related set of templates from which incoming signals about reality may be compared, tested, accepted, or rejected.

The construction of self is a socially-moderated affair. As individuals, we seek a general view of our fellow humans and seek to be 'normal' within our own society. As we gain evidence about our surroundings, both social and physical, this informs the sets of inner templates with which we set about explaining and accounting for the ways of the world. This personal constructivism will be determined by the unique or massed (for instance, that gained through formal education) set of experiences that are encountered, including the beliefs, values, and opinions of those around us.

Self-reflection induces the ability to symbolise and manipulate systems (including language, mathematics, and science) in order to use mark-making as a way to externalise thought, both as semiotic (drawing) and symbolic (writing) templates. This Big Leap Forward of 50,000 years ago for human conceptual capability beyond know how/know that is what underpins the human technological capacity by imbuing action with meaning and imputing order to chaos.

Cognitive archaeologists such as Mithen (1994) assert that this integration of conceptual domains within the brain was the evolutionary turning point in the unique development of our species and which radically shifted the relationship between humans and the material environment. Greek neurologist Agrafiotis (2004) comments: "There are two ways to think: one can assume that society shapes technology or that technology shapes society" (p. 100).

THE NATURE OF CONCEPTUAL KNOWLEDGE

Human conceptual knowledge is essentially symbolic and systemic. The human brain expects to encounter systems that it can comprehend, manipulate, and act upon. Bruner (1979) quotes Weston's assertion that we convert new problems into known puzzle forms in order to solve them.

Humans have the ability to abstract and generalise in order to take an overview of inter-related connectives that we can mentally manipulate for specific applications by creating a stream of inner models that fit with our constantly updating perceptions of a problem or opportunity.

Being able to control these mental models in imagery and symbols (both linguistic and graphic) and to be able to move easily and fluidly between the inner and outer worlds is the essential power of the human mind. Our ability to name the parts and to verbalise the actions makes us tool-makers par excellence. However, our self-reflective capabilities enable us to verbalise the abstract world as enactive and predictive and make such statements as "What you could do is …" in relation to other people's ideas that do not yet exist in the real world but are conveyed only through descriptive language (Hope, 2002).

The use of plans, diagrams, maquettes, and other means of making inner thoughts external and viewable not only enables communication to others but powerfully supports self-reflection. In a dispute over whether something will work or not, the challenge is frequently given: Draw it. Drawing and physical model-making enables what is known to be portrayed and shared; in addition, the doing may in itself generate new knowledge or deeper understanding. The use of iconic, semiotic, or symbolic support systems as externalisations of the inner image is an essential part of technological capability, one of the building blocks of technological literacy.

The tantalisingly clear – yet at the same time ambiguous – properties of a model, especially a drawing, that promises new possibilities whilst supporting the definition of half-conceived thoughts and ideas, allows us to symbolically create new meanings and learn as we create. Such external models enable learning to be examined, both by the creator and by others, as well as enabling a discussion in which the model takes the role of a dialogic partner (Hope, 2008). This symbolically-created conceptual framework feeds directly into the perceptual framework. The tools are selected and the materials are viewed as if the product already exists.

To be literate is to be able to understand and communicate in a symbolic system. It is to be able to use that system creatively to express ideas. Knowing about the system is not enough. True literacy involves being able to use the symbol system purposefully and creatively, to have a fluency in capability that allows easy use of the tools, and to be able to choose appropriate options in techniques and materials (Hope, 2002). Technological literacy as an educational goal needs to fit young people across the globe for a world that includes small-scale production alongside advanced computer-controlled industrial systems; supply essential needs such as clean water, good food, and adequate shelter, as well as designer labels and must-have fashion items; and utilise the earth's resources whilst preserving the ecological environment. Simple answers fitted to a simpler industrialised age will no longer suffice. Not only science, but ethics and ecology need to be part of technological educational package (Aduriz-Bravo, 2002).

The characteristics of technology, and therefore that for which we should be educating young people, are not just the ability to perform certain actions effectively and efficiently so as to effect desired changes in the material world, nor even to have a clear inner image of the goal of the activity. Rather, it requires the ability to analyse, generalise, and synthesise perceptions, observations, actions, and knowledge; as well as the ability to make leaps of the imagination, to see possibilities, to take risks, to try out new ideas, and to be innovative. Such analogical leaps involve "as if" or "possibility thinking" (Craft, 2001) and "seeing as" and perceiving "family resemblances" (Wittgenstein, 1969) between situations that will lead to new and novel applications of prior knowledge, skills, and understanding.

All higher vertebrates play – but most reserve this joyous phase only for infanthood – to practice the skills they will need as adults. Humans engage in symbolic play from infanthood, making a 'house' under a clothes airer, for instance. Symbolism lifts play into new dimensions that can be continued throughout life. Lieberman's (1977) research in the United States indicated that children who played with ideas imaginatively (e.g., being able to suggest multiple ideas for improving toys) continued to show such playful inventiveness as adults.

The parallel dialogic nature of both play and invention can be seen by juxtaposing Winnicott's (1991) typification of play as the interaction between the inner world of the imagination and the outer reality of the external world with the views of the French cognitive psychologist Leroi-Gourhan (1945). He described technology as a dialogue between the artisan and the material, which is situated at the junction between the internal and external milieu, implicating the natural determination of the *tendance* with the cultural idiosyncrasy of the *fait*. These inner and outer milieux he saw as forming a *chaîne opératoire* (operational chain) that involve the social and cultural world as well as the corporeal and bodily aspects of human existence. Schlanger (1997) regards these *chaînes opératoires* as a means of expressing the interplay between the fixed and the flexible within any technological solution.

This capability implies the perception of mirror image symmetry between the abstract and symbolic generalisation that constitutes the conceptual knowledge held in the brain, and the unique physical reality of the particular problem or specific situation (van der Leeuw, 1994). The innovative use of this interaction depends on the ability to reflect on the topological symmetry between the mirror image of the inner concept and the external world to initiate a search for a perspective with sufficient symmetries to clarify the problem, and to define the dimensions which capture these symmetries in order to identify a solution.

Such ability to key into an inner conceptual fluidity is essential for the capacity to model technological ideas. Roberts (2001a) asserted that there were two forms of modelling used by designers: 'modelling of' (to clarify their ideas or explain them to others) and 'modelling for' (which is the essence of design process). Such externalisation through drawing or making prototypes supports the development of ideas whilst making them available for scrutiny, evaluation, modification, and so on. The greatest design thinkers are those with the broadest conceptual stores, not necessarily the greatest factual knowledge, as their understanding of general principles cascades into the specific design problem or opportunity. Leonardo da Vinci, for instance, left sheets of drawings of engineering ideas that never came to fruition. The greatest thinkers are always on to the 'next new thing'.

Human-built environments are inevitably landscapes of change, constantly re-invented and adapted to new purposes and in response to human needs and desires. Technology education, therefore, cannot duck moral and social issues but must embrace the difficult and frequently contradictory mess of ethics and responsibility. Kimbell and Perry (2001) claim that

> ... design and technology empowers us to change the made world ... design is not just about change, it is about improvement. ... Any designed object is the manifestation of a set of values. And the concept of improvement is essentially value-laden. ... The pupil is transformed from passive recipient into active participant. (pp. 3, 7)

This is important because it grants to the pupil a sense of personal empowerment within a culture which increasing dictates the choices to be made through market forces and advertising media. Technology education can attempt to redress the balance by encouraging pupils to critically examine products for fitness of purpose. It endorses the personal creativity of each individual and raises issues of responsibility to society, to the environment, and to the safety of the planet. Humans, uniquely of all species, have the self-reflective capacities to question the consequences of their actions and products.

LEARNING TECHNOLOGICAL CONCEPTS

As discussed so far, conceptually-based technological knowledge is complex, being informed by perceptual input in practical experience and by the internalisation of prior learning in intra-personal, scientific, and symbolic realms. These interactions in the mind of the learner are further supported by aspects of learning theory involving the testing of hypotheses (Gholson, 1980) and the use of mental models or frameworks from which to make comparisons or judgements about the world and to form new ideas.

There are parallels here with Piaget's (1952) schemata, the cognitive building blocks, which develop through the accommodation and assimilation of experiences that contribute to the maintenance and revision of mental models. This is allied to personal construct theory (Kelly, 1955), 'constructivism' (Driver, Guesne, & Tiberghien, 1985; Valanides & Angeli, 2001), or 'constructionism' (Papert, 1991). The chief tenets of these views of learning divide into two main strands, the one operating at an individualised level (personal constructivism) and the other, wider form linked to the idea of communities of practice (social constructivism) where learning groups may be seen to be adapted to reside and share values within particular conceptual habitats. The personal construction of technological understanding involves the inner modelling of ideas and possibilities, whereas the social construction of technological understanding includes both the sharing and passing on of knowledge, skills, and understanding from teacher to learner, and also the joint construction of solutions amongst peers. Constructionism can be seen as an extension of social constructivism whereby virtual and constructed forms themselves carry meaning and enable the sharing of ideas. One could envisage a post-constructionalism, parallel to Foucault's (1969) post-structualism, which could be applied to the study of technological capability within a social context; an *Archaeology of Technicacity* to parallel his *L'Archéologie du Savior*.

Engagement with and interactions between the phenomena of the physical, social, and mental worlds may lead to strands of action that involve creative question-raising, the re-structuring of problems, or the embracing of emerging opportunities. Following Piaget's thought, these actions will necessarily involve comparisons with past exposures to the phenomenon encountered, embedded within our experiential vaults. Conceptual models or frameworks provide the structuring which forms the basis from which to compare and contrast a question or explore a novel situation. Such conceptual templates can be seen as cognitive structures which permit the testing of ideas against the application of reason and analogy. Questions and problems both make conceptual demands and require a cognitive response.

The conceptual template allows new experiences to be gauged against what has gone before, yet is itself constantly and fluidly reflexive in the light of new experiences, perceptions, and questions. For if these provide fresh evidence, unavailable before, that serves to challenge the very basis of the template itself as a viable means of explaining experience, the conceptual template may be revised and adapted, or even demolished and a new one erected in its place. Where these are inconsistent with generally shared notions (e.g., with regard to scientific explanations of physical phenomena), they may be carried as 'alternative frameworks' or 'alternative conceptions' (Wandersee, Mintzes, & Novak, 1994). Learners frequently carry a number of alternative frameworks, or even multiple conceptual templates, and these are used to explain, exploit and employ materials, techniques and opportunities in different ways and in different settings.

Constructivist explanations of conceptual templates are inherent in the way in which all knowledge is socially constructed, not just that of young learners. For example, Roberts (2001b) describes the tensions between differing theoretical positions and differing existential persuasions (and hence curriculum practices) which derive from groups of practitioners who think their version is rationally or self-evidently well-founded. His comment – that "paradigmatic change inevitably brings disturbance in the taken-for-granted perspectives and rationales of normal practices" (p. 13) – could be applied equally to the 6 or 60 year old, to the novice craftsperson or the expert industrialist. The self-reflection of the human mind allows for this fluidity of rational thought and perception of analogical symmetry.

TECHNOLOGY, KNOWLEDGE, AND THE GLOBAL ECONOMY

On a global scale, the conceptual component of technology education in the modern world has been driven by factors such as the forces of competitive economics and the instrumentalist goal of either becoming or remaining in positions of strength in relation to other nations. The technology education agenda, therefore, has frequently been seen as moving citizens away from traditional forms of knowledge and skills into those that are relevant to the global marketplace.

Beyond the agenda of national politics and economic development, however, the job of delivering the technology education curriculum is passed on to teachers and pedagogues who frequently have a somewhat different perspective (Ginestié, 2005). In their hands, the form and direction of technology education has been driven by matters such as values, relevance, learning styles, and debates concerning the balances and conflicts between the 'content' (the substance and outcomes of technology-based actions) and 'process' (the 'doing' of technology). However, teachers are frequently constrained in curriculum delivery by government policies and centralised curricular and examination/assessment regimes, despite the individuality and personal freedom that democracy implies (as reported by individual contributors from the eight countries represented in Ginestié's (2005) handbook).

In economic terms alone, education that does not develop children's creativity and innovative capacities will not promote prosperity within a knowledge society (Hargeaves, 2003). The shift from the industrial age to the conceptual age, which Pink (2005) suggests as vital to the economic survival of the USA, demands the capabilities of design, story, symphony (synthesis), empathy, play, and meaning. In contrast, educational agendas frequently focus on instrumentalist goals, which may as Toffler (1970) suggests, result in "... cranking out Industrial Man when Johnny lives in the hurricane's eye of change" (p. 371). Even worse would be cranking out a modern-day version of *Homo neanderthalensis*: Knowing that 'knowing how' is advantageous to personal survival, but lacking the conceptual capacities to reflect, evaluate, analyse, re-appraise, consider, invent, and innovate, that underlie the defining technological capability of *Homo sapiens* in the face of the ever-increasing industrial and global changes that comprise our present and future habitat.

REFERENCES

Aduriz-Bravo, G. (2002). Design and implementation of a course on the Philosophy of Science within the degree of science teaching. Report of experiences from the University of Buenos Aires (Argentina) and the Universitat Autonoma de Barcelona. In S. Gatt (Ed.), *Proceedings of conference: Linking science, technology and mathematics education and their social relevance* (pp. 96–104). Malta: University of Malta.

Agrafiotis, D. (2004). Statement by Professor Demosthenes. *Proceedings and synthesis report "Connecting brains and society: The present and future of brain science: What is possible, what is desirable?" European Workshop April 2004.* Amsterdam: King Baudouin Foundation (Belgium) and Rathenau (Netherlands).

Balfanz, R. (1991). Local knowledge, academic skills and individual productivity: An alternative view. *Educational Policy, 5*(4), 343–370.

Broks, P. (2003). *Into the silent land: Travels in neuropsychology*. London: Atlantic Books.

Bruner, J. (1979). *On knowing: Essays for the left hand.* Cambridge, MA: Harvard University Press.

Craft, A. (2001). Little c creativity. In A. Craft, B. Jeffrey, & M. Leibling (Eds.), *Creativity in education* (pp. 45–61). London: Continuum.

de Vries, M. J., & Tamir, A. (1997). Shaping concepts of technology. *International Journal of Technology and Design Education, 7*(1–2), 3–10.

Driver, R., Guesne, E., & Tiberghien, A. (1985). *Children's ideas in science*. Milton Keynes, Buckinghamshire: Open University Press.

Foucault, M. (1969). *L'Archéologie du Savior*. Paris: Editions Gallimard.

Gholson, B. (1980). *The cognitive-developmental basis of human learning. Studies in hypothesis testing*. London: Academic Press.

Ginestie, J. (2005). *Teacher training, preparing young people for their future lives: An international study in technology education*. Aix-Marseille, France: ALFA/IUFM Aix-Marseille.

Hamlyn, D. W. (1970). *The theory of knowledge*. Basingstoke and London: Macmillan.

Hargreaves, A. (2003). *Teaching in the knowledge society*. New York: Teachers College Press.

Hope, G. (2000). Beyond 'Draw one & make it'. In R. Kimbell (Ed.), *Conference proceedings: Design and technology international millennium conference* (pp. 82–89). Wellesbourne, Warwickshire: Design & Technology Association.

Hope, G. (2002). Learning something new: Problem solving and the development of knowledge. In S. Gatt (Ed.), *Proceedings of conference "Linking science, technology and mathematics education and their social relevance"* (pp. 77–89). Malta: University of Malta.

Hope, G. (2008). Beyond knowing how to make it work. In E. W. L. Norman & D. Spendlove (Eds.), *Conference proceedings, education and research conference 2008* (pp. 105–112). Wellesbourne, Warwickshire: Design and Technology Association.

Ihde, D. (1979). *Technics and praxis*. Dordrecht, The Netherlands: D. Reidel.

Jones, A. (1997). Recent research in learning technological concepts and processes. *International Journal of Technology and Design Education, 7*(1–2), 83–96.

Karlin, C., & Julien, M. (1994). Prehistoric technology: A cognitive science. In C. Renfrew & E. B. W. Zubrow (Eds.), *The ancient mind. Elements of cognitive archaeology* (pp. 152–164). Cambridge, MA: Cambridge University Press.

Keirl, S. (2000). An episode in technology curriculum refinement: It's only another design brief… In P. Roberts & E. W. L. Norman (Eds.), *IDATER2000 international conference on design & technology educational research and curriculum development* (pp. 113–117). Loughbourough University: IDATER.

Kelly, G. A. (1955). *The psychology of personal constructs. Vol. 1. A theory of personality*. New York: W.W. Norton and Company Inc.

Kimbell, R., & Perry, D. (2001). *Design & technology in a knowledge economy*. London: The Engineering Council.

Leroi-Gourhan, A. (1945). *Milieu et techniques*. Paris: Albain Michel.

Lieberman, J. N. (1977). *Playfulness*. New York: Academic Press.

Mithen, S. (1994). From domain specific to generalised intelligence: A cognitive interpretation of the Middle/Upper Palaeolithic transition. In C. Renfrew & E. B. W. Zubrow (Eds.), *The ancient mind. Elements of cognitive archaeology* (pp. 29–39). Cambridge, MA: Cambridge University Press.

Papert, S. (1991). Situating constructionism. In I. Harel & S. Papert (Eds.), *Constructionism* (pp. 1–12). Norwood, NJ: Ablex.

Pelegrin, R. (1991). Les savoir-faire: une très longue histoire. *Terrain, 16*, 106–113.

Piaget, J. (1952). *The origins of intelligence in children*. Madison, USA: International Universities Press.

Pink, D. (2005). *A whole new mind: Moving from the information age to the conceptual age*. New York: Riverhead.

Pitts, M. (2000). *Hengeworld*. London: Century.

Roberts, P. (2001a). An invitation to IDATER99. In E. W. L. Norman & P. Roberts (Eds.), *Design & technology educational research and curriculum development: The emerging international research agenda* (pp. 3–9). Loughborough University. IDATER.

Roberts, P. (2001b). Opening address to IDATER2000. In E. W. L. Norman & P. Roberts (Eds.), *Design & technology educational research and curriculum development: The emerging international research agenda* (pp. 10–30). Loughborough University. IDATER.

Ryle, G. (1949). *The concept of mind*. Harmondsworth: Penguin.

Schlanger, N. (1997). Mindful technology: Unleashing the chaîne opératoire for an archaeology of mind. In C. Renfrew & E. B. W. Zubrow (Eds.), *The ancient mind. Elements of cognitive archaeology* (pp. 143–151). Cambridge, UK: Cambridge University Press.

Toffler, A. (1970). *Future shock*. London: Pan Books.

Valanides, N., & Angeli, C. (2001). Constructivism in the classroom. In N. Valanides (Ed.), *Science and technology education: Preparing future citizens* (Vol. 2, pp. 157–164). International Organisation for Science and Technology Education. Nicosia: University of Cyprus/Cyprus Ministry of Education and Culture/IOSTE.

van der Leeuw, S. E. (1994). Cognitive aspects of "technique". In C. Renfrew & E. B. W. Zubrow (Eds.), *The ancient mind. Elements of cognitive archaeology* (pp. 135–142). Cambridge, UK: Cambridge University Press.
Wandersee, J. H., Mintzes, J. J., & Novak, J. D. (1994). Research on alternative conceptions in science. In D. L Gabel (Ed.), *Handbook of research on science teaching and learning* (pp. 177–210). New York: Macmillan.
Winnicott, D. W. (1991). *Playing and reality* (2nd ed.). London: Routledge.
Wittgenstein, L. (1969). *Philosophical investigations; the blue book; the brown book.* Oxford, UK: Basil Blackwell.

Eric Parkinson
Gill Hope
Faculty of Education
Canterbury Christ Church University
UK

SECTION 4

TECHNOLOGY AND SCIENCE-TECHNOLOGY-SOCIETY-ENVIRONMENT (STSE) EDUCATION

DEREK HODSON

23. TECHNOLOGY IN SCIENCE-TECHNOLOGY-SOCIETY-ENVIRONMENT (STSE) EDUCATION: INTRODUCTORY REMARKS

Traditional forms of science education aim to socialise those intending to study science beyond the school level and those contemplating a career in science, medicine, and engineering into scientific ways of thinking, speaking, acting, and believing. Because their role is primarily a form of pre-professional training, these programmes succeed in serving the needs of only a particular elite group of students. A more broadly based alternative approach promotes the practical utility of scientific knowledge and a connectedness with personal and social values in pursuit of developing every student's capacity to function effectively as a knowledgeable and responsible citizen in a world increasingly impacted by science and technology. The drive to equip students to understand science in its social, cultural, economic, and political contexts is the underpinning rationale of the so-called science-technology-society (STS) approach – more recently expanded to STSE (where E stands for environment). James Gallagher (1971), one of the pioneers of STS education, captures its overall flavour particularly well:

> For future citizens in a democracy, understanding the interrelations of science, technology, and society may be as important as understanding the concepts and processes of science. An awareness of the interrelations between science, technology, and society may be a prerequisite to intelligent action on the part of a future electorate and their chosen leaders. (p. 337)

STS has always been a purposefully ill-defined field that leaves ample scope for varying interpretations and approaches, and much has changed over its 35+ year lifetime in terms of its priorities and relative emphases (Aikenhead, 2003; Kumar & Chubin, 2000; Pedretti, 2003; Solomon & Aikenhead, 1994; Yager, 1996). Aikenhead (2005, 2006) describes how the early emphasis on values and social responsibility was systematised by utilising a theoretical framework deriving from sociology of science: (i) the interactions of science and scientists with social aspects, issues, and institutions *external* to the community of scientists, and (ii) the social interactions of scientists *within* the scientific community. Both emphases have remained strong, though much has changed with respect to the socio-political and economic contexts in which educators and scientists work; in our understanding of key issues in the history, philosophy, and sociology of science; and in our theoretical knowledge concerning concept acquisition and development. Interestingly, as consideration of the nature of science has become a much more prominent part of regular science curricula, even a central part in many educational jurisdictions (American Association for the Advancement of Science [AAAS], 1993; Council of Ministers of Education, Canada, 1997; Department of Education, Republic of South Africa, 2002; Goodrum, Hackling, & Rennie, 2000; Millar & Osborne 1998; National Research Council, 1996; OECD, 1999; UNESCO 1993), so emphasis in STSE education has shifted much more towards confrontation of socioscientific issues (SSI).

A.T. Jones and M.J. de Vries (eds.), International Handbook of Research and Development in Technology Education, 267–273.
© 2009 Sense Publishers. All rights reserved.

Zeidler, Sadler, Simmons, and Howes (2005) contrast SSI-oriented teaching with STS or STSE education in terms of its emphasis on developing habits of mind (specifically, developing skepticism, maintaining open-mindedness, acquiring the capacity for critical thinking, recognising that there are multiple forms of inquiry, accepting ambiguity, and searching for data-driven knowledge) and "empowering students to consider how science-based issues reflect, in part, moral principles and elements of virtue that encompass their own lives, as well as the physical and social world around them" (p. 357). They argue that while STS education emphasises the impact of scientific and technological development on society, it does not focus explicitly on the moral-ethical issues embedded in decision-making:

> STS(E) education as currently practiced ... only 'points out' ethical dilemmas or controversies, but does not necessarily exploit the inherent pedagogical power of discourse, reasoned argumentation, explicit NOS considerations, emotive, developmental, cultural or epistemological connections within the issues themselves ... nor does it consider the moral or character development of students. (p. 359)

Bingle and Gaskell (1994) had earlier noted that STS education tends to emphasise what Bruno Latour (1987) calls "ready made science" (with all its attendant implicit messages about certainty) rather than "science-in-the-making" (with its emphasis on social construction). Simmons and Zeidler (2003) argue that it is the priority given to science-in-the-making through consideration of *controversial* SSI that gives the SSI approach its special character: "Using controversial socioscientific issues as a foundation for individual consideration and group interaction provides an environment where students can and *will* develop their critical thinking and moral reasoning" (p. 83, emphasis added). In a further attempt at delineation, Zeidler, Walker, Ackett, and Simmons (2002) claim that the SSI approach has much broader scope in that it "subsumes all that STS has to offer, while also considering the ethical dimensions of science, the moral reasoning of the child, and the emotional development of the student" (p. 344).

In the Western contemporary world, technology is all pervasive; its social and environmental impact is clear; its disconcerting social implications and disturbing moral-ethical dilemmas are made apparent almost every day in popular newspapers, TV news bulletins, and Internet postings. In many ways, it is much easier to recognise how technology is determined by the sociocultural context in which it is located than to see how science is driven by such factors. It is also often much easier to see the environmental impact of technology than to see the ways in which science impacts on society and environment. For these kinds of reasons, it makes good sense to use problems and issues in technology and engineering as the major vehicles for contextualising the science curriculum. This is categorically not an argument against teaching science; rather, it is an argument for teaching the science that informs an understanding of everyday technological issues and problems and may assist students in reaching tentative solutions about where they stand on key SSI. I have argued elsewhere (Hodson, 1994, 2003) for an SSI-oriented approach comprising four levels of sophistication:

- *Level 1*: Appreciating the societal impact of scientific and technological change, and recognising that science and technology are, to some extent, culturally determined.
- *Level 2*: Recognising that decisions about scientific and technological development are taken in pursuit of particular interests, and that benefits accruing to some may be at the expense of others. Recognising that scientific and technological development are inextricably linked with the distribution of wealth and power.
- *Level 3*: Developing one's own views and establishing one's own underlying value positions.
- *Level 4*: Preparing for and taking action on socioscientific and environmental issues.

Much of STS and STSE education is located at Level 1, with decision-making in science and technology seen simply as a matter of reaching consensus or affecting a compromise. In contrast, the intention at Level 2 is to enable students to recognise that scientific and technological decisions are taken in pursuit of particular interests, justified by particular values, and sometimes implemented by those with sufficient economic or political power to override the needs and interests of others. As a consequence, the advantages and disadvantages of scientific and technological developments often impact differentially within society and between countries and regions. Level 3 is concerned primarily with supporting students in their attempts to formulate their own opinions on important issues and establish their own value positions, rather than with promoting the 'official' or textbook view. It focuses much more overtly on values clarification, developing strong feelings about issues, and actively thinking about what it means to act wisely, justly, and 'rightly' in particular social, political, and environmental contexts. Level 4 is concerned with helping students to prepare for and engage in responsible action, and assisting them in developing the skills, attitudes, and values that will enable them to work with others to bring about change. In short, this form of SSI-oriented education aims to politicise students and to promote student empowerment through civic participation (Pedretti, 1999; Roth & Désautels, 2002). It sees scientific literacy as the driving force for socio-political action, and the science curriculum as a key element in education for social reconstruction. Roth and Barton (2004) make essentially the same point: "critical scientific literacy is inextricably linked with social and political literacy in the service of social responsibility" (p. 10). In common with Lee and Roth (2002) and Roth and Lee (2004), they recognise that significant impact on SSI decision-making is more likely through collective action than individual efforts, thus shifting the ultimate focus of education for scientific literacy towards effective public practice, summed up by the increasingly popular notion of 'enhanced public engagement with science'.

If students are to get to grips with SSI at any level beyond the merely superficial they need relevant scientific and technological knowledge. It is not surprising that Keselman, Kaufman, and Patel (2004) found that the capacity of grade 7 and grade 9 students to reason about information concerning HIV-AIDS and formulate responsible courses of action was directly related to the nature and extent of their knowledge of biology, particularly the nature of viruses, mechanisms of infection, and the immune system. Similarly, Lewis and Leach (2006) show that the quality of discussion of issues relating to genetic engineering is substantially enhanced by basic understanding of genetics, and note that the necessary level of understanding is 'relatively modest' and fairly easily achieved. Those wishing to assess likely risks from the proximity of overhead power lines or nuclear power stations, the dumping of toxic waste, and the frequent use of mobile phones will need considerable relevant technological knowledge, the capacity to engage in risk-benefit analyses, and the ability to evaluate the views of 'experts' and officials. It is often the case that scientific knowledge has to be re-worked and reorganised to make it more readily applicable to real practical situations – what Layton, Jenkins, MacGill, and Davey (1993) refer to as "practical knowledge for action". Addressing SSI in class may require students to engage in a similar kind of re-working of scientific knowledge to that of engineers and lay people confronting science-related dilemmas in daily life:

> Adjusting the level of abstraction, 'repackaging' knowledge to bring together components of scientific knowledge that pedagogical and disciplinary considerations have uncoupled, and 'recontextualizing' scientific knowledge to reassimilate the messy realities that have been idealized in order to shape and address a problem with the rigour deemed necessary to move towards a scientific solution. (Jenkins, 1994, p. 601)

In Chapter 24 of this section of the *Handbook*, Catherine Buntting and Alister Jones explore the range of relationships between science and technology, including science as a precursor, knowledge source and enabler of technology, technology as a precursor and enabler of science, independent development, and situations in which science and technology grow and develop together. They conclude that its location at the intersection of science and technology make biotechnology an ideal vehicle for illustrating the ways in which society's beliefs, needs, interests, and values shape and are shaped by scientific development and technological innovation. Three examples of teaching and learning initiatives in biotechnology are described: the first concerns ways of dealing with brushtail possums – an introduced species in New Zealand that threatens indigenous flora and fauna and has detrimental impact of the cattle and deer industry through the spread of bovine tuberculosis; the second reports on a partnership between a research organisation (The Cawthron Institute) and a New Zealand high school, during which students collected data on the impact of environmental variations on the size and weight of green lipped mussels; the third is an example of a 'low-tech' activity in which students explored the role of yeast in the production of fizzy drinks and looked at ways of changing the flavour of a drink to meet consumer preferences.

Because SSI are often located in disputed frontier science (or science-in-the-making) rather than in established textbook science, knowledge and understanding *about* science is crucial. Students need to be able to interpret reports; make sense of disagreements and evaluate knowledge claims; scrutinise arguments; distinguish among facts, arguments, and opinions; make judgements about good science and bad science; detect bias and vested interest; and so on. This is the focus of Jim Ryder's argument in Chapter 43 that nature of science (NOS) knowledge should be learned in specific disciplinary contexts (emphasising differences among the sciences and between science and technology) and as 'knowledge in action' (that is, in the context of particular scientific concerns, technological problems, or socioscientific issues). Clearly, students' NOS knowledge and views will impact on the way they address SSI, but not always in a simple, straightforward, and predictable way (Zeidler et al., 2002). There is a complex, reflexive interaction: more sophisticated NOS views open up new possibilities for scrutinising SSI; engagement with important and personally significant SSI enhances and refines NOS understanding. Thus, it could be argued that an SSI-oriented curriculum both demands and justifies a high curriculum priority for NOS. A closely related concern is the need to develop students' argumentation skills (see Chapter 36 by Jonathan Osborne).

Of course, many SSI are controversial. Some teachers are alarmed by the prospect of teaching about controversial issues and seek strenuously to avoid it. Justin Dillon in Chapter 26 argues strongly that we should not shelter students from controversy (whether in science or SSI); controversy is a key element of life and all citizens need to learn how to recognise it and deal with it. In rejecting the once fashionable notion of the teacher as neutral chair of classroom discussions, Dillon argues for a position in which students focus on the nature of controversy itself and the reasons why individuals may disagree on a particular issue. The thrust of the approach is to encourage and support students in subjecting their own views and the views of others to rigorous critical scrutiny, develop their capacity to identify bias and faulty reasoning, promote open-mindedness and a willingness to change one's view when appropriate, and avoid premature closure. This approach demands that teachers share their views with students and make explicit the ways in which they arrived at those views.

Of course many SSI have both explicit and implicit ethical dimensions and so any treatment of them in class will inevitably raise questions about what is the *right* decision and what *ought* we to do? For example, should any individual with a disabling genetic condition be able to have gene therapy using material from embryonic stem cells? Should prenatal genetic testing be readily available and selective abortion of a foetus with a severe genetic disorder be permitted? Asking questions about what we *ought* to do raises questions and concerns about morality (what it is right or wrong to do) and ethics (the reasons and justifications for judging these things to be right or wrong). After considering arguments for the teaching of ethics, and arguments against (both sets of arguments focusing on NOS issues and the consequences of teaching or not teaching ethics), Michael Reiss in Chapter 27 proceeds to draw distinctions among different kinds of values pertinent to education in science and technology: moral values, intellectual values, political values, and aesthetic values. After identifying four general aims of teaching ethics and values in STSE education (enhancing moral sensitivity, enlarging students' stock of moral knowledge, improving their moral judgement, and making them more virtuous), he focuses on three specific goals related to consideration of SSI: autonomy, well-being, and justice. Moral-ethical and values issues are likely to surface in consideration of most SSI, and are particularly prominent in Dillon's discussion of "soft disasters"(Chapter 26) and Buntting and Jones' discussion of methods used for possum control and eradication (Chapter 24).

Lamenting the current preoccupation of teachers and educational administrators with curriculum specifications, assessment targets, and reporting mechanisms, Steve Alsop describes in Chapter 28 a radical alternative to the usual classroom concern with science and technology that is distant, unfamiliar, uninteresting, and unimportant (to students). He urges teachers to look both *outward* "to the world in which they live and the communities in which they work" and *inward* "to the moral and ethical commitments that lured them into teaching" and to reverse the usual order of priorities – that is, to begin with exploration of learners, communities, and environments and then consider the implications for curriculum, teaching, and schooling. Drawing on an approach that features strongly in Aboriginal education and environmental education, Alsop advocates a form of "place-based" education – in this case, a focus on the immediate physical, social, political, cultural, and economic environment in which the school is located. His approach is illustrated by two case studies: one, a description of how a group of three students worked with a severely ill 5 year old (sister to one of the participants), particularly their efforts to gain familiarity with ECGs and treatment of sleep disorders in order to provide both emotional support and science-based advice; the other, an account of two students investigating the local fish market, seeking answers to questions such as: Where do the fish come from? How nutritious are they? Which species are in decline? Emphasis in both projects (and others in the programme Alsop describes) is on ownership, personalisation, and relevance to the community in which the students live. It is an inspiring chapter showing that we *can* re-envision and re-direct STSE education, at least in part, without any need for substantial investment in fancy equipment and facilities.

In Chapter 29, Angela Calabrese Barton and Edna Tan document a similar shift of power, authority, and control from teacher to students in their description of how a group of 11 year olds investigating the phenomenon of urban heat islands in a Midwestern city in the United States chose to design and produce an 8 minute 40 second multimedia documentary titled *Where da Heat Go?* In pursuit of its goal of educating the local community about the phenomenon of urban heat islands, the documentary incorporates a scientific-technological explanation of the phenomenon, locally generated temperature data and heat island profiles, and interviews with residents. The significance of the work described in this chapter and the one by Alsop lies in the way they shift definitions of scientific and technological literacy from what one knows, and how one knows it, to engagement with SSI in specific, local contexts of action.

All the authors contributing to this section of the *Handbook* make the point that many teachers are ill-equipped and ill-prepared to adopt an STS, STSE, or SSI-oriented approach. It is inevitable that some teachers will lack confidence and expertise in handling unstructured, open-ended discussions. It is not surprising that teachers new to SSI-oriented teaching commonly express concern bordering on anxiety that they will be accused of bias or lay themselves open to charges of indoctrination. Some teachers remain uncommitted to the approach because, they argue, it takes too much time and is likely to divert attention away from content aspects of the curriculum. Some are concerned that a shift from the supposed 'certainties' of science to the uncertainties of SSI-oriented teaching will constitute a threat to their classroom authority and may even estrange them from their specific discipline-based community – see Pedretti, Bencze, Hewitt, Romkey, and Jivraj (2008) for a discussion of the range of concerns expressed by preservice teachers. Even those who are committed to the approach may still not implement it, citing lack of time to plan lessons and prepare materials capable of integrating coverage of content with social concerns, economic considerations, and moral-ethical dilemmas. Others cite difficulties associated with design of assessment and evaluation strategies. The message is clear: we need to make a substantial investment in teacher education at both the preservice and inservice levels and in the development of user-friendly curriculum resources.

In Chapter 30, Deborah Corrigan discusses the enormously beneficial role that short industrial placements for teachers can play in helping them acquire the background necessary for teaching about science in the real world. She also points out some key differences between the highly formal, structured, targeted, and teacher-led learning of the conventional school classroom and the informal, unstructured, open-ended, and learner-centred ways in which knowledge and understanding are acquired outside school, noting that students need to be able to bridge that divide if they are to address SSI in a critical and productive way. Fieldwork, museums, and site visits have considerable potential for assisting students in this task, as Corrigan illustrates through two contrasting case studies of site visits – one teacher attempted to meet the interests of individual students, the other focused on an industry of significance to the whole local community in terms of employment opportunities.

Taken together, the initiatives advocated by the seven chapters contributing to this section of the *Handbook* can go a long way towards achieving Oxfam's (1997) definition of a global citizen as someone who

> … is aware of the wider world and has a sense of his or her own role as a world citizen; respects and values diversity; has an understanding of how the world works economically, politically, socially, culturally, technologically and environmentally; is outraged by social injustice, participates in and contributes to the community at a range of levels from the local to the global; is willing to act to make the world a more equitable and sustainable place; takes responsibility for his or her actions. (p. 2)

REFERENCES

Aikenhead, G. S. (2003). STS education: A rose by any other name. In R. Cross (Ed.), *A vision for science education: Responding to the work of Peter Fensham* (pp. 59–75). New York: RoutledgeFalmer.

Aikenhead, G. S. (2005). Research into STS science education. *Educación Quimica, 16*(3), 384–397.

Aikenhead, G. S. (2006). *Science education for everyday life: Evidence-based practice*. New York: Teachers College Press.

American Association for the Advancement of Science [AAAS]. (1993). *Benchmarks for scientific literacy*. New York: Oxford University Press.

Bingle, W. H., & Gaskell, P. J. (1994). Scientific literacy for decision making and the social construction of scientific knowledge. *Science Education, 78*(2), 185–201.

Council of Ministers of Education, Canada. (1997). *Common framework of science learning outcomes*. Toronto, ON: Author.

Department of Education, Republic of South Africa. (2002). *Revised national curriculum statement for grades 8-9 schools*. Pretoria: Author.

Gallagher, J. J. (1971). A broader base for science education. *Science Education, 55*(3), 329–338.

Goodrum, D., Hackling, M., & Rennie, L. (2000). *The status and quality of teaching and learning of science in Australian schools: A research report.* Canberra, ACT: Department of Education, Training and Youth Affairs.

Hodson, D. (1994). Seeking directions for change: The personalisation and politicisation of science education. *Curriculum Studies, 2*, 71–98.

Hodson, D. (2003). Time for action: Science education for an alternative future. *International Journal of Science Education, 25*(6), 645–670.

Jenkins, E. W. (1994). Public understanding and science education for action. *Journal of Curriculum Studies, 26*, 601–611.

Keselman, A., Kaufman, D. R., & Patel, V. L. (2004). 'You can exercise your way out of HIV' and other stories: The role of biological knowledge in adolescents' evaluation of myths. *Science Education, 88*(4), 548–573.

Kumar, D. D., & Chubin, D. E. (Eds.). (2000). *Science, technology, and society: A sourcebook of research on research and practice*. New York: Kluwer/Plenum.

Latour, B. (1987). *Science in action: How to follow scientists and engineers through society.* Cambridge, MA: Harvard University Press.

Layton, D., Jenkins, E., MacGill, S., & Davey, A. (1993). *Inarticulate science?* Driffield, UK: Studies in Education.

Lee, S., & Roth, W.-M. (2002). Learning science in the community. In W.-M. Roth & J. Désautels (Eds.), *Science education as/for sociopolitical action* (pp. 37–66). New York: Peter Lang.

Lewis, J., & Leach, J. (2006). Discussion of socio-scientific issues: The role of science knowledge. *International Journal of Science Education, 28*(11), 1267–1287.

Millar, R., & Osborne, J. (Eds.). (1998). *Beyond 2000: Science education for the future.* London: King's College.

National Research Council. (1996). *National science education standards.* Washington, DC: National Academy Press.

OECD. (1999). *Measuring student knowledge and skills.* Paris: Author.

Pedretti, E. (1999). Decision-making and STS education: Exploring scientific knowledge and social responsibility in schools and science centers through an issues-based approach. *School Science & Mathematics, 99*(4), 174–181.

Pedretti, E. (2003). Teaching science, technology, society and environment (STSE) education: Preservice teachers' philosophical and pedagogical landscapes. In D. L. Zeidler (Ed.), *The role of moral reasoning on socioscientific issues and discourse in science education* (pp. 219–239). Dordrecht, The Netherlands: Kluwer.

Pedretti, E. G., Bencze, L., Hewitt, J., Romkey, L., & Jivraj, A. (2008). Promoting issues-based STSE perspectives in science teacher education: Problems of identity and ideology. *Science & Education, 17*(8&9), 941–960.

Roth, W.-M., & Barton, A. C. (2004). *Rethinking scientific literacy.* New York: RoutledgeFalmer.

Roth, W.-M., & Désautels, J. (2002). *Science education as/for sociopolitical action.* New York: Peter Lang.

Roth, W.-M., & Lee, S. (2004). Science education as/for participation in the community. *Science Education, 88*(2), 263–291.

Simmons, M. L., & Zeidler, D. (2003). Beliefs in the nature of science and responses to socioscientific issues. In D. L. Zeidler (Ed.), *The role of moral reasoning on socioscientific issues and discourse in science education* (pp. 81–94). Dordrecht, The Netherlands: Kluwer.

Solomon, J., & Aikenhead, G. (Eds.). (1994). *STS education: International perspectives on reform.* New York: Teachers College Press.

UNESCO. (1993). *International forum on scientific and technological literacy for all. Final report.* Paris: Author.

Yager, R. E. (Ed.). (1996). *Science/technology/society as reform in science education.* Albany, NY: State University of New York Press.

Zeidler, D. L., Walker, K. A., Ackett, W. A., & Simmons, M. L. (2002). Tangled up in views: Beliefs in the nature of science and responses to socioscientific dilemmas. *Science Education, 86*(3), 343–367.

Zeidler, D. L., Sadler, T. D., Simmons, M. L., & Howes, E. V. (2005). Beyond STS: A research-based framework for socioscientific issues education. *Science Education, 89*(3), 357–377.

CATHERINE BUNTTING AND ALISTER JONES

24. UNPACKING THE INTERFACE BETWEEN SCIENCE, TECHNOLOGY, AND THE ENVIRONMENT: BIOTECHNOLOGY AS AN EXAMPLE

INTRODUCTION

This chapter argues that recognising the interdependence of scientific and technological literacy has potential to enhance learning of both science and technology, although there are few reports in the education literature of cases where this is successfully achieved. More often, technology is included in science curricula in the context of 'technology as applied science'; and technology curricula tend largely to ignore concepts associated with the relevant science (see Fensham's (1988) analysis of STS courses in chapter 30 by Deborah Corrigan).

The emerging field of biotechnology lies at the intersection of different disciplines and provides a powerful example of how knowledge and products evolve through the ongoing interaction between both scientific and technological communities. In addition, biotechnology is often strongly linked with environmental issues and, being inherently value-laden, it provides a vehicle for exploration of values in relation to science, technology and the environment. Biotechnology contexts create opportunities to explore both science and technological concepts, develop science and technological skills and consider interactions between science, technology, society and the environment. Moreover, such explorations need not be limited to senior secondary school.

This chapter examines the relationship between STS and STSE using biotechnology as an example, and illustrate how biotechnology learning contexts can be used to enhance both scientific and technological literacy.

BACKGROUND

Although early twentieth century scholars divided the study of natural phenomena into physics, chemistry and biology as distinct disciplines, the reality of modern science and technology research and/or development is that it tends to be carried out by a transdisciplinary collage of scientists, technologists, engineers and funding agencies (Latour, 1987). In addition, current funding regimes dictate a focus on the functional aspects of science and technology, particularly as they relate to economic development, social progress, and human and environmental wellbeing (DeHart Hurd, 1998). This interdependence between science, technology, the environment and the social and cultural context, is not only worthy of exploration within compulsory education, but necessary if students are to develop into scientifically and technologically literate citizens, as advocated by contemporary school curricula.

A.T. Jones and M.J. de Vries (eds.), International Handbook of Research and Development in Technology Education, 275–285.
© 2009 Sense Publishers. All rights reserved.

TECHNOLOGICAL LITERACY WITHIN SCIENCE

The interdependence of scientific and technological literacy is becoming much more prominent in science education research literature, with major science education journals publishing special issues focusing on technology education, for example, *Journal of Research in Science Teaching* (2001), *Research in Science Education* (2001) and *Canadian Journal of Science, Mathematics and Technology Education* (2003). Within the first of these publications, Cajas (2001) specifically recommends that scientific literacy includes a better understanding of technology. Indeed, there has been a general move to broaden science curricula to include concepts associated with the nature of science, as well as interactions between science, technology, society and the environment. However, a narrow view of technology as applied science seems all too often to apply (e.g., Jones, 2007).

This treatment of technology as embedded in science and resulting from science is cause for concern because of the limited view of technology it portrays (Bell, Jones & Carr, 1995). For example, such a view assumes a linear relationship in which science generates technology. Typically, the story of a technological development is projected through the lens of science (Gardner, 1995). The development of penicillin as a result of Sir Alexander Fleming's serendipitous discovery of the effect of *Penicillium* moulds on bacteria is a typical example (France-Farmer, 1997). Although providing a fruitful illustration of certain views of 'scientific process', such an approach does not accurately reflect the strategic and interdisciplinary nature of modern scientific and technological endeavours. As such, it does not allow full development of the learner's scientific *or* technological literacies.

De Vries (2001) suggests that the history of industrial research laboratories provides an opportunity for studying the complex relationships between science and technology. He describes three different interaction patterns: science as an enabler for technology; science as a forerunner of technology; and science as a knowledge resource for technology. All focus on the role of science in technological development, and an important element missing from this framework seems to be technology as an enabler of scientific discovery (Jones, 2007). Technological innovations and developments that have no direct links to science as a discipline are also ignored (Jones, in press).

Gardner (1994, 1995) explores the additional viewpoints of science and technology as independent communities, and science and technology as equal and interacting communities. Roth (2001) describes the co-evolution of scientific knowledge and technological development, arguing that it should not be surprising that scientific knowledge is often the by-product of technological activities and, conversely, technological artefacts are the by-products of scientific activities.

An alternative approach to using examples of technological developments to illustrate scientific principles is the incorporation of technological problems in science lessons. Examples include bridge-building (Cajas, 2001) and the design of simple machines (Roth, 2001). However, in order to make the science-technology link successful teachers need to help students make salient the concepts that are relevant to science (Roth, Tobin & Ritchie, 2001).

A further limitation to incorporating technology concepts meaningfully in science curricula appears to be the difficulty many science teachers report experiencing when required to explore concepts associated with ethics and values (e.g., Levinson & Turner, 2001; Michael, Grinyer, & Turner, 1997; Oulton, 2004), concepts that are key to scientific *and* technological literacy (see Chapter 27 by Michael Reiss). As Solomon (1998) warns:

> Our students need to see technology as the application of knowledge, scientific and other, for social purposes. They will have to recognise some of the pressures for innovation (once called the 'technological imperative') and its cultural dependence. Technology may be either

small scale or large, but it is always the process of producing social changes as well as the result of social needs (p. 381).

SCIENTIFIC LITERACY WITHIN TECHNOLOGY EDUCATION

Historically, technology subjects such as woodwork, metalwork, and home economics (sewing and cooking) have focused on vocational outcomes and emphasised the design and production of artefactss. Even recently introduced subjects like information and communication technology (ICT) tend to focus on using digital technologies to construct a product. In contrast, curriculum reforms in the 1990s emphasised the need to include concepts associated with the nature of technology in order to reflect new learning needs in a new technological age. For example, Cajas (2001) suggests that the goal of technology studies is "to have citizens who understand the nature of technology and its interaction with science and society" (p. 718) and Zuga (1991) suggests that enabling students to become citizens who make wise personal and collective choices from a growing number of options requires technology educators to recognise the need to incorporate information about science as well as society.

In spite of these goals for 'technological literacy', many of the contexts traditionally used in the technology classroom do not involve the exploration of specific scientific concepts, and curriculum reform continues to be heavily influenced by tensions between vocational skill development for a select group of students, and technological literacy for all (see Section 1, international developments). In contrast, Cajas (2001) argues that the introduction of technological ideas, such as design and failure, frame technological tasks not only in the construction of the artefact, but also – and mainly – in understanding how the artefact is designed, and how and why it could fail. This, he says, requires students to use both scientific and technological ideas. De Vries and Tamir (1997) similarly argue that the role of scientific knowledge in the development of new technologies and products is a necessary component of a realistic perception of technology, and Zuga (1991) suggests that theory without practice, or practice without theory, cannot be technology. In consequence, these and other authors advocate the view that it is an 'old conception' of technology education that values the construction as most important, rather than emphasising an understanding of the underpinning concepts.

Several reports demonstrate how relevant scientific concepts can be effectively incorporated in technology classroom programmes. For example, Cox (2007) describes an electronics unit in which students were required to understand circuits and how to construct them before building their own 'Robobugs', and Maley (1987) describes how windmills can be used as a context for blending science and mathematics with a traditional technology project, with students unpacking scientific concepts such as friction and energy conservation.

SCIENCE/TECHNOLOGICAL DETERMINISM VERSUS SOCIAL DETERMINISM

The STS interface entails understanding not only the influence science and technology have on society (scientific or technological determinism), but also the influence society has on science and technology (social determinism). The two-way nature of these interactions – of science/technology shaping society, and society shaping science/technology – is complex but important. In particular, a critical understanding of the latter (how society can and does shape science/technology) is crucial to developing in students a sense of empowerment. This is because one cannot initiate change without understanding the frameworks upon which any change is dependent. As argued elsewhere:

> Understanding the relationship between technology [or science] and society is about not only learning the 'rules of the game' but being in a position to critique these rules and feel empowered to change them should you find them to be inappropriate. (Jones, 2007, p. 93)

Historically, however, it seems that science and technology education curricula have emphasised the impact of science/technology on society, rather than focusing on ways in which society's needs, values, and beliefs have shaped science/technological developments. In contrast, we advocate a view where technological determinism and social determinism are seen as neither dichotomous nor mutually exclusive. Educational experiences that enhance technological literacy need to provide scope for students to critically consider long-term impacts prior to intervening in the world in ways that would fundamentally change it (Keulartz, Korthals, Schermer, & Swierstra, 2002). Modern biotechnologies provide fruitful, real-world contexts in which this curricular vision can be effectively implemented.

BIOTECHNOLOGY AS TECHNOLOGY AND SCIENCE

For the purpose of this discussion, biotechnology is broadly defined as any technique that uses biological processes to create products or solve problems for the benefit of people and/or the environment. Biotechnology has been part of everyday life for thousands of years, for example, using bacteria to making yoghurt and cheese, fermenting grapes to make wine, and using yeast in bread-making. However, greater understanding of the complex systems inside living organisms has led to new and different ways in which biological knowledge can be used. For example, targeted genetic modification of any organism is now technically possible using laboratory procedures. Although such genetic manipulation can circumvent the time and organism numbers required by conventional breeding to achieve the same effect, it can also be used to produce outcomes previously impossible with conventional breeding. For this reason, modern biotechnologies involving genetic modification have become a source of major public controversy.

Biotechnology contexts, and particularly 'modern' biotechnologies (which may or may not involve genetic modification), have the potential to provide powerful learning opportunities to unpack the interactions between science and technology, as well as social and environmental dimensions. Additionally, the increasing pervasiveness of biotechnologies means that examples can be found that link to many of the controversial topics often selected by teachers for classroom exploration. To illustrate, Justin Dillon's 'soft disasters' in Chapter 26 include BSE, GM crops, HIV-AIDS, and global climate change. Biotechnology contexts associated with each of these could include, for example, the development of technologies to diagnose diseases such as BSE, use of terminator technologies to prevent genetically modified crops from being self-seeding (ostensibly to control their environmental distribution, but with significant financial implications for farmers), the development of drug cocktails to reduce the transfer of HIV-AIDS from mother to child, and the creation of new varieties of drought-resistant crops.

Each of these examples fulfils a clear 'need' or opportunity, has economic and societal implications, requires an understanding of both science and technology concepts, and could be used to engage senior students in argumentation (see Chapter 36 by Jonathan Osborne). However, not all biotechnology contexts need to be heavily dependent on advanced biological/chemical concepts such as cellular microstructures. More traditional technologies, such as fermentation, can be embraced by young students, and have the added advantage of greater opportunity for hands-on activities and the development of scientific and technological skills. (In Chapter 25, Jim Ryder demonstrates how mobile phone technologies can be explored as an alternative science/technology context for unpacking STSE interactions.)

The section below provides three illustrations of biotechnology learning contexts developed and used in New Zealand. The first demonstrates how science and technology can work together to create a biotechnological solution to an environmental problem; in the second, an effective school-industry link showcases the role of authentic tasks; and the third provides an example of a low-tech scenario suitable for all ages. To make these and other examples accessible to schools across the country, the New Zealand Biotechnology Learning Hub was developed by the authors of this paper in conjunction with a national team of educators and industry experts.

The Hub was initiated in 2004 by New Zealand's Ministry of Research, Science and Technology to connect the education and industry sectors, primarily via an online portal. Content is rich in multimedia and includes up-to-date examples of biotechnology practice, computer-mediated interactive activities, virtual tours, and stories of people working as biotechnologists. Teachers can also access resources for planning and implementing classroom studies related to biotechnology; teaching units include both science and technology learning outcomes, with values and ethics integrated throughout. The examples below have been selected to illustrate the breadth of possible contexts that can foster meaningful learning in this area.

Science and Technology Work Together to Address an Environmental Problem

Brushtail possums (*Trichosurus vulpecula*) were introduced to New Zealand from Australia in 1837 in order to establish a fur industry but have since become the dominant economic and ecological vertebrate pest. This is because they tolerate a wide range of natural habitats and have no natural predators or competitors. As well as spreading bovine tuberculosis, at great cost to the dairy, beef and deer industries, they are a serious conservation threat, altering the composition of native forests, competing with native fauna for food, and preying directly on native birds and insects.

The possum population is currently controlled by shooting, trapping and poisoning. Challenges to trapping include designing traps that are light weight, capture efficient, meet animal welfare standards and don't catch non-target fauna. A range of poisons are available, with 1080 (sodium monofluoroacetate) being the most widely used, despite significant public controversy: it can be applied at both bait stations and via aerial drops, and is highly effective. However, poor quality bait containing small pieces can cause bird death and there is a risk of secondary poisoning from carcasses, especially to dogs. It can also generate bait-shyness if a sub-lethal dose is ingested.

Shooting, trapping and poisoning are all time-intensive, costly, and need to be ongoing in order for possum numbers to remain controlled. This is because surviving possums breed better and live longer due to reduced competition. In addition, there is increasing concern about public acceptability and animal welfare aspects of current control methods (Landcare Research, n.d.). (Michael Reiss provides a more complete discussion of ethical issues in STSE contexts in Chapter 27).

Biological control, or biocontrol, is the control of a pest population by the introduction or augmentation of a natural predator or other biological agent, such as a micro-organism or parasite. This definition can also be expanded to include the use of biological organisms (e.g., parasites or viruses) as vectors to distribute technologies such as immuno-contraceptive agents and toxins, and it is this strategy that is currently the focus of active research in New Zealand. The National Research Centre for Possum Biocontrol was established to this end, a research platform formed through partnership between two major science providers and key end-user organisations such as the Department of Conservation and Animal Health Board. It is largely government-funded.

One promising area of research is the genetic modification of a possum-specific parasitic worm to produce a toxin to interfere with possum fertility. Video clips on the Biotechnology Learning Hub showcase how and why the parasitic worm will be modified, as well as unpacking factors that need to be considered and understood when selecting and implementing a successful biocontrol programme.

Key science concepts include conservation, pests, reproduction, fertility, parasitism, vectors, species-specificity and genetic modification. An interactive activity on the Biotechnology Learning Hub also focuses on the concept of fair testing in the context of determining the amount of toxin required to affect possum fertility – tested by counting the cells in the female possum's urine, causing humorous fascination amongst students and teachers alike.

Key technology concepts include (i) testing efficacy of the control solution (such as how effectively the modified worm affects possum fertility, and how efficiently the modified worm is spread within the possum population); (ii) operating within constraints (e.g., species-specificity, animal welfare, public acceptance, and technical and economic viability); (iii) selecting from multiple options; (iv) operating within regulatory parameters (the introduction of all new organisms is controlled in New Zealand by the Environmental Risk Management Authority); and (v) collaboration within multidisciplinary teams of experts working within different organisations.

The role of public engagement is recognised by the presence of a bioethics panel on the National Research Centre for Possum Biocontrol. A 2000 report, '*Caught in the Headlights: New Zealanders' reflections on possums, control options, and genetic engineering*' (Parliamentary Commissioner for the Environment, 2000) is further demonstration of this, and can be used to initiate student exploration in this area. As stated in the preface of this report:

> If science fails to engage with society, to listen to communities' concerns, and to jointly develop the needed research questions, there may be a high probability that any resultant technologies would be rejected. The expectations for public involvement are only heightened by the introduction of genetic technologies into the possible toolbox of possum control methods. (Commissioner's preface)

The development of control options to address New Zealand's possum problem therefore includes a strong mandate for exploration of the societal aspects, and this is incorporated in the unit plan written to support the teaching resources published on the Biotechnology Learning Hub (BLH, 2005):

> Biological solutions to pest control are complex and require input from a diverse range of sources. Beliefs, values and ethics of individuals and groups promote or constrain technological developments. The development of an effective system must take into account environmental pressures and societal attitudes in order for it to gain acceptance by the community.

In summary, the development of a biocontrol solution to New Zealand's possum problem provides a context that is real, relevant and important to the country's students. It demonstrates how science and technology can interact to address environmental issues; the science and technology concepts can be identified and explored; and social aspects can be investigated. By undertaking learning within this particular context, students might be expected to have a sense of ownership of the problem and possible solutions, understand key science and technology concepts (such as pests, parasites, vectors, species-specificity, genetic modification, and the implications of economic, social and environmental constraints), and be able to explain how science and technology are working together to address a serious environmental issue.

School-Industry Links Enhance Learning Opportunities

The second example of biotechnology as a promising context for learning about STSE interactions focuses on a partnership between a New Zealand secondary school, Queen Charlotte College, and a privately-funded research organisation, The Cawthron Institute. Cawthron provides high quality research, advice and analytical services to support sustainable management of the coastal and freshwater environment, and New Zealand's seafood industry. One of its research projects is selective breeding of Greenshell™ mussels with commercial goals. The programme combines traditional breeding techniques with industry-focused data analysis. Associated research includes molecular analysis to genetically identify key characteristics for selective breeding purposes and for forensic traceability.

The school offers a specialised aquaculture programme which includes structured workplace learning. The programme was initiated in 2002 in partnership with the New Zealand Marine Farming Association. Students are introduced to aquaculture as part of their compulsory science classes during their second year of high school (14 year olds). In the following year, an optional course focuses on Greenshell™ mussels and students also obtain qualifications in diving and boating. This supports additional optional courses in subsequent years, with successful students receiving an industry-recognised National Certificate in Seafood (Aquaculture).

A key component of all the courses is the engagement of students through hands-on learning. To this end, the school has established a research line of Greenshell™ mussels in a nearby bay. The mussels are provided from Cawthron's selective breeding programme and the school is involved in tagging the mussels (including working out the most effective, long term, economic way of doing this) and harvesting them. Key parameters, such as shell length and body weight, are entered into a database alongside information about site conditions. These data are contributing to knowledge about how specific genetic lines respond in different environmental conditions.

As part of this project, the school identified the presence of a sea squirt, *Didemnum vexillum*, smothering the mussels. In order to produce an ongoing record of *D. vexillum* abundance on the research line, the school is using underwater digital video equipment and undertaking regular monitoring, and hopes to become involved with a worldwide study of invasive marine species based on protocols developed by the Smithsonian Institute in the US. Students are also testing the anti-fouling effectiveness of five different marine paints, as well as a *Didemnum* extract.

By accessing marine research expertise from Cawthron and other research organisations and using donated equipment and facilities, students are able to carry out meaningful aquaculture research both on land and *in situ*. Key science concepts introduced during the introductory year of study include reproduction, mussel lifecycles, selective breeding, marine ecology, sustainability, fair testing, and the use of technology for scientific purposes. Key technology concepts include operating within constraints (e.g., an underwater environment, equipment limitations, cost); liaising with key stakeholders (e.g., The Cawthron Institute, the local aquaculture industry); selecting from multiple options (e.g., working out the most effective way to tag mussels so that their genetic heritage is known); developing methodologies to accurately collect and store data; and taking a systems approach to their work.

Environmental issues are a core component of the learning, since aquaculture by its very nature affects the aquatic ecosystem. By way of example, students participate hands-on in the 'Adopt a Beach' clean-up scheme established by the New Zealand Marine Farming Association, including underwater clean-ups of offshore seabeds. Students also have the opportunity to participate in relevant public events, such as regional development conferences and the annual meetings of relevant aquaculture bodies. This provides them with scope to engage directly with key stakeholders.

This example demonstrates how an effective school-industry link has been used to foster learning of STSE interactions in the context of aquaculture research and development projects. Access to expertise and equipment provides students with opportunities for authentic hands-on learning, and key science and technology concepts are effectively integrated throughout. Attention is also given to relevant environmental and social issues to ensure that the science and technology is not explored in isolation of these other components.

Low-Tech Biotech

Not all biotech projects require access to sophisticated equipment and facilities, as demonstrated by fermentation and cheese-making studies already being carried out in many classrooms around the world. By carefully selecting and focusing on key concepts, meaningful learning can occur in both science and technology, as demonstrated in a case study by Wine and Moreland (2007).

The focus of the learning, in this case by 11 and 12 year olds, was on adapting a traditional fermented drink recipe by considering both consumer feedback and how changes impact on the activity of the yeast. The tasks included following a recipe to make a traditional drink, designing and implementing science experiments to identify parameters that affect yeast growth, and designing and carrying out a consumer survey to identify consumer preferences. The traditional recipes were subsequently adapted to take into account taste tests, survey results, and science experiments. This was an iterative process, and provides an illustration of how new technologies often build on existing ones. Conceptual learning outcomes related to understanding that yeast comprises living organisms, and identifying significant components of the fermentation process and how these can be manipulated. Procedural learning outcomes related to accurate measuring and the making of fermented drinks; designing, undertaking and analysing a consumer survey; carrying out meaningful science experiments; developing specifications for the adapted fermented drink (incorporating survey data, results from science experiments, and constraints in the manufacturing process); and making the adapted drink. The unit culminated in a product launch during which students presented their modified drinks and justified the adaptations they had made. Societal learning outcomes related to both historical aspects (fermented drinks have a long history and are influenced by the needs and culture of the time) and economic factors (consumer preferences influence product development). The teacher also created several opportunities for students to explore their understanding of biotechnology, which developed to include notions of meeting human needs.

Although a variety of solutions were expected, students needed to include results from taste testing, surveys, and the science experiments. In other words, the problem was not so open that it allowed for any solution, or so closed that students would lose interest. The teacher also needed to help students make key science and technology concepts salient to each step. In addition, she purposefully created opportunities for students to form connections between the different components of work, creating what Moreland and Cowie (2007) call continuity, connection and coherence. The task met criteria reported elsewhere (e.g., Crismond, 2001) as being important – that is, it used familiar, easy-to-work with materials; had clearly defined outcomes that allowed for multiple solution pathways; allowed for multiple design iterations; and had clear links to a limited number of science and technology concepts.

This case study with 11 and 12 year olds exploring the role of yeast in the production of fizzy drinks demonstrates that a fun and hands-on unit helps engage students in biotechnology learning. By the end of the unit, students demonstrated an understanding that when yeast is 'fed' sugar under the right conditions, it grows and produces carbon dioxide. This can be used to make drinks fizzy. Any changes to conditions might affect the growth of the yeast, and therefore the final product. This needs to be considered when modifying a recipe in order to take account of consumer feedback.

Space for Biotechnology in School Curricula

A major review of the positioning of biotechnology in school curricula internationally demonstrates its prominence in senior science and biology, where it tends to appear in the form of technology as applied science (France, 2007). This positioning underplays not only the 'technology' component, but also limits opportunities for students to develop a greater understanding of the relationships between science and technology, and the values inherent in these relationships. Further, by focusing on examples aligned with public controversy, such as genetic modification or whole-animal cloning, students may develop a distorted view of the area rather than seeing it in its fuller context (Jones, 2006). As noted earlier, some science teachers may also resist teaching what they see as 'social science' components. The overlapping scientific, technological and social dimensions of biotechnology thus present something of a challenge for the science curriculum and science teachers.

The New Zealand curriculum provides additional scope for teaching biotechnology within the Technology curriculum (Ministry of Education, 1995; 2007). Elements of biotechnology are also visible in other New Zealand school curriculum and guideline documents, including science, biology and environmental education. Indeed, there is a plethora of opportunities to implement biotechnology education. This is in line with key government goals, as outlined in the New Zealand Biotechnology Strategy (Minsitry of Research, Science and Technology, 2003). However, as France (2007) points out, such a positioning of biotechnology - as either science or technology - has implications for what students experience as 'biotechnology' in their schooling, and who will teach it.

For teachers to implement appropriate forms of biotechnology education, they need to be aware of their own perceptions of what biotechnology is, and the ways in which biotechnology is perceived within the curriculum area in which they are working. As France (2007) points out, this perception of biotechnology will firstly determine the curriculum requirements for biotechnology education, and secondly influence their teaching. Professional development and high quality teaching resources are also needed.

CONCLUSION

This chapter has used the example of biotechnology to demonstrate how interactions between science, technology, society and the environment might be unpacked and explored within a classroom setting. To this end, a broad definition of biotechnology was adopted. Three examples were used to illustrate the range of contexts that are possible. However, curriculum requirements and teacher perceptions remain a dominating influence in what biotechnology is taught, and where and how students are exposed to it during their schooling.

We argue that the teaching of technology only as a subset of, or subservient to, science is detrimental for students' development of a clear understanding of the interactions between science, technology, society and the environment. An expanded view, consistent with curriculum goals for science and technological literacy, means that students can be introduced to a wider range of ideas, and consider the values that both constrain and promote scientific and technological developments. These include economic, political, social and environmental aspects. Because it lies at the intersection of different disciplines, biotechnology offers a fruitful context in which interactions between science and technology can be explored. In addition, recent advances present society with ethical issues and dilemmas that require informed citizens capable of contributing meaningfully to public debate.

ACKNOWLEDGEMENTS

We thank Associate Professor Bev France for her research and leadership within the biotechnology education community in New Zealand.

REFERENCES

Bell, B., Jones, A., & Carr, M. (1995). The development of the recent national New Zealand Science Curriculum. Invited paper for a special issue on policy. *Studies in Science Education, 26*, 73–105.

Biotechnology Learning Hub [BLH]. (2005). *Making a case for possum biocontrol.* Retrieved March 1, 2009, from http://www.biotechlearn.org.nz/focus_stories/biological_control_of_possums/making_a_case_for_possum_biocontrol

Cajas, F. (2001). The science/technology interaction: Implications for science literacy. *Journal of Research in Science Teaching, 38*(7), 715–729.

Chapman, B. (1991). The overselling of science education in the eighties. *School Science Review, 72*(260), 47–62.

Cox, T. (2007). Electronics. In M. de Vries, R. Custer, J. Dakers, & G. Martin (Eds.), *Analyzing best practices in technology education* (pp. 59–67). Rotterdam, The Netherlands: Sense.

Crismond, D. (2001). Learning and using science ideas when doing investigate-and-redesign tasks: A study of naive, novice, and expert doing constrained and scaffolded design work. *Journal of Research in Science Teaching, 38*(7), 791–820.

DeHart Hurd, P. (1998). Scientific literacy: New minds for a changing world. *Science Education, 82*(3), 402–416.

de Vries, M. (2001). The history of industrial research laboratories as a resource for teaching about science-technology relationships. *Research in Science Education, 31*, 15–28.

de Vries, M., & Tamir, A. (1997). Shaping concepts of technology: What concepts and how to shape them. In M. J. de Vries & A. Tamir (Eds.), *Shaping concepts of technology: From philosophical perspectives to mental images* (pp. 3–10). Dordrecht, The Netherlands: Kluwer.

Dreyfus, A. (1995). Biological knowledge as a prerequisite for the development of values and attitudes. *Journal of Biological Education, 29*(3), 215–219.

France, B. (2007). Location, location, location: Positioning biotechnology education for the 21st century. *Studies in Science Education, 43*, 88–122.

France-Farmer, B. (1997). Biotechnology: Applied science or technology? In B. Bell & R. Baker (Eds.), *Developing the Science curriculum in Aotearoa New Zealand* (pp. 242–255). Auckland, NZ: Addison Wesley Longman.

Gardner, P. (1995). Representations of the relationship between science and technology in the curriculum. *Studies in Science Education, 24*, 1–28.

Gardner, P. (1995). The relationship between technology and science: Some historical and philosophical reflections: Part 2. *International Journal of Technology and Design Education, 5*(1), 1–33.

Jones, A. (2007). The valuing of technology in the science curriculum. In D. Corrigan, J. Dillon, & R. Gunston (Eds.), *The re-emergence of values in science education* (pp. 89–100). Rotterdam, The Netherlands and Taipei, Taiwan: Sense.

Jones, A. (in press). Exploring the tension and synergies between science and technology in science education. In S. M. Ritchie (Ed.), *The world of science education: Australasia*. Rotterdam, The Netherlands: Sense.

Hickman, L. A. (1990). *John Dewey's pragmatic technology*. Bloomington, IN: Indiana University Press.

Keulartz, J., Korthals, M., Schermer, M., & Swierstra, T. (2002). Ethics in a technological culture: A proposal for a pragmatist approach. In J. Keulartz, M. Korthals, M. Schermer, & T. Swierstra (Eds.), *Pragmatist ethics for a technological culture*. Dordrecht, The Netherlands: Kluwer.

Landcare Research (n.d.). *Possum information sheets and manuals.* Retrieved March 1, 2009, from http://www.landcareresearch.co.nz/publications/infosheets/possums/

Latour, B. (1987). *Science in action.* Cambridge, MA: Harvard University.

Levinson, R., & Turner, S. (2001). *Valuable lessons: Engaging with the social context of science in schools. Recommendations and summary of research findings.* London: Wellcome Trust. Retrieved March 1, 2008, from http://www.wellcome.ac.uk/doc_WTD003435.html.

Maley, D. (1987). Integrating math and science into technology education. *The Technology Teacher, 46*(8), 9–12.

Michael, M., Grinyer, A., & Turner, J. (1997). Teaching biotechnology: Identity in the context of ignorance and knoweldgeability. *Public Understanding of Science, 6*, 1–17.

Ministry of Education. (1995). *Technology in the New Zealand Curriculum*. Wellington, NZ: Learning Media

Ministry of Education. (2007). *The New Zealand Curriculum*. Wellington, NZ: Learning Media.

Ministry of Research, Science and Technology. (2003). *New Zealand biotechnology strategy*. Wellington, NZ: Author.

Moreland, J., & Cowie, B. (2007). Teaching approaches. In M. de Vries, R. Custer, J. Dakers, & G. Martin (Eds.), *Analyzing best practices in technology education* (pp. 213–219). Rotterdam, The Netherlands: Sense.

Oulton, C. (2004). Reconceptualizing the teaching of controversial issues. *International Journal of Science Education, 26*(4), 411–423.

Parliamentary Commissioner for the Environment (2000). *Caught in the headlights: New Zealanders' reflections on possums, control options, and genetic engineering*. Wellington, NZ: Office of the Parliamentary Commissioner for the Environment. Retrieved March 1, 2009, from http://www.pce.govt.nz/work_programme/reports_by_subject/all_reports/science__and__technology/caught_in_the_headlights

Roth, W.-M. (2001). Learning science through technological design. *Journal of Research in Science Teaching, 38*(7), 768–790.

Roth, W.-M., Tobin, K., & Ritchie, S., (2001). *Re/constructing elementary science*. New York: Peter Lang.

Solomon, J. (1988). Science technology and society courses: Tools for thinking about social issues. *International Journal of Science Education, 10*(4), 379–397.

Wine, P., & Moreland, J. (2007). Teaching biotechnology. In M. de Vries, R. Custer, J. Dakers, & G. Martin (Eds.), *Analyzing best practices in technology education* (pp. 13–25). Rotterdam, The Netherlands: Sense.

Cathy Buntting
Alister Jones
School of Education
The University of Waikato
Hamilton, New Zealand

JIM RYDER

25. ENHANCING ENGAGEMENT WITH SCIENCE/TECHNOLOGY-RELATED ISSUES

The Role of Students' Understandings about the Nature of Science

INTRODUCTION

A key aim of science-technology-society-environment (STSE) education is to prepare young people to engage effectively with science/technology-related issues as they encounter them outside of school science (American Association for the Advancement of Science, 1995; European Commission, 2007; Osborne & Dillon, 2008; Tytler, 2007). Such an aim raises many questions. What does *effective* engagement mean, for the people involved and for science educators promoting this aim? To what extent does knowledge of science and technology have a role in influencing people's engagement, alongside other understandings and commitments? What can school education realistically contribute to such future engagement? How might science and technology education be designed to maximise its impact on students' interactions with science/ technology-related issues as adults?

Drawing upon published case studies, this chapter begins with an examination of how people engage with science/technology-related issues. It considers what effective engagement might mean in specific contexts, and the extent to which science knowledge and technological understandings feature in such cases. A particular focus is placed on the role of understanding the nature of science. The chapter then draws together analyses of how people engage with science/technology-related issues outside of formal education, with discussions about effective teaching and learning about the nature of science in school settings.

HOW DO PEOPLE ENGAGE WITH SCIENCE/TECHNOLOGY-RELATED ISSUES?

There have been many studies examining the experiences of those not professionally involved in science engaging with science/technology-related issues outside of formal educational settings. Published collections of case studies include Irwin (1995), Irwin and Wynne (1996), and Layton, Jenkins, Macgill and Davey (1993). Examples of the contexts in which these studies are conducted include people working on a community advisory forum reviewing options for the management of local waste (Petts, 1997), the experiences of parents of children with Down's syndrome (Layton et al., 1993), the incidence of BSE in cattle and human health concerns (Jasanoff, 1997) and how AIDS activists engaged with scientists investigating HIV/AIDS (Epstein, 1995). Given the focus here on implications for the school curriculum, it is significant that many of these case studies involve issues in which scientific findings are uncertain and perhaps disputed. This contrasts with the canonical, secure science knowledge typical of the school science classroom: 'Science in the making' versus 'ready made science' (Latour, 1987).

A.T. Jones and M.J. de Vries (eds.), International Handbook of Research and Development in Technology Education, 287–296.
© 2009 Sense Publishers. All rights reserved.

The following case study of people's engagement with mobile phone technology illustrates the range of knowledge necessary for effective engagement with a specific science/technology-related issue. This case study is chosen because it highlights the significance of both science knowledge and technological understanding for people's engagement. Following Layton (1993), the case study shows how relevant technological understandings do not simply reflect applied science, but are understandings of a different kind and purpose.

Case Study: People's Engagement with Mobile Phone Technology

This section draws heavily on two published studies. The first of these is a wide ranging review, from a sociological perspective, of people's responses to perceived health hazards associated with mobile phone use (Burgess, 2004). This is supplemented by a detailed case study of a local protest into the siting of a mobile phone mast (Drake, 2006). Perspectives influencing people's decisions about mobile phone technology identified in these analyses are summarised in Table 1 and explored below.

Relevant *science knowledge* includes the source-receiver model of radiation, the distinction between ionising and non-ionising radiation, and the place of microwaves within a broader set of related electromagnetic radiations including, for example, radio waves. Important *technological understandings* include the use and purpose of a cellular network of base stations and the nature of communication between mobile phones and base stations. Also relevant is recognising that radiation intensity will decrease with distance from the source. However, Burgess provides an example of how this basic *science* knowledge was not directly or simply applicable, given the *technology* involved. Due to the directivity of the microwave beam from a typical mobile phone mast, the maximum intensity at ground level occurs around 100 metres or more from the mast and diminishes closer to the mast itself (Burgess, 2004, p. 138), thus highlighting the importance of considering context-specific technological issues when examining the relevance of basic scientific understanding.

Table 1. Perspectives influencing people's engagement with mobile phone technology.

Understanding of science concepts
Understanding of mobile phone technology
Consideration of risks and associated benefits, including the extent to which risks can be controlled by the individual
Trust in sources of information (from scientists, government bodies, and business representatives)
Knowledge of public policy guidelines
Understandings about the nature of science

Burgess (2004) identifies three misconceptions that hamper effective engagement with mobile phone technology: (i) all 'radiation' is ionising and therefore harmful; (ii) man-made radiation as necessarily more harmful than naturally occurring radiation; and (iii) if a large dose of radiation is harmful then a little must also have some harmful effect (p. 123). Drake (2006) found that people often draw upon analogies with more familiar phenomenon to understand the science and technology issues involved. For example, people linked heating food in a microwave oven with the potential heating effects of mobile phone use on body tissue. Similarly, individuals drew an analogy with epileptic fits being triggered by flashing lights to appreciate the potential for microwaves to cause non-thermal ill-health effects. Tackling the misconceptions identified by Burgess, perhaps by drawing upon the explanatory analogies above, is likely to be a key aim of a school science and technology education that promotes effective engagement with mobile phone technology.

Risk-benefit considerations and *risk control* are also significant issues for people's engagement with science/technology-related issues such as mobile phone technology. For example, people who recognise the benefits of mobile phone use (e.g., in enhancing personal safety) may be willing to accept a degree of uncertainty about potential risks to their health. The extent to which individuals can control risk is also a major theme. Drake (2006) shows that people are willing to accept a risk that they can control (e.g., the extent of their personal use of mobile phones, whether or not they use a receiver away from their ear, etc.) but not a related risk that is beyond their control (e.g., the positioning of a mobile phone mast near to their home).

The studies also show that people are concerned about the extent to which they can *trust information sources*. For example, people ask 'Who has funded this research?' or 'What does the mobile phone company/government stand to gain from the building of these mobile phone masts?' Many individuals also engage with *public policy guidelines*. For example, in the UK, the Health Protection Agency Radiation Protection Division (formerly the National Radiological Protection Board) publishes guidelines indicating what they consider to be acceptable levels of microwave radiation exposure. These guidelines include the need to keep exposure 'as low as reasonably achievable' (ALARA). An enquiry in the UK into the health effects of mobile phone use suggested that in certain circumstances (e.g., for children) the use of mobile phones should be limited in case there are harmful health effects that have yet to be identified; the so-called 'precautionary principle' (Stewart, 2000). Similarly, in different engagement contexts, whether or not to trust the views of official policy experts was an issue for individuals deciding whether to allow their child to take the combined measles-mumps-rubella (MMR) vaccine in the UK (McMurray, Cheater, Weighall, Nelson, Schweiger, & Mukherjee, 2004), and people's responses to the dangers of Severe Acute Respiratory Syndrome (SARS) (Lee, 2008; Wong, Kwan, Hodson, & Yung, 2009).

Understandings about the *nature of science* were also significant for those engaging with mobile phone technology. Here I use 'nature of science' as an overarching term that brings together insights into the nature of scientific knowledge and the activities of scientists and scientific communities. This includes consideration of the purposes of science, the design of scientific investigations, the nature and status of scientific knowledge, how conflicts of ideas are resolved within the scientific community, and the social and institutional organisation of scientific activity (McComas, 1998; Osborne, Collins, Ratcliffe, Millar, & Duschl, 2003). For example, a prominent member of a mobile phone mast protest group recognised that there was a great deal of uncertainty about the possible longer-term non-thermal effects of mobile phone use, given the long time horizons involved and the complexities involved in undertaking such studies (Drake, 2006). Such an insight requires understandings about how science investigations are conducted in complex contexts outside of the laboratory. Burgess (2004) describes the case of a parent who was convinced that moving her child away from a school sited near to a mobile phone mast was the cause of a dramatic improvement in her child's health: "She's a different child now – it's all the proof I need to convince me there is a link between those wretched masts and the health of children" (p. 1). This individual is using a single case linking two factors as a *necessary* indication of a causal link. Such a misunderstanding of how data from individual cases can be used appropriately as evidence for a causal link has been identified in several studies of young people's ideas about science (Donovan & Brandsford, 2005; Ryder, 2007). For example, as part of an interview-based study of students' reasoning skills, students (aged 11–16 years) were asked to interpret data about the impact of different foods on whether or not children get colds (Kuhn, Amsel, & O'Loughlin, 1988). They found that around a quarter of the students inferred a causal link between food type and 'getting a cold' based on a single co-occurrence.

By deploying a more developed understanding of how data from individual cases can be used appropriately as evidence for a causal link, the parent cited above could have made a more effective case for her concerns about the siting of a mobile phone mast near her child's school. Instead, her claim was easily challenged on the basis of sample size (i.e. generalising from a single case) and a lack of distinction between potential correlation and cause. With a more developed understanding of the relationship between co-occurrences, correlation, and cause, this parent could have drawn upon evidence relating to clusters of cases in a range of geographical locations, along with evidence of a potential mechanism linking the two suspect factors (e.g., thermal heating of human tissue by microwaves). In this context, a more developed understanding of data interpretation may have supported the parent, and those with whom she interacted, in engaging more effectively with the available data and the sources of these data in addressing her concerns.

Other Studies of Public Engagement with Science/Technology-Related Issues

The mobile phone technology case study provides one example of the range of issues that are relevant to individuals engaging with science and technology in a specific context. However, to what extent is the analysis summarised in Table 1 reflected in a broader range of engagement contexts? Aikenhead (2005) provides a comparable list of issues in a review of how students make decisions in a number of contexts in which science experts provide conflicting advice. Similarly, Ryder (2001) reviewed 31 case studies of people's engagement with science/technology-related issues to identify what knowledge of science was relevant to the individuals involved. This analysis generated a framework of science understandings featuring in the studies. In addition to an understanding of the facts and concepts of science, a significant part of this framework focused on understandings about the nature of science, for example, how scientists assess the quality of data, multiple interpretations of a single dataset, and the distinction between models and data in science. This analysis suggests that understandings about the nature of science have an important role to play in many contexts in which people engage with science/technology-related issues outside of formal science education.

A further outcome of these analyses is recognition that the understandings identified in Table 1 are only a part of the broad range of influences that guide people's engagement with science/technology-related issues. Referring back to the mobile phone technology case study: "Evidence has only a 'walk on' part into a drama that is actually propelled by feelings and beliefs derived from experience of society, rather than newfound scientific knowledge about biological processes" (Burgess, 2004, p. 127).

In addition to understandings about science, technology, and the nature of science, broader issues such as religious commitments, ethical perspectives, political ideologies, and social networks can have a powerful influence on people's engagement with science/technology-related issues (Aikenhead, 1985; Lee, 2008).

The studies referred to above do not explore how people's engagement with science/technology-related issues might be influenced by, for example, an increased focus on such engagement in school science and technology education; they are descriptive analyses rather than intervention studies. There are many case studies of individuals who do not draw upon science and technology when engaging with science/technology-related issues (e.g., in making decisions about the MMR vaccine, see McMurray et al., 2004). However, that does not mean that this is desirable, or a situation that cannot be changed through formal and informal science education. Given the emphasis placed on STSE education within schools it is striking how little we know about how such school science innovations impact upon the ways in which people engage with science issues outside of the school science context.

TEACHING AND LEARNING ABOUT THE NATURE OF SCIENCE

This section turns to the design and implementation of teaching activities that aim to develop school-age students' understandings of the nature of science *for the purpose of* supporting their engagement with science/technology-related issues. The focus is on formal schooling of young people. I do not address here the important role that informal learning contexts have in supporting effective engagement with socioscientific issues.

Various approaches to teaching about the nature of science have been advocated: Using episodes from historical and philosophical studies of science (Matthews, 1994; Monk & Osborne, 1997; Shortland & Warwick, 1989); immersing students in 'authentic' science practice (Chinn & Malhotra, 2002; Roth, 1995); and exploring the nature of science alongside the teaching of traditional school science concepts (Cartier & Stewart, 2000; Leach, Hind, & Ryder, 2003). These published studies show that *all* such approaches can develop students' understandings of the nature of science. A closely related issue is whether or not messages about the nature of science should be communicated to students implicitly or explicitly (Khishfe & Abd-El-Khalick, 2002; Khishfe & Lederman, 2007). Again, published studies show that *both* approaches can result in the development of students' ideas about the nature of science. For example, my own work has examined teaching and learning contexts in which messages about the epistemology of science have been communicated implicitly (Ryder, Leach, & Driver, 1999) and explicitly (Leach et al., 2003; Ryder & Leach, 2008). Here I suggest that identifying the significant factors underlying the effectiveness of these teaching/ learning approaches involves going deeper to look at the detail of each teaching sequence and how it is enacted in the classroom. Two arguments are developed here. Firstly, teaching about the nature of science needs to relate these ideas to a range of science disciplinary contexts. Secondly, given the goal of supporting current and future engagement with science/technology-related issues, teaching needs to emphasise the role of understandings about the nature of science as 'knowledge in action'. Elsewhere, following analyses of effective teaching/learning about science concepts, I have also argued that providing opportunities for students to communicate their ideas about the nature of science *prior to teaching* and having these challenged and developed by the teacher, and their peers, is also a key factor influencing the success of teaching in this area (Ryder & Leach).

Disciplinary Context

Teaching about the nature of science involves recognising the range of distinctive practices within and between the science disciplines (Collins & Pinch, 1994; Giere, 1995; Knorr-Cetina, 1999; Schwartz & Lederman, 2008). Scientists deal with phenomena, measurements, correlations, causal mechanisms, explanatory models, each of which has a distinctive epistemic status. Thus it has been claimed that there is no single 'nature of science' (Rudolph, 2000). Rather, it is important to recognise the differing natures of the sciences, whilst recognising that at a broad level there are features of these differing activities that identify them as 'science'. For example, the ways in which theoretical ideas interact with data can be very different in astrophysics and condensed matter physics. Observations of natural phenomena in the laboratory under controlled conditions in which experiments can be set up and key variables controlled is rarely possible in astrophysics, whereas such experiments are the norm in condensed matter physics. Thus, although it is possible to design teaching that provides students with definitions of terms such as 'theory', 'law', and 'model', this decontextualised approach does not reflect the differing ways in which aspects of the nature of science feature within specific science subject areas (Samarapungavan, Westby, & Bodner, 2006). This makes such learning aims distinct from many existing school science learning aims that involve specific science contexts, for example, 'pupils should be taught the properties and uses of the noble gases'. Providing clear and meaningful definitions of the terms 'theory', 'law' and 'model', and how these interact, is inherently difficult. Thus, communicating an appropriate view of the nature of science involves recognising this disciplinary context dependence rather than communicating generic statements about aspects of the nature of science. Indeed, there is evidence that advanced science learners do draw upon distinctive positions about the nature of science depending on the subject domain (Brickhouse, Dagher, Letts, & Shipman, 2000; Leach, Millar, Ryder, & Séré, 2000).

Many studies have described and evaluated teaching/learning sequences in which ideas about the nature of science have been communicated within specific science and technology contexts (Bellomo, 2005; De Andrade & Silva, 2001; Kolstoe, 2001; Roth & Lucas, 1997; Zeidler et al., 2002). For example, the mobile phones context explored earlier has been used as an effective context for teaching about the validity and reliability of data and the distinction between correlation and cause (Albe, 2008; Hind, Leach, & Ryder, 2000). Cartier & Stewart (2000) describe a contextual approach to teaching about the nature of science in high school biology in which students examine epistemic ideas within the context of inquiry activities in genetics – an approach that helped students to gain a coherent view of the nature of distinctive knowledge claims within the discipline of genetics. In each of these studies, learning about the nature of science involves introducing students to a wide range of science contexts and building up a student's 'toolkit' of contextualised examples of the nature of science (Mortimer, 1995; Ryder et al., 1999; Southerland et al., 2005).

The Nature of Science as 'Knowledge in Action'

A further argument against a predominantly decontextualised teaching approach arises from consideration of the educational purposes of teaching about the nature of science. A commonly-cited purpose is to support students in making decisions and engaging in debates concerning scientific and technological issues; the democratic argument for teaching science (Durant, 1990; Roth & Désautels, 2002). Here the focus is on students' actions in the science/technology-related contexts in which they are likely to be engaged as adults, rather than their ability to articulate a view about the nature of science. This reflects the character of such knowledge as 'knowledge-in-action'. For example, in the context of newspaper reports concerning a possible link between deep vein thrombosis (DVT) and long distance air travel (the so-called 'economy-class syndrome') students need to be encouraged to ask questions about the reliability and validity of scientific findings, sample size and any potential sampling bias, whether any causal mechanism has been established, and the possibility of DVT resulting from other factors not considered in the newspaper report.

To support the development of effective engagement with science, students need to gain the confidence to see themselves as capable of raising such questions (an affective outcome) and experience a range of relevant contexts in which to use and develop their understanding of the nature of science. This is not to say that such teaching should not address aspects of the nature of science explicitly. However, such explicit discussions might be most appropriately set in science contexts from the outset, rather than through a de-contextualised presentation on the nature of science followed by illustrative examples of how such ideas might be applied in a citizen engagement context of contemporary concern.

In this chapter I have discussed teaching/learning about the nature of science as a way of supporting young people's future engagement with science/technology-related issues as adults. However, there are many studies that have taken a more immediate stance on action within a school science setting (Kolstoe, 2000; Sutti, 1991). For example, one study has involved students in taking measurements of the levels of pollutants in a nearby stream and presenting their findings to the local community (Lee & Roth, 2002). Jenkins (2002) suggests that such teaching contexts provide motivation for students since the context involves immediate action by the students and is of personal relevance to them. He goes on to suggest (citing Posch, 1993) that this motivation can be lost if the teaching does not involve direct action but, for example, uses role play to discuss or enact a case study, the activity becoming vicarious rather than personal.

REFLECTIONS

An enhanced focus on science education for citizenship and the nature of science within school science and technology is a significant curriculum policy shift. There are many unanswered questions concerning the impact of such a change on students. Will an enhanced focus on nature of science and citizen engagement contexts change the ways in which these students interact with science and technology outside of formal education? Traditionally science education has focused on 'hard facts'. Indeed, such reliable and powerful knowledge is perhaps the defining feature of the scientific enterprise. An increased focus on the social and cultural aspects of science, and disputes and opposing views in science, softens this 'hard edge'. How will this 'softening' impact on the career aspirations of school students who are attracted to the 'hard edge' of science and would traditionally go on to pursue post-compulsory science courses, and perhaps enter the science professions? Will the impact of these curriculum innovations on student participation in post-compulsory science courses be different for boys and girls? Current curriculum innovations, and associated evaluations, are providing insights into some of these questions (e.g., Millar, 2006). However, there is an urgent need for large-scale longitudinal evaluation studies of the impact of this curriculum policy shift on the student experience.

These curriculum policy developments also present significant challenges for the teaching profession. Of course there is a need for classroom teaching resources and associated professional development activities to support teachers. However, more fundamentally, how will the teaching profession respond to this shift in curriculum focus? In England the science curriculum for 14–16 year olds experienced a major reform in 2006 (Qualifications & Curriculum Authority, 2005). One feature of this reform is an enhanced focus on citizenship issues and teaching about the nature of science. My own experience of talking to teachers about these developments indicates a wide range of initial responses. Many teachers have embraced teaching about the nature of science and its relevance to people's engagement with contemporary science/technology-related issues. Such teachers often cite the positive impact the new courses are having on students previously disaffected from school science. Other teachers have responded negatively, and vociferously, to what they perceive as a 'dumbing down' of the science curriculum. These responses suggest that this shift in science curriculum content and purpose is striking at the heart of many teachers' identities and sense of professionalism. Many teachers feel that they are being asked to see themselves, and their subject, differently. We have much to learn about the differing ways in which this shift in curriculum policy is being enacted by the teaching profession.

Finally, the significance of summative student assessment within schools is an issue that is often not given sufficient emphasis in discussions about teaching the nature of science within the educational research literature. In many countries the use and significance of national testing is growing. Such tests are used to create league tables of school performance and can even guide decisions about whether schools should be closed (e.g., Curtis, 2008). Understandably, school management is concerned about the performance of science departments within such testing regimes and will apply pressure on science staff to ensure that their students 'perform' well in such tests. How does teaching about citizen engagement and the nature of science feature within this context? For example, if understandings about the nature of science do not appear on national tests then it is likely that in many school contexts such teaching will not be pursued. However, assessment instruments addressing ideas about the nature of science have been developed mainly for research purposes rather than for use in the classroom. There is an urgent need for the design and evaluation of nature of science assessment instruments that can be used by teachers within the school science classroom and in the formal examination and accreditation of students.

REFERENCES

Aikenhead, G. S. (1985). Collective decision making in the social context of science. *Science Education, 69(4)*, 453–475.

Aikenhead, G. S. (2005). *Science education for everyday life*. New York: Teachers' College Press.

Albe, V. (2008). When scientific knowledge, daily life experience, epistemological and social considerations intersect: Students' argumentation in group discussions on a socio-scientific issue. *Research in Science Education, 38*(1), 67–90.

American Association for the Advancement of Science. (1995). *Science for all Americans*. New York: Oxford University Press.

Bellomo, K. (2005). Exploring the nature of science: Reinterpreting the Burgess Shale fossils. In S. Alsop, L. Bencze, & E. Pedretti (Eds.), *Analysing exemplary science teaching: Theoretical lenses and a spectrum of possibilities for practice* (pp. 46–52). Maidenhead, UK: Open University Press.

Brickhouse, N. W., Dagher, Z. R., Letts, W. J., & Shipman, H. L. (2000). Diversity of students' views about evidence, theory, and the interface between science and religion in an Astronomy course. *Journal of Research in Science Teaching, 37*(4), 340–362.

Burgess, A. (2004). *Cellular phones, public fears and a culture of precaution*. Cambridge, UK: Cambridge University Press.

Cartier, J. L., & Stewart, J. (2000). Teaching the nature of inquiry: Further developments in a high school genetics curriculum. *Science and Education, 9*, 247–267.

Chinn, C. A., & Malhotra, B. A. (2002). Epistemologically authentic inquiry in schools: A theoretical framework for evaluating inquiry tasks. *Science Education, 86(2)*, 175–218.

Collins, H. M., & Pinch, T. (1994). *The Golem: What everyone should know about science*. Cambridge, UK: Cambridge University Press.

Curtis, P. (2008, March 13). Struggling schools given more money and less time to hit their targets. *The Guardian*. Retrieved April 1, 2008, from http://www.guardian.co.uk/education/2008/mar/13/ schools.uk1#

De Andrade, M. R., & Silva, C. C. (2001). Newton and colour: The complex interplay of theory and experiment. *Science and Education, 10*, 287–305.

Donovan, M. S., & Brandsford, J. D. (Eds.). (2005). *How students learn: Science in the classroom*. Washington DC: The National Academies Press.

Drake, F. (2006). Mobile phone masts: Protesting the scientific evidence. *Public Understanding of Science, 15*, 387–410.

Durant, J. R. (1990). What is scientific literacy? *European Review, 2*(1), 83–89.

Epstein, S. (1995). The construction of lay expertise: AIDS activism and the forging of credibility in the reform of clinical trials. *Science, Technology and Human Values, 20*(4), 408–437.

European Commission. (2007). *Science education NOW: A renewed pedagogy for the future of Europe*. Luxembourg: Office for Official Publications of the European Communities.

Giere, R. N. (1995). *Understanding scientific reasoning* (3rd ed.). New York and London: Holt, Rinehart and Winston.

Hind, A., Leach, J., & Ryder, J. (2000). *Nuffield teaching about science: Resources for AS/A level*. Retrieved April 1, 2008, from http://www.nuffieldcurriculumcentre.org/go/minisite/Post16TeachingAboutScience/Introduction

Irwin, A. (1995). *Citizen science*. London and New York: Routledge.

Irwin, A., & Wynne, B. (Eds.). (1996). *Misunderstanding science? The public reconstruction of science and technology*. Cambridge, UK: Cambridge University Press.

Jasanoff, S. (1997). Civilisation and madness: The great BSE scare of 1996. *Public Understanding of Science, 6*, 221–232.

Jenkins, E. W. (2002). Linking school science education with action. In W.-M. Roth & J. Desautels (Eds.), *Science education as/for sociopolitical action* (pp. 17–34). New York: Peter Lang.

Khishfe, R., & Abd-El-Khalick, F. (2002). Influence of explicit and reflective versus implicit inquiry-oriented instruction on sixth graders' views of nature of science. *Journal of Research in Science Teaching, 39*(7), 551–578.

Khishfe, R., & Lederman, N. G. (2007). Relationship between instructional context and views of nature of science. *International Journal of Science Education, 29*(8), 939–961.

Knorr-Cetina, K. D. (1999). *Epistemic cultures: How the sciences make knowledge*. Cambridge, MA: Harvard University Press.

Kolstoe, S. D. (2000). Consensus projects: Teaching science for citizenship. *International Journal of Science Education, 22*(6), 645–664.

Kolstoe, S. D. (2001). 'To trust or not to trust...' – pupils' ways of judging information encountered in a socio-scientific issue. *International Journal of Science Education, 23*(9), 877–901.

Kuhn, D., Amsel, E., & O'Loughlin, M. (1988). *The development of scientific thinking skills.* San Diego, CA: Academic Press.

Latour, B. (1987). *Science in action.* Milton Keynes: Open University Press.

Layton, D. (1993). *Technology's challenge to science education.* Milton Keynes: Open University Press.

Layton, D., Jenkins, E., Macgill, S., & Davey, A. (1993). *Inarticulate science.* Driffield, UK: Studies in Education.

Leach, J., Hind, A., & Ryder, J. (2003). Designing and evaluating short teaching interventions about the epistemology of science in high school classrooms. *Science Education, 87(3)*, 831–848.

Leach, J., Millar, R., Ryder, J., & Séré, M. G. (2000). Epistemological understanding in science learning: The consistency of representations across contexts. *Learning and Instruction, 10*(6), 497–527.

Lee, Y. C. (2008). Exploring the roles and nature of science: A case study of severe acute respiratory syndrome. *International Journal of Science Education, 30*(4), 515–541.

Lee, S., & Roth, W.-M. (2002). Learning science in the community. In W.-M. Roth & J. Desautels (Eds.), *Science education as/for sociopolitical action* (pp. 37–66). New York: Peter Lang.

Matthews, M. R. (1994). *Science teaching: The role of history and philosophy of science.* New York and London: Routledge.

McComas, W. F. (Ed.). (1998). *The nature of science in science education: Rationales and strategies.* Dordrecht, The Netherlands; Boston and London: Kluwer.

McMurray, R., Cheater, F., Weighall, A., Nelson, C., Schweiger, M., & Mukherjee, S. (2004). Managing controversy through consultation: A qualitative study of communication and trust around MMR vaccination decisions. *British Journal of General Practice, 54*, 520–525.

Millar, R. (2006). Twenty first century science: Insights from the design and implementation of a scientific literacy approach in school science. *International Journal of Science Education, 28*(13), 1499–1521.

Monk, M., & Osborne, J. (1997). Placing the history and philosophy of science on the curriculum: A model for the development of pedagogy. *Science Education, 81*(4), 405–424.

Mortimer, E. F. (1995). Conceptual change or conceptual profile change? *Science and Education, 4*, 267–285.

Osborne, J., Collins, S., Ratcliffe, M., Millar, R., & Duschl, R. (2003). What 'ideas-about-science' should be taught in school science? *Journal of Research in Science Teaching, 40*(7), 692–720.

Osborne, J., & Dillon, J. (2008). *Science education in Europe: Critical reflections.* London: The Nuffield Foundation.

Petts, J. (1997). The public-expert interface in local waste management decisions: Expertise, credibility and process. *Public Understanding of Science, 6*, 359–381.

Posch, P. (1993). Research issues in environmental education. *Studies in Science Education, 21*, 21–48.

Qualifications & Curriculum Authority. (2005). *New key stage 4 programme of study from 2006.* Retrieved April 1, 2008, from http://www.qca.org.uk/default.aspx

Roth, W.-M., & Lucas, K. B. (1997). From "truth" to "invented reality": A discourse analysis of high school physics students' talk about scientific knowledge. *Journal of Research in Science Teaching, 34*(2), 145–179.

Roth, W.-M., & Désautels, J. (Eds.). (2002). *Science education as/for sociopolitical action.* New York: Peter Lang.

Roth, W.-M. (1995). *Authentic school science. Knowing and learning in open-inquiry science laboratories.* Dordrecht, The Netherlands and London: Kluwer.

Rudolph, J. (2000). Reconsidering the nature of science as a curriculum component. *Journal of Curriculum Studies, 32*, 403–419.

Ryder, J. (2001). Identifying science understanding for functional scientific literacy. *Studies in Science Education, 36*, 1–44.

Ryder, J. (2007, August). *The development of secondary school students' understandings of 'ideas about science'.* Paper presented at the European Science Education Research Association Conference, Malmo, Sweden.

Ryder, J., & Leach, J. (2008). Teaching about the epistemology of science in upper secondary schools: An analysis of teachers' classroom talk. *Science and Education, 17*, 289–315.

Ryder, J., Leach, J., & Driver, R. (1999). Undergraduate science students' images of science. *Journal of Research in Science Teaching, 36*(2), 201–219.

Samarapungavan, A., Westby, E. L., & Bodner, G. M. (2006). Contextual epistemic development in science: A comparison of chemistry students and research chemists. *Science Education, 90*(3), 468–495.

Schwartz, R., & Lederman, N. (2008). What scientists say: Scientists' views of nature of science and relation to science context. *International Journal of Science Education, 30(6)*, 727–771.

Shortland, M., & Warwick, A. (Eds.). (1989). *Teaching the history of science*. Oxford, UK: The British Society for the History of Science.

Southerland, S. A., Johnston, A., Sowell, S., & Settlage, J. (2005, April). *Perhaps triangulation isn't enough: A call for crystallization as a methodological referent in NOS research.* Paper presented at the American Education Research Association, Montreal, Canada.

Stewart, W. (2000). *Mobile phones and health.* Report of the Independent Expert Group on Mobile phones (National Radiological Protection Board). Retrieved February 1, 2009, from http://www.iegmp.org.uk/report/text.htm

Sutti, A. (1991). The water analysis project: An alternative model for environmental study. In *Environment, schools and active learning* (pp. 59–65). Paris: OECD.

Tytler, R. (2007). *Re-imagining science education: Engaging students in science for Australia's future.* Victoria: Australian Council for Education Research.

Wong, S., Kwan, J., Hodson, D., & Yung, B. (2009). Turning crisis into opportunity: Nature of science and scientific inquiry as illustrated in the scientific research on Severe Acute Respiratory Syndrome. *Science & Education, 18*(1), 95–118.

Zeidler, D. L, Walker, K. A, Ackett, W. A, & Simmons, M. L. (2002). Tangled up in views: Beliefs in the nature of science and responses to socioscientific dilemmas. *Science Education, 86*(3), 343–367.

Jim Ryder
Centre for Studies in Science and Mathematics Education
School of Education
University of Leeds, UK

JUSTIN DILLON

26. APPROACHING 'SOFT DISASTERS' IN THE CLASSROOM: TEACHING ABOUT CONTROVERSIAL ISSUES IN SCIENCE, TECHNOLOGY, SOCIETY, AND ENVIRONMENT EDUCATION

INTRODUCTION

Today's citizens are confronted by a range of what are called 'soft disasters' – "environmental and political crises that emerge only slowly but at high cost to society, not least the erosion of public confidence and legitimacy" (ESRC Global Environmental Change Programme, 2000, p. 3). Such soft disasters include socio-scientific issues such as BSE, the GM food debate, HIV-AIDS and global climate change. To understand such issues, the public needs an education that is broad and deep enough to unravel the complex inter-relationships among science, technology, society and the environment (STSE). However, many of these topics are controversial and teachers face many challenges in attempting to address them in their classrooms.

Not surprisingly, the teaching of controversial STSE issues is itself somewhat controversial. In 2004, I gave a talk entitled 'Teaching Controversial Issues' as part of a session on 'Teaching Science for Responsibility' at the British Association Festival of Science in Exeter. My aim was to disseminate the findings of research I had been involved in with colleagues from two other universities (Oulton, Day, Dillon, & Grace, 2001). The press release issued by King's College to publicise the event contained a comment from me that summarised one of the key points emerging from our research:

> Teachers should be open with their own biases rather than pretend to be neutral and students should be asked to take any bias into account when making up their minds on a topic. Taking a neutral stance is not a good strategy for teaching children how society works.

The press release contained another statement attributed to me:

> The traditional approach to leading a discussion on a controversial science subject is for the teacher to take a neutral role. ... We believe that this strategy is wrong and that it is unethical to pretend to pupils that teachers have no opinion.

The story was picked up by Mark Crow (2004) of the *Guardian*, one of the UK's more thoughtful newspapers. Crow opened his story by stating that "New research has lambasted science teaching in the UK as 'unethical' for its failure to acknowledge the importance of personal bias" (first paragraph). Crow had sought the views of a pressure group, Save British Science, whose spokesperson, Rosemary Davies, was quoted as saying: "Whilst it may be inappropriate for teachers to claim to be dispassionate about controversial science issues, students should still be trained to aspire towards objectivity". 'Aspire' maybe, but – as I will attempt to demonstrate later – students need to be aware that values play a key role in many of the decisions made by themselves, by politicians, by scientists, and by many others. Science and technology education

A.T. Jones and M.J. de Vries (eds.), International Handbook of Research and Development in Technology Education, 297–306.
© 2009 Sense Publishers. All rights reserved.

that does not leave students with an awareness of the complexity of the STSE issues facing society is not fit for its purpose. Davies was also quoted in the article as saying:

> Just as subjects within science such as biology or physics are best taught by fully trained specialists, 'science for citizenship' type courses also need to be taught by those with adequate training and a thorough understanding of policy making and ethics. We can't invest in the skills of our pupils without investing in the people who are going to teach them. (Crow, 2004, final paragraph)

I would agree with the sentiment Davies expresses here. One of the findings of the research on which I was reporting is that science teachers often state that they are not trained to teach about socioscientific issues. Davies' quote masks a concern examined by Cross and Price (1996) about teachers who present science as unproblematic and characterised by content and certainty. As Kibble (1998) points out, there is a danger of an "over-simplistic presentation of moral dilemmas" (p. 54). Reporting on their work with pre-service and in-service teachers, Cross and Price found that science teachers were very loyal to their subject discipline, that is to say, a loyalty to science rather than science education. Partly as a result of this divided loyalty, Cross and Price reported that students lacked opportunities to discuss the nature of science or the construction of knowledge. A similar point has been made by Camino and Calcagno (1995):

> The most delicate matter still has to be faced, that is encouraging [science] teachers to change their ways of thinking. This implies, first, trying to persuade teachers to abandon the safety of viewing science as an objective and neutral discipline in favour of the idea that knowledge is 'a-disciplinary', transitory and loaded with values. (p. 72)

The British Association talk was also picked up by Anna Salleh (2004) of the Australian Broadcasting Corporation, whose online piece was entitled 'Teachers told to spice up lessons'. Salleh had contacted Deborah Crossing, Executive Director of the Australian Science Teachers Association, who said that "controversies such as immunisation, Australia's relationship to the Kyoto Protocol and genetic engineering were regularly discussed in Australian schools". She added that "teachers in Australia were probably not encouraged to express their own point of view when teaching controversies, and this might be because of concerns about litigation. ... Instead," she said, "they were encouraged to facilitate debate". Crossing continued by providing what has become the orthodox view of teacher as neutral chair: "A good teacher will engage debate, provide the students with different points of scientific view, get them to question, get them to test and get them to come up with their own solution." She went on to make a point with which I wholeheartedly disagree: "It's kind of irrelevant whether they put their point of view". I would, however, agree with her subsequent statement that: "The important thing is that they are encouraging their students to come up with their point of view based on good scientific evidence and [showing them] how to assess whether it's good scientific evidence." Interestingly, Crossing seemed to contradict her support for teachers adopting a neutral stance when she commented that "students often did not question their teachers' views and this was counterproductive". Unless teachers give their views, how can students question them?

THE NATURE OF CONTROVERSY AND CONTROVERSIAL ISSUES

Controversy seems to be a common feature of many cultures. New 'soft disasters' appear daily somewhere in the world. Over time, controversies also disappear; what might be thought controversial one year might not be so the next. Issues are not intrinsically controversial – if it takes two to tango then it takes two or more for something to be controversial. Indeed, I am using the phrase 'controversial issues' in this chapter to mean that "significant numbers of people argue about them without reaching a conclusion" (Oulton, Dillon, & Grace, 2004, p. 411). Dearden (1981), writing more than 25 years ago, opined that:

> A matter is controversial if contrary views can be held on it without those views being contrary to reason. This can be the case, for example, where insufficient evidence is held in order to decide the controversy, or, where the outcomes depend on future events that cannot be predicted with certainty, and where judgement about the issue depends on how to weigh or give value to the various information that is known about the issue. (p. 38)

Using this conceptualisation, one can see how the increase in the use of mobile phones (see Chapter 25), which led to some people raising concerns about their safety, can be considered controversial because there was insufficient evidence about links between phone use and users' health. The point here is that people, when presented with the same evidence, whether it be scientific or otherwise, can come to radically different conclusions depending on their personal values (Lynch & McKenna, 1990). So Crossing's point, referred to above, that the "important thing is that [teachers] are encouraging their students to come up with their point of view based on good scientific evidence" seems oblivious to the impact of students' values on their interpretation of 'good scientific evidence'.

Writing a few years later than Dearden, Stradling (1985) defined controversial issues as "those issues on which our society is clearly divided and significant groups within society advocate conflicting explanations or solutions based on alternative values" (p. 9). Thus, nanotechnologies can be seen as offering great value or great peril depending on the values of people debating their use. Of course there are also degrees of being controversial, and what might be controversial in one situation is not controversial in another. For as long as I can remember, the use of nuclear power has been far more controversial in, say, the UK than it has been in France.

TEACHING ABOUT CONTROVERSIAL ISSUES

> Education should not attempt to shelter our nation's children from even the harsher controversies of adult life, but should prepare them to deal with such controversies knowledgeably, sensibly, tolerantly and morally. (Qualifications and Curriculum Authority [QCA], 1998, p. 56)

Thus, the UK Qualifications and Curriculum Authority sees teachers as having a duty to prepare students to "deal with" controversial issues. Such an argument is supported by writers such as Dewhurst (1992), who argued that teachers have "to help their students to handle questions of value, to learn to make judgements which are truly their own" (p. 153); and Leib (1998), who argued that "it is [the educators'] duty not to shy away from tackling controversial topics" (p. 230). In order to carry out this "duty", Geddis (1991) asserts that teachers need to teach in such a way that students:

> … uncover how particular knowledge claims may serve the interests of different claimants. If they are to be able to take other points of view into account in developing their own positions on issues, they need to attempt to unravel the interplay of interests that underlie these other points of view. (p. 171)

The point is that controversial issues cannot be taught in a value-free way. Extending Geddis' points further, Oulton et al. (2004) argue that teachers need to teach about controversial issues in such a way that the following points are made:

- Groups within society hold differing views about them;
- groups base their views on either different sets of information or they interpret the same information in different ways;
- the interpretations may occur because of the different ways in which individuals or groups understand or 'see' the world (i.e. their worldview);
- differing worldviews can occur because the individuals adhere to different value systems;
- controversial issues cannot always be resolved by recourse to reason, logic or experiment; and
- controversial issues may be resolved as more information becomes available.

A point here is that students may well benefit from studying the nature of controversy, as well as developing better understanding of particular controversial issues. Science and technology teachers are, *potentially*, in a strong position to teach about the nature of controversy in their subject areas.

Research into pedagogical approaches to the teaching of controversial issues has developed relatively recently. Oulton et al. (2004, p. 415), summarising recent research, note that teachers often suggest that their pedagogy is underpinned by their stance on three issues:

- The focus should be on rationality, reasoning and sticking to the facts;
- a balanced view on the issue must be presented; and
- the teacher should remain neutral.

The neutral stance is a pedagogic strategy that has been advocated by, among others, Lawrence Stenhouse (1983). 'Procedural neutrality', as Stenhouse called it, is easier to advocate than to sustain in the classroom. Indeed, Stradling (1985) reports that teachers find procedural neutrality difficult to sustain because it threatens the rapport they have built up with the students and because the approach appears to cast doubt on their personal credibility. More realistically, Kelly (1986) suggests that teachers show 'committed impartiality'. With this approach, teachers attempt to provide different sides of an argument but do, at some stage, share their own views with the class.

In 1998, the UK Qualifications and Curriculum Authority concluded that teachers need to be flexible rather than show allegiance to one particular stance. Noting that it was not their "business or intention to try to tell teachers how to teach", they did, however, comment in a disturbingly vague manner that "a 'common sense' approach has much to recommend it" (QCA, 1998, p. 60). Three years later, the UK Department for Education and Employment (DfEE) published a non-statutory framework for Personal, Social and Health Education (PSHE) and Citizenship to accompany the recently implemented statutory National Curriculum for secondary (high) schools. Interestingly, the Citizenship documentation (DfEE, 1999) implied that certain values were to be promoted, for example:

> Education for sustainable development, through developing pupils' skills in, and commitment to, effective participation in the democratic and other decision making processes that affect the quality, structure and health of environments and society and exploring values that determine people's actions within society, the economy and the environment. (pp. 7–8)

So what might teachers do to promote such values should they so wish? Oulton et al. (2004) argue that there is a need for pedagogical approaches that:

- focus on the nature of controversy and controversial issues – that is, that people disagree and have different worldviews; values and limitations of science; political understanding; power; etc.;

- motivate pupils to recognize that a person's stance on an issue will be affected by their worldview;
- emphasize the importance of teachers and learners reflecting critically on their own stance and recognize the need to avoid the prejudice that comes from a lack of critical reflection;
- give pupils the skills and abilities to identify bias for themselves, encouraging them to take a critical stance towards claims of neutrality, a lack of bias and claims to offer a balanced view;
- promote open mindedness, a thirst for more information and more sources of information and a willingness to change one's view as appropriate, and avoid strategies that encourage pupils to actually make up their minds on an issue too hastily; and
- motivate teachers, as much as possible, to share their views with pupils and make explicit the way in which they arrive at their own stance on an issue. (p. 420)

Such a pedagogy, however, is rarely found in classrooms. Indeed, as suggested by Deborah Crossing, teachers might be reluctant to express their own point of view because of the risk of litigation (in Salleh, 2004, referred to above).

In recent years, curriculum changes in the UK have encouraged teachers to develop their students' awareness and understanding of controversial issues. For example, ever since the introduction of the National Curriculum in England, teachers have been encouraged by statute to devote some time to the nature of science. At its best, such an approach to science teaching can illuminate the values that underpin science and the use of science by society. In its latest incarnation, the science curriculum for 14-16-year-olds indicates that pupils should be taught about 'How Science Works'. Under the heading 'Applications and implications of science', teachers are instructed that pupils should be taught:

- about the use of contemporary scientific and technological developments and their benefits, drawbacks and risks
- to consider how and why decisions about science and technology are made, including those that raise ethical issues, and about the social, economic and environmental effects of such decisions
- how uncertainties in scientific knowledge and scientific ideas change over time and about the role of the scientific community in validating these changes. (QCA, 2007, p. 4)

A short series of 'Explanatory Notes' expands on what the learning outcomes should be:

> During key stage 4, pupils learn about the way science and scientists work within society. They consider the relationships between data, evidence, theories and explanations, and develop their practical, problem-solving and enquiry skills, working individually and in groups. They evaluate enquiry methods and conclusions both qualitatively and quantitatively, and communicate their ideas with clarity and precision.
>
> All pupils develop their ability to relate their understanding of science to their own and others' decisions about lifestyles, and to scientific and technological developments in society. (ibid.)

Few teachers have substantial experience of working as a scientist other than during their own education. As a result, they depend for their knowledge of how science works on a range of secondary sources, including television and textbooks. Some teachers arrange for scientists to come into the classroom, but they are in a minority. Perhaps teachers should spend some time shadowing scientists and technologists for a short time.

RECENT RESEARCH IN THE TEACHING OF CONTROVERSIAL ISSUES

Although teaching about STSE issues has been advocated for many years (see, for example, Aikenhead, 2003; Solomon, 1993; van Weelie & Wals, 2002), it is only relatively recently that it has generated much research interest. For example, a growing body of research has examined how teachers deal with socioscientific issues. Albe (2008) carried out a study in a French 11th-grade science class (n=12) in the context of vocational secondary education. Albe's participants were 16-18-year-olds specialising in sciences and technologies for agronomy and the environment. The students were divided into two groups to debate whether or not mobile phones were dangerous to human health. Both groups contained equal numbers of boys and girls. The teaching intervention was based on a module designed by science education researchers and teachers (Hind, Leach, Ryder, & Prideaux, 2001). The purpose of the lesson was to develop the students' understanding of how to assess the quality of scientific data.

> In the first part of the lesson, students examined the issues raised by the controversy, the technological and scientific concepts involved and were trained to evaluate the validity and reliability of research results. Students were then asked to review a number of pieces of research and to select those that support a particular point of view. To do so, students in group discussions studied seven research extracts on mobile phone effects. Each extract was an approximately 10-lines of text describing the research protocol and the results obtained by a team of scientists. They are related to the occurrence of illnesses in animals which had been subjected to experiments but also on epidemiological surveys and memory tests. (Albe, 2008, p. 69)

The teacher then set up a court case scenario. Students took on the roles of expert witnesses in a trial in which an employer was taken to court by an employee whose health was poor. The employee's argument was that it was excessive mobile phone use that had led to health problems. Students had to present coherent arguments to support the employer or the employee. Having presented their case, they then had to question the arguments put forward by the other group. The teacher's role was that of a magistrate presiding over the case. After the role playing was over, the students were given an opportunity to reflect on their opinions about the controversy. Albe used a "discourse analytic approach grounded in sociolinguistics and supported by an ethnographic perspective" (p. 86), reporting that:

> The activity of reviewing research results and elaborating arguments in a discussion group, with a view to participating in a role-play on the controversial issue of the danger of using cellular telephones, is very demanding in social and epistemological terms. (p. 85)

In addition:

> It has been observed that collaborative argumentation was developed within a group when a student acts as a helper collaborating with a student assuming the leadership of group discussions. Students' requests addressed to others to explain their claims or to develop justifications create opportunities for collaborative argumentation. These results tend to indicate that group discussions on a socio-scientific issue should be organised to regulate students' interactions on the social plane. Teachers should attempt to arrange group discussions to avoid student conflicts. (p. 86)

Following on from Grace (2005), who suggested ground rules for discussions of socioscientific arguments, Albe (2008) suggests that "guides could be provided to students within group discussions to explain their views and consider others. A discussion at the end of the activity may also be valuable to enable students to reflect on their views and appreciate the value of group discussion" (p. 86). Importantly, Albe argues that as well as looking at controversial issues themselves, students should also examine:

> … the way in which scientific knowledge is produced within a community and, in particular, the role of controversy in this process. For instance, the inclusion of considerations of the

> nature and limitations of science, the status, role and limits of evidence, the interests involved and the ways science communities operate to establish knowledge are very important when dealing with contemporary science controversies. (p. 86)

One wonders how many teachers are *au fait* with 'the ways science communities operate to establish knowledge'. (See also Chapter 25 for a discussion of the importance of the nature of science in relation to socioscientific issues.) Albe recognises that teachers need help, noting:

> Teachers can be uncomfortable with such issues that could question their own epistemology and teaching practices. A way of assisting teacher professional development can be the co-operative design of teaching sequences on socio-scientific issues. This can be an occasion to explore with teachers ideas about the nature of controversy and elaboration modes of scientific knowledge, issues about teaching approaches appropriate to socio-scientific issues and the question of teacher position in class. (p. 86)

Writing two years earlier, Levinson (2006) pointed out that "the teaching of controversial issues needs a stronger theoretical base" (p. 1201) and proposed a theoretical framework for the teaching of socio-scientific controversial issues:

> Drawing on a liberal democratic conception of possible sources of conflict, three strands are developed that provide a framework for teachers when teaching socio-scientific issues: these are categories of reasonable disagreement, the communicative virtues, and modes of thought. (ibid.)

Levinson also provides 'an epistemological model of controversy' that, he argues, is: "a framework for approaching the teaching of controversial issues in schools and a tool for analysing the ways in which elements of a controversial issue are deployed in the classroom" (p. 1217):

- Categories of reasonable disagreement.
- Communicative virtues when conducting a discussion with conflicting views
- Patience
- Tolerance
- Respect for differences
- Attentive and thoughtful listening
- Openness
- Honest self-expression
- Adherence to agreed procedures
- Freedom of expression
- Equality
- Modes of thought
- Narrative mode which involves the voices of the participants. For lay people addressing socioscientific issues the substantive science is often reconstructed, marginalized or inert.
- Logico-scientific mode, based on scientific evidence. (ibid.)

Drawing on various authors' work (for example, Rawls, 1983), Levinson provides a secure theoretical framework that complements and develops other work in the field of argumentation (e.g., Newton, Driver, & Osborne, 1999; Zeidler, Osborne, Erduran, Simon, & Monk, 2003; and Chapter 36 by Osborne).

CONTROVERSIAL ISSUES AND THE ENVIRONMENT

With regard to research specifically focused on the teaching of controversial environmental issues, Cotton (2006) noted that "there has been little research into the strategies that teachers adopt" (p. 225). Adding that much existing research involved large-scale surveys of teachers' attitudes, students' attitudes, or classroom interventions aimed at attitudinal change, Cotton argues that they have provided little insight into teachers' beliefs and normal classroom behaviours:

> The few recent studies which have investigated teachers' beliefs about teaching controversial issues in other curriculum areas have largely neglected to incorporate classroom observation, being reliant instead on post hoc reports of behaviour. (p. 225)

Cotton's study "aimed to provide an alternative perspective by investigating the beliefs and practices of three experienced teachers in English secondary schools, drawing upon both interview data and transcripts of classroom interaction" (ibid.). Cotton focused on a geography course for students aged 16–19 (n=11-18) and studied three cases, each involving observations over a 5-6 week period at three different schools. The topics covered were: indigenous people's land rights in the rainforest, the role of NGOs in governing Antarctica, and reconciling the needs of conservation and tourism in National Parks. Cotton reported that all the teachers adapted their teaching strategies "to encourage student involvement in discussions and producing lively and interesting lessons" (p. 237). However, he went on to note:

> All of the teachers studied experienced great difficulty in implementing their beliefs about balance and neutrality, and the classroom data suggest that the influence of the teachers' own environmental attitudes was greater than they either intended or, in all probability, realized. Moreover, in line with the findings of Geddis (1991), there were occasions where the teachers' intent appeared to be persuasion rather than instruction. Their desire not to express their own views frequently led to the situation where these views were expressed indirectly in the form of questions, or by control of students' turns in discussion. Whilst these strategies enabled the teachers to avoid explicitly stating their views, such an indirect expression of attitudes may have been harder for the students to challenge than a direct argument presented by the teacher. (ibid.)

Cotton's findings seem to provide support for the more transparent approach advocated by Oulton et al. (2004) (see above).

In a similar vein, Tal and Kedmi (2006), reporting on the implementation of an innovative Science and Technology in Society (STiS) curriculum in Israel, described a study involving observations of six classes and interviews with four teachers and 20 grade 10 and 11 students. The unit taught was entitled 'Treasures in the Sea: Use and Abuse'. Noting that the "use of relevant and authentic topics that encourage and allow open discussions within and between communities reinforces the idea of citizens-science" (p. 640), they caution that "teachers cannot dramatically change their practice unless they go through a longer process of reviewing their own perceptions of science and teaching philosophy, and gradually practicing different teaching and assessment culture in class" (p. 641).

CONCLUDING COMMENTS

The long-haul journey to broaden the school curriculum in order to address STSE issues has been a bumpy ride. While many countries now promote the discussion of such issues, the evidence of any substantial impact on teachers' practices and students' understandings, attitudes, or behaviours remains weak. It is easy to advocate their inclusion in the curriculum; it is much more difficult to impact on the dominant pedagogies of science and technology education in an assessment climate that values rote learning of atomised concepts. This raises fundamental questions about how science teachers see science and technology and, thus, education in these areas. I worry that many science teachers see themselves as scientists first and teachers second when, in actuality, they have little experience of contemporary science in the making. Worse still, few science teachers appear to have had an adequate training in teaching socioscientific issues and so find it difficult to shift their pedagogical practices.

The focus of recent international comparisons, such as PISA, towards a focus on scientific literacy, might promote the cause of STSE education but it will take a major shift in teacher understanding and pedagogy before schools adequately prepare students for the complexity of life in the 21st Century.

ACKNOWLEDGEMENTS

I wish to thank two colleagues, Alex Manning and Melissa Glackin, for their insightful comments on an earlier draft of this chapter. I also acknowledge the contribution to my thinking about the teaching of controversial issues made by Marcus Grace (University of Southampton) and Chris Oulton (formerly of the University of Gloucester).

REFERENCES

Aikenhead, G. (2003). STS education: A rose by any other name. In R. Cross (Ed.), *A vision for science education: Responding to the work of Peter Fensham* (pp. 59–75). London: RoutledgeFalmer.

Albe, V. (2008). When scientific knowledge, daily life experience, epistemological and social considerations intersect: Students' argumentation in group discussions on a socio-scientific issue. *Research in Science Education, 38*(1), 67–90.

Camino, E., & Calcagno, C. (1995). An interactive methodology for 'empowering' students to deal with controversial environmental problems. *Environmental Education Research, 1*, 59–74.

Cotton, D. R. E. (2006). Teaching controversial environmental issues: Neutrality and balance in the reality of the classroom. *Educational Research, 48*(2), 223–241.

Cross, R., & Price, R. (1996). Science teachers' social conscience and the role of teaching controversial issues in the teaching of science. *Journal of Research in Science Teaching, 33*(3), 319–333.

Crow, M. (2004). Call for a new approach to science teaching. *Guardian*. Retrieved from http://guardian.co.uk/education

Dearden, R. F. (1981). Controversial issues in the curriculum. *Journal of Curriculum Studies, 13*, 37–44.

Department for Education and Employment [DfEE]. (1999). *The national curriculum for England: Citizenship*. London: Department for Education and Employment and the Qualifications and Curriculum Authority.

Dewhurst, D. (1992). The teaching of controversial issues. *Journal of Philosophy of Education, 26*(2), 153–163.

ESRC Global Environmental Change Programme. (2000). *Risky choices, soft disasters: Environmental decision making under uncertainty*. Brighton, Brighton and Hove: University of Sussex. Retrieved from www.gecko.ac.uk

Geddis, A. (1991). Improving the quality of science classroom discourse on controversial issues. *Science Education, 75*(2), 169–183.

Grace, M. M. (2005, August/September). *Developing personal values and argumentation skills through decision-making discussions about biological conservation*. Paper presented at the European Science Education Research Association (ESERA) conference, Barcelona, Spain.

Hind, A., Leach, J., Ryder, J., & Prideaux, N. (2001). *Teaching about the nature of scientific knowledge and investigation on AS/A level science courses*. Leeds, UK: Centre for Studies in Science and Mathematics Education.

Kelly, T. (1986). Discussing controversial issues: Four perspectives on the teacher's role. *Theory and Research in Social Education, 14*(2), 113–138.

Kibble, D. (1998). Moral education dilemmas for the teacher. *Curriculum Journal, 9*, 51–61.

Leib, J. (1998). Teaching controversial topics: Iconography and the Confederate Battle Flag in the South. *Journal of Geography, 97*, 229–240.

Levinson, R. (2006). Towards a theoretical framework for teaching controversial socio-scientific Issues. *International Journal of Science Education, 28*(10), 1201–1224.
Lynch, D., & McKenna, M. (1990). Teaching controversial material: New issues for teachers. *Social Education, 54*, 317–319.
Newton, P., Driver, R., & Osborne, J. (1999). The place of argumentation in the pedagogy of school science. *International Journal of Science Education, 21*(5), 553–576.
Oulton, C. R., Day, V., Dillon, J., & Grace, M. (2001). *Unlocking controversial issues: A report to the Countryside Foundation for Education.* Worcester, UK: University College Worcester.
Oulton, C., Dillon, J., & Grace, M. (2004). Reconceptualizing the teaching of controversial issues. *International Journal of Science Education, 26*(4), 411–423.
Qualifications and Curriculum Authority [QCA]. (1998). Education for citizenship and the teaching of democracy in schools. London: Author.
QCA. (2007). *Science. Programme of study for key stage 4.* London: Author.
Rawls, J. (1993). *Political liberalism.* New York: Columbia University Press.
Salleh, A. (2004). Teachers told to spice up lessons. *ABC Science Online.* Retrieved from September 13. Retrieved from http://www.abc.net.au/science/news/
Solomon, J. (1993). *Teaching science, technology and society.* Buckingham, UK: Open University Press.
Stenhouse, L. (1983). *Authority, education and emancipation.* London: Heinemann.
Stradling, B. (1985). Controversial issues in the curriculum. *Bulletin of Environmental Education, 170*, 9–13.
Tal, T., & Kedmi, Y. (2006). Teaching socioscientific issues: Classroom culture and students' performances. *Cultural Studies of Science Education, 1*(4), 615–644.
Van Weelie, D., & Wals, A. (2002). Making biodiversity meaningful through environmental education. *International Journal of Science Education, 24*, 1143–1156.
Zeidler, D. L., Osborne, J., Erduran, S., Simon, S., & Monk, M. (2003). The role of argument during discourse about socioscientific issues. In D. L. Zeidler (Ed.), *The role of moral reasoning on socioscientific issues and discourse in science education* (pp. 97–116). Boston: Kluwer.

Justin Dillon
Science and Technology Education Group
King's College London, UK

MICHAEL J. REISS

27. ETHICAL REASONING AND ACTION IN STSE EDUCATION

INTRODUCTION

That ethics is a part of technology education is fairly uncontroversial, but the extent to which ethics should be taught within science education is more contested and has been the subject of debate for some considerable time. Indeed, there are many who remain suspicious about the place of ethics in science education. Here I explore what teaching about ethics and values might hope to accomplish in STSE (Science-Technology-Society-Environment) education. I relate this to the broader aims of education and, with reference to particular instances, such as teaching about climate change, discuss the extent to which a valid education of young people can expect action or whether it should concern itself solely with cognitive and attitudinal aims.

TEACHING ABOUT ETHICS IN SCIENCE EDUCATION

Broadly speaking, there are two main families of arguments against the teaching of ethics in science. One is to do with the nature of science and the other to do with the consequences of teaching ethics in science (Reiss, 1999).

The argument against the teaching of ethics in science that stems from a consideration of the nature of science has its roots in epistemological distinctions between forms of knowledge. It was David Hume who first pointed out the difference between what he termed 'relations of ideas' and 'matters of fact'. Mathematics concerns itself with relations of ideas. As Hume explained, it doesn't matter how many [planar] triangles you investigate, the sum of the internal angles of each of them is always 180°. Contrast this with science where the veracity of matters of fact is subject to empirical investigation. As Popper so famously pointed out, science is therefore primarily about the possible refutation of hypotheses, through the comparison of their predictions with reality. (There is no need here to introduce subsequent more historically – and sociologically-situated – accounts of the nature of the scientific enterprise, cf. Kuhn (1970), Lakatos (1978), and Yearley (2005). The point is that different domains of knowledge have different ways of having their truth claims established.)

It was Hume, too, who pointed out that there is no logical connection between what is and what ought to be – the contradiction of which has subsequently become known as the naturalistic fallacy. Thus is stated the essence of the argument that there should be no ethics in science: science concerns itself with matters of fact, with what is, including how we arrive at such knowledge; whereas ethics concerns itself with what ought to be. In other words, the two disciplines of science and ethics occupy separate spheres of knowledge. In claiming that ethics should be taught in science one might as well claim that science teachers should teach aesthetics. The job of a physics teacher is to explain why we get rainbows, not to pontificate on whether they are beautiful or to urge us what we should do on seeing one.

A.T. Jones and M.J. de Vries (eds.), International Handbook of Research and Development in Technology Education, 307–318.
© 2009 Sense Publishers. All rights reserved.

This point of view has been forcefully presented by Donnelly (2002) who, following Ogborn (1995) and others, argues in science for 'the refusal of the personal', 'the demarcation from ethics', and 'the absence of reflexivity'. In the section on the demarcation of science from ethics, Donnelly maintains that while pragmatic attempts have been made to insert ethics in a thoroughgoing way into the practice of science, the claim that this has epistemological or ontic consequences is an extreme one. Similarly, in a protest against a perceived widening of the science curriculum to include moral and ethical considerations, Hall (1999) pleads:

> Science is a discipline concerned exclusively with the reliability that can be attributed to factual ('is') statements as a result of empirical investigation. It is widely recognised that 'is' statements in science cannot be turned into the 'ought' statements of moral discourse. For example, science can fairly accurately judge the consequences of bringing together a number of sub-critical masses of U_{235} above a densely populated geographical area. It can say absolutely nothing, however, about whether such an action would be right or wrong. The answer to the latter question lies outside the domain of science, but within the remit of moral discourse. The domains of scientific and moral discourse are fundamentally different; they have different core concepts (space, time, energy and good, right, ought), different procedural ground rules and different tests for truth. (p. 15)

The arguments against the teaching of ethics in science that stem from a consideration of the consequences that would follow were such a practice to become common are more speculative, but might go something like the following: Science teachers are generally educated in science not in moral philosophy. It is therefore unrealistic and unfair to expect them to teach ethics. If such teaching is required it would/might (a) decrease the time they have available to teach science; (b) lead to lower quality teaching, since science teachers will be teaching outside their sphere of competence; (c) lead to lower levels of professional satisfaction amongst existing science teachers; and (d) result in fewer science graduates wanting to enter teaching, thus exacerbating the shortage of specialists science teachers that exists in many countries.

There is evidence that school science teachers are less confident than humanities teachers about teaching social and ethical issues generally (e.g., Levinson, Douglas, Evans, & Turner, 1999). This lack of confidence can be for various reasons. As one science teacher put it:

> When we talk about the ethics of anything you're going to give an opinion rather than something that's fact based. Once you start giving an opinion then you express disagreement. Then they treat the whole of the subject in the same way that they treat your opinion in that they disagree with it personally. So they might end up treating your fact based stuff in the same manner. (p. 4)

The argument from consequences is therefore a pragmatic argument, a 'the straw that broke the camel's back' argument. As Matthews (1998), writing about the teaching of the nature of science, has put it:

> Previously, the science teacher had to master his or her subject matter and the techniques of making it interesting and intelligible to students (Shulman's pedagogical content knowledge). Good professional teachers ... tried to master something of the orthodox history and philosophy of their subject. All of this was demanding enough. Now, in addition, teachers need to understand and evaluate the postmodern challenge to orthodoxy. (p. 163)

Are we now to require science teachers to add ethics to the list of expectations placed on them? In response, there are, broadly speaking, two main families of arguments in favour of teaching ethics in science. As with arguments *against* teaching ethics, one is to do with the nature of science and the other to do with the consequences of teaching ethics in science.

The argument in favour of the teaching of ethics in science that stems from a consideration of the nature of science is an argument about the source and purpose of scientific knowledge. It begins by asserting that, even if we accept a characterisation of science as open-minded, universalist, disinterested, and communal (Merton, 1973), all scientific knowledge is formulated within particular social contexts (e.g., Fuller, 1997). At the very least this means that the topics on which scientists (not to mention technologists and engineers) work – and so the subject matter of science itself – to some extent reflect the interests, motivations, and aspirations both of the scientists who carry out such work and those who fund it. There is no doubt that the majority, almost certainly the great majority, of the funding provided for scientists, both currently and for some considerable time past, has been provided with the hope/expectation that particular applied ends will be met. These might be the production of a new vaccine, the development of a new variety of crop, the synthesis of a new chemical dye, the construction of a better missile detection system, and so on.

The point is that it can be argued that ethics is inevitably and inexorably conflated with science in most cases. Both the scientists and those who fund them hope that production of a new vaccine will lead to more lives being saved (presumed to be a good thing); that the development of a new variety of crop will lead to increased food yields (presumed to be a good thing); that the synthesis of a new chemical dye will lead to greater cash flows, increased profits, improved customer satisfaction, or increased employment (all presumed to be good things); and that the construction of a better missile detection system will lead to increased military security (presumed to be a good thing). In each of these cases, the science is carried out for a purpose. Purposes can be judged normatively, that is they may be good or bad. Indeed, just beginning to spell out some of the intended or presumed goods (increased crop yields, increased military security, etc.) alerts us to the fact that perhaps there are other ways of meeting these ends or, indeed, perhaps these ends are not unquestionably the 'good' that may have been assumed.

Further, it can be argued that the separation of science from values in general, and ethical considerations in particular, is a relatively recent, Western and secular phenomenon (cf. Cobern, 1998). The history of the secularisation of Western science is a complicated one, with the relationship between science and religion seeming rarely to fall into the simple, opposing camps of 'conflict' and 'harmony' (Barbour, 1990; Brooke, 1991). Also worthy of note is the frequent finding that in many countries school science teachers mostly have non-secular presuppositions (e.g., Haidar, 1997; Ogunniyi, Jegede, Ogawa, Yandila, & Oladele, 1995). For example, Islamic science has being described as a science whose processes and methodologies incorporate the spirit of Islamic values (Sardar, 1989). Early classifications of the elements within Islamic science included metaphysics, within which was knowledge of noncorporeal beings, leading finally to knowledge of the Truth, that is, of God, one of whose names is the Truth (Nasr, 1987). To this day Islamic science "takes upon a more holistic human-centred approach that is grounded in values that promote social justice, public welfare and responsibility towards the environment" (Loo, 1996, p. 285).

It is not to be supposed that only in societies where religious views are still important is ethics held to be a central part of science. In modern-day secular Japan a number of authors have written about the importance of a love of nature for science and science education. Put rather simplistically, Japanese science teaching puts value on pupils gazing on objects (an activity which overlaps with but does not equate to Western scientific 'observation') and encourages pupils to commune with and to love nature (Ogawa, 1998).

Whereas this argument for the teaching of ethics in science is contentious, the arguments in favour of the teaching of ethics in science that stem from a consideration of the consequences that would follow are perhaps likely to be accepted by a greater proportion of science educators. These arguments stem from a consideration of what school students would prefer science lessons to consist of. It is widely recognised that in a number of countries students enter their secondary schooling (around the age of 11 years) with high expectations of science and a positive attitude towards it. Over the succeeding years, though, interest in science generally wanes, especially in chemistry and physics (see Cerini, Murray, & Reiss, 2003; Lindahl, 2003; Osborne & Collins, 2001). Discussions with both students and their parents suggest that one of the principal reasons for this is that much of what students learn is not perceived to be 'relevant', more particularly that school science fails to be meaningful or to engage with students' developing sense of their identities (Lindahl, 2003; Reiss, 2000; Schreiner & Sjøberg, 2007). 'Relevance' encompasses a number of things, but for many young people it is the ethical issues raised by science that too often seem to be absent from their science lessons (e.g., Cerini et al., 2003).

VALUES IN SCIENCE AND TECHNOLOGY EDUCATION

Having examined the arguments for and against the place of ethics in science education, I shall now examine the place of values in science and technology education. There is still much disagreement about the term 'values'. People disagree, for example, on the extent to which values hold universally or only apply within particular cultures or at particular times, whether values must be shared or are simply a matter of personal preference, whether there is a difference between private and public values, and whether there are any overarching principles by which conflicts between values may be resolved (Halstead & Reiss, 2003).

Values include fairness, equality, freedom, love, justice, happiness, security, peace of mind, and truth. They can be distinguished, at least to a certain extent, from related and sometimes overlapping terms such as 'virtues' (which are personal qualities or dispositions like truthfulness, generosity, courage, loyalty, and kindness) and, again to a certain extent, though there is some overlap, from 'attitudes' (which are acquired tendencies or predispositions to make judgments and behave in a predictable manner, for example, openness, tolerance, respect, acceptance, and freedom from prejudice). Furthermore, there are different types of values. There are clearly fundamental distinctions between, for example, moral values (e.g., justice), intellectual values (e.g., criticality, respect for logical argument), political values (which are diverse), and aesthetic values (which are perhaps even more diverse than political values). Science education in general, and STSE education specifically, has much to learn from the tradition in technology education of taking values seriously. Marc de Vries (2007) argues that philosophers of technology have established that there are four different ways of reflecting on the nature of technology:

- technology as artefacts;
- technology as knowledge;
- technology as processes; and
- technology as a property of humans.

If we think of biotechnology as an example of a school curriculum topic that draws on both science and technology, it is clear how values, particularly moral values, are implicated – whether the technology is seen as an artefact, knowledge, a process, or as the property of humans – to the extent that there is an on-going battle between scientists, technologists, and industrialists seeking to conduct trials on genetically modified crops and those seeking to prevent such trials from going ahead (e.g., Figure 1).

WHAT MIGHT TEACHING ABOUT ETHICS AND VALUES HOPE TO ACCOMPLISH IN STSE EDUCATION?

I do not think it is worthwhile for science and technology teachers to spend a great deal of time in schools discussing with students the fact that scientists and technologists are morally beholden to be truthful, to give credit to those to whom it is due, and so on. Very appropriately, journals such as *Science and Engineering Ethics* devote a great deal of space to such issues as fraud in science and technology, the contributions of co-authors and the appropriate citation of the works of others. However, there is little here that is specific to science or technology and essential for someone still in compulsory education consciously to reflect on or learn about. So, more positively, what might be the aims of teaching ethics and values in science or technology? Based on Davis (1999), at least four can be suggested.

Figure 1. Protestors destroy a genetically modified crop trial in Oxfordshire, UK. Photograph from http://www.telegraph.co.uk/earth/graphics/2008/07/29/scigm129.jpg

First, teaching ethics and values in science or technology might intend to heighten the moral *sensitivity* of participants. For example, a chemistry teacher might encourage students to consider what happens to any heavy metal ions, such as those of lead, that are poured down school laboratory sinks. (I had never thought about this issue until, in a secondary school in Japan, I was shown some equipment, apparently widely installed in Japanese school laboratories, that enables such ions to be retrieved.)

Secondly, teaching ethics and values in science or technology might increase the moral *knowledge* of students. The arguments in favour of this are much the same as the arguments in favour of teaching any knowledge – in part that such knowledge is intrinsically worth possessing, in part that possession of such knowledge has useful consequences. For example, appropriate teaching about the issue of rights might help students to distinguish between legal and moral rights and to understand something of the connections between rights and duties.

Thirdly, teaching ethics and values in science or technology might improve the moral *judgement* of students. This would help them to reach what most morally well educated people would agree are 'right' responses when faced with new ethical dilemmas in science or technology.

Fourthly, and perhaps most ambitiously, teaching ethics and values in science or technology might make students morally *better people* in the sense of making them more virtuous or otherwise more likely to take action by the implementation of normatively right choices. To give a simple and relatively non-controversial example, a unit on renewable and non-renewable resources might lead students to greater re-use and recycling of materials. Such teaching might lead to students seeing certain artefacts, such as cars, differently so that the criteria of 'good design' shift towards such features as a high mileage per unit of fossil fuel and away from the time it takes to go from 0 to 100 mph. There is, within the field of moral education, a substantial literature both on ways of teaching people to 'be good' and on evaluating how efficacious such attempts are (e.g., Carr, 1991; Wilson, 1990). Here it suffices to note that while care needs to be taken to distinguish between education and indoctrination, there is considerable evidence that moral education programmes can achieve intended and appropriate results.

THE AIMS OF EDUCATION

More generally then, how might these various aims of teaching ethics and values in science or technology relate to the broader aims of education? (See also Reiss, 2007.) Some philosophers of education have argued that education would do well to have no aims. This somewhat startling position is well illustrated by asking what the aims of Aberdeen are (Standish, 1999). The point (in part) is that too tight a focus on aims may be stifling; we might do best simply to strive to make Aberdeen a 'good' city and need not even have widespread agreement on what precisely is meant by 'good' in this context (which might be taken, depending on one's perspective, as a high level of employment, adequate housing, low crime rates, easy access to public services, a balance between residential accommodation, industry, shops and entertainment, etc.).

Nevertheless, it is difficult to defend the notion that education in general, and science and technology education in particular, should be aimless. However, at the least, Standish's question helps us to ask whether it is realistic (helpful, sensible) to envisage a single aim or whether, in a fragmented world, we need to embrace a plurality of aims, not all of which we may be able fully to articulate. Kress (2000), for example, has argued that whereas the previous era required an education for stability, the coming era requires an education for instability. Such a scenario would require considerable provisionality and open-endedness about any precise, subject-specific set of aims.

We can begin with the obvious point that even a cursory historical survey shows that education has had diverse aims over the years. Nevertheless, we can discern two broad groupings. First, those where the intention is to develop the individual for her/his own benefit; secondly where the intention is to develop individuals so that they may collectively contribute to making the world a better place. We may note that this is typical of much social policy in many countries. So, for example, in the West under-age pregnancy, illicit drug misuse, and speeding are generally seen as bad both for the individuals concerned (loss of opportunities, mental and physical harm, risk of injury or death) and for the rest of society (financial cost, more burglaries, harm and upset caused to families and friends).

Granted these various caveats as to the worth of attempting to discern the aims of education, and acknowledging the historical and cultural contingency of any attempt to do so, contemporary philosophers of education have examined in some depth what these aims might be. Chief among the suggestions are autonomy, well-being, and justice.

Autonomy

To be autonomous is to decide things for oneself. It may immediately be objected that no man (or woman) is an island; that we make our decisions influenced by our historical past and social present. This is true, but the argument for autonomy still holds. If you choose to listen to Schoenberg and Stockhausen rather than the Bee Gees and Dolly Parton, the reasons are likely to be due in large part to your parents, schooling, and friends; but that doesn't mean that you lack agency or act with diminished autonomy in listening to them.

That education should intend pupils to achieve autonomy is widely accepted within liberal traditions of education. Of course, there are degrees of autonomy and different conceptions of autonomy. The concept of rational autonomy has, after all, a long history, being informed by Plato and other classical philosophers and developed further by philosophers with such different analyses as Kant and Mill. It is generally held (though nowadays we are more sensitive to the Eastern and feminist positions that too strong an emphasis on rationality may be both discriminatory and illusory) that rational autonomy is displayed by individuals who act intentionally, with understanding, and without external controlling influences that determine their actions (e.g., Haworth, 1986). An optimistic belief in rational autonomy (it is not rational autonomously to choose routinely to deceive and steal, as even a cursory analysis should convince me that this is neither fair to others nor likely to lead to a flourishing society) would mean that a society in which a sufficient proportion (hopefully not necessarily all) of its citizens had achieved rational autonomy would function successfully.

There are other terms that overlap with autonomy but have their own emphases. One can speak of a learner being self-determined or authentic, that is, true to herself (e.g., Walker, 1999). However, both self-determination and authenticity have, though the notion may be paradoxical to some, to be acquired. We are not born with a self that waits to be unveiled, nor is authenticity a state isolated from all others. Furthermore, as with autonomy, both self-determination and authenticity immediately cause us to ask about the relationship between what is good for an individual and what is good for society. It is just possible that I might validly lead a self-determined or authentic life by devoting myself to a life in which I spend all my free time consuming chocolates or memorising e-mail addresses (though this is to accept a rather narrow conception of self-determination and authenticity). But we should not accept that individuals can validly achieve self-determination or authenticity by amassing well-catalogued collections of child pornography or painstakingly researching ways to steal and torture cats.

Well-being

The idea that education should try to enhance people's well-being (sometimes expressed as enabling human flourishing), whether through good teaching, conventionally understood, or more recent educational conceptions of care, is closely allied to the notion that it should enable people to act autonomously; it is generally the case that the two go hand-in-hand. However, we can imagine situations (and introductory textbooks on medical ethics have them aplenty) when they are opposed. Suppose, for example, an adult injured in an accident needs urgently some life-saving medical treatment that can only be delivered by injection, but attempts to reject it because they are terrified of needles. Most of us would favour trying to persuade the person to accept the treatment, but if that proved unsuccessful should we consider giving the treatment without the person's consent (i.e., refusing to allow the person to act autonomously) on the grounds that there is a greater good than autonomy in this case, namely the preservation of life?

This example illustrates the point of view that there may be values that exist outside of an individual's preferences. However, this does not necessarily mean that there are absolute values that exist outside the sphere of human preferences. It might be that to save someone's life even when this means overriding their preferences is not to accede to some absolute, Platonic-like notion of the good but simply to hold that, consciously or otherwise, we (i.e., groups of people) decide what is the right course of action in such problematic cases.

As White (1999) points out, those who hold that it is people who decide what is right or wrong can be labelled as subjectivists, a subjectivist being "someone who holds that ethical values, including personal well-being, are not locatable in a reality outside human beings' desire-structures" (p. 193). The subjectivist position in ethics is put particularly strongly in a book whose title describes its main stance, *Ethics: Inventing Right and Wrong* (Mackie, 1977). The attraction of the non-subjectivist position is that one's conception of what is right and what is wrong is tied to something more durable than human preference. The principal difficulty, though, with the non-subjectivist position is where to locate these transcendental values. Once the majority of people might have found them in their scriptures but, at least in liberal societies such as those in Western Europe, only a minority of people hold strong religious convictions, or even if they do, see scripture as a place where definitive moral values are to be found solely by virtue of them being there (e.g., we don't need the Ten Commandments to tell us that theft is wrong).

Justice

A further approach is to see education as striving for justice. There are various conceptions of justice but they have in common an emphasis not only on the actions of individuals but also on the consequences of individuals' actions and social structures on relationships between individuals and on the distributions of resources between individuals.

Social justice is about the right treatment of others (what Gewirtz (1998) characterises as the relational dimension of social justice) and the fair distribution of resources or opportunities (the distributional dimension). Of course, considerable disagreement exists about what precisely counts as right treatment and a fair distribution of resources. For example, some people accept that an unequal distribution of certain resources may be fair provided certain other criteria are satisfied (e.g., the resources are purchased with money earned, inherited, or obtained in some other socially sanctioned way – such as gambling in some, but not all, cultures). At the other extreme, it can be argued either that we should ensure that all resources are distributed equally or that all people have what they need. Such distributions might be achieved through legislative coercion, social customs, or altruism on the part of those who would otherwise end up with more than the average.

TEACHING ABOUT CLIMATE CHANGE AT SECONDARY LEVEL

For a number of reasons, climate change is a particularly fruitful case for considering the importance of ethics and values. For a start, there are uncertainties and controversies about the extent, importance, and causes of global warming so that it is a good topic for getting students to think about how scientific knowledge is produced and how the establishment of reliable scientific data relies on technology. And then, of course, there is the question of how we should respond to it, which broadens the issue from science education (understood narrowly) to STSE education.

Global warming is held by many to be potentially extremely serious both for humans and for other organisms. The topic is therefore one that allows, even demands, first, that ethical aspects of the issue be addressed and, secondly, that the possibility that science education could lead to action be raised (e.g., Roth & Désautels, 2002). Furthermore, there is a political edge to the debates that surround the existence (or otherwise) and importance of global warming (Dessler & Parson, 2006). The topic can therefore be used to help students to develop their understanding of the extent to which scientific knowledge – and the use made of such knowledge – is objective and unproblematic.

Finally, the issue is increasingly in the news and of interest to a high proportion of school students in many countries. It is therefore 'relevant', an important consideration given that much school science teaching is criticised by students for being irrelevant and uninteresting (Aikenhead, 2007; Schreiner & Sjøberg, 2007). Indeed, recent years have seen a growth in a number of countries in the importance attached in schools to socioscientific issues education (SSI education). As Zeidler, Sadler, Simmons, and Howes (2005) put it:

> ... the SSI movement focuses on empowering students to consider how science-based issues reflect, in part, moral principles and elements of virtue that encompass their own lives, as well as the physical and social world around them. (p. 357)

Equally, Ratcliffe and Grace (2003) argue that the teaching of socioscientific issues can enable science education for citizenship and enhance scientific literacy, and cite with approval (p. 22) Oxfam's understanding of a global citizen as someone who:

- Is aware of the wider world and has a sense of his or her own role as a world citizen;
- Respects and values diversity;
- Has an understanding of how the world works economically, politically, socially, culturally, technologically and environmentally;
- Is outraged by social injustice;
- Participates in and contributes to the community at a range of levels from the local to the global;
- Is willing to act to make the world a more equitable and sustainable place;
- Takes responsibility for his or her actions. (Oxfam, 1997, p. 2)

Education about global warming, potentially the most important aspect of climate change, might include within its scope:

- Discussing whether we have duties only to humans, or to non-human sentient animals or to the environment more broadly. For example, to what extent are the atmosphere, biosphere, and hydrosphere common human heritage (Miller, 2005)?
- Examining the evidence that changes in human activities might reduce the rate at which the Earth is getting warmer (Houghton, 2004).
- Examining the costs (financial and otherwise) of reducing the rate at which the Earth is getting warmer (Lomborg, 2001).
- Situating the issue of global warming within broader considerations. For example, what is the optimal number of people on the Earth (Cunningham, Cunningham, & Saigo, 2005) and how great can/should be the inequalities between people in their patterns of consumption?
- Discussing the kinds of changes that might permit transition to an ecologically sustainable, peaceful, and equitable society (e.g., Westra, 1998). As Erhlich and Kennedy (2005) point out "... society has no prominent international forum in which such issues (like how we treat our environment and each other) are publicly discussed" (p. 562).
- Helping students to realise that some actions to increase sustainability can be taken by individuals (e.g., using less energy, buying energy from renewable sources), some locally (e.g., tree planting on land owned by a church or local council), some nationally (e.g., increasing the proportion of a nation's energy supplies from renewable sources), and some internationally (e.g., agreeing and then implementing agreements on carbon dioxide emissions).

CONCLUDING THOUGHTS – ACTION IN STSE EDUCATION

An increasing number of authors have argued that a good science and technology education includes the possibility of students taking action (Hammond, 2001; Hodson, 1994, 2003; Roth & Calabrese Barton, 2004; Roth & Désautels, 2002). For example, Hammond worked with pre-service teachers based in elementary schools in a variety of urban school districts in California's Central Valley. In one project, the school in which Hammond and her collaborators worked had children fairly evenly split between transient first-language English-speaking poor people who populate motels, trailer parks, and small homes in the neighbourhood, and recently arrived families from overseas, mostly refugees from Southeast Asia. The majority of these Southeast Asia refugee families were Mien. The Mien language has no secular written form and Mien families had typically never attended school prior to arriving in the United States. There is no word for 'science' in the Mien language. However, Mien adults come to the United States with a broad technological base, deriving from their lives as almost completely self-sufficient gatherer-hunters and slash-and-burn agriculturalists in the highland jungles of Laos. Therefore, they know how to grow food, hunt for food, preserve food, build their own houses, produce intricate fabrics, and use metallurgy to produce both tools and jewellery. The contrast between their knowledge and school knowledge is a notable one which, as Hammond says, "provides a rich tapestry through which to examine more general questions about cross-cultural science education" (p. 985).

One of the activities that Hammond and those working with her undertook was to develop urban garden plots (Figure 2). Science work in gardens not only allows a large number of conventional school science activities to be undertaken, it also allows children to learn, practically, about sustainability. The garden project was aided by the award of a USDA food security grant entitled 'Project FIELD: Food Independence through Economic Literacy Development'. This helped 80 families to use the plots to grow their own food. Some of this food was then bought back by the school for use in its cafeteria.

Figure 2. Student teachers working with Mien children in an urban garden as part of a school science activity. Photograph provided by Lorie Hammond.

Such examples of action in STSE education can inspire. However, I have a number of cautions about *expecting* STSE education to entail action. A mundane, but important, one is that to teach in this way is to take most science and technology teachers beyond their comfort zone. This does not mean that such teaching should never happen, but reminds us that high quality initial teacher education and continuing professional development will be required. Secondly, I doubt that actually taking action will ever be a primary aim of science and technology education. Rather, as Hodson (2003) points out, *preparing for* taking action may be a key element of the science and technology curriculum. In much the same way, the artefacts made in school technology courses and the research projects undertaken in school science courses are not the ultimate ends; rather, such exercises (and that is what they are) serve in part to help students appreciate what is entailed in producing artefacts or undertaking scientific research, and in part to help students should they ever subsequently produce artefacts or undertake scientific research. In fact, only a small minority of students will do either of these for career purposes. However, all of us have the potential to take morally valid actions as a result of the science and technology education we receive. A science and technology education that encompasses ethics and values can contribute to this goal.

REFERENCES

Aikenhead, G. S. (2007). Humanistic perspectives in the science curriculum. In S. K. Abell & N. G. Lederman (Eds.), *Handbook of research on science education* (pp. 881–910). Mahwah, NJ: Lawrence Erlbaum.

Barbour, I. G. (1990). *Religion in an age of science. The Gifford lectures 1989-1991* (Vol. 1). London: SCM.

Brooke, J. H. (1991). *Science and religion: Some historical perspectives.* Cambridge, MA: Cambridge University Press.

Carr, D. (1991). *Educating the virtues: An essay on the philosophical psychology of moral development and education.* London: Routledge.

Cerini, B., Murray, I., & Reiss, M. (2003). *Student review of the science curriculum: Major findings.* London: Planet Science.

Cobern, W. W. (Ed.). (1998). *Socio-cultural perspectives on science education: An international dialogue.* Dordrecht, The Netherlands: Kluwer.

Cunningham, W, Cunningham, M., & Saigo, B. (2005). *Environmental science: A global concern* (8th ed.). Boston: McGraw-Hill.

Davis, M. (1999). *Ethics and the university.* London: Routledge.

Dessler, A. E., & Parson, E. A. (2006). *The science and politics of global climate change: A guide to the debate.* Cambridge, MA: Cambridge University Press.

de Vries, M. J. (2007). Philosophical reflections on the nature of design & technology. In D. Barlex (Ed.), *Design & technology for the next generation* (pp. 20–33). Whitchurch: Cliffe & Company.

Donnelly, J. (2002). Instrumentality, hermeneutics and the place of science in the school curriculum. *Science and Education, 11*, 135–153.

Ehrlich, P., & Kennedy, D. (2005). Millennium assessment of human behavior. *Science, 309*, 562–563.

Fuller, S. (1997). *Science.* Buckingham, UK: Open University Press.

Gewirtz, S. (1998). Conceptualizing social justice in education: Mapping the territory. *Journal of Education Policy, 13*, 469–484.

Haidar, A. H. (1997). Arab prospective science teachers' world view: Presuppositions towards nature. *International Journal of Science Teaching, 19*, 1093–1109.

Hall, E. G. (1999). Science education and social responsibility. *School Science Review, 81*(295), 14–16.

Halstead, J. M., & Reiss, M. J. (2003). *Values in sex education: From principles to practice.* London: RoutledgeFalmer.

Hammond, L. (2001). Notes from California: An anthropological approach to urban science education for language minority families. *Journal of Research in Science Teaching, 38*, 983–999.

Haworth, L. (1986). *Autonomy: An essay in philosophical psychology and ethics.* New Haven, CT: Yale University Press.

Hodson, D. (1994). Seeking directions for change: The personalisation and politicisation of science education. *Curriculum Studies, 2*, 71–98.

Hodson, D. (2003). Time for action: Science education for an alternative future. *International Journal of Science Education, 25*(6), 645–670.

Houghton, J. (2004). *Global warming: The complete briefing* (3rd ed.). Cambridge, MA: Cambridge University Press.

Kress, G. (2000). A curriculum for the future. *Cambridge Journal of Education, 30*, 133–145.

Kuhn, T. S. (1970). *The structure of scientific revolutions* (2nd ed.). Chicago, IL: University of Chicago Press.

Lakatos, I. (1978). *The methodology of scientific research programmes*. Cambridge, MA: Cambridge University Press.
Levinson, R., Douglas, A., Evans, J., & Turner, S. (1999). *The teaching of social and ethical issues arising from biomedical research: Interim report to The Wellcome Trust*. London: Institute of Education, University of London.
Lindahl, B. (2003). *Pupils' responses to school science and technology? A longitudinal study of pathways to upper secondary school*. Göteborg: Acta Universitatis, Gothorburgensis.
Lomborg, B. (2001). *The skeptical environmentalist: Measuring the real state of the world*. Cambridge, UK: Cambridge University Press.
Loo, S. P. (1996). The four horsemen of Islamic science: A critical analysis. *International Journal of Science Education, 18*, 285–294.
Mackie, J. L. (1977). *Ethics: Inventing right and wrong*. London: Penguin.
Matthews, M. R. (1998). In defense of modest goals when teaching about the nature of science. *Journal of Research in Science Teaching, 35*, 161–174.
Merton, R. (1973). *The sociology of science*. Chicago: University of Chicago Press.
Miller, T., Jr. (2005). *Living in the environment* (14th ed.). London: Thomson Learning.
Nasr, S. H. (1987). *Science and civilisation in Islam* (2nd ed.). Cambridge, UK: Islamic Texts Society.
Ogborn, J. (1995). Recovering reality. *Studies in Science Education, 25*, 3–38.
Ogawa, M. (1998). A cultural history of science education in Japan: An epic description. In W. W. Cobern (Ed.), *Socio-cultural perspectives on science education: An international dialogue* (pp. 139–161). Dordrecht, The Netherlands: Kluwer.
Ogunniyi, M., Jegede, O., Ogawa, M., Yandila, C., & Oladele, F. (1995). Nature of world view presuppositions among science teachers in Botswana, Indonesia, Japan, Nigeria, and Philippines. *Journal of Research in Science Teaching, 32*, 817–831.
Osborne, J., & Collins, S. (2001). Pupils' views of the role and value of the science curriculum: A focus-group study. *International Journal of Science Education, 23*, 441–467.
Oxfam. (1997). *A curriculum for global citizenship*. London: Author.
Ratcliffe, M., & Grace, M. (2003). *Science education for citizenship*. Maidenhead, UK: Open University Press.
Reiss, M. J. (1999). Teaching ethics in science. *Studies in Science Education, 34*, 115–140.
Reiss, M. J. (2000). *Understanding science lessons: Five years of science teaching*. Buckingham, UK: Open University Press.
Reiss, M. J. (2007). What should be the aim(s) of school science education? In D. Corrigan, J. Dillon, & R. Gunstone (Eds.), *The re-emergence of values in science education* (pp. 13–28). Rotterdam, The Netherlands: Sense.
Roth, W.-M., & Calabrese Barton, A. (2004). *Rethinking scientific literacy*. New York: RoutledgeFalmer.
Roth, W.-M., & Désautels, J. (Eds.). (2002). *Science education as/for sociopolitical action*. New York: Peter Lang.
Sardar, Z. (1989). *Explorations in Islamic science*. London: Mansell.
Schreiner, C., & Sjøberg, S. (2007). Science education and youth's identity construction – two incompatible projects? In D. Corrigan, J. Dillon, & R. Gunstone (Eds.), *The re-emergence of values in science education* (pp. 231–247). Rotterdam, The Netherlands: Sense.
Standish, P. (1999). Education without aims? In R. Marples (Ed.), *The aims of education* (pp. 35–49). London: Routledge.
Walker, J. C. (1999). Self-determination as an educational aim. In R. Marples (Ed.), *The aims of education* (pp. 112–123). London: Routledge.
Westra, L. (1998). *Living in integrity: A global ethic to restore a fragmented Earth*. Lanham, MD: Rowman & Littlefield.
White, J. (1999). In defence of liberal aims in education. In R. Marples (Ed.), *The aims of education* (pp. 185–200). London: Routledge.
Wilson, J. (1990). *A new introduction to moral education*. London: Cassell.
Yearley, S. (2005). *Making sense of science: Understanding the social study of science*. London: SAGE.
Zeidler, D. L., Sadler, T. D., Simmons, M. L., & Howes, E. V. (2005). Beyond STS: A research-based framework for socioscientific education. *Science Education, 89*, 357–377.

Michael J. Reiss
Institute of Education
University of London
UK

STEVE ALSOP

28. NOT QUITE THE REVOLUTION: SCIENCE AND TECHNOLOGY EDUCATION IN A WORLD THAT CHANGED

INTRODUCTION

There is whisper of a bigger picture, a growing awareness that science and technology education is much more than raising achievement levels, meeting standards, and providing a favourable return on state investment. In some circles there is even rumour of new beginnings, of the emergence of a revitalised pedagogical sense of humanity, democracy, beauty, and justice. For far too long our practices have become lost in an overwhelming fixation on improving quality through standardisation and increased accountability. Loud and powerful voices have spoken of tradition and the importance of covering the basics, teaching children to master certain kinds of academic knowledge – the established, 'objective' truths of the world. What should 16-year olds be expected to know for our new knowledge-economy? How can we raise school achievement to be truly globally competitive? How can we meet the needs of industry? Such questions have given rise to a contemporary reality in which daily practices have become driven by economies of performance, measured by centralised assessment outcomes and standardised tests. The 'order of things', to use the language of Foucault (1973), is now *covering* the curriculum, *meeting* assessment targets, *completing* reporting mechanisms, *following* Ministry funding formulas, and *satisfying* terms and conditions of employment. Framed in this way, 'business as usual' seems to be compliance – a culture in which teachers and learners, for their own benefit, adhere to the latest set of policy guidelines and implement mandates and training courses of the relevant Ministry.

In the meantime, however, the world changed. Images of war, starvation, disease, and looming ecological disasters continue to fill television news reports, but now there is growing anxiety about the future of humanity. Over the past 20 years, ecosystems have declined more rapidly and more extensively than in any comparable period of time in human history. Savage injustices have affected the lives of millions and although the rise of India and China has brought about global economic relief to some, the depressing reality is that nearly half the world's population still lives without basic facilities such as access to regular fresh water and functioning sanitation (Millenium Ecosystem Assessment [MEA], 2005). This is all widely known: population growth, poverty, climate change, declining air quality, peak oil, and peak water have become part of the public psyche, embedded in a growing consciousness of environmentalism, a green movement fuelled with the idea that we are living way beyond our means. Coffee houses, wine-bars, working-clubs, and playgrounds are filled with talk of different 'Rs' - not the educational basics of reading, writing, and arithmetic; but reducing, reusing, recycling, and rethinking. Nowadays even transnational and multinational corporations are holding regular retreats and debating sustainability, carbon neutrality, corporate social responsibility, and the triple bottom line, attempting to combine traditional economics with social and ecological values (although I should confess that I continue to be sceptical about whether this is possible in practice).

A.T. Jones and M.J. de Vries (eds.), International Handbook of Research and Development in Technology Education, 319–328.
© 2009 Sense Publishers. All rights reserved.

In the same 20-year period, science and technology education has changed. The curriculum in many jurisdictions has grown substantially; there has been considerable debate about teaching and learning, and didactic methods and deficit models have given way to desires of more learner-centred and active constructivist practices. Information technology has increasingly permeated our practices and the science-technology and society (STS) movement has accentuated the social, political, and ethical dimensions of knowledge. Through policies and curriculum reform we have grappled with equal opportunities and the purposeful concept of education-for-all. But is this enough? Are we really grappling with the pressing societal and ecological transformations that now seem necessary? Our response to global warming is perhaps a pause for reflection. Described by *Scientific American* as "arguably the most imposing scientific and technological challenge that humanity has ever faced" (Stix, 2006, p. 46), it seems surprising that there has not been a more rapid mobilisation. We have mostly waited for distant governments to sanction action through incremental curriculum revision and reform. Why didn't we follow our own convictions and suspend the curriculum for a week (or even longer), cancelling all subjects to come together and organise for collective action? At the very least, we might have forged alliances with other teachers, parents, and students, and changed our teaching practices.

According to United Nations statistics there are over 60 million teachers in the world (and a significant percentage of these are, of course, science and technology teachers). A group of this size, with this expertise, offers hope of responding to today's truly global social and ecological concerns. Of course we cannot work in isolation; schools alone cannot address the entrenched double bind of the crisis of justice and crisis of nature (Sachs, 1999). But as science and technology educators, we can respond to them. And we can do this because we have an important voice and want to 'make a difference'. This is what attracts so many of us to the profession; frequently articulated in pre-service applications.

As I write this chapter I sense a growing wave of optimism; many of my local schools, for example, have established environmental groups and community events are taking place with greater frequency, albeit for the most part in the educational margins of after-school programmes and contexts. Under consideration in this chapter is a way in which we might bring this educational coalition into the mainstream.

CHALLENGING THE ESTABLISHED ORDER OF THINGS

When we speak passionately of 'science and technology for all', we need to speak of science and technology *for whom* and *for what*? When we raise educational standards, we must question standards *for whom*, as well as standards *for what*? We cannot overlook broader questions of educational purpose for fear of drifting aimlessly without point of anchorage. As Egan (2003) persuasively argues, school systems have a habit of embracing and expressing contradictory aims and purposes in, for example, mission statements that include commitments to employability skills, academic excellence, citizenship, and personal development. They often conflate desires of (a) initiation and induction into an academic discipline, (b) personal and social development of children, and (c) preparing future citizens and workers. Egan's point is that these goals are more often than not conflicting and incompatible.

My recent conversations with teachers certainly highlight the dominance of a particular kind of curriculum thinking. I have been part of numerous in-service meetings in which teachers and administrators repeatedly frame their practice in terms of covering subject content and meeting prescribed curriculum expectations. If changes are suggested to see things differently – to push curriculum and report cards aside for a brief moment and engage in education activities with local communities, or go on a field trip to sites of scientific interest, or spend extended time building relationships with an uneasy class – allegiances become very clear: "In my school we have to do it this way. I am already behind and have to cover the earth and space unit before the end of the term"; "There are examinations at the end of the year and I have to make sure that I cover all the content"; "I don't think the parents would like it, neither would my principal, to be honest"; and "There is so little time to think during the week nowadays because of all the admin and reporting".

If we want to break through "limits of the conventional" (Greene, 1994, p. 1) and the taken-for-granted assumptions of everyday practices, we need to question this dominance. Unless careful, our ideals and hopes for education might evaporate as our daily practices become homogenised through assumptions in conflict with our pedagogical values and wishes. Inspired by the writing of Adrent (1968), Dewey (1938), Freire (1970), Illich (1970), Greene (1986), and Postman (1995), much of my recent work has been searching for conversations that enable teachers to look outward to the world in which they live and the communities in which they work, and inward to the moral and ethical commitments that lured them into teaching in the first place. My hope is that this juxtaposition helps formulate a clear and coherent vision of legitimate aims and purposes to inform collaborative practices. In a broad sense, I am interested in encouraging teachers, learners, and citizens to re-envision science and technology education in relation to global (but not globalised), local (but not parochial), and personal (but not individualistic) contexts and commitments.

IN SEARCH OF CONVERSATIONS

In search of such a conversation, this chapter briefly outlines one response in the form of a small-scale case study. Despite its size, I have found it to be a source of some inspiration and critical reflection. In the remainder of this chapter, I describe and explain this study and then turn to some of the questions that I believe it provokes. I do not presume to offer any definitive answers, but rather view my collaborative work as part of an ongoing conversation.

There has been much recent discussion of the importance of contextualised and situated learning. Science, technology, society and environment education (STSE) is an established movement with a growing number of high profile advocates (Blades, 2006; Hodson, 2003; Pedretti & Little, 2008; Roth & Barton, 2004; Solomon & Aikenhead, 1994). Associated with a broad church of pedagogical methods, STSE education uses historical, political, and sociological framing to understand relationships between science, technology, societies, and environments.

In my work I have been exploring a heavily contextualised and localised practice in which science and technology emerge from local social and ecological contexts, rather than an orientation in which science and technology content is related, in a passive way, to non-specific contexts. This thinking has been greatly influenced by growing literature on what has been called a contemporary social movement of 'new localism' (Gruenewald, 2003; Gruenewald & Smith, 2008; Smith 2007). It is also influenced by a long-standing ecological and social concern about the decontextualised, extemporised, and fractured nature of much academic knowledge (see Capra, 1985; Suzuki, 2006).

My referent for praxis is that a responsibility for knowing well and doing well is born out of a sense of responsibility for understanding and acting in one's own neighbourhood. This is an argument persuasively made by Postman (1995) in his suitably provocative text entitled *The End of Education.* It is also, in part, embodied in the familiar mantra of thinking globally and acting locally. If we are interested in the wellbeing of communities and ecologies and the responsibility that knowledge brings, we must begin by paying close attention to the relationships between communities and ecological wellbeing in our schools and local neighbourhoods. To this end, I take a particular approach that seeks to separate education and schooling, suggesting that most of the dilemmas that we face with science and technology education today are not subject issues *per se* but issues concerning the ways in which our subject is schooled. In response, I propose that science and technology education might usefully embrace a two-phase process of 'de-schooling' and 're-envisioning'. This thinking has been greatly influenced by Gruenewald's (2003) positioning of 'critical-place-based-education' as a process of 'decolonisation' and 'reinhabitation' and my continued reflection on Illich's (1970) ever-contemporary work on deschooling society. I briefly illustrate my approach here through a shift in pedagogical narratives.

In Figure 1, science and technology education is presented as a dialectic-of-triads. The first triad represents the nature of school (the traditional narratives of schooling, the curriculum, and teaching methods). The second triad represents the culture, language, and traditions of local places (the narratives of learners and their communities and connections with environments). I draw this separation for purposes of orientation and emphasis. In the context of formal education, the narratives of schooling, curriculum, and teaching (triad 1) frame the ways in which learners, for the most part, experience science and technology. Although STSE education is often seen as a progressive reform, these dominant narratives also shape its practices and as a consequence greater emphasis is usually given to science and technology than to society and environments. In this subject-focused manner, common practices of STSE education connect, or relate, science and technology to distant, often non-specific, real world exemplars (such exemplars are usually selected by teachers on the basis of subject content not on the basis of society or environments) - compare Fensham's (1988) analysis with Corrigan's Chapter 30 in this volume.

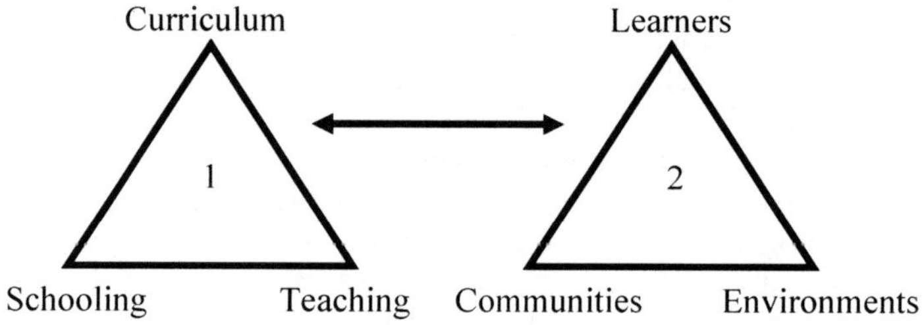

Figure 1. Dialectic of pedagogical triads.

The pedagogy that I am proposing here aims to reverse this approach. It starts with society and environments, as framed by narratives in the second triad (children and their local communities and environments), and then through research and action seeks to reconnect these narratives with school science and technology in response to local needs and desires. To this end, colleagues and I have purposefully created pedagogical interventions that seek to resist dominant narratives of schooling and start by immersing education in local places (de-schooling). Having explored and legitimised these experiences, we then return to dominant narratives of schooling in search of new ways in which science and technology schooling might respond (re-envisioning) to understand and make local places better.

Much of the emphasis of contemporary education, I posit, tacitly assumes a school-centred model in which parents and communities respond to the needs and agendas of school subjects (through, for example, supporting homework and school-parent meetings). In contrast, the orientation advocated here hopes to reverse this emphasis and poses the question: How can schools (and science and technology subjects, in particular) better serve learners' local communities and environments?

PICTURES, PLACES, AND POSSIBILITIES

Over the past three years, teachers, graduate students and I have been exploring the implications of this reversal. Our participatory action research team, funded by the Social Science and Humanities Research Council of Canada, comprises five science teachers working with three science education researchers and 25 grade 6 students (aged 10–11 years) and 11 grade 7/8 students (aged 12–13 years). We are all indebted to Sheliza Ibrahim, who took a leading role in organising our work.

Our emergent pedagogy uses visual images as a way of reconnecting to local contexts and concerns. As part of this process, teachers and students are provided with cameras to photograph images of themselves, their communities and local environments, and how they might relate in some way to science and technology. Such images are then discussed and offer points of departure for student-led science and technology action research (see also Chapter 40).

We start by looking at inequalities and injustices in the world through large-scale colour photographs depicting communities and environments under threat (many of these pictures are available online at www.focusonyourworld.com). We discuss images of poverty, pollution, hunger, and environmental destruction and come together, in the language of Ayers (2004), to "feel the weight of the world" (p. 103) as a precursor to looking at *our* local places and ways in which we might use STSE education to better understand and respond to them. We follow this with a neighbourhood walk, escaping the classroom to wander around the local market, shops, and parks, as well as where the students and their relatives live. To understand each other better we share stories of these experiences and then take cameras home (research funds were used to purchase disposable cameras) and bring back images of science and technology in our contexts – creating opportunities to express our thoughts and hopes of change in photographic format. We then discuss these images and how they might offer places of learning, starting points for collaborative research, and action. To illustrate this further, I now focus on two projects (Alsop & Ibrahim, 2007, 2008).

Place 1: Learning to Help Your Sister

Nested in the mosaic of photographs displayed on the classroom wall was an image of a child in a hospital bed wearing an oxygen mask, clearly suffering from a severe illness. We learned that a sister of one of the students was undergoing a host of tests to investigate breathing patterns and other health complications. Up until this point, nobody (including the teacher and all the other students) was even aware of Mary's[1] sister's illness and there was an intense silence in the classroom as Mary spoke clearly and precisely. Later, she joined with other students (Jenny, Olivia, and Brianna) and they decided to investigate the science behind this illness and its profound, often debilitating effects. Mary and her friends *wanted* to learn science to *help* her sister; they wanted to offer some advice and understand what it was like to live with illness. A lengthy planning discussion led to a series of questions about an aspect of the condition. While some information was available on the Internet, the students wanted to meet Shania (the sister) and talk about her life experiences. Permission was sought and the four students and researcher (Sheliza) visited the 5- year old in her home. They sat patiently on the sofa next to her and asked how it felt to have the syndrome, and what was it like having tubes in your mouth while in hospital – Brianna asked, "Were you sad when you knew you had a problem?" At the students' request, a trip was then organised to a sleep rehabilitation laboratory located in a research institute in a downtown hospital. Shania makes regular visits to a similar laboratory. After some organisational hurdles, the four students, their teacher (Chris), and Sheliza headed to the lab, located a few blocks from their school. They interviewed the technician and doctor in order to better understand the science and technology involved. The technician took them on a tour of the facility and showed them how computers capture changes in sleep patterns. The wavelengths that were generated from the electrodes that are securely fastened to the sleeping patient showed characteristic peaks for several common sleep problems, and the technician showed the students how to interpret the electrode data. They tried on oxygen masks and the ECG electrode cap that measures brain activity while sleeping. They learned about the international electrode placement map and how different probes detect brain activity and different sleep stages, including stage one, two, low wave, and REM sleep.

After their visit, the students wanted to act, to use their knowledge and experience to make a difference, and so an opportunity was created for them to return to share their expertise with Shania. They sat huddled together on the sofa with Shania and her mother and discussed the syndrome and how they experienced the tests and as a result now saw things differently. They gently offered advice, warning Shania about sleeping on her stomach. "You mustn't do this … you promise, don't you?", one of the students implored. At this moment, schooling changes: The children, the emergent scientists, become teachers, caring citizens, as they offer advice and listen intensely and carefully to each other's responses. Sheliza, the researcher and Chris, the teacher, became learners as they listened, often with a lump in their throats, to the family's and the children's intimate exchanges.

Place 2: Learning with Fishmongers

Surrounding the school and the homes of many of the students is a large urban market. It is a distinctively high-density, inner-city multicultural space spanning a few blocks as it borders three popular, high traffic-density streets. This is where Dale and Rich live. Dale's parents were born in Canada and Rich's parents moved to Canada quite recently from South America. During the neighbourhood walk, the students took photographs of the local fishmongers. One fishmonger picked up two fish from the icy display case and posed them for their photo-op as if the fish were kissing. The students seemed genuinely fascinated, and excitedly asked questions about locality, anatomy, reproduction, water systems, transport, and habitat. Rich was the only student brave enough to touch the fish; others were squeamish and held their noses because of the smell. The students were encouraged to direct their questions to the fishmonger behind the counter, but they became shy and reluctant and left with many unanswered questions. Dale and Rich wanted to pursue some of these questions and decided to return to the fish store. In their action plan, which they developed closely with the teacher as a 'critical friend' (Carr & Kemmis, 1986), they hoped to revisit the fish market, meet the storeowners, and further investigate the fishing industry and aquatic ecosystems more generally. They had an abundance of questions: Where do these fish come from? How nutritious are they? How can you tell a boy fish from a girl fish? How much fish gets shipped to you in a day? Year? Week? Why are they kept on ice? Even when dead, why do the eyes still look so glossy? What's the most common fish to be found in the ocean? What are the rarest? And many other questions.

Upon returning to the store, they video-documented as many marine species as they could. Some items did not have a sign with details of the fish, so they asked the owner for the common names and origins of each fish they pointed at. They became keen to understand more about the geographical habitats of different marine life forms and wanted to know where the seafood came from, how it got to the fish store, and what this might mean for our environment and aquatic ecosystems. They sought information about the practices of the local store. As the United Nations Millennium Ecosystem Exercise (2005) so clearly indicates, aquatic ecosystems are our most vulnerable systems with over 65% of their services now in decline (the Millennium Ecosystem Assessment exercise adopted the term Ecosystem Services to measure the human demand for natural assets). In a context of increasing demand, Dale and Rich became curious about food and ecological footprints, the difference between fresh fish and packaged fish, and the geographic locations/understanding of fish-farms or aquaculture plants linked to the local store.

After a brief conversation, it became clear that the practices of the fishmongers have adapted – that more abundant species such as cod have declined and new species such as tilapia have taken their places on the iced counters. With assistance from their teacher, Dale and Rich compiled a poster of this information to be displayed, with the owner's permission, in the store. A clear message in the display was that changing aquatic ecosystems impact practices geographically dispersed, even including a local store far away from salt water.

As the students interviewed the storeowners, the depth and sophistication of their knowledge became apparent. Local knowledge built up over generations offered a commentary supplementing and extending the teacher's and researcher's expertise (although at times the fishmongers refused to believe in their expertise). In this way, Rich and Dale's research connected global and local issues with science and technology and the desire to bring attention to declining global ecosystem services and the impact that this has on all those connected with the fish industry. In so doing, their work raises important questions for science and technology education, including ways in which we might acknowledge knowledge created outside of the academy, and how local actions have global consequences. Such discussions offer an interesting comment on what it might mean, and should mean, to be called an ichthyologist?

NOT QUITE THE REVOLUTION

There has been a long-standing debate about aims of education. Whether we educate for social replication and control – a process of producing passive, obedient workers and consumers – or whether we educate for critical consciousness, liberty, participatory democracy, and freedom. While such binaries are, of course, oversimplified, I believe that when presented with these stark choices, most science and technology teachers will identify with the latter goal rather than former. In this light, the practices outlined in this chapter are not particularly revolutionary nor, for that matter, especially new. Their distinctiveness, as is in most case studies, resides in specificity and methods – ways in which emphasis is given to children's agency and community involvement. Like many before us, we have relinquished some instructional power and listened to and acted upon learners' voices interests and desires. To use the words of Smith (2007), our work "reflects the thoughtful practice of adults concerned about connecting children in more meaningful ways to their locales rather than about allegiances to a set of educational principles promulgated by regional or national institutions" (p. 191).

There is a growing scepticism, I believe, in the intensity of recent curriculum reforms, and citizens around the world are looking to science and technology education to respond in an era of social and ecological uncertainty. But in such an era, how are we to understand science and technology education? To what and to whom should we be responsive? The projects previously outlined here sought (albeit for a fleeting instance) to escape the normal flow of things and offer opportunities to reflect upon responsibilities as educators and learners beyond the ingrained day-to-day enactments of covering the curriculum, adhering to school rules, and meeting subject expectations. The choices of actions roll back dominating propaedeutic tendencies (seeing school subjects as preparation for future studies) and necessitate a reinvestigation of how we understand science and technology schooling, whether defined by place, tradition, intention, practice, or spirit.

We would all agree that young people need a science and technology education that nurtures their imagination and capacity to research and act – an education that confirms their role as caring citizens and community members. In this light, there is much to admire in these two projects. A genuine attempt was made by elementary students (children aged 10–13), teachers, and researchers to participate with families and local communities and to further what it means to be an actively knowledgeable member of society. In the course of the students' work, knowledge of science and technology emerged not only as a wonderful way of seeing and explaining the world, but also as a way of recognising local practices and responding to local needs.

In post-project interviews, the students and teachers spoke of their impressions of learning in this way and an abundant sense of enthusiasm was evident. As students reflected on their projects it was clear that many had rediscovered a genuine motive to learn outside of academic performance. Science and technology was for all; in contrast to the familiar scenario of academic winners and losers, achievers and underachievers, associated with many school activities. Here was a group of children coming together and being together in ways that made a difference, exploring common concerns, and becoming cognisant of their own agency and capacities to contribute to collective action. For just a moment, education became much more than following the state controlled curriculum; indeed, this curriculum for a short time took a backseat along with many ingrained conventions and practices of schooling. As Chris (the teacher) and Sheliza (the researcher) listened to Shania's description of her illness and the ways in which her parents continue to offer all the support and love they can, a deep sense of humanity emerged. As students learned with fishmongers, the explanatory power of local knowledges became self-evident; storeowners – community ichthyologists – displayed considerable out-of-school expertise and students saw local knowledge, and scientific and technological knowledge, in a new light, as worthy of extended investigation and study.

There are, of course, many ways of framing this pedagogy. Freire (1970), for instance, writes of the need for learners to reflect upon their own 'situationality' to the extent that they are challenged to act. Dewey (1938) writes of progressive practices that aim to integrate schools and communities into patterns of civic engagement essential to democratic participation. Adrent (1968) writes of education as searching for a fuller sense of humanity, as a point between the past and the future, and as the renewal of a common world. I suggest, in a very modest way, that the students, teachers, and researchers involved in the projects described above encountered an opportunity to learn; learning for *my* family, *my* community, *my* environment is demonstrably different than learning for *my* grades, *my* teacher, and *my* school. In many ways, perhaps the most profound change was simply a shift in roles and relationships: learners became teachers and teachers became learners. We are too often deprived of this much talked about pedagogical ideal.

My hope for this research was not to create a false dichotomy between schools and their local communities and environments, or to separate science and technology for schools and science and technology for communities. Rather, it is necessary to reconnect the triads (as previously outlined), represented as a dynamic dialectic, a third space of school-community hybridity. For purposes of emphasis, the triads are depicted in this way to cast light on the dominance of the first triad (as illustrated by the conversations recorded in in-service teacher education settings). I am also not claiming that all science and technology education can or should be taught in this way. However, I am reasserting the need to reconsider our practices in response to global realities, and the need to be cognisant of multiple referent points for our practices and the ever-present dangers of self-referencing (we teach this way because we have always taught this way; we teach science and technology for science and technology).

In terms of teacher education, student-led participatory action research offered a powerful vantage point to reconsider practices. The teachers involved in the study described the experience as an opportunity to connect with students in new ways and find out more about their communities. (As is the case in many inner-city schools, teachers rarely live in the local neighbourhoods; they drive through them on their way to work.) During the course of the three-year project, teaching did change, our conversations did escape some of the more traditional ways of seeing our practice, and, in many ways, science and technology was de-schooled and then re-envisioned.

Nevertheless, dominant forces step in and actions become swiftly subsumed into established orders. Science and technology education has always been tied to national agendas, linked in some magic way to economic prosperity and global domination. In an era of diminishing political influence, schooling has become a political game of 'raising standards' and 'out-performing others' in regional, provincial, and international tests (such as TIMSS and PISA). In contrast, the very success of this project perched not on some distant mantle of economics and GDPs, but on *local* expertise and concerns. Regrettably it seems that to take up 'the local' necessitates practitioners holding at bay 'the national' (curriculum expectations and testing) and this is only possible for a short period of time. In this light, Jenny, Olivia, Brianna, Mary, Dale, and Rich's projects existed only within the margins, within cracks in consent. The normal flow of things is commercialised textbooks and accompanying worksheets – the conveyer belt of our daily practices over which they (and we) seem to have little control. Such practices continue to be structured around desires for universal solutions, and for teachers and learners to focus on standardised content and methods (the first triad). This practice, although often not in opposition to the pedagogy that I am proposing, seems so pervasive and extensive that it extinguishes other pedagogical possibilities grounded in diversity and unique social and ecological concerns (phased simply, it masks sight of the second triad).

To return to the two questions above: In contemporary times how are we to understand science and technology education? To what and to whom should we be responsive? Perhaps the simple message from this project is that we cannot lose sight of the world and our ability to mobilise for action. We must not be passively reduced to covering a subject area without concomitant consideration of the ways in which our subject expertise might react to local and global needs and desires. We must not be overrun by loud voices from the academy that continue to bemoan declining academic standards, because school science and technology is much more than catering for the minority of students who will continue with higher studies. Moreover, we must not be seduced by universal answers to economic prosperity and global dominance (this is much more complex than educational policy likes to assume). Instead, I believe, our primary referents for school science and technology education are learners, communities, and environments (the *whom*) and the ways that our knowledge and expertise can bring about change (the *what*). Let's not lose vision of the real power of our praxis, residing in nurturing the personal and social growth and development of children in our care and supporting a collective struggle – mobilising for a better world. Such an education stands in opposition to systems that perpetuate personal, social, and ecological injustices and instead nurtures solidarity, epistemological responsibility, and action to heal and celebrate relationships between people, communities, and life on earth.

I have little doubt that for some this approach will be challenging. In a recent talk, for example, I was approached by an indignant colleague who asked "how I can get away with this work", suggesting that I "must be working in inner city schools that out of desperation are forced do something different". In this regard, as Gruenewald and Smith (2008) lament, it seems likely that innovations capable of addressing the types of changes that I and many other science and technology educators believe necessary (see, for example, Bencze, 2001; Hodson, 2003; Roth & Barton, 2004) will come from the margins of power rather than the centre – from inner-cities and isolated rural communities. But, in the same breath, it is important to note that there is a whisper of change and it is within our powers to respond. Maxine Greene (1986) writes so beautifully and inspiringly of the "shadow of silent majorities" and the ways in which education can "inspire hitherto unheard voices" (p. 45).

NOTES

[1] Pseudonyms are used to refer to the children in these case descriptions.

REFERENCES

Adrent, H. (1968). *Between past and future: Eight exercises in political thought*. New York: Penguin.

Alsop, S., & Ibrahim, S. (2007). Searching for a science motive: Community, imagery and agency. *Alberta Science Education Journal, 38*(2), 17–24.

Alsop, S., & Ibrahim, S. (2008). Visual journeys into critical place-based science education. In Y.-J. Lee & A.-L. Tan (Eds.), *Science education at the nexus of theory and practice* (pp. 289–300). Rotterdam, The Netherlands: Sense.

Ayers, W. (2004). *Teaching toward freedom: Moral commitment and ethical action in the classroom*. Boston: Beacon Press.

Bencze, J. L. (2001). Subverting corporatism in school science. *Canadian Journal of Science, Mathematics and Technology Education, 1*(3), 349–355.

Blades, D. (2006). Levinas and an ethics for science education. *Educational Philosophy and Theory, 38*(5), 647–664.

Capra, F. (1985). *The turning point: Science, society and the rising culture*. New York: Simon & Schuster.

Carr, W., & Kemmis, S. (1986). *Becoming critical: Education, knowledge and action research*. London: Falmer.

Dewey, J. (1938). *Experience and education*. New York: Simon & Schuster.

Egan, K. (2003). The ideas that run schools. *Education Canada, 43*(2), 23–24.

Fensham, P. J. (1988). Approaches to the teaching of STS in science education. *International Journal of Science Education, 10*(4), 346–356.

Foucault, M. (1973). *The order of things*. New York: Vintage.

Freire, P. (1970). *Pedagogy of the oppressed*. New York: Continuum.

Greene, M. (1986). *Releasing the imagination: Essays on education, the arts and social change*. San Francisco: Jossey-Bass.

Greene, M. (1994). Teaching for openings: Pedagogy as dialectic. In P. A. Sullivan & D. J. Qualley (Eds.), *Pedagogy in the age of politics* (pp. 1–12). Urbana, IL: NCTE.

Gruenewald, D. (2003). The best of both worlds: A critical pedagogy of place. *Education Researcher, 32*(4), 3–12.

Gruenewald, D., & Smith, G. (2008). *Place-based education in the global age*. New York: Lawrence Erlbaum & Associates.

Hodson, D. (2003). Time for action: Science education for an alternative future. *International Journal of Science Education, 25*(6), 645–670.

Illich, I. (1970). *Deschooling society*. New York: Harper & Row.

Millennnium Ecosystem Assessment. (2005). *Ecosystems and human well-being: Synthesis*. Washington, DC: Island Press.

Pedretti, E., & Little, C. (2008). *From engagement to empowerment: Reflections on science education for Ontario*. Toronto, ON: Pearson.

Postman, N. (1995). *The end of education: Redefining the value of school*. New York: Vintage.

Roth, W.-M., & Barton, A. (2004). *Rethinking scientific literacy*. London: RouthledgeFalmer.

Sachs, W. (1999). *Planet dialectics: Explorations in environment & development*. London: Zed Books.

Smith, G. (2007). Place-based education: Breaking through the constraining regularities of public school. *Environmental Education Research, 13*(2), 189–207.

Solomon, J., & Aikenhead, G. (Eds.). (1994). *STS education: International perspectives on reform*. New York: Teachers College Press.

Stix, G. (2006). A climate repair manual. *Scientific American, 295*(3), 46–49.

Suzuki, D. (2006). *The David Suzuki reader: A lifetime of ideas from a leading activist and thinker*. Quebec: Greystone Books.

Steve Alsop
Faculty of Education
York University
Toronto, Canada

ANGELA CALABRESE BARTON AND EDNA TAN

29. THE EVOLUTION OF DA HEAT

Making a Case for Scientific and Technology Literacy as Robust Participation

INTRODUCTION

It is July in the summer of 2007. Naomi, Kathy, Shernice, and Jeremy are reflecting on the mini science documentary on urban heat islands – *Where da Heat Go* – that they had directed and premiered before an audience including their peers at the *Boys and Girls Club*, adult administrators, science education professors, doctoral students, and engineers. In their documentary, they had gone out into the downtown neighbourhood of their community on a hot, 90°F summer day to interview the general public on their opinions of the urban heat island phenomenon. There were several scientific and technological elements to their movie, including temperature taking, laser tape measuring, digital photography, interviewing, and note taking. The final cut of *Where da Heat Go?* is a kaleidoscopic mix of science discourse and youth culture, featuring scientific and technological information, hip hop songs, more serious science footage, and blooper reels all blended seamlessly into an engaging story. During a reflection conversation with Naomi, Kathy, Shernice, and Jeremy three months after their premiere, they noted certain recommendations they would like to offer the state's Governor regarding urban heat islands:

> Naomi: If I had a chance to talk to the Governor, I won't be shy, I would tell her what a urban heat island is, and for all those who don't know what it is, it's just a place that doesn't have a lot of trees and has a lot of buildings and it's very very hot. Thank you.
>
> Jeremy: Um, I would tell the Governor that urban heat islands is a place with very little trees … and a lot of buildings, the black tar absorbs a lot of heat so it gets hotter in a heat island … I would want her to put rooftop gardens on the buildings so that it absorb carbon dioxide or make there be less cars driving around in the city so it's not as hot.

Throughout the documentary, the general lack of awareness of the urban heat island phenomenon among adult community members was playfully emphasised, with the youth positioned as the community experts. Instead of assuming the customary role of recipients of scientific and technological knowledge who then seek to apply such knowledge locally in their own lives, the youth recast the meaning of scientific and technological literacy by positioning themselves as the catalysts and harbingers of a serious socioscientific issue in their community. The manner of their engagement with this issue potentially disrupts the entrenched hierarchies and orderly segregations of scientific and technological literacy. In this chapter, we analyse the youth's engagement process and purposes with the urban heat island phenomenon through the lens of scientific and technological literacy guided by the following questions:

- How did the youth manifest scientific and technological literacy through making the documentary?

A.T. Jones and M.J. de Vries (eds.), International Handbook of Research and Development in Technology Education, 329–346.
© 2009 Sense Publishers. All rights reserved.

- How did the form and function of their manifested scientific and technological literacy differ from the prescribed notions of scientific and technological literacy?

In elucidating these questions, we trace the participation of the four students as they engaged in the summer *GET City* programme, drawing from video data, field notes, student work, and interviews to make our claims. We also rely heavily on the youth's video documentary *Where da Heat Go?* – the culmination of the students' work in this summer program.

SCIENTIFIC LITERACY: A BRIEF OVERVIEW

Scientific literacy has been an overarching goal of science education since it was formally introduced in the 1950s (Hurd, 1958; McCurdy, 1958). Since then, it has eluded precise definition and has become a broad concept, taking on different meanings as the historically significant educational themes have shifted over time.

In his review of scientific literacy, Laugksch (2000) noted various 'interest groups', each with their unique perspective on the concepts, purposes, and measurement of scientific literacy. Laugksch postulates three gradated interpretations of scientific literacy, ranging from "literate as learned" through "literate as competent" to "literate as able to function minimally in society as consumers or citizens" (p. 82). He also argues that the goals of scientific literacy can be grouped into two categories – macro and micro. The former is concerned with the alleged benefits that scientific literacy will bestow on the nation, science or society; the latter relates to the enhancement of individual lives.

While there is general agreement that scientific literacy refers to what individuals ought to know and why it matters socially, scholars have recently attempted to further clarify the notion of scientific literacy along three strands of thought:

- clarifying what individuals ought to know and why;
- defining the relationship between knowing science and civic participation; and
- the sociocultural dimensions of science literacy as collective praxis.

What Should a Scientifically Literate Person Know?

Scientific literacy has ranged from "what the general public ought to know about science" (Durant, 1993, p. 129) to a universal slogan or rallying call for science educators to improve science education (Bybee, 1997). Arguments for why we need scientific literacy have taken various forms: from lauding the virtues of learning science as a pure intellectual pursuit, to strengthening science education as a response to World War II and the launch of Sputnik (National Society for the Study of Education [NSSE], 1947, 1960). The term finally took deep root in the United States when the American Association for the Advancement of Science (AAAS) used the idea to make a case for reforming mathematics, science, and technology education. While the vision of scientific literacy put forth by the AAAS has ranged from what individuals ought to know to how they might participate in society, the subsequent Benchmarks and the National Science Education Standards (produced by the National Research Council) have focused such benchmarks and standards primarily on *what* individuals should know in order to be regarded as scientifically literate.

Anderson (2005) argues that framing scientific literacy around deep conceptual understanding of particular content – what he refers to as scientific literacy for conceptual understanding – is arguably the tradition with the "longest history and the most influence within the science education community" (p. 7). At the heart of this stance is the idea that learning science involves developing deep conceptual understandings of the big ideas of science. Undergirding this position is the notion that science is inherently conceptual and theoretical, and that students are proto-scientists. It is not surprising, then, that both the AAAS and the National Research Council put forth a set of recommendations "spelling out the knowledge, skills, and attitudes all students should acquire as a consequence of their total school experience" (AAAS, 1989, p. 3) if they are to be regarded as scientifically literate.

More recently, some have argued that the process of becoming scientifically literate ought to be framed in cultural terms because it is not just *what* one knows that matters, but also how one gains access to that knowledge. Wallace (2004), for example, argues that for students to develop authentic scientific literacy, they need to be afforded opportunities to engage in scientific discourse, to negotiate among multiple discourses, and to cooperate with teachers and peers in ascribing meaning to scientific language in a "third space" setting. Wallace contends that it is only through "engagement in hearing, writing or reading an argument that a learner would become most familiar with those reasoning forms that are particular to science, including questioning, interpreting data, making claims, and providing evidence for those claims" (p. 902). Wallace's argument is important because it pushes the discourse around 'what students should know' to consideration of how such knowledge acquisition is socio-culturally situated. She asserts that a student has achieved authentic communication when she uses scientific language in recounting her experiences. Thus, successful science learning (and a mark of scientific literacy) is when the child's everyday language appropriates scientific language (defined as that used by academic scientists) into authentic communication.

Regardless of whether the focus is on what students should learn or how access might be achieved, there is a presupposition that such understandings matter in how individuals learn to engage in a democratic society. However, little attention has been given to how this knowledge ought to transform participation in the larger global society.

Science and Civic Participation

Another side of the science literacy debate that has received little attention in the United States, but has been at the centre of the science education debate in the UK and Canada, is the extent to which scientific and technological literacy should explicitly incorporate notions of citizenship and democratic participation. For example, Hobson (2003) contends that in "industrial nations, many of the most crucial decisions concern science and technology, and in democracies, citizens decide" (p. 109).

While those who argue for 'science literacy for citizenship' are quite clear that school science should be concerned with fostering understanding of basic foundational ideas (similar to AAAS's 'big ideas') and the nature of scientific inquiry, they also argue that it is the responsibility of school science to foster a regular practice of consuming science information through reading about and evaluating scientific ideas in the public domain (Miller, 2002). Hobson (2003) argues that this can be achieved by teaching topics like energy resources and the environment within the context of content-domain courses like physics. These topics, because of their social relevance, can provide teaching and learning opportunities that foster critical engagement with scientific ideas in the forefront of public debate.

Hodson (1999) also argues that science for civic participation should involve the teaching and learning of domains in science that are culturally relevant. However, he pushes other researchers engaged in designing and defending civic science literacy to take into account the positions that science, as culture and practice, hold in society. From this vantage point, Hodson argues that because science, like schooling, is not neutral, the goals of scientific literacy ought to be focused, in part, on exposing the culturally and politically located aspects of what it means to become scientifically and technologically literate. Affording opportunities to develop such literacy means fostering real opportunities to engage with the subject, along with opportunities to work through four levels or challenges around the political dimension of knowing/doing that include developing critical awareness of the societal impact of scientific/technological change, recognising that decisions about scientific and technological development are always socially situated and biased, making sense of one's own assumptions, and preparing for and taking action.

Civic scientific literacy is, at its core, not really all that different from those ideas promoted by Project 2061 when this project first introduced scientific literacy to the science education community's lexicon in the 1980's (AAAS, 1993). After all, one of the fundamental arguments of the vision advocated by the AAAS was to advance citizen participation in scientific debate. However, civic science literacy foregrounds the relevance and importance of engaging with specific science knowledge (i.e., 'why' this knowledge is important and ought to be taken up in daily life) in addition to the science knowledge itself (i.e., 'what' one should know).

The Sociocultural Dimensions of Science Literacy as Collective Praxis

A small but growing group of science educators has been more far-reaching in their view of civic science literacy or science literacy for citizenship by arguing that what is important is not just 'what' individuals know or how they come to know it, but how such knowledge is actually brought to bear in the decision-making process, and for what purposes. For example, 'real' literacy in science involves deep understandings of the human and social aspects of the production of scientific knowledge and a critical assessment of knowledge and sources of knowledge, so that individuals can be thoughtful decision-makers on socioscientific issues that dominate life in the 21st century, such as gene therapy and cloning (Kolstoe, 2000).

In a similar fashion, Backstrand (2003) argues that citizens (and hence civil society) need to be brought into more robust participation in community science through "participatory practices in the conduct of science" (p. 28). The point here is that civic science literacy involves the public in both the production and the use of scientific knowledge, changing relationships between science, expert knowledge, and citizens in democratic societies. Others push us even further to consider how the process of coming to make decisions on larger socioscientific issues involves collective praxis and situated knowing/doing – thus taking science literacy out of the individual realm and into public practice (Roth & Lee, 2004). Nonetheless, cutting across these arguments for science literacy for citizenship is the concern that reform documents like Project 2061 do not go 'far enough' in delineating 'how' individuals might use what they know in participation in democratic society.

While much of the work around science literacy for citizenship has focused on how scientific knowledge and habits of mind might be brought to bear on public decision-making, what has garnered almost no attention is science literacy for *everyday practice*. Roth (2009) raises the question of how science and science education might assist everyday folk in, and with, the problematic situations they face in their lives. Here, science literacy involves not just what people need to know, but how that knowledge is fashioned by location.

To frame science literacy in this more personal, highly contextual way demands that science be accessible to all individuals, and that knowing science be positioned as situated. This stands in stark contrast to the positivistic history upon which the growth and concentration of power in the sciences has been built. Positioning science literacy in the practice of everyday life transforms science literacy from being about what one must 'be able to know and do' to an account of how such knowing and doing intertwine epistemological, methodological, and political activity. It shows us not only how becoming scientifically and technologically literate may be about developing understandings and habits of mind and using them in democratic participation, but also about the personal process of coming to know in a highly contested world.

We would be remiss if we did not clearly state that deeply empowering scientific literacy includes dimensions from all three stances described above: conceptual understandings of the big ideas and practices in science, knowledge of how to use these ideas in ways that matter in democratic society, and knowledge of how to use these ideas in critical awareness of one's worlds. However, we want to expose the limitations of these views (including our own) and argue for a more deeply embedded stance on scientific and technological literacy by examining closely the work of four young people in a summer science programme. What did they learn? How did they use their knowledge? How did they position themselves *vis à vis* science, technology, community, and their peers as they enacted and performed with/in the science club?

Advancing Science Literacy as Collective Praxis through Technological Literacy

With regard to understanding the role of technology in learning science (or any other subject), part of the challenge in recent years, with the emergence of a range of advanced information technologies, is a focus on the technology itself rather than how it is used (Mishra & Koehler, 2006). This approach to framing how teachers teach and how students learn with and about technology is myopic and unsystematic. The real focus should be on developing a conceptual framework that takes into account not only how technology is used but also how it is framed by the broader social, cultural, and educational contexts which give rise to such use (Selfe, 1990). Mishra and Koehler have recently introduced the term 'technology content knowledge' to refer to "knowledge about the manner in which technology and content are reciprocally related" (p. 1028).

The idea of technology content knowledge helps to focus the importance of framing questions of technology literacy in the greater context of teaching and learning science. It is in understanding how students learn to use technology so as to afford "newer and more varied representations" of scientific ideas and to "navigate across a range of representations" (Mishra & Koehler, 2006, p. 1028) that we learn how technology literacy can make a difference to science literacy. Thus it is in how youth learn to engage technology to support the use of their knowledge to position themselves *vis à vis* science, their peers, and their communities that can open up new and varied venues for becoming scientifically literate. These intersections between knowing science, representing knowledge through identity, practice, and the role of technology in doing so, have been at the centre of our own efforts to understand more deeply how youth seek to represent themselves as scientifically literate and engaged people.

GET CITY IN SUMMER 2007

Green Energy Technologies in the City

GET City (Green Energy Technologies in the City) is an after-school, voluntary science/technology/social club for youth aged 11–15 years. Housed in the *Boys and Girls Club* (BCG) of a large Midwestern city (named River City for the purposes of this chapter), the programme began in Summer 2007 with 15 students, and currently enrols 40 youth. *GET City* is funded by a federal grant and holds as its dual goals to foster deep and meaningful learning among urban youth in the areas of advanced information technologies (including data acquisition, management, and analysis tools; and communication tools) and the science and engineering of green energy issues.

GET City is housed in the "club room" at the BGC, a multipurpose room serving as cafeteria, auditorium, and classroom space. With movable walls, it can be split into three smaller rooms, although *GET City* regularly uses two-thirds of the space for meeting times. The room is rather spartan, but includes foldable tables and chairs, a white board, three flags (United States, Michigan, and BGC), and a piano. The room is located off the main social area where youth of all ages (5–18) hang out or line up at the canteen to buy candy or hotdogs. This social space opens up to the games room, the bathrooms, the computer room, the canteen, the hallway that leads to the gym, the teen room, and the children's room. To state that it is always bustling with activity would be an understatement. The net result is that *GET City* is positioned in the centre of activity. Noise from the social space leaks into the room and there is constant movement between these spaces as youth excuse themselves momentarily to get food from the canteen, use the bathroom, or leave for a few moments to see friends not in *GET City*. At the same time, the location also gives *GET City* status, with youth (mainly younger ones) knocking on the door to ask when can they join.

Another important dimension to the location of *GET City* is that by virtue of being housed at the BGC and not in a school building, the programme operates on the students' turf. The BGC is where they rule – socially – and arguably may be one of the most important other-figured worlds of their lives outside school. The youths' identities as prominent BGC club members are very important to them. For example, a recent survey we conducted shows that 92% of the youth in *GET City* have participated in the BGC since they were in the 6th grade, with most of them having participated in the club since 3rd or 4th grade. Nearly all of the students, on average, attend the club four days a week (Tuesday – Friday), for three to four hours a day (4pm – 7pm), although many of the teens stay until closing time (9pm). The main reasons youth state for joining the club are summed up in the following questionnaire responses by Camden, Le'Don, and Amy:[1]

> [I] have been coming here since i was six years old. What i like about the club is that it keeps you from getting into trouble. (Camden, 13)

> I have been coming here for 3 years and I enjoy being around kids where I can be myself. (Le'Don, 14)

> [I have been coming to the club] since 5th grade. What I like about the club is that it's a good atmosphere to be in to be with your friends after school and the activities and programs you do like getcity (Amy, 13)

Studying Urban Heat Islands

The urban heat island phenomenon was explored with the *GET City* youth during a five-week unit in June and July 2007. The youth met three times a week (Monday, Wednesday, and Friday) for two and a half hours a day. *GET City* is technology-focused and the youth work on laptop computers at every session. The *GET City* mobile learning lab is a huge attraction for the youth, most of whom do not have computer access beyond what is available to them in school and at the club.

The urban heat island unit was first introduced to students by asking them to consider whether River City is an urban heat island, and how they might find out. Urban heat islands are urban and suburban regions that are 1-6°C hotter than nearby rural areas. Elevated temperatures can impact communities by increasing peak energy demand, levels of air pollution such as ozone, and heat-related illness (Environmental Protection Agency [EPA], 2007). In our curricular unit, we intended for the youth to learn about the causes of heat islands and how they are related to energy usage and health quality indicators such as ozone levels. We also intended that the project would build on their Geographic Information Systems (GIS) skills by producing and managing data and drawing upon EPA databases to study the relationship between environments and local air temperatures (e.g., GIS analysis of thermal images and aerial photos of various land-cover types in Michigan will predict surface and air temperatures for different developed environments). The participants would map temperature and ozone levels using both thermal and chemical probes to relate heat islands to air quality, and would explore the impact of urbanization on heat island formation by relating various surfaces (concrete, asphalt, vegetation) to surface temperature under identical conditions.

Because none of the students had heard of urban heat islands before (as we anticipated) we asked them to consider the question: Where would you rather be on a hot summer day? Standing in the middle of the mall parking lot or under a shady tree? An animated debate followed, raising a set of relevant issues: A mall parking lot would be hot and uncomfortable but a desired location because it meant you were heading to an air conditioned mall to shop. A shady tree would be cooler and relaxing but possibly boring with no proximity to air conditioning, shopping, and cold drinks. Thus, the youth immediately pushed beyond the intended science question of which decontextualised space would be cooler and thus more comfortable, to a more complex consideration of how these spaces are situated physically and socially with everyday desires and practices.

After conducting some controlled experiments for which they built different coloured houses out of construction paper and measured their indoor and outdoor temperatures, both in the sun and shade, the youth eschewed any other ideas we posed (e.g., building model landscapes that examined the relationship between land cover and surface temperature) and elected to create a documentary on the community's understanding of the urban heat island phenomenon. The youth were very enthusiastic and excited about interviewing community members. When an adult expert, an engineer from the local state university, visited one of the *GET City* sessions, the youth took turns to interview him about the urban heat island phenomenon. They used GIS technology (Google earth) to investigate the land cover of River City in order to select a location where the urban heat island effect would be most pronounced. Downtown River City was selected as the site for the filming of the youth's science documentaries. Three documentaries were produced by the youth. We focus on one of them – *Where da Heat Go?* – created by Naomi, Kathy, Shernice and Jeremy (see Figure 1).

Snapshots of da Heat *members: Naomi, Kathy, Shernice and Jeremy*

Naomi, Kathy, Shernice, and Jeremy were part of the pilot summer cohort of students who enrolled in *GET City*. All four were regular attendees of the Boys and Girls Club and were all ascending into the 6th grade at the time of the programme.

Naomi. Naomi is an 11-year old girl of African American and Native American descent who attends a mathematics and science focused magnet school. When we first advertised the programme to the youth in the club, she stood up from the crowd, asking: "Do I have to use my brain?" (should she join GET City). In spite of her seeming lack of desire to engage in cognitive endeavours, Naomi is very concerned about her performance at GET City. She constantly asked us if we thought she was doing OK in the programme, and would register looks of disappointment or anxiety when she deemed a task too difficult. With adult help, she willingly and eagerly tackled any activity on hand.

Figure 1. Self group portrait used in the Heat's documentary on urban heat islands.

Naomi was very excited to have picked one of the two roles of "reporter" for her group's production of the documentary. During the practice interview with the engineer, Dr. Scott, she was the last reporter to interview him. Both interviewers before her had read their questions from slips of paper held in their hands. Just before it was her turn, Naomi asked if we would hold the paper for her just out of the camera's range so that she could position herself well and look "real professional, like a real reporter." During the interview, she role-played a "real reporter" with her body language, posture, and speech to the audience behind the camera. As the reporter for her group in the documentary, Naomi was confident and excited about interviewing people in the community.

Kathy. Kathy is an 11-year old African American girl who is also the club president's daughter. She attends a performance arts focused magnet school. A bright girl who quickly catches onto ideas, Kathy was the group leader of *da Heat*. As leader, Kathy was authoritative and decisive. When she was dissatisfied with the way Naomi and Jeremy (the reporters of the group) delivered their lines, she would insist that the footage be shot again. During the documentary editing process, she had very definite ideas about *what* their documentary should look *like*, and held the group to high standards.

She also went home and shared what she had learnt from *GET City* with her mother and grandmother. When a neighbour cut down one of his trees, Kathy's mom told us how distraught she was and how she argued with her mom that they should plant something to compensate for the neighbour's trees because of heat island effects.

Shernice. Shernice is an African American girl who attends a regular middle school. She is interested in science although she expressed disappointment at how science is being taught at her school. She also said, passionately, that she thinks science is for nerds. Shernice professes to love technology but, in her own opinion, she types "real slow". Shernice was very interested in urban heat islands and was one of the few youth who, with our help, carried out thorough online research of the phenomenon. Shernice recited her research in the role of a science expert for the documentary; it ended up as one of the blooper reels. Shernice was also the photographer for her team during the movie-making process. Shernice was very proud of their movie, saying that it made her "feel really good that [she] could do a lot of work and that [she] is not lazy."

Jeremy. Jeremy is an 11-year old boy of mixed ethnicities. He is very bright and one of the top students in his mathematics and science focused magnet school. Together with Naomi, he was a reporter for the group. Although he had not previously used the computer program i-movie, Jeremy was very quick in picking up i-movie skills. He did a lot of the editing work for the group, even though the group negotiated the edits collaboratively. At the last minute, it was Jeremy who decided to incorporate aspects of the coloured house experiment into the movie. At the end of the summer session, when we gave the youth a quiz on urban heat islands, Jeremy scored the highest out of all the youth with 22/20 (including correct answers for a bonus question).

Where Da Heat go? – *A student produced and directed mini-documentary on urban heat islands*

This 8 minute 41 second mini science documentary opens with a black screen and blue text, "*GET City: Where da Heat Go?*".[2] In the background, Chris Brown's "Run It" is playing. At 18 seconds, the group's photo appears with the text "The Heat" layered on top. At 23 seconds, a new title flashes across a black background. It states: "Look at what people think… Do they really know?????" Then a second title appears: "Is downtown River City an Urban Heat Island?". Beginning at 38 seconds, the documentary moves the viewer through a series of scenes that bring in the voices of community members as they are interviewed on their views about whether downtown River City is an urban heat island, including interviews with a tour guide of the capital building, two men visiting the capital building, and a woman walking to work. As the scenes build, the theme that no-one knows what an urban heat island is becomes stronger and stronger as the youth select scenes to show the interviewees either offering inaccurate statements on urban heat islands or statements of confusion or "I don't know". For example, in the very first of these scenes, the youth interview a tour guide at the capital building:

> Youth: Is River City an urban heat island?
> Adult: No. I wouldn't consider that, no.
> Youth: Why not?
> Adult: Why? Um just because we're in the city, we have more businesses down here which would mean, yes we would have more heat but, we get a lot of rain in Michigan and we're right in the middle of two huge lakes... No I don't think we are heated.

Or later when interviewing a man visiting the capital building:

> Naomi: What would you say an urban heat island is?
> Adult: Urban heat island?? I have no idea!
> Naomi: Okay …. Would you consider downtown River City an urban heat island?
> Adult: I don't know what it is!! *laughs*

Each of these interview scenes is separated from the other with brief musical interludes, with snippets from songs like M&M's "This is Why I'm Hot" and Buster Pointdexter's "Hot Hot Hot" and captions like "I don't know".

After these scenes, the youth then pan in on themselves and share some of their initial findings about the temperatures recorded in the sun and shade at various points on the capital property. After another brief musical interlude (again, Buster Pointdexter's "Hot Hot Hot"), the scene cuts to a series of pictures of the model houses the students built in the very early sessions. In the background, Jeremy is offering an explanation of the experiments and what they tell us about urban heat islands. He shows a graph of temperature measurements of the house in indoor and outdoor conditions. After the scene is over, an image of Jeremy punching himself in the face while making a funny face is shown with the subtitle "Expert Reporting: Jeremy".

The next scenes take about two minutes (3:40 – 5:40) and provide the viewer with a deeper explanation of what urban heat islands are and their impact on environmental and personal health.

> Shernice: An urban heat island is where urban heat air and surfaces have temperatures that are up to 10 degrees Fahrenheit higher than the surrounding natural landcovers. Downtown River City is an example of an urban heat island.
>
> Kathy: Consequences of an urban heat island: Cities experience higher rates of illnesses and death due to dehydration. Approximately 7,421 deaths occurred between 1979 through 1998 due to the urban heat island effect.
>
> Shernice: Consequences of heat islands to the environment: Heat islands produce ozone pollution, which provides great damage to vegetation and local ecosystems.
>
> Together: HEAT ISLANDS ARE A BIG NO NO! WHAT CAN YOU DO ABOUT IT?

The images accompanying the voice-over include snapshots of major urban centres in the United States where the urban heat island effect has been officially documented by the EPA (e.g., Atlanta and Los Angeles), and other cities that matter to the youth (i.e., Detroit and River City). In the background, Rihanna's "Umbrella", a current hip hop song that is very popular with the youth, is playing. While it does not come across explicitly in the video documentary, the youth selected this song because some heat islands, like Houston and Atlanta, have been documented to create their own weather patterns, causing more rain to fall on the eastern edges of these cities, as the youth learned by exploring data provided by the EPA (http://www.epa.gov/hiri). They also used graphical representations to demonstrate the relationship between the temperatures of the urban heat islands versus the surrounding areas, as shown in Figure 2.

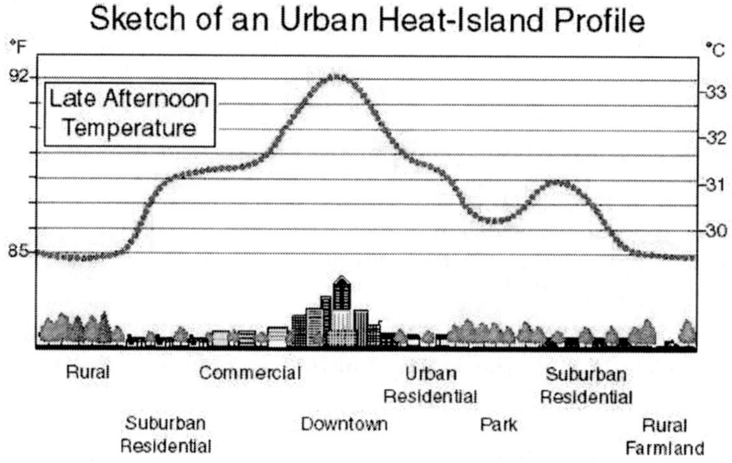

Figure 2. Representation used in da Heat*'s documentary on urban heat islands. (Source: http://www.epa.gov/heatisland/index.htm)*

The documentary then moves on to pictures of polar bear families and the voice-over focuses more intently on the relationship between urban heat islands and global warming, and what this is doing to global biodiversity. The text subtitles state, "Polar bears are drowning because the Arctic ice is melting". The song transitions to "Where is the Love" (Black Eyed Peas). The middle part of the song was selected because it focuses on how something is wrong with the world and how we need to act to turn it around:

Ask yourself
Where is the love?
Where is the love?
Where is the love?
Where is the love?
Father, Father, Father help us
Send some guidance from above
'Cause people got me got me questioning
Where is the love?
Sing with me yo
one world one world
we only got
one world one world
dats all we got
one world one world
and something wrong with
yeah
somethings wrong with
yeah
somethings wrong with the w w world
yeah
we only got
one world one world
dat all we got
one world one world

The next few scenes offer a series of suggestions for "what can we do about it" and the song transitions to R. Kelly's song, "World's Greatest", as if to suggest that young people can and will make this needed difference. A series of still photographs presented with the 'Ken Burns effect'[3] is shown, together with the subtitles presented in Table 1.

The final three minutes of the documentary are dedicated to "bloopers" and "wacky scenes", including Naomi and Jeremy reporting on camera before being interrupted by some other youth; Jeremy dissolving into fits of laughter when Naomi suddenly burps in the middle of reporting; Naomi trying to smile at the camera and direct two ladies she was interviewing to look into the camera at the same time, with one of the ladies exclaiming, "Are we supposed to be looking there?"; and Shernice sticking out her tongue after she had made a mistake with her 'expert' recording. Rolling credits accompanied by the youth themselves singing R. Kelly's "World's Greatest" follow the blooper reel. The documentary ends with a still photograph of earth taken from space, with the subtitles "Walk to Work. Save our planet. The End" accompanied by the last strains of Black Eyed Peas' "Where is the Love":

Table 1. Juxtaposition of lyrics, figures, and subtitles in "How we can help?" segment of da Heat

Still Photograph	Subtitles	Song Lyrics
Still photograph of a girl planting a tree with kids watching	Plant trees to help your city!!!	Something's wrong wit it, yeah Something's wrong wit it, yeah Something's wrong wit it – world world world
Still photograph of building scaffolding, black and white	Help cities restore buildings	Yeah, we only got One world, one world That's all we got One world, one world
Still photograph of cans in a trash bin	Help recycle so your city looks better	Instrumental beginning bars to R. Kelly's "World Greatest"
Still photograph of a boy whispering to a girl	Draw awareness to other people so they help to[o]	Continuation of R. Kelly's "World's Greatest"
Still photograph of a working man in shirt and tie riding a bike	Walk or ride your bike to school or work	Continuation of R. Kelly's "World's Greatest"
Still photograph of roof top gardens in a city		Continuation R. Kelly's "World's Greatest"

we only got
one world one world
dats all we got
one world one world
and something wrong with
yeah
somethings wrong with
yeah
somethings wrong with the w w world
yeah
we only got
one world one world
dat all we got
one world one world

Throughout the documentary, the youth sprinkled their scene transitions with diverse i-movie editing effects such as "far far away", "earthquake", "unscramble", "flying words", and "dissolve" (i-movie HD, Macintosh).

Scientific and Technological Literacy in da Heat

Throughout the process of producing this documentary, from conceptualising the movie and conducting the interview and filming the sequences, to final editing and premiering of the documentary, the group manifested scientific and technological literacy in ways that challenged current constructions. We take up two claims in the following discussion:

- Authoring alternative routes and outcomes: science and technological literacy for robust participation; and
- the youth authoured new identities that positioned them with power in respect of science and technology literacy.

Authoring alternative routes and outcomes. Scientific and technological literacy for robust participation. While the goals of scientific literacy have been expanded over the years, achieving scientific literacy has been reduced to a set of ideas that individuals should know and that teachers can teach if they just follow best practices. It is assumed by most that once students have imbibed the required compendium of academic scientific knowledge and practices, they are equipped to draw upon these resources in their lives for decision making on socioscientific issues. It seems, then, that students, through mirroring academic scientific discourse and practices, are supposed to incrementally accrue elements of scientific literacy so that when they graduate from high school they will exhibit the level of scientific understanding existing among adults, as DeBoer (1991) described. Thus, students are positioned as novices and learners who need to acquire a prescribed repertoire of highly specialised knowledge and skills before they can act as informed citizens. Such a prescribed route towards scientific literacy tends to reduce students to mere empty vessels and evokes the banking notion of education that Friere (1970) cautioned against. Such an approach also disempowers students and narrows opportunities for them to develop a sense of agency with and in science.

With *da Heat*, the youth forged an alternative route towards scientific literacy that was not grounded only in the essential route of "hearing, writing or reading an argument" (Wallace, 2004, p. 902). After conducting the controlled experiments focused on coloured houses, the youth elected to engage in an ethnographic investigation of the urban heat island phenomenon instead of performing more science experiments. Their elected mode of engagement with the subject matter – investigating urban heat islands – diverged from the forms of engagement usually espoused for achieving scientific literacy (see also Chapter 28). In lieu of interacting with text-based sources, the youth opted for live, face-to-face encounters through interviews, first with a science expert and then later with community members.

The youth were also purposeful in the ways they negotiated specific information. For example, one of the first interviews they conducted was with Dr. Scott, an 'energy engineer'. Before Dr. Scott arrived the youth crafted specific questions on the urban heat island phenomenon they were interested in asking him. Through their purposeful questioning and interviewing, the youth ensured that the content on urban heat islands was negotiated, between themselves and the expert, rather than being dispensed by an authority figure. In the interview session, the youth concentrated their questions on deciphering what the word 'island' in the term 'urban heat island' really means. Many of the youth were puzzled and imagined that it implied a geographical island, a land mass surrounded by water, an image discordant with the general concept of an urban heat island. Since he lives in a large Northeastern city well known for the phenomenon, they specifically asked Dr. Scott for examples of urban heat islands. The ensuing conversation led to discussions of how an underground subway system is an urban heat island and to the youth realising that their kitchen at home can be considered a local urban heat island.

Through this dialogue with a science expert, the youth addressed aspects of urban heat islands that perturbed them, and gained a more robust understanding of the metaphoric use of the word 'island' in describing the physical effects of the phenomenon. We believe that this 'negotiated interview session' between the youth and an adult science expert constitutes a third space experience, as suggested by Wallace (2004) in her scientific literacy framework. However, we argue that the direction of negotiations leaned more heavily towards the youth than the adult authority, who was careful and open to the youth's purposeful inquiry and held no ultimate agenda to steer them towards any specific scientific goals.

Dialectical exploration of scientific concepts continued when the youth interviewed residents and workers in downtown River City. When Jeremy interviewed two women and asked them if they knew of any urban heat islands in River City, one lady hesitantly offered "Eagle island", a wooded park with a small lake that is a popular weekend venue for families. Jeremy and his peers concluded that the lady was most probably thrown off by the word 'island' and latched onto the most familiar venue in River City that also bore that word in its name. Other adult community members interviewed by Jeremy and Naomi either did not know or had not heard of the urban heat island phenomenon, and were hard pressed to name an example. This led the youth to highlight this lack of awareness within the community in their documentary video. Through their interviews, they were also made more aware of what the community needs to know about the urban heat island phenomenon. This is evident in the way *da Heat* includes diagrams and pictures of real urban heat island examples to illustrate that urban heat islands are not literal islands (Figure 3) and that the phenomenon is widespread and exhibited in many cities (Figure 4).

The documentary *Where da Heat Go?* featured both science discourse and youth discourse without elevating one over the other. In fact, youth discourse and culture as exemplified by the hip hop music, dramatic transitions, cartoons, and snazzy subtitles emphasised the youth's determination not to yield to a complete appropriation of scientific discourse. Instead, they displayed mastery of the discourse in the truest sense of the word when they were able to use the knowledge of scientific concepts to further their aims of communicating a message.

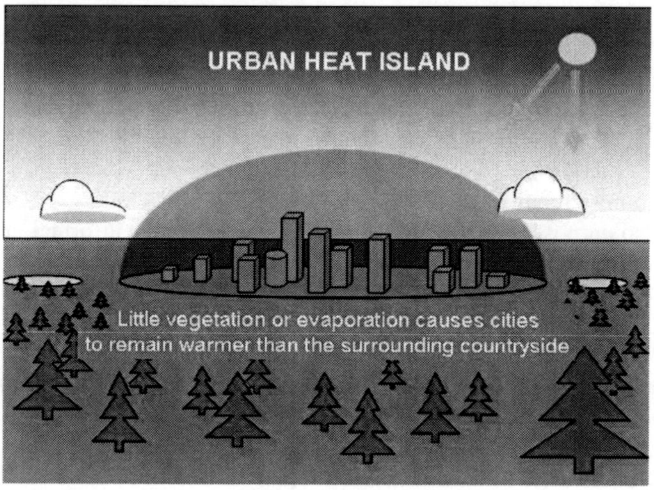

Figure 3. Scientific representation of an urban heat island, used in da Heat.
(source: http://www.sustainableskylines.org/Dallas/heatisland.html)

Figure 4. Atlanta, an example of urban heat island, used in da Heat.

While Wallace (2004) stresses the importance of multiple discourses, she asserts that the negotiation between these diverse discourses is ultimately intended to enable the student to appropriate a full scientific discourse as evidence of achieving scientific literacy. We argue that the hybrid discourse exemplified in the documentary does not detract from the fact that the students have achieved and demonstrated scientific literacy. The hybrid discourse allowed the students to define a serious science topic in their own terms while, at the same time, share sound scientific information with an audience they determined to include both peers and adults. We believe the students were purposeful in creating a final "signature science artefact" (Calabrese Barton, Tan, & Rivet, 2008, p. 82) heavily infused with their identities as urban youth, but at the same time showcasing their expertise in regard to an important and serious socioscientific issue.

In addition to our focus on how youth appropriated highly contested scientific ideas in ways that were situationally meaningful, we also wish to draw attention to the goals of these appropriations. We argued earlier that scientific literacy involves not just *what* one knows or *how* one knows it, but also how such knowledge actively positions one to engage more robustly in scientific practice and discourse as it intersects with civil society. The fact that the youth chose to present their findings to the public in the form of a documentary rather than a science report is also indicative of the youth's desires to redefine what scientific literacy means and looks like to them. While we, as teachers and researchers, worked to make the summer science programme as relevant and authentic as possible – we were engaged in an exploration of whether River City was an urban heat island – the youth in the programme used this investigation to produce a scientific documentary to meet a need discerned in their ethnography of the urban heat island phenomenon. The level of alarm exhibited by the youth regarding the lack of community knowledge of this phenomenon was more than palpable. As one student, Ron, exclaimed:

> I am surprised that people [in River City] don't think this is an urban heat island. Right now you can actually see the beads of heat-induced sweat. Do you see it? [Ron points to his forehead where he is visibly sweating.] They are beads. Not little droplets. *Beads*! I cannot believe this! The people around here are so *unknowledged* [drawn out]. We should really do something about this. Have a heat island awareness day. Yah.

The "unknowledged" became a focal point of *da Heat*'s efforts, and of nearly all of the students in the summer science programme. That the outcome of the youth's summer work was to create a science story intended to engage multiple audiences with differing levels of scientific expertise speaks strongly to how much the students value the localised context of their scientific efforts and their desire to share their knowledge with members of their community. Using evidence to demonstrate the urban heat island effect in River City, and then using their findings to educate others, reflects considerable effort on the youth's part to reposition the nature of knowing and learning and science, and to make becoming scientifically literate an effort to participate in science/society to its fullest extent. Instead of being students and consumers of ideas, they took up the role of citizens who have a clear and compelling stake "in the science-politics interface, which can no longer be viewed as an exclusive domain for scientific experts and policy-makers only" (Backstrand, 2003, p. 24). From issues of life, death, and extinction (i.e., polar bears), to issues of personal health and practices (i.e., consequences of urban heat islands and what can you do about it), the youth made the juncture of science and civic society their place to make a case for robust participation.

In short, throughout their engagement, the youth did not allow themselves to be positioned as *tabula rasa* to be inscribed with the foreign knowledge of experts. Instead, they exhibited agency in how they wanted to investigate, as well as how they wanted to act on what they discovered about the urban heat island phenomenon. While they were actively co-constructing their knowledge with community members, the youth were simultaneously engaging in active acts of citizenship (i.e., with the creation of their science mini-documentary). Again, this is in contradiction to the prescribed route to scientific literacy, where students are encouraged to first gain sound knowledge before they utilise such knowledge in acts of citizenship as adults.

The youth authored new identities that positioned them with power in regard to scientific literacy. By simultaneously engaging in acts of citizenship, the youth both reversed the traditional route to scientific literacy and challenged the notion that children who are still engaging in science learning (i.e., have not yet graduated from high school) typically do not have the political power to raise public awareness of such issues. By choosing to investigate the science concepts through interviews, instead of via text-based sources, the youth authored agentic citizenship identities from the onset. By engaging in dialogic discussion with the science expert and community members, the youth collaboratively negotiated with others regarding what the urban heat island phenomenon might mean to the public and how best to share information about the phenomenon with the larger community. In initiating this investigation through their interviews, and then through their mini-documentary, the youth rejected the traditional position of learners and consumers of science knowledge ascribed to students. Instead, they positioned themselves as the source of information and the epicentre of this urban heat island socioscientific issue within their community. In so doing, they expanded the notion of active citizenship from the individual engaged in personal decision-making to that of educating the larger community. In the process, the youth also continued to build on their increasing knowledge of urban heat islands. The dual gain in knowledge and citizenship afforded by these agentic positions led them to feel confident about sharing their thoughts with the Governor, should such an opportunity arise, as Naomi and Jeremy noted in their reflection conversations.

With their mini-documentary, the youth not only changed the usual direction in which science knowledge is shared (i.e., from adult experts to student learners), they also transformed the nature of the conduit itself, from text to media and human agents. With their science documentary, the youth shared the phenomenon of the urban heat island across settings with their peers at the club, the adult administrators, science educators, doctoral students, and potential donors to the *Boys and*

Girls Club. These donors were so impressed with the documentary that they funded an additional 100% increase in the number of laptops in the *GET City* mobile learning lab.

Because the mini-documentary *Where da Heat Go?* is an atypical science product (i.e., unlike artefacts such as a science report or text) it enabled the youth to display different facets of scientific literacy. Not only were they community science experts with a robust understanding of the urban heat island phenomenon, they were also educators and active citizens who brought awareness of the issue to other youth and adults in the community. This signature science artefact also acted as an advertisement for *GET City* to the other youth in the club in drawing them to future *GET City* sessions. This fulfills the underlying philosophy of scientific literacy espoused by scholars such as DeBoer (2000): That students should find science interesting so that they will be inclined to continue learning science both formally and informally.

CONCLUSION

Conceptions of science literacy must move beyond notions of "what one must know and be able to do" to account for how such knowing and doing intertwine epistemological, methodological, and political activity. Scientific literacy should involve learning about the big ideas in science, the nature of scientific knowledge, and the practices of science, but it should do so in ways that account for the personal process of coming to know in a highly contested world. It should also do so in ways that help individuals to author pathways to more robust participation at the nexus of science and society.

Da Heat's efforts to author a space of participation at the intersection of science and society work to advance current constructions of science literacy. *Da Heat*'s story also challenges us to remember just how much of the development of scientific literacy is situated in context, and that learning to engage in science and society together nurtures a scientific literacy with authentic credence because it is grounded in relevance. Their decision to investigate 'real' urban heat islands (rather than only models of them) positioned the students to engage with real community members in dialogue about the urban heat island phenomenon. Students talked with legislators, police officers, capital workers, and residents, and were surprised that, with few exceptions, almost no-one in their city knew much about heat islands. By producing a high quality science documentary intended for audiences broader than themselves, the youth ultimately positioned themselves to speak to a wide range of communities that have influence over scientific, political, and educational work. By such means, they can 'make a difference' in the real world efforts to understand and minimise the heat island phenomenon.

NOTES

1. All names used in this chapter are pseudonyms.
2. To view the complete documentary, see: getcity.org/UHI.html
3. An Apple i-movie photo editing feature which allows the zooming and panning of still photographs when embedded in motion pictures named after American documentarian Ken Burns, who uses this technique extensively.

REFERENCES

American Association for the Advancement of Science [AAAS]. (1993). *Benchmarks for science literacy*. New York: Oxford University Press.

Anderson, C. (2005). Perspectives on science learning. In S. Abell & N. Lederman (Eds.), *Handbook of research in science teaching* (pp. 3–30). Mahwah, NJ: Lawrence Erlbaum.

Backstrand, K. (2003). Civic science for sustainability: Reframing the role of experts, policy-makers and citizens in environmental governance. *Global Environmental Politics, 3*(4), 24–41.
Bybee, R. (1997). Toward an understanding of scientific literacy. In W. Graber & C. Bolte (Eds.), *Scientific literacy* (pp. 37–68). Kiel, Germany: Institute for Science Education (IPN).
Calabrese Barton, A., Tan, E., & Rivet, A. (2008). Creating hybrid spaces for engaging school science among urban middle school girls. *American Educational Research Journal, 45*(1), 68–103.
DeBoer, G. E. (1991). *A history of ideas in science education: Implications for practice*. New York: Teachers College Press.
Durant, J. R. (1993). What is scientific literacy? In J. R. Durant & J. Gregory (Eds.), *Science and culture in Europe* (pp. 129–137). London: Science Museum.
Environmental Protection Agency. (2007). *Heat island effects*. Retrieved from http://www.epa.gov/heatisland/
Freire, P. (1970). *Pedagogy of the oppressed*. New York: Continuum.
Hobson, A. (2003). Physics literacy, energpy and the environment. *Physics Education, 38*, 109–114.
Hodson, D. (1999). Going beyond cultural pluralism: Science education for sociopolitical action. *Science Education, 83*(6), 775–796.
Hurd, P. (1958). Science literacy: Its meaning for American schools. *Educational Leadership, 16*, 13–16.
Kolstoe, S. (2000). Consensus projects: Teaching science for citizenship. *International Journal of Science Education, 22*, 645–664.
Laugksch, R. (2000). Scientific literacy: A conceptual overview. *Science Education, 84*(1), 71–94.
McCurdy, R. (1958). Toward a population literate in science. *The Science Teacher, 25*, 366–368.
Miller, J. (2002). Civic scientific literacy: A necessity in the 21st century. *Federation of American Scientists Public Interest Report, 55*, 3–6.
Mishra, P., & Koehler, M. (2006). Technological pedagogical content knowledge: A framework for teacher knowledge. *Teachers College Record, 108*(6), 1017–1054.
National Society for the Study of Education [NSSE]. (1947). *Science education in American schools: Forty-sixth yearbook of the NSSE*. Chicago: University of Chicago Press.
NSSE. (1960). *Rethinking science education: Fifty-ninth yearbook of the NSSE*. Chicago: University of Chicago Press.
Roth, W.-M. (Ed.). (2009). *Science education from people for people: Taking a stand(point)*. New York: Routledge.
Roth, W.-M., & Lee, S. (2004). Science education as/for participation in the community. *Science Education, 88*(2), 263–291.
Selfe, C. (1990). Technology in the english classroom: Computers through the lens of feminist pedagogy. In C. Handa (Ed.), *Computers and community: Teaching composition in the twenty first century* (pp. 118–139). Portsmouth, NH: Boyton/Cook.
Wallace, C. S. (2004). Framing new research in science literacy and language use: Authenticity, multiple discourses and the "third space". *Science Education, 88*(6), 901–914.

Angela Calabrese Barton
Edna Tan
Department of Teacher Education
Michigan State University
USA

DEBORAH CORRIGAN

30. AN APPROACH TO CHEMISTRY EDUCATION THAT BRINGS SCIENCE AND TECHNOLOGY TOGETHER

INTRODUCTION

My purpose in this chapter is to consider different perspectives for learning science or, in this particular case, chemistry. In considering curriculum from the viewpoint of the student, De Vos, Bulte, and Pilot (2002) have suggested there are a number of roles a student needs to 'play': They need to have an idea of what it means to be a scientific researcher, a consumer of science, and a responsible citizen making decisions based on science. In learning these roles, students need to be provided with authentic experiences, which De Vos et al. suggest can be developed by thinking about the four faces of chemistry: chemistry as a science; chemistry as technology; chemistry as a craft; and chemistry as magic. Each of these faces requires a different type of thinking about chemistry. Traditionally, curriculum has focussed only on the first. In this chapter, I consider an example of an intended, implemented, and to some extent experienced curriculum of chemistry from the perspective of chemistry as technology.

The chapter highlights the difficulties for teachers planning and implementing such a curriculum resulting from their own experiences in learning chemistry. Some insights on student learning from this perspective are also presented, with particular attention to providing more authentic learning experiences for students and integrating formal and informal contexts for learning chemistry. The specific example of industrial site visits as an informal learning context is used to provide some insights into how such authentic learning experiences might be achieved. By considering different perspectives to learning chemistry, students may gain greater appreciation of the roles they need to adopt as adults when considering chemistry in their future lives.

BACKGROUND

Until the 1980s, science education was largely designed to provide a future science workforce, with little emphasis on science education for the general population. In 1985, Fensham's paper titled *Science for All* echoed the voices of a number of science educators (including Aikenhead, Solomon, Layton, and Jenkins) who were more concerned with science education that was embedded in its cultural context and that made specific links with technology and society in the movement now called STS (Science-Technology-Society) and more recently STSE, where E is for Environment. As highlighted by Aikenhead (1980), the general goals of STS included:

- the characteristics of science; including its aims and values, its human character, and its strategies for decision making and extending knowledge;
- the limitations of scientific knowledge, values, strategies and techniques; including recognition that science is but one knowledge system among many in society, and an examination of boundaries between science and politics, economics, religion, technology and ethics;

A.T. Jones and M.J. de Vries (eds.), International Handbook of Research and Development in Technology Education, 347–359.
© 2009 Sense Publishers. All rights reserved.

– the characteristics of science and its place in society; including case studies of science-related problems, personal interpretation of one's community by making decisions as a consumer, voter, in career planning and so on. (p. 13)

The tension that has existed since the 1980s between producing a science curriculum for a future science workforce and a science curriculum for everyone is reflected in more recent discussions centred on 'scientific literacy'. While Roberts (2007) points out that there is no one definition of scientific literacy, or even consistent use of the term, there do appear to be *types* of scientific literacy apparent in the science education literature, extending along a continuum. Roberts uses the terms Vision I and Vision II to describe the extremes of this continuum. Vision I represents a scientist-centred view that focuses on decontextualised science subject matter and aims to enculturate students into scientific disciplines – a type of pre-professional training. This more traditional view of scientific literacy focuses on the products and process of canonical science. In contrast, Vision II is student-centred and focuses on science as it exists within a particular context, and aiming to enculturate students into their local, national, and global communities. Situations that students are likely to encounter, and the science embedded in such situations, characterise this view and consequently reinforce the notion that science education takes place within a technological and social context.

Just as there are multiple interpretations of scientific literacy, there is also no definitive view on STS. For example, Fensham (1988) identified three approaches to determining the 'knowledge of worth' embedded in STS curricula: science-determined, technology-determined, and society-determined (see Table 1). Categories A, B, and C represent knowledge of worth determined by traditional logic and the sequencing of science topics. The concepts and principles are similar to those in the 1960s curriculum reforms; the additional STS material is often optional and consequently the depth of learning associated with this additional material varies. Categories D, E, and F represent knowledge of worth determined by technology. Here, science knowledge is included via the technological or socio-technological issue being studied. Such an approach results in this knowledge being associated with other science knowledge of a type that has *not* traditionally been part of school science. Categories G and H represent knowledge of worth determined by the social significance or topicality of the material. The science and technology learnt arise directly from the study of the relevant societal knowledge. Curricula adopting this latter approach tend to be more general than traditional science courses and often serve as 'add-on' or complementary courses of study in science.

Aikenhead (1994) and Holman (1987) have presented other approaches to thinking about STS curricula. For example, Aikenhead proposed eight categories of STS along a continuum ranging from motivation by STS content (traditional school science with mention of STS content for creating interest), casual infusion of STS content (traditional school science with a short STS content study), purposeful infusion of STS content (traditional school science with a series of short STS content studies that form cohesive themes), single discipline through STS content (STS content serves as an organiser for the science content and its sequence within a single discipline, for example, chemistry), science through STS content (STS content serves as an organiser for the science content and its sequence where the content is multidisciplinary), science along with STS content (STS content is the focus of instruction and relevant science content enriches the learning), infusion of science into STS content (STS content is the focus of instruction and relevant science content is mentioned but not systematically taught), to STS content (a major technology or social issue is studied; science content is mentioned but only to indicate an existing link to science). In many ways Aikenhead's categories and Fensham's 'knowledge of worth' approach have many similarities to Roberts' (2007) continuum of scientific literacy. Holman's threefold classification of STS science curriculum is somewhat different:

Table 1. Categories of STS curriculum materials that generate knowledge of worth (Fensham, 1988, p. 354).

A. Motivational use
Descriptive accounts of technology in action are used at the beginning and throughout the learning of a science topic as a motivational aid to learning the science. There is little or no expectation that any systematic learning about the technology will occur.

B. Factual references to randomly-chosen examples
These materials are studded with examples of applications of the concepts of the science topic. The examples are not treated or intended to be studied in detail, nor are they chosen because they are a type or category of technology. There is some expectation of rote learning of some of the factual information.

C. Factual references to selected examples of related technology
In these materials a succession of examples, related both to the science concepts and to each other, are presented. More systematic learning about these technologies is intended that is not only descriptive but also comparative and critical in relation to the societal functions the technologies perform, or to aspects of socioscientific issues.

D. Thematic or topical
In these materials a broad STS theme or a technology is chosen as the object of study and it becomes the context that determines the science that will be studied. Some systematic learning about the theme or the technology is possible in parallel with the science that is to be learnt, but this is not developed as conceptually as is the science.

E. Constructional technology
The construction of a technological model or the practice of a technological process is used as the context for learning scientific facts and principles. Learning of scientific skills and practical knowledge associated with the construction and use of the models or process is intended.

F. Social and scientific concepts related to a technology
A few concepts are chosen for study that are significant to the operation in society of technologies that are based on the science topic. These concepts are then studied in increasing depth as the unit progresses in a manner that parallels the sequential learning of the science concepts.

G. Scientific aspects of technology or of socio-technical topics
These materials include some of the detailed science that is important in the construction and operation of a technology or understanding of a socioscientific issue. There is emphasis on the broader scientific principles involved, together with details of its social and economic context.

H. Science and society
In these materials, priority is given to systematic learning about one or more societal aspects of a major technology or a broad field of applied science. A general appreciation of the possibilities and limitations of the underlying science is intended, but no detailed learning is expected.

- The utilitarian perspective or science for work. In this perspective, the primary focus of the curriculum is the translation of scientific principles into technological innovation and efficiency.
- The idealistic perspective or science for citizenship. The focus of this curriculum is on helping the general public to understand the place of science in their world. Social issues of science and the application of science are stressed.
- The pragmatic perspective or science for interest. In this perspective, importance is placed on familiarity and relevance of the material and whether it lies within everyday applications or social issues.

What is evident in each of these analyses of STS or STSE curricula is that a student-centred approach allows the science content to arise from real life situations and that traditional science content can be embedded in a socio-technological context, which can be chosen for its meaningfulness to students or its socioeconomic or sociopolitical importance, as determined by the teacher.

IMPLEMENTING STS STYLE CURRICULUM: ISSUES FOR TEACHERS

Teachers play a crucial role in translating and implementing the intentions of any curriculum. In this section, I consider influences on curriculum implementation focusing on teacher knowledge of chemistry, student-centred pedagogical strategies, teachers' and students' values, and differences between formal and informal learning.

Content Issues and Pedagogical Strategies

Most science teachers' own science education has largely been of a Vision 1 type. The science they learnt (successfully) centred on reasoning that has established "what is the case" (Roberts & MacDonald, 1991, p. 4) – in particular, mastery of laws, theories, and generalisations (the products of science) and understanding how these products were developed and tested (the processes of science). With the introduction of STS curricula and the consequent curriculum emphasis on both the *external* aspects of science (such as pollution, acid rain, and conservation) and the *internal* aspects (such as the nature of scientific theories, the ways in which knowledge or concepts are generated, and characteristics of the scientific community), teachers are being asked to introduce science with 'practical reasoning' to their students. This goes beyond establishing "what is the case" to include the development of values and consideration of "what *can* and what *should* be done" (Roberts & MacDonald, p. 4). In STS curricula, such practical reasoning is established through the introduction of technological and societal aspects of science – aspects outside the realm of many teachers' own educational experience.

Research conducted by myself and Peter Fensham (Corrigan, Fensham, Sheed, & Hutchinson, 1992; Fensham & Corrigan, 1995; Fensham, Corrigan, Sheed, & Hutchinson, 1994) into the dissemination of information and perception of VCE[1] Chemistry (a new STS-style senior secondary chemistry curriculum) found that establishing content unfamiliar to most science teachers, with their narrow academic backgrounds, as learning of worth was more difficult than introducing new pedagogical strategies. However, the experience of teaching a particular curriculum seemed to assist teachers' construction of curricular meaning. Networking with other teachers played a key role in the clarification of messages and ideas contained within the curriculum. This was particularly the case if the networking was not based on social grounds and 'like-mindedness', but on such grounds as geographical location – especially important for teachers in rural areas.

Values Issues: The Intended, Implemented, and Learned Curriculum

Our research also confirmed what research by Mitchiner and Anderson (1989) and Aikenhead (1990) had previously found – that is, teachers remain the key figures in determining the success or failure of a new curriculum. Teachers' beliefs and values critically influence whether a curriculum with a particular set of curriculum emphases will be adopted. Hildebrand (2007) argues that in the process of going from the *intended* curriculum to the *implemented* and *experienced* curriculum, there are a number of mediation points for the values embedded within the curriculum.

While the intended curriculum is often mono-vocal, the implemented and experienced curriculum provide opportunities to be pluri-vocal: "There are many different values that teachers draw on to construct their worldview, that students bring to the classroom, and that the broader community, education jurisdictions and science areas voice" (p. 49). She points out that

> . . . if we desire our school science programs to enable all students to develop lifelong scientific capability then we need to design ways in which diverse values positions – of scientists, science educators, teachers and students – can be embedded in our curriculum. (p. 45)

Further, the diversity of values positions needs to be thought of in multiple ways and from a range of worldviews:

> scientists ... those who integrate knowledge in action (e.g. engineers, biotechnologists, pharmacologists, agronomists); education experts with a range of values stances (such as feminist, special educators, problem-based learning experts) and students whose sociocultural contexts vary (e.g. rural students, girls, boys, indigenous students, Muslim Australians, Chinese Australians, and so on. (p. 46)

Although chemistry curriculum developers may put emphasis in the intended curriculum on understanding chemistry across a broad range of contexts, there is no guarantee that such an emphasis will persist into the translated curriculum implemented by teachers. Teachers will respond in diverse ways: some will implement the curriculum as intended, others will not. In other words, the curriculum that students experience depends on the degree of teacher filtering. If the intentions of the curriculum have been well articulated and understood, and such intentions are within the experiences of the teacher, then relatively little filtering will take place. More commonly, the filtering process will be substantial.

Filtering processes are dependent upon value positions of teachers – and of students, too, since they will also engage in filtering. For example, students will engage in chemistry learning experiences they regard as 'proper chemistry' – subject to assessment and consequently valued by others such as curriculum developers, and relevant to the 'real world'. As Ratcliffe (2007) points out, the explicit discussion of values in science classes is not common. Such an emphasis on values requires teachers and students to reflect on their own values and experiences and to view the purposes of learning experiences in science to be directed towards decision-making and the evaluation of evidence. Such a shift in pedagogy by teachers, and expectations of science learning experiences by students, may not be comfortable. If, as Hildebrand (2007) suggests, we are to consider a variety of value positions, from a range of worldviews, it is essential that some effort be made to explore, critique, and (if necessary) modify the value positions and worldviews currently existing in chemistry learning experiences. If not, extensive filtering will persist.

Formal and Informal Learning of Science

A curriculum that requires a broad contextual understanding of chemistry or science also recognises the importance of both the formal (i.e. in the classroom) and informal, (e.g., via the mass media) procedures students adopt in their learning. The key differences between formal and informal learning as outlined by Wellington (1991) are presented in Table 2.

If links between the formal and informal learning of science are not established, students may continue to operate with two sets of science: their public science (the science they practise in formal settings, such as classrooms with a theoretical base) and their private science (the science they practice in everyday settings, with a practical base). As Solomon (1987) points out, contexts seen as relevant by students may obscure the acquisition of science content and the solution of scientific problems, and both students and science teachers may experience difficulty in relating theoretical concepts to the everyday world. However, Rennie (2007) points out that learning is ongoing and cumulative, and is a result of multiple experiences in a number of settings. Therefore, she says, "It does not make sense to try to distinguish it [learning] as formal or informal . . . it is helpful to distinguish between the formal and informal contexts where learning opportunities arise" (p. 126). Hein (1998) provides a simple distinction: "the terms 'formal' and 'informal' [refer to] a description of settings and the presences or absence of a formal curriculum" (p. 7).

Table 2. Features of informal and formal learning in science (Wellington, 1991, p. 365).

Informal learning	**Formal learning**
Voluntary	Compulsory
Haphazard, unstructured, unsequenced	Structured and sequenced
Non-assessed, non-certificated	Assessed, certificated
Open-ended	More closed
Learner-led, learner-centred	Teacher-led, teacher-centred
Outside of formal settings	Classroom & institution based
Unplanned	Planned
Many unintended outcomes (& outcomes are more difficult to measure)	Fewer intended outcomes
Social aspect central eg. social interactions, between visitors	Social aspect less central
Low "currency"	High "currency"
Un-directed, not legislated for	Legislated and directed (controlled)

Museums, interactive science exhibits, and industrial plants are all sites where links can be established between formal and informal learning. Price and Hein (1991) identify some factors that improve learning possibilities for students undertaking fieldtrips and museum visits:

- Multiple visits help students learn by reducing the novelty of the setting. Students appear less able to concentrate on the learning task in new environments, and they need unstructured time to make sense out of the confusion of a novel setting.
- Teachers need to be actively involved in the planning and implementation of visits in order to prepare students adequately for the experience. Such planning assists in utilising the rich resources available in the field but not in schools.
- Students should be encouraged to develop questions in order to identify student interest, ignorance and knowledge, and initiate conversation. Questions also help to remind students that there is often no 'right' answer.
- Activities should be varied, interactive wherever possible, and provide first-hand experiences and observations in order to develop divergent thinking, questioning, and observation skills.

In her review of learning science outside of school, Rennie (2007) states:

> The evidence for learning from well-structured field trips is so strong that they should be integral to each year's science program. Research of field trips consistently concludes that effective field trips complement the in-school curriculum. Guidelines . . . [include] the need for teachers to plan jointly with students and venue staff, incorporate pre-visit instruction,

> conduct a flexible but focused visit where students have independence and responsibility, and make use of organised post-visit reflection. This structure has the best chance of effective learning because it exploits the personal, contextual, and cumulative nature of learning. (p. 155)

Industrial site visits are particularly valuable in chemistry education and can enable students to experience something that cannot be encountered in the classroom. There are many occasions when providing such first-hand experiences may be more effective in promoting learning than vicarious experiences. Understanding a gold mine or polymer plant is enormously enhanced by going to those places and directly observing such things as safety requirements and problems of large-scale operations.

Visits to industry are of two main types: visits arranged primarily for student learning, although teachers may also learn a great deal from these experiences, and visits or placements for teachers that are designed to improve their understanding of the industrial setting. In the following discussion, both these types of visits are considered.

Hofstein (1996) identifies the following goals for including consideration of industrial chemistry in the curriculum: the application of chemical principles, the importance of the industry to society and the economy, the factors involved in establishing a particular industry, the specific problems faced by the local chemical industry, and the need to address the common unfavourable image of the chemical industry. Murdoch (1996) notes that an industrial chemistry context can provide opportunities for developing students' problem solving skills and practical abilities, and can foster positive attitudes towards chemistry. Hofstein and Kesner (2006) studied the effects of including industrial chemistry in chemistry courses on students' perceptions and attitudes towards chemistry as well as the effectiveness of the training provided for teachers to support its inclusion. They found that students with no exposure to industrial chemistry found their study of chemistry less relevant and less applicable, and that student motivation was greatly enhanced by the inclusion of industrial chemistry *if* it was supported by teacher training encouraging a wide range of instructional techniques. A similar study by Kay (1996) reported that most A-level chemistry students found industrial site visits useful and enjoyable. Over half of the teachers in this study reported that they had found the visit useful and that it connected well to the chemistry course. Kay reports a correlation between students' and teachers' judgements about the site visits.

Between 60 and 75% of the teachers in the study of learning science at industrial sites reported by Parvin and Stephenson (2004) said that they had learnt new knowledge, developed a 'new' image of industry, gained more specific information with respect to science jobs and processes, and learnt how to teach about industry. A smaller number of teachers (40%) said that they had learnt to teach science more effectively and 31% said they had learnt about using industry contexts to teach science. This research also found that 70% of teachers reported increased student motivation and enthusiasm for learning science deriving from this experience. All the teachers in this study had received professional development support prior to undertaking the visits.

Dexter (1996) suggests that site visits are examples of good industrial-educational links because they provide a definite curricular focus, with clear aims and objectives, and provide students with a relevant, modern, interesting, and high quality experience. Site visits can demonstrate an enthusiasm and commitment by the school and the company to provide contextually-based learning opportunities. They can also be an opportunity for opening up communication channels between students and the company.

Thus, the inclusion of industrial contexts for students studying science has some substantial support in research. However, the ways in which teachers manage the inclusion of industrial contexts in science curricula has been less well documented. It is clear that teachers with limited experiences of industry may experience difficulty in making industrial contexts a meaningful experience for their students. Mulroney (1996) provides eight case studies of Victorian science teachers participating in a programme that places teachers in industry for a period of 40 weeks (Teacher Release to Industry Programme – TRIP). Mulroney suggests that this industry-based professional development of teachers produced the following outcomes. First, there appeared to be changes in the teachers' knowledge and philosophy of science, which included a broadening of their knowledge base, application of science in 'real world' situations, establishment of links between science and economic endeavour, an increased awareness of the way science work is carried out, and a broader sense of careers in science. The second major outcome comprised appreciation of management from a wide range of perspectives, including business issues, personnel management and quality assurance, together with development of interpersonal skills related to communication and team work, enhanced understanding of the difficulties of learning as an adult, and improved computing skills. The third outcome related to changes in classroom practice: the teachers placed greater emphasis on practical applications, and had more confidence in doing so, and also emphasised the ways in which scientific work is conducted as well as possible careers and vocational paths in science. Similar findings are also found in research from industry sources, where the knowledge and skills identified as important aspects of an industrial scientific workforce include improved scientific, technical, and economic knowledge; social skills; teamwork; problem-solving skills; and a commitment to life long learning (Unger, 1996). Similarly, Coles (1996) identifies the key aspects of 'scientific capability' as practical skills, design of experimental procedures, decision-making based on evidence, scientific habits of mind, and knowledge of scientific principles.

To demonstrate how more authentic learning experiences can be provided, a specific example of the inclusion of an industrial site visit into a senior chemistry curriculum will be explored.

CASE STUDIES OF SITE VISITS IN VCE CHEMISTRY

In 1991, the Victorian Certificate of Education (VCE) chemistry course, a senior secondary two-year post compulsory course for students aged 16–18 years, was introduced in Victoria, Australia (Victorian Board of Studies, 1986). The curriculum reflected a major shift in emphasis to a more STS style that could be viewed as largely technology-determined (Fensham, 1988) and was intended to link the chemistry classroom with the 'real world'. As such, it represented a significant change for chemistry teachers, most of whom had never worked in industry and had no experience of industrial site visits.

The purpose of the site visit was to provide a real world context that was meaningful to students and enabled them to explore some chemistry in depth. The course required students to write a report of the visit, focusing on the history and use of the particular chemical(s) investigated, the production process (including the treatment of wastes and monitoring of quality), the work environment, and possible careers for people working within this industry. The site visit has proved problematic: the purposes were not well recognised, links between the site visit and the learners world were not apparent, and the context was often unfamiliar to both students and teachers (Corrigan, 1999, 2003, 2004). Between 1991 and 2008 a number of revisions to the curriculum have occurred, although the current curriculum (Victorian Curriculum and Assessment Authority, 2005) still has links with industry and technology through the requirement for a report on an aspect of industrial chemistry or emerging areas such as nanotechnology and green chemistry. The course also involves students in a range of investigations designed to promote self-management, team-work, problem solving, initiative, enterprise, and communication.

Using questionnaires and interviews, Corrigan (1999) gathered information on the ways in which 17 teachers and their students (228 in total) prepared for, and responded to, site visits. The visits were seen as a type of fieldwork that enabled students to experience chemistry first hand and more authentically than could be captured through the use of media, although teachers admitted to a lack of experience in site visit settings. Significant preparation by both teachers and students was needed in terms of developing the relevant chemistry knowledge, understanding the site, and establishing a clear understanding of the purposes of the visit. Students and teachers agreed that the visit brought about a better understanding of the characteristics of the industry and possible career prospects within the industry. In exploring the relationship between chemistry, technology, and society, students generally felt that their chemistry knowledge was adequate to understand a variety of social contexts for chemistry. Chemistry teachers were not so confident, indicating that they felt their own experiences were inadequate in helping them to cope with industrial and technological as well as societal applications of chemical knowledge.

The data from this research confirms much of what Rennie (2007) and Price and Hein (1991) have identified as important if site visits and field trips are to be beneficial learning experiences. Much care should be taken over the choice of the site so that visitors have the opportunity to become familiar with the site and find it relevant to their own personal experiences. It should provide visual appeal to the visitor and, if possible, allow for active participation. The knowledge of the site visit leader is critical in supporting learning for the visitors so that they feel sufficiently comfortable to ask questions. Preparation is crucial so that the site is familiar and able to provide an authentic example of a real life use of chemical knowledge. Preparation also includes the development of an understanding of the nature of chemistry, particularly in applied settings. For example, teachers often had difficulty seeing that chemistry was more than just the chemical reaction involved. The conditions under which such a reaction takes place, whether the reaction generates or absorbs heat, the material of construction for the reaction vessels, whether the vessels are open or closed, and so on, were often not viewed as part of chemistry by teachers.

Hofstein and Kesner (2006) and Parvin and Stephenson (2004) also strongly support the notion that teachers need to be well prepared for such visits and receive some professional development in helping them to understand this new 'applied' chemistry knowledge. Such professional development was not available for the VCE Chemistry teachers. However, there were examples of good practice and two successful examples are described below. While the teachers adopted very different approaches to the task, there are some important common elements.

Adrian's Story

In his approach to investigating a chemical in everyday use, Adrian (a pseudonym), who teaches in an all girls' school, invited his students to identify an area of personal interest to them. Choices included chemicals used in fire-fighting, the dangers of fluoride overdose, and the role of benzene in the Coode Island fires. In each case, the students brought some personal experience to the topic. For example, the investigation of the chemicals used in fire-fighting was selected by a student who lived in an area that had been devastated by fire ten years earlier (when the student was eight). Her father was also a volunteer fire-fighter. The student investigating the dangers of fluoride overdose was concerned on a personal level because she had fluoride staining on her teeth resulting from the high doses of fluoride her mother had taken during pregnancy. The third example focused on a huge chemical fire that had occurred in the previous year in the Coode Island chemical facility, located close to the centre of Melbourne. The facility stored a number of chemicals, but primarily benzene. The intense chemical fire covered large areas of Melbourne with black smoke and areas of the central business district were evacuated.

Each student was expected to organise their own investigation, including site visits, with Adrian acting as a resource for the students. Interviews with Adrian revealed that he found the role of 'learning resource' for students quite demanding because it required knowledge of a variety of issues and associated chemistry, as well as a sound knowledge of industrial practices. For example, students needed to write to appropriate industry sites with well-defined requests and questions in order to obtain information and seek permission for a visit. In this, his experience as an industrial chemist had been invaluable. It gave him a broad understanding of chemistry and the necessary background to help his students approach the various industries, enabling him to cope with the wide-ranging issues that interested them. While this approach had taken huge amounts of time, the students completed the task and remained interested and enthusiastic about their investigation throughout. Adrian comments:

> ... it was a lot of work, but it was worth it to see them [the students] actually engaged in investigating a chemical in such an enthusiastic way. A number of students have said that they felt like real scientists, really investigating something. (Corrigan, 1999, p. 238)

Adrian's students echoed his thoughts through comments such as:

> ... it was really important to thoroughly investigate my topic, but it was interesting as I thought I was really investigating it and organizing everything myself. Seeing the industry setting helped me clarify the research I was doing and made me understand better. (ibid.)

In their questionnaire responses, Adrian's students made positive comments about what they had learnt and about the experience generally. They were able to articulate their ideas and understanding very well in response to questions and, while they recognised the extra effort required by this approach, felt it was well worth it.

Paul's Story

Paul (a pseudonym) is a chemistry teacher in a country secondary school in a town in the far north-west of Victoria, some 5.5 hours drive from Melbourne. Paul had lived in this country town for 20 years: 10 years working as a winemaker in a local winery and 10 years teaching chemistry at the school. For the investigation of a chemical in everyday use, Paul chose to focus the work of all his students on the local winery. While Paul had worked in this winery, this was not the only factor in his decision. The winery is an integral part of the community and, therefore, is well known to the students. It is the largest industry in the community and one in which his students are likely be involved in some way; many of the local families grow grapes for the winery or are employed at grape harvesting time as pickers. Paul felt that using the winery as the focus for the investigation would give his students a sense of ownership.

Paul was very positive about the investigation. He said that the willingness of the students to question and organise their information was a particular highlight. He commented on the high standard of work produced by his students and the motivation they maintained throughout the investigation.

> ... it was a fantastic experience for any teacher to see his students attack a piece of work with such enthusiasm and maintain it all the way through. They [the students] deserve a lot of credit for what they learnt and the way they approached the investigation. They produced a first rate product at the end too. (Corrigan, 1999, p. 239)

Paul's students also expressed positive feelings about the investigation, with comments such as the following:

> I learnt about pH levels, organic chemistry, decomposition and fermentation, and acidity involved in the winemaking process. I never thought I'd use any of this stuff, but I did and it's right here, almost in my own backyard. (ibid.)

> The best part was when the winemaker let us use our own judgement and chemistry to blend some wines. It was fantastic to realize that I did know something and could use it. (ibid.)

These case studies highlight features of fieldwork, museum trips, and site visits that are essential for successful learning. First, the choice of the site to visit is very important: it needs to build on previous experience and provide a context in which students can interact with others. In the first example, students had ownership over what they chose to explore; in the second, although the site was chosen by the teacher, due attention was paid to the personal experience base of the students. It is noteworthy that both these teachers had previous experience within an industrial setting, and so had a good 'applied chemistry' knowledge base. These teachers were able to assist their students in developing an understanding of the industry and creating new knowledge of chemistry within an industrial setting; they could also offer students some insights into career prospects and the skills required for such a career.

SOME IMPLICATIONS

This chapter has traversed aspects of the intended, implemented, and experienced curriculum using the specific curriculum example of VCE Chemistry. The intended curriculum shifted from a Vision I type curriculum, with a focus on canonical chemistry content that is largely decontextualised and intended to provide a pre-professional training for future chemists, to a curriculum of Vision II type, with a focus on enculturating chemistry students into their local, national, and global communities using chemistry knowledge that is embedded in technological and social contexts. The learning orientation shifted from teacher-centred to student-centred.

In implementing this new curriculum, there have been some problems for teachers, both in planning to teach the curriculum and in their own understanding of chemistry. From a planning perspective, teachers had been asked to include a diversity of worldviews in their teaching of chemistry, not just those of scientists. In presenting a more pluri-vocal voice, teachers need to consider different value positions that they and their students bring to the classroom, and pay attention to the voices of the local community, educational jurisdiction, and community of science. The representation of such diverse worldviews is often outside the experiences of teachers whose own successful science career has usually been based on a mono-vocal and mono-cultural worldview of science.

In addition, teachers often struggle to appreciate 'applied' contexts such as chemical industry sites and to access the knowledge required for teaching and learning in such contexts. There has been limited opportunity for teachers to gain an appreciation for the context of work in such settings, including the different ways skills 'play out'. A consistent message is that, like their students, teachers need to be given learning opportunities to consolidate their new knowledge and view of industry. They need to develop strategies to teach about science and industry using industrial settings in ways that facilitate and motivate students' learning. Learning in these settings needs to be based on prior experiences, contextual, cumulative in nature, and should aim to reduce the distinctions between formal and informal learning contexts for both teachers and their students.

[1] VCE stands for the Victorian Certificate of Education, the post-compulsory school qualification in Victoria, Australia.

REFERENCES

Aikenhead, G. S. (1980). *Science in social issues: Implications for teaching. A discussion paper.* Ottawa, ON: Science Council of Canada.

Aikenhead, G. S. (1990, September). *Consequences to learning science through STS: A research perspective.* Paper presented to the British Council's "Science, Technology and Society Education" course, Department of Education Studies, Oxford University.

Aikenhead, G. (1994). What is STS science teaching? In J. Solomon & G. Aikenhead (Eds.), *STS education: International perspectives on reform* (pp. 47–59). New York: Teachers College Press.

Coles, M. (1996). Learning in the context of industry's need: A research project. In J. N. Lazonby & D. J. Waddington (Eds.), *Partners in chemical education. Proceedings of the international conference on industry-education initiatives in chemistry* (pp. 120–121). Heslington, York: University of York.

Corrigan, D. J. (1999). *Industry and technology links with chemistry curriculum.* Unpublished doctoral dissertation, Monash University, Australia.

Corrigan, D. J. (2003, July). *What is real world chemistry?* Invited keynote at the biennial conference of Chemistry Education New Zealand, Wellington.

Corrigan, D. J. (2004, February). *Real world chemistry – Myth or reality?* Invited keynote at Royal Australian Chemical Institute – Chemistry Education Conference, University of Tasmania, Hobart.

Corrigan, D. J., Fensham, P. J., Sheed, J., & Hutchinson, R. (1992, July). *Perceptions of VCE chemistry as a curriculum innovation.* Paper presented at the annual conference of the Australasian Science Education Research Association, Hamilton, New Zealand.

De Vos, W., Bulte, A., & Pilot, A. (2002). Chemistry curricula for general education: Analysis and elements of a design. In J. Gilbert, O. De Jong, R. Justi, D. Treagust, & J. Van Driel (Eds.), *Chemical education: Towards research-based practice* (pp. 101–124). Dordrecht, The Netherlands: Kluwer.

Dexter, J. (1996). How can a visit to an industrial site effectively support chemistry teaching? In J. N. Lazonby & D. J. Waddington (Eds.), *Partners in chemical education. Proceedings of the international conference on industry-education initiatives in chemistry* (pp. 77–78). Heslington, York: University of York.

Fensham, P. J. (1985). Science for all: A reflective essay. *Journal of Curriculum Studies, 17*(4), 415–435.

Fensham, P. J. (1988). Approaches to the teaching of STS in science education. *International Journal of Science Education, 10*(4), 346–356.

Fensham, P. J., & Corrigan, D. J. (1995). The implementation of an STS chemistry course in Australia: A research perspective. In J. Soloman & G. Aikenhead (Eds), *Science, technology and society education* (pp. 194–204). New York: Teachers College Press.

Fensham, P., Corrigan, D., Sheed, J., & Hutchinson, R. (1994). The construction of meaning about VCE chemistry: Research approach to the sharing of curricular knowledge. *Curriculum Perspectives, 14*(3), 25–32.

Hein, G. E. (1998). *Learning in the museum.* London: Routledge.

Hildebrand, G. (2007). Diversity, values and the science curriculum. In D. J. Corrigan, J. Dillon, & R. F. Gunstone (Eds.), *The re-emergence of values in science education* (pp. 45–60). Rotterdam, The Netherlands: Sense.

Hofstein, A, (1996). How to make chemistry more relevant to students? In J. N. Lazonby & D. J. Waddington (Eds.), *Partners in chemical education. Proceedings of the international conference on industry-education initiatives in chemistry* (pp. 80–81). Heslington, York: University of York.

Hofstein, A., & Kesner, M. (2006). Industrial chemistry and school chemistry: Making chemistry studies more relevant. *International Journal of Science Education, 28*(9), 1017–1039.

Holman, J. (1987). Resources or courses? Contrasting approaches to the introduction of industry and technology to the secondary curriculum. *School Science Review, 68*(244), 432–438.

Kay, M. B. (1996). How do A-level (age 16–18) students perceive the chemical industry and how might their perceptions and views be influenced by their A-level chemistry course and the teaching of it? In J. N. Lazonby & D. J. Waddington (Eds.), *Partners in chemical education. Proceedings of the international conference on industry-education initiatives in chemistry* (pp. 78–79). Heslington, York: University of York.

Mitchiner, C. P., & Anderson, R. D. (1989). Teachers' perspectives: Developing and implementing an STS curriculum. *Journal of Research in Science Teaching, 26*(4), 351–369.

Mulroney, G. (1996, August). *Teacher internships for science and technology education*. Paper presented at the 8th Symposium of the International Organization for Science and Technology Education, Edmonton, Canada.

Murdoch, J. (1996). Industrial topics in the school curriculum. In J. N. Lazonby & D. J. Waddington (Eds.), *Partners in chemical education. Proceedings of the international conference on industry-education initiatives in chemistry* (pp. 126–129). Heslington, York: University of York.

Parvin, J., & Stephenson, M. (2004). Learning science at industrial sites. In M. Braund & M. J. Reiss (Eds.), *Learning science outside the classroom* (pp. 129–149). London: Routledge.

Price, S., & Hein, G. E. (1991). More than a field trip: Science programmes for elementary school groups at museums. *International Journal of Science Education, 13*(5), 505–519.

Ratcliffe, M., (2007). Values in the science curriculum. In D. J. Corrigan, J. Dillon, & R. F. Gunstone (Eds.), *The re-emergence of values in science education* (pp. 119–132). Rotterdam, The Netherlands: Sense.

Rennie, L. J. (2007). Learning science outside of school. In S. K. Abell & N. G. Lederman (Eds.), *Handbook of research on science education* (pp. 125–167). New Jersey: Lawrence Erlbaum.

Roberts, D. A. (2007). Scientific literacy/Science literacy. In S. K. Abell & N. G. Lederman (Eds.), *Handbook of research on science education* (pp. 729–780). New Jersey: Lawrence Erlbaum.

Roberts, D. A., & MacDonald, D. (1991, April). *Keeping the science proper in school classrooms: Some conceptual challenges inherent in STS teaching*. Paper presented at the annual conference of the American Educational Research Association, Chicago.

Solomon, J. (1987). Social influences on the construction of pupils' understanding of science. *Studies in Science Education, 14*, 63–82.

Unger, M. (1996). Changes in the chemical industry. In J. N. Lazonby & D. J. Waddington (Eds.), *Partners in chemical education. Proceedings of the international conference on industry-education initiatives in chemistry* (pp. 117–119). Heslington, York: University of York.

Victorian Board of Studies. (1986). *Chemistry study design*. Carlton: Author.

Victorian Curriculum and Assessment Authority. (2005). *Chemistry Victorian Certificate of Education study design*. East Melbourne, Victoria: Author.

Wellington, J. (1991). Newspaper science, school science: Friends or enemies. *International Journal of Science Education, 13*(4), 363–372.

Ziman, J. (1994). The rationale of STS education is in the approach. In J. Solomon & G. Aikenhead (Eds.), *STS education: International perspectives on reform* (pp. 21–31). New York: Teachers College Press.

Deborah Corrigan
Centre for Science, Mathematics and Technology Education
Faculty of Education
Monash University
Victoria, Australia

SECTION 5

LEARNING AND TEACHING

ROBERT McCORMICK

31. LEARNING AND TEACHING: AN INTRODUCTION

LOOKING BACK

When I last considered the state of the field in relation to learning and teaching technology (McCormick, 2004a), I reviewed research that I had been involved in over the years in relation to both 'content' and 'processes' of technology. The headings I used included the 'nature of technological knowledge'; 'procedural knowledge', in particular 'problem solving'; and 'technological conceptual knowledge'. These remain sensible ways of at least dealing with the 'learning' element that is the concern of this section of the *Handbook*. Frank Banks, in the first article in this section (Chapter 32), provides an overview of the research connected with learning and teaching technology, and he adopts a more comprehensive set of headings within two sections: 'pupils' learning of technology' and 'teachers teaching technology'. Within his set of headings appear some of the ones I adopted in 2004.

In the early days of writing in technology education the focus was on the curriculum (another heading Banks uses) and on prescriptions for teaching technology. Much of the work lacked accounts of what actually occurred in the classroom or workshop; rather, it addressed issues of curriculum design and development. These issues were concerned with new approaches based on design activities and problem solving, in contrast to the craft-based 'making' approaches. Design-based activities took their lead from what professional designers or design teachers did (starting with design briefs and the like), while problem-solving activities started with posing problems with criteria that required students to seek solutions to situations they had never encountered before. The first accounts of such activities in action tended to be based on algorithms or stages that sought to capture design or problem-solving processes.

Technological conceptual knowledge in these early days was hardly evident, though the selection of content for a technology curriculum was a central concern. It was not until our colleagues from the science education tradition took an interest in technological conceptual knowledge that the situation changed. Science education has had a long and almost obsessive concern with how students learn science concepts, and some of this has found its way into technology, either as a concern with science concepts and their use in technology or, on rare occasions, technological concepts. Stephen Norton and Stephen Ritchie in Chapter 35 give us an insight into this world of conceptual learning (including that of mathematical educators). However, the area of technological conceptual knowledge has not been taken up by technology educators with the same enthusiasm as that of 'design' and 'problem solving'.

Science educators, and to a lesser extent mathematics educators, challenged technology education to first of all increase the amount of empirical work done in researching technology classrooms and workshops and to consider some of the understandings of learning and teaching science. My own work with colleagues at the Open University in the UK was influenced by what, for example, science educators knew about problem solving, although we were later to diverge greatly from their work as a result of our own classroom research findings. No doubt there is an

A.T. Jones and M.J. de Vries (eds.), International Handbook of Research and Development in Technology Education, 363–371.
© 2009 Sense Publishers. All rights reserved.

interesting study to be done on the influences of these science education researchers who saw a rich field in technology classrooms for understanding both ideas on learning and teaching in general, and science and technology in particular.

I ended my article in 2004 with the inevitable issues for research in technology education that formed a research agenda:

- further clarify student experience of procedural knowledge, such as problem solving and design;
- delineate other procedural knowledge relevant to technology (e.g. systems analysis), perhaps through studies of experts in industry;
- investigate how students use both scientific and technological knowledge in technology classrooms;
- investigate student understanding of systems ideas and how they relate them to their use in a variety of areas of technology;
- investigate the nature of qualitative knowledge particularly as used by expert technologists;
- investigate the kinds and uses of qualitative knowledge that occur in the classroom and how it can be encouraged and developed by teachers and students. (McCormick, 2004a, p. 41)

Some of these are issues that perhaps reflect a personal agenda (e.g., qualitative knowledge), but most are found in what I see as the ways we have come to think about learning and teaching technology. Banks deals directly with many of them in Chapter 32 in his reflection of what makes up the research in the field. In my examination of the field, for example, in looking at the *International Journal of Technology and Design Education* [IJTDE] since 2004 it is possible to see the developments that have taken place in research on teaching and learning.[1] There are around 70-80 articles published in the journal since that date and around ten deal with 'processes' and four 'concepts', thus reflecting the initial emphasis of empirical research and classroom practice. But two 'new' categories are evident: concerns with teachers and their pedagogy, and with learning processes *per se* (i.e. independent of what in technology was being learned). The 12 or so articles concerned with teachers focus largely on teacher pedagogical knowledge, with the number being 'boosted' by a 2008 special issue on the topic (volume 18, issue 3). These articles interestingly bring 'pedagogy' into the lexicon of the technology educator. The articles concerned with 'processes' deal with design and problem solving (including that of trainee teachers), but these have been refined with concerns for modelling (both 2-D and 3-D) and 'design without making' (a heresy at one point!). Those concerned with 'concepts' are interesting in that they include 'systems', one of my 2004 'agenda items', and the kind of mathematics and science concept work represented by Norton and Ritchie in Chapter 35 of the *Handbook*. For some in the technological world, 'systems' is both a conceptual base and a collection of processes (T274 course team, 1992), but the 'systems approach' has yet to be subjected to any research, not least because it rarely features in the school technology classroom. Lest you think my view is too idiosyncratic, Ginns, Norton, and McRobbie (2005) come to similar conclusions, although in the context of primary education:

> A relatively small quantity of research has been conducted into primary school students' understandings of concepts related to technology practice and materials ... Cajas (1999) and Jones and Moreland (2000) suggest that one of the foci of research in technology needs to be on students' learning of key technological ideas and processes at more abstract levels ... Even less research has been conducted into students' understandings of concepts embodied in systems in design and technology. (p. 49)

The new area of 'learning processes' in technology education research, as reflected in journals such as the *International Journal of Technology and Design Education* (IJTDE), is a welcome and interesting development in concerns for learning and teaching technology. It includes the familiar topic of creativity, long advocated as both an outcome and as driver of technology education. Importantly, there is study that is empirically based rather than just the eloquent arguments for technology education enabling creativity to be developed. Newer topics include 'authenticity', again something that technology educators have used as a selling point, but in the IJTDE it is again manifest for the first time as drawing on empirical studies. This signals an important underlying strand in the work over the last few years: a concern for basing studies or accounts of learning and teaching on *views of learning*. Banks, in Chapter 32, indicates the two dominant approaches: symbol-processing and sociocultural traditions. Those who advocate the latter tend to make this explicit in their accounts, in particular to lay out a territory that may be unfamiliar to many in technology education.[2] On the other hand, those who take a symbol-processing or a cognitivist stance (within which I would include those who might describe their work as using a Piagetian approach) do not make this explicit, but it becomes evident in their language and assumptions. Sadly, there are still some empirical articles on teaching and learning in the IJTDE that give no attention to any learning literature, perhaps because they consider it 'obvious' or (more worryingly) because they may not be familiar with any. In laying out these two stances on learning Banks does the field a good service. We have, however, come a long way in the ten years since Rowell (1999) argued:

> Perhaps one reason why research in technology education is diffuse and lacking focus is because we are lacking theoretical frameworks in which to question classroom practices. We need to develop ways of talking about specific dimensions of teaching and learning in technology classrooms in such a way that we can begin to interrogate the nature of classroom interactions. Because of the interplay of cognitive and manipulative demands in technological problem-solving, we need a perspective which enables us to view learning as more than an acquisition of knowledge of concepts and procedures. Moreover, we need a perspective which foregrounds the social and cultural dimensions of technological problem solving. Such theoretical frameworks will hopefully enable us to ground the findings of ethnographic and case study research and, in so doing, contribute to an elaboration of such frameworks.

Another way of looking back, but over a longer timescale than the above, is to consider a recent review of the last twenty years of technology education (de Vries & Mottier, 2006). Of the 38 chapters, some 40% (16) deal with learning or pedagogy in some form, though only seven refer substantially to empirical work; this is not surprising given the 'review-like' nature of the book. Taking into account all of those that have some concern for learning or pedagogy, then almost three-quarters of the chapters deal with learning and teaching. This signals to me a healthy state of where we have arrived at, particularly when we look back at where we have come from.

WHERE WE ARE

The learning and teaching section of the *Handbook* presents one view of where we are today in technology education, given the inevitable limitations of space. Following Bank's overview of the field of research on learning and teaching (Chapter 32), David Mioduser revisits the familiar process of problem solving in Chapter 33. He makes clear his theoretical view of learning, which is more closely aligned with symbol-processing, but he combines this with a view of technology as being central to human endeavour. He turns a clear view of learning technological problem solving into a pedagogy that provides a model for supporting young children's technological thinking and learning of such problem solving. Rather than starting with a view of how problem solving can be conceived, this model is built on empirical work in classrooms. Interestingly, he ends having reversed the relationship we have had up to now in technology education by, as indicated above, seeing it as an opportunity to understand more about the "cognitive foundations of young children's technological thinking and their performance within (and perhaps beyond?) developmental affordances and constraints ...".

The development of empirical studies in learning and teaching research that I referred to earlier with some satisfaction has now reached the area of work that most occupied early technology curriculum developers: how to build in a progression of student work. Alister Jones in Chapter 34 bases his work on a clear view of learning "as a process of construction through guided participation in social activity" and on an extensive set of empirical studies. This enables him to ground the overall categories of student learning in technology, and the details of the specific aspects in each category that show progression, in what students do in classroom activities and their resulting work. This research is not just about establishing a curriculum; rather, it is part of the pedagogy that teachers need to employ in identifying where students come from and what they may be able to achieve in their learning. Thus we have the intimate and important link of pedagogy and learning; the use of the word 'pedagogy' rather than 'teaching' captures just such a relationship.

Jones brought his formidable research experience to technology education from his earlier and continuing work in science education, bringing not only an empirical research base but an understanding of related learning issues in science. Norton and Ritchie (Chapter 35), as indicated earlier, bring with them a similar richness of experience and insight. It has long been the case that science educators have looked to technology as a means to bring to life science and to base it in what can be seen as authentic activity. This approach does not just see technology as a *means* to learning science, but as an important activity *in itself* that puts different demands on learners. When this is combined with the use of mathematics, albeit less researched than the use of science in technological activity, learners can experience – in the words of Norton and Ritchie – a 'multiplier effect'. It can also go some way to reducing fragmentation in the curriculum, though that is as much an artefact of the way teachers and schools organise themselves as how students engage with the curriculum. Indeed, Norton and Ritchie point out the challenges for teachers seeking to integrate learning and plan activities that use science and mathematics in technological activity. The pedagogical traditions of each of the areas are as much a barrier to this integration as the conceptual and procedural knowledge of the subject – something to which I will return later.

Having noted the way in which science education has brought a concern for processes like problem solving, as well as the difficulties for learners of understanding and using concepts, Chapter 36 by Jonathan Osborne contributes yet more innovative thinking about learning. In this chapter, Osborne discusses the role of argument in teaching and learning science and technology. This approach to pedagogy offers another challenge to technology education to extend the envelope of concern. Argumentation has the potential to add greater rigour to student design thinking. Osborne argues that technology teachers' awareness and skill in developing argument with students will help improve their pedagogy, and e presents a list of epistemic discourse activities that can become a repertoire for a teacher. This raised for me the intriguing prospect of how designing might stretch the discourse of description when students are dealing with objects that don't yet exist. The role of graphical representations offer a rich seam for further research. From a sociocultural perspective, such discourses and their use are seen as part of being a community of practice (of say designers), and teachers need to involve students in these different discourses and their varied uses in science and technology. This is part of the challenge that Norton and Ritchie saw for teacher and student understanding of the demands of the different language, expectations, and classroom practices associated with different domains.

The chapters in this section on learning and teaching present, of course, only snapshots into particular areas of work, and into the range of work that is being done on learning and teaching.

They show a vibrant concern on a range of fronts, indicating a healthy field. But they show there is much to do.

LOOKING FORWARD

The various attempts at laying out agendas for the future rarely act as directors for the path ahead, not least because there is no central body that can do this. One of the strong messages for me in looking back over the research on learning and teaching technology is that there exist a number of exceptional researchers who continually feature in the publication lists and who are contributing considerably to the collective endeavour. Predicting what they might find intriguing, interesting, and challenging to research and develop is impossible, as the case of the introduction of argumentation into the pedagogic repertoire illustrates. My attempts in 2004 show that a concern for qualitative knowledge, for example, though finding some resonance in subsequent research, has not itself been a central part of published work.

All this, of course, doesn't prevent people such as me making predictions! I remain convinced that our understanding of learning and teaching technology will only be progressed if we continue our efforts to understand the nature of student learning in its most general sense, and to seek to think about this in the context of technology. For me the most productive avenue is to explore the sociocultural views on learning, but it is clear that there will remain those who take other perspectives, and some who will, in Sfard's (1998) terms, mix their learning metaphors. Starting from clear perspectives on learning, we can be more secure in looking at various elements of pedagogy. Interestingly we have relatively few studies that investigate more general pedagogic strategies such as group work or problem-based learning (but they do exist, particularly with respect to collaboration, as Banks reports in Chapter 32). The dominance of design-and-make activities have led to a limited range of strategies, mainly based on individual work, and this may have restricted technology pedagogy.

Whatever common understanding of learning we may have as teachers, there is no doubt that pedagogy has a strong 'subject' orientation, and we cannot simply import general ideas from elsewhere without a great deal of development work. For example, there is interesting evidence of domain differences in pedagogy in the two important areas of formative assessment (also known as assessment for learning) and promoting communities of learners. Neither of these has featured much in technology education research on teaching and learning, so I will spend a little time exploring the ways in which each must be seen from a subject perspective.

Formative Assessment

Formative assessment is a pedagogy that is considered to be a generic approach and involves a variety of classroom practices that are drawn from a range of research (Black & Wiliam, 1998) and that were trialled in a project with a limited range of secondary school subjects (Black, Harrison, Lee, Marshall, & Wiliam, 2003). These practices involve:

- teacher questions that require students to think (and time to allow them to do so);
- feedback to students that focuses on how they can improve their performance;
- sharing criteria with students and giving them examples of good quality work; and
- self and peer assessment.

This work was further developed in a larger-scale project investigating how teachers might learn to develop and use these practices in both primary and secondary schools and across the whole curriculum (James, McCormick, Black, Carmichael, Drummond, Fox, et al., 2007).

In a comparison of formative assessment in two subjects, mathematics and English, Hogden and Marshall (2005) examined the extent to which there was indeed a subject specific-ness of formative assessment. They indicate that although formative assessment has generic ideas (in the four areas identified above – questioning, feedback, sharing criteria, and self and peer assessment), they are manifest differently in different subjects. These differences reflect the guild differences[3] that students are introduced to in each subject. For example, as Marshall (2004) argued, criteria for assessing work in English reflect English as a subject. Indeed, she concludes that "Much of the practice which appears to be formative on the surface arises more out of the subject discipline than it does out of a specific desire to be formative" (p. 112). Hogden and Marshall (2005) similarly discuss some of the problems for mathematics in implementing formative assessment:

> Despite, or perhaps because of, the differences in the nature of the disciplines, we feel that the two subjects have much to learn from each other. In our view the formative assessment within the mathematics lesson described here is good and considerably better than the current practice in the majority of mathematics classrooms. However, we feel that this practice is itself limited in the particularities of school mathematics in England. School mathematics tends to be taught as formal and abstract concepts … (pp. 172–173)

We could construct a parallel in technology through, for example, teachers' comments on design ideas or students in a group discussing each other's work.

Technology teachers who have used formative assessment in the ways discussed here may be able to spell out the particular issues for implementing it in technology education (see Section 6, particularly Chapters 38 and 39). In research started a decade ago, Jones and Moreland (2005) argue for the importance of the teacher's pedagogical content knowledge. They show how teachers need "a clear understanding of the nature of technology and the conceptual and procedural aspects of the different technological areas" (p. 195). In addition, any understanding of the conceptual terrain teachers are exploring needs to be combined with "a pedagogical sense of the likely understandings the students will bring to the domain" (p. 195). They also pick up the theme I drew from Marshall (2004), that is, of the need for teachers to be interpreters of their observations and student outcomes so that they can provide the kind of feedback that is central to a formative assessment approach. Any lack of critique of students' conceptual and procedural understanding and skills will affect the quality of the guidance teachers give to students to inform their subsequent learning. In their work with schools, Jones and Moreland saw how technology teachers became an example for those elsewhere in the school, but that science teachers, for example, had also to focus on the nature of the science activity to implement formative practices in their classrooms.

Any implementation of formative assessment will become even more complex when technology teachers join with their science and mathematics colleagues to work on joint technology, science, and mathematics activities. This will add greatly to the problems Norton and Ritchie anticipate in their desire to move forward on science and mathematics through technology practice.

Communities of Learners

Fostering 'communities of learners' was a pedagogy devised by Ann Brown and developed in conjunction with her partner Joseph Campione (Brown, 1997; Brown & Campione, 1996) and involves students trying to reach a deep understanding of disciplinary content.[4] A 'big idea' of a subject is chosen and split into topics, each of which is assigned to a student group to research and become 'expert' in that topic. Later, new groups are formed containing one student from each of the expert groups and the students share their understandings of the individual topics to develop their understanding of the big idea (jigsawing). This then enables them to complete a task that is dependent on the students all understanding the ideas they have been learning. In essence, this is a generic pedagogy that Brown and others were able to show improved student performance in some selected areas of the curriculum, premised upon the organisation of knowledge as disciplinary knowledge.

Recently, a group of researchers explored the way in which this generic pedagogy is developed across a range of subjects: social studies (Mintrop, 2004), biology (Rico & Shulman, 2004), English (Whitcomb, 2004), and mathematics (Sherin, Mendez, & Louis, 2004). This group were working with Lee Shulmam and hence were informed by his ideas on pedagogic content knowledge and therefore some of the ideas that underlie the formulations of Banks (2008) and his colleagues who have investigated technology teachers' knowledge. What emerges from the studies undertaken by Shulman and his colleagues is that the details of how to Foster Communities of Learners (FCL) do depend on the particular domain, and there are sometimes conflicts between, on the one hand, what the discipline would encourage and the principles of FCL and, on the other, the 'default' pedagogy of teachers of a particular subject (much as the quote from Hogden and Marshall, 2005, indicated above). Thus, one of the English teachers resorted to her more familiar school subject teacher-directed approach rather than involving the students. Science teachers could create the big ideas, but in the research phase – despite the traditions of science – they resorted to the usual science education approach of focusing on the grasping of a body of facts, undermining the FCL approach. The social studies teachers had difficulty creating the big ideas because they didn't have sufficient discipline knowledge across the range of social sciences and, although they could set up the jigsaw groups, they could not maintain students' clear focus on the big ideas. Another English teacher, anxious to create discourse in the jigsaw process, had to go back to first principles for this to create dialogue and was unable to rely on generic procedural approaches.[5] As one team of investigators put it: "Our experiences suggest that it is possible to teach mathematics through FCL pedagogy, but that doing so requires a reconceptualization of mathematics instruction as well as some rethinking of essential FCL features" (Sherin et al., 2004, p. 229). This shows that it is not just that there has to be 'subject-specific' interpretations of the pedagogy, but that there is mutual change in the subject pedagogy and that of the 'generic' pedagogy.

We can only speculate on what issues of implementation this approach might give to technology teachers, but they would no doubt find themselves in the same situation as the social studies teachers in the study by Mintorp (2004), faced with all the possible disciplines that feed into technology. Certainly the experience of formative assessment reported by Jones and Moreland (2005) encountered a problem of teachers' pedagogical content knowledge not being sufficiently refined to enable the identification of key subject areas. The disciplines that feed into technology include science and mathematics, which are the subject of Chapter 35 by Norton and Ritchie, as well as social studies, when we consider how people interact with technologies. Even for technology teachers focusing on what they see as technological concepts, the evidence from areas such as systems would lead us to anticipate problems for them in securing the pedagogical content knowledge required. All this argues for extending our research in such areas.

If I am to speculate on an area of technology learning and teaching for the future, then it would be around ICT, an area in which I have personally developed some interest. The focus of my work in ICT in technology education was in collaborative work (McCormick, 2004b), but computer-aided design (CAD) is the most common application found in technology classrooms. Despite the growth of ICT in society and the use of CAD in schools, we have relatively few studies in the technology education literature – only one since 2004, and focusing on undergraduates (Chester,

2007).[6] A sociocultural view of learning offers rich interpretations of CAD when compared to traditional drawing techniques as modelling media in terms of the way the 'tool' mediates the learning and thinking during designing. Although there has been some recent work on drawing and practical (3-D) modelling, little has been done to explore different modelling environments, and in particular what CAD offers to student thinking and learning.

Whatever topics are on our collective agenda for research in learning and teaching technology, we are dependent on new researchers coming forward to build on the work of those who are currently holding the flame. This section of the *Handbook* will hopefully act as an inspiration for these new researchers. Section 8, on theoretical and practical approaches to research, will aid them, and it is also good to see that there are now books that specifically address the development of the skills such technology researchers need (Middleton, 2008).

NOTES

1 In what follows, I have avoided naming articles to prevent invidious comparisons being made between what are all worthwhile studies.

2 This lack of familiarity can not primarily be because it has only recently been written about, as I drew on situated approaches (which is a sociocultural approach) in mainstream learning when I wrote about learning through apprenticeship some 15 years ago (McCormick, 1994).

3 Marshall (2004) indicates the importance of the development of guild knowledge and the role of the community of interpreters in this development. Such a community is experienced through the participation of teachers in public examination marking, where they argue about what grades should be awarded to student work. These teachers develop constructs of what constitutes the various grades through a process of citing work and discussing why it is worth a particular grade.

4 Note that the term 'communities of learners' as used by Brown is not the same as that used by Wenger (1998), though there may be some similarities in terms of a subject community and its discourse(s).

5 This is identical to the findings of the large-scale assessment for learning project (LHTL) mentioned earlier, where Marshall and Drummond (2006) showed how those who were able to practice successful assessment for learning practices in their classroom had developed them from a deep understanding of the principles, taking the spirit of the approach rather than following the 'letter' of it; that is, only adopting procedures without any basic change in the culture of the classroom.

6 There have been studies of computer use as a technology in its own right (e.g., Barron, Martin, & Roberts, 2007).

REFERENCES

Banks, F. (2008). Learning in DEPTH: Developing a graphical tool for professional thinking for technology teachers. *International Journal of Technology and Design Education, 18*(3), 221–229.

Barron, B., Martin, C. K., & Roberts, E. (2007). Sparking self-sustaining learning: Report on a design experiment to build technological fluency and bridge divides. *International Journal of Technology and Design Education, 17*(1), 75–105.

Black, P., & Wiliam, D. (1998). Assessment and classroom learning. *Assessment in Education: Principles, Policy and Practice, 5*(1), 7–71.

Black, P., Harrison, C., Lee, C., Marshall, B., & Wiliam, D. (2003). *Assessment for learning: Putting it into practice.* Maidenhead, UK: Open University Press.

Brown, A. L. (1997). Transforming schools into communities of thinking and learning about serious matters. *American Psychologist, 52*(4), 399–413.

Brown, A. L., & Campione, J. C. (1996). Psychological theory and the design of innovative learning environments: On procedures, principles, and systems. In L. Schauble & R. Glaser (Eds.), *Innovations in learning: New environments for education* (pp. 289–325). Mahwah, NJ: Erlbaum.

Cajas, F. (1999). *Overview of the conference.* Proceedings of the American Association for the Advancement of Science Technology Education Research Conference, Washington, DC. Retrieved February 17, 2009, from http://www.project2061.org/events/meetings/technology/Overview.htm

Chester, I. (2007). Teaching for CAD expertise. *International Journal of Technology and Design Education, 17*(1), 23–35.
de Vries, M. J., & Mottier, I. (2006). *International handbook of technology education: Reviewing the past twenty years.* Rotterdam, The Netherlands and Taipei, Taiwan: Sense.
Ginns, I. S., Norton, S. J., & McRobbie, C. J. (2005). Adding value to the teaching and learning of design and technology. *International Journal of Technology and Design Education, 15*(1), 47–60.
Hogden, J., & Marshall, B. (2005). Assessment for learning in English and mathematics: A comparison. *The Curriculum Journal, 16*(2), 153–176.
James, M., McCormick, R., Black, P., Carmichael, P., Drummond, M.-J., Fox, A., et al. (2007). *Improving learning how to learn in classrooms, schools and networks.* TLRP Improving Learning Series. A Gateway Book. London: Routledge.
Jones, A., & Moreland, J. (2000, July). *The role and place of teacher knowledge in enhancing classroom formative interactions.* Paper presented at the Annual Meeting of the Australasian Science Education Research Association, Freemantle, Western Australia.
Jones, A., & Moreland, J. (2005). The importance of pedagogical content knowledge in assessment for learning practices: A case study of a whole-school approach. *The Curriculum Journal, 16*(2), 193–206.
Marshall, B. (2004). Goals or horizons – The conundrum of progression in English: Or a possible way of understanding formative assessment in English. *The Curriculum Journal, 15*(2), 101–113.
Marshall, B., & Drummond, M.-J. (2006). How teachers engage with assessment for learning: Lessons from the classroom. *Research Papers in Education, 21*(2), 133–149.
McCormick, R. (1994). Learning through apprenticeship. In D. Blandow & M. Dyrenfurth (Eds.), *Technology education in school and industry* (pp. 16–36). Berlin: Springer-Verlag.
McCormick, R. (2004a). Issues of learning and knowledge in technology education. *International Journal of Technology and Design Education, 14*(1), 22–44.
McCormick, R. (2004b). Collaboration: The challenge of ICT. *International Journal of Technology and Design Education, 14*(2), 159–176.
Middleton, H. (Ed.). (2008). *Researching technology education: Methods and techniques.* Rotterdam/Taipei: Sense.
Mintrop, H. (2004). Fostering constructivist communities of learners in the amalgamated multi-discipline of social studies. *Journal of Curriculum Studies, 36*(2), 141–158.
Rico, S. A., & Shulman, J. H. (2004). Invertebrates and organ systems: Science instruction and 'fostering a community of learners'. *Journal of Curriculum Studies, 36*(2), 159–181.
Rowell, P. M. (1999). *Looking back, looking forward: Reflections on the technology education research conference.* Proceedings of the American Association for the Advancement of Science Technology Education Research Conference, Washington, DC. Retrieved February 17, 2009, from http://www.project2061.org/events/meetings/ technology/ papers/ Rowell.htm
Sfard, A. (1998). On two metaphors for learning and the dangers of choosing just one. *Educational Researcher, 27*(2), 4–13.
Sherin, M. G., Mendez, E. P., & Louis, D. A. (2004). A discipline apart: The challenges of 'Fostering a Community of Learners' in a mathematics classroom. *Journal of Curriculum Studies, 36*(2), 207–232.
T274 course team. (1992). A systems approach to food provision. In R. McCormick, C. Newey, & J. Sparkes (Eds.), *Technology for technology education* (pp. 142–155). Wokingham, Berkshire: Addison-Wesley.
Wenger, E. (1998). *Communities of practice: Learning, meaning and identity.* Cambridge, UK: Cambridge University Press.
Whitcomb, J. A. (2004). Dilemmas of design and predicaments of practice: Adapting the 'Fostering a Community of Learners' model in secondary school English language arts classrooms. *Journal of Curriculum Studies, 36*(2), 183–206.

Robert McCormick
Faculty of Education and Language Studies
The Open University
Milton Keynes, UK

FRANK BANKS

32. RESEARCH ON TEACHING AND LEARNING IN TECHNOLOGY EDUCATION

INTRODUCTION

What do we know about how pupils learn, and what is effective teaching, in the school subject of technology? This chapter looks at research that has been carried out into teaching and learning in school technology, and how research studies in other areas of education have implications for technology teachers' day to day work.

In writing about this topic in the past, I referred to technology as being a "relatively new curriculum area" (Banks, 2002, p. 299). In comparison to subjects such as science or mathematics, school technology is indeed relatively new, but I was surprised recently when I took down my old copies of the materials written for England's Schools Council's Project Technology to realise that the project ran from 1969-1971. It was a full twenty years later that England and Wales eventually made technology a compulsory subject, part of a new national curriculum for all pupils' general education from the ages of 5 to 16 years. In the early 1990s, many countries, including those as diverse as The Netherlands and Colombia, also introduced the study of technology into the general curriculum (see Pena, 1992; Van der Velde, 1992). As technology as a school subject has now been around in some schools and in some countries for a very long time, surely there must be considerable empirical evidence of what school technology should be, how pupils learn, and what appear to be effective teaching strategies. But as de Vries (2006) says: "Of course one cannot reasonably expect a new or drastically reformed school subject to result in concrete evidence of success in just 20 years. Yet, for several countries the fate of technology education depends on that" (p. 5).

In considering research on teaching and learning in technology education there is therefore a need for some boundaries. How far back in time is legitimate? How widely should more general research in teaching and learning be considered when, naturally, it too has implications for technology lessons? To attempt to answer these questions, this chapter considers the types and approaches of research studies that have been conducted in the period where technology has gained in popularity across the world as a general subject for all pupils. It looks at empirical and theoretical works, and some which are offering a polemic, addressing the following areas:

- The development of the subject of school technology curriculum;
- pupils' learning in technology; and
- teachers teaching technology.

For convenience I have chosen to address research in these areas separately, but naturally learning is related to teaching, which is linked in turn to what is specified to be taught.

A.T. Jones and M.J. de Vries (eds.), International Handbook of Research and Development in Technology Education, 373–389.
© 2009 Sense Publishers. All rights reserved.

THE DEVELOPMENT OF SCHOOL TECHNOLOGY CURRICULUM

Technological Literacy

Research studies that are offering a polemic are arguably most common when investigating and critiquing the creation of the differing technology curricula in various countries. International comparative studies of technology curricula of the 1990s tried to answer questions related to the content and purpose for general technology education for pupils, largely through a comparison of curriculum documents, syllabi, and State orders – to set out a rationale for 'technology for all' (see McCormick, 1992). As Pavlova (2006) put it: "People involved in the development of technology education were looking around the world for ideas" (p. 20). Those ideas assumed that there was a general common goal for technology education which was, at least at the broad level, vocational in order to create a flexible and adaptable workforce. As Pavlova summarised:

> In the UK, the former Secretary of State for Education, Kenneth Baker, announced that technology as a subject was considered to be "of great significance for the economic well-being of this country" (cited in Barnett, 1992, p. 85). A Statement on technology for Australian Schools explained: "Technology programs prepare students for living and working in an increasingly technological world and equip them for innovative and productive activity" (Curriculum Corporation, 1997, p. 4). In the USA it was announced that technology education was "vital to human welfare and economic prosperity" (ITEA [International Technology Education Association], 1996, p. 1). (p. 20)

Such international surveys and comparisons continue, seeking to find solutions to practical issues of implementation and teaching organisation. In planning the new technology curriculum for Finland, for example, Rasinen (2003) conducted an analysis of the 'theory and practice' underlying technology education across six countries: Australia, England, France, The Netherlands, Sweden, and the United States. Rasinen concluded:

> Technological literacy is a universal goal. Principal objectives include understanding the role of science and technology in society, the balance between technology and the environment, the development of technological literacy, and the development of skills such as planning, making, evaluating, social/moral/ethical thinking, innovativeness, awareness, flexibility, and entrepreneurship. The prominent methods focus on experiences for students that engage them in planning, analysing, inventing, innovating, making, and evaluating. The most significant content includes the systems and structures of technology, professions in technology and industry, safety practices, ergonomics, design, construction techniques, assessment practices, the role and history of technological development, problem-solving strategies, and evaluating and valuing the relationship between society and nature. The list of content included in the curricula of the six countries was quite broad and extensive, making it very difficult to condense it. The long standing argument of breadth versus depth was clearly evident across all of the curricula, with the former being more prevalent than the latter. (p. 45)

Comparative research has shown that the development of technology curricula across the world has been slow and implementation restricted, even when the new subject is 'compulsory'. For example, Ginner (2007) noted similarities between the technology curricula of Sweden and New Zealand and counselled that just because it is prescribed for schools, does not mean it is taught:

> It has been and still is a challenge to implement Technology in the Swedish schools. That is not to say that nothing has happened since 1994, but given the fact that it is a compulsory subject the number of schools neglecting the area or doing too little is surprisingly high. …

> The situation is of course much better now than ten years ago. The coming years will be decisive – at least in Sweden. It could either finally take its place and be accepted as a natural and vital part of the curriculum or gradually evaporate and/or merge into science. (p. 2)

Ginner is clear that just specifying technology in a document is not enough to ensure implementation. A related issue is how much the assessment regime complementing the specification constrains its implementation.

In September 2008, the latest version of the national Design and Technology (D&T) curriculum for 11-14 year olds in England came into force. Although there is mention of innovation and consideration of 'industrial issues', in contrast to the original aims of the politicians there is now little that is explicit about 'economic well being' or 'economic prosperity' in the new statement, which suggests that D&T is important because:

> In design and technology pupils combine practical and technological skills with creative thinking to design and make products and systems that meet human needs. They learn to use current technologies and consider the impact of future technological developments. They learn to think creatively and intervene to improve the quality of life, solving problems as individuals and members of a team.
>
> Working in stimulating contexts that provide a range of opportunities and draw on the local ethos, community and wider world, pupils identify needs and opportunities. They respond with ideas, products and systems, challenging expectations where appropriate. They combine practical and intellectual skills with an understanding of aesthetic, technical, cultural, health, social, emotional, economic, industrial and environmental issues. As they do so, they evaluate present and past design and technology, and its uses and effects. Through design and technology pupils develop confidence in using practical skills and become discriminating users of products. They apply their creative thinking and learn to innovate. (Qualifications and Curriculum Authority [QCA], 2007, p. 51)

This builds on a succession of previous curriculum documents and consultations in England and, unlike some past curricula, was received well by teachers and other stakeholders with few political problems. Such a view of the subject D&T in England seems now to be accepted. This is very different from the reception of the original and the world's first compulsory Technology curriculum for all pupils in 1990, which was very severely criticised by The Engineering Council which baldly stated "Technology in the national curriculum is a mess" (Smithers & Robinson, 1992, p. 1) and led to a number of rapid changes. However, despite its quiescent reception, the new statement is as political as the original curriculum both in what it says about D&T education and what it does not say. It highlights 'creative thinking', 'current technologies', 'problem solving', and 'design' (a form of problem-solving), and suggests that "pupils develop confidence in using practical skills ...as individuals and members of a team" (p. 51). These terms are loaded. For example, there has been much criticism as to whether, under a rigid assessment process, pupils can truly be 'creative', and whether in reality team work is compatible with the making of individual products. There is also some disquiet about the balance between traditional craft skills and current technologies, and if in practice teachers really facilitate problem-solving or merely 'play it safe' and direct their pupils closely to ensure they gain high marks. We will look at some research evidence in these areas later.

Some also criticise what is missing from such a definition of technology education. Keirl (2007) and Petrina (1998, 2007) point out that technology education should consider the politics involved in making choices of products and materials. They maintain that consideration of 'who wins, who loses' in the manufacture of, for example, footballs in China or fashion garments in India or

Bangladesh should be part of every pupil's education. Similarly, Elshof (2005, 2006) and Hill (2006) make a case for environmental sustainability as being a much more important driver of 'technology education for all' than just design-and-make assignments, with or without the extensive use of CAD/CAM and other new technologies (see also Chapter 21 by Leo Elshof).

Creativity

The investigation of the place of 'creativity' in schools – much emphasised in the technology curricula of a number of countries – provides an example of research into the enactment of school technology. Does the rhetoric match the reality? Taking the particular case of England, following the publication of the so-called Robinson Report (National Advisory Committee on Creative and Cultural Education [NACCCE], 1999) concerning 'Creativity, Culture and Education', the place of 'creativity' in schools has received considerable attention. Kimbell (2000) noted that in the former national curriculum for England more than half of the subjects claimed creativity as their core concern, but questioned if a climate of rigid testing, league tables, and monitoring by school inspectors was conducive to promoting a vibrant risk-taking environment for pupils:

> Teachers know that naming and shaming is the order of the day. To hell with trust, faith and supportive risk-taking environments. ... We should not be surprised if teachers play it safe rather than take creative risks. It is not sensible for them to do so. (p. 209)

Kimbell drew on school inspection reports (with their low interest in creativity) to support his claims. A series of investigations followed, similarly concerned with 'Creativity in Crisis' (Barlex 2003; Barlex 2007; Barlex & Rutland, 2008; Spendlove, 2007), all suggesting that due to the assessment regime in England pupils were given too few opportunities for truly design-based tasks. Barlex and Rutland offered suggestions for improving the situation:

> In order to enhance teaching for creativity in design and technology for pupils aged 11–14 years and beyond, secondary design and technology teachers can learn from art and design teachers in terms of an overall approach which would encourage more conceptual design i.e. ideas for different sorts of product and greater aesthetic awareness in developing the appearance of products. (p. 163)

So here is an example of how research indicates that, in the case of creativity at least, pupils are not being developed according to the desired declarations of the curriculum nor to the desired priorities of the investigating researchers, who express their findings as 'disappointing'. Further, the way that pupils are taught design or problem-solving (see below) sometimes tends to be rather ritualistic, as what gains marks in examinations is the illustration of the 'technological process' and sometime pupils and teachers follow a fixed routine – what McCormick (2004) has termed an 'algorithm' – to ensure all the relevant elements of the 'process' have been recorded. We turn now to wider investigations of pupil learning in technology.

PUPILS' LEARNING IN TECHNOLOGY

A consideration of technology curricula, therefore, suggests there is as yet little common agreement as to the subject domain of technology, its key purposes, and to what is valid content. However, certain assumptions are made as to what pupils will *want* to study and *how* they will learn (see Rasinen's (2003) comparative study introduced above). Teachers around the world naturally see it as their duty to work with their pupils for examination success and, as we have seen (and will also see later), what pupils actually learn in technology is heavily influenced by the prevailing assessment regime of the class, school, and at national levels.

How should we try to understand learning, knowledge, and the technology curriculum that we have just explored?

Ideas about Learning

"Learning in technology education is under-theorised, and could draw extensively from contemporary learning theory" (Stevenson, 2004, p. 5). However, in a chapter such as this there is only space to outline some aspects of 'contemporary learning theory' that Stevenson advocates, and references are made to other work where a fuller discussion takes place.

There are broadly two theoretical traditions about what is termed 'mind' and how the mind works. On the one hand is the idea of mind as a computational device which engages in symbol-processing:

- Thinking and learning are individual processes;
- social influences come from outside;
- reality is mirrored by symbols and language;
- people passively process information;
- thinking is separated from sensory input and precedes activity; and
- people act on the environment.

On the other hand is an idea of mind formed by the use of human culture, a so-called 'sociocultural' view:

- Individual change is not separable from social change;
- individual understanding is always distributed in its nature;
- language only has meaning in the context of activity when words are being used in a particular way;
- people are agents in solving problems;
- perception and action arise together and co-construct each other; and
- people act with the environment (Adapted from Open University, 2008, pp. 57–60; see also Bruner, 1996).

In the first instance – symbol-processing – knowledge is viewed as being 'out there', external to people and independent of when and how it is acquired. In this tradition, the job of a teacher is seen as being a 'transmitter' of information. In contrast, a sociocultural view of mind holds that words (spoken or written) gain meaning in use, and language and reality are not separate. So there are those who see learning as primarily individual and those who see learning as primarily social.

Between these two extreme positions are some important ideas about how people learn. In subjects such as technology and science there has been a long tradition of 'learning by doing'. In interacting with the environment, a learner can construct meaning and experiment with phenomena. In this way, babies and small children have been described as acting like 'scientists'. This is termed 'constructivism'. The learning is still considered to be centred on the individual but the mind is 'agentive' and actively seeking meaning, not just learning by receiving. An extension of constructivism would be 'social constructivism', where individual learning is socially influenced. In addition to tools, materials, and other physical resources, other people's ideas are an additional resource to inform individual learning. For example, working in groups might raise common ideas which can then be explored through actual experiments and interactions with the real world for individual learning. That is not the same thing as a sociocultural perspective, where the interaction with others is to develop a common frame of mind.

McCormick (2004, 2006) looks to contemporary views of learning to help consider what the nature of technological knowledge should be. He is particularly concerned about how we define and think about that knowledge, especially in the context of how students learn and use knowledge in technology education.

Types of Knowledge

Stevenson (2004) suggests drawing on a consideration of activity theory to help us understand the construction of knowledge in technology (see Engeström, 1987, for a discussion of this, and also Chapter 54 by Michael Roth). He suggests technology education is well placed to develop rich connections between different ways of constructing meaning because technology is a combination of concrete experiential meanings developed in interaction with materials and equipment, with meanings constructed through talk with peers, teachers, and others. Zuga (2004) agrees with Stevenson in the view that much of what is done in technology education is a-theoretical:

> For example, problem solving becomes the catch phrase used in the field, rather than constructivism, cognition, or some other related learning theory. What kind of research is generated when 'problem solving' is used as a key construct? Questions such as: Is using problem solving better than traditional teaching? What do students learn from problem solving? And Do students exposed to technology education problem solving activities do better on mathematics and science achievement tests? Often, studies such as these are conducted in absence of a theoretical perspective with little discussion of the learning theories that underpin the act of using problem solving as a method of teaching. Basing research about problem solving in a learning theory such as constructivism could result in a detailed look at how students construct their own reality and what concepts they are learning via technology education design activities as they make sense of the world. (p. 82)

Zuga takes a constructivist view. She rejects the view of the learner as an empty vessel into which we can pour knowledge, and the view of good teachers as 'careful pourers' (see Driver, 1983) but bemoans the fact that so little empirical work has been done in technology education, and that what has been done is not well grounded in learning theory. This is not the case in science education, for example, where it was discovered that practical work sometimes reinforced incorrect concepts because the learner already had some prior ideas about the concept which had been built up from a very young age; pupils are not 'empty vessels'. For instance, many children (and some adults!) think heavy objects fall faster than light objects – after all, a cannon ball falls faster than a feather! A pupil already has certain ideas about gravity and the teacher's idea that all objects fall together (provided we ignore air resistance) just seems to contradict common sense.

The research suggests that pupils construct their own meaning of the world around them by incorporating new ideas into their current understandings. In learning technological concepts too, therefore, pupils are often required to modify or abandon their current understanding and, perhaps not surprisingly, they do so reluctantly. Scott (1987) has summarised the following points, taking a social-constructivist view of learning:

- Learning outcomes depend not only on the learning environment but also on the prior knowledge, attitudes and goals of the learner. What is already in the learners' mind matters.
- Learning involves the construction of knowledge through experience with the physical environment and through social interaction. Individuals construct their own meaning.

- Constructing links with prior knowledge is an active process involving the generation, checking and restructuring of ideas or hypotheses. The construction of meaning is a continuous and active process.
- Learning science is not simply a matter of adding to and extending existing concepts, but may involve their radical re-organisation. Learning may involve conceptual change.
- Meanings, once constructed, can be accepted or rejected. The construction of meaning does not always lead to belief
- Learning is not passive. Individuals are purposive beings who set their own goals and control their own learning. Learners have the final responsibility for their learning.
- Students frequently bring similar ideas about natural phenomena to the classroom. This is hardly surprising when one considers the extent of their shared experiences – school life, hobbies, clubs, television, magazines, music etc. Some constructed meanings are shared. (pp. 7-8)

Let us now turn to 'problem-solving', which Zuga (2004) considers is in danger of being merely a 'catch phrase' in technology education. As we will see, in contrast to the learning of some concepts in science, in technology the real-world context of the problem cannot be ignored.

Problem-Solving

Zuga (2004) is right in cautioning that discussions about technology as a vehicle for the teaching of problem-solving sometimes become emotionally charged rather than theoretically sound, as those who have argued for 'technology for all' have considered problem-solving as central to technological activity and quite different from the craft tradition, with its emphasis on promoting just making. Over the years, those proposing different technology curricula have used this argument as a principal way of advocating that technology should have an enhanced status in the school curriculum because a general ability to solve problems is central to satisfying human needs. What is the research evidence for this? Glaser (1984), Hennessy and McCormick (2002), and Layton (1993) all suggest that learning is heavily influenced by the context in which it occurs. McCormick (2006) in particular suggests that this is to be expected if one takes a sociocultural view of learning, where knowledge is the result of the social interactions in which it occurs and is inseparable for them.

Pupils do not easily transfer their ability in a particular activity from one learning 'domain' to another. Technology teachers have assumed that if pupils are taught to investigate the factors influencing the design decisions for making one product, for example a moisture sensor, then they will be able to transfer those techniques to consider the different design decisions for, say, batch food production. The evidence is that pupils do not easily transfer their understanding across these different contexts and require considerable support from their teacher to help them do so (McCormick, 2006). Barak (2007) agrees that no all-purpose problem-solving method exists, but has set out a set of series of what he calls 'strategies, schemes and heuristics' that would help teachers and their pupils to start with a framework for considering various possible problem-solving techniques. Murphy and McCormick (1997), however, caution that such strategies become an 'algorithm' which sometimes teachers and pupils follow rather slavishly. Problem-solving within a specific context is not confined to learning in technology; many people are able to add up effectively when shopping in a supermarket but find a similar sum set in a maths lesson very difficult (Lave, 1988).

Researchers investigating how people tackle problem-solving see a close association in a particular context between the conceptual knowledge associated with the particular problem and an understanding of what action needs to be done to tackle that problem (procedural knowledge). These researchers claim that people think within the context in which they find themselves –

'situated-cognition', as discussed above. Murphy (2006) and Murphy and McCormick (1997) suggest that when pupils are presented with problems in unfamiliar contexts they tend to use everyday knowledge to tackle them. For example, a group of Year 9 pupils was set the problem of investigating the thermal insulation properties of various textiles in a science lesson. The investigation suggested by many pupils (particularly the boys) was that they should lag identical beakers with the different materials and plot comparative cooling curves. When the same problem was set within the context of choosing the best material for a mountaineer's jacket, however, the problem became more difficult, with girls in particular focusing on the context. Bringing everyday knowledge to bear, they rejected materials unsuitable for manufacture (such as cotton wool) and modelled the situation by making the material into a small jacket (Open University, 1990). This research has important implications for how we teach both procedural and conceptual knowledge in design and technology.

Designing

Although problem-solving in its different manifestations and techniques is seen as central to the teaching of technology, 'designing' is sometimes considered as so important that it is separated out – as in 'Technology and Design' – perhaps for extra emphasis as in most of the school technology curricula around the world pupils engage in designing to some extent. Mawson (2003) working in New Zealand sets out the particular emphasis on the 'design-make-evaluate' process there and in many other countries. He also notes the widespread criticism of how such an artificially linear 'design process' is taught in schools, drawing on a wide range of research studies in Australia , Canada, and England and going back very many years. For example, Archer (1973) advocated Design to be developed to a level which merited scholarly consideration, and Eggleston (1992) agreed about its importance:

> At the heart of the matter is the design process. This is the process of problem-solving which begins with a detailed preliminary identification of a problem and a diagnosis of needs that have to be met by a solution, and goes through a series of stages in which various solutions are conceived, explored and evaluated until an optimum answer is found that appears to satisfy the necessary criteria as fully as possible within the limits and opportunities available. (p. 18)

Eggleston, therefore, sees Design as a special form of problem-solving, and just as is the case with problem-solving discussed above, many researchers have criticised the simplistic models that were promoted when technology (or *Design* and Technology!) was first introduced into schools as a more scholarly activity than the former craft-based subjects. Initially such criticisms manifested themselves in the search for alternative models which better described what people engaged in a design activity actually did. This search for the 'holy grail' of a supposedly correct description of the design process might have been seen as imposing order on what is necessarily a complicated and iterative process. That some countries wish to do this is to be able to assess and give pupils credit for the *process* of technology rather than just the end product that they make. However, this desire to assess builds in a level of unfortunate artificiality – even game playing – that is unacceptable to many learners. For example, when evidence of ideas is judged through a portfolio of drawings and notes, it is not unheard of for a pupil to be advised (after they have completed their final made artefact) to go back and invent some more 'initial ideas'!

The research into the teaching of design in schools has moved away from attempts to articulate a 'true' model of the design process that should be taught, and considers instead how pupils learn within context. For example, Mawson (2003) advocates that prior to any introduction of a perceived need, pupils need to be exposed to the context within which the task will be based:

> During this exploration of the general knowledge, relevant information, and social attitudes relating to the particular context, children should also be given an opportunity to explore the range of materials available to them when working towards their solution. (p. 123)

As we will see later when we consider the pedagogy associated with teaching technology, motivation for learning is obviously important, and the authenticity of the proposed task is crucial to such motivation for most pupils. Not only does the context shape their ideas and thoughts about their emerging design, so does the opportunity to engage in their work with others. The opportunities for such collaborative work in a sociocultural sense are often not offered to pupils in the individualistic common 'design-and-make' paradigm common in many countries.

Modelling and Systems Thinking

Much is made about the wide knowledge base needed for technology. For example, a pupil in technology needs skills of making, communication of ideas, and notions of production that might involve calculations of costs and economies of scale. Are there any concepts that are principally associated with technology education? Although certainly not unique to technology, two areas that might claim that prize are modelling and systems thinking. There are many forms of modelling in technology lessons: two-dimensional, three-dimensional, and conceptual models such as offered by adapting ideas from science and mathematics. Models are seen as ways in which pupils can communicate their ideas to others, and also generate and develop those ideas for themselves. Such 'thinking on paper', especially exploring initial ideas, has been investigated by a number of researchers (e.g., Garner, 2001; Welch, Barlex, & Lim, 2000).

Anning (1992) illustrates the way preliminary drawing on paper can become divorced from the eventual activity of creating in three dimensions. Young children in particular find the translation between two and three dimensions difficult. The research indicates that they appear to think in different ways in the two modelling contexts. The aspects of the problem they consider when using modelling kits, for example, are different from those they bring into play when drawing out their ideas. Parkinson (2007) takes this further and suggests that for young pupils, concrete models may be used as hypotheses from which to test ideas about the nature of the world, and so provide crucial platforms for learning.

Using the examples of twenty case studies of professional designers, such as architects, engineers, and theatre designers, Garner (1994) showed how drawing helped these people when thinking about their designs. A practice that appeared to have been common among the different designers was the use of sketches to stimulate their own ideas and to clarify more carefully the problem they were trying to solve. This 'homing in' extends the contribution of drawing from a problem-solving to a problem-finding strategy.

A different notion of modelling is offered by the use of conceptual models in technology education (see Barlex, 1994). For example, heat flow in science is often conceptualised using the kinetic theory of molecular motion. This is of limited value in technology where heat flow related to conductivity or even 'U values' and temperature difference is usually much more useful in practical situations.

In order to use a particular idea for practical action, it is sometimes the case that a full scientific explanation is unnecessary and too abstract to be useful:

> [Reconstruction of knowledge] involves creating or inventing new ‘concepts’ which are more appropriate than the scientific ones to the practical task being worked upon. . . . Science frequently advances by the simplification of complex real-life situations; its beams in elementary physics are perfectly rigid; its levers rarely bend; balls rolling down inclined planes are truly spherical and unhampered by air resistance and friction. Decontextualisation, the separation of general knowledge from particular experience, is one of its most successful strategies. Solving technological problems necessitates building back into the situation all the complications of ‘real life’, reversing the process of reductionism by recontextualising knowledge. What results may be applicable in a particular context or set of circumstances only. (Layton, 1993, p. 59)

Again the importance of the particular context for learning is stressed. The message for teachers from this research is to think carefully about the purpose of teaching a particular scientific concept for use in design and technology. Will a robust scientific explanation help the work in hand?

Considering devices and processes in systems terms, although certainly not unique to technology, can be considered as a common conceptual tool. For example, systems thinking – such as considering electronic devices in terms of input, process, and output blocks – can simplify the learning of electronics. A technologist does not need to know about the detailed working of an integrated circuit, or even a transistor, in terms of the physics involved, just how to use it in a range of circumstances. However, systems thinking can sometime make simple ideas more complex. Consider the example of a flush cistern in a toilet where the ballcock regulates the level of the tank. If a variety of technologists are asked to draw a systems diagram of a cistern, they will probably produce different diagrams. Similarly, when asked to identify the input to a simple ‘burglar alarm’ of a switch-relay-buzzer circuit, pupils often identify the input as electricity or the battery (McCormick & Banks, 1994). As McCormick (2004) notes:

> This is an area for research, to explore across what range of situations systems concepts can be used, and to see how understanding of complexity might vary according to age. Although in technology education it is thought to be important to teach students about systems, as a central concept, little is understood about how students come to understand the concepts involved. (p. 35)

TEACHERS TEACHING TECHNOLOGY

At the start of this chapter I made the point that, for convenience, I was separating curriculum, learning, and teaching. It is an artificial division and particularly so in this section where we consider the research done on technology teaching. What teachers do in the technology workroom relates directly to the curriculum they are required to teach and to their view of mind and learning in choosing pedagogic strategies.

Pupil Interaction and Team Work

As part of my personal research work, every year I teach and observe both science and technology lessons across the UK, in India, and in Bangladesh (Banks, 2008). There are surprising similarities across these subjects despite the different international contexts. Both subjects make much of ‘hands-on’ learning; both promote problem-solving and other ‘processes’; both try to explicitly link school tasks to useful learning for everyday life and the needs of the work-place. There are differences, too, particularly a difference in the classroom culture of the two subjects. Students in science classrooms usually engage in practical work in groups. This is partly driven by the need to share apparatus but is also due to a tradition in science classes of discussing ideas and concepts as part of 'social constructivist' view of learning as discussed above. This is true even in classroom contexts where there is little practical work and pupils instead watch a demonstration.

Despite the rhetoric of technology lessons being an opportunity for pupils to work 'as individuals and members of a team', my observation of practical work is that the reality is pupils are more solitary in their learning. Although pupils are often given common design briefs and engage in similar preparatory activities, most of the time work is done on their own, and the later making stage is almost always done individually. However, to explore the values, perspectives, and opinions of others, discussion is essential (see Coles & Norman, 2005; Conway, 2000; Pavlova, 2005). Empirical research in technology classrooms has highlighted the benefits of collaboration if the teacher is convinced that it is of value:

> The Young Foresight curriculum initiative, funded by industry and government departments (Barlex, 2000), highlights the role of collaboration in students' problem-solving in Design and Technology. An evaluation of the implementation of the programme with students aged 14–15 years found that teachers who recognised the need for agency in learning were well placed to develop their practice to include pedagogic strategies to support collaboration . . . Furthermore we argue that the strategies to support students to learn how to collaborate need to be developed alongside those for supporting their learning through collaboration. As teachers develop and implement these strategies they can use the feedback from practice to refine their views of learning and knowledge. The Young Foresight initiative has shown that teachers are willing to undertake this challenge because of the increased motivation and learning of their students that collaboration achieves. Students were found to enjoy the opportunity to work together and understood the cognitive benefits that accrued from it. However the evaluation also found that teachers who view learners as passive in the learning process see no benefits in changing their practice to support collaboration between students. The case for collaboration has to be made for these teachers. (Murphy, Davidson, & Lunn, 2001, pp. 235–236)

Similarly, Dakers (2005) argues that teachers should move from what he terms a behaviourist paradigm as:

> Learners bring with them into the technology classroom, a variety of beliefs, opinions and values. These beliefs, opinions and values are shaped through interaction with their fellow human beings as well as their particular environment. Their formation is consequently sociocultural. (p. 119)

Dakers thus sets out his view that a transmission model of teaching, grounded in a local symbol-processing view of mind, is by its nature a monologue, and interaction between teacher and pupil is a one-way process that leads to a rejection of technology with unfortunate consequences for society at large. Dow (2006), from her work as part of a European-wide research group, suggests that this view of the teachers' role is widespread:

> Despite cultural differences, curricular reforms, policy developments, scientific and technological advances and developments in theories of what constitutes effective learning, the prevailing model of pedagogy in schools across Europe has clearly remained in many ways essentially unchanged. Although a number of countries have new and innovative curricula for technology education, in terms of pedagogy, these are clearly too often subverted by teachers' reluctance to change. (p. 312)

In particular, where countries still have a rationale for technology as the acquisition of only practical craft skills, an apprentice-based 'show, copy, and practice' teaching model is likely to be prevalent. This, however, is a very restricted view of the possible technology curriculum, and not widely prevalent as shown by Rasinen (2003) in his overview of technology curricula internationally.

Gender

Concern has been expressed about the lack of women in engineering and the physical sciences for many years. In continental Western Europe girls are under-represented in technological subjects and a number of research projects have investigated why this may be so, looking at biological differences, the way that parents treat boys differently from girls in terms of access to different types of toys, and differing teacher expectations between the sexes (see Mammes, 2004; Murphy, 2007).

Before the introduction of technology as a subject for all pupils, the upper years of its predecessors were predominantly single-sex. Entrants for wood and metal craft subjects were almost exclusively male and for home economics subjects almost exclusively female, and still the optional elements within technology reflect these traditional gender-specific subjects (see Harding, 2002). Research has focused on a number of issues which have been considered to be the cause of the gender imbalance that result in women being under-represented in the designing and making area (Mammes, 2004). Although women are out-numbered by men in the work of product design, they are the principal consumers of technological products. Researchers have looked at:

- The attractiveness of design and technology as a subject for women and strategies for improving their participation;
- gender bias in the classroom or workshop; and
- the potential impact of women's views on the nature of technology.

The final point is extremely important as many technological objects are inappropriate for the needs and preferences of most female consumers and users. Male designers take account of women through anthropometric studies and market research, but do not necessarily translate this knowledge into products commensurate with women's requirements and needs (Bruce, 1986). However, increasing the number of female designers necessitates encouraging the involvement and retention of more girls across all aspects of technology education at school. The first two points are therefore of crucial importance to teachers.

The work done in Germany and Britain illustrates gender differences in a number of areas. Girls have less experience with certain tools and measuring instruments than boys. Their hobbies are less likely to include using construction sets and they tend not to tinker with items around the home (Mammes, 2004). Murphy (2006) also noticed that the school experience of practical work is different for girls and boys, and teachers engaged in different levels of discourse especially in mixed sex classes. Contrary to what is thought about boys monopolising classroom time, in the workshop girls had more frequent and longer teacher contacts than boys. Randall (1987) found, however, that the requests were for help and encouragement, and the teachers rather accepted this position, reinforcing the girls' sense of inadequacy. Single-sex classes should significantly affect higher levels of engagement by girls but, as mentioned above, the selection of appropriate tasks is very important for motivation to learn.

Work conducted by Hackett (1989) gave a number of suggestions for making practical work more 'girl friendly':

- Organise practical work in mixed groups from the first day. If pupils have already chosen to work in single-sex groups, or pupils object, and the result is less work achieved, ensure that each bench has some boys' groups and some girls' groups.

- Address questions to specific pupils by name rather than choosing the first hand up.
- Choose girls by name to come out to help with demonstrations.
- Praise girls for good ideas as well as neat work; praise boys for neat work as well as good ideas.
- Never ask a boy who understands something to help a girl who does not.
- Encourage girls to be self-reliant and think things out for themselves. [...]
- Ensure fair distribution [of tools, materials and equipment]. (pp. 21–22)

The research conducted in this area would suggest that teachers' (and society's) attitudes to girls' engagement in technological subjects is deep-seated, and a much more profound response is needed than a simple check-list such as this. However, even such a simple list of 'dos and don'ts' causes us to pause and think through day-to-day classroom realities.

Assessment

The examination system, external to the school and independent of government, is arguably more extensive in the United Kingdom than anywhere else. Competitive examinations have been a feature of selection for university entrance and employment since the nineteenth century. It is also the case that the period from the 1960s through to the 1980s saw a revolution in examination strategies across all school subjects, possibly none more profoundly than in school technology (Kimbell, 1997). The separation of examinations in 'theory' from 'practical' was replaced by an integrated assessment regime that attempted to assess the technological process as well as the outcome of that process. This was done by pupils not only producing an individual artefact, but also a project portfolio showing the process of design and construction that they went through. This portfolio is internally marked by the teacher and the score moderated by a visiting assessor. However, assessing a number of pupils each making a different artefact, within an assessment regime where the number of pupils achieving high grades is very important to the teacher and the prestige of the school, leads to teachers developing strategies that will maximise pupil performance. This, of course, is very praiseworthy. A good teacher helps their pupils attain good results. A consequence, however, is that teachers adopt a rather routine approach which helps to guarantee 'examination success' but potentially minimises creativity and flair. For example, the curriculum statement suggests pupils undertaking design and technology courses:

- learn to use current technologies
- learn to think creatively and intervene
- [work as] members of a team.
- consider the impact of future technological developments
- apply their creative thinking and learn to innovate. (QCA, 2007, p. 51)

Like McCormick (2004), Kimbell (2002) suggests that in reality the teacher continues to follow an 'algorithm' both in terms of the portfolio and for the final project. The outcome is that:

- it is (typically) not creative – but formulaic;
- it is (typically) not using new technology – but traditional making;
- it is (typically) not in teams – but individual;
- it is (typically) not dealing with social/environmental issues – but largely technical and (sometimes) economic; and
- it is (typically) not innovative – but 'safe' [predictable and dull]. (pp. 172–173)

Kimbell suggests that a way out of this unfortunate consequence of the examination process is to re-think what has become, in England at least, the tyranny of 'long-tem project assessment' and instead look at short-term structured response activity. Kimbell, Wheeler, Miller, and Pollit (2007) developed this idea further to see if the project portfolio could be used as a more 'active' tool for thinking rather than a passive reporting device, and also if new technologies (such as a PDA) could engender a more ephemeral approach to recording ideas, by engaging students in short (six hour) assessment tasks. In the context of this research project, at least, the outcomes have been very encouraging.

These assessment issues have resonances internationally (see also Section 6 of the *Handbook*). McLaren (2007) and Williams (2007) set out respectively the need to clarify the purposes of assessment in different curriculum regimes internationally and the advantages of a global curriculum which, while recognising the importance of local cultures, could set an international benchmark for performance while navigating a path through conflicting tensions of postmodernism, rationalism and globalisation. In the United States, too, great care has gone into the establishment of benchmark standards for technology, distinct from the well-accepted benchmark standards for science, as a possible precursor to State testing of the subject (ITEA, 2007).

One of the key points brought out by McLaren (2007) in her international overview is the distinction between 'assessment *of* learning' and 'assessment *for* learning'. (See also Chapter 38 by Christine Harrison and Chapter 39 by Judy Moreland and Bronwen Cowie.) Formative assessment, assessment *for* learning in technology, is the central feature of the work of Atkinson and Black (2007). They point out that:

> A key point here is that effective learning demands interaction, started by the teacher to evoke the pupil's ideas, leading to feedback, from pupil to teacher and from pupil to pupil and then back to the pupil, along lines and at levels determined, not by some pre-ordained plan, but by the needs revealed through the pupil's response. The teacher has to use this feedback to modify the teaching plan, so that new vocabulary and structures can be introduced through challenging activities that extend pupils' learning. (p. 200)

Atkinson and Black make the point that teachers taking on the development of formative assessment need to work in a team for mutual support in sharing ideas and resources.

CONCLUSION

The point made by Atkinson and Black (2007) above is key – that teachers need to support each other in sharing ideas and resources. Even forty years is a short time to establish a new curriculum area such as technology and design and, as this chapter has highlighted, teachers have been instrumental in developing this curriculum with its different and changing emphases, by working through the difficulties of different aspects of learning that are intended for pupils and often being innovative in adopting new pedagogies using new resources. Although research has highlighted some difficulties on the journey, there is much to celebrate.

Teachers do, however, need mutual support in sharing ideas and resources. A series of international studies (Banks, Barlex, Jarvinen, O'Sullivan, Owen-Jackson, & Rutland, 2004; Fox-Turnbull, 2006; Stein, Ginns, & McDonald, 2007) have indicated that developing teachers' subject knowledge, pedagogical strategies, and understanding of how technology in school (school knowledge) differs from technology outside school, are all important. Most significant, however, is acknowledging a teacher's personal subject construct. Banks et al. and Stein et al. all put this at the centre of their ideas about technology teacher development. A teacher's past experience of learning technology, their personal view of what constitutes 'good' teaching, and a belief in the purpose of technology as a subject underpins a teacher's professional knowledge. This 'personal subject construct' will have a key impact on the way in which he or she responds to a professional development activity. Technology education, as part of general education for all, even with roots in the 1960s and earlier, is still relatively new in most curricula around the world. The work of researchers in teaching and learning have found that continual professional development, in new pedagogies as well as content knowledge, provides technology teachers with mutual support and a way of articulating what they know, understand, and can do, and – most of all – what they believe about teaching technology.

REFERENCES

Anning, A. (1992). Learning design and technology in primary schools. In R. McCormick, P. Murphy, & M. Harrison. (Eds.), *Teaching and learning technology* (pp. 176–187). Wokingham, UK: Addison-Wesley.

Archer, L. B. (1973). *The need for design education.* Paper presented to DES conference N850, Horncastle. Mimeo and London: Royal College of Art.

Atkinson, S., & Black, P. (2007). Useful assessment for design & technology: Formative assessment, learning & technology. In D. Barlex (Ed.), *Design and technology for the next generation* (pp. 198–215). Whitchurch, UK: Cliffe & Company.

Banks, F. (2002). Research in design and technology education. In G. Owen-Jackson (Ed.), *Teaching design and technology in secondary schools* (pp. 299–317). London: RoutledgeFalmer.

Banks, F. (2008). Teaching design and technology. In G. Owen-Jackson (Ed.), *Learning to teach design and technology in the secondary school* (2nd ed., pp. 151–169). London: RoutledgeFalmer.

Banks, F., Barlex, D., Jarvinen, E.-M., O'Sullivan, G., Owen-Jackson, & Rutland, M. (2004). DEPTH—Developing professional thinking for technology teachers: An international study. *International Journal of Technology and Design Education, 14*(2), 141–157.

Barak, M. (2007). Problem-solving in technology education: The role of strategies, schemes & heuristics. In D. Barlex (Ed.), *Design and technology for the next generation* (pp. 154–169). Whitchurch, UK: Cliffe & Company.

Barlex, D. (1994). A comparison between the nature of modelling in science and design and technology. In F. Banks (Ed.), *Teaching technology* (pp. 74–81). London: Routledge.

Barlex, D. (2000). *Young Foresight: Shaping things to come. Handbook for teachers and mentors.* London: Software Production Enterprises.

Barlex, D. (2003). *Creativity in crisis? Design & technology at KS3 and KS4.* Wellesbourne, UK: DATA.

Barlex, D. (2007). Creativity in school design & technology in England: A discussion of influences. *International Journal of Technology & Design Education, 17*(2), 149–162.

Barlex, D., & Rutland, M. (2008). Perspectives on pupil creativity in design and technology in the lower secondary curriculum in England. *International Journal of Technology & Design Education, 18*(2), 139–165.

Barnett, M. (1992). Technology, within the National Curriculum and elsewhere. In J. Byenon & H. Mackay (Eds.), *Technological literacy and the curriculum* (pp. 84–104). London: Falmer Press.

Bruce, M. (1986). A missing link: Women and industrial design. In A. Cross & R. McCormick (Eds.), *Technology in schools* (pp. 167–178). Milton Keynes, UK: Open University Press.

Bruner, J. (1996). *The culture of education.* Cambridge, MA: Harvard University Press.

Coles, R., & Norman, E. (2005). An exploration of the role values play in design decision-making. *International Journal of Technology & Design Education, 15*(2), 155–171.

Conway, R. (2000). Ethical judgements in genetic engineering: The implications for technology education. *International Journal of Technology & Design Education, 10*(4), 239–254.

Curriculum Corporation. (1997). *A statement on technology for Australian schools.* Carlton, Victoria: Author.

Dakers, J. R. (2005). The hegemonic behaviorist cycle. *International Journal of Technology & Design Education, 15*(2), 111–126.

de Vries, M. J. (2006). Two decades of technology education in retrospect. In M. J. de Vries & I. Mottier (Eds.), *International handbook of technology education: Reviewing the past twenty years* (pp. 3–11). Rotterdam, The Netherlands: Sense.

Dow, W. (2006). The need to change pedagogies in science and technology subjects: A European perspective. *International Journal of Technology & Design Education, 16*(4), 307–321.

Driver, R. (1983). *The pupil as scientist?* Milton Keynes, UK: Open University Press.

Eggleston, J. (1992). *Teaching design and technology.* Buckingham, UK: Open University Press.

Elshof, L. (2005). Teacher's interpretation of sustainable development. *International Journal of Technology & Design Education, 15*(2), 173–178.

Elshof, L. (2006). Technological education and environmental sustainability, a critical examination of twenty years of Canadian practices and polices. In M. J. de Vries & I. Mottier (Eds.), *International handbook of technology education: Reviewing the past twenty years* (pp. 179–182). Rotterdam, The Netherlands: Sense.

Engeström, Y. (1987). *Learning by expanding: An activity theoretical approach to developmental research.* Helsinki: Orienta-Konsultit Oy.

Fox-Turnbull, W. (2006). The influences of teacher knowledge and authentic formative assessment on student learning in technology education. *International Journal of Technology & Design Education, 16*(1), 53–77.

Garner, S. (1994). The importance of graphic modelling in design activity. In F. Banks (Ed.), *Teaching Technology* (pp. 68–73). London: Routledge.

Garner, S. (2001). Comparing graphic actions between remote and proximal design teams. *Design Studies, 22*(4), 365–376.

Ginner, T. (2007, October). *Implementing technology education: Not just a question of excellent steering documents. Key note lecture*. New Zealand Technology Education Conference, Auckland, New Zealand.

Glaser, R. (1984). Education and thinking: The role of knowledge. *American Psychologist, 39*(2), 93–104.

Hackett, G. (1989). Equal opportunities project—starting at the top. *Change in Focus. Towards Better Science, 2*, 12–14.

Harding, J. (2002). Gender and design and technology education. In G. Owen-Jackson (Ed.), *Teaching design and technology in secondary schools* (pp. 237–248). London: Routledge/Falmer.

Hennessy, S., & McCormick, R. (2002). The general problem-solving process in technology education. Myth or reality. In G. Owen-Jackson (Ed.), *Teaching design and technology in secondary schools* (pp. 109–123). London: Routledge/Falmer.

Hill, A. M. (2006). Reflections on twenty years of wandering through the pathways and forest of technological education in Ontario, Canada. In M. J. de Vries & I. Mottier (Eds.), *International handbook of technology education: Reviewing the past twenty years* (pp. 133–159). Rotterdam, The Netherlands: Sense.

International Technology Education Association [ITEA]. (1996). *Technology for all Americans. A rationale and structure for the study of technology.* Reston, VA: Author.

ITEA. (2007). *Standards for technological literacy* (3rd ed.). Reston, VA: Author.

Kimbell, R. (1997). *Assessing technology: International trends in curriculum and assessment.* Buckingham, UK: Open University Press.

Kimbell, R. (2000). Creativity in crisis. *The Journal of Design and Technology Education, 5*(3), 206–211.

Kimbell, R. (2002). Assessing design innovation: The famous five and the terrible two. *The Journal of Design and Technology Education, 7*(3), 172–180.

Kimbell, R., Wheeler, T., Miller, S., & Pollit, A. (2007). *E-scapes portfolio assessment phase 2. Report for DfES and QCA.* London: TERU, Goldsmiths College.

Keirl, S. (2007). The politics of technology curriculum. In D. Barlex (Ed.), *Design and technology for the next generation* (pp. 60–73). Whitchurch, UK: Cliffe & Company.

Lave, J. (1988). *Cognition in practice.* Cambridge, MA: Cambridge University Press.

Layton, D. (1993). Science education and praxis: The relationship of school science to practical action (II). *Studies in Science Education, 19*, 63–75.

Mammes, I. (2004). Promoting girls' interest in technology through technology education: A research study. *International Journal of Technology & Design Education, 14*(2), 89–100.

Mawson, B. (2003). Beyond 'The design process': An alternative pedagogy for technology education. *International Journal of Technology & Design Education, 13*(2), 117–128.

McCormick, R. (1992). The evolution of current practice in technology education. Part 1. *Journal of Epsilon Pi Tau, 18*(2), 19–28.

McCormick, R. (2004). Issues of learning and knowledge in technology education. *International Journal of Technology & Design Education, 14*(1), 21–44.

McCormick, R. (2006). Technology and knowledge. In J. Dakers (Ed.), *Defining technological literacy* (pp. 31–47). New York: Palgrave.

McCormick, R., & Banks F. (1994). *Study guide E650: Design and technology in the secondary curriculum.* Milton Keynes, UK: The Open University.

McLaren, S. V. (2007). An international overview of assessment issues in technology education: Disentangling the influences, confusion and complexities. *The Journal of Design and Technology Education, 12*(2), 10–24.

Murphy, P. (2006). Gender and technology: Gender mediation in school knowledge construction. In J. Dakers (Ed.), *Defining technological literacy* (pp. 219–237). New York: Palgrave.

Murphy, P. (2007). Gender & pedagogy. In D. Barlex (Ed.), *Design and technology for the next generation* (pp. 236–261). Whitchurch, UK: Cliffe & Company.

Murphy, P., & McCormick, R. (1997). Problem solving in science and technology education. *Research in Science Education, 27*(3), 461–481.

Murphy, P., Davidson, M., & Lunn, S. (2001). *Young foresight phase 1 evaluation: Interim summary report.* London: Software Production Enterprises.

Murphy, P., & Hennessy, S. (2001). Realising the potential—and lost opportunities—for peer collaboration in a D&T setting. *International Journal of Technology & Design Education, 11*(4), 203–237.

National Advisory Committee on Creative and Cultural Education [NACCCE]. (1999). *All our futures: Creativity, culture and education.* London: DfEE.

Open University. (1990). *Study guide pE819. Curriculum, learning and assessment.* Milton Keynes, UK: The Open University.

Open University. (2008). *Study guide-E846. Curriculum, learning and society: Investigating practice.* Milton Keynes, UK: The Open University.

Parkinson, E. (2007). Practical modelling and hypothesis testing in primary design and technology education. *International Journal of Technology & Design Education, 17*(3), 233–251.

Pavlova, M. (2005). Knowledge and values in technology education. *International Journal of Technology & Design Education,* 15(2), 127–147.

Pavlova, M. (2006). Comparing perspectives: Comparative research in technology education. In M. J. de Vries & I. Mottier (Eds.), *International handbook of technology education: Reviewing the past twenty years* (pp. 19–32). Rotterdam, The Netherlands: Sense.

Pena, M. (1992). Technology in the general curriculum: A Latin American perspective. In D. Blandow & M. Dyrenfurth (Eds.), *Technology literacy competence and innovation in human resource development. Proceedings of the first international conference on technology education* (pp. 147–152). Weimar, Germany.

Petrina, S. (1998). The politics of research in technology education: A critical content and discourse analysis of the *Journal of Technology Education, Volumes 1–8. Journal of Technology Education, 10*(1), 27–57.

Petrina, S. (2007). 2020 vision – on the politics of technology. In D. Barlex (Ed.), *Design and technology for the next generation* (pp. 34–43). Whitchurch, UK: Cliffe & Company.

Qualifications and Curriculum Authority [QCA]. (2007). *Design and technology programme of study: KS3.* London: Author.

Randall, G. J. (1987). Gender differences in pupil-teacher interactions in workshops and laboratories. In G. Weiner & M. Arnot. (Eds.), *Gender under scrutiny* (pp. 163–172). Milton Keynes, UK: Open University Press.

Rasinen, A. (2003). An analysis of the technology education curriculum of six countries. *Journal of Technology Education, 15*(1), 31–47.

Scott, P. (1987). *A constructivist view of learning and teaching science.* Leeds, UK: University of Leeds.

Smithers, A., & Robinson, P. (1992). *Technology in the National Curriculum: Getting it right.* London: The Engineering Council.

Spendlove, D. (2007). The locating of emotion within a creative, learning and product orientated design and technology experience: Person, process, product. *International Journal of Technology & Design Education, 18*(1), 45–57.

Stein, S. J., Ginns, I. S., & McDonald, C. V. (2007). Teachers learning about technology and technology education: Insights from a professional development experience. *International Journal of Technology & Design Education, 17*(2), 179–195.

Stevenson, J. (2004). Developing technological knowledge. *International Journal of Technology & Design Education, 14*(1), 5–19.

Van der Velde, J. (1992). Technology in basic education. In D. Blandow & M. Dyrenfurth (Eds.), *Technology literacy competence and innovation in human resource development. Proceedings of the first international conference on technology education* (pp. 159–160). Weimar, Germany.

Welch, M., Barlex, D., & Lim, H. S. (2000). Sketching: Friend or foe to the novice designer? *International Journal of Technology & Design Education, 10*(3), 125–148.

Williams, P. J. (2007). A global curriculum: Design technology in the International Baccalaureate. *The Journal of Design and Technology Education, 12*(3), 47–57.

Zuga, K. F. (2004). Improving technology education research on cognition. *International Journal of Technology & Design Education, 14*(1), 79–87.

Frank Banks
Centre for Research in Education and Educational Technology
The Open University

DAVID MIODUSER

33. LEARNING TECHNOLOGICAL PROBLEM SOLVING – A COGNITIVE/EPISTEMOLOGICAL PERSPECTIVE

INTRODUCTION

> Life in the zoological sense consists of such actions as are necessary for existence in nature. But Man arranges things so that the claims for this life are reduced to a minimum. In the vacuum arising after he has left behind his animal life he devotes himself to a series of nonbiological occupations which are not imposed by nature but invented by himself. This invented life … Man calls "human life", "well being". (Ortega y Gasset, 1941, p. 107)
>
> To understand is to invent. (Piaget, 1973)

The need to supply all citizens with core knowledge and skills concerning technology and the artificial world is today widely consensual, as expressed in numerous policy documents and reports published in many countries (e.g., Pearson & Young, 2002). The justifications for this need are rooted in a perspective on modern (post-modern?) life emphasising that: (a) for the last century, more than in previous times, we live immersed in a technology-saturated environment continuously evolving and growing in sophistication and complexity; and (b) living in such an environment demands, in correspondence, the capability to implement sophisticated knowledge and skills, and to make decisions, based on both technological and social, moral, and ethical considerations as well. From this perspective – a socio/technological perspective – learning technology means acquiring knowledge and skills considered necessary for functioning appropriately (e.g., as citizens, workers, students, members of a community) in 21st century technology-saturated societies.

Without neglecting or diminishing the importance of the socio/technological perspective, I will elaborate in this chapter on the need to teach and learn technology from a different standpoint: the cognitive/epistemological perspective. My main claim will be that we should learn technology because the making of the artificial world has always been a defining characteristic of human beings' thinking and intellectual development (Kirlik, 2005; Nye, 2006; Sternberg & Preiss, 2005). Learning technology is learning about thinking – about learning – about the (outer) invented world, vis-à-vis the (inner) inventors' world, and about the results of the co-evolution – both phylogenetic and ontogenetic – of the designed and the designers' worlds. Or in Cole and Derry's (2005) conclusive words: "We have met technology and it is us" (p. 209).

Within the complex and continuously evolving body of knowledge of technology, I focus in this chapter on the learning of problem solving and design. The study of these processes has been part of a long term research agenda focusing on technological thinking: *DESIGNING MINDS*$^{(dm)}$, named in twofold reference to minds *doing design* as well as *being designed* while doing design.

In the following sections I present a brief overview of the rationale for this work, followed by an elaboration on two main claims concerning the learning of technological problem solving (TPS), based on the results of our studies:

A.T. Jones and M.J. de Vries (eds.), International Handbook of Research and Development in Technology Education, 391–406.
© 2009 Sense Publishers. All rights reserved.

- Claim 1: Teaching/learning TPS should be integral to the early stages of children's intellectual development; and
- Claim 2: Teaching/Learning TPS should be based on students' evolving cognitive models of the design process.

CONCEPTUAL FRAMEWORK

In this section I present a brief account of the conceptual framework of the research, focusing on a cognitive/epistemological approach to technology in general, to TPS in particular, and to learning TPS more specifically.

Technology as a Cognitive/Epistemological Human Endeavour

Technology is a defining characteristic of humankind. In Ortega y Gasset's (1941) words:

> Technology, in contrast to the adaptation of the individual to the medium, is the adaptation of the medium to the individual. ... To react upon his environment, not to resign himself to the world as it is – that is the token of man. ... Man without technology – that is, without reaction upon his medium – is not man. (p. 96)

This line of thought emphasises the crucial role of technology as humankinds' way to continue the evolutionary path from the stage at which we became *Homo sapiens* by means of cultural evolution, that is, by growing exosomatic instead of endosomatic organs (Medawar & Medawar, 1977), inventing cultural organs, and accommodating the natural world to our needs. This is in fact a co-evolutionary path in which technology stands at the meeting point between the 'inventor's mind' and the 'invented world', serving both the phylogeny and ontogeny of human knowledge, perceptions, beliefs, thinking processes, and skills (Cole & Derry, 2005; Nye, 2006; Ortega y Gasset, 1941).

Here, technology is defined in its broadest sense as human knowledge generated and applied for solving perceived problems and needs. Artefacts (whether concrete or abstract, material or symbolic) are the resulting instantiations of this knowledge within specific contexts and conjunctures. In different contexts (e.g., environmental conditions, states of knowledge, available resources, cultural circumstances) different solutions might be generated. This is how we can refer across time to the co-evolution of technologies and the technologies-generating mind, and across space to the co-existence of different technological realities within the same historical time (e.g., Fanta, 2007).

Technological Problem Solving

Technological problem solving (TPS), as performed by both experts and novices, has been a subject of study for several years, including theoretical considerations (e.g., Chen & Strupp, 1993; Ihde, 1990, 1993; Mitcham, 1994; Simon, 1981) as well as empirical research (e.g., McCormick, Murphy, & Hennessy 1994; Mioduser & Dagan, 2007; Mioduser, Venezky, & Gong, 1996; Wu, Custer, & Dyrenfurth 1996). Part of this work examines TPS processes by professional practitioners (e.g., Bucciarelli, 1996; Dym, Agogino, Eris, Frey, & Leifer, 2005; Vincenti, 1990), while other parts focus on TPS skills to be learned by any technologically literate person (e.g., Dyrenfurth, 1991; Pearson & Young, 2002).

The development of technological solutions may take a variety of forms: from a completely open-ended process (e.g., no *a-priory* plan exists), to highly structured ones (e.g., strictly following the course of action commonly mentioned in the literature as 'the design process'). In all cases it involves a wide range of cognitive and metacognitive processes, for example, exploration, experimentation, and reflection-in-action (Pacey, 1999; Rowe, 1987; Waks, 2001). It can be characterised as a highly creative and multi-faceted course of action in which informed doing-and-evaluation loops gradually advance the generation of a sound solution.

A key methodological construct in TPS is the 'design process' – a compound of procedures, intellectual means, and formally defined courses of action used by experts to generate technological solutions. In our work with young learners, our focus on design is on the way the human mind generates and implements solutions for bridging perceived gaps between *current* and *desired* states of affairs. Design is about thinking, learning, making decisions, and forming solutions. Thus, design encompasses a whole set of cognitive resources and processes, for example: perceiving the gap (analysing the current state to identify what desired aspects are not covered and anticipating the desired state to identify what aspects might be solved); formally defining the need to be satisfied; elaborating on the resources and constraints contributing to or affecting the construction of the appropriate solution (e.g., materials; processes; funding; human resources; and social, cultural, esthetical, ethical, or environmental considerations and implications); gradual construction of the solution (both as concrete and abstract constructs, including recruiting the required resources, for example, knowledge, expertise, materials, tools, and time); and debugging, while exercising continuous control (evaluation) over the process.

Design is obviously about *making*, but moreover it is about *generating knowledge on how to make*, how to solve a problem, how to improve ways to solve it, and how to transform the problem solving procedures themselves into general intellectual tools. Much controversy exists as to whether problem solving skills can be viewed as general purpose constructs applicable across areas or domains (e.g., McCormick, 2004). Mostly, the question is focused on a particular skill acquired in one domain and applied (successfully or unsuccessfully) in another domain. The perspective presented here regarding generalisable tools has a different focus: the *process* of their development, and consolidation as a result of recurrent situated experiences. Students at an early age are capable of identifying commonalities, regularities, and patterns not only in observed processes and phenomena but in the way they deal with these (e.g., Levy & Mioduser, 2008). I will suggest later on that appropriate pedagogies addressing these metacognitions in every situated experience might support the gradual construction of generalisable tools and the gradual widening of their application space.

Learning TPS

Technological problem solving knowledge and skills have become primary components of every curricular endeavour in technology education in the last decades (e.g., DES, 1990; New Mabat, 2008; Nuffield, 1995; Science & Technology Education Syllabus, 1996). However, in contrast with the malleable and many-sided process depicted in the research regarding real-life problem solving, most curricular versions of this process have resulted in overly structured and rigid didactic models (Hill, 1998; Mioduser, 1998; Mioduser & Dagan, 2007). These are mostly linear models comprising a variable number of phases, all having in common at least four stages: problem identification, exploration of alternative solutions, implementation of the chosen solution, and its evaluation (e.g., Hutchinson & Karsnitz, 1994; Johnsey, 1995). In some cases feedback paths among stages are suggested. However, due to implementation constraints (e.g., time, logistics, clarity of teaching goals, teachers' linear perceptions of the process), these cyclical models rarely transcend the textbook to be actually implemented in the classroom. Moreover, most curricular solutions seem to ignore current pedagogical approaches that encourage student-controlled knowledge-construction processes within unstructured and resourceful learning environments (Fleer, 2000; Hennessy & Murphy, 1999; Hill, 1998; Resnick & Ocko, 1990). This gap between the highly creative and flexible nature of the design process of technological solutions as characterised in the research literature, and the highly structured and prescriptive models of this process as incorporated in teaching materials, is an evident source of learning and motivation difficulties in design activities (Baynes, 1992; de Vries, 1997; Hill & Anning, 2001).

Recent research results indicate that students' problem solving resembles the way expert designers or technology practitioners approach the process (Hill & Anning, 2001). Students proceed in a variety of solution paths (e.g., they may begin with exploration of materials, then building; or begin building, then evaluating the design and reworking it; or begin planning and drawing, then building and evaluating the results) and approach in different ways the lack of correspondence between actual and expected results (Fleer, 2000). Students seem also to engage in a constant reformulation of intermediate goals while developing a technological solution. A great deal of tacit knowledge, which can be observed in students' actions but cannot be easily verbalised or formalised by them, seems to develop during the work (Mioduser & Kipperman, 2002). And the whole process, rather than being linear and ordered, appears to be iterative and cyclical, 'evaluation' being a driving factor for the initiation or ending of design phases.

In light of the above research results, there is clear need to devise pedagogical solutions addressing the cognitive/epistemological needs of learners that foster the gradual construction and formalisation of design knowledge and skills as powerful intellectual tools.

ON TEACHING/LEARNING TECHNOLOGICAL PROBLEM SOLVING

Claim 1: Teaching/Learning TPS Should Be Integral to the Early Stages of Children's Intellectual Development

We grow and live in technology-saturated environments. From the beginning of our lives we interact with a wide range of artefacts, we consume products synthesised in artificial processes, we live immersed in the 'second order' environment we have created – the artificial world. As a consequence, we face the need to construct gradually the appropriate cognitive and metacognitive resources required for interacting with, and even contributing to the making of, the artificial world. Children, as a result of the intense interaction with complex-artefact- saturated environments (e.g., sophisticated toys, microprocessor-embedded domestic devices, remote-controlled music or TV sets, automatic doors and elevators, and many other 'behaving' devices) develop these resources intuitively over the years. I firmly believe that we should build on these intuitive constructs, supporting their further refinement and the construction of new ones, from the very early stages of children's intellectual development.

Fostering this goal has been the *Designing Minds$^{(dm)}$* model for supporting young children's technological thinking and learning of TPS. The model is built upon the confluence of three main conceptual corpuses. The first, introduced above, is the perception of *technology* in both its phylogenetic and ontogenetic nature as means for developing powerful intellectual tools.

The second conceptual component behind the model is the perception of *learning* as a complex process amalgamating several approaches: constructivism, social-constructivism, and constructionism (Kafai & Resnick, 1996; Papert, 1991; Vygotzky, 1978); a situated perspective (Collins, Brown, & Newman, 1989); and learning-by-design (and its close pedagogical associates, problem- and project-based learning and case-based learning, see Kolodner, Camp, Crismond, Fasse, Gray, Hollbrook, et al., 2003; Williams & Williams, 1997). As a whole, we rely on the robust body of theoretical and practical knowledge that locates the learner (individuals and groups) at the centre of the learning process, in the dual role of main engine and main beneficiary of this process.

The third component relates to the perception of children's age level as developmentally appropriate for promoting the construction of key technological-thinking schemas and skills. The literature concerning developmental changes throughout the ages 5–7 pinpoint major changes taking place at the cognitive, neuropsychological, emotional, and social levels, for example, changes in attentional capacity, working memory, executive functions, and concepts acquisition (Case, 1992; Rueda, Fan, McCandliss, Halparin, Gruber, Lercari, et al., 2004; Siegler, 1996); neuroanatomical or organic changes in the frontal and prefrontal areas of the brain, to which high cognitive functions are attributed (Anderson, Northam, Hendy & Wrennall, 2001; Fuster, 2001; Kanemura, Aihara, Aoki, Araki, & Nakazawa, 2003); changes in moral development, in self's structure and organisation, in understanding the balance of positive and negative attributes (Case, 1992; Harter, 1996); and obviously the beginning of formal schooling. All these changes are indicative of a crucial transition stage at which conceptions, schemas, concepts, attitudes, and intellectual tools evolve and become consolidated, forming the child's cognitive/affective stance towards the world – including the designed world. This transitional stage represents a fertile cognitive and emotional space for the development of the intellectual tools compromised in the creation of technological solutions.

The *Designing Minds*$^{(dm)}$ model has been implemented and studied for the last three years in regular kindergartens in Israel. A brief account of the model components and examples of its actual implementation are presented in the following sections. A wide range of tasks are carried out daily in the kindergarten in five main strands (S1 to S5), either by individuals, by small groups, or as plenary activities.

S1 – The designed/artificial world. In this strand, a twofold purpose is pursued, fostering children's acquaintance with (a) the designed world (its components, functioning, characteristics), and (b) the thinking, knowledge, and considerations involved in its design.

At the *content level* the children deal, in every task, with questions relating to an artefact's 'identity' (e.g., What is it? What is it for? What name should we give it? Does it work 'by itself' or do we need to operate it?) as well as its structural and causal configuration (e.g., What parts is it made of? How do the parts work together?). Dealing with these questions is frequently accompanied by classification activities, aiming towards the formation of categories on the basis of structural or functional properties.

At the *processes level* the children are actively involved in diverse forms of interaction with artefacts. The artefacts are physically handled, taken apart both manually (analysing causal interrelations) and conceptually (e.g., constructing a simple functional map). Information about artefacts is searched for, and alternative solutions are discussed, and often built.

Examples of activities carried out in one of the kindergartens in our study (Mioduser, Levy, Kuperman, & Rudolf, 2008) in an ongoing fashion are:

– 'The artefacts corner', for which the children are asked to bring in artefacts (discarded at home) for the kindergarten's collection (see Figure 1). They are asked to elaborate at home (with a family member) on the characteristics of the artefact (e.g., purpose, materials, functioning), and bring in a written summary of their ideas to share with the group during the plenary discussion. On occasions, the discussion focuses on the 'artefact of the week'.
– 'Exploring the artificial world' are tasks aimed to support the evolving perception of artefacts in relation to the context for their invention and the people who have built and use them. Figure 2 relates to a collaborative analysis performed with a seemingly simple artefact – a clothespin.

Several questions were explored, including: What is it made from? What are all those little 'things' in its shape? What is a spring? What is it for? Can we think of other ways to attach the clothes to the clothesline? Figure 3 shows the results of a classification activity guided by the question "Does it work 'by itself' or do we need to operate it?" (which introduced later discussions and activities on 'smart artefacts', see S5 below). So far, the electric kettle that "knows when the water is boiling and stops by itself" stands alone in its category.

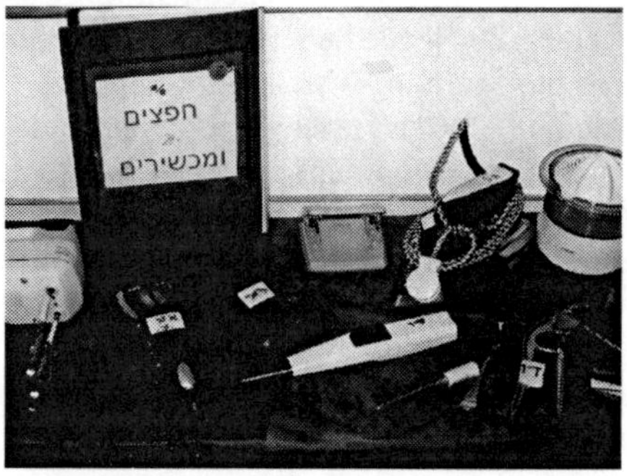

Figure 1. The artefacts corner.

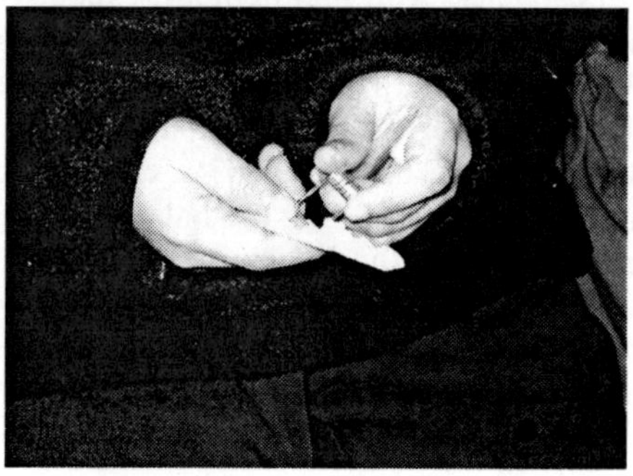

Figure 2. Analysing a clothespin.

Figure 3. A classifying activity.

S2 – Problem solving. While working on this strand's activities, mostly emerging spontaneously in relevant situations of the kindergarten's daily life, a problem solving culture is gradually co-constructed by children and teachers. The trigger for the process is usually a declaration – by a staff member or by one of the children – that "we do have a problem". Iconic signs are used to portray a scheme of the process (the troubled face sign in Figure 4a is used to raise attention and start a discussion). After generating and representing (e.g., drawing) several alternative solutions and discussing their pros and cons, the children cast a vote, selecting the solution they choose to pursue (Figure 4b).

Although the situations in which children are involved in problem solving processes relate at first to the artificial realm, very rapidly the approach and use of strategies propagate into other areas of the kindergarten's life (e.g., conflicts among children, dilemmas related to contents being learned), thus becoming a more general tool. In all cases the discussions are supported by the construction of a representational chart of the problem solving process (by itself an artificial/invented construct).

S3 – Design ('the Bauhaus kids'). Design is the heart of technology. As a highly structured or mindful process, as a completely intuitive and unplanned/on-the-fly process, or in any combination of these two modalities, we design every time we attempt to close a perceived gap between an existing and a desired state-of-affairs by means of an invented solution.

Working in this strand, the children get involved in a long journey, which starts with very basic experiences with materials and building kits and intuitive attempts to construct objects and devices. Along the journey, several activities support the construction of a cognitive toolkit for design: reflection on what is being decided and done, verbalisation and formalisation of procedures, gathering relevant information (e.g., on materials, procedures, existing or past solutions), using representations and notations for either documentation or planning (linkage to S4, see below), and addressing moral and ethical questions.

Figure 4a. A problem: How can we fix the juice maker?

Figure 4b. Alternative solutions and children's voting.

The way the learning proceeds can be depicted as an evolving journey (inspired in a way by the Bauhaus' conceptual approach as embedded in its syllabus (Itten, 1964): starting from a preliminary acquaintance with materials, tools and processes and the very idea of constructing, via systematic work with specific building kits and design situations, up to the integration of knowledge and abilities in individual, group and whole kindergarten design projects. Expanding the children's intuitive and creative encounter with the rich repertoire of building kits that normally populate any kindergarten, the children are requested to accomplish various exploration and conceptualisation tasks, for example, composing an ID card for the building parts or blocks, naming them, hypothesising about their functions, classifying them into meaningful groups, and exploring possible assemblages of parts forming functional compounds (see Figure 5).

In addition, drawing is used as reflection and thinking aid before, while, and after designing and building. First in the progression is drawing-after-building, for the purposes of documentation and communication (Figure 6a). In this modality, drawing promotes (and even demands) reflection on the object constructed, for example, on structural aspects, on mechanisms (e.g., types of transmissions), or representational issues (e.g., scale, 3D to 2D translation, level of detail).

Gradually, drawing is introduced for planning purposes, fostering anticipatory thinking, envisioning, and concrete representation of the structure and mechanisms required to materialise the object (Figure 6b).

Figure 5a. ID of part of a magnetic building kit.

Figure 5b. Classification and naming activity.

Figure 5c. Investigating functional compounds.

S4 – Notations (from conventional signs to computer programs). Symbolic behaviour is a defining characteristic of humans, and the invention of notational systems one of the essential achievements of the human mind (Barton & Hamilton, 1999). The generation and manipulation of representations have been part of human intellectual and cultural development since forever, taking countless forms and fulfilling diverse functional goals:

> From buying and selling to the most holy rituals, most social acts involve the creation and interpretation of external representations. Besides communicative referential tools, external representations are also epistemic tools, *objects to think with.* (Tolchinsky, 2007, p. 2)

Notational systems constitute a unique strand in our model for two reasons. First, because of its centrality as an intellectual tool serving all layers of technological thinking and making – thoughts get objectified, allowing cognitive operations on them, for example, reflection, analysis, decomposition, comparison, grouping, arguing for or against. Second, by being involved in the construction of notations (e.g., defining conventions, selecting representational units and symbolic entities, establishing representational rules), children deal with both the represented content and the representing means – the 'epistemic tools' – and with the very process of inventing conventional representation and communication systems. In addition, the work within this strand naturally blends with the kindergarten's curricular requirements in the areas of literacy and numeracy.

Figure 6a. Drawing as documenting.

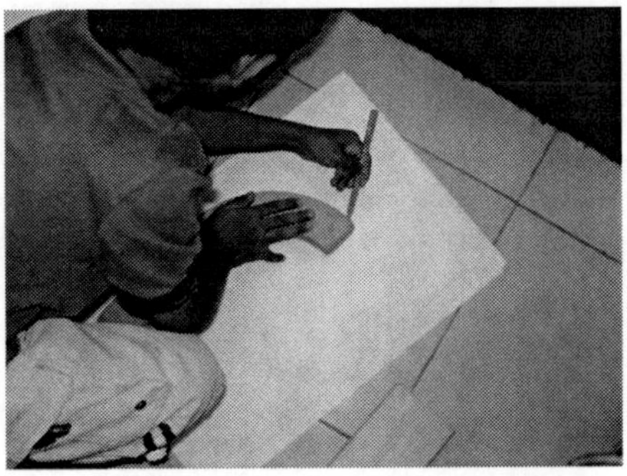

Figure 6b. Drawing as planning.

All activities are built as progressions from concrete involvement in situations in which action and representation are intertwined, via different stages of conventionalisation and symbolic translation, to work on formal representations (Figure 7). One example is the devising of conventions for "telling someone how to navigate a labyrinth" (Figure 8). The progression starts as actual navigation in a path, while trying to formalise the description of the required set of actions (e.g., two steps forward, right turn, one step forward, etc.). Next will be the generation of a description for navigating the labyrinth – and its formal representation – to be communicated to another 'traveller'. At the end of this progression stand the programming activities by which the children construct the program required for a robot to navigate the path (linkage with strand 5).

Figure 7. From enacted path building, to 'floor representations', to notational descriptions using symbols and numbers.

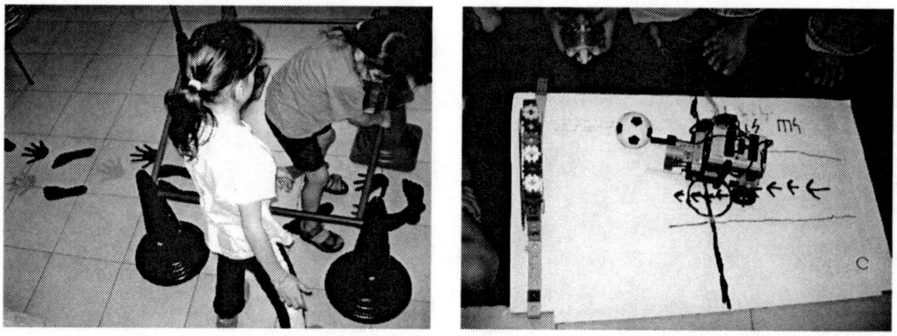

Figure 8. From navigating the labyrinth to programming the robot.

S5 – Understanding and constructing artificial-behaviour (from playing, analysing, and formalising enacted behaviours to the design of artificial adaptive behaviours in robots). Today, controlled systems (e.g., automatic doors, domestic devices, programmable toys, communication and computational devices, robotic systems) have become central to everyday life. With the widespread development of microprocessor-based devices, a new category of artefacts infiltrated the traditional and intuitive distinctions between the alive and not-alive, animate and inanimate, human-operated and autonomous – the realm of 'behaving artefacts'. While pertaining to the artificial world, these devices are conceived as endowed with traits such as purposeful functioning capabilities (behaviour), programmability and knowledge accumulation capabilities (learning), autonomous decision-making, and adaptive behaviour. The interaction with this controlled-artefact-saturated environment (either as a mere user, trouble-shooter, or generator of controlled processes at home, work, or in learning settings) demands a new cognitive approach, including the development of new conceptual schemes, a different mental modelling of devices (their structural and functional configurations, as well as overall behaviour patterns), and a new set of skills associated with the design and implementation of artefacts' behaviour.

In this strand's activities, children are engaged in discussions on the nature of devices "which make decisions" (e.g., an electric kettle or an automatic door), on the physical resources required to construct behaving artefacts (e.g., for these to be able to move or to sense), and on the logical constructs governing their behaviour (e.g., routines, If…Then… rules). Gradually, they enter the process of constructing a robot's behaviour by means of a specially developed programming interface (Mioduser, Levy, & Talis, 2009). In the programming activities, the children are engaged in recursive cycles of construction/reflexion, concrete-experimentation/abstract-symbolisation, and analysis/synthesis, while solving a progression of tasks of increasing complexity.

Figure 9 shows the children's attempts to define If … Then … rules for a situation under discussion. They do this using the If …Then … cards (in Hebrew in the picture), and completing on their own the condition and action part of the rule by drawing and/or annotating the additional cards. Normally, in everyday life, they perform this exercise not only when working on the robot tasks, but eventually whenever an interesting situation arises and children or teachers think that its reformulation in terms of rules may help in the discussion.

Figure 9. From definition of If... Then... behaviour units (rules), to programming the robot.

Coda. Three years of implementation of the model have been enlightening, on a wide range of learning issues emerging from the encounter between children and the artificial. Examples are: perceptions and concept development concerning the artefacts' worlds; situated and transferred problem solving; the development of representational and notational abilities; recursive cycles between the concrete and the abstract while constructing complex artificial behaviours; and the generalisation of intellectual constructs consolidated in technological tasks for facing other subjects and even social conflicts and situations.

Claim 2: Teaching/Learning TPS Should Build on Students' Evolving Cognitive Models of the Design Process

Curricular solutions for teaching TPS reflect the conflicting perceptions of design as a highly creative, branching, and cyclical process; and at the same time as a process that has to proceed in stages, to follow a well defined and logical path, to meet schedules. Overall, most curricular proposals follow two main approaches: (a) the structural (stage-by-stage) approach, or (b) the functional approach.

The structural approach emphasises the need for an ordered learning of the *stages* of the design process (Hutchinson & Karsnitz, 1994; Waetjen, 1989). Different models have been developed all over the world for teaching design as an organised and systematic process (e.g., in UK: DES, 1990; in the US: ITEA, 2000; in Australia: Kimbell, 1997; in Israel: Science and Technology Education Syllabus, 1996).

The functional approach emphasises the teaching and study of design *functions* (rather than stages), for example, the identification and definition of problems or gaps to be bridged, exploration and investigation, decision making, planning, making, and evaluation. At every stage of the process the problem solver may use more than one of the design functions (e.g., investigation and evaluation) depending on the specific context and requirements of the particular stage. For example, the *investigation* function will take a different form and goals depending on whether it is applied for defining the design goal or for scanning alternative solution paths. In this approach, the process of problem solving is flexible and cyclical. Correspondingly, the instructional plan is based on the teaching of the different design functions (Chidgey, 1994; Lawson, 1997; McCormick et al., 1994; Mioduser, 1998) so that the students use them in the way that best matches the problem, the situation, and their own learning and working styles.

The structural approach is the one more commonly implemented in curricular materials, and many studies have focused on it. These studies' results raise doubts about students' ability to achieve a holistic view of the process by this instructional approach (de Vries, 1997; Hennesy & McCormick, 1994; Johnson, 1994; Jones, 2002; Mawson, 2003; McCormick, 1997). In contrast, only very few attempts have been made to develop instructional materials for the functional approach (Johnsey, 1995; Mawson, 2003) or study their contribution to TPS learning (Hill, 1998; Mawson, 2003).

In a recent study (Mioduser & Dagan, 2007) we tested the claim that learning by the functional approach towards design fits the students' cognitive model of the process and brings about a more efficient construction of design knowledge and skills. Two groups of students were taught by one of the approaches (structural or functional) and worked on a wide range of design projects. In the functional approach, the teachers supplied a 'functional tool-kit' and the overall goals and guidelines of a process expected to evolve dynamically. This modality of work demanded a continuous mental switching between the systemic level (the perception of the process as a whole) and the specific level (the particular design activity at any given stage), much like experts' cognitive modelling of the design process (DeKleer & Brown, 1981; Hegarty, 1991; Norman, 1983). In contrast, in the structural approach, the students were taught the stage-wise model, ignoring any alternative processes they might possibly have followed (Chidgey, 1994; Hennessy & McCormick, 1994).

During the learning process and at the end of it, significant differences were observed between the students' mental models in both groups and in all sets of variables examined (i.e. type of cognitive model of design, changes in type of model along the learning/design process, and qualitative characteristics of the models). Specifically, we found that students in the functional group developed mental models that resembled those of expert designers in characteristics such as flexibility, change in the course of the process, and recurrent use of design functions (Bucciarelli, 1996; Cross, 2002; Lawson, 1997). In each and every lesson, students had to develop their own image or scenario of the particular stage they were at, which obliged them to revise and reconstruct their models of the whole process. The significant and continuous changes observed in the group's models in the course of the learning process were indicative of their ability to adapt these in light of incoming data. These features were far less frequent among the structural group students, who constructed to a large extent partial, localised, and inflexible mental models of the design process.

Investigation and *exploration* functions were among the most frequent functions used by students in the functional group. This clearly indicates that these students became aware of the importance of information gathering and analysis in the various stages of the design process. For example, the *exploration* function was instantiated for identifying the issue or problem to be solved; for defining requirements and constraints for the solution; for reviewing previous solutions for similar problems; or for comparing alternative materials or procedures. In contrast, the scarce use of the *investigation* and *exploration* functions by the structural group students is indicative of their perception of information gathering as a specific and ad-hoc stage in the process (i.e. search for alternative solutions). In most other stages, an almost exclusive use of intuitive knowledge by these students was observed.

Overall, we might conclude that the functional approach towards design instruction was more effective than the traditional structural approach for supporting the construction of holistic, flexible, and effective mental models for designing technological solutions. By demanding from the students the active revision of their models at each and every stage and the making of decisions about the next step to be taken, learning became a highly constructivist process. In addition, the evident overlap between the characteristics of the functional approach and the students' cognitive modelling of the design process contributed to the quality of their designs (products) as well as of the concepts and skills acquired and implemented during their work.

CONCLUDING REMARKS

In this chapter I have elaborated on the need to teach and learn technological problem solving from a cognitive/epistemological perspective. By this perspective, the designed (outer) world and the designer's (inner) world are co-constructed and evolve as a result of the recurrent and mutually-affecting interaction between them.

Learning technology in this way is learning about the wide range of mind capabilities involved in inventing the artificial world – knowledge, skills, methods and strategies, epistemic perspectives, beliefs, and moral judgements. Learning about technological solutions in context (e.g., historical, social, cultural, moral, scientific and technological, economic) is learning about the evolution of ideas and their adaptation to fit affordances and constraints. And learning about TPS methods and processes is learning about the functional knowledge and skills serving as specific technology-related tools as well as general intellectual tools.

This perspective has led to the implementation of the above ideas in different learning settings and with different age levels. A sample of the focal ideas pursued in the work was briefly surveyed in this chapter: the belief that TPS should be integrated at the very early stages of schooling and intellectual development, that TPS learning should build on learners' conceptions and cognitive modelling of the solution-generation process as the basis for further formalisation of methods and strategies, and that active cognitive involvement in doing TPS is the *sine-qua-non* of design knowledge and skills learning.

The world of young children's learning and doing TPS is highly intriguing, and not well explored. This world is currently the main focus of work within the research group with which I am associated, and we have already begun to navigate its fascinating alleys. These studies will give us the opportunity to understand better the cognitive foundations of young children's technological thinking and their performance within (and perhaps beyond?) developmental affordances and constraints *vis-à-vis* this unexplored domain, and identify the cognitive platform upon which further development of technology-related skills should rely.

REFERENCES

Anderson, V., Northam, E., Hendy, J., & Wrennall, J. (2001). *Developmental neuropsychology: A clinical approach.* Philadelphia: Psychology Press.

Barton, D., & Hamilton, M. (1999). Social and cognitive factors in the historical elaboration of writing. In A. Lock & C. Peters (Eds.), *Handbook of human symbolic evolution* (pp. 793–858). Malden, MA: Blackwell.

Baynes, K. (1992). *Children designing—Learning design.* Occasional Paper No. 1. Loughborough University.

Bucciarelli, L. (1996). *Designing engineers.* Cambridge, MA: MIT Press.

Case, R. (1992). The role of frontal lobes in the regulation of cognitive development. *Brain and Cognition, 20*, 57–63.

Chen, D., & Strupp, W. (1993). General systems theory: Towards a conceptual framework for science and technology education for all. *Journal of Science Education and Technology, 2*(3), 447–459.

Chidgey, J. (1994). A critique of the design process. In F. Banks (Ed.), *Teaching technology* (pp. 89–93). London: Routledge.

Cole, M., & Derry, J. (2005). We have met technology and it is us. In R. Sternberg & D. Preiss (Eds.), *Intelligence and technology* (pp. 209–227). Mahwah, NJ: Erlbaum.

Collins, A., Brown, J. S., & Newman, S. E. (1989). Cognitive apprenticeship: Teaching the crafts of reading, writing, and mathematics. In L. B. Resnick (Ed.), *Knowing, learning, and instruction: Essays in honor of Robert Glaser* (pp. 453–494). Hillsdale, NJ: Lawrence Erlbaum.

Cross, N. (2002). The nature and nurture of design ability. In G., Owen-Jackson (Ed.), *Teaching design and technology in secondary schools* (pp. 124–139). London: The Open University.

de Vries, M. J. (1997). Science, technology and society: A methodological perspective. *International Journal of Technology and Design Education, 7*(1–2), 21–31.

DeKleer, J., & Brown, J. S. (1981). Mental models of physical mechanisms and their acquisition. In J. R. Anderson (Ed.), *Cognitive skills and their acquisition* (pp. 285–310). Hillsdale, NJ: Lawrence Erlbaum.

DES. (1990). *Technology in the national curriculum*. London: Department of Education and Science and the Welsh Office, HMSO.

Dym, C., Agogino, A., Eris, O., Frey, D., & Leifer, L. (2005). Engineering design thinking, teaching and learning. *Journal of Engineering Education, 94*(1), 103–120.

Dyrenfurth, M. J. (1991). Technology literacy synthesized. In M. J. Dyrenfurth & M. Kozak (Eds.), *Technological literacy, 40th Yearbook of the Council on Technology Teacher Education* (pp. 138–183). Peoria, IL: Glencoe Division, Macmillan/McGraw-Hill.

Fanta, Y. (2007). *Technological thinking among illiterate Ethiopian immigrants in Israel*. Unpublished doctoral dissertation, School of Education, Tel Aviv University.

Fleer, M. (2000). Working technologically: Investigations into how young children design and make during technology education. *International Journal of Technology and Design Education, 10*(1), 43–59.

Fuster, J. M. (2001). The prefrontal cortex – an update: Time is of the essence. *Neuron, 30*, 319–333.

Harter, S. (1996). Developmental changes in self-understanding across the 5 to 7 shift. In A. Sammeroff & M. Haith (Eds.), *The five to seven year shift* (pp. 207–236). Chicago, IL: The University of Chicago Press.

Hegarty, M. (1991). Knowledge and processes in mechnical problem solving. In R. J. Sternberg & P. A. Frensch (Eds.), *Complex problem solving* (pp. 253–285). Hillsdale, NJ: Lawrence Erlbaum.

Hennessy, S., & McCormick, R. (1994). The general problem-solving process in technology education: Myth or reality? In F. Banks (Ed.), *Teaching technology* (pp. 94–108). London: Routledge.

Hennesy, S., & Murphy, P. (1999). The potential for collaborative problem solving in design and technology. *International Journal of Technology and Design Education, 9*(1), 1–36.

Hill, A. (1998). Problem solving in real-life contexts: An alternative for design in technology education. *International Journal of Technology and Design Education, 8*, 203–220.

Hill, A., & Anning, A. (2001). Primary teachers' and students' understanding of school situated design in Canada and England. *Research in Science Education, 31*, 117–135.

Hutchinson, J., & Karsnitz, J. (1994). *Design and problem solving in technology*. Albany, NY: Delmar.

Ihde, D. (1990). *Technology and the lifeworld—from garden to earth*. Bloomington, IN: Indiana University Press.

Ihde, D. (1993). *Philosophy of technology: An introduction*. New York: Paragon House.

ITEA. (2000). *Standards for technological literacy—Content for the study of technology*. Reston, VA: Author.

Itten, J. (1964). *Design and form—The basic course at the Bauhaus*. New York: Reinhold.

Johnsey, R. (1995). The design process—Does it exist? A critical review of published models for the design process in England and Wales. *International Journal of Technology and Design Education, 5*(3), 199–217.

Johnson, S. D. (1994). Research on problem solving instruction: What works, what doesn't? *The Technology Teacher, May/June*, 27–29.

Jones, A. (2002). Research in learning technological concepts and processes. In G. Owen-Jackson, (Ed.), *Teaching design and technology in secondary schools* (pp. 79–92). London: The Open University.

Kafai, Y., & Resnick, M. (Eds.). (1996). *Constructionism in practice: Designing, thinking and learning in a digital world*. Mahwah, NJ: Lawrence Erlbaum.

Kanemura, H., Aihara, M., Aoki, S., Araki, T., & Nakazawa, S. (2003). Development of the prefrontal lobe in infants and children: A three-dimensional magnetic resonance volumetric study. *Brain & Development, 25*, 195–199.

Kimbell, R. (1997). *Assessing technology*. Buckingham, UK: Open University Press.

Kirlik, A. (2005). Work in progress: Reinventing intelligence for an invented world. In R. Sternberg & D. Preiss (Eds.), *Intelligence and technology* (pp. 105–134). Mahwah, NJ: Erlbaum.

Kolodner, J., Camp, P., Crismond, D., Fasse, B., Gray, J., Hollbrook, J., et al. (2003). Problem-based learning meets case-based reasoning in the middle-school science classroom: Putting Learning by Design™ into practice. *The Journal of the Learning Sciences, 12*(4), 495–547.

Lawson, B. (1997). *How designers think – The design process demystified* (2nd ed.). London: The Architectural Press.

Levy, S. T., & Mioduser, D. (2008). Does it "want" or "was it programmed to ..."? Kindergarten children's explanations of an autonomous robot's adaptive functioning. *International Journal of Technology and Design Education, 18*(4), 337–359.

Mawson, B. (2003). Beyond "the design process": An alternative pedagogy for technology education. *International Journal of Technology and Design Education, 13*(2), 117–128.

McCormick, R. (1997). Conceptual and procedural knowledge. *International Journal of Technology and Design Education, 7*(1–2), 141–159.

McCormick, R. (2004). Issues of learning and knowledge in technology education. *International Journal of Technology and Design Education, 14*(1), 21–44.

McCormick, R., Murphy, P., & Henessy, S. (1994). Problem-solving process in technology education: A pilot study. *International Journal of Technology and Design Education, 4*(1), 5–34.

Medawar, P., & Medawar, J. (1977). *The life science*. London: Wildwood House.

Mioduser, D. (1998). Framework for the study of the cognitive nature and architecture of technological problem solving. *International Journal of Technology and Design Education, 8*(2), 167–184.

Mioduser, D., & Dagan, O. (2007). The effect of alternative approaches to design instruction (structural or functional) on students' mental models of technological design processes. *International Journal of Technology and Design Education, 17*(2), 135–148.

Mioduser, D., & Kipperman, D. (2002). Evaluation/modification cycles in junior high students' technological problem solving. *International Journal of Technology and Design Education, 12*(2), 123–138.

Mioduser, D., Levy, S. T., & Talis, V. (2009). Episodes to scripts to rules: Concrete-abstractions in kindergarten children's explanations of a robot's behavior. *International Journal of Technology and Design Education, 19*(1), 15–36.

Mioduser, D., Levy, S. T., Kuperman, A., & Rudolf, L. (2008). Design and learning in the kindergarten. In D. Kiperman, O. Dagan, & M. de Vries (Eds.), *PATT 20 proceedings* (pp. 262–274). Tel Aviv: Ort Israel.

Mioduser, D., Venezky, R. L., & Gong, B. (1996). Student's perceptions and designs of simple control systems. *Computers in Human Behavior, 12*(3), 363–388.

Mitcham, C. (1994). *Thinking through technology*. Chicago, IL: University of Chicago Press.

New Mabat. (2008). *Science & technology education series—primary school 1 to 6*. Tel Aviv: Ramot.

Norman, D. A. (1983). Some observations on mental models. In D. Genter & A. L. Stevens (Eds.), *Mental models* (pp. 7–14). Hillsdale, NJ: Lawrence Erlbaum.

Nuffield. (1995). *Nuffield design and technology—students book*. London: Longman.

Nye, D. (2006). *Technology matters—questions to live with*. Cambridge, MA: MIT Press.

Ortega y Gasset, J. (1941). *Toward a philosophy of history*. New York: W.W. Norton.

Pacey, A. (1999). *Meaning in technology*. Cambridge, MA: MIT Press.

Papert, S. (1991). Situating constructionism. In I. Harel & S. Papert. (Eds.), *Constructionism* (pp. 1–12). Norwood, NJ: Ablex.

Pearson, G., & Young, T. (Eds.). (2002). *Technically speaking*. Washington DC: Academy Press.

Piaget, J. (1973). *To understand is to invent: The future of education*. New York: Grossman.

Resnik, M., & Ocko, S. (1990). *Constructionist learning. LEGO/LOGO: Learning through and about design*. Boston, MA: MIT Media Laboratory.

Rowe, P. (1987). *Design thinking*. Cambridge, MA: MIT Press.

Rueda, M. R., Fan, J., McCandliss, B. D., Halparin, J. D., Gruber, D. B., Lercari, L. P., & Posner, M. I. (2004). Developmenat of attentional networks in childhood. *Neuropsychologia, 42*, 1029–1040.

Science & Technology Education Syllabus. (1996). Jerusalem: Ministry of Education.

Siegler, R. S. (1996). Unidimensional thinking, multidimensional thinking and characteristics tendencies of thought. In A. Sammeroff & M. Haith (Eds.), *The five to seven year shift* (pp. 63–84). Chicago, IL: The University of Chicago Press.

Simon, H. (1981). *The sciences of the artificial*. Cambridge, MA: MIT Press.

Sternberg, R., & Preiss, D. (2005). *Intelligence and technology*. Mahwah, NJ: Erlbaum.

Tolchinsky, L. (2007). The multiple functions of external representations: Introduction. In E. Teuval, J. Dockrell, & L. Tolchinsky (Eds.), *Notational knowledge, historical and developmental perspectives* (pp. 1–10). Rotterdam, The Netherlands: Sense.

Vincenti, W. (1990). *What engineers know and how they know it*. Baltimore: John Hopkins University Press.

Vygotsky, L. S. (1978). *Mind in society: The development of higher psychological processes*. Cambridge, MA: Harvard University Press.

Waetjen, W. (1989). *Technological problem solving: A proposal*. Reston, VA: International Technology Education Association.

Waks, L. (2001). Donald Schon's philosophy of design and design education. *International Journal of Technology and Design Education, 11*(1), 37–51.

Williams, A., & Williams, J. (1997). Problem-based learning: An appropriate methodology for technology education. *Research in Science & Technological Education, 15*(1), 91–103.

Wu, T., Custer, R., & Dyrenfurth, M. (1996). Technological and personal problem solving styles: Is there a difference? *Journal of Technology Education, 7*(2), 55–71.

David Mioduser
School of Education
Tel Aviv University
Israel

ALISTER JONES

34. TOWARDS AN ARTICULATION OF STUDENTS MAKING PROGRESS IN LEARNING TECHNOLOGICAL CONCEPTS AND PROCESSES

INTRODUCTION

Technology education is concerned with developing students' technological literacy through exploring and solving complex and interrelated technological problems that involve multiple conceptual, procedural, societal, and technical variables. The researching of learning in technology therefore needs to be multi-dimensional. As research in this area develops, students' learning of technological concepts and processes is becoming more clearly articulated.

Consideration of the nature of technology indicates that technological knowledge and processes are socially constructed and context dependent and where human mental processes are situated within their historical, cultural, and institutional setting (Wertsch, 1991). (The nature of technology and its historical and philosophical roots are discussed in Section 2 of this *Handbook.*) Technology therefore is essentially an activity that involves not just the social context, but also the physical context, with thinking being associated with and structured by the objects and tools of action. This means research in student learning has to consider not just the conceptual and procedural understanding alone but also the way technological tools and objects influence and interact with student thinking and doing. This chapter will discuss the importance of thinking about student and teacher concepts of technology and technological practice, and ways to think about assisting students to make progress in learning technology.

STUDENT AND TEACHER CONCEPTS

Existing concepts that a learner holds provide a framework from which other concepts may be constructed, as well as determine courses of action. This means that existing concepts that students may have will have a direct impact on their learning in technological education. Not only is it students' existing concepts of technological knowledge and practices but also their concepts of technology, and specifically the nature of technology, which may influence their learning and action in technology.

Concepts may also be broadly taken to include dispositions that include inclinations, sensitivity, and abilities (Perkins, Jay, & Tishman, 1993). These terms are defined in the following way: "*Inclination* refers to the person's felt tendency toward behaviour *X* ... *sensitivity* refers to the person being alert to *X* occasions … *Ability* refers to the actual ability to follow through with the *X* behaviour" (p. 4). McCormick (1997) indicates that conceptual knowledge is not just a series of isolated items of knowledge, but the relationships amongst that knowledge. Procedural knowledge is defined in terms of 'how to' knowledge. As McCormick highlights, in technology conceptual and procedural knowledge are strongly linked. For example, in terms of device knowledge, "as the

A.T. Jones and M.J. de Vries (eds.), International Handbook of Research and Development in Technology Education, 407–417.
© 2009 Sense Publishers. All rights reserved.

complexity of devices increases so does the importance of interaction of device knowledge and procedural knowledge" (p. 149).

The wealth of research data from science education on students' existing concepts indicates that conceptual knowledge significantly affects students' predictions, explanations, and perceptions of phenomena and problems. Students' concepts of technology, and what it means to be learning in a technology classroom, will influence student learning of technological concepts and processes. Students' concepts of technology or concepts of the nature of technology range from seeing technology as being just 'hi-tech' devices through to broader notions of technology to be the wider made environment.

Not only have students' concepts of technology been found to influence their actions in technology; so has the nature of the classroom environment. Student concepts of what it means to be a learner, teachers' concepts of technology and technology education, as well as curriculum directions, will also all influence student learning in technology. This means, therefore, that students' and teachers' concepts of technology, including concepts of technological knowledge and processes, will impact on the way in which technological practice is undertaken in the classroom. Conversely, students' technological practice and limited concepts of technology can affect the teachers' own perceptions of technology education, especially where the teacher's concept of technology is somewhat fragile in nature. It is therefore crucial to explore teachers' and students' concepts of technology to understand the way these might influence the learning of technological concepts and processes.

Research findings (Jones, 1997) show strong links between students' concepts of technology and their technological practice, which includes the learning and use of technological concepts and processes. The interaction between concepts and future learning is complex and dependent on a number of factors. For example, where students' concepts of the nature of technology are broad then students are more likely to undertake their technological activities in a holistic fashion that shows links between the various stages in the process. Where students' concepts of technology are narrow, this constrains their technological practice and limits the potential for learning technological concepts and processes. This was exemplified in Jones' research by a group of students who wanted to make their product 'technological' by putting flashing lights on it. Flashing lights were not in any way related to the solution, but the students' view was that a waterproof container could not be technology and therefore additions had to be made to it. The students' existing concepts of technology thus had a greater impact on their technological practice than the associated teaching strategies. If students have limited concepts of the nature of technology, then it can be assumed that this will limit their learning of technological concepts and processes.

For teachers, the teaching and learning of technology is often bound up with their initiation and the socialisation of teachers into particular subject subcultural settings (Goodson, 1985). These subcultures, according to Paechter (1991), represent reasonably consistent views about the role of the teacher, the nature of their subject, the way it should be taught, and expectations of the students' learning. Paechter also points out that the teachers' beliefs about what is important for students to learn in their existing subject areas (such as craft, design and technology, home economics, or art) were transferred to technology education. For example, art teachers may make the technology curriculum into a design course and science teachers into a laboratory course; science teachers may discuss technology in terms of applied science, social studies teachers tend to focus on the effect technology has on society, and technical teachers' view of technology tends to be primarily focused on specific skills and design and making artefacts (Jones & Carr, 1992; Lindblad, 1990). Subject backgrounds thus have a direct influence on the way teachers structure lessons and develop classroom strategies to teach technological concepts and processes, and teachers tend to develop strategies to allow for learning outcomes that were often more closely related to their particular teaching subject than to technology. This can be particularly noticeable when teachers enter areas of uncertainty, causing them to revert to their traditional teaching approaches and learning outcomes. McCormick, Murphy, and Hennessy (1994) also identify the importance of the teachers' concepts of technology in relation to the skills and knowledge that students consider necessary to operationalise.

Student learning technological concepts and processes will also be influenced by approaches supported by particular curricula, both at the national, school, and classroom level. Curricula provide the framework for possible learning and assessment outcomes and this, influencing the ways in which students undertake technological problems. Curricula also influence student learning of technological concepts and processes. In particular, a separation between knowledge and processes causes problems in student and teacher understanding of the interrelationship between processes and concepts.

Learning is often enhanced when students are involved in authentic activities. However, most classroom approaches do not reflect actual technological problem solving, which involves mainly adaptation and modification rather than starting from scratch (Hennessy, 1993). This then limits the potential learning of technological concepts and processes. For example, where students expect just to make a model to present to the rest of the class, the students may not undertake planning and design or take into account appropriate measurements in solving their problems, nor do they see the need to carry out appropriate research. Although they may be introduced to technological principles, their expectations mean these are largely irrelevant. The students' expectations of what an appropriate outcome is in a particular class thus influence the approaches they adopt, and what knowledge and skills they identify and operationalise in a particular activity (Jones, 1997).

Assessment approaches and the definition of the technological task can also affect student technological practice and learning. Emphasis on particular summative assessment procedures in a classroom may mean students omit unsuccessful designs and ideas from portfolios and do not appreciate particular technological principles involved in the development and design of technological solutions. The openness of the activity can also affect student learning of technological concepts and processes. For example, if a task is too closely defined, then this can limit students' approaches and affect the potential to learn and incorporate technological principles. This is often found where students are constrained by a defined design cycle, with resources and outcomes clearly identified. This can also limit innovative approaches that students might adopt, and limit the learning of technological principles, such as modifications and adaptations.

This chapter argues that a framework of technological knowledge and processes needs to be developed, in which both concepts of the nature of technology and technological concepts are explicitly identified and understood. In addition, there needs to be an understanding of the complexity of learning in technology; that the use and application of appropriate knowledge and skills does not necessarily occur in a simplistic fashion. Learning is characterised more as a process of construction through guided participation in social activity. (This is explored by Frank Banks in Chapter 32.) It is therefore essential that strategies are developed that broaden students' concepts of nature of technology, technological concepts, and technological practice.

THINKING ABOUT MAKING PROGRESS IN TECHNOLOGY

Whilst there is published research on what students do when involved in technological activities, there is a need for more research on students' learning in technology and ways in which this learning can be enhanced. In thinking about this we need to construct a notion of progression as a verb rather than a noun – that is, we need to think about students making progress, rather than progression. Recently there has been increased research in examining how students make progress in technology, as well as influences on their progress. For example, Mawson (2007) explored the way in which personal factors and school context possibly influence students' progression in technological literacy. Compton and Harwood (2005) provide illustrative examples of student work against a possible progression matrix that was being trialed with teachers.

In a review of 600 research articles on curriculum, assessment, and pedagogy, Carr, McGee, Jones, McKinley, Bell, Barr, et al. (2000) argued that teacher knowledge of their subject (content knowledge), how students learn (pedagogical content knowledge), and the interaction of these two factors are essential to support learning. Good teacher knowledge of the content of their subject was found to have a positive effect on decisions to change pedagogical strategies to create learning opportunities. In addition, sound content knowledge seems to have a positive effect on curriculum planning, assessment, implementation, and development. For a new curriculum area such as technology, this presents particular challenges for teachers as they search to construct a coherent, technological content base and appropriate assessment practices. (For more on assessment, see Section 6 of this *Handbook*.)

Learning in technology is more than simply the sum of its parts; making progress can not be seen in terms of a linear development of atomistic individual skills or knowledge or bits of procedural quality (Kimbell, 1997). Rather, progression is seen in the constructs of holistic judgements as an encapsulation of the wholeness of student learning. Holistic judgements backed by detail are both more reliable and valid. For example, if you judge it as yes or no, you are operating in a world of certainties and absolutes and there is nothing in between the yes and no. The reality is that no one just can, and no one just cannot. The implications of this view are:

- progression cannot be defined in absolute terms, it only has meaning within the context of an individual learner's development;
- progression is about managing progress and therefore requires an understanding of the learner's current level of technological literacy as well as a clear view of how this might be further developed; and
- to make judgements about progression, teachers need a sound understanding not only of their students' needs but also of the technology involved in the tasks and projects their students undertake (Jones & Moreland, 2003).

Research in progression shows the need to be careful and clear about specifying what is being talked about (Black, Brown, Simon, & Blondel, 1996). For example, progression may be shown in the following ways:

- hierarchical progression through the subject, for example, the order in which a child's learning takes place or the sequential difficulty within a subject;
- developing cognitive capabilities, for example, a child's ability to process increasingly more complex ideas;
- progression in increasing conceptual understandings;
- progression in procedural aspects, including the number of independent variables in a task, the nature of the variables (continuous, discrete, categoric), and measurements needed to judge dependent variables;
- progression as use of knowledge in different contexts;
- progression related to task variables, for example, the number, complexity, sophistication level, order, and discreteness; and
- progression influenced by factors of task specificity or openness.

Progression does not mean that in order to make progress one simply does something extra and different. Often, it is doing the same thing to a higher level of quality. Improvement may amount to doing the same thing but in progressively richer ways. Based on extensive classroom data (Jones &

Moreland, 2003; Moreland, 2003; Moreland, Jones, & Northover, 2001) we define progression in learning technology in the following categories:
- the nature of technology;
- student technological practice;
- generic conceptual, procedural, societal and technical aspects; and
- specific conceptual, procedural, societal and technical aspects.

Understanding the nature of technology includes a broad understanding of technology and the relationships between technology and society. 'Student technological practice' includes the ways in which students bring together the different aspects of the technological task, that is, the ways in which they bring knowledge and skills together to solve a technological problem. Student technological practice can therefore be defined as the operationalisation and integration of conceptual, procedural, societal, and technical variables in technological activity to solve a technological problem. 'Conceptual aspects' include knowledge and understanding of relevant technological concepts and procedures. 'Procedural aspects' include the strategic application of procedures and processes. 'Societal aspects' are related to the interrelationship between technology and groups of people. 'Technical skills' are related to practical techniques and tool use. A further categorisation occurs with conceptual, procedural, societal, and technical aspects, with these aspects being defined as generic or specific. 'Generic' is taken to mean those aspects common to more than one technological area. 'Specific' is defined as those aspects related to a particular technological area, such as biotechnology or materials. Within each of these categorisations there is also the complexity of the variables that are involved as well as the number of variables that the students are asked to consider or do consider and operationalise.

This categorisation forms the basis for our thinking about progression. Thus, progression in understanding the nature of technology relates to:
- developing a broader, more sophisticated view of technology and its relationship to society and the key characteristics of technology.

Progression in student technological practice relates to being more cognisant of, and more able to:
- operationalise inter-related conceptual, procedural, societal, and technical variables;
- operationalise an increasing number of conceptual, procedural, societal, and technical variables; and
- operationalise an increasing complexity and sophistication of conceptual, procedural, societal, and technical variables.

Progression in generic aspects relates to a number of technological areas or domains and an increasing ability to:
- make links between generic aspects in different technological areas or domains;
- develop broader understandings across a range of technologies;
- understand an increasing number of generic conceptual, procedural, societal, and technical aspects; and
- understand increasingly complex and sophisticated generic conceptual, procedural, societal, and technical aspects.

Progression in specific aspects relates to each of the technological areas and an increasing ability to:
- develop more detailed understandings within different technologies;
- understand an increasing number of specific conceptual, procedural, societal, and technical aspects; and
- understand increasingly complex and sophisticated specific conceptual, procedural, societal, and technical aspects.

Jones and Moreland (2003), from a data set of over 300 students, have identified the following characteristics of progression in children's learning from Year 1 to Year 6 (5–11 years of age). The analysis of the student work is shown in Table 1.

Table 1. Characteristics of progression in learning technology (Years 1–6) (Jones & Moreland, 2003, pp. 61–62).

Characteristic	Early primary	Middle primary	Senior primary
Nature of technology	Distinguish between made and natural world and that people use tools and techniques to help them do things. Notions of betterment and enhancement and initial understanding of the way technology impacts on them.	Knowing about a greater range of technologies and the differences between them. Better understanding of the relationship between technology and other areas.	Knowing that technology has altered the way we think about the made world. Knowing how technology has evolved over time. Aware of both the positive and negative aspects of technology.
Technological practice	Usually solving simple technological tasks incorporating beginning understandings of a limited number of technological principles, concepts and procedures. Beginning development of ability to connect stages to each other.	Developing ability to solve more complex technology tasks, which may include accounting for simultaneous task variables and incorporating understandings of an expanding number of technological principles, concepts and procedures. Beginning iteration between stages.	Developing ability to solve a variety of more complex technology tasks incorporating several inter-related task demands. These include, e.g., the users, product analysis, manufacturing requirements, and construction techniques. Beginning prioritisation of conflicting demands. Stronger iteration between stages.
Conceptual aspects	Usually dichotomous (e.g., bends/doesn't bend, structural stability – balanced/not balanced, material strength – strong/not strong, user friendliness – easy to use/not easy to use).	Often able to identify technological characteristics of the concepts (e.g., portable – easy to carry and store, durable – strong joins and will stand repeated use). Beginning understanding of mock-ups and prototypes.	Able to identify characteristics of technological concepts (e.g., stability, malleability). These concepts are related directly to the context. Developing classification. Awareness of market analysis of consumer choices. Increasing understanding of mock-ups and prototypes and their purposes.

			Starting to identify relationship between moving and non moving parts.
Procedural aspects	Ideas often generated through simple shaping, assembling and rearranging of materials and components. Judgements often yes/no about the outcomes of work. Usually select from a limited range of materials, tools and techniques. Usually assemble and join materials in limited ways. Usually use pictures and words and some beginning use of drawing conventions to convey ideas – simple technological communication.	Usually undertake product analysis through direct observation and comparison using agreed criteria to generate ideas. Often develop initial ideas as 2D/3D sketches, link these to annotated 2D/3D drawings for planning construction with increasing ability to visualise solutions. Beginning use of mock-ups and prototypes. Construct and modify using simple techniques and simple tools with reference to both drawings and agreed criteria. Usually make decisions on when to do things, reflect on processes and make justifiable changes.	Identification of conflicting demands between product development and users. Increasing accuracy in designing, making and testing, paying attention to quality of finish. Developing own criteria related to product and users. Increasing range of formal drawing methods. Working from plans, generating ideas from an understanding of familiar products and from other sources (e.g., survey results). Wider incorporation of mock-ups and prototypes. Increasing ability to test and modify plans and products. Usually evaluate against product, design intentions and user.
Societal aspects	Often related to self and immediate others (e.g., I like/don't like, Mum likes/doesn't like). Often dichotomous views of effects of technology (e.g., good/bad).	More likely to include own, and others' viewpoints for technological choices and decisions made.	Includes own and others' viewpoints for technological choices and decisions made. Aware of costs and benefits. Beginning understanding of risk analysis. Environmental considerations begin to be raised.
Technical aspects	Usually inexact and imprecise. Beginning use of some hand tools.	Beginning precision and accuracy (e.g., some measurements included). Greater range of hand tools used.	Increasing precision and accuracy (e.g., greater range of finer measurements included). Use of more sophisticated tools.

The trends in this table indicate that as students develop an understanding of the nature of technology they move from thinking about made and natural artefacts to include seeing that technology has evolved over time and that societies' values and beliefs impact on and are impacted by technology. As students' technological practice develops they are able to deal with an increasing number and complexity of variables. As their conceptual understanding develops they can use and define concepts such as prototypes, malleability, etc. In terms of the relationship between technology and society they move from thinking about the impact of technology as being either good or bad to consider technology as a value-laden activity. Students develop a greater understanding of costs and benefits as well as environmental aspects. They develop a greater understanding of the processes of technology, and incorporate these processes into their technological practice. In terms of technical aspects students increase their precision and accuracy and use more sophisticated and appropriate tools.

This model proposed by Jones and Moreland (2003) was applied to a national sample of student work from Years 4 and 8 (8- and 12-year olds, respectively) from the *National Educational Monitoring Project* (Crooks & Flockton, 2000). This project assesses students' understanding through a variety of tasks and a 3% national sample is reported on every four years. In exploring student progression from this national sample, nine tasks were selected and for each task a sample of thirty student responses were chosen, representing 10 low, 10 medium, and 10 high performing students (giving a total of 270 responses). In comparing data between the year groups, several trends emerged, mostly in terms of the complexity and interrelatedness of the variables that the students identified and operationalised in terms of the nature of technology; technological practice; and conceptual, procedural, and societal aspects.

Compared with Year 4 students, Year 8 students were more likely to consider a greater number and more functional alternatives, more conflicting demands, and relationships between variables and the end user. They were also more able to identify and operationalise a greater range of task variables and demonstrated a greater use of technological language. Though hardly surprising, the trends have the potential to assist us in understanding what students can do, and allow us to interact with students in more effective ways. Understanding capabilities students bring to tasks allows us to think about fast-forwarding particular aspects. In addition, it assists in designing appropriate tasks to maximise task affordance, which may also enhance learning.

The analysis of the nine tasks in relation to student technological practice is illustrated through students designing and making a pet house through thinking about designing and making with particular materials. Year 8 students were more likely to consider problems with cardboard as a material for use as a pet house than Year 4 students. Only a few Year 8 students suggested more than one variable. High performers at both year levels were more likely to identify the weaknesses of cardboard. In terms of functionality, more students in both year levels considered just one variable in this part of the task. Year 8 students were more likely to consider more than one variable than Year 4 students. High performers at both year levels were more likely to identify more than one variable. Students tended to suggest problems with either the material or the design – not a combination of both. Year 8 students were more likely than Year 4 students to refer to both, and higher performing students at both year levels were more likely to refer to both factors.

In exploring student understanding, student drawings at both year levels elicited more information than written answers about pet needs, structural requirements, and accessibility. High performing students in both year levels were more likely to consider these factors. Students in the lower level of Year 4, and 70% of the low group of Year 8, did not incorporate any of the design elements they had written about in their drawings. All students in the high groups at both year levels incorporated their written ideas into their drawings. More students considered the conflicting demands in their drawings than in their written answers, although awareness of structural requirements of the pet house did not improve. Students in the high performing groups at both year levels were more likely to consider the conflicting demands. Year 8 students were more likely to consider conflicting demands and the high performing students in both year levels were more likely to consider more than one conflicting demand. Half of the high performing group of Year 4 students (17% of the Year 4 group) added a closing mechanism on the opening in their drawing, whereas none had written about the need for such a mechanism. Thirty-seven percent of the Year 8 group drew a closing mechanism compared to 3% who wrote about such a mechanism. Year 8 students were more likely to show an understanding of how a given material performs a function for its intended use. High performers at both year levels were more likely to identify the weaknesses of the given material. In terms of functionality, or showing an understanding of how an environment performs a function for its intended user, Year 8 students were more likely to consider more than one variable. High performers at both year levels were more likely to identify more than one variable. Year 8 students were more likely to consider the structural needs of a pet house. Year 8 students were more likely to consider the conflicting demands of each variable, and to consider a greater number of variables. High performing students at both year levels were more likely to consider a greater number of conflicting demands.

In terms of the technical aspects of drawing, 53% of Year 8 students drew an annotated drawing using one viewpoint – these students tended to be in the low or medium groups and half of these students incorporated all their written ideas into the drawing. Forty three percent of Year 8 students drew a 3D isometric projection – these students tended to be in the high and medium groups and half of these students incorporated all their written ideas into the drawing.

In thinking about materials for an outdoor solution, Year 4 students tended to base their selection of a material on one (38%) or two (35%) variables, whereas 50% of Year 8 students based their selection on three or more variables. However, on closer inspection, analysis revealed a difference between the high, medium, and low performing groups in both year levels. Year 4 students offered less detail and were more likely to offer simple descriptions and explanations than Year 8 students. Although Year 4 students were able to describe the functions and qualities of materials, they were less likely to compare and contrast materials than the Year 8 students. Within the Year 4 level, high performing students were able to consider 3+ variables in their choices, and 1-2 conflicting demands. Medium students considered 1-2 variables and 1-2 conflicting demands. Low students considered 0-1 variable and 0-1 conflicting demands. Within the Year 8 level, high performing students were able to consider 3+ variables in their choices and 2-3 conflicting demands. Low students considered 1 variable and 1 conflicting demand.

Other tasks allowed for the exploration of student understanding of the relationship between technology and society. One such task was about the placement of a timber processing plant in a particular community. In thinking about the relationship between technology and society, and in particular the impact of a technology on the community, students in the high and medium performing groups were more likely to identify more than one stakeholder and students in the high and medium performing groups were more likely to identify those who might be affected indirectly by the new factory. In thinking about progression, there was an increasing ability to examine a greater number of positive and negative effects of technology on a community, such as impacts on the environment and on the social structure of the society. From the low performers in this task to the high performers there was a trend towards an increasing number of positive issues being identified, with higher performers tending to identify more than one stakeholder. There was an increasing ability to show empathy with different groups within the community. High performers were more likely to identify more than one concern than the lower performers in terms of environmental effects, issues for the nearby town, and issues for business people including farmers. There was an increasing concern with more complex issues surrounding change in society and a trend towards consideration of more complex issues by the students in the higher groups, who suggested not only ideas such as better services being available to the town if a factory was built nearby, but also the concern that there would be increased traffic flow, in particular logging trucks, that could cause more accidents and other safety issues related to the security of the site.

Using this Evidence with Teachers

These models of progression should not be interpreted as absolute and fine-grained; rather, they represent a coarse-grained analytical tool to help teachers think about the characteristics of student learning in technology. They can assist teachers to think about what students may bring to the classroom, and how this might be built upon. Jones and Moreland (2003) highlight the following changes for both teachers and students when these areas of thinking about technology are used: First, they found that when teachers were more aware of possible student learning progression then they were more able to identify specific learning intentions and develop more appropriate tasks for students to develop particular learning outcomes. Second, there was evidence that identifying and clarifying possible technological learning outcomes enhanced student-teacher interactions around technology ideas. Third, as a result of more appropriate planning taking into account emerging evidence of student learning there appeared to be greater student understanding of the nature of technology and an increasing complexity and sophistication of technological conceptual and procedural understanding. Jones and Moreland also provide evidence of more appropriate and prolific use of technological vocabulary, and increased teacher and student motivation and interest in technology.

CONCLUSION

This chapter has argued that students' and teachers' existing concepts of technology can limit their learning of technological concepts and processes, including their understanding of the nature of technology. The identification of the characteristics of student learning in technology as defined in four categories – nature of technology; student technological practice; and generic and specific conceptual, procedural, societal, and technical aspects – has been useful in exploring student learning, thinking about their progression, and examining task affordance. In thinking about student learning in technology and their responses to assessment tasks it was usually the number and type of variables identified and operationalised by the students that indicated the different levels of performance. This increased understanding of the capabilities that students bring to tasks allows us to think about fast-forwarding particular aspects of student learning. Such awareness also assists in designing appropriate tasks to maximise student technological activities, which may also enhance learning. However, there is no guarantee that once a view of progression is defined learners will make progress in learning. To make progress, teachers and students need to be intimately involved in the learning process.

Further research needs to be taken along the areas identified in this chapter, that is, learning in technology for both teachers and students and developing enhanced notions of progression.

ACKNOWLEDGEMENTS

I am grateful to Dr Judy Moreland, with whom I have worked on these studies, and the significant contributions she has made to our thinking in technology education.

REFERENCES

Black, P., Brown, M., Simon, S., & Blondel, J., (1996). Progression in learning: Issues in evidence in mathematics and science. In M. Hughes (Series Ed.), *Teaching and learning in changing times.* Oxford, UK: Blackwell.

Carr, M., McGee, C., Jones, A., McKinley, E., Bell, B., Barr, H., & Simpson, T. (2000). *The effects of curricula and assessment on pedagogical approaches and on education outcomes.* Wellington, NZ: Ministry of Education.

Compton, V., & Harwood, C. (2005). Progression in technology education in New Zealand: Components of practice as a way forward. *International Journal of Technology and Design Education, 15*(3), 253–287.

Crooks, T., & Flockton, L. (2000). *Technology assessment results. National Education Monitoring Report.* Dunedin, NZ: University of Otago.

Goodson, I. F. (1985). Subjects for study. In I. F. Goodson (Ed.), *Social histories of the secondary curriculum. Subjects for study* (pp. 343–367). Lewes, UK: Falmer.

Hennessy, S. (1993). Situated cognition and cognitive apprenticeship: Implications for classroom learning. *Studies in Science Education, 22,* 1–41.

Jones, A. (1997). Recent research in student learning of technological concepts and processes. *International Journal of Technology and Design Education, 7*(1–2), 83–96.

Jones, A., & Carr, M. (1992). Teachers' perceptions of technology education - implications for curriculum innovation. *Research in Science Education, 22,* 230–239.

Jones, A., & Moreland, J. (2003). Developing classroom-focused research in technology education. *Canadian Journal of Science, Mathematics and Technology Education, 5,* 1–66.

Kimbell, R. (1997). *Assessing technology: International trends in curriculum and assessment.* Buckingham, UK: Open University Press.

Lindblad, S. (1990). From technology to craft: On teachers' experimental adoption of technology as a new subject in the Swedish primary school. *Journal of Curriculum Studies, 22*(2), 165–175.

Mawson, B. (2007). Factors affecting learning in technology in the early years at school. *International Journal of Technology and Design Education, 17*(3), 253–269.

McCormick, R. (1997). Conceptual and procedural knowledge. *International Journal of Technology and Design Education, 7*(1–2), 141–159.

McCormick, R., Murphy, P., & Hennessy, S. (1994). Problem solving processes in technology education: A pilot study. *International Journal Technology and Design Education, 4*(1), 5–34.

Moreland, J. (2003). Becoming effective technology teachers: Enhancing assessment practices in primary classroom. Unpublished doctoral thesis, University of Waikato, Hamilton, New Zealand.

Moreland, J., Jones, A., & Northover, A. (2001). Enhancing teachers' technological knowledge and assessment practices to enhance student learning in technology: A two-year classroom study. *Research in Science Education, 31*(1), 155–176.

Paechter, C. (1991, September). *Subject subcultures and the negotiation of open work: Conflict and co-operation in cross-curricular.* Paper presented at St. Hilda's conference, Warwick University.

Perkins, D., Jay, E., & Tishman, S. (1993). Beyond abilities: A dispositional theory of thinking. *Merrill-Palmer Quarterly, 39*(1), 1–21.

Wertsch, J. V. (1991). *Voices of the mind: A sociocultural approach to mediated action.* Cambridge, MA: Harvard University Press.

Alister Jones
School of Education
University of Waikato
Hamilton, New Zealand

STEPHEN NORTON AND STEPHEN M. RITCHIE

35. TEACHING AND LEARNING SCIENCE AND MATHEMATICS THROUGH TECHNOLOGY PRACTICE

INTRODUCTION

In this chapter we review studies of the engagement of students in design projects that emphasise integration of technology practice and the enabling sciences, which include physics and mathematics. We give special attention to affective and conceptual outcomes from innovative interventions of design projects.

This is important work because of growing international concern that demand for professionals with technological expertise is increasing rapidly, while the supply of students willing to undertake the rigors of study in the enabling sciences is proportionally reducing (e.g., Barringtion, 2006; Hannover & Kessels, 2004; Yurtseven, 2002). The net effect is that the shortage in qualified workers is having a detrimental effect upon economic and social potential in Westernised countries (e.g., Department of Education, Science and Training [DEST], 2003; National Numeracy Review Panel and National Numeracy Review Secretarial, 2007; Yurtseven, 2002). Interestingly, this trend is reversed in developing economies including China and India (Anderson & Gilbride, 2003).

The ever-increasing technological demands of the 21st Century require all citizens to become mathematically, scientifically, and technologically literate. Those without these capacities will become disadvantaged economically and socially within society (Australian Academy of Technological Science and Engineering [ASTE], 2008). The changing technological demands of society and the falling proportion of participating students in the enabling sciences have implications for pedagogy and curriculum design.

Past learning models where mathematics and science concepts are learned quite separately from each other have been under challenge for some time. Criticisms of this curriculum models come from the reported limited transfer of mathematics and science concepts to applications in real-world contexts (Wilhelm, Sherrod, & Walters, 2008). Traditional mathematics teaching is particularly criticised for fostering a view that mathematics is a decontextualised domain of study. The teaching of the enabling sciences through design processes associated with technology practice is worth examining as a potential solution to increasing student participation in the study of the enabling sciences, in part because it potentially offers a common context with which to study technology, mathematics, and science.

THE CHANGING NATURE OF TECHNOLOGY EDUCATION

In 1987 the Australian Educational Council (AEC) began a series of initiatives that led to the 1994 nationally agreed curriculum statements that established technology as one of eight key learning areas. Similar approaches have been adopted in the USA with their *Standards of technological literacy: Content for the study of technology* (International Technology Education Association

A.T. Jones and M.J. de Vries (eds.), International Handbook of Research and Development in Technology Education, 419–429.
© 2009 Sense Publishers. All rights reserved.

[ITEA], 2007, 2008) and other nations, all responding to the recognition of the increasingly important role of technology in society.

Technology is broadly defined as the application of knowledge and resources that are used to extend human capabilities (Australian Education Council [AEC], 1994; ITEA, 2007). Lewis (2006) cites a number of authors to define the processes of technology in which 'generating a design', 'making a plan', and 'evaluating a design' are the central processes of technology learning and practice. Technological problem solving essentially combines principles of design with the material aspects of production to satisfy a material need. Other authors note that the design process is part of procedural knowledge, or 'knowing how to do it', that is highly dependent upon contextual knowledge (e.g., McCormick, 1997).

The major change in emphasis in technology education has been a shift from skill development to an emphasis on generic problem solving and ways of thinking (Custer, 2003). Today a technically literate person is someone who understands technology, how it shapes society, and how to 'do' technology that includes designing things to solve practical problems (e.g., ITEA, 2008; National Academy of Engineering, 2008a).

Technology curriculum documents attempt to provide thinking frameworks that encourage students to be creative and innovative. For example, the 'technology practice' model of Queensland Studies Authority (QSA, 2003) in Australia lists key processes of investigation (examining, analysing, researching…); ideation (generating, planning, designing…); production (creating, developing, managing, making…), and evaluating (testing, judging, reflecting, and comparing…). Such rubrics imply that problem solving, including solving problems associated with design, can be taught as a generic set of skills within the domain of technology, and that these could even be transferred to such other domains as mathematics and science. Yet, McCormick (1997) argued that such assumptions have little empirical support.

CURRICULUM CONSIDERATIONS FOR LEARNING SCIENCE AND MATHEMATICS THROUGH TECHNOLOGY PRACTICE

The idea of teaching science and mathematics through the practice of design became a topic of discussion from the 1960s. Lewis, Barlex, and Chapman (2007) reported that in the UK efforts were made to establish science and technology as a single examinable subject. Similar science and technology integration curriculum models were described in other Western nations including Canada (e.g., Lewis, Barlex, & Chapman, 2007), The Netherlands (e.g., de Beurs, 2000), and the USA (e.g., National Academy of Engineering, 2008b) throughout the 1980s and 1990s.

Design projects offer a rich context in which students can transform their science and mathematics understandings. Learning science and mathematics through design projects can be viewed as 'situated learning' (Wicklein & Schell, 1995). From a situated learning perspective, students working collaboratively on design projects might demonstrate increased use of knowledge in different situations (potentially across subject domains), and improved problem solving and social outcomes. The degree to which technology practice might offer increased understanding, develop general problem solving skills, and engage students in the learning of the enabling sciences is examined below.

Several authors have noted that classroom design projects can give a task 'authentic purpose' that leads to high student engagement (e.g., American Association for the Advancement of Science [AAAS], 1993; Dow, 2006; ITEA, 2007; Lewis, 2006; McNair & Clarke, 2007; Mettas & Constantinou, 2007; Norton, McRobbie, & Ginns, 2007). Authenticity has two aspects: personal authenticity and cultural authenticity (McCormick, 2004). Personal authenticity means an activity has meaning and value for the learner. In contrast, a task with cultural authenticity has value beyond the individual and even the classroom; that is, the task is worthy of completion and/or the solution to a problem is likely to benefit the wider community.

Roth (1998) reported that technology-design projects (also called engineering-based projects) offer students the opportunity to develop rich technology and science language by engaging in activity that was personally relevant. Lewis, Barlex and Chapman (2007) cited the Engineering Council's (1990) definition of an engineer as "one who acquires and uses scientific, technical and other pertinent knowledge and skills to create …" (p. 10). Roth found that engaging children in engineering and design projects changed the nature of classroom learning from children attempting to mimic the teacher's definitions and language to children using the language of science in activities they perceived as personally authentic.

The critical importance of students developing language from their personal engagement with the phenomena of science is a recurrent theme in the literature on authentic learning in science (e.g., Roth, 2005). Earlier, Lemke (1990) commented that learning in science means learning to 'talk science'. Implicit in 'talking' science is the engagement of students in the practices of science. While language is central to thinking, it is accompanied by other attributes – including purpose, procedures, and conceptual knowledge – that may overlap to varying degrees. The extent of overlap determines the extent to which students can engage with the practices of science and technology simultaneously (McCormick, 2004). Yet, Roth (1998) argued that the study of technology and science are mutually supportive practices since engaging in technology-based activities overlaps with the practices of science.

Even though Lewis (2006) found considerable convergence between science and technology reasoning, mental models, testing processes, and dependence upon overlapping content knowledge, he also reported differences in purpose, the role of constraints, the role of failure, and the greater importance of context in technology. Overall, he asserted that science and technology were potentially mutually supportive discourses, concluding: "The act of designing introduces problems and challenges in classrooms that can be resolved by knowledge of science" (p. 276). In contrast, McCormick (2004) noted that in traditional approaches to science, teachers have tended to focus on abstraction or the stripping away of context, a practice not shared by most technology educators. Yet, science educators who advocate context-based approaches to science do not support a decontextualised treatment of science concepts either (see King & Ritchie, in press).

Another distinction between science and technology teaching McCormick (2004) makes is that cognitive processes differ according to the domain of thinking, and frequently those domains are distinctly different in technology and science. Similarly, Barlex and Pitt (2000) noted that the thinking demands of science and technology are quite distinct. For example, they considered that the essence of science is the production and testing of knowledge while technology practice is primarily concerned with "transforming this and other sorts of knowledge into techniques and artefacts for which there is a human demand" (p. 5). Benson (2001) supported this argument with his contention that since science and technology are so different it is illogical to subsume them under one "science and technology label" (p. 6).

A further concern was made by Paechter (1995) who noted that, in secondary schools, teachers have such divergent 'subject cultures' that agreement on common learning goals would be difficult to reach. The divergent subject cultures develop from different values placed upon abstraction and context, with technology tending to focus on relevance to an application, and science (and mathematics) upon theories and abstractions. As noted above, these differences in subject culture are manifested in the language and processes involved in teaching as well as assessment models used to grade and reward students (McCormick, 2004). In technology, students are generally assessed according to how the product performs, while in science students are more likely to be assessed on the quality of their explanations of underlying principles.

As with science, learning mathematics through design activities has the potential to give purpose to and generate positive attitudes towards mathematics and perceptions about the nature and value of mathematical endeavour (Norton, 2006a, 2006b; Norton & Cooper, 2008; Wicklein & Schell, 1995). However, relatively little research has been conducted to determine the extent to which mathematical concepts can be effectively developed through integrated mathematics and technology activities. McCormick and Evans (1998) argued that different classroom practices in mathematics and technology classes could hinder the simultaneous learning of technology and mathematics in integrated tasks. They found, for example, that in technology classrooms the teacher did the mathematical work to ensure production progress. Furthermore, they asserted that students' fragile and varied understanding of mathematics was frequently not sufficient to make sense of the mathematics concepts when they were applied in authentic contexts.

A different outcome was reported following an evaluation of a large-scale intervention of an integrated program in primary schools – known as the *New Basics* – enacted by Education Queensland (2003) in Australia. This curricular innovation emphasises student engagement in integrated projects known as 'rich tasks'. These tasks provided opportunities for students to learn mathematical concepts in the context of real-world problems (e.g., travel itineraries, Pi in the sky, and international trade). The State's report on the project (Education Queensland, 2004) indicated that students whose schools adopted this approach were not disadvantaged on standardised tests. One plausible reason for this anomalous result is that, in most cases, the schools also sustained a supplementary program of mathematics independent of the rich tasks.

An independent report on the *New Basics* by Cooper, Nuyen, and Baturo (2003) found that rich tasks were inadequate on their own to teach, assess, and report on the learning of mathematics, and that there were major components of mathematics that needed to be taught outside of the tasks. This suggests that without the parallel curriculum of mathematics, student learning may have been minimal for many students, as was the case for learning science in the robotics study reported by Norton et al. (2007).

TEACHING AND LEARNING SCIENCE AND MATHEMATICS THROUGH TECHNOLOGY PRACTICE

From the perspective of the classroom teacher, a critical question is which comes first in the process of integrated science and technology learning? Is it better to teach the science and then apply it in a technological context, or to learn the science through engagement with technology practice? One consideration is that the application of science in subsequent technology contexts runs the risk of technology being seen as 'applied science', a description not favoured by some authors not least because science is cast in a 'a gatekeeper' role that potentially limits entry to technology (e.g., Gardner, Penna, & Brass, 1990).

Several design projects involving children from Year 2 to Year 7 across the US, Canada, and Australia reported students' appropriation of canonically accurate science language as the artefacts took shape (Roth, Tobin, & Ritchie, 2001). This was particularly pronounced when the teacher purposefully introduced the language of science 'just in time' for the students to learn the related scientific concept in-the-making of the artefact (Ritchie, 1999; Ritchie & Hampson, 1996). These studies assumed the situated and contingent nature of design activities. As Roth et al. argued:

> Designing is not a cognitive activity that can be reduced to individual mental processes. Decisions taken by principal actors, or owners of projects, cannot be understood apart from the tools, materials, artefacts, teacher interdictions and constraints, and the emergent properties of collective, discursive and material activity, and design artefacts. (p. 29)

Furthermore, they recognised that emerging artefacts provide a context for subsequent design moves that lead to additional opportunities for learning science concepts.

Norton et al. (2007) reported on students' engagement in integrated science and technology projects making logo-controlled robots in two classes. They found that most of the science remained implicit. More specifically, the few students who had completed a prior physics unit took account of concepts such as friction and leverage, and used the language of these phenomena, yet the designs of the majority of students paid scant regard to the physics of mechanics. This was hardly surprising since almost all aspects of the classroom environment did not support the abstraction of science. The assessment focused on a working product, most teacher support was related to helping students to get their product to work, and most student time was spent designing and particularly building working robots. Analysis of classroom discourse indicated that compared to a 'science only' focused classroom, the teacher's use of science language was negligible and student-to-student discourse was essentially bare of science language and concepts for most groups. A manifestation of the different views that teachers had of the purpose of technology and science was seen in their planning of the unit: They did not have a clear goal to make explicit the learning of science concepts. In addition, in one of the classes there was not a suitable forum for discussing the underpinning science, and in the classes most students simply did not have access to the linguistic resources with which to engage in discussions about the science underpinning the design project. McCormick (2004) and McCormick, Davidson, and Levinson (1995) reported similar outcomes from other technology projects.

In another technology-based learning study in Years 3 to 7, students were asked to design a theme park ride (Norton, 2006a). The first step was to experience the forces first-hand by going to a theme park where an onsite engineer guided the students by explaining much of the physics underpinning the rides. Some Year 4 and 5 boys subsequently chose to build a model of the *Centrifugal Ride*. The concept of centripetal force made sense to them after their visit: It was the centre-seeking force that kept them moving in a circle when they were pinned to the wall of the ride despite the floor falling beneath them. The release of centripetal force also spun the water out of a lettuce colander, and they adapted this to make their model of the ride. In this instance, the science was experienced and a language linked to that experience such that the students could think about and communicate about the science phenomena when designing and making their ride. Analysis of the discourse describing their projects indicated that these young students fluently used the language of science, including terms such as force, push, magnets, magnetism, speed, going faster, slowing down (acceleration), electricity, minutes, seconds, watts, kilowatts, power, and so on. In this study, the focus on classroom activity was upon making science explicit, applying the laws of physics, and using measurement and geometry aspects of mathematics to design and build the rides.

In a more recent study reported by Norton and Cooper (2008), two case studies had an explicit focus on science, one on insect ecology and the other on the sun, including seasons and telling the time by the sun. In regard to the insect ecology study the design brief was to construct a 'bug catcher'. Opportunities for learning science occurred as a result of collecting insects and making observations about numbers, types, and locations of the insects. The Year 3–4 students asked questions such as "Are there more bugs in some parts of the school?", "Do different ants come out at night?", "Do all ants eat honey?", and "Do more butterflies come out after rain?" In this instance, the design of the bug catcher was an integral part of the inquiry process of learning science, as recommended in the contemporary science education literature where students seek to answer their own questions (e.g., Duschl, 2008). A scientific need provided a rationale for constructing a bug catcher, a reason to engage in the practice of science, and a context to learn the language of science. It could also be seen that the need to solve a science question provided a rationale to learn technology and design principles.

The classroom teacher in both the bug catcher and the sundial projects reported high levels of student motivation and engagement. However, while the learning of science concepts associated with insect ecology and data presentation seemed to be well catered for in the bug collection project, success in meeting technology outcomes was less clear. The design practices of investigate, ideate, produce, and evaluate were not realised, much of the production stage was craft orientated and the evaluation was largely coincidental. The students applied the mathematics of scale drawing to sketching designs, but there was no evidence to suggest that these diagrams guided construction. As in the previous study (i.e., Norton, 2006a), construction tended to be dictated by the practicalities of cutting and fixing materials. It appeared that much of the 'thinking' processes of technology had not been abstracted during the project, possibly due to the teacher's limited familiarity with the technology curriculum (see also de Beurs, 2000), but also because of the different demands and ways of working in the various subject domains.

In the sundial-based integrated technology/mathematics/science project described by Norton and Cooper (2008), somewhat different integration processes were observed. The composite class of Year 4–7 students wanted to make a sundial and sundial mount because the school's old sundial had been eaten away by termites and the students wished to replace it with a robust artefact. This need provided a reason to research various sundial designs as well as to assess various materials. The teacher recognised considerable overlap between the "natural and processed materials" strand in science and the "materials" strand in the technology curriculum. In this project, the students evaluated various sundial designs and built prototypes with cardboard. They then contracted the local metal industry to manufacture the sundial to their specifications.

The observations of student technology-based projects in the contexts of robots, theme park rides, bug catchers, and sundials support Roth's (2005) findings that student engagement in design-based activity had the effect of transforming the classroom discourse from teacher-centred to child-centred learning communities. Learning mathematics through technology practice has potential to create learning environments that are rich in mathematics learning and application opportunities (e.g., Norton, 2005; Norton & Cooper, 2008). Realising this potential in typical classroom settings is a challenge for the same reasons as is the case with science learning. Some of these challenges, faced by teachers implementing engaging technology projects that enhance the likelihood for science and mathematics learning outcomes, are now considered.

CHALLENGES THAT CONFRONT SCIENCE AND MATHEMATICS TEACHERS IMPLEMENTING TECHNOLOGY-BASED DESIGN ACTIVITIES

An ongoing challenge for science and mathematics educators has been to support student learning in ways that assist students to apply the concepts beyond the settings in which they are learnt. The criticism of traditional textbook-based learning is that many students do not recognise and apply science and mathematics concepts outside of their decontextualised classroom activities. The studies reviewed in this chapter suggest that while students may well have been engaged in the manipulation of materials to produce artefacts in design activities, without skilful teacher intervention (e.g., Ritchie, 1999), many students do not independently develop deep understanding of science (and mathematical) concepts. Not surprisingly, teacher support is also a likely requisite for successful implementation of design projects (Roth et al., 2001).

There are two ways for teachers to scaffold technology practice to enhance mathematical and scientific understanding. The first is a 'just in case' approach. This is typified by introducing the concept in mathematics or science class before it is needed in the design project. For example, proportion was taught using models of coloured counters, different quantities of mixtures, and scale in diagrams and maps before introducing it in the form of ratio in gearing within a design setting (Norton, 2006b). In this model, the design activity becomes an opportunity to apply a mathematical concept in an authentic setting. One advantage of this approach to the integrated learning of technology and mathematics is that the teacher can plan ahead to address specific aspects of the curriculum purposefully, a significant factor considering the necessity imposed by most education authorities that require mathematical concepts, for example, to be taught sequentially (e.g., multiplication before division, and differentiation before integration).

The second approach can be called 'just in time'. In this approach, mathematics or science concepts relevant to the unfolding artefact are taught as the need arises. This gives an immediate purpose for learning the mathematics or science concept. Potentially, this form of integration is powerful and has the theoretical support of those who value situated learning (e.g., Brown, Collins, & Duguid, 1989; Lave & Wenger, 1990). While this just in time approach has not been observed regularly in mathematics education studies, it is more commonly reported in the science education literature (see Roth et al., 2001).

Teachers attempting to implement either form of scaffolding face formidable challenges associated with the use of different language, different procedures, having different backgrounds or contextual knowledge, and different classroom management processes in technology, science, and mathematics classrooms. In addition, the students become increasingly cognisant of the particular cultures associated with learning in each subject, and contribute to the maintenance of conforming practices (i.e., cultural reproduction). Considering the divergent activity associated with the different enabling sciences, a 'just in time' approach places very high demand upon teachers' domain and pedagogic knowledge.

Given the reports of difficulties finding teachers with adequate breadth and depth of mathematical and science knowledge (e.g., National Numeracy Review Panel & National Numeracy Review Secretariat, 2007), the expectation that teachers would also have widespread experience with technology practice may be unrealistic. While support for integrated learning has been reported from primary to tertiary levels (e.g., Wilhelm et al., 2008), clearly concerns about depth of knowledge become increasingly significant as children progress from primary school settings to secondary schools, with distinct language, procedural and conceptual knowledge, and staff subcultures in each subject domain.

CONCLUDING REMARKS

The use of technology practice as a framework for the learning of mathematics and science offers considerable potential for students to develop deep understanding of phenomena in relevant contexts. As such, design projects can have a multiplier effect on learning science and mathematics. There are possibly three mechanisms for effect. The first is additional time. Students can learn mathematics and science in their allocated time in the school timetable and have additional time to apply scientific and mathematical thinking while engaged in a technology-based project. Second, students are afforded additional representations. For example, Norton (2006b) described students learning ratio with varying representations in mathematics and applying those understandings in gearing contexts as part of a technology project. The third multiplier effect is frequency of engagement. Students have the potential of engaging with mathematical and scientific concepts over a number of subject domains in a teaching week.

Some authors suggest that specific knowledge of mathematics and science concepts mutually supports the design, production, and evaluation processes of technology (e.g., Roth, 1995; Wicklein & Schell, 1995). Technology practice can be seen as a mechanism to integrate the learning of the enabling sciences, and has considerable potential to address many of the problems in engaging students in learning mathematics and science, including those related to a fragmented and disconnected curriculum and teacher-centred approaches. However, realising this potential is not without its challenges, and mounting evidence suggests that these challenges become increasingly important as students pass through the school system and each subject domain develops disparate language conventions, contextual knowledge, and procedures including problem solving procedures.

Primary and secondary teachers face different obstacles in implementing design projects. In primary schools, one teacher generally teaches all key learning areas and has considerable flexibility in planning the curriculum. Primary teachers might overcome some obstacles by integrating discipline learning in design projects through the 'just in case' model. This model might be enacted more easily because there is a gradual transition towards increased integration, supported by appropriate curriculum material and professional assistance. The relatively low assumptions of student contextual, subject, and procedural knowledge also facilitate a 'just in case' model of scaffolding.

The successful implementation of the 'just in time' model is likely to require greater discipline knowledge and familiarity with technology practice both in terms of procedural and conceptual knowledge (McCormick, 1997). While secondary teachers are more likely to have the requisite specialist background in a particular discipline and access to relevant resources, they are usually less experienced in designing integrated activities. In contrast, primary teachers have indeed used this approach successfully in curricular innovations studied (e.g., Roth et al., 2001). Furthermore, few secondary teachers have the timetable flexibility and breadth of discipline and pedagogic knowledge to attempt integration of this nature. Secondary teachers might, however, aim for more mutually supportive teaching relationships across discipline boundaries with their colleagues. Through co-ordination, teachers of technology, mathematics, and science might become informed of the teaching goals of the other disciplines and plan their curricula such that the different subjects support each other. For example, a physics unit on forces and vectors might be planned to coincide with teaching rates of change and modelling polynomials in mathematics, and Newtonian physics in science such that students' conceptual development could be reinforced through the design of machines with a purpose.

A deeper collaborative relationship between secondary teachers of technology, mathematics, and science could be developed with adequate resources, including time for in-depth planning. Collaboration would include the co-ordination of curricula to the extent of sharing some common tasks. Research into the effects of different organisational and leadership structures could illuminate how teachers might initiate such curricular initiatives (see Ritchie, Tobin, Roth, & Carambo, 2007).

REFERENCES

American Association for the Advancement of Science [AAAS]. (1993). *Benchmarks for science literacy: Project 2061.* New York: Oxford University Press.

Anderson, L., & Gilbride, K. (2003). Pre-university outreach: Encouraging students to consider engineering courses. *Global Journal of Engineering Education, 7*(1), 87–93.

Australian Academy of Technology Science and Engineering [ATSE]. (2008). *Technology education essential.* Retrieved November 8, 2008, from http://www.atse.org.au/index.php?sectionid=1226

Australian Education Council [AEC]. (1994). *A statement on technology for Australian schools.* Carlton, Victoria: Curriculum Cooperation.

Barrington, F. (2006). *Participation in year 12 mathematics across Australia 1995–2004.* Melbourne, Victoria: ICEME/AMSI, University of Melbourne.

Barlex, D., & Pitt, J. (2000). *Interaction, the relationship between science and design technology in the secondary school curriculum.* London: Engineering Council.

Benson, G. (2001). The unrealized potential of everyday technology as a context for learning. *Journal of Research in Science Teaching. 28*(7), 730–745.

Brown, J. S., Collins, A., & Duguid, S. (1989). Situated cognition and the culture of learning. *Educational Researcher, 18*(1), 32–42.

Cooper, T., Nuyen, A., & Baturo, A. (2003). *Rich task project: Integrated mathematics outcomes with rich tasks within a productive pedagogies framework*. Queensland University of Technology. Retrieved September 1, 2008, from http://education.qld.gov.au/corporate/newbasics/pdfs/richtaskrep3-final-8.4.03.pdf

Custer, R. (2003). Technology education in the United States – A status report. In G. Martin & H. Middleton (Eds.), *Initiatives in technology education: Comparative perspectives* (pp. 16–29). Brisbane, Queensland: Technical Foundation of America/CTER, Griffith University.

de Beurs, C. (2000). Technology for 15+. Integration of technology in science curriculum of upper level secondary education in the Netherlands. In R. Kimbell (Ed.), *Design and technology international millennium conference, England*. Loughborough University Institutional Repository. IDATER Archive. Available from https://dspace.lboro.ac.uk/dspace-jspui/

Department of Education, Science and Training [DEST]. (2003). *Australia's teachers: Australia's future. Advancing innovation, science, technology and mathematics. Background data and analysis*. Canberra, ACT: Commonwealth of Australia.

Dow, W. (2006). The need to change pedagogies in science and technology subjects: European perspective. *International Journal of Technology and Design Education, 16*(3), 307–321.

Duschl, R. (2008). Science education in three-part harmony: Balancing conceptual, epistemic, and social learning goals. *Review of Research in Education, 32*(1), 268–291.

Education Queensland. (2003). *The new basics project*. Retrieved September 1, 2008, from http://education.qld.gov.au/corporate/newbasics/

Education Queensland. (2004). *New basics research report*. Retrieved September 1, 2008, from http://education.qld.gov.au/corporate/newbasics/html/research/research.html

Engineering Council. (1990). *Policy statement: Standards and routes to registration* (2nd ed.). London: Author.

Gardner, P., Penna, C., & Brass, K. (1990). Technology and science: Meanings and educational implications. *The Australian Science Teachers Journal, 36*(3), 22–28.

Hannover, B., & Kessels, U. (2004). Self-to-prototype matching as a strategy for making academic choices. Why high school students do not like math and science. *Learning and Instruction, 14*, 51–67.

International Technology Education Association [ITEA]. (2007). *Standards for technological literacy: Content for the study of Technology*. Reston, VA: Author.

ITEA. (2008). *Technologically literate citizens*. Retrieved September 1, 2008, from http://www.iteaconnect.org/TAA/TAA_Literacy.html

King, D., & Ritchie, S. M. (in press). Learning science through real-world contexts. In B. Fraser, K. Tobin, & C. McRobbie (Eds.), *Second international handbook of science education*. Dordrecht, The Netherlands: Springer.

Lave, J., & Wenger, E. (1990). *Situated learning: Legitimate peripheral participation*. Cambridge, UK: Cambridge University Press.

Lemke, J. L., (1990). *Talking science: Language, learning and values*. Norwood, NJ: Ablex.

Lewis, T. (2006). Design and inquiry: Bases for an accommodation between science and technology education and curriculum? *Journal of Research in Science Teaching, 43*(3), 255–281.

Lewis, T., Barlex, D., & Chapman, C. (2007). Investigating the interaction between science and design & technology (D&T) in the secondary school—A case study approach. *Research in Science and Technology Education, 25*(1), 37–58.

McCormick, R. (1997). Conceptual and procedural knowledge. *International Journal of Technology and Design Education, 7*(1–2), 141–159.

McCormick, R. (2004). Issues of learning and knowledge in technology education. *International Journal of Technology and Design Education, 14*(1), 22–44.

McCormick, R., Davidson, M., & Levinson, R. (1995). Making connections: Students using science understanding of electric circuits. In J. S. Smith (Ed.), *IDATER 95 – International conference on design and technology educational research and curriculum development* (pp. 63–67). Loughborough, UK: University of Loughborough.

McCormick, R., & Evans, H. (1998, August). *The use of mathematics in secondary school d&t*. Paper presented at the British Educational Research Association Annual Conference, Belfast, August 27–30.

McNair, V., & Clarke, R. (2007). Effective technology and design teaching: Getting it right in the classroom. *International Journal of Technology and Design Education, 17*(3), 271–290.

Mettas, A., & Constantinou, C. (2007). The technology fair: A project-based learning approach for enhancing problem solving skills and interest in design and technology education. *International Journal of Technology and Design Education, 18*(1), 79–100.

National Academy of Engineering. (2008a). *Technically speaking. What is tech lit?* Retrieved August 28, 2008, from http://www.nae.edu/nae/techlithome.nsf/weblinks/KGRG-569LNP?OpenDocument

National Academy of Engineering. (2008b). *Technically speaking. What's been done?* Retrieved November 10, 2008, from http://www.nae.edu/nae/techlithome.nsf/weblinks/KGRG-55X6N7?OpenDocument

National Numeracy Review Panel & National Numeracy Review Secretariat. (2007). *National numeracy review: Background paper, June 2007*. Retrieved January 1, 2009, from: http://www.dest.gov.au/NR/rdonlyres/A6687654-F732-44CF-B39F-16511ADFBE2/18249/NumeracyReviewBackgroundPaper.pdf

Norton, S. J. (2005). The constructional of proportional reasoning. In H. Chick & J. Vincent (Eds.), *Proceedings of the 29th conference of the International Group for the Psychology of Mathematics Education* (pp. 17–23). Department of Science and Mathematics Education, University of Melbourne.

Norton, S. J. (2006a). The use of design practice to teach mathematics and science. *International Journal of Technology and Design Education, 18*(1), 19–44.

Norton, S. J. (2006b). Pedagogies for the engagement of girls in the learning of proportional reasoning through technology practice. *Mathematics Education Research Journal, 18*(3), 69–99.

Norton, S. J., & Cooper, T. J. (2008). Teaching mathematics and technology through design practice. In M. Goos, R. Brown, & K. Makar (Eds.), *Navigating currents and charting directions. Proceedings of the 31th annual conference of the Mathematics Education Research Group of Australasia* (pp. 377–384). Brisbane, Queensland: The University of Queensland, St Lucia.

Norton, S. J., McRobbie, C. J., & Ginns, I. S. (2007). Problem solving in a middle school robotics design classroom. *Research in Science Education, 37*(3), 261–277.

Paechter, C. (1995). Subcultural retreat: Negotiating the design and technology curriculum. *British Educational Research Journal, 21*(1), 75–87.

Queensland Studies Authority [QSA]. (2003). *Technology: Years 1–10 syllabus*. Brisbane, Queensland: Author.

Ritchie, S. M. (1999). The craft of intervention: The construction of a teacher's personal practical theory for science learning and teaching during teacher-student interactions. *Science Education, 83*(2), 213–231.

Ritchie, S. M., & Hampson, B. (1996). Learning in-the-making: A case study of science and technology projects in a year six classroom. *Research in Science Education, 26*(4), 391–407.

Ritchie, S. M., Tobin, K., Roth, W.-M., & Carambo, C. (2007). Transforming an academy through the enactment of collective curriculum leadership. *Journal of Curriculum Studies, 39*(2), 151–175.

Roth, W.-M. (1995). *Authentic school science*. Dordrecht, The Netherlands: Kluwer.

Roth, W.-M. (1998). *Designing communities*. Dordrecht, The Netherlands: Kluwer.

Roth, W.-M. (2005). *Talking science: Language and learning in science classrooms*. Oxford, UK: Rowman & Littlefield.

Roth, W.-M., Tobin, K., & Ritchie, S. (2001). *Re/Constructing elementary science*. New York: Peter Lang.

Wicklein, R., & Schell, J. (1995). Case studies of multidisciplinary approaches to integrating mathematics, science and technology education. *Journal of Technology Education, 6*(2), 59–76.

Wilhelm, J., Sherrod, S., & Walters, K. (2008). Project-based learning environments: Challenging preservice teachers to act in the moment. *Journal of Educational Research, 101*(4), 220–235.

Yurtseven, H. (2002). How does the image of engineering affect student recruitment and retention? A perspective from USA. *Global Journal of Engineering Education, 6*(1), 17–23.

Stephen Norton
School of Curriculum Teaching and Learning
Griffith University, Mount Gravatt Campus
Queensland, Australia

Stephen M. Ritchie
School of Mathematics, Science & Technology Education
Queensland University of Technology
Brisbane, Australia

JONATHAN OSBORNE

36. LEARNING TO ARGUE, ARGUING TO LEARN

The Role of Argument in Teaching and Learning Science and Technology

THE ROLE OF ARGUMENTATION IN SCIENCE AND TECHNOLOGY

In the case of science, the arguments for the centrality of argument have been fully explored and well-established (Driver, Newton, & Osborne, 2000). In essence, that:

> Science is a social practice and scientific knowledge the product of a community. New knowledge does not become public knowledge in science until it has been checked through the various institutions of science. Papers are reviewed by peers before being published in journals. Claims made in published papers are scrutinized and criticized by the wider community of scientists; sometimes experiments are repeated, checked, and alternative interpretations are put forward. In this process of critical scrutiny argument plays a central role. In presenting and evaluating arguments, scientists are influenced by factors beyond those internal to science, factors such as scientists' social commitments, values, and by the wider culture of ideas and technological capabilities in society at the time. (p. 296)

Translated into the context of science education, this means that there has been a shift from seeing science as a body of objective 'facts' about the world, to presenting the subject as one where ideas are tested and evaluated against a body of evidence and data derived from the material world (Duschl, 2007). Argumentation can then be used, for instance, to establish how to interpret ambiguous data sets; to decide how to classify rocks or specimens; to make a prediction about the outcome of an experiment; or to explore the implications of different ways of generating energy.

The first reaction to the concept that argument and argumentation has anything to do with technology, however, is perhaps puzzlement. At the core of technology is the manipulation of the material world – the production of new objects – be it a new carbon fibre bicycle or the development of a new food product. Technology is fundamentally a human response to problems and their solutions, such as climate change, providing clean water, or developing more efficient petrol engines. The history of technology is the history of humanities' ability to provide creative and inventive material solutions to the problems that confront it daily (Pacey, 1983). Thus, technology, whilst it draws on scientific knowledge, is different. What then, moreover, has this discipline, which has the manipulation and exploitation of the material world very much at its core, got to do with a discursive practice such as argumentation?

The answer comes from an examination of the nature of technology. In the past decade, there has been a shift from seeing technology education as a process of undertaking set-piece tasks to one which sees it as a process of solving real and complex problems (Black & Atkin, 1996). Admittedly, within the context of the technology classroom, the boundaries defining the problem and its complexity are often constrained. Nevertheless, the same set of issues arises as those that confront the professional technologist. The core difference between technology and science is that

A.T. Jones and M.J. de Vries (eds.), International Handbook of Research and Development in Technology Education, 431–442.
© 2009 Sense Publishers. All rights reserved.

the goal of technology is the production of a solution to a given problem or the design of a product or system to meet a specific need. The goal of science, in contrast, is the production of new knowledge. Technological solutions are invariably interdisciplinary, drawing on a range of bodies of knowledge. Scientific understanding is required to know, for example, both what materials are capable of and the causal mechanisms that are inherent to any design. Mathematical knowledge may well be required to make predictions about how the artefact may respond in different contexts. Economic considerations must be used to decide whether it is commercially feasible, and aesthetic considerations are required to decide whether it will appeal to consumers. Unlike science, there is no consensually agreed answer. Rather, there are invariably a plethora of solutions and deciding which is the most appropriate is a process of considering a range of alternatives and evaluating their competing merits using a variety of types of criteria and areas of knowledge (e.g., technology, science, mathematics, and economics).

Technological solutions are invariably the product of a design cycle where competing solutions are offered and modeled. Careful evaluation and analysis of their competing merits is dependent on a reasoned and rational presentation – that is, argued – and attempts to identify weaknesses. Argument is not some peripheral feature of this process – it is *the* central discursive process. There is then the process of constructing the artefact – a process that itself may be the subject of considerable debate, followed by the task of evaluating critically the outcome and its fitness for purpose. At any of the steps in the process, students can be asked to engage in small-group discussions. For instance, groups could be asked to develop alternative solutions and present these to each other. After every presentation, each group could then be asked to discuss and indentify one positive feature and one weakness of the design. After all presentations, groups could be asked to revisit their designs, and in the light of what they have heard, re-evaluate and reconsider their proposal. A similar process could be undertaken when the alternative artefacts have been produced and students need to evaluate their solutions. Being able to explain why a proposed solution is inappropriate not only helps to develop a critical sense of what constitutes a better solution, but also makes higher order cognitive demands of synthesis and evaluation (Bloom, 1956). Such a process is, for instance, a key element of architectural education where students have to present their ideas and defend them publically.

Thus, technical answers to the plethora of problems that confront humanity are constructed through a social process, in which discursive activity is central (Solomon, 1993). If technology education is to provide its students with an authentic experience of what it means to do technology, then providing students with opportunities to engage in argumentation should be a prominent feature of the teaching of technology.

Nevertheless, this argument for argumentation does not rest solely on the need for 'authentic experiences'. After all, the practice of technology is distinct from the learning of technology and confusion arises when the two activities are elided. Rather, the case for argumentation in technology lies in the growing body of evidence that such processes enhance not only students' learning but also their engagement with learning. In short, the adoption of a more discursive or dialogic approach to the teaching of technology can enhance the quality of its pedagogic practice. This chapter offers a summary of the research in the field of argumentation and a rationale for its value.

WHAT IS MEANT BY ARGUMENT?

Argument in everyday language commonly carries with it a pejorative meaning associated with confrontation and feelings of discomfort. This is the notion of argument as war (Cohen, 1995). For this reason, some prefer to talk about the process of discussion, reasoning, or debate. Whilst this is understandable, what such terms fail to capture is that the resolution of problems or issues comes from contradiction or, as Bachelard (1940) so elegantly phrased it: "Two people must first contradict each other if they really wish to understand each other. Truth is the child of argument, not of fond affinity" (p. 5). From such a perspective argument is seen as something which is essential to the resolution of disputes – an idea captured by Van Eemeren and Grootendorst (2004) when they defined argumentation to be: "verbal, social and rational activity aimed at convincing a reasonable critic of the acceptability of a standpoint by putting forward a constellation of propositions justifying or refuting the proposition expressed in the standpoint" (p. 1).

Thus, argumentation is a term which both refers to the product – the argument and its justification – and the process. Within education, however, argumentation has a twofold goal – one as a means to develop or aid thinking, and second, as a means to render belief rational. From this perspective, one which sees argument as the furtherance of enquiry, a more appropriate metaphor for argument would be to see it as part of the process of brainstorming – a process that results in an exchange and evaluation of ideas rather than their imposition by one side on another. As Cohen (1995) suggests, this means that argumentation can be seen as a process of 'arguing for something' – in the case of technology the optimal nature of a given design – without necessarily arguing against anybody – a much more positive perspective.

Argumentation has three generally recognised forms: analytical, dialectical, and rhetorical (Van Eemeren, 1996). The application of analytical arguments (e.g., formal logic) to evaluate science claims has been extensive and pervasive. The capstone event of applying argumentation to the sciences was perhaps Hemple-Oppenheimer's Deductive-Nomological Explanation Model (Hemple, 1965), wherein deduction was used as an account to establish the objectivity of scientific explanations. Toulmin's (1958) examination of argumentation was, however, one of the first to challenge the 'truth' seeking role of argument and instead push us to consider the rhetorical elements of argumentation. From Toulmin's perspective, arguments consist of claims about the world that are advanced as statements of truth. These claims are supported by data whose connection to the claim is articulated through a warrant that demonstrates the significance of the evidence. Such warrants often rest on theoretical suppositions or backings that may be explicit but that are often implicit. Claims can be qualified to show the extent of the domain in which they hold true and are often subject to rebuttals or counter-arguments that may attempt to show why either the warrant, or the data, or the qualifiers are fallacious.

For Toulmin, arguments are field dependent as, in practice, the warrants and backings used to make claims are shaped by the guiding conceptions and values of the field. In science, what counts as evidence, and the theoretical assumptions driving the interpretations of that evidence, are consensually and socially agreed by the community. Technology is both similar and different. It is similar in that there are socially agreed criteria about the material solutions which will work. For instance, a building must satisfy a range of engineering specifications which are a product of our knowledge of the material world. Technology is different from science, however, in that the design must satisfy at least some elements of the design specification – a specification which is socially determined. Argumentation is central to the process of establishing the extent to which any proposed design satisfies (or not) the given design criteria. More fundamentally, what counts as 'good' design is rooted in the values associated with aesthetic judgement, which have a much greater degree of social and cultural dependence. In this respect, there is less commonality of shared standards and more potential for innovative and creative solutions. And, given that there is less social consensus about the normative criteria of 'good' design, there are many more claims that can be advanced – all of which must be evaluated through a process of argument and counter-argument.

What Toulmin offers then is twofold. First of all, his achievement is to bring argumentation out of the reified context of academia – an activity which is the sole preserve of logicians and philosophers – into the realm of the everyday. In this context, argument is a universal and daily occurrence which occurs in discussions about the best route to get from A to B as much as it does in academic discussions about the relative merits of competing explanations about climate change or what proposals might best meet the challenge of what to do. In this manner argumentation is a normative rather than exceptional dialogic process that only a few are required to undertake. Its pre-eminent role is to serve a rhetorical function of persuading the listener of the validity of the speaker's world-view. All technologists – be they engineers, textile designers, or software designers – have to engage in this process when confronted with the need to persuade their clients that their solution not only matches the specifications but offers the best solution.

Toulmin's second contribution is that his framework provides a meta-linguistic vocabulary for describing the features and elements of an argument. This is not to say that the individual concepts such as data, reasons, and justifications did not have an *a priori* existence, but rather that Toulmin's perspective on argument synthesises these elements in such a way that the significance and function become transparent. Providing such a vocabulary is important as it offers us both a schema for the analysis of argumentative discourse and – more importantly – a means of describing the function of the elements of an argument. Commonly, these are linguistic, but in the case of technology, which is heavily reliant on drawings and models, these may be visual as well. Thus, Toulmin's work provides us with a meta-level understanding that is an essential requirement to develop both teachers' theoretical understanding of argumentation and a professional language to describe and evaluate its function and purpose.

THE ROLE OF ARGUMENTATION IN THE CLASSROOM

Within an educational context, the function of argumentation must be to provide an opportunity for exposing the reasons for belief both of the ideas propounded by the teacher and the ideas held by the students. In this sense, it offers the opportunity to lay bare the rationality that lies at the heart of science and technology – essentially a commitment to evidence and rational argument as the basis of belief and action (Siegel, 1989). But in order to make judgements about the basis of belief, engaging in argumentation demands the use of the higher order cognitive processes of analysis, evaluation, and synthesis (Bloom, 1956). These in turn require students to clarify their prior knowledge (Aufschnaiter, Erduran, Osborne, & Simon, 2008). Thus, learning to argue is both a process of *learning to think* and a process of *arguing to learn* (Andriessen, 2006).

The evidence for its role in assisting conceptual learning comes from a number of studies. For instance, Hynd and Alverman (1986) found that if students were given texts that not only explained why the right answer was right *but also explained why the wrong answer was wrong*, and students were provided with an opportunity to deliberate about the merits of the competing explanations, then students' conceptual learning, compared to a control group, was enhanced. Likewise, Howe, Rodgers, and Tolmie (1990) and Howe, Tolmie, and Rodgers (1992) found in studies based on a comparison of learning of groups that held differing preconceptions compared to those who held similar preconceptions, that the groups consisting of those who held different ideas consistently made greater progress whilst those who worked in groups who held similar ideas made no progress whatsoever.

More evidence comes from the research conducted by Zohar and Nemet (2002) working with a two classes of 16–17 year old students studying genetics. Their approach required the teacher to permit students opportunities to engage in argumentative discourse about the appropriate answer to specific problems. They then compared the outcomes of their work with similar control classes who were taught in a more traditional manner. Three statistically significant findings emerged from their work: first, the frequency of the students who did not consider biological knowledge was higher in the comparison group as compared with the experimental group (30.4% versus 11.3%, respectively); second, the frequency of students who used false considerations of biological knowledge was higher in the comparison group as compared with the experimental group (16.1% verses 4.8%, respectively); and, finally, the frequency of students who correctly considered specific biological knowledge was higher in the experimental group as compared with the comparison group (53.2% vs. 8.9%, respectively).

In our own work (Aufscheiter et al., 2008) we have found that providing students the opportunity to engage in argumentation about scientific concepts was an opportunity not so much to develop new knowledge but an opportunity to consolidate old knowledge. In the case of technology, the specific concepts to be learnt are undoubtedly more nebulous – an understanding of the strengths and weaknesses of any given materials; how they perform under load; the criteria that constitute 'good' design, etc. Moreover, there is little research that has been conducted on argumentation in this context. But what is common to all the above examples is not so much the discipline but the process and the positive outcomes. Hence, it is reasonable to predict that offering students of technology the opportunity to engage in small-group discussion and argumentation would provide a means by which individual thinking can be held up for examination by others and, if necessary, challenged. Through this process, both parties would gain a deeper understanding of some of the core ideas of the discipline – a primary educational goal of technology education.

HOW DOES ARGUMENTATION WORK?

Why does engaging in argumentation have these effects? The answer comes from a range of work but basically is explained by the demands that it makes on a range of higher order skills and, as put by Billig (1996), "learning to argue may be a crucial phase in learning to think" (p. 141).

Argumentation requires the participant to a) listen to what the other person has to say, which, in and of itself, is a social skill that many young children lack; and b) to compare and contrast the idea that they have just heard with their own thinking, which requires a student to reflect not only on their own ideas and cognition but also on the ideas of others. This process of juxtaposing ideas develops an *evaluative* epistemology where the individual begins to see their ideas for what they are – as claims that need to be considered in the light of competing alternatives or ambivalent evidence. Individuals who do not adopt such a stance hold personal epistemologies which are *absolute* where they take their theories for granted – essentially as statements about the way the world is. Alternatively, many young people hold a *multiplist* epistemology that sees all ideas as being of equal worth and with no criteria, which could establish the relative merits of competing claims – essentially a highly subjective and individualist view of life. More than any other school discipline, technology is the one field where it is patently obvious that such thinking is inadequate. Clearly there are criteria that make some designs and solutions better than others, making the multiplist position untenable. Moreover, these criteria and what they should be are the subject of debate and argumentation. Technology education therefore has a vital role to play in advancing student thinking towards becoming *evaluativist*.

Kuhn (1999) argues for a developmental model of critical thinking where young people start out as absolutists. Then, in adolescence, recognising that such a position is untenable, many shift to becoming multiplists – a position typified in lay terms by the response of 'whatever' to any challenge – a response which means that what you think is fine for you just as what I think is fine for me. However, such a position is ultimately untenable as it denies the possibility that some solutions are better than others and gives free rein to those who might wish to hold ideas which might be considered racist or offensive in other ways. As Kuhn argues:

> To be competent and motivated to 'know how you know' puts one in charge of one's own knowing, of deciding what to believe and why and of updating and revising those beliefs as one deems warranted. To achieve this control of their own thinking is arguably the most important way in which people both individually and collectively take control of their own lives. (p. 23)

It is only by considering alternatives – by seeking to identify what is not – that one can begin to achieve any certainty about what is. However, to know why the wrong answer is 'wrong', or in the case of technology to understand why one solution is more optimal than another, requires calls on the higher order cognitive skills of analysis, synthesis, and evaluation. Some insights into why this is so, and how argument 'works', comes from the work of Ohlsson (1996). He suggests that it is possible to distinguish two principal forms of learning – learning to do and learning to understand. Learning to understand requires engagement in a variety of discourse acts, which he defines as shown in Table 1. All of these discourse acts are to be found in the kind of dialogic discourse promoted by asking students to engage in argumentation. Clearly in the context of science, the processes of predicting, explaining, and explicating might predominate; in technology, one might expect arguing and critiquing to dominate. As Ohlsson argues, however:

> This tentative taxonomy of epistemically relevant activities is short but surprisingly complete. Whatever else do we ever do when we talk or write, over and above describe, explain, predict, argue, explicate and define? Although there are many other types of speech acts (e.g. to promise, request and threaten) no epistemically relevant extensions to the taxonomy comes to mind.

It is important to note that Ohlsson makes no distinction between the importance afforded to each of these categories. They are not, for instance, to be seen as a hierarchy reflecting increasing cognitive demand. Moreover, it is reasonable to suppose that some of them may be more epistemically demanding than others. Tentatively, I would wish to argue, following Kuhn (1991) and Miller (1987), that some of these skills are more indicative of higher order cognitive thinking than others. For instance, description is a relatively undemanding task that is often reliant on a mastery of language simply to convey the major perceptual features of an object. The task becomes more difficult when the object is not easily visible, that is, when it is too small to be seen or too large to be imagined. In such cases, individuals have to resort to the use of analogy or metaphor to construct meaning for the individual (Harré, 1986). Nevertheless, description is perceptually driven and rarely requires abstraction. However, describing an object not yet in existence, that is, one that is being imagined requires the ability to visualise and forms of representations other than linguistic.

Explanation, predicting, and arguing, in contrast, are essentially all forms of warranted justification. The essential difference between explanation and argument is that in an explanation we reason backwards from a known fact to the cause or reason for that fact using theoretical ideas to justify why a particular event has occurred. In an argument or a prediction, we reason forward from a premise to that conclusion. However, and this is the primary point, all of these discourse acts make varying degrees of cognitive demand on students, a demand that may generate conceptual conflict requiring either assimilation or accommodation. Without such demand, conceptual change and restructuring is heavily reliant on intra-mental processes to generate cognitive conflict and the intrinsic motivation of an individual, making it less likely to occur.

Table 1. Epistemic discourse activities (Ohlsson, 1996).

Describing	To describe is to fashion a discourse referring to an object or an event such that a person who partakes of that discourse acquires an accurate conception of that object or event.
Explaining	In the canonical explanation task, the explainer is faced with an event of some sort (e.g., the sinking of the Titanic, the demise of the dinosaurs) and fashions a discourse such that a person who partakes of that discourse understands why that event happened.
Predicting	To make a prediction is to fashion a discourse such that a person who partakes of that discourse becomes convinced that such and such an event will happen (under such and such circumstances).
Arguing	To argue is to state reasons for (or against) a particular position on some issue, thereby increasing (or decreasing) the recipient's confidence that the position is right.
Critiquing (evaluating)	To critique a cultural product is to fashion a discourse such that a person who partakes of that discourse becomes aware of the good and bad points of that product.
Explicating	To explicate a concept is to fashion a discourse such that the person who partakes of that discourse acquires a clearer understanding of its meaning.
Defining	To define a term is to propose a usage for it. When the term already exists in the language, the boundary between defining and explicating is blurred.

Evidence for the epistemic function of Ohlsson's (1996) set of discourse activities comes from the work of Chi, Bassok, Lewis, Reimann, and Glaser (1989) where eight college students were asked to explain to one another what they understood from reading statements from three examples taken from a physics text. The four students who were more successful at solving problems at the end of the chapter (averaging 82% correct in the post-test) were the ones who had spontaneously generated a greater number of self-explanations (15.3 explanations per example) while studying the chosen examples. In contrast, the four students who were less successful achieved only 46% on the post-test and had generated only 2.8 explanations per example. Similar findings emerged from a later study with 14 eighth grade students where the focus this time was on declarative knowledge of the functioning of the valves in the heart (Chi, De Leeuw, Chiu, & Lavancher, 1994). Notably, the improvement for the students required to generate explanations was more significant for the more complex questions used to test understanding. Chi et al. postulate the reason such explanation is effective is that, having articulated an incorrect explanation, further reading of ensuing sentences (which always present the correct information) or counter arguments offered by another will ultimately lead to the generation of a contradiction that will generate conflict. The outcome of such conflict will require metacognitive reflection and self-repair by the student if resolution is to occur. Thus, argumentation 'works' because it enhances the chances of demonstrating cognitive conflict.

The exercise that Chi et al. (1994) conducted was essentially a decontextualised psychological experiment and not a process that lends itself easily to the classroom. Small-group work and argumentation, however, provide a naturalistic context in which self-explanation is both required and challenged – the very activities that Chi et al. have found to be so effective at enhancing student learning. Evidence that this is so comes from a review conducted by Noreen Webb (1989) of 19 published studies on learning mathematics and computer science. Her finding was that the level of elaboration of students' interaction with other students was related to achievement: those that gave high-level elaboration to other members of the group achieved more highly whereas offering only low-level explanations had no effect. Her explanation for this effect was that:

> In the process of clarifying and reorganizing the material, the helper may discover gaps in his or her own understanding or discrepancies with others' work or previous work. To resolve those discrepancies, the helper may search for new information and subsequently resolve those inconsistencies, thereby learning the material better than before ... Furthermore, when an explanation given to a team-mate is not successful, the helper is forced to formulate the explanation in new or different ways. (p. 29)

Clearly technology education offers good opportunities for group work. Groups of students might explore different design alternatives, make their presentations arguing why they believe their solution is optimal, and defend it against criticism that their fellow students and teachers may offer.

Another important insight into the value or argumentation comes from the work of Hatano and Inagaki (1991). The data they draw from many experimental studies of small groups working collaboratively on problems led them to draw three conclusions. First, that students often produce knowledge that can seldom be acquired without such interaction – a point which was recognised by the Headmaster of Eton, a leading English private school when he stated that "We have always recognised two principles in education, which are first that young people teach each other more than adults think they teach them, and, secondly, that at least as much learning goes on outside the classroom as within it" (Eyres, 2008). Hatano and Inagaki's second finding was that the nature of the knowledge acquired by the majority of the participants is very context-specific to the group, even when the structure and nature of the activities are very similar; that is, that the variance between groups is large. Finally, they also found that the knowledge acquired within the groups by individual members differed considerably; that is, that the intra-group variance is large. What this suggests is that the nature of the group dynamics has a considerable effect on the potential outcomes of any small-group discussions or argumentation and, thus, the organisation and selection of groups requires attention by any teacher.

IMPLEMENTING ARGUMENTATION IN THE CLASSROOM

For the teacher, the issue of how to establish argumentative discourse of high quality within their classroom is akin to an engineering challenge. Much effort has been expended by those working on computer supported collaborative learning exploring how this might be achieved (Andriessen, Baker, & Suthers, 2003; Linn & Bell, 2000; Scardamalia, Bereiter, & Lamon, 1996). However, many classrooms do not have ready access to such technology and their needs are both more prosaic and more complex: More prosaic in that the issue is simply one of how can argumentative discourse be implemented without any computer technology; and more complex in that there is more to using such an approach than can be captured by any technology.

The approach taken by the author and his co-workers was to develop a pack of materials to support the professional development and learning of teachers of science (Osborne, Erduran, & Simon, 2004b) in collaboration with a group of eight teachers. This pack focused on six themes considered to be essential theoretical and pedagogical knowledge for any teacher wishing to adopt the use of argumentation in their classroom. Whilst these themes were developed in the context of science, they are no different for the teacher of English (Nystrand, Gamoran, Kachur, & Prendegarst, 1997) or the teacher of technology. They are:

Introducing argument Underlying the use of argument is a set of theoretical ideas about the elements that constitute an argument, why it is central to the science, and how it might assist student learning. Such theoretical understandings are essential knowledge without which teacher pedagogy becomes simply the performance of a set of practices whose value and intent is obscure (Adey, Landau, Hewitt, & Hewitt, 2003).

Managing small group discussion Small-group discussion is a central element of using argumentation in the classroom. Yet its use is relatively rare in the science classroom (Newton, Driver, & Osborne, 1999; Osborne & Collins, 2000; Sands, 1981) and standard pedagogical techniques of pairs, envoys, listening triads and jigsawing (Johnson, Johnson, & Johnson-Houlbec, 2002) are unfamiliar to many teachers of science and technology.

Teaching argumentation Work on our earlier project with 12 teachers (Osborne, Erduran, & Simon, 2004a) had established that there are number of pragmatic issues that confront teachers. For instance, teachers need to be able to define and value goals other than conceptual learning. Such goals and their value must be communicated to students. What are the different structures and lesson types that will promote argumentation? In addition, the rhetorical project of the teacher is to persuade students of the superiority of the scientific explanation (Osborne, 2001). Providing space to discuss erroneous ideas or misconceptions would seem, at least in the first instance, to undermine such a goal. Furthermore, teachers need to have some understanding of the scaffolds that will stimulate argument in small groups (Simon, Erduran, & Osborne, 2006).

Resources for argumentation Teachers need exemplars of materials and strategies that can be used to develop argumentation with their students. In essence, a starter pack of ideas. In this case, a set of materials that had been developed by the teachers we had worked with initially was incorporated in the pack.

Evaluating argument Once students have begun to engage in argumentation, teachers have to facilitate and scaffold the process. Doing so requires judgements to be made about the quality of argumentation. Many frameworks exist for evaluating argument but most are reliant on Toulmin's (1958) framework and an appreciation of the need to identify which of these elements are present or absent in a given argument. Teachers need opportunities to analyse arguments and share their understanding of the relative merits of a range of arguments on a common topic.

Modelling argument Part of the pedagogical knowledge associated with argumentation is an understanding of how to represent and communicate what the elements of an argument are to students using appropriate language. Exploring ways of doing this is an important part of the professional learning required to engage in argumentation.

Crucial to teachers' attempts at innovative strategies, however, are the insights that come from observation of another teacher engaging in the practice in a context similar to their own. This enables critical reflection on the strengths and weaknesses of any novel practice and reflection on how it might be adapted to their own students and context. For this reason, these materials were supplemented by 28 video excerpts drawn from lessons of experienced science teachers using argumentation. As Von Glaserfeld (1989) points out, such video extracts are "a far more powerful didactic tool" (p. 3) than oral or written materials which are often, at best, only half understood. Video, he argues, has been responsible for the major advances in the teaching of sports in the past two decades. Likewise, we would contend, its value for teacher professional development.

Least but not last is the capacity for the use of argumentation as a pedagogic approach to enhance student engagement. Some direct empirical evidence for this comes from the work of Nolen (2003). In a study of 377 students in 22 introductory science classes in the ninth grade year, she showed that perceptions shared amongst the students that classrooms focused on understanding and independent thinking positively predicted students' self-reported satisfaction with learning. The conclusion she drew was that "in these classrooms where students perceive their science teacher as interested in student understanding and independent thinking, rather than in the speedy recitation of correct answers, students were more likely to have productive and satisfying learning experiences" (p. 365).

CONCLUSION

In conclusion, clearly the data here are limited, but asking students to engage in argumentation is a recognition that what students think and their reasoning is something which is both valued and valuable. As such, argumentation offers a means of transforming the deep grammar of pedagogy in school science and technology from one rooted in a view of science and technology as well-established bodies of knowledge, skills, and processes to be transmitted; to one which sees science and technology as a set of beliefs that must be justified by an appeal to evidence – a transformation which is long overdue.

REFERENCES

Adey, P. S., Landau, N., Hewitt, G., & Hewitt, J. (2003). *The professional development of teachers: Practice and theory.* Dordrecht, The Netherlands: Kluwer.

Andriessen, J. (2006). Arguing to learn. In K. Sawyer (Ed.), *Handbook for the learning sciences* (pp. 443–460). Cambridge, UK: Cambridge University Press.

Andriessen, J., Baker, M., & Suthers, D. (Eds.). (2003). *Arguing to learn: Confronting cognitions in computer-supported collaborative learning environments.* Dordrecht, The Netherlands: Kluwer Academic Publishers.

Aufschnaiter, C., Erduran, S., Osborne, J. F., & Simon, S. (2008). Arguing to learn and learning to argue: Case studies of how students' argumentation relates to their scientific knowledge. *Journal of Research in Science Teaching, 45*(1), 101–131.

Bachelard, G. (1940). *The philosophy of no.* Paris: Paris University Press.

Billig, M. (1996). *Arguing and thinking* (2nd ed.). Cambridge, UK: Cambridge University Press.

Black, P., & Atkin, J. M. (Eds.). (1996). *Changing the subject: Innovations in science, mathematics and technology education.* London: Routledge.

Bloom, B. S. (Ed.). (1956). *Taxonomy of educational objectives: The classification of educational goals. Handbook 1, Cognitive domain.* London: Longmans.

Chi, M., Bassok, M., Lewis, M. W., Reimann, P., & Glaser, R. (1989). Self-explanations: How students study and use examples in learning to solve problems. *Cognitive Science, 13*, 145–182.

Chi, M., De Leeuw, N., Chiu, M. H., & Lavancher, C. (1994). Eliciting self-explanations improves understanding. *Cognitive Science, 18*, 439–477.

Cohen, D. (1995). Argument is war....and war is Hell: Philosophy, education, and metaphors for argumentation. *Informal Logic, 17*(2), 177–188.

Driver, R., Newton, P., & Osborne, J. (2000). Establishing the norms of scientific argumentation in classrooms. *Science Education, 84*(3), 287–312.

Duschl, R. (2007). Science education in three-part harmony: Balancing conceptual, epistemic and social learning goals. *Review of Research in Education, 32*, 268–291.

Eyres, H. (2008, May 23). Bold Etonians. *Financial Times*, p. 1.

Harré, R. (1986). *Varieties of realism: A rationale for the natural sciences.* Oxford, UK: Basil Blackwell.

Hatano, G., & Inagaki, K. (1991). Sharing cognition through collective comprehension activity. In L. Resnick, J. M. Levine, & S. D. Teasley (Eds.), *Perspectives on socially shared cognition* (pp. 331–348). Washington, DC: American Psychological Association.

Hemple, C. G. (1965). *Aspects of scientific explanation.* New York: Free Press.

Howe, C. J., Rodgers, C., & Tolmie, A. (1990). Physics in the primary school: Peer interaction and the understanding of floating and sinking. *European Journal of Psychology of Education, V*, 459–475.

Howe, C. J., Tolmie, A., & Rodgers, C. (1992). The acquisition of conceptual knowledge in science by primary school children: Group interaction and the understanding of motion down an inclined plane. *British Journal of Developmental Psychology, 10*, 113–130.

Hynd, C., & Alvermann, D. E. (1986). The role of refutation text in overcoming difficulty with science concepts. *Journal of Reading, 29*(5), 440–446.

Johnson, D. W., Johnson, R. T., & Johnson-Houlbec, E. (2002). *Circles of learning: Cooperation in the classroom* (5th ed.). Minnesota: Interaction Book Company.

Kuhn, D. (1991). *The skills of argument.* Cambridge, UK: Cambridge University Press.

Kuhn, D. (1999). A developmental model of critical thinking. *Educational Researcher, 28*(2), 16–46.

Linn, M., & Bell, P. (2000). Designing the knowledge integration environment. *International Journal of Science Education, 22*(8), 781–796.

Miller, M. (1987). Argumentation and cognition. In M. Hickmann (Ed.), *Social and functional approaches to language and thought* (pp. 225–249). Orlando, FL: Academic Press.

Newton, P., Driver, R., & Osborne, J. (1999). The place of argumentation in the pedagogy of school science. *International Journal of Science Education, 21*(5), 553–576.

Nolen, S. B. (2003). Learning environment, motivation and achievement in high school science. *Journal of Research in Science Teaching, 40*(4), 347–368.

Nystrand, M., Gamoran, A., Kachur, R., & Prendegarst, C. (1997). *Opening dialogue: Understanding the dynamics of language and learning in the english classroom.* New York: Teachers College Press.

Ohlsson, S. (1996). Learning to do and learning to understand? A lesson and a challenge for cognitive modelling. In P. Reimann & H. Spada (Eds.), *Learning in humans and machines* (pp. 37–62). Oxford, UK: Elsevier.

Osborne, J. F. (2001). Promoting argument in the science classroom: A rhetorical perspective. *Canadian Journal of Science, Mathematics and Technology Education, 1*(3), 271–290.

Osborne, J. F., & Collins, S. (2000). Pupils' and parents' views of the school science curriculum. *School Science Review, 82*(298), 23–31.

Osborne, J. F., Erduran, S., & Simon, S. (2004a). Enhancing the quality of argument in school science. *Journal of Research in Science Teaching, 41*(10), 994–1020.

Osborne, J. F., Erduran, S., & Simon, S. (2004b). *The IDEAS Project.* London: King's College London.

Pacey, A. (1983). *The culture of technology.* Oxford, UK: Basil Blackwell.

Sands, M. K. (1981). Group work in science: Myth and reality. *School Science Review, 62*(221), 765–769.

Scardamalia, M., Bereiter, C., & Lamon, M. (1996). The CSILE Project: Trying to bring the classroom into world 3. In K. McGill (Ed.), *Classroom lessons: Integrating cognitive theory and classroom practice* (pp. 201–228). Cambridge, MA: MIT Press.

Siegel, H. (1989). The rationality of science, critical thinking and science education. *Synthese, 80*(1), 9–42.

Simon, S., Erduran, S., & Osborne, J. F. (2006). Learning to teach argumentation: Research and development in the science classroom. *International Journal of Science Education, 28*(2–3), 235–260.

Solomon, J. (1993). *Teaching science, technology and society.* Buckingham, UK: Open University Press.

Toulmin, S. (1958). *The uses of argument.* Cambridge, UK: Cambridge University Press.

Van Eemeren, F. H. (1996). *Fundamentals of argumentation theory: A handbook of historical developments and contemporary developments.* Mawhah, NJ: Lawrence Erlbaum.

Van Eemeren, F. H., & Grootendorst, R. (2004). *A systematic theory of argumentation: The pragma-dialetical approach*. Cambridge, UK: Cambridge University Press.
Von Glaserfeld, E. (Ed.). (1989). *Constructivism in education*. Oxford, UK: Pergamon.
Webb, N. M. (1989). Peer interaction and learning in small groups. *International Journal of Education Research, 13*, 21–39.
Zohar, A., & Nemet, F. (2002). Fostering students' knowledge and argumentation skills through dilemmas in human genetics. *Journal of Research in Science Teaching, 39*(1), 35–62.

Jonathan Osborne
Graduate School of Education
Stanford University
Palo Alto, California
USA

SECTION 6

ASSESSMENT IN TECHNOLOGY EDUCATION

JUDY MORELAND

37. ASSESSMENT: FOCUSING ON THE LEARNER AND THE SUBJECT

Although for many years assessment in technology education has focused on far more than marks of attainment, the development of assessment practices has been inextricably tied to what has been considered important in technology education itself. For assessment to be meaningful and authentic to technology the main learning purposes of technology education need to be defined. Technology education has evolved over the years to be a subject far more than that focusing only on student replication of a narrow range of technical aspects.

Whilst there is no one universally accepted view of technology education there is increasing international agreement that it is a subject requiring both the intellectual and practical involvement of students. It generally includes knowing about, and using, technological knowledge and processes; developing technological artefacts, systems and environments; developing technological competencies and skills; and knowledge of the relationships between technology and groups of people, cultures, societies, environment, and economies. Creative and critical thinking are required; innovation, problem solving, and designing are at the subject's heart.

At the same time as developing all-round technological literacy in students, teachers have the added responsibility of encouraging students to become life-long learners who are intimately involved in their own learning progress. The challenge for technology teachers then becomes one of juggling juxtaposing demands – on the one hand they need to help students become creative, innovate, and critical thinkers in technology; on the other they need to help students achieve particular technology learning goals. They are required to help students become life-long learners (a long-term and open-ended goal) yet at the same time they need to provide evidence of technology learning in a unit (a short-term and definite goal). So the question of authentic assessment arises – what is authentic for the subject and what is authentic for the learner. In technology, authentic assessment involves examining the processes students undertake to resolve a design task and how they use knowledge and skills as a resource for action. The interactive processes of thinking and decision-making are more important than the products that students create. The interest is in how and why students do things rather than what they choose to do.

The chapters in this section of the *Handbook* explore assessment from several perspectives and are a somewhat eclectic grouping. A range of insights is presented and these are drawn from a variety of contexts and nations. This diversity means that a rich set of chapters is provided, and collectively they highlight many issues and ideas that arise when technology assessment is put under the microscope. Both formative assessment, now often referred to as assessment for learning (AfL), and summative assessment are explored, as is assessment in primary and secondary schools.

In Chapter 38, Christine Harrison considers assessment for learning practices and the contribution that effective AfL makes to learning. She takes a broad overview of AfL and describes the key principles underpinning it. She especially focuses on the work undertaken by the team in assessment at Kings College, London and draws on work published in the *Black Box*

A.T. Jones and M.J. de Vries (eds.), International Handbook of Research and Development in Technology Education, 445–447.
© 2009 Sense Publishers. All rights reserved.

series. She highlights areas teachers may need to consider if they wish to strengthen their formative practice, and explains the role of talk in this process. For example, teachers need to orchestrate class opportunities to discuss ideas, encourage students to express tentative ideas more openly, and plan for generating effective questions. Examples from technology classrooms are used to illustrate these points. The chapter concludes with a discussion of the importance of teacher professional learning to help teachers think about AfL. Such professional learning should be undertaken in a continuous, iterative process and include regular and sustained times for professional dialogue, teacher reflection and learning, and teacher evaluation of their progress over time and with colleagues.

Chapter 39, written by myself and Bronwen Cowie, is situated within a New Zealand primary school context. The chapter explores the development of AfL strategies to support teachers with learning in technology, and ways to develop teacher knowledge to better deal with learning in the classroom. The idea of pedagogical content knowledge (PCK), a specialised form of knowledge about a subject, is raised in the context of technology teachers becoming more knowledgeable for teaching technology effectively. The chapter goes on to highlight the impact of a technology-specific planner. Using the planner helped teachers focus on technology learning as multidimensional, coherent, and connected. It also helped them think of how they could employ AfL practices to help their students develop as independent, creative, and critical designers. Examples are also provided of primary teachers employing AfL in their technology classrooms to help students learn effectively in technology. These examples provide a case for teachers' AfL practices to encompass the multidimensional nature of technology; to focus on building connections, continuity, and coherence between ideas and actions over time; and to use multimodal approaches. Related assessment issues and ideas in technology are raised.

Chapter 40, written by Marilyn Fleer and Gloria Quiñones, deals with assessment from a funds of knowledge perspective. Like Moreland and Cowie in Chapter 39, they agree that teachers need to be knowledgeable about the students they teach if they wish for their interactions with them to be productive. They argue that in educational settings teachers often know little of the breadth and depth of knowledge their students have, and therefore have a restricted base from which to create worthwhile experiences to help students learn more effectively. A funds of knowledge theoretical approach frames assessment beyond the classroom and bridges into students' homes and community, assisting teachers to gain insights into the lived experiences of their students. The authors go on to demonstrate auto-photography as a method to assist students to research and learn about their own technological knowledge and capability within and beyond school. To this end, they describe a small-scale research project set in two primary classrooms in southern Australia where students took technology photographs and talked about them. They conclude that the insights teachers can gain from involvement with students in such a project can deepen and broaden their knowledge about their students, and therefore impact on the experiences and conversations they subsequently have with their students in the classroom.

Chapter 41 by Kay Stables and Jenny Bain adds to the discussion on assessment for learning from yet another perspective – by considering ways to shift towards an integrated system of assessment in the context of technology education. They examine the role summative assessment methods can play as part of effective formative practice, providing explicit examples from their research in English schools relating to performance based, authentic assessment. The potential of the 'unpickled portfolio' to help learners reflect on their progress – and teachers to gain insights into the effectiveness of their teaching – is described. Three case studies are used to illustrate how the unpickled portfolio can provide scope for broader assessment rather than it being just a narrow, explicitly summative assessment instrument. The case studies also demonstrate how the unpickled portfolio approach to assessment has the potential to have a positive effect on learning, to develop sustainable assessment practices, and to provide comprehensive assessment data. The authors conclude by discussing the central place of the learner in the development of sustainable assessment and their belief that to achieve such assessment requires democratising. Even in high stakes summative assessment, Kay and Jenny see a role for learners *and* teachers, as a shared involvement has the potential to maximise learning because all parties need to work together to develop mutual understandings of the holistic qualities being assessed.

Building on the work presented in Chapter 41, Richard Kimbell reports in Chapter 42 on the development of performance portfolios at the secondary level in the UK and how they can be used to demonstrate students' thinking, their ability to tackle tasks, and the ways they bring them to effective resolution. The Technology Education Research Unit (TERU) at Goldsmiths was commissioned to find ways to overcome the problem of schools working to upgrade their league table positions through 'doctoring' students design portfolios to match the mark scheme and thus gain higher marks. Richard describes the evolution of portfolio assessment to now include e-scape portfolios, where real-time performance portfolios focus on students generating direct evidence of process-centered capability. He describes the steps in the creation of an *e-scape* portfolio and illustrates the richness of the data generated by the students who were involved. He concludes that the use of *e-scape* portfolios assists students to better describe their work and the circumstances surrounding it; to leave a real-time evidence trail; and that their approach to the task is enriched as they are more motivated, take more risks, and think more deeply about what they are doing. Finally, he discusses the three-round judging process for assessing *e*-scape portfolios. The positive features of the e-scape system have been acknowledged and will be used from September 2009 by Awarding Bodies, and that real awards (GCSEs and Diplomas) will be given as a result.

The important role of peer and self assessment runs through all the chapters. Although teachers and external bodies are also necessarily involved in assessment, for students to become independent and innovative technology learners they must take responsibility for placing value on their ideas and capabilities. To do this, they need to understand those technology practices and ideas viewed as central to the subject and especially how they bring those practices and ideas to light.

The chapters in this section serve to focus on what is important for technology, technology teachers, and technology learners. Interlaced throughout are ideas about how assessment in technology is increasingly focused on the learner and is more technologically authentic.

Judy Moreland
Wilf Malcolm Institute of Educational Research
University of Waikato
Hamilton, New Zealand

CHRIS HARRISON

38. ASSESSMENT FOR LEARNING – A FORMATIVE APPROACH TO CLASSROOM PRACTICE

INTRODUCTION

Since the publication of the review *Inside the Black Box* by Paul Black and Dylan Wiliam (1998b), Assessment for Learning (AfL) has been a key initiative for schools in the UK and elsewhere to improve the learning experiences of their students and to raise standards. This chapter outlines some of the factors involved with this initiative and highlights areas that teachers may need to consider if they wish to strengthen their formative practice. The first section looks at the main ideas that underpin Assessment for Learning, while the second section focuses on how classroom talk helps drive the formative approach. The final section highlights the importance of professional learning, including how and why a collaborative and long-term approach is needed to establish and sustain the changes needed for Assessment for Learning to be effectively developed. While the ideas discussed are generic in that formative assessment underpins all teaching-learning scenarios, the ideas will be explained and illustrated with a view to how such practice can work effectively in a design and technology classroom.

Classroom assessment has an essential role to play in all curricular areas and technology is no exception to this. Technological areas, contexts, and the way related variables are identified and operationalised, affect the complexity of the task, which, in turn, will affect the approach to assessment in deciding both what and how to assess. To develop strategies for assessment of technology tasks and activities, teachers require clear understandings of the nature of technology and the conceptual and procedural aspects of different technological areas (Jones, Hawe, & Mather, 1994). It is important that teachers have a view of technology consistent with the technology curriculum they are teaching because when there is any gap or friction between these, assessment becomes problematic (Jones & Moreland, 2004). At the same time, students need to be involved with and assessed across a variety of areas, contexts, and variables (Jones et al.) so it is important for technology teachers to not only focus on assessment within technological areas in which they lead the teaching, but also to consider the learning experiences students engage with across the technology curriculum. This is particularly so for schools who deliver the technology curriculum through a carousel approach with students engaging with different technological areas on a termly, half-termly, or monthly basis. The assessment processes for each part of this carousel approach need to link between and progress with each block of technological learning.

It is important for assessment processes to also consider both procedural and conceptual aspects, since the learning outcomes of technology education involve both knowing and doing. Formative assessment, in particular, helps teachers determine next steps for learning by enabling them to filter the rich data that arise when students undertake technology. Feedback, including peer and self-assessment, has an important role to play in this process and, utilised properly, formative assessment can result in large learning gains (Black & Wiliam, 1998a,b; Wiliam, Lee, Harrison, & Black, 2004).

A.T. Jones and M.J. de Vries (eds.), International Handbook of Research and Development in Technology Education, 449–459.
© 2009 Sense Publishers. All rights reserved.

THE PRINCIPLES

Formative practice is about providing guidance for improvement. Learners need feedback during the learning process so that they can reshape what they do and what they think. This means that teachers need to look for evidence in students' work that exemplifies how well it is understood. This helps teachers to more carefully pitch the next events in the learning, be they in the discussion that follows or in follow-on activities in the same or subsequent lessons.

At first it seems a simple task to decide on next steps in learning, but when teachers start to explore children's understanding more carefully, they become more aware of the misunderstandings that need to be unravelled, the differentiated approach that needs to be adopted with different groups of learners, and how the pace of future learning is dependent on confidence in the current learning. Teachers also need to provide opportunities for the learner to respond to feedback and to give dedicated time to reviewing their work. In this way, the 'partly known' ideas are sorted out as well and the correct ideas consolidated.

Sometimes the starter activity for the lesson helps the teacher identify the level of understanding, but it is more likely for this to occur as the lesson progresses and students begin to develop their ideas and designs. When students are engaged with drawing, designing, and making activities it is important for the teacher to working alongside them so that they can interact with the students' emerging ideas. Leaving the students to work unassisted conveys the message that design and technology is a self-help activity, which can deter students from exploring their ideas with others. Through discussion, students are able to help each other by clarifying and developing ideas, communicating alternatives and solutions, and critiquing their own and others' ideas in a supportive environment (Moreland, Cowie, & Jones, 2007). Class and peer discussion provides the teacher with opportunities to notice, recognise, and respond to students' ideas as they emerge and so make it easier for them to take formative action (Cowie, 2000). One technology teacher described this way of working as:

> My formative interactions have definitely changed. Before I'd go over and see how they're getting on and let them go. But now I'm getting them thinking all the time. I used to accept things whereas now I look to see if the conceptual, procedural, technical and societal stuff is there. My questions got them to think about things, so they could change them on the way. They could improve it while they're going. (Moreland, Jones, & Barlex, 2008, p. 17)

Two of the main ideas developed by the research in this area are that new learning starts from a learner's existing understanding and that the learner should be actively involved in the learning process. Assessment for Learning activities are productive because they exemplify these principles, and also because they put into effect two other research findings: That teachers should develop the learner's overview of quality of achievement and self-assessment (i.e. what is a good technology design and product for the age group); and how does my attempt compare social learning (i.e. learning through discussion with one's peers) (Wood, 1988).

Many of the strategies that teachers use to teach in a formative way are now well known by teachers. For example, teachers might decide to use 'Think-pair-share', a method to encourage active, productive classroom talk where students turn to the person sitting next to them to discuss the question being asked. This allows for exchange of ideas, sharing of vocabulary, and an opportunity to formulate an answer through collaborative peer talk before sharing with the whole class. What is important here is how the teacher orchestrates this strategy in the classroom so that it gives opportunity for learners to consider their current understanding, to share this with others, and to map this against the developing understanding of the group as the learning activity progresses. This requires teachers to be both proactive, in setting up activities that foster this co-construction of knowledge, and reactive to the learners' needs, in scaffolding the activity as it progresses to keep all learners actively on board.

Thus, it is not simply a matter of teachers adopting Assessment for Learning strategies but rather thinking through how they can create learning scenarios that offer opportunities to cue students into learning intentions and expectations that provide feedback as the learning is taking place. Through this process, learners begin to understand what constitutes quality (Sadler, 1989) and how they might improve their current standard of work. It is the messages that learners receive about their teacher's expectations and aspirations for their learning that matter, rather than the Assessment for Learning strategy itself. The most important of these messages is that children should eventually be able to self-regulate their own learning, and many of the skills that the teacher is trying to develop in their learners lead to this ultimate goal.

In many classrooms there is still a considerable journey to be made before learners can work independently and so self-regulate their learning. Rather, many learners are dependent on their teachers and expect their teachers to frame, direct, and schedule what they do in their technology work. While there may be a need for teachers to control the focus, pace, and direction of work in the early stages of learning new skills, space must be made for students to make decisions and take control of their own learning if they are to develop technological competency and confidence. Table 1 shows some of the stages involved in moving towards self-regulation. This progression is aided by formative practice, which moves the learning process from directed to guided and eventually to being independent. The elements of Assessment for Learning help teachers and learners develop the skills and behaviours needed for the self-regulated classroom to evolve and function.

Formative feedback is about moving learning forward. Feedback should be detailed, pointing out where progress has been made and setting two or three targets for improvement. This can be done orally or in writing. In order to develop the right tenor and emphasis, it is important for teachers to realise and believe that learning is not just a cognitive exercise but that it involves the whole child. Teachers know how important it is to motivate learners, but it is often assumed that this is best done by offering such extrinsic rewards as merits, grades, and prizes. There is ample evidence that challenges this assumption (Deci, 1971; Kohn, 1993; Ryan & Deci, 1996; Strong, Silver, & Robinson, 1995). Students will only invest effort in a task if they believe that they can achieve something. If a learning exercise is seen as a competition, then everyone is aware that there will be losers as well as winners; those who have a track record as losers will see little point in trying. The emotional effect leads these learners to question their learning capabilities (Stiggins, 2005). Thus, the solution is to find ways to motivate everyone even though some are bound to achieve less than others, and so the focus needs to move from comparison with others to comparison with one's own earlier attempts – a so-called ipsative approach.

Teachers need to have the mindset that all can improve, and to make this explicit through the ways they interact with their learners. In tackling this problem the type of feedback is very important (Dweck, 2000). In general, feedback given as rewards or grades enhances 'ego-involvement' rather than 'task-involvement'. It can focus students' attention on their 'ability' rather than on the importance of effort. This damages the self-esteem of low attainers, and leads some high-attainers to avoid tasks that challenge because of the fear that they will turn out to be not so bright after all. Feedback that focuses on what needs to be done can encourage all to believe that they can improve. Such feedback can enhance learning, both directly through the effort that can ensue, and indirectly by supporting the motivation to invest such effort.

Table 1. Using formative practice to move students from being directed to being independent learners.

Pedagogic style:	Directed	Guided	Independent
Student perspective	Education is done to me.	Education is done for me.	I am responsible for my own learning.
Content	Controlled by teacher. No negotiation.	Teacher controls topics and frames of reference but these will be discussed and shared with learners.	Students select relevant information and focus.
Student's role	Acceptance of what the teacher asks them to do.	Teachers apprentice learners into the learning landscape.	Students select approach and methods.
Classroom talk	Teacher talk, where teacher generates questions and summarises.	Group and class discussions, where teacher teases out understanding and misunderstandings.	Students investigate questions that interest them, making use of peers and teacher as they see the need.
Classroom activities	Dictated notes and individual exercises.	Group activities that promote discussion.	Individual and team projects.

Marks or grades on work have no place in a formative environment for a number of reasons. The first of these is because the quality aspects of the work are coded within the grade or mark and so learners do not get specific advice on how the learning can be improved. This means that learners might know what they have achieved but do not understand how, and in which ways, the task could be done better. Secondly, marks and grades emphasise competition between learners rather than highlighting personal improvement. A knock-on effect of this is that it discourages collaborative learning and so mitigates against the very environment in which Assessment for Learning flourishes, limiting the sharing of any tentative current thinking. Thirdly, marks and grades reduce the likelihood that learners will take risks. All learners in the competitive classroom understand the culture and know the risk of failure. For many low attainers, the emotional turmoil of continual failure leads them to opt out the learning environment – a syndrome that Dweck (2000) refers to as 'learned helplessness'. It is less painful for these individuals to be thought of as 'unable to do something' than as 'failing'. For high attainers, a competitive environment also creates risks and these learners are stifled in what they do, limiting themselves to challenges that they recognise they can achieve. Thus the competitive classroom gives no challenge to high attainers who are reluctant to take risks in case they are incorrect.

In the formative classroom, the environment is much more enabling for all learners as was reported by Roger Higton, a science and technology teacher, after working on changing his Assessment for Learning practices over an 18 month period: "Here the classroom ceased to be a habitat where only the brightest survived and flourished, but one where, with careful grouping and good questioning, every student could feel themselves making progress through the lesson" (Black, Harrison, Lee, Marshall, & Wiliam 2003, p. 90).

FEEDBACK

Feedback provides guidance for how to improve (Butler, 1987) and is beneficial when it focuses on the strengths and weaknesses of student work as this focus enables students to take more control of their own learning. The following examples, adapted from Moreland (2003), demonstrate how formative interactions assisted students to become aware of and attempt to close some technology gaps. They also show how teachers noticed the students' technological strengths and weaknesses and used these to foster student learning. The first is between a teacher and a 6–year-old student and happened when the student was planning how to build a hanging birdfeeder. The teacher was not sure whether the student had thought about stability (a targeted principle). The brief conversation shows the teacher helping her student to think about stability while she was still planning:

> **T**: I see in your drawing that you've got three cords for hanging your birdfeeder. Can you tell me how you will keep your birdfeeder even and balanced and stable?
> **P**: Well, when I make it I could try to make it level by putting all the string together in my hand. And if one is longer than the others, I will cut it to the same as the others.
> **T**: That should work. (p. 226)

The second example lists some of the questions a teacher asked her 7– and 8–year-old students when they were creating the patterns for making bird masks. Her questions focused student thinking on aspects they may not have included.

> Have you thought about having lots of space for gluing on bits?
> Where will it be fastened?
> How is it fastened?
> How will you add the beak?
> What is your mask for? Will it be strong enough to last through the production?
> Do you think this material you have chosen will be flexible? What could you choose? How will you make it flexible? (p. 245)

Feedback will only become formative when it is used to take future learning forward. Therefore, if learners are given feedback on their work, they need also to be given opportunity to react to and act on advice. This is essential both for the learning at that point in time and also for instilling in youngsters a willingness to improve. It means providing opportunities within lesson time for learners to read or hear comments on their work and to identify with their peers and teacher the specific improvements that are needed. Learners need to recognise that the teacher really does want a piece of work improved and that, together, the teacher and learner can ensure that progress is made. It is useful to allocate time for the improvements to be made in class at regular intervals. A cycle of review, reflection, target setting, and formative action is a useful route to take.

If students rely solely on teachers' feedback about their work, this limits the ability of the student to understand quality in a variety of contexts. Instead, students must become intimately involved in assessing their own progress if they are truly to become self-regulated learners; the more involved they are in the assessment process, the greater will be their understanding of how to extend their learning. This is not easy to implement, however, as it requires the student to have a picture of the targets in the learning trajectory ahead of them and a means of moving forward to close the learning gap. Peer assessment has a massive role to play in this development since learners are able to see faults in a peer's work that may be common to those in their own attempts, but which are masked by an inability to reflect fully on one's own attempts. Discussion about other people's work can thus provide insights for learners engaging with quality aspects of their own work.

Teachers are increasingly using digital cameras to record work in progress, which students can refer to and compare their efforts against while still in the development stage. This has been particularly useful in practical subjects like design and technology where products can be seen during the making process rather than by comparing completed objects at the end. One technology teacher I know simply produced two copies of each of the students' work during the lesson and each child took home photographs of their work and the work of one of their peers. The homework was to explain what was similar and what was different about the two pieces of work. The results from this homework were used as a starter activity in the next lesson, where discussion of the details of the different pieces of work fostered the development of collaborative understanding both of quality aspects and how to achieve these. Such an approach is particularly useful when the students themselves have identified the stages involved in achieving quality, and have used the technology themselves to capture the images.

THE ROLE OF TALK IN ASSESSMENT FOR LEARNING

As was explained earlier in this chapter, discussing ideas with others is a vital part of Assessment for Learning practice. Many studies have mapped the type of talk that happens in classrooms (e.g., Alexander, 2004). If we look first at who does the most talking, it is clear that in most classrooms the teacher is responsible for the majority of what is said. In the King's-Medway-Oxford-Formative-Assessment-Project KMOFAP study (Black, Harrison, Lee, Marshall, & Wiliam, 2004), we looked at how teachers generally started their lessons. We found that teachers often began lessons with question and answer sessions intended to link the lesson with previous learning experiences. At the start of the KMOFAP project, we found that on average teachers dominated talk in most of theses lesson starters by an average word count of between 10:1 and 12:1. While teachers did try and engage learners by asking questions, the required answers tended to be limited to one or two words. This approach to the start of lessons tipped the dominance of talk in favour of the teacher, limiting opportunities for learners to express their ideas and making it difficult for the teacher to collect evidence of strengths and weaknesses in student understanding. With support from the King's College research team, the project teachers began to address this imbalance in the classroom talk.

By the end of the project, most of the KMOFAP teachers had introduced techniques that helped students to verbalise their ideas, reducing the dominance of teacher talk and giving students a voice. This began by consciously increasing 'wait time' (Rowe, 1974), the time a teacher takes between asking a question and accepting an answer. It was also enhanced by many teachers permitting students to rehearse and construct answers in groups prior to the whole class discussion, and working on techniques that encouraged the continuation of themes and ideas within the talk. This involved teachers planning scenarios that the class could discuss rather than using the classroom talk as a series of questions to check whether some students know the answers or not.

Improving classroom talk also involved the teachers working on how they could encourage learners to develop the tentative ideas they expressed in the comfort of a small group in the larger, more public arena of whole class discussion. To help them with this the teachers envisaged the talk as something like a jigsaw. Within the groups, individuals find and present pieces of the picture. Some pieces will be identified as important for the whole picture, while others will be discarded or set aside temporarily when their part in the whole idea is not yet clear. Sometimes jigsaw pieces can be joined to others or reshaped to fit with others such that the part of the whole picture that any group in the class possesses is likely to be clearer than the single pieces held by any individual within the group. In the next stage of whole class discussion, these group ideas can be held up and examined alongside those of other groups.

Class discussion here has a different role to play than the group dialogue. Discussion, etymologically, is derived from the Latin meaning "smash to pieces" (Isaacs, 1999) and it is through the whole class discussion that ideas are judged and reshaped, having emerged and been moulded through group dialogue. On the KMOFAP project, the teachers saw the class discussion as being more of a sieving process, where some ideas were sufficiently refined to pass on to the next stage whereas others faltered and stuck, making little progress. The way that the teacher handles class discussion has a great deal to do with how effective this is for learning as well as it being a sharing experience. So, choosing which group speaks first is important and decisions about how much of the group dialogue needs to be revealed before a second group is allowed to add, contrast, or compare their ideas needs to be considered. The skilful teacher helps draw together the themes and ideas that emerge from the group dialogues while at the same time holding up these experiences for scrutiny, challenge, and consolidation through the whole class discussion.

The teacher not only has a facilitation role to play in this, but also an active role. Learning can benefit greatly from the talk that is generated from good questions and teachers need to put planning time aside to generate these questions and to share effective questions with colleagues. Formulating good questions that make students think and motivate them to want to discuss ideas is an effective starting point. For example, questions such as, "Is it always true that metals rust?" are better at generating talk than "Which types of metals rust?" Questions that require students to predict or consider alternatives are better than those that lead to a set answer. For example, "What might the manufacturer have done to improve the packaging?" or "How would the packaging work differently if you used thinner cardboard?" are far better questions for promoting active discussion and thought than "How has the item been packaged?" and "Can you use thinner cardboard?".

Grouping, categorising, deciding on alternatives, and recognising exceptions in technology provide useful focal points for discussion. These processes encourage students to articulate their ideas and teachers to use this information to understand both students' existing knowledge and how to use that information to develop and refine students' ideas. Examples include: "Which materials would make the apron easy to clean?", "Would changing the type of fastenings or their position make it easier to put on and take off?", and "Why might buckle fasteners be a problem for aprons?".

Prediction activities provide a useful mechanism through which students can test and extend their ideas, by applying their knowledge and understanding to future situations. Through encouraging learners to bring their ideas to the fore, the teacher creates opportunities to gauge student understanding. For example, if the problem is reasoning whether a design solution is feasible or predicting if a product will be safe, the teacher may ask questions like:

- How do you know if your design will work/is safe?
- How might you check if your design will work/is safe?
- Which of your design features is key to ensuring this works/is safe?

Teachers also need to try and draw out meanings and explanations about the ideas that students suggest. The questions and statements below (adapted from Harrison & Howard, 2008) have proved useful in probing ideas further and encouraging students to talk.

For clarification:
What do you mean by that? Say a bit more. Can you give us an example?

For probing assumptions:
Why would someone say that? What do you think happened? Might someone else use it differently? Is that always what happens?

For probing reasons and evidence:
What are your reasons for saying that? Are you saying ...? Which means ...? Which part of you design fits with that idea?

For probing implications and consequences:
What might happen if you did this ... not that ...?
What other reasons might there be for that happening?
What might you do to strengthen that feature?
How might your design work if it were used with younger children/in different weather conditions/for several weeks?

For gaining viewpoints or perspectives:
What would be another way of saying that? How do Sian's ideas differ from Gashan's? What is similar and what is different about all these ideas? Would the user look at the design in the same way?

For focusing:
What other questions might be useful? Explain how that question is going to help us.

Teachers also help scaffold the emerging ideas as they arise within groups by intervention. Through careful eavesdropping of student conversations the teacher comes to understand what learners know, what they partly know, and what they do not yet know (Black & Harrison, 2004). This allows learners to work at their 'leading edge of learning' as their knowledge base is continually being challenged and upgraded through this process.

The teacher's selection of groups and the questions that she asks other groups in response to what the first group has said are also part of the teacher's intervention repertoire. A searching question or a variation on an idea from a second group can cause the first group, who has already spoken in the whole class discussion, to reconsider their ideas or to re-establish and reconfirm what they meant in their original presentation of the idea. At the same time, other groups, who have not had opportunity to share their thoughts, look at ideas already raised and begin to identify where their own thinking fits and varies. So the teacher intervention holds up a mirror for all the groups to investigate, re-establish, and evolve their ideas.

Sometimes teachers need to work on strategies to prevent them 'cutting off' classroom talk and this mostly involves them withholding from the students their judgement about an answer so that the students can compare answers given with the answers they hold inside their own heads. For example, if a teacher asks, "What's similar and what's different about the fastenings?" she can collect ideas before intervening by using a technique called 'Pose-Pause-Pounce-Bounce'. First, the teacher poses the question. Then there is a pause for thinking (while the children discuss in pairs or small groups) before she selects a student to answer. On hearing the first answer, the teacher can immediately bounce the question to a second student. The 'pounce-bounce' action prevents the teacher reacting to the first answer, possibly cutting off the class talk. The second student might give the same or a different answer to the first student and may or may not respond to what was said in the first answer. Whatever happens, the point is that the classroom talk has started to move away from the teacher judging publicly whether the student is correct, to the voicing of more student talk and ideas. The ultimate aim is to achieve several 'bounces' before they express their reaction to the ideas. This pushes the talk in the direction of the learners, which in itself is beneficial to learning, and also gives the teacher lessential 'thinking time' to plan appropriate intervention to help drive the learning forward.

PROFESSIONAL LEARNING

Change in practice entails much more than introducing teachers to new ideas and strategies. It requires teachers evolving new practice from their existing pedagogy in the light of their aspirations of how these new ideas might work with their learners. Assessment for Learning requires changes in the ways teachers work with learners, which some may find risky, and which may be challenging at times. The process is like a voyage of discovery, a journey into new territories of teaching and learning (Harrison & Howard, 2008) and requires teachers who are willing to consider their practice and what they might do to make Assessment for Learning work better in their classes. For many teachers, this is a very different type of professional development programme to ones they have previously encountered; the intention is not to show teachers what to do, nor is it designed to simply fit in to what they already do, but rather it requires them to think about the current learning behaviours of their classes and consider what pedagogic decisions they need to make to create and nurture opportunities for learning. This means a continuously iterative process of focusing on the learning of a particular class and deciding how to develop a selected Assessment for Learning strategy to best purpose. Such a voyage takes considerable time to complete and, in fact, is never really 'finished' because as the teaching changes, the learners also begin to change and this requires further tweaks to the developing pedagogy.

As teachers begin to strengthen their formative practice, it is not only their teaching that changes but also the way they react to and describe events on their professional voyage of discovery. One of the striking features of the KMOFAP project introduced earlier was the way in which, in the early stages of professional learning, many of the teachers described the new approach as 'scary' because they were concerned that they were going to lose control of their classes. Towards the end of the project, they described this same process not as a loss of control, but as one of sharing responsibility for the learning with the class – exactly the same process, but viewed from two very different perspectives. In one perspective, the teachers and students are in a delivery-recipient relationship; in the other they are partners in pursuit of a shared goal. As one of the teachers reported: "What formative assessment has done for me is made me focus less on myself but more on the children. I have had the confidence to empower the students to take it forward" (Black et al., 2004, p. 20). Another of the teachers recognised the change that he made in the locus of control:

> There was a definite transition at some point, focusing on what I was putting into the process and what the students were contributing. It became obvious that one way to make a significant sustainable change was to get the students to do more of the thinking. I then began to search for ways to make the learning process more transparent to the students. Indeed I now spend my time looking for ways to get students to take responsibility for their learning at the same time making the learning more collaborative. (p. 19)

There are no recipes for success – teachers need to take decisions about the approach they take with their formative practice. What matters here is that the approach is tailored to particular groups of learners because it is the teacher who manages and facilitates the formative classroom. It is how teachers provide guidance to help students learn and how teachers structure and facilitate the activities that enable learning. Creating an Assessment for Learning environment requires teachers to weigh up and assess how effective talk is employed in their classroom, how confident and enthusiastic their students are to 'have a go' at learning, and how well peers support and challenge one another when engaged in collaborative learning.

Teachers need to create opportunities in their classrooms for formative practice to emerge but, at the same time, openly examine this with colleagues to give credence to the information that it reveals. This is not an easy balance to manage, since change in practice is often uncomfortable and possibly threatening to individuals. Thus, such practices need support and encouragement from middle and senior management in schools, and ultimately from government. This move is unlikely to occur without extensive training through professional development programmes. Such programmes need to provide regular and sustained opportunity for professional dialogue in order to promote teacher reflection and learning, so that new practices can be evolved, moulded, and honed from existing classroom practice. Evaluation is a vital part of any plan. This should be on-going, in terms of mutual observation and of sharing of ideas and resources to support professional learning.

REFERENCES

Alexander, R. (2004). *Towards dialogic teaching: Rethinking classroom talk.* Cambridge, UK: Dialogos.

Black, P., & Harrison, C. (2004). *Science inside the black box: Assessment for learning in the science classroom.* London: NFERNelson.

Black, P., & Wiliam, D. (1998a). Assessment and classroom learning. *Assessment in Education: Principles Policy and Practice, 5*(1), 7–73.

Black, P., & Wiliam, D. (1998b). *Inside the black box.* London: GL Assessment.

Black, P., Harrison, C., Lee, C., Marshall, B., & Dylan, W. (2004). *Working inside the black box: Assessment for learning in the classroom.* London: GL Assessment.

Black, P., Harrison, C., Lee, C., Marshall, B., & Wiliam, D. (2003). *Assessment for learning: Putting it into practice.* London: OUP.

Butler, R. (1987). Task-involving and ego-involving properties of evaluation: Effects of different feedback conditions on motivational perceptions, interest and performance. *Journal of Educational Psychology, 79*(4), 474–482.

Cowie, B. (2000). *Formative assessment in science classrooms.* Unpublished doctoral thesis, University of Waikato, New Zealand.

Deci, E. (1971). Effects of externally mediated rewards on intrinsic motivation. *Journal of Personality and Social Psychology, 18,* 105–115.

Dweck, C. S. (2000). *Self theories: Their role in motivation, personality and development.* London: Taylor and Francis.

Harrison, C., & Howard, S. (2009). *Inside the primary black box.* London: GL Assessment.

Isaacs, W. (1999). *Dialogue and the art of thinking together.* New York: Doubleday.

Jones, A., Hawe, E., & Mather, V. (1994). Assessment in technology education. In *Issues in the practice of technology education. Vol. 3: Working papers from the third phase of the Learning in Technology Project.* Hamilton, NZ: Centre for Science, Mathematics and Technology Education Research, University of Waikato.

Jones, A., & Moreland. J. (2004). Enhancing practising primary school teachers' pedagogical content knowledge. *International Journal of Technology and Design Education, 14*(2), 121–140.

Kohn, A. (1993). Rewards verses learning: A response to Paul Chance. *Phi Delta Kappan, 74,* 783–787.

Moreland, J. P. (2003). *Becoming effective technology teachers: Enhancing assessment practices in primary classrooms.* Unpublished doctoral thesis. University of Waikato, New Zealand.

Moreland, J., Cowie, B., & Jones, A. (2007). Assessment for learning practices in primary technology classrooms. *International Journal of Design and Technology Education, 12*(2), 37–48.

Moreland, J., Jones, A., & Barlex, D. (2008). *Design and technology inside the black box: Assessment for learning in the design and technology classroom.* London: GL Assessment.

Ryan, R., & Deci, E. (1996). When paradigms clash: Comments on Cameron and Pierce's claim that rewards do not undermine intrinsic motivation. *Review of Educational Research, 66,* 33–38.

Rowe, M. B. (1974). Wait time and rewards as instructional variables, their influence on language, logic and fate control. *Journal of Research in Science Teaching, 11,* 81–94.

Sadler, R. (1989). Formative assessment and the design of instructional systems. *Instructional Science, 18,* 119–144.

Stiggins, R. (2005). From formative assessment to assessment FOR learning: A path to success in standards-based schools. *Phi Delta Kappan, 87*(4), 324–328.

Strong, R., Silver, H., & Robinson, A. (1995). What do students want (and what really motivates them)? *Educational Leadership, 53,* 8–12.

Wiliam, D., Lee, C., Harrison, C., & Black, P. J. (2004). Teachers developing assessment for learning: Impact on student achievement. *Assessment in Education: Principles Policy and Practice*, *11*(1), 49–65.
Wood, D. (1988). *How children think and learn* (2nd ed.). Oxford, UK: Blackwell.

Chris Harrison
Department of Education and Professional Studies
King's College
London, UK

JUDY MORELAND AND BRONWEN COWIE

39. MAKING MEANING IN PRIMARY TECHNOLOGY CLASSROOMS THROUGH ASSESSMENT FOR LEARNING

INTRODUCTION

Emerging international research suggests that enhancing teachers' – and students' – assessment for learning (AfL) practices are key to effective student learning. AfL in primary technology classrooms is complex, but effective AfL contributes to students' technology learning. When students are involved with meaningful technology tasks they develop, and deploy, designerly thinking and action; they conceive and take new ideas from the minds-eye into the made world.

Designing requires creativity and problem-solving abilities and involves concepttual, procedural, societal, and technical dimensions. For teachers to teach well in technology, they need to help their students combine their designing skills with knowledge, skill, and understanding in order to design and make products, processes, and systems. For this, teachers require rich pedagogical content knowledge (PCK), a specialised form of knowledge about a subject, its particular ideas and skills, and effective ways to teach the subject ideas and skills. This chapter highlights the impact of a technology-specific planner on teacher PCK and AfL. Using the planner helped teachers focus on technology learning as multidimensional, coherent, and connected. Further, the use of the planner helped them think of how they could employ AfL practices to help their students develop as independent, creative, critical designers. The chapter also provides examples of primary teachers employing AfL in their technology classrooms, helping students learn effectively. Their AfL practices encompassed the multidimensional nature of technology; helped students build connections, continuity, and coherence between ideas and actions over time; and were multimodal.

TECHNOLOGY EDUCATION

People use technology to expand their possibilities, intervening in the world through the development of products, systems, and environments. This requires the application of intellectual and practical resources. Design, innovation, invention, form, and function are features of technological thinking and action. Any technology design task requires that students think and work reflectively and protectively so that the task can be accomplished in an iterative and holistic manner, rather than by using a step-by-step, unconnected, and linear approach. Because young students often find it difficult to work iteratively when designing, making, refining, and testing, however, they require teacher assistance to maintain a focus on the overall goal and the overarching design task (Kimbell, Stables, Wheeler, Wozniak, & Kelly, 1991; Stables, 1997). Without teacher assistance, design processes may become ritualised and steplike and students may design in a mechanical, step-wise fashion creating a veneer of accomplishment. Student portfolios that provide evidence of technological processes and progress, for example, may be revamped retrospectively to ensure that evidence of all steps they needed to address are included, even if those steps were not actually undertaken (Hennessy, McCormick, & Murphy, 1993; Lave, 1988).

A.T. Jones and M.J. de Vries (eds.), International Handbook of Research and Development in Technology Education, 461–476.
© 2009 Sense Publishers. All rights reserved.

Innovation, creativity, and genuine design problem solving may be sacrificed for performativity (Nicholl & McLellan, 2008).

Additionally, when students undertake complex technology tasks, problems and issues may arise that are unforeseen by them or their teacher. These problems and issues often need to be dealt with on the spot and frequently require teacher assistance for students to continue. For teachers to respond productively to students' emerging technological problems and issues, and to help students develop their ideas and skills, they require an in-depth knowledge of the multidimensional and integrated nature of technology and of effective assessment for learning (AfL) principles. For students to have ownership of their technology learning and work in independent ways, they require design tasks that are relevant to them, provide creative learning opportunities, and afford some leeway to depart from strictly adhering to fixed design specifications in a rigid order.

TEACHERS AND ASSESSMENT FOR LEARNING

Students learn effectively when teachers utilise a variety of appropriate approaches, such as providing models to be replicated, orchestrating tasks and opportunities, providing feedback and guidance, and providing explicit explanations of principles and procedures (Wells & Claxton, 2002). Using such a variety of approaches when teaching technology poses a number of challenges given that technology tasks need to be multidimensional, complex, integrated, and undertaken over time. However, the use of AfL provides a pedagogical approach for teachers who are seeking to meet these challenges, since AfL is undertaken during learning and is about recognising and responding to learning for the purposes of enhancing it (Bell & Cowie, 2001). AfL emphasises the crucial role of formative assessment in the learning process, is an essential feature of classroom learning, and its development can raise achievement (Black & Wiliam, 1998). To promote learning, AfL encourages authenticity and complexity in student learning and assessment rather than emphasising the reproduction of knowledge, it promotes formative assessment rather than summative assessment as the main driver for teaching and learning, it enables students to build confidence and capabilities, it promotes informal and formal feedback on learning, and it develops student autonomy as learners. Student involvement plays a crucial role in AfL, where the aim is to foster student development through encouraging them to develop their ability to evaluate and improve their own learning. Such self-assessment is at the heart of autonomous learning and of students developing as life-long learners (Boud & Falchikov, 2006).

AfL may be accomplished through interactions between teachers and students. Productive interactions call on and utilise all of the modes of expression and communication that are available in the classroom, including speech, drawing, modelling, demonstrative action, etc. Productive interactions also involve both descriptive feedback focused on key subject ideas and learning goals, and discussion with students about their next steps in the learning process (Tunstall & Gipps, 1996). For teachers to undertake AfL effectively they require particular knowledge and skills, some of which is encompassed in pedagogical content knowledge (PCK). PCK is concerned with ways teachers blend content/subject matter knowledge and pedagogical knowledge so that the idea/skill to be taught can be learned by their students. Yet, at the same time, the integrity of the idea/skill must be maintained (Shulman, 1987). In other words, teachers need to translate what they know about a subject's ideas and skills into forms learnable for their students, and transform generic pedagogical practices to help their students learn specific subject ideas and practices. Teachers need to understand subject matter and be able to clarify it "in new ways, reorganise and partition it, clothe it in activities and emotions, in metaphors and exercises, and in examples and demonstrations, so that it can be grasped by students' (Shulman, p. 13). When teachers also account for their students' preconceptions, understandings, and prior experiences, they are able to deploy pedagogically powerful ways to represent content.

Primary teachers generally have a deep knowledge of their students and a broad repertoire of pedagogical practices, including AfL. However, they often lack in-depth technology content knowledge and tend to have a limited repertoire of technology-specific pedagogies. In sum, their technology PCK is restricted. This exacerbates the challenges inherent in any technology-focused AfL interactions (Moreland & Jones, 2000). AfL interactions require that teachers confront simultaneously issues to do with content, pedagogy, student skills, interests, and understandings. Black and Wiliam (1998) identified distortions to teacher AfL interactions where there is minimal subject knowledge and/or gaps in knowing how that subject knowledge is best represented and developed for students. When teachers lack knowledge and understanding of what is important in technology they are not able to structure AfL around the key technology curriculum learning goals. Instead, they tend to focus AfL on managerial and social aspects, such as completing tasks on time, co-operative group work, and working independently, rather than on improving student understanding of subject ideas and the development of subject skills and practices (Black & Wiliam, 1998; Jones & Moreland, 2004; Moreland, 2003).

In this chapter it is argued that using a technology-specific planner helps teachers distil out the salient technology learning outcomes their students will learn and the ways they will teach these learning outcomes for effective student learning. It is also argued that teachers with sound technology PCK are able to undertake AfL to help students understand, and integrate, multidimensional technology learning goals. Because technology units are undertaken over several weeks, effective technology teachers plan for and deploy AfL practices that help students make connections and build coherent learning experiences across ideas, skills, lessons, and days. The chapter also illustrates that when teachers actively involve students in their technology learning through interactions that use multiple modes evocative of technology – such as drawing and modelling, not just talk – students are engaged in their learning and develop technology ideas and skills. Consideration of multimodal AfL is therefore required.

FINDINGS RELATED TO ASSESSMENT FOR LEARNING IN PRIMARY TECHNOLOGY CLASSROOMS

The *Interactions in Science an Technology Education* (InSiTE) project, through researching with teachers their AfL interactions, highlighted several factors as useful for teachers to think about (Cowie, Moreland, Jones, & Otrel-Cass, 2008). These include the importance of teachers planning to promote technology learning by being clear about and focusing on the multiple dimensions of technology as well as their interconnections; structuring and deploying AfL conversations to build connections, continuity, and coherence, including helping students understand and experience technology as multidimensional; and the use of multiple modes to develop a broad range of technology ideas and skills.

Planning to Promote Technology Learning: Being Clear about and Focusing on the Multiple Dimensions of Technology and the Interconnections

The InSiTE project highlighted the importance of working with primary teachers to help them identify, articulate, and enhance their technology PCK. Teachers need to be able to translate their own technology content knowledge into forms that their students can learn, and be able to change generic pedagogical practices to foster their students' technology learning. AfL also needs to become technology specific.

To help students develop a sophisticated understanding of the multiple dimensions of technology it is important for teachers to first identify and include the relevant multiple dimensions for their teaching and students' learning in their planning. This identification assists teachers to prepare for their technology AfL interactions before they enter the classroom since part of their planning involves clarifying the inter-related technological conceptual (knowledge and understanding of relevant technological concepts and procedures), procedural (knowing how to do something, what to do, and when to do it), societal (the interrelationship between technology and groups of people), and technical (skills related to manual/practical techniques) aspects (Jones & Moreland, 2003). In addition, teachers need to consider and plan for how they will establish and sustain a dynamic interaction between these aspects. The use of a two-part planner (described below) supported this process by helping teachers to plan for clear goal setting; plan for activities to meet those goals; and plan for how to make connections between days, lessons, activities, and ideas.

The first part of the planner helped to focus the teachers on analysing their own content knowledge in order to derive appropriate student learning goals (see Figure 1). The second part of the planner focused the teachers on how they would help students work towards the learning outcomes in an appropriate technological and student-centred manner. The attention paid by teachers to identifying technology learning goals as their first step in planning was important. Previous research in technology has highlighted that teachers just plan activities without focusing on the technology-specific learning outcomes – they plan the activities they want students to do, rather than the ideas, design processes, or skills they would like students to learn. The effect of not articulating the technological conceptual, procedural, societal, and technical outcomes within tasks at the planning stage means that teachers subsequently find it difficult in the classroom to communicate the technology learning outcomes to students and/or help the students work towards them (Jones & Moreland, 2004).

Having to complete the first part of the technology planner focused the teachers on analysing their own technology knowledge and skills that they thought they would need for their teaching. From this analysis, their knowledge of their students' interests, prior knowledge and skills, and a scrutiny of whether they thought that the technology foci had the capacity to enhance students' technology learning, they articulated the main/macro task the students would undertake. They also identified the technological area/s, the overall dimensions of technology, and the possible conceptual, procedural, societal, and technical student learning goals.

Main/ Macro task:		Technological Area/s:	
Overall dimensions of technological practice:			
Conceptual learning outcomes:	Procedural learning outcomes:	Societal learning outcomes:	Technical learning outcomes:

Figure 1. Technology planning framework.

The teachers thus began their planning by deciding on a main/macro design task within a particular technological area/s that was appropriate for their students, for example, to use materials found on a farm to make musical instruments for a public performance. They expanded out from the main task to the overall dimensions of technological practice, and then even further out to the more specific learning outcomes. The effect was to tease out from the main task the relevant concepts, procedures, and skills they thought important for their students to learn. The detailing of the overall dimensions of technological practice helped teachers thinking about how they would build in connections and coherency in student technological practice. Thinking about connections and coherency moved teachers away from teaching learning outcomes as individual and/or discrete. They thought about the range of specific learning outcomes inherent and nested in the main task and within particular technological area/s. They also thought about how the learning outcomes in the four categories – conceptual, procedural, societal, and technical – might coalesce and work together when students were undertaking technological practice. When teachers were completing this first part they worked iteratively, back and forth, between defining the main task, the associated technological practice, and the intended learning outcomes in the four categories. The iteration helped them to think about the links between the categories. The in-depth articulation also helped the teachers to clarify what *they* needed to know to teach the particular task. At the same time, teachers also began to think about ways their students might work towards the learning outcomes.

Figure 2 shows the first part of Jenny's planning for her Year 4/5 (9– and 10–year old) students' percussion-instrument-making unit. (All names are pseudonyms.) The students were involved in a school production where they needed to produce a musical item. This context provided an authentic rationale for the creation of percussion instruments. Because her students were studying dairy farming in social studies, Jenny also decided to use the farm visit for the students to gather material from which to make their instruments. The use of found/recycled materials was aimed at developing her students' environmental consciousness.

Jenny defined the technology learning in detail and with precision. She focused on the key ideas and skills of percussion instrument-making technology that she wanted her students to learn. She defined the main/macro task and this provided the overall direction for student learning, i.e. "Make a percussion instrument from farm/recycled materials for the school production". Below the macro task she specified three broad technology learning outcomes as indicative of the integrated dimensions of technological practice: (1) develop an understanding of appropriate materials and structures that make different sounds in a percussion instrument; (2) design and produce a percussion instrument using recycled materials according to a plan; and (3) know that musical instruments are used by people for a variety of purposes. These have clarified the macro task and identified broad technological concepts and procedures. More specific intended learning outcomes related to conceptual, procedural, societal, and technical aspects were then defined.

The list of outcomes demonstrates Jenny's understanding of technology as multidimensional and illustrates her PCK, for she has configured the learning outcomes to suit her students. For example, she has defined drawings in ways her Year 4/5 students would understand – "conceptual drawings can show general ideas and/or finished products" and "drawings can be plans for identifying materials and construction steps". Her expectation was that her students could learn to understand design drawings from these two perspectives. By configuring the learning outcomes in language matched to the level of their students, other teachers in the project also demonstrated that they were thinking about students' metacognitive functioning in the classroom. They used language that would help their students develop technology ideas and skills. The teachers had their students' technological pathways in mind when they approached planning. The first part of the planner thus prompted teachers to map-out the pertinent technology ideas and practices they wanted their students to learn before they planned the related lessons and subtasks. It assisted them to clarify the directions for their teaching, and student directions for their learning.

<table>
<tr><td colspan="2">Macro task: Make a percussion instrument from recycled/farm materials for the school production</td><td colspan="2">Technological Area/s:
Materials

Year 4/5 students</td></tr>
<tr><td colspan="4">Integrated dimensions of technological practice:
1. Develop an understanding of appropriate materials and structures that make different sounds in a percussion instrument.
2. Design and produce a percussion instrument using recycled materials according to a plan.
3. Know that people use musical instruments for a variety of purposes.</td></tr>
<tr><th>Conceptual Learning Outcomes</th><th>Procedural Learning Outcomes</th><th>Societal Learning Outcomes</th><th>Technical Learning Outcomes</th></tr>
<tr><td>• Understand how percussion instruments are constructed and work.
• Understand that materials used to make percussion instruments need specific properties.
• Understand that conceptual drawings can show general ideas and/or finished products.
• Understand that drawings can be plans for identifying materials and construction steps.</td><td>• Research and examine different percussion instruments to identify requirements of design and materials used.
• Examine and test appropriate materials for percussion instruments and examine other instruments to generate ideas.
• Develop a 2D plan to identify:
– Recycled materials used in construction
– Structure and shape.
• Construct a percussion instrument with reference to design.</td><td>• Understand that people use music and instruments for enjoyment, work, and sharing information.</td><td>• Develop 2D drawing skills with annotations, e.g.. labelling.
• Develop technical skills for cutting, gluing, punching, joining, folding, lacing, tying.</td></tr>
</table>

Figure 2. Jenny's planning.

The second part of the planner prompted the teachers to plan the technology-specific teaching approaches and tasks that would assist students to meet the learning goals. They outlined in sequence the meso and micro tasks that would help students achieve the main/macro task and help them to work towards the possible learning outcomes (Moreland & Cowie, 2007). This segmentation and sequencing allowed teachers to structure the tasks in a nested and interrelated manner. As a result, teachers were viewing technology as an integrated and complex knowing and doing activity. The macro-meso-micro task structure helped them think about how to connect the various tasks together coherently so that students could learn particular technology ideas and skills in a holistic manner. Considering which learning outcomes could be made manifest in the nested subtasks was essential for helping teachers incorporate AfL strategies in their planning, including what might be appropriate feedback, and how to encourage students to be active learners and self-assess their learning. By knowing the ideas and skills inherent in the tasks teachers were clearer about the focus for their AfL interactions.

Teachers also itemised the resources, or focal artefacts (Roth, McGinn, Woszczyna, & Boutonne, 1999), they would use. The focal artefacts were the teaching materials to be used to help structure lessons, to focus student attention, to introduce and develop student ideas. Alongside each focal artefact, teachers detailed some of the interactions they expected and considered some likely responses from their students. The detailing of some planned interactions encouraged the teachers to predict how they might interact with their students and what they might say when introducing and working through tasks. Teachers also anticipated student actions and the process encouraged them to undertake a dress rehearsal/dry run before classroom teaching.

Technology units are usually undertaken over several lessons. This long-term and multidimensional nature of technology tasks means students need help to see the connections between ideas, tasks, lessons, and days, and to see that the design process is a dynamic, iterative process. Teachers therefore need effective technology PCK in order to help students make connections and build continuity and coherence. One way to help students build connections and develop continuity and coherency is for teachers to think about these aspects when planning. Breaking the macro task into meso and micro components helped the InSiTE teachers to design nested and related subtasks that contributed to the achievement of the main task. This organisation helped to provide structure, support, and direction for student learning. For example, Ellie (a pseudonym) planned for her students to make masks to wear in their school production as the main task. She developed a series of eight sub tasks. The first related to building on students' existing ideas about masks. Within this first subtask, she devised four smaller, nested tasks: class discussion about masks students had seen, listening to the school production story to think about the attributes of the characters, in pairs examining books to find out how masks are made and their component parts, and a class discussion about the features their character masks require for their production. Other subsequent subtasks helped students build their skills and develop their understandings as they progressed through the unit, with the final subtask being a product evaluation.

In their planning, the teachers also allocated time for students to reflect and think through the activities. These times were to help students appreciate the connections between what they did last time, what they were currently doing, and what they would do next. Such sharing helps students to undertake the design process in an iterative manner. For example, Grant planned for his class to share in small groups their designs for outdoor signs. This was to give his students the opportunity to discuss and demonstrate how they were going to make their signs so that they could seek suggestions from their peers on how to improve. All the teachers planned for students to share their design progress with their peers and the teacher to provide them with genuine opportunities to be actively engaged in their learning and to build on their ideas and skills.

Cycles of using the planner for planning and then implementing the plan in the classroom over the three years enabled our teachers to better anticipate students' learning difficulties, have multiple opportunities to translate their technology content knowledge into outcomes that were learnable for their students, and design and prepare authentic tasks that supported student learning. Planning with a subject-specific planner set up a coherent framework for teachers to think about how they would teach technology and support students' technology learning at the same time. Its use helped our teachers identify the technology ideas and skills appropriate for their students, the knowledge they needed in order to teach the ideas and skills, and the appropriate pedagogical approaches for effective teaching. Teachers developed and refined their PCK prior to, and in preparation for, teaching in the classroom. The framework also structured and supported in-depth collaborative discussion and analysis of what might be required to teach and learn a particular technology task. The teachers saw that their rehearsal of tasks before they began teaching as useful for developing their PCK. Such 'dry runs' enabled them to explore the knowledge, understandings, and skills most salient in the tasks; to assess the appropriateness of the tasks for their students; and to foreshadow potential problems:

> It was good that I tried out how to make cheese at home before I had the children try this. I found how long it would take and I thought about easier ways they could do it. (Brenda)

Teachers planned for the multiple dimensions of technological learning and designed activities to support student learning of these multiple dimensions. They reported that this planning changed their classroom interactions; they now focused more clearly on the multiple conceptual, procedural, societal, and technical aspects within the tasks. They also commented on the benefits of thinking about and planning for classroom interactions, including the sorts of responses their students might make. Such forethought meant that they could optimise learning within their AfL interactions with students:

> It made me think through each stage thoroughly instead of always having to think on my toes, which can often mean missing good opportunities, or not choosing the best way to do something. (Jenny)

The teachers' technology PCK was also enhanced when they used their plans for teaching in the classroom. It was then that their PCK was brought to light, especially in their AfL interactions when they were supporting student technology learning. The teachers said their plans were an important reference, acting as a backup for their actions and interactions in the classroom.

> I often had the plan beside me when I was teaching. I referred to it a lot. It clearly marked where I wanted to take the children with their learning and what I would need to do to get them there. (Lois)

Teachers' technology PCK was further refined when they reflected on what had occurred in the classroom in relationship to what they had anticipated in their planning. However, the teachers were emphatic that their detailed planning did not lead to them to constraining student learning possibilities but rather allowed them to respond flexibly to student ideas, interests and skills.

The Need for AfL Conversations to Build Connections, Continuity, and Coherence

The long-term and multidimensional nature of technology tasks means that students may find it difficult to see the connections between days, activities, and ideas. Teachers therefore need to help students build connections, continuity, and coherence. The technology planner focused the InSiTE teachers on developing sequences of nested and linked tasks. This form of planning ensured that the teachers were clear about the links and connections between the ideas, tasks, and lessons before they began teaching. The benefits of this planning process were that when the teachers interacted with their students they were able to make decisions in the moment whether or not to pursue particular student ideas and practices; they were able to ascertain the worth of students pursuing learning pathways against the parameters of the overall planned technology task. This allowed them to be responsive to their students' interests and abilities, yet at the same maintain the integrity of the students' technology learning.

One way that teachers assist students to build continuity and coherence is by structuring lessons so that everyone has time and opportunities to think about and discuss the links between lessons, tasks, and ideas. When teachers focus on actual technology aspects during these times and opportunities, students are more likely to make effective technology connections. An example showing how a teacher helped her students make connections is from Ellie and her Year 3 and 4 students when they created masks for their school production. For each of the five days that Ellie and her students worked on this technology task, Ellie opened and closed every lesson with a whole-class discussion. Beginning lesson conversations focused on two aspects: first, she orchestrated class conversations to look back at what had already been achieved; and second, she introduced the tasks and ideas for that day. End of lesson conversations also had two aspects. First, Ellie orchestrated discussion reviewing the tasks and ideas for that day. Second, Ellie and the students talked about what might happen the next day. Phrases like 'Today we're going to go on from there' helped make visible to her students the connections between tasks and days. Her students were helped to see how the tasks they were undertaking linked to each other and contributed to the overall macro, or main, task of making a mask for the school production. Because she encouraged her students to think ahead to next steps and to think back to what had happened and what they learned, students also had opportunities to self-monitor their technology ideas and progress over the course of the five days of the unit.

Success criteria are often advocated within AfL as a means to support student self-assessment. Just as importantly, they can serve as a mechanism to develop coherence in student technology learning and understanding. In technology, product specifications are encompassed in the brief that technologists work to. The InSiTE teachers acted to help students understand multidimensional technology learning goals through developing multiple product criteria. These product criteria were the same as the multidimensional technology learning goals. All teachers developed product criteria with their students. As teachers and students developed the multiple criteria, teachers recorded the criteria on whiteboards and charts as a visible record. This ensured that students had easy access to the criteria and the ready record helped them keep the criteria in mind for the duration of the unit. Sometimes the criteria were written in the form of questions. Criteria written in this way challenged students to think and talk about their ideas, helping the teachers find out not only what their students could do in technology, but also what they understood. For example, Figure 3 shows the criteria Tayla and her Year 7 and 8 students (12– and 13–year olds) developed for their outdoor game. The criteria detailed that the students had to make their game fun, consider the weight and size of the board and pieces in relation to who was playing it and how, choose colours for their attractiveness and defining playing areas, ensure their game was easy to pack up and move, and so on. The list was constantly referred to and was used as a self-assessment tool by the students. It helped them critically consider their game and make decisions on how well their design was meeting the criteria. The open nature of the design task fostered creative design thinking but the criteria ensured a functional product could be developed.

Is it fun?
The weight and size of the board and pieces in relation to who's playing it and how?
The colour – for attractiveness and defining playing areas.
Easy to pack up and move. Where and how you will store it? Permanently outside, or do you put parts of it away each day? Weather durable?
Is it challenging? Does it make us think?
Is it cost effective?

Figure 3. Criteria for outdoor game.

Conversations between teachers and students throughout the units focused on the multiple criteria, thus helping to focus student learning on technology as multidimensional and to also help them connect their ideas. In addition, the focus on criteria assisted students to work independently of their teacher. Criteria were often listed on students design drawings to focus their attention, and the same criteria were also used as evaluative criteria for appraising their products. Figure 4 shows two Year 1–3 students (5– to 7–year olds) completing a taste test survey. Using a three-point scale and their product criteria as the evaluative criteria they coloured the appropriate scale after tasting the product. The students accomplished this task with little teacher intervention since they clearly understood the criteria they were using for the taste test evaluation. The same criteria had been continuously used throughout the unit so that by the time the students came to this evaluative task they were fully conversant with the criteria and were able to deploy them to justify their choices.

A key to successful self- and peer-assessment was that teachers and students talked together about the technology task and the inherent criteria, and worked alongside each other. Under these circumstances students gained the language and skills they needed for discussing and developing technological concepts and procedures. In addition, when the teachers focused on all the technological learning outcomes in a holistic manner, students could see the connections and the task in its entirety and were able to look back over their work and undertake self-assessment focused on the technological aspects. Students thus experienced assessment as an authentic activity connected to their learning.

Using Multiple Modes to Develop Technology Ideas and Skills

Classroom studies in technology education are beginning to explore teacher and student use of multiple and multimodal resources for meaning-making and communication. In technology education, McCormick (2004) illustrates the way the specifics of the technology task along with the available resources shapes what students learn and do. Technologists draw and make models and prototypes to develop and test ideas. In technology, teachers typically engage students in tasks that have a practical aspect. As teachers and students work together on technology tasks, meaning is developed through the interplay of linguistic, pictorial, material, and gestural resources (Kress, Jewitt, Ogborn, & Tsatsarelius, 2001; Roth, 2005; see also Chapter 53 by Roth). This provides students with access to multiple modes for developing, representing, evaluating, and communicating their technological ideas. An analysis of the teacher-student interactions in the InSiTE study revealed that the teachers deployed a range of strategies when interacting with students to assess and come to a shared understanding of the task and related ideas. The multimodal nature of interaction emerged as a key focus for the research. Teacher and student talk played a pivotal role in interaction, but talk was invariably anchored and augmented by other modes.

The InSiTE teachers made active use of tasks that provided students with many opportunities to make and express meaning through a combination of modes –drawing, talking, writing, action as modeling, gesture and dramatisation, and the use and production of artefacts. The teachers and students made substantial use of visual materials and the visual mode in teacher-whole class interactions, teacher-group interactions, and individual student interactions. For example, Brenda and her Year 5/6 students (10– and 11–year olds) were involved in designing and making a light tower for a model stage. She used students' drawings – and their difficulties in interpreting each other's drawings – to develop students' understandings of the need to construct a complete circuit for a workable light tower and to generate a shared means for representing electric circuits. The students' drawings served to bridge and mediate their learning about conventions for representing electric circuits and then being able to incorporate these conventions in their subsequent light tower designs.

Figure 4. Taste testing a healthy snack.

Drawing played a role in developing student understanding and abilities in the practices of representation in technology. Students produced drawings as an initial representation of their ideas as part of the design process in all of the technology units observed in the InSiTE project. These drawings provided a forum for discussion and teacher feedback. Students were not left to draw unassisted or without conversation. For example, Tayla had explained to her Year 7/8 students that they needed to develop concise construction steps and include simple annotations in their design plan of their musical instrument to be made from recycled materials; this plan was to be their blueprint for construction. When she examined one group's plan she saw two things: first, that no detailed construction steps had been included and, second, that they had decided to only glue the joints of the wooden rack. The following dialogue ensued:

Tayla: Remember you need to outline your equipment and your construction steps. Talk me through your steps bit by bit.
Tim: First off we're going to cut this piece of wood, so it's the same size as that [points to Step 1 drawing].
Tayla: Right, that's your first step.
Garth: Then once it's like that, we're going to take out all the nails [points to Step 2 drawing].
Tayla: So that's your second step – take out the nails. ... We're cutting the wood. We're taking out the nails; that's the second step.
Tim: [Pointing to Step 3 drawing] Once it's like that, we're going to get a hot glue gun to join it.
Tayla: A hot glue gun! Why a hot glue gun?
David: We're going to ask Mr. T (specialist technology teacher) about what might be the best connection.
Tayla: The best connection. Just one thing, what do you think makes the best connection for wood and wood?
Garth: Nail the ends.
Tayla: Nails.
David: Oh, so we're going to put glue in there, as well as the nails.
Tim: It's like so it will stick together better.
Tayla: Make it stronger, robust, really strong, really, robust.

Tim: Yeah, we'll put in the glue; then nail it. [The boys then drew step four and annotated glue and nails on their plan.]

In this example, the combination of talk, drawing, and gesture were pivotal to Tayla and the boys coming to understand ideas about a robust joint being 'more than glue'. The drawing process provided a forum for the boys' discussion about how to proceed with the production of their xylophone. The drawings were a record of the boys' thinking in Tayla's absence and they provided her with an entry when she came to interact with them. She was able to probe what would be 'the best connection for wood and wood'. The drawings then served as a tool that contributed to developing the meaning of robustness, which came to be taken-as-shared by Tayla and the boys. In this case it was 'glue plus nails' so that the wood 'will stick together better'. This example illustrates the potential for drawing to anchor and augment talk to support the development of a shared understanding about the intent and potential limitations of a particular design. It also illustrates how student drawings can provide teachers with opportunities for meaningful (and useful) feedback about technology ideas that can then lead to the expansion of their students' technology approaches and solutions. The necessity for a conversation that involved more than talk is evident, as the interpretation and subsequent development of the boys' plan led to a more feasible design. It was one of many occasions where the InSiTE teachers augmented talk by drawing themselves and/or asking their students to draw and explain ideas.

Action as a mode for expressing, developing, and evaluating meaning includes the use of gesture, the use of space, modelling, demonstrative action, and dramatisation. The InSiTE teachers made use of all of these hands-on activities. Teacher action and talk provided a model for how to act and speak technologically. Modelling and demonstrations helped students gain a sense of how to carry out a task and what their contribution would be. Sometimes students were asked to act out ideas and processes before they undertook them for real. These activities were undertaken to enhance the introduction of new ideas and to provide an alternative means for students to engage with and express ideas. Dramatisation provided a stimulus to, and a forum for, talk. For example, when Lois' Year 1–3 students were producing healthy snacks she introduced the notion of batch production by dramatising the process.

Technological development may involve the analysis and refinement of existing technologies. With appropriate teacher guidance, the use of artefacts is a source of ideas and feedback for students. Artefacts – those brought into the classroom from home (sample objects, such as kites or lunch boxes), and those designed by teachers (such as design-prompt worksheets or posters) – provided scenarios and resources to underpin the InSiTE teachers' interactions with their students. Teachers made substantial use of pictures and images accessed from books, newspapers, and the Internet, and of digital photographs they had taken themselves. They used images to stimulate student curiosity, to gain student interest, to foster student design creativity, to develop student design evaluative abilities, and to co-ordinate and inform talk. For example, Ellie used books with illustrations and photographs of masks with her students when they worked in pairs to develop design ideas about different mask styles. Tayla used an interactive whiteboard to develop student understandings when her class classified and grouped images of musical instruments. Grant's students made outdoor signs for their school. Initially they went on a 'discovery walk' around the neighbourhood to examine signs *in situ* to discern their shape, their visual impact, and fitness for purpose. He took photographs of signs around the town and placed them in a PowerPoint presentation. He made printouts of the photographs. His students initially examined the PowerPoint presentation as a class and discussed the salient features of the depicted signs. With this information and the knowledge they had gleaned from their walk they created the specifications for their own signs. They independently referred to the photographic printouts over time for more particular detail, usually as individuals. Criteria such as suitable materials, weatherproofing, structural stability, and sign and lettering size and colour were evident in the students' designs and final products. Incorporating these criteria was critical to the feasibility of their designs given that the signs were to be actually used in the school grounds.

In whole-class settings, publicly visible and readily accessible artefacts provided direction, guidance, and support for interactions between students across different locations and time. Artefacts helped students to work together in small-group settings, providing guidance in the absence of the teacher. However, the influence of an artefact on interaction was not a given; rather this depended on how it was introduced to the students and how it came to be integrated into the interactions. Over the course of the project the teachers, as they came to appreciate the effect of artefacts, paid more attention to their form, fit, and function.

The use of real objects in particular helped teachers to elicit prior knowledge and experiences that could then be used as a resource to expand the pool of ideas available for whole-class discussion. Real objects were also used to help students analyse and talk about the strengths and weaknesses in their own designs and products through directly comparing their design or product features with real artefacts. Discussing what was the same, what was different, is it better or not, and why, helped students further critique, refine, and adjust their own designs and products. The analysis of existing artefacts, systems, and environments provided a source of ideas for students when they designed and produced their own artefacts, systems, and environments. Artefacts provided a vehicle for students to think about technology and a forum for teachers and students to talk about technology. However, teacher guidance was important in helping students to analyse, transform, and translate the salient aspects of real artefacts into something that was useful and creative for their particular context and purpose. Since effective technology learning is not promoted when students only produce highly imitative copies, teachers needed to encourage students to take some risks with their ideas. Assessment for learning was important in these instances for providing feedback to students so they could manage the degree of risk against the feasibility of their ideas (Barlex, 2003).

There was evidence from the InSiTE project that student use of each of the modes was an activity that required knowledge. While the teachers were always careful to introduce and explain the language of technology they came to realise the need to also make technological practices explicit. For example, the use of drawing to enhance talk was not without challenge. This was illustrated when Jane asked her Year 1 students to draw an electric circuit as part of planning the lighting for a model theatre. The students were required to draw a circuit diagram on blank paper. She provided no instruction on how to do this and in consequence the students struggled to realise the connection between their drawing and the circuit they were to make. Jane discussed her students' difficulties with Martha, the Year 2 teacher at the same school, because she knew Martha was going to do a similar activity with her students. As a result, Martha had her students draw their plan for wiring an electric circuit for a front door light for their model house on the model house itself, rather than on a separate piece of paper. Martha then asked her students to trace their circuit with their finger and explain to her why they thought the light would glow. The practical issues that arose included whether the circuit was complete, whether someone would be able to reach the light switch, whether a person could open the door if the circuit wiring was positioned as it was on their plan, and whether the placement of the light would efficiently light the room. Through this process a number of students realised that they had not considered these ideas. In this instance, student drawings provided a useful avenue for teacher-student talk about working circuits and the need for the technological solution to be practical and useful. Martha's forethought on having the students draw on the house, tracing and talking about their circuit, mitigated some of the problems Jane encountered with her students only drawing.

The InSITE project demonstrated that not only did students enjoy and value hands-on tasks, teachers also enjoyed and valued them. However, student use of artefacts was not unproblematic. Students needed to appreciate the pertinent physical, conceptual, and procedural aspects of an artefact, and the relationships between these aspects, for them to use an artefact as the teacher intended. This was just as much the case for real objects as it was for teacher-designed artefacts. The InSiTE teachers spent considerable time collecting and preparing resources to support their students' learning, but the influence of an artefact on their interactions was not a given. This depended on how the teacher introduced an artefact to the students and how it came to be integrated into interactions. Providing students with multiple opportunities to learn through the use of a combination of modes helped them develop as independent, creative, and critical designers.

DISCUSSION AND SUMMARY

In this chapter some evidence has been provided of the ways that teachers can set up and sustain AfL interactions around the important ideas and practices of technology. Teacher knowledge and understanding of technology is important in structuring formative interactions around key learning goals. The role teacher pedagogical content knowledge plays in assisting teachers to plan for student learning and to think about how they might interact with students around those ideas and practices has been highlighted. The use over an extended timeframe of a subject-specific planner proved to be a powerful tool for helping the InSiTE teachers clarify the intended technology learning goals, deepen and extend their pedagogical content knowledge, design appropriate technology learning activities, and anticipate AfL interactions. The teachers indicated that the use of the two-part planner had changed their formative interactions to focus more on the conceptual, procedural, societal, and technical aspects of tasks. Focusing on the coalescence of these aspects helped teachers provide coherent technology learning experiences for their students. The teachers also reported that detailed planning increased their confidence and ability to formatively interact with students' technology learning, whilst at the same time providing opportunities for students to pursue divergent ideas and interests.

When teacher AfL practices focused on the multidimensional nature of technology and multiple product criteria, students were able to connect technology ideas and experiences over time and design in an authentic manner. Teachers were able to provide descriptive feedback where key technology ideas and learning goals were the focus. Teachers and students were able to ascertain the next steps for improvement in the learning process and students had opportunities to act on this advice. Students were active in their learning and were able to self assess their work as they progressed through a technology unit.

Talk was a concise and efficient means for negotiating understandings and communicating ideas and the InSiTE teachers explicitly focused on developing their students' technology language proficiency in their interactions with them. However, almost invariably talk was augmented by writing, action, and the visual mode. Technology provides authentic contexts to use the visual in conjunction with, and as an alternative to, oral and written text. Like writing, images and artefacts are available for examination and use all at once and as a whole. Images and artefacts were used by the InSiTE teachers and students to develop and represent ideas, products, and past and future events that could not have been expressed or represented by words and written text alone. Teacher and student drawings and their making of models and prototypes helped them express, communicate, and develop their ideas; it was in these modes that students' technology ideas were brought to light from the minds-eye. Teachers and students also used actions to demonstrate and illustrate their understandings, skills, and practices. Knowing how and being able to do are critical to the development of students' technological literacy.

A focus for the InSiTE project was to help the teachers enhance their interactions around technology ideas and practices. Over the course of the project the teachers came to appreciate and value the multimodal nature of interactions, and the fact that technology is inherently multimodal in nature – technology involves technologists in drawing and talking, and the manipulation and production of products, systems, and processes; it is about practical involvement in our made world and creating change in the made world. Multimodal conversations therefore provided rich entry points into students developing technological capability. Teachers used different modes singly and in combination to help students customise the technological task, develop shared understandings, and extend their technological capabilities. When multiple modes were co-ordinated, or used in conjunction with one another, teacher and student opportunities for expressing ideas and developing shared understandings were enhanced and enriched.

Effective AfL interactions in technology in the InSiTE project focused on the multidimensional nature of technology. Such a focus helped students to build a sense of connectedness, continuity, and coherence between ideas and actions over time. Success criteria linked to specifications served as a mechanism for developing such connections, continuity, and coherence in student learning and understanding.

Effective AfL interactions lead to effective student learning when the teacher has the confidence, knowledge, and ability to explore and extend student ideas, and when students are involved with realising the learning goals. As technology is multidimensional and multimodal the challenges faced by teachers and students in AfL are complex. Teachers need support and guidance to develop their own understandings of technology and technology-specific AfL. However, because of the nature of technology itself, teachers have rich opportunities to find out about, and engage with, student learning and help students become independent, creative, and critical designers.

REFERENCES

Barlex, D. (2003). *Creativity in crisis? Design and technology at KS3 and KS4.* Wellesbourne, UK: Design and technology association.

Bell, B., & Cowie, B. (2001). The characteristics of formative assessment in science education. *Science Education, 85*(5), 536–553.

Black, P., & Wiliam, D. (1998). Assessment and classroom learning. *Assessment in Education, 5*(1), 7–74.

Boud, D., & Falchikov, N. (2006). Aligning assessment with long-term learning. *Assessment and Evaluation in Higher Education, 31*(4), 399–413.

Cowie, M., Moreland, J., Jones, A., & Otrel-Cass, K. (2008). *The classroom InSiTE project: Understanding classroom interactions to enhance teaching and learning in science and technology.* Teaching and Learning Research Initiative, Final Report. Wellington, NZ: New Zealand Council for Education Research.

Denzin, N., & Lincoln, Y. (Eds.). (2000). *Handbook of qualitative research.* Thousand Oaks, CA: Sage.

Erickson, F. (1998). Qualitative research methods for science education. In B. J. Fraser & K. G. Tobin (Eds.), *International handbook of science education* (pp. 1155–1173). Dordrecht, The Netherlands: Kluwer.

Hennessy, S., McCormick, R., & Murphy, P. (1993). The myth of general problem solving capability: Design and technology as an example. *The Curriculum Journal, 4*(1), 73–89.

Jones, A., & Moreland, J. (2004). Enhancing practicing primary school teachers' pedagogical content knowledge. *International Journal of Technology and Design Education, 14*(2), 121–140.

Jones, A., & Moreland, J. (2003). Developing classroom-focused research in technology education. *Canadian Journal of Science, Mathematics and Technology Education, 3*(1), 37–50.

Kimbell, R., Stables, K., Wheeler, T., Wozniak, A., & Kelly, V. (1991). *The assessment of performance in design and technology—The final report of the APU design and technology project.* London: SEAC/COI.

Kress, G., Jewitt, C., Ogborn, J., & Tsatsarelius, C. (2001). *Multimodal teaching and learning: The rhetorics of the science classroom.* London: Continuum.

Lave, J. (1988). *Cognition in practice: Mind, mathematics and culture in everyday life.* Cambridge, UK: Cambridge University Press.

Lemke, J. (2000). Across the scales of time: Artefacts, activities, and meanings in ecosocial systems. *Mind, Culture and Activity*, *7*(4), 273–290.

McCormick, R. (2004). Issues of learning and knowledge in technology education. *International Journal of Technology and Design Education*, *14*(1), 21–44.

Mercer, N. (1995). *Guided construction of knowledge: Talk amongst teachers and students*. Clevedon, UK: Multilingual Matters Ltd.

Ministry of Education. (1995). *Technology in the New Zealand curriculum*. Wellington, NZ: Learning Media.

Mitchell, J. (1984). Typicality and the case study. In R. F. Ellen (Ed.), *Ethnographic research: A guide to general conduct* (pp. 238–241). San Diego, CA: Academic.

Moreland, J. (2003). *Becoming effective technology teachers: Enhancing assessment practices in primary classrooms*. Unpublished doctoral thesis, University of Waikato, New Zealand.

Moreland, J., & Cowie, B. (2007). Making masks for Maui: Keeping the macro task in mind. *Australian Research in Early Childhood Education*, *14*(1), 81–93.

Moreland, J., France, B., Cowie, B., & Milne, L. (2004). Case studies of biotechnology in the classroom. In A. Jones (Ed.), *Biotechnology in the New Zealand Curriculum* (pp. 92–120). Hamilton, NZ: CSTER, University of Waikato.

Moreland, J., & Jones, A. (2000). Emerging assessment practices in an emergent curriculum: Implications for technology. *International Journal of Technology and Design Education*, *10*(3), 283–305.

Nicholl, B., & McLellan, R. (2008). 'We're all in this game whether we like it or not to get a number of As to Cs.' Design and technology teachers' struggles to implement the creativity and performativity policies. *British Educational Research Journal*, *34*(5), 585–600.

Roth, W.-M. (2005). *Talking science: Language and learning in science classrooms*. Lanham, MD: Rowman and Littlefield.

Roth, W.-M., McGinn, M., Woszczyna, C., & Boutonne, S. (1999). Differential participation during science conversations: The interaction of focal artifacts, social configuration and physical arrangements. *Journal of Learning Sciences*, *8*, 293–347.

Shulman, L. (1987). Knowledge and teaching: Foundations of the new reform. *Harvard Educational Review*, *57*(1), 1–22.

Stables, K. (1997). Critical issues to consider when introducing technology education into the curriculum of young learners. *Journal of Technology Education*, *8*(2), 1–17.

Tunstall, P., & Gipps, C. (1996). Teacher feedback to young students in formative assessment: A typology. *British Educational Research Journal*, *22*, 389–404.

Wells, G., & Claxton, G. (Eds.). (2002). *Learning for life in the 21st century: Sociocultural perspectives on the future of education*. Oxford, UK: Blackwell.

Judy Moreland
Bronwen Cowie
Wilf Malcolm Institute for Educational Research
University of Waikato
Hamilton, New Zealand

MARILYN FLEER AND GLORIA QUIÑONES

40. ASSESSMENT OF CHILDREN'S TECHNOLOGICAL FUNDS OF KNOWLEDGE AS EMBEDDED COMMUNITY PRACTICES

INTRODUCTION

Cultural-historical theory has provided a dynamic approach to framing research and has drawn scholars' attention to community activities for building children's capabilities (Vygotsky, 1998). Within this context 'funds of knowledge' best describes how learning in communities generates social, economic, and cognitive capacities within the home and community (see Moll & Greenberg, 1990). In this chapter, assessment is framed as developing an understanding of a collective cultural-historical enterprise through positioning children as researchers of their own technological knowledge and capability, as located within their home and community practices in the form of funds of knowledge. A study of young children engaged in researching their own home, school, and community environments in relation to their technological activities is outlined in order to make visible the conceptual argument being put forward in this chapter.

THEORETICAL PERSPECTIVES OF CONCEPT FORMATION

Moreland and Jones (2000, 2004), Moreland (2003), and Rohaan, Taconis, and Jochems (2008) have argued that teachers' perceptions and understandings of technological concepts directly influence their formative interactions during technology education in classrooms. For example, Moreland's research showed that teachers' growing understandings about the nature of technology and knowledge in different technology areas greatly "enhanced understandings about ways to plan, teach and assess technology", particularly in relation to giving descriptive ongoing feedback to children (p. 303). Jones and Moreland, in drawing upon sociocultural theory, argued for the importance of building teacher conceptual knowledge in technology education not just for more effective assessment, but for enhanced technological practice in schools.

Vygotsky (1987) argued that if we are to understand concept formation, then we should study concepts "in the process of change" (pp. 64–65) rather than use methods which decompose "the whole into its elements" through "partitioning the whole into its units" (pp. 46–47). Yet the examination of dynamic conceptual contexts, and therefore attempting to understand concepts which are ever changing during pedagogical activity, is most problematic (Vygotsky & Luria, 1994), particularly in relation to assessment practices within early childhood settings (Anning, 1994; 1997) and beyond (Black, Harrison, Lee, & Wiliam, 2003; Black & Wiliam, 1998; Middleton, 2008).

In studying the dynamics of concept formation, it has also been noted by Sakharov (1994) that "One cannot be limited to the study of ready made concepts" (p. 82). That is, teachers should not assess children's technological capabilities at the conclusion of their project work. Rather, Sakharov suggests that the "process of formation of new concepts is important" (p. 82) and that a concept must be studied in its functional context. In classrooms and early childhood settings,

A.T. Jones and M.J. de Vries (eds.), International Handbook of Research and Development in Technology Education, 477–491.
© 2009 Sense Publishers. All rights reserved.

children have many opportunities in which they can explore technological concepts and develop technological capabilities. Moreland (2003) has shown that teachers who have conceptual knowledge in technology can and do assess children in action, providing feedback to children as they work. Her research shows that teachers with professional knowledge of technological concepts go beyond reinforcing 'task completion' or focussing only on 'working effectively in teams'. The latter approach, by teachers without high levels of conceptual knowledge in technology, was noted for most technology teachers Moreland studied in New Zealand.

Langermeyer and Roth (2006) have stated when examining Vygotsky's (1987) methodological writings that a "multi-dimensional object of study . . . cannot be investigated by looking at isolated elements or components", but rather through "determining relevant 'units'" (pp. 26–27). That is, teachers cannot make judgements about a multi-dimensional and complicated activity, such as occurs in technology education, by only examining isolated elements. Assessment should be multi-faceted, dynamic and fluid, and encompass many different dimensions simultaneously. This dialectical approach to research is one of the cornerstones of Vygotsky's cultural-historical theory and provides a useful framework for thinking about technology assessment in early childhood settings. This is consistent with other studies, undertaken with older pupils, where the dependency culture on teacher assessment was removed and more sustainable assessment practices were encouraged, such as time spent on learner engagement in reflection and evaluation (Bjorklund, 2008).

We draw upon this body of theoretical work to inform the assessment of children's technological knowledge and capabilities in the early years of learning. The chapter begins with a theoretical discussion of the funds of knowledge as the guiding body of literature used to frame how assessment practices were conceptualised and enacted. This is followed by an overview of the study context and design in relation to a review of the international literature in this area. In the latter part of the chapter, the findings and their implications are discussed in relation to building a collective community oriented approach to thinking about assessment in early years' technology education.

FUNDS OF KNOWLEDGE TO FRAME ASSESSMENT PRACTICES

In order to understand a child's growing knowledge of technology, we argue in this chapter that teachers must also understand the social relations and contexts in which a child is located. As such, it is also important to examine children's technological contexts outside of the school environment, such as children's home and community environments. The conceptual tool which best supported the directions of this approach is 'funds of knowledge'. This theory has been framed from an anthropologic perspective (see Velez Ibáñez, 1988) and also from an educational perspective (e.g., González, Moll, Floyd-Tenery, Rivers, Rendon, González, et al., 1993; Moll & Greenberg, 1990) and, as such, is useful for guiding the approach taken in this chapter.

We begin in this section by discussing the origins of the term, funds of knowledge, and then move forward to examine how researchers have used this concept in their studies. We are particularly interested in the work of Moll, González, and colleagues, and seek to examine and discuss how the particular nuances put forward by scholars have helped inform how assessment in technology education may be conceptualised.

Moll and colleagues (González, Andrade, Civil, & Moll, 2001; González et al., 1993; Moll, Amanti, Neff, & González, 1992; Moll & Greenberg, 1990) and other researchers (Monzo & Rueda, 2003) have used the term 'funds of knowledge' to refer to "the historically accumulated and culturally developed bodies of knowledge and skills essential for household or individual functioning and well-being" (Moll, et al., 2001, p. 133). Vélez-Ibáñez and Greenberg (1992) describe the term as "strategic and cultural resources that households contain" (p. 313), and which are often ignored in schools as important resources for classrooms. One of the many challenges that educational institutions face is the assumption that working class, minority students do not emerge from a household rich in social and intellectual resources and that they are seen as poor (González et al., 1993; Moll et al., 1992). How schools and early childhood teachers build understandings or assess beyond the classroom or centre in order to capture children's knowledge has been actively considered by cross-cultural researchers using cultural-historical theory. For example, Vélez-Ibáñez (1988) noted important networks of knowledge exchange within communities, consisting of "very thick and dense reciprocal relationships" (p. 28).

These reciprocal relationships serve as mediating a variety of functions; and, from very early in life, children replicate them. In Vélez-Ibáñez's (1988) study, Mexican children were introduced to these social and cultural expectations through mutual trust, termed *confianza*, which supported how these exchange relations were enacted. The knowledge transmission from adults to children is also based on transmission from generation to generation. As Vélez-Ibáñez and Greenberg (1992) note:

> The probability that these funds of knowledge that such clusters [household knowledge] contain will be transmitted to the following generation rests not only on an appropriate economic and social context for their application, but also on the early expectation for learning that children gain in such contexts. (p. 324)

In a sense, the confianza that was noted by Vélez-Ibáñez (1988) is the glue that holds community knowledge together between families, communities, and generations. Moll and Greenberg (1990) point out that funds of knowledge are "manifested through events or activities" (p. 326). In many of the research studies which have examined funds of knowledge, the objects and events are representative of important technological knowledge and capabilities that communities value and pass on to children through mutual engagement. Furthermore, Moll and Greenberg point out that "children are active in creating their own activities within tasks created by adults and . . . knowledge is obtained by children, not imposed by adults" (ibid.).

Thus, children are familiar with the bodies of knowledge learned in their households, and children are active in this process. In our study, we wanted to know both the children's funds of knowledge, and how teachers were acknowledging these as important dimensions of assessing technological learning. As argued by Moll et al. (1992), using a funds of knowledge framework enables teachers to see children in classrooms as a 'whole' person rather than simply as the 'student'. That is, their technological knowledge and capability must be understood in the context of its use – which goes beyond the walls of the classroom or centre. In using a funds of knowledge framework for assessment, Moll et al. argue that the relationship that teachers have is not the typical teacher-student relationship that is based on performance; rather, it is based on taking into account children's active knowledge construction both inside the school and in the community.

Research which has used the conceptual lens of funds of knowledge for investigating working-class Latino households has tended to be designed from an ethnographic perspective (González et al., 1993, 2001; Moll et al., 1992). In these studies, teachers carry out home visits in order to know about the household's funds of knowledge. The role of the teacher in these home contexts is seen as building knowledge about the learning and knowing that takes place for the child in the community. Through gaining knowledge beyond the walls of the classroom, teachers begin to know more about their children's household knowledge and can plan activities in their classrooms that draw upon this knowledge. Similarly, teachers can make better judgements about children's learning if they understand how their homes and communities operationalise this knowledge for children. This represents a positive "potential utility" (González et al., 1993, p. 134) for classroom curricula for technology education.

A funds of knowledge approach not only focuses on gaining greater knowledge of children, but this theoretical lens also locates teachers within the community contexts in which their children live. Research done by Monzo and Rueda (2003) found that teachers not only gain insights into children's funds of knowledge, but these funds of knowledge can be drawn upon to create meaningful ways of knowledge instruction in their classroom. González, Moll, and Amanti (2005), in an extensive study of working class Latino students, focussed on the idea of the teacher as a researcher. Through this project they found that the teachers formed new relationships with their children and, importantly, the teachers valued the students' funds of knowledge. This was particularly significant as the children were from poor communities. Further research by González et al. (2001), Andrews and Ching Yee (2006), and Hughes and Pollard (2006) has also shown positive assessment practices in mathematics when teacher researchers specifically investigated children's funds of knowledge and used this in their classrooms. Work by Riojas-Cortez (2001) has shown similar outcomes for preschool-aged children in the context of socio-dramatic play. These studies suggest that schools should pay more attention to children's outside worlds.

From Australia, Fuller and Hood (2005) and Fuller (2006) studied primary school children in their third and fourth year who were located in an Aboriginal community, demonstrating how an everyday experience of being in a camp could be used by teachers to develop children's literacy learning in schools. Findings from the UK (Andrews & Ching Yee, 2006; Hughes & Pollard, 2006) and the US (De Jesus & Antrop-González, 2006) also support the importance of using understandings of funds of knowledge to inform pedagogical practices and make curriculum relevant to children.

Whilst this overview suggests that research into funds of knowledge has much potential as a theoretical tool for informing assessment in technology education, there has been limited research directed towards generating knowledge about the funds of knowledge in technological learning outside schools and preschools in Australia.

Investigating Technology in the Everyday Lives of Children

When we examined the international literature on the relations between home and school contexts for technological learning and assessment, we found that both teachers and researchers predominantly concentrate upon what happens directly in classrooms (see Bain & McLaren, 2006; Braund & Reiss, 2006; Kimbell, 1997). With the exception of studies in museums (see Aronso & Gilbert, 2006; Griffin, 1994, 1998, 2004; Griffin, Mehan & Jay, 2003; Griffin & Symington, 1997; Luehmann & Markowitz, 2007), very few studies have focussed on children's learning in contexts other than schools. In drawing upon research by de Vries (2003a), Solomonidou and Tassios (2007) argue that research which seeks to find out children's technological concepts is very rare.

In examining the literature, we noted that some researchers have investigated community engagement of some kind in relation to technology education, but have not specifically discussed children's funds of knowledge. For instance, Moreland (2005) provided primary-aged children with cameras and invited them to photograph their environments in order to produce a children's book on their scientific and technological world. She found that children were able to easily photograph many everyday practices, recognising their significance for science and technology. In line with Rose (2007), Moreland's work points to some important methodological directions and these findings are significant for technology and design education. However, very few studies have used photographs in their research (Pozzer-Ardenghi & Roth, 2005).

Research in the past using either picture cards, drawings, or lists of words (de Vries, 1996; Jarvis & Rennie, 1994; Rennie & Jarvis, 1995; Symington, 1987) into how children conceptualise technology has tended to show that children only think about the technology in their world in relation to 'high-tech' technologies such as computers and televisions. Interestingly, more recent research has also found technology to be associated with 'products' rather than 'processes' (see de Vries, 2003b; Tasios & Solomonidou, 2002). For example, Tasios and Solomonidou invited children aged 9–12 years to respond to open-ended questions in relation to technology in their daily life by drawing or writing their responses. As with earlier research, they found that children mostly associated technology with communications technology (computers), domestic technology (electrical appliances), and transport (cars, planes). In a follow up study, Solomonidou and Tassios (2007) investigated 60 Greek primary school children's views on everyday technologies. Children were invited to view a series of 20 pictures which showed technological objects, such as tools, appliances, and some technological process and environments, for example, medical examinations, repair work, and natural environments, such as a forest. In line with previous research in this area, the children were asked to sort the pictures into those that represented technology and those that did not. They were then interviewed in relation to the criteria used for sorting. Using a discussion approach, the researchers asked questions about daily-life technologies, technological change, and the impact of technological use. The study found that children tended to categorise and discuss the pictures in relation to thinking about daily-life technology and human-oriented criteria. That is, children who saw technology as artefacts as separate from people tended to think that they had no influence over the technology (i.e. they thought about technology as following a manual), whilst those who categorised and discussed the technology in relation to people indicated that users are responsible for any technological effect. They found that the majority of students did not relate the everyday technologies to human activities and needs. Although students were able to discuss environmental problems caused by the wide use of some technologies, one third had no idea about technological change. In line with longstanding research, and in contrast to Moreland's (2005) research, their work uses predetermined photographic/line drawing images for children to sort and discuss.

Solomonidou and Tassios (2007) in reviewing the literature in this area (see de Vries, 1996; Hill 1998; McCormick, 1997; McCormick, Murphy, & Hennessy, 1994) state that "in the 'real world' the requirements of handling a situation in which technology is involved cannot be anticipated and the knowledge and skills that are needed may have not been acquired through formal instruction. The learning of technology in authentic environments – and not in an abstract and highly specified level of knowledge – by trying to solve real-life problems seems to be a key factor for teaching and learning technology" (pp. 115–116).

In a related study which investigated children's design questions, Fleer (1999) capitalised on children's home experiences of building cubbies (cosy spaces created by children) to generate cubby building experiences in two classrooms (5–year olds and 11–year olds). In this work, she showed how children's everyday home technology could be used in schools to promote technological learning in the early and later years of primary education. Her work demonstrates how everyday home contexts easily generate children's technological questions, such as: everyday design questions (how to get running water into a tree house?); structural questions (how to join paddle pop sticks in cubby model making?); materials (deciding upon material use for lining cubbies); historical focus (who built the first cubby?); fantasy focus (designing space-age cubbies); and social elements (do all kids like cubbies?).

Carr (2000), using a theoretically similar approach (sociocultural theory and affordance theory), examined how young children construct themselves as learners (individually and collectively) during their explorations and use of technology whilst in preschool. In her study, she documented children's construction of 51 hats for family members, birthday parties, princesses (hats/tiaras), or more complex and functional purposes (e.g., movement, sun visors). Her research found that children constructed a narrative to inform their technological work, which related to the children's everyday lives – for example, personal elements such as 'being nearly five' and 'being a friend', or professional roles, such as 'being a hat maker'. Carr argues that both the physical affordances (materials and context) and the social practices (being a friend/hat maker) illustrate a transactional model for learning in technology education. She notes that her research provides a case example of one technological activity and is not representative of the children's everyday lives. However, her study does demonstrate the importance of a personal narrative for children when engaged in technological activity during their preschool time.

In another paper by Carr (2001), where she examined the relationship between learners and technology, she foregrounded the importance of the learning narrative. She argued that the learning narrative that is generated by the child gives meaning to the technological activity. Importantly to the content of this chapter, she discusses the privileging of a discourse or social intent. She cites Hatch and Gardner (1993): "We argue that we need to expand the scope of such terms as 'cognition' and 'intelligence' to include the conditions under which problems are discovered and solved and within which skills are developed" (p. 165).

Although few scholars have explicitly examined the everyday technological lives of learners outside of the classroom, the studies reviewed in this section show the importance of including in the research design the social context of young learners, particularly the documentation of what matters to children in their everyday lives.

RESEARCH DESIGN

Research Question

Our study sought to document and understand the children's everyday and academic concepts of technology with a view to gaining insights into the funds of knowledge of technology available within the children's home/community. In particular, the study examined how everyday concepts of technology were drawn upon by the teachers to bring together everyday concepts and academic concepts within classroom programmes.

Sample

Two classrooms from a school located in a low socioeconomic community within the southern part of Australia were invited to participate in this study. This community is located on the Mornington Peninsula, fronting the western shore of the Peninsula. The community has an extensive foreshore area consisting of large and grassy open spaces, native bush, and mangroves along the water's edge. The community has a modern commercial centre and heavy industry, including an oil refinery and a steel plant. According to the *Australian Early Development Index: Building Better Communities for Children* (2004/2005), children in this community are 37.4% more vulnerable than the comparison communities on the Mornington Peninsula.

Fifteen children aged 5 years and 6 months (age range 4 years, 9 months to 6 years, 11 months) from the first year of school participated in this study. Their teacher had recently graduated from university and had taught in the school for four years. Twenty-five children from the third and fourth classes, aged 9 years (age range 7 years, 9 months to 10 years, 8 months) also participated in the study. Their teacher had spent 22 years teaching in the school and was well known to the families within the community. The principal of the school was new to the position, but had grown up in the community and was well known to the families.

Procedure

The study design positioned the children as researchers over the six-month period of data gathering. The first activity was for the children to create a "researcher" badge using their own drawings. This experience gave them an opportunity to display their initial understandings of technology by illustrating their badge with examples they had chosen to represent these themes.

Teachers and researchers planned a set of environmental walks during the course of the project. The children took digital photographs and video images of their walk, which were later discussed in class with the researchers. The aim of the environmental walks was for children to find examples of technology and technological activities around their school and community environments. Specifically, the children researched the school and community for local knowledge about technology in order to make these understandings visible to teachers. The first environmental walk was in the children's school followed by walks to community places, such as the "Backtracks" (a familiar play space for the children).

Photographic Documentation and Interviews

Researchers and classroom teachers were given 20 disposable cameras to pass on to the children and their families (ten per classroom). The disposable cameras were accompanied with a letter that invited the children to photograph examples of technology around their home and community (adopting the same procedure conducted in the school grounds). Classroom teachers selected the children, who were given one week to return the camera back to the school. The teachers and researchers subsequently interviewed the children during school times about the photos they had taken. We based the interview questions around three broad areas:

- Is this (photograph) science, technology or both? Why?
- Why do you think it is science/technology?
- What makes you think it is science/technology?

The children's comments were immediately transcribed onto a computer next to a digital copy of their photograph.

Analysis

In keeping with a cultural-historical approach through generating dynamic conceptual contexts, the data were clustered in relation to both the organised environmental walks (as linked through photographs, transcribed group discussions and video), and also the photographic journeys documented by individual children about their homes and communities. This allowed the data to include concepts which are ever changing during pedagogical activity (Vygotsky & Luria, 1994).

In drawing upon the theoretical framework of funds of knowledge, the dynamic scripts of data were analysed in relation to investigating the children's funds of knowledge and how everyday and academic concepts were being interlaced within their home, school, and community environments.

We based our research on the premise that by getting involved, teachers will understand more about households' knowledge (in this case, with relation to technology) and that this can be used to plan classroom activities that represent a positive "potential utility" (González et al., 1993, p. 134) for classroom curriculum. In this study, the disposable cameras and photographs became a mediating tool. They served as a way for teachers to understand where children came from in relation to their understandings about technology, thus supporting a more broadly-framed view of assessment in technology education. We were also interested in see what teachers gained from this project and how they utilised the children's funds of knowledge.

FINDINGS

Children's Focus of Attention

An analysis of the data generated through this study shows a broad range of everyday and community knowledge about technology. Through children researching their own communities for evidence of technology in their environment, it was possible to gain an understanding of the funds of knowledge held in this particular community and to generate a broader assessment context for the learners.

A range of photographs was taken by the children. The photographs were categorised as electrical goods, environment, materials around the households, and community images. The dominant category was electrical goods (total number, 144), followed by artefacts that the children believed to be of great importance within their home (total number of photographs 56). Although the children went on environmental walks within the school grounds and surrounding community areas, the children's photographs indicate that the focus of their attention was on artefacts within the home (total of 200). Only 28 photographs were taken of the environment, and only 18 photographs were directed towards the community rather than within the home.

What Are the Funds of Knowledge in This Community?

One example has been chosen to illustrate the children's funds of knowledge in relation to the everyday artefacts displayed in the children's home photographs and noted in their interview responses. Table 1 illustrates one child's keen attention to the way in which her father organises and operates his equipment to service and repair car engines. She understands not only the full activity, but has noted the order for particular processes. She notes in her interview the way in which the problem solving activity for identifying engine problems is computerised within this particular workshop. She has conceptualised her father's activity in the workshop as a problem-solving task. This finding is consistent with the principal's understanding of the capacity of the children within this particular community, having reported that the children keenly participate in problem-solving activities with everyday artefacts. In drawing upon the children's funds of knowledge, the principal had set up within the school grounds a 'Tinkering Club':

> *... where a group of tradesmen and retired tradies come in and work with the kids. Doing stuff like pulling apart lawnmowers or building billy carts or fixing appliances or whatever. But it's good ... The Tinkering Club was already established in the school and the Rotary Club and tradesmen works with kids.*

The problem-solving capacity of the children and their resourcefulness in utilising everyday materials were noted by the school, as explained by the principal:

> . . . [Student X] really became involved in the Tinkering Club. And he enjoyed it – so much that on the weekends he was pulling things apart and what not. And we had a whipper snipper that a neighbour had given me – these kids ranging from about 9 or 8 years old through to about 12 – and, cleaned it right up and put it back together and – And, I'm sitting

> there one day pulling a bike apart or something and I hear this nanananananing. "Hey, Mr. W, we've got it going and started whipper snippering.

These young children's capacity for problem solving with everyday artefacts within the tinkering club and within their own home and community was not only noted by the principal, but was used by the school to support the children's learning during school hours. The principal recognised the children's funds of knowledge.

The principal of the school comes from this particular community. He has grown up within the community, and is therefore well known to the families. The principal describes the special relationship he has with the community, as linked to how he feels. He feels as though he is a part of the community, where he is trusted by the families. Velez-Ibáñez's (1998) suggests the importance of *confianza* (mutual trust) in supporting how these exchange relationships are enacted within the community of parents.

Table 1. Mechanical and problem solving funds of knowledge (Ka, Year 3 girl).

Photograph		Interview
Photo 1		That's my dad's wheel machine. It get plug in they go on to the wheels it computerise what is wrong with the cars. Maybe science and technology.
Photo 2		You can't do all cars at once, so checking the car, place the wheels up. He brings it up and check underneath the car and plugs the car. Both, science and technology.

Valuing Funds of Knowledge

Because the community under study occupies the first place amongst the more disadvantaged communities within the Mornington Shire and in local government in Victoria (SEIFA index of disadvantage, 2001), most attention is directed towards the limited literacy competence of the children. In contrast, the advanced problem-solving skills and resourcefulness of the children that were noted in this study are not always reflected in how the school is portrayed within the education system. Unfortunately, recognition is not always valued, as noted by the Year 3–4 teacher when viewing the children's photographs and interview transcripts:

> And when you take this environment and this school and this community, technology is huge, you know, it's what we've got, what we buy, what we say as, as successful – science would be very much on their agenda ... And if you were in a different community – I think you'd have totally different results. You wouldn't have so much – you know, saturation with the

> technology that's in their home because that's part of their "upness" (climbing the social ladder) you know. [reflective tone] Yeah, the vast majority is something that they plugged in with a remote to it so, um yeah – they do have a lot of that stuff.

This teacher has carefully noted the dominant focus of attention by the children on electrical and household artefacts. She also observed that the children had taken less photos of the environment (a total of 28). This surprised her because the children had on their school environmental walks taken many photos of the plants and trees. What is interesting here is the teacher's analysis of the photographs. She interpreted the children's focus on technology as problematic – it was seen as evidence of social climbing (she used the term 'upness'). Her comment that the photographs would be very different in another community was also framed negatively.

Noting the technological focus of the children was something all the participants found. However, how this finding was valued varied between staff. This finding tends to suggest that even when the technological world of children outside the classroom is captured, it does not in itself guarantee that teachers will interpret this knowledge positively and use this information to inform how they frame learning for children.

Methodological writings on the use of visual techniques, such as those of Rose (2007), have shown how some viewers of visual images adopt a 'visual connoisseurship'. That is, the viewer looks at images in ways that are not methodologically or theoretically explicit. Rose argues that "compositional interpretation claims to look at images for 'what they are', rather than for, say, what they do or how they were or are used" (p. 36). In our study, the teachers' past experiences of the children in this particular community provided the analytical frame for looking at and interpreting the technological images the children generated. For the principal, this meant 'technological resourcefulness', and for the Year 3–4 teacher this meant 'technological upness'. Pozzer-Ardenghi and Roth (2005) have also noted these methodological challenges when using photographs, stating that photographs have a "dialectical character" because they "simultaneously lack determinacy and exhibit an excess of meaning" (p. 219). For instance, they argue that photographs:

> … give rise to innumerable, different interpretations because … their meaning emerges from the dialectic relation between the photographer's way of seeing and the perception of the reader. It is the reader's work of reading, the viewer's perception of the narrative and perceptual order of the photographic image and the surrounding text, and the meaning-making resources available to the reader that allows a specific interpretation of a photograph to arise … (p. 220)

NEW PERSPECTIVES FOR ASSESSMENT PRACTICE

In framing assessment as located beyond the school or centre walls, it has been possible to gain insights into the lived experiences of the children who participated in this study. The children were located in a poor community where money was limited, and where children and communities needed to be resourceful in order to maintain what they had, rather than throw it away. The children showed real interest in problem solving and working with existing materials. Although their focus of attention tended to be on artefacts and their continued use and repair, they were motivated to participate in generating new creations in the school tinkering programme. Roth (1998) has argued that "design is the epitome of a context in which elementary students can learn to solve open-ended problems" (p. 18).

In our study, problem-solving in the tinkering club simulated the world that the children experienced outside of their classroom and provided an affirming and positive experience for them (see Figure 1). They created an important and real environment for finding out more about the children's technological knowledge and capability.

For example, the children's interest in materials, particularly mechanical and electrical artefacts, was capitalised upon by the school through the principal. In this way, the principal recognised the funds of knowledge the children had and valued their extraordinary capacity and ingenuity when manipulating mechanical and electrical artefacts. The tinkering club provided an intellectual and practical space for the children to bring to their school environment the funds of knowledge they held in relation to problem solving to fix technological artefacts. The tinkering club allowed for a meaningful learning context to be generated for these children for technology education, and provided the potential for building new and different understandings about technological learning. Specifically, the tinkering club provided the possibility for exploring design, systems, and information, as three other examples of learning in technology education.

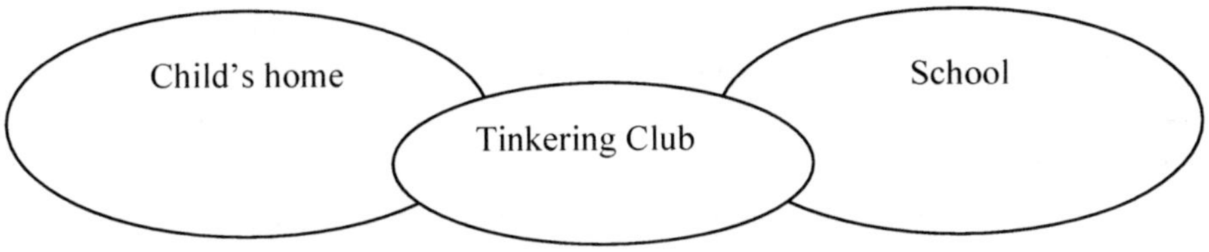

Figure 1. Simulating the child's home environment.

Roth (1998) has suggested that "design has also been described as a paradigmatic learning environment for newcomers and old-timers in a domain" (p. 19). With expert knowledge from Rotary Club members in the tinkering club, a pathway from old-timers to newcomers is possible.

It was through this study that the staff in the school gained additional access to assessing the knowledge and capabilities of the children in technology education. As a result of the children photographing their school, community, and home environments, it was possible to gain a collective perspective on the technological knowledge held in this particular community. The environmental walks and the use of disposable cameras collectively provided a vehicle for teachers and researchers to gain insights into what mattered for children in relation to their technological world. The photographic images allowed teachers and researchers to find out from the children what their home and community environment looked like. Through this, the teachers gained useful contexts and artefacts that they could discuss with the children.

The photographs thus acted as a tool for mediating the world of the child at home and the world of the child at school (see Figure 2). The teacher could gain insights into what technological activities and understandings were enacted in the home and the community, and draw upon this knowledge for generating more meaningful learning. For example, the teachers organised opportunities for the children to look at the photographs, discuss potential questions and investigation, and produce a set of 'challenge cards'. They also organised a circus of experiences which focused on the themes that were generated by the photographs:

- What is inside it (electrical goods)?
- How does it work (mechanical objects)?
- How does it move (community wind turbine)?

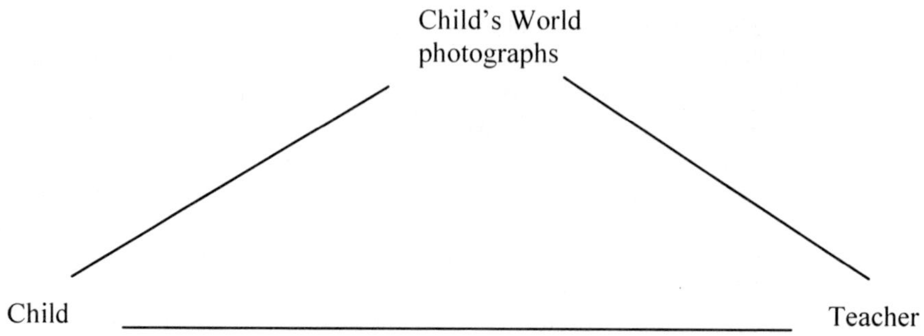

Figure 2. Mediating tools.

The circus and the challenge cards were designed to support the children in exploring technological ideas that they had photographed and asked questions about. Roth (1998) suggests:

> Design is the unfolding conversation in which the designer interacts with the materials of his work; these material and the emerging artefacts continuously talk back to her ... As she assesses unanticipated problems and potentials, the designer evaluates the move that have created them, thus linking past and present and future actions. (p. 19)

Although the worthiness of these technological activities may be open for discussion, the important point here is how the mediating tool of the photographs and the tinkering club provided a mechanism for more closely aligning the school with the children's home and community.

The dialectical relations between the home and the school were foregrounded in this particular study, and point towards the need for generating processes or structures, such as the tinkering club, or the photographs, for finding out the children's funds of knowledge. More authentic assessment is possible when the real world of the child is understood. As noted by Lund (2008) "It would seem that much work remains in order to establish principled practices where assessment includes peer support and the use of cultural tools instead of withholding such resources" (p. 49) as traditionally occurs when children are assessed individually and frequently out of the context of their everyday lives.

It should also be noted that the study highlighted how teacher beliefs and prejudices strongly influence how the photographs and the interviews of the children could be viewed. The perspective held by one teacher that "the children were surrounded by unnecessary and numerically too many electrical artefacts" also featured. This finding is worthy of further analysis, and due to space limitations has not been fully considered in this chapter. However, this finding does highlight the significance of teacher beliefs in relation to how judgments or assessments of children's learning may be being framed. It also points to the importance of the legitimacy of assessment, as noted by Lund (2008):

> One implication is that (sic) possible inconsistency between what counts as competence in and out of school ... However, if we look beyond schooling we see the increasing importance of capacity to engage in collectives, in diverse contexts, and in new practices and the capacity to appropriate the cultural tools that accompany them. (p. 49)

Although it has been beyond the scope of this chapter to give details of teacher assessment in the classroom, the discussions presented highlight the significance of moving the assessment lens beyond the walls of the school in order to find structures and process that allow teachers and families to work together so that important technological funds of knowledge can be factored into the assessment process. The three models presented above in Figures 1–3 provide a theoretical base for furthering this important work.

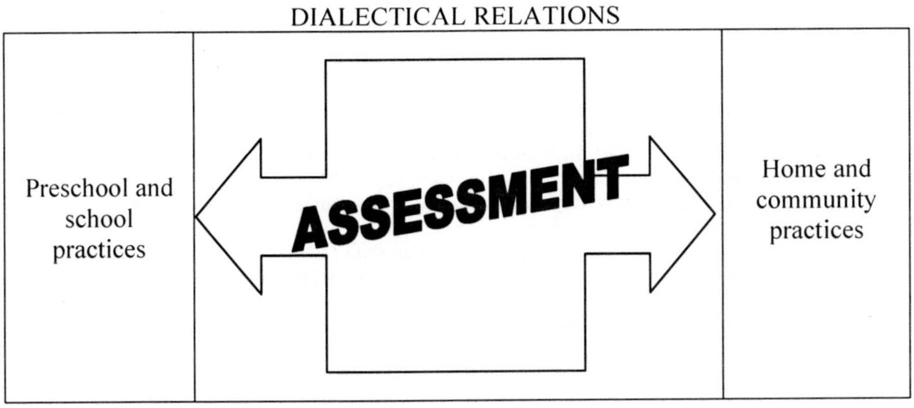

Figure 3. Assessment as a dialectical relation between the home and the school.

ACKNOWLEDGMENTS

We wish to thank the children, families, and staff of the schools who participated in this study. Monash student teachers, both those undertaking work towards their studies and volunteers, were important members of the research team: John Gipps, Dr. Robyn Zink, Isabella Lablanche, Marni-Lee Dickson, Emma Kean, Emma E. Jackson, Emma Wilson, Marie Castelli, Melinda Junginger, Denise Rado-Lynch, and Gee Gee Liew. We would also like to acknowledge the expertise of Carol Linney who transcribed the interview and classroom videotapes. Finally, we wish to thank the Australian Government for providing research funds to support this important study through their ASISTM scheme. All methodological and theoretical discussions on technology education reported here and any limitations noted by the reader are the sole responsibility of the first author.

REFERENCES

Andrews, J., & Ching Yee, W. (2006). Children's 'funds of knowledge' and their real life activities: Two minority ethnic children learning in out-of-school contexts in the UK. *Educational Review*, *58*(4), 435–449.

Anning, A. (1994). Dilemmas and opportunities of a new curriculum: Design and technology with young children. *International Journal of Technology and Design Education*, *4*(2), 155–177.

Anning, A. (1997). Drawing out ideas: Graphicacy and young children. *International Journal of Technology and Design Education*, *7*(3), 219–239.

Aronso, A. S., & Gilbert, J. K. (2006). The use of memories in understanding interactive science and technology exhibits. *International Journal of Science Education*, *28*(13), 1523–1544.

Australian Early Development Index: Building Better Communities for Children. (2004/2005). *Mornington Peninsula Shire community profile. SEIFA Index of disadvantage 2001.* Retrieved February 29, 2008, from http://203.84.234.220/profile06/Default.aspx?id=118&pg=240&gid=140&type=enum

Bain, J., & McLaren, S. V. (2006). Sustainable assessment: Exploring a learner centred approach to practice. In H. Middleton, M. Pavlova, & D. Roebuck (Eds.), *Values in technology education. Fourth Biennial International Conference on Technology Education Research* (pp. 1–7). Queensland: Griffith University.

Bjorklund, B. (2008). The repertory grid technique. Making tacit knowledge explicit: Assessing creative work and problem solving skills. In H. Middleton (Ed.), *Researching technology education. Methods and techniques* (pp. 46–69). Rotterdam, The Netherlands: Sense.

Black, P., Harrison, C., Lee, C., & Wiliam, D. (2003). *Assessment for learning. Putting it into practice*. London: Open University Press.

Black, P., & Wiliam, D. (1998). *Inside the black box*. London: School of Education, Kings College.

Braund, M., & Reiss, M. (2006). Towards a more authentic science curriculum: The contributions of out-of-school learning. *International Journal of Science Education, 28*(12), 1373–1388.

Carr, M. (2000). Technological affordance, social practice and learning narratives in an early childhood setting. *International Journal of Technology and Design Education, 10*(1), 61–79.

Carr, M. (2001). Let me count the ways. Analysing the relationship between the learner and everyday technology in early childhood. *Research in Science Education, 3*(1), 29–47.

De Jesus, A., & Antrop-Gonzalez, R. (2006). Instrumental relationships and high expectations: Exploring critical care in two Latino community-based schools. *Intercultural Education, 17*(3), 281–299.

de Vries, M. J. (1996). Technology education: Beyond the 'technology is applied science' paradigm. *Journal of Technology Education, 8*(1), 7–15.

de Vries, M. J. (2003a). Editorial. *International Journal of Technology and Design Education, 13*(3), 199–205.

de Vries, M. J. (2003b). The nature of technological knowledge: Extending empirically informed studies into what engineers know. *Techné, 6*(3), 1–15.

Fleer, M. (1999). The science of technology: Young children working technologically. *International Journal of Technology and Design Education, 9*(3), 269–291.

Fuller, R. (2006, October). Stella: Star turn for learning. *Teacher, 17*, 38–43.

Fuller, R., & Hood, D. (2005). Utilising community funds of knowledge as resources for school learning. In B. Kambler & B. Comber (Eds.), *Turn around pedagogies: Improving the education of at-risk students* (pp. 63–76). Newton, NSW: Primary English Teachers Association.

González, N., Andrade, R., Civil, M., & Moll, L. (2001). Bridging funds of distributed knowledge: Creating zones of practices in mathematics. *Journal of Education for Students placed at Risk, 6*(1 and 2), 115–132.

González, N., Moll, L., & Amanti, C. (2005). *Funds of knowledge. Theorizing practices in households, communities and classrooms*. New Jersey: Lawrence Erlbaum Associates.

González, N., Moll, L., Floyd-Tenery, M., Rivers, A., Rendon, P., González, R., & Amanti, C. (1993). *Teacher research on funds of knowledge: Learning from households*. National Center for Research on cultural diversity and second language learning Educational Practice Report: 6. Santa Cruz: University of California.

Griffin, J. (1994). Learning to learn in informal settings. *Research in Science Education, 2*, 121–128.

Griffin, J. (1998). *School-museum integrated learning experiences in science: A learning journey*. Unpublished doctoral thesis, University of Technology, Sydney.

Griffin, J. (2004). Research on students and museums: Looking more closely at the students in school groups. *Science Education, 88*(1), 59–70.

Griffin, J., Mehan, C., & Jay, D. (2003). *The other side of evaluating student learning in museums: Separating the how from the what*. Paper presented at the Museums Australia Conference, Perth, Western Australia.

Griffin, J., & Symington, D. (1997). Moving from task-oriented to learning-oriented strategies on school excursions to museums. *Science Education, 81*(6), 763–779.

Hatch, T., & Gardner, H. (1993). Finding cognition in the classroom: An expanded view of human intelligence. In G. Salomon (Ed.), *Distributed cognitions: Psychological and educational considerations* (pp. 167–187). Cambridge, UK: Cambridge University Press.

Hill, A.-M. (1998). Problem solving in real-life contexts: An alternative for design in technology education. *International Journal of Technology and Design Education, 8*(3), 203–220.

Hughes, M., & Greenhough, P. (2006). Boxes, bags and videotape: Enhancing home-school communication through knowledge exchange activities. *Educational Review, 58*(4), 471–487.

Hughes, M., & Pollard, A. (2006). Home-school knowledge exchange in context. *Educational Review, 58*(4), 385–395.

Jarvis, T., & Rennie, L. (1994, March). *Children's perceptions about technology: An international comparison*. Paper presented at the Annual Meeting of the National Association for Research in Science Teaching. Anaheim, California.

Jones, A., & Moreland, J. (2004). Enhancing practicing primary school teachers' pedagogical content knowledge in technology. *International Journal of Technology and Design Education, 14*(2), 121–140.

Kimbell, R. (1997). *Assessing technology. International trends in curriculum and assessment*. Buckingham, UK: Open University Press.

Langermeyer, I., & Roth, W.-M. (2006). Is cultural-historical activity theory threatened to fall short of its own principles and possibilities as a dialectical social science? *Outlines, 2*, 20–42.

Luehmann, A. L., & Markowitz, D. (2007). Science teachers' perceived benefits of an out-of-school enrichment programme: Identity needs and university affordances. *International Journal of Science Education, 29*(6), 643–661.

Lund, A. (2008). Assessment made visible: Individual and collective practices. *Mind, Culture, and Activity, 15*(1), 32–51.

McCormick, R. (1997). Conceptual and procedural knowledge. *International Journal of Technology and Design Education, 7*(1 and 2), 141–159.

McCormick, R., Murphy, P., & Hennessy, S. (1994). Problem-solving processes in technology education: A pilot study. *International Journal of Technology and Design Education, 4*(1), 5–34.

Middleton, H. (Ed.). (2008). *Researching technology education. Methods and techniques.* Rotterdam, The Netherlands: Sense.

Moll, L., Amanti, C., Neff, D., & González, N. (1992). Funds of knowledge for teaching: Using a qualitative approach to connect homes and classrooms. *Theory into Practice, 31*(2), 132–141.

Moll, L., & Greenberg, J. (1990). Creating zones of possibilities: Combining social contexts for instruction. In L. C. Moll (Ed.), *Vygotsky and education: Instructional implications and applications of sociohistorical psychology* (pp. 319–348). Cambridge, UK: Cambridge University Press.

Monzo, L. D., & Rueda, R. (2003). Shaping education through diverse funds of knowledge: A look at one Latina paraeducator's lived experiences, beliefs, and teaching practice. *Anthropology and Education Quarterly, 34*(1), 72–95.

Moreland, J. P. (2003). *Becoming effective technology teachers: Enhancing assessment practices in primary classrooms.* Unpublished doctoral thesis, University of Waikato, New Zealand.

Moreland, J. P. (2005). *Using a digital methodological approach in science and technology education.* Presentation at Monash University, Faculty of Education, Peninsula campus.

Moreland, J., & Jones, A. (2000). Emerging assessment practices in an emergent curriculum: Implications for technology. *International Journal of Technology and Design Education, 10*(3), 283–305.

Pozzer-Ardenghi, L., & Roth, W.-M. (2005). Making sense of photographs. *Science Education, 89*(2), 219–241.

Rennie, L., & Jarvis, T. (1995). Children's choice of drawing to communicate their ideas about technology. *Research in Science Education, 25*(3), 239–252.

Riojas-Cortez, M. (2001). Preschoolers' funds of knowledge displayed through sociodramatic play episodes in a bilingual classroom. *Early Childhood Education Journal, 29*(1), 35–40.

Rohaan, E. J., Taconis, R., & Jochems, W. M. G. (2008). Reviewing the relations between teachers' knowledge and pupils' attitude in the field of primary technology education. *International Journal of Technology and Design Education.* Advance online publication. DOI: 10.1007/210798-008-9055-7

Rose, G. (2007). *Visual methodologies. An introduction to the interpretation of visual materials* (2nd ed.). California: Sage.

Roth, W.-M. (1998). *Communities. Science and technology education library.* Boston: Kluwer.

Sakharov, L. (1994). Methods of investigating concepts. In R. Van Der Veer & J. Valsiner (Eds.), *The Vygotsky reader* (pp. 73–98). Oxford, UK: Blackwell.

Solomonidou, C., & Tassios, A. (2007). A phenomenographic study of Greek primary school students' representations concerning technology in daily life. *International Journal of Technology and Design Education, 17*(2), 113–133.

Symington, D. J. (1987). Technology in the primary school curriculum: Teacher ideas. *Research in Science and Technology Education, 5*(20), 167–172.

Tassios, A., & Solomonidou, C. (2002, September). *Students' (aged 9–12 years old) representations about everyday life technologies and the place of ICT in those representations.* Paper presented at the 3rd Pan-Hellenic Conference on information and communication technologies in education (presented in Greek), Athens, Greece.

Vélez-Ibáñez, C. G. (1988). Networks of exchange among Mexicans in the U.S. and Mexico. Local level mediating responses to national and international transformations. *Urban Anthropology, 17*(1), 27–51.

Vélez-Ibáñez, C. G., & Greenberg, J. (1992). Formation and transformation of funds of knowledge among U.S. – Mexican Households. *Urban Anthropology, 23*(4), 313–335.

Vygotsky, L. S. (1987). Thinking and speech (N. Minick, Trans.). In R. W. Rieber & A. S. Carton (Eds.), *The collected works of L.S. Vygotsky* (Vol. 1, pp. 37–285). New York: Plenum.

Vygotsky, L. S. (1998). Child psychology (M. J. Hall, Trans.). In R. W. Rieber (Ed.), *The collected works of L.S. Vygotsky* (Vol. 5). New York: Plenum.

Vygotsky, L., & Luria, A. (1994). Tool and symbol in child development. In R. Van Der Veer & J. Valsiner (Eds.), *The Vygotsky Reader* (pp. 99–174). Oxford, UK: Blackwell.

Marilyn Fleer
Gloria Quiñones
Faculty of Education
Monash University
Australia

KAY STABLES AND JENNY BAIN

41. TRANSFORMING ASSESSMENT: ASSESSMENT OF LEARNING BECOMES ASSESSMENT FOR LEARNING

CONTEXT

Increasingly, international school systems are focusing on how assessment can be embedded in curriculum and, as a result, on the questions this raises about the purposes of assessment. Growing concern about the impact of summative assessment on learning has led to major attempts to re-focus debate from a summative approach (or assessment *of* learning) to more formative approaches (or assessment *for* learning). This need to re-focus assessment practice – led by research including that of Black and Wiliam (1998), Harlen and Deakin-Crick (2003), and Jones and Moreland (2005) – is emphasised in a global monitoring report, *Education for All: The Quality Imperative* (United Nations Educational, Scientific and Cultural Organisation [UNESCO], 2005). The report, an independent publication commissioned by UNESCO on behalf of the international community, concludes:

> For governments seeking to improve education quality, a sound assessment policy is crucial. For school-level assessment to be influential, it should be consistent, regular and reliable, part of an overall school development policy and reconcile both formative and summative assessments with a strong focus on providing feedback to the learner and teacher. (p. 157)

The report goes on to acknowledge that, despite international trends to include formative assessment practice, ''continuous and formative assessment" often amounts to "repeated summative assessment" (p. 158).

Accordingly, this chapter will consider ways of making the shift to a more integrated system of assessment in the context of technology education and will examine the role that summative assessment methods can play as part of effective formative assessment practice. It will draw on case studies of research of performance based, authentic assessment in design and technology and will take as its backcloth the context of English schools where the need to re-think assessment has become an acute issue.

A SHIFT TO ASSESSMENT 'FIT FOR PURPOSE'

Through the 1980 Education Act in England and Wales that required secondary schools to publish their examination results in 'league tables' and through National Curriculum reforms, summative assessment was introduced at ages 7, 11, 14, 16, 17, and 18. These assessments have increasingly been driven by a 'standards' and 'audit' culture. All this has been despite evidence that league tables have little value either for supporting learning and teaching or for informing parental choice (Ball & Vincent 1998; Lawton 1992; Nuttall 1990; Wilson, Croxson, & Atkinson, 2006). The negative impacts of this are becoming all too apparent and a distinct shift in approach is being felt. Indeed, as we write, the UK Government has announced that the current compulsory national tests

A.T. Jones and M.J. de Vries (eds.), International Handbook of Research and Development in Technology Education, 493–508.
© 2009 Sense Publishers. All rights reserved.

for 14-year olds will be replaced by improved classroom assessment by teachers (Department for Children, Schools and Families [DCSF], 2008). At the forefront of this shift is the Government body responsible for curriculum and assessment, the Quality and Curriculum Authority (QCA), who have undertaken a review of the National Curriculum. This review has been informed by a series of thematic studies conducted as part of the *International Review of Curriculum and Assessment Frameworks* (INCA) project carried out by the National Foundation for Educational Research in England and Wales (NFER, 2002 to present) and, as such, the impetus for curriculum change may be considered to reflect best practice in international assessment trends. This has led the QCA to lay down an outline for what they refer to as the 'big picture' of a curriculum for the 21st Century:

> If the curriculum is to be inspiring and challenging and fit for the 21st century, it has to be dynamic, responsive to change, relevant and engaging. [To support this concept] QCA has developed a curriculum big picture to reinforce the concept of curriculum as the entire planned learning experience of a young person. (QCA, 2008)

This 'big picture' of the National Curriculum is not a policy document but a tool for discussion aimed at helping shape curriculum design. Its introduction aims to reinforce a change of focus for assessment, with an assessment system 'fit for purpose' that has formative assessment at its heart. Such an assessment system might provide opportunities for both formative and summative assessment and be designed to serve both purposes well (Newton, 2003; Wiliam, 2003).

In explaining the 'big picture', Mick Waters, Director of Curriculum at QCA, accounted for the problematic situation that has developed in respect of summative assessment and offered a vision of what an assessment system fit for purpose might focus on:

> One of our challenges in England is to make assessment fit for purpose. What we've got at the moment is a whole range of possibilities but we focus on one more than any other. We focus on testing at the end of key stages. We've become obsessed with it as a country. We can't get out of it now. We're testing at the end of key stages to the extent that becomes so important because of all the league tables, that beyond that, teachers start doing testing to check that children are good at testing. And so we practice testing so that we know whether we're good at testing. If we're not, we do more tests to make sure we are. And the testing machine is starting to eat itself, and we've somehow got to redress that back into peer assessment, children assessing with each other, children assessing with parents, assessing against what you used to be able to do. Assessing against what you think you want to do. Assessing against things you've never even heard about yet. (Waters/QCA, 2007, p. 1)

This more comprehensive view of 'assessment fit for purpose' is summarised in the extract from the QCA Big Picture of the curriculum (QCA, 2008) presented in Table 1. Whilst this recent shift is useful in guiding the future focus of assessment practice, the reality has been – as indicated in Waters' quote above – a major emphasis on summative assessment, led directly by the policy makers and enforcers. Coupled with the political agenda on standards outlined earlier, this has led to a greater commissioning of research around summative rather than formative assessment. Although this has had a distinctive impact on the focus of assessment research in design and technology education conducted within the Technology Education Research Unit (TERU) at Goldsmiths, it has become apparent that through the development of design and technology assessment activities, primarily summative in focus yet designed to support and record progress, an integrated and flexible assessment instrument has emerged. It is the emergence of this assessment instrument, the 'unpickled portfolio', that provides focus for an examination of an assessment system that that might be fit for purpose in the ways described above.

SUMMATIVE ASSESSMENT DEVELOPMENTS IN TERU

Through a series of research projects conducted over a span of 20 years, researchers within TERU have created and developed an approach to facilitating dynamic, intensive, focused, authentic design activities that provide evidence for summative assessment of design and technological performance. We have come to term these evidence-based activities 'unpickled portfolios' (Stables & Kimbell, 2000) because of the short time frame (between 90 minutes and 6 hours) in which an on-task design portfolio is created.

Table 1. Extract from A Big Picture of the Curriculum working draft *(QCA, 2008).*

To make learning and teaching more effective so that learners understand quality and how to improve										
Assess-ment fit for purpose	Is integral to effective teaching and learning	Draws on a wide range of evidence of pupils' learning	Promotes a broad and engaging curricu-lum	Maxim-ises pupils' progress	Gives helpful feed-back for the learner and other stake-holders	Helps identify clear targets for improve-ment	Links to national standards which are consistently interpreted	Informs future planning and teaching	Uses tests and tasks appro-priately	Embraces peer- and self-assess-ment

From the outset (Kimbell, Stables, Wheeler, Wozniak, & Kelly, 1991) there has been anecdotal evidence to suggest that, while the primary purpose of the activities has been summative assessment, learners have found engaging in the tasks insightful in terms of their own designing. In addition, teachers have identified ways in which they could draw on the unpickled portfolio approach for more formative learning and teaching practices.

More recently, this approach has been actively developed as a diagnostic tool for learners to identify their own strengths and areas for development in design and technology activities (McLaren, Stables, & Bain, 2007). This has involved on-task peer assessment and learner self-reflection; post-task learner self-reflection and target setting; and teaching and learning interventions.

In what follows we outline the key features of this approach to performance assessment and explore its potential for both summative and formative assessment – for 'assessment fit for purpose' – and the issues and ideas raised for learning and teaching in design and technology education.

THE POTENTIAL OF THE 'UNPICKLED PORTFOLIO' FOR ASSESSMENT FOR LEARNING

The concept of the 'unpickled portfolio' was developed originally in response to the need for short, authentic design and technology assessment activities for the Assessment of Performance Unit in the late 1980s (Kimbell et al., 1991). In essence, the original activities, and all that have been developed since, have been short, structured activities that engage learners in assessment activities that both support their performance and provide evidence of their design and technological capability (Kimbell & Stables, 2007). Critical within these activities is a positive tension between the assessment intentions of the assessors (e.g. to assess how well a learner develops design ideas) and the more instrumental intentions of the learner (i.e. to get on with their designing). Most recently, Stables and Kimbell (2007) have described this relationship in terms of mirrors, suggesting that providing authentic activity structures and prompts that create evidence of capability is like providing the teacher with a two-way mirror that they can observe the learner through. When the pedagogic balance is good and the activity is sufficiently authentic, when the learner is enabled to 'see' the evidence that they have just created, it also provides a looking glass – reflecting back to the learner insights into their progress and their learning and to the teacher insights into the effectiveness of their teaching actions (see Figure 1).

The unpickled portfolio approach has the potential to go beyond helping the learner modify and develop the design they are working on, and allows the teacher to gain insights into this. In supporting the learner to make their thinking explicit, it increases the learner's metacognitive grasp on their own (more generic) designing processes (Stables & Kimbell, 2007). In doing so, it has the potential for high consequential validity (Boud, 1995; Messick, 1989) as part of an assessment system that seeks to maximise learner progress whilst supporting learners in developing sustainable assessment practice.

Sustainable assessment might be defined as a consequence of assessment, rather than a method or approach (Boud, 2000). Here, 'sustainable assessment' centres around assessment practice that promotes, for example, active engagement with learning tasks, use of feedback and access to peer support. It is similar to what Mentkowski (2000) terms "learning that lasts" (p. 216). Sustainable assessment may be considered, along with formative assessment, as features of assessment that present a view of independent learners able to, as Waters (Waters/QCA, 2007, p. 3) suggests: "Assess against things you've never even heard about yet". Consequential validity refers to the effect of assessment on learning and other educational matters (Linn, Baker, & Dunbar, 1991; Messick, 1989). It prompts the question, "What are the broader consequences of a given assessment activity beyond those which are immediately evident?" (Boud, 1995, p. 39). Accordingly, the consequential validity of assessment is considered high when there is a positive effect on learning and low when it encourages ways of learning that are counter to that desired.

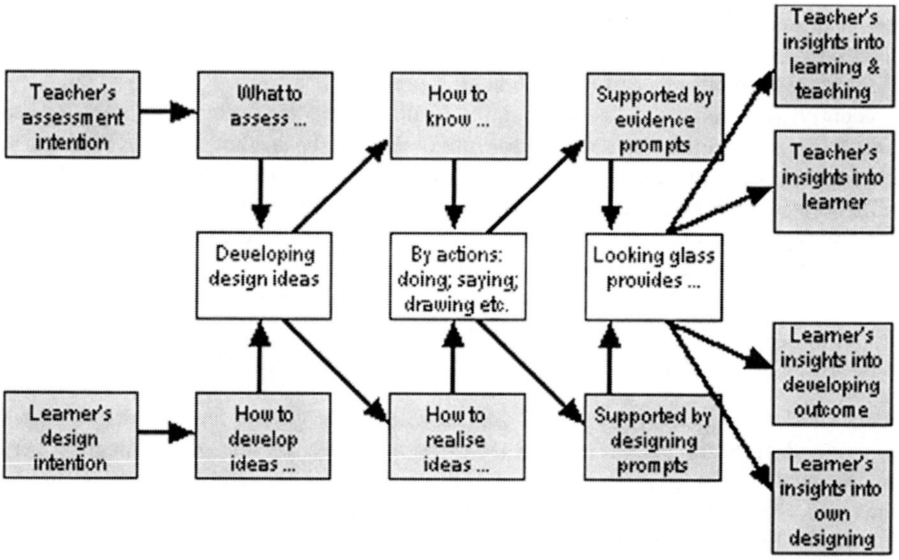

Figure 1. Assessment activity as a looking glass.

Through an examination of three case studies undertaken using the 'unpickled portfolio' approach, we illustrate below how this approach to assessment might be seen to provide scope for broader assessment 'fit for purpose' rather than as a narrower, explicitly summative assessment instrument. We also illustrate how, particularly when used with supplementary instruments such as learner evaluation questionnaires and an holistic assessment rubric, the unpickled portfolio approach to assessment has the potential to have a positive effect on learning, to develop sustainable assessment practice through peer and self assessment opportunities, and to provide a comprehensive set of assessment data that can be used by learners and teachers to achieve high consequential validity.

THE CASE STUDIES

The case studies are linked projects in assessing design and technology: *Assessing Design Innovation; Creativity and Progression in Transition through Assessment for Learning in Design and Technology (CAPITTALDT)*, and *e-solutions for creative assessment in portfolio environments (e-scape).*

Assessing Design Innovation (2003–2004)

In late 2002, TERU was commissioned to develop approaches to summative assessment that would focus on creativity, and specifically creativity in the context of public examinations at 16+ in design and technology (the GCSE examination). Concern had been expressed about the lack emphasis on creativity in design and technology, particularly in the context of assessment (e.g., Atkinson 1994, 2000; Davies, 2000; Kimbell 2000). In addition, an independent review of the design and technology National Curriculum documentation (Prest, 2002) showed a mismatch between the vision for design and technology as supporting learners to 'intervene creatively' and become 'creative problem solvers' and 'innovators', and the legislated Programmes of Study and Attainment Target (DfES/QCA, 1999).

With a clear recognition of the way in which summative assessment, such as that embodied in the GCSE, influenced the local development of curriculum at the school level, TERU were taxed with developing assessment activities that would promote and assess creativity and, in addition, have the double value of providing summative assessment and a model for incorporating creativity more broadly into design and technology project work. The resulting project, *Assessing Design Innovation* (Kimbell, Miller, Bain, Wright, Wheeler, & Stables, 2004) developed a series of 6-hour design activities through which individual learners produced their own design solutions to a task, whilst working in a group of three 'critical friends' (Costa & Kallick, 1993). Learners were further supported by a handling collection of inspiration objects, a set of materials that allowed for the rapid 3D modelling of design ideas, and, to capture the dynamic nature of their modelling, snapshot photographs of their developing work, taken roughly once an hour during the working session. Figure 2 indicates the overall structure of the activity.

For summative assessment purposes, the activities worked well and have now been incorporated into one Examination Awarding Body's national suite of assessments. This has been reported elsewhere (Kimbell et al., 2004; Kimbell & Stables, 2007) and is not of direct interest here. What is significant from a formative assessment viewpoint are the following findings:

- the reactions of the learners to the support provided for their creative responses, including their reactions to being able to 'see' the evidence of progress in their designing (quite literally through the photo storyboard – see Bain, 2005 – created through the snapshots of their design), the impact of the 'critical friends', and explicit prompts to regularly review their work; and
- the reactions of the teachers to what was, in effect, an examination.

Learner feedback was collected through a questionnaire that was structured to include a series of questions with responses on a 4-point Likert scale and also a free response section on the best and worst aspects of the activity. A key overarching finding from the questionnaire was that the learners rated the activity highly in terms of it being a good way to assess design and technology capability (3.5 mean - strongly agree being 4 and strongly disagree being 1). Seen from this perspective of the learners validating the activity for assessment, the examples from the free response section illustrate further aspects that indicate high consequential validity in terms of engagement, opportunities to be creative, opportunities for self-reflection, and the value of collaboration.

'LIGHT FANTASTIC' TASK

A light-bulb company wants to minimise packaging waste and extend the product range they offer. They want a new range of light-bulb packaging that people won't throw away.

Your task is to come up with exciting ideas for light-bulb packaging that people won't throw away because it transforms into interesting lighting features & structures.

By the end of the activity you must have produced

- a working light-bulb package containing everything for the lighting feature;
- an assembled lighting feature:
- a persuasive argument for your product to attract purchasers.

Outline structure

1. read task to the group and establish what is involved
2. explore a series of 'idea-objects' on an 'inspiration table' and in a handling collection designed to promote ideas for transformation
3. put down first ideas in a designated box in the booklet
4. swop work within team - for further development by team mates
5. work returned to 'owner' to consider which ideas to pursue
6. teacher introduces the modelling/resource kit
7. learners develop their ideas through drawing – and/or through 3D modelling.
8. learners reflect on the *user* of the end product and the *context* of use, before continuing with development
9. at set intervals, learners pause and throw a 'questions' dice, e.g. "how would your ideas change if you had to make 100?". Answers recorded in their booklet
10. approximately every hour photos of modelling taken to develop *visual story line* of evolution of design ideas
11. end of 1st morning, learners reflect on own and team members work
12. 2nd morning starts with celebration of work from day 1 using 'post-it' notes to highlight 'best' idea, 'wackiest idea' biggest problem' and 'next steps'.
13. prototype development continues
14. hourly photos and pauses for reflective thought continue
15. final team reflections on each others' ideas and progress
16. learners 'fast-forward' their idea - what it will look like when finished

Figure 2. Outline structure of an Assessing Design Innovation *task (a 6-hour assessment activity).*

The best things about the activity:

I liked letting my imagination go wild (learner number AB4111)

It was a very good way of finding how many skills you have (learner number AB4110)

I could show people my ability (learner number AB2116)

The photos was the best way to show the steps, which I think was good (learner number AB326)

It's easier to understand when you try it out for yourself (learner number AB1117)

The pictures help, cos I could always look back and see how good I'd done (learner number AB1119)

Working in groups but having the ability to work independently (learner number AB2115)

These have been the best two days in school (learner number AB1111)

In terms of engagement, the following comment from one of the teachers is particularly telling:

One of the remarks that I recall from reading the self assessment sheet was "it shows what I can do in a positive way" – this was written by a pupil who is a school phobic and finds school work difficult. (Kimbell et al., 2004, p. 34)

Feedback from the teachers, through post activity semi-structured interviews and questionnaires, revealed that they were generally impressed by both the positive way in which the learners engaged with the task and with the quality of the development work, including through the interactions within the smaller groups. Their comments made little, if any, reference to the summative assessment potential of the tasks but attest to their authenticity as creative design and technology learning activities, highlighting the ways in which the activities give learners "permission to take a

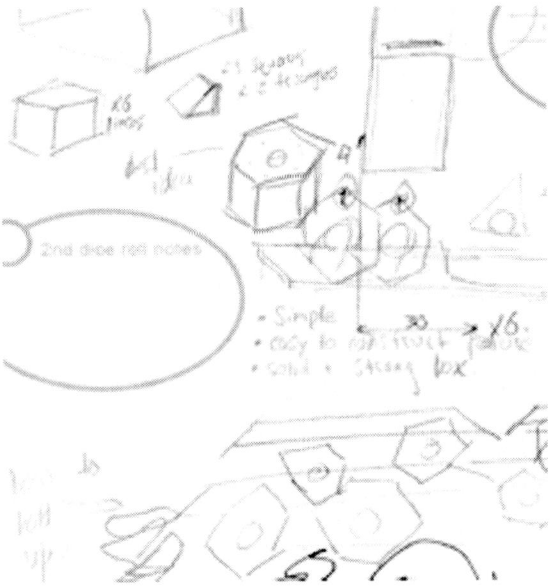

Figure 3. Reflective annotations (from Kimbell et al., 2004, p. 53).

risk" through a process which is "pacey", in which "nothing becomes overworked or laboured" and in which the learners become "aware that other people's opinions could help and more importantly they function as a team" (Kimbell et al., 2004, pp. 34–35). There is a certain irony that this summative assessment activity has, for the teachers involved, encouraged them to think freshly, not about assessment, but about learning and teaching. This was illustrated by a comment from an extremely experienced teacher, who commented: "This trial has had a real effect on my teaching. It has reinforced things I do, reminded me of things that I have done, and prodded me to think of things I have never done" (p. 37).

But the fact of the matter remains that the activities were targeted at assessment, and in particular on gathering evidence of performance. The research brief we were working to emphasised creativity and innovation, and early in the process we analysed sets of work provided by teachers and the examination Awarding Bodies that had achieved high marks in the standard summative assessment processes for GCSE and that we asked to be identified as holistically (taking the whole work into account) as either innovative or not innovative, using key words to describe the work. From this we created a holistic continuum from 'wow' (highly innovative) to 'yawn' (limited/no innovation) and then analysed each piece to identify evidence of creative and innovative performance in order to derive criteria that could be developed into an assessment rubric. What quickly became clear was that innovative work was driven by creative ideas developed through an iterative process of having (sparkiness), growing (modelling), and proving (criticality) the ideas as they emerged. By analysing a range of work (96 pieces) we were able to characterise different levels of 'having', 'growing', and 'proving' ideas. (For further details see Kimbell et al., 2004, pp. 59–60).

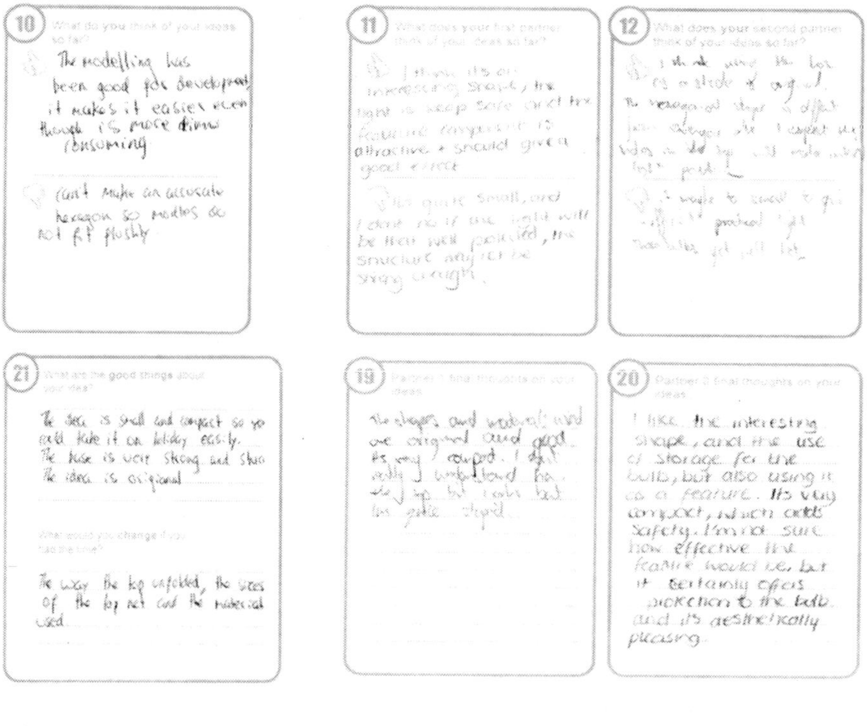

Figure 4. Prompted critique: Peer and self reflection (from Kimbell et al., 2004, p. 54).

The analysis drew heavily on the visual evidence from the photo storyboard of the development of the ideas, supported by the learner's ongoing reflection on their work. It is the combination of these two facets that provide the win-win assessment prize outlined earlier in Table 1. Figure 3 illustrates ad hoc reflection through annotations and Figure 4 illustrates the prompted critique – showing the comments of both the learner and their two 'critical friends'.

So while the main thrust of Assessing Design Innovation was summative, inbuilt in its structure there was feedback for learners that had a formative use. It was this implicit potential of Assessing Design Innovation that we exploited in the next project – *CAPPITTAL-DT.*

THE *CAPITTAL-DT* PROJECT

The *CAPITTAL-DT* pilot study (McLaren, Stables, & Bain, 2006) was a direct 'spin-out' from *Assessing Design Innovation.* The focus of the project was creativity and innovative thinking in the context of design activity at the transition between primary and secondary schooling, which has been identified as problematic and as having negative impact on progression (SOED, 1996, 1998). Within this, the study focused on the development of learner-centred sustainable assessment approaches as they relate to progression in creativity and enterprising design thinking (Bain & McLaren, 2006). The methodology adopted for the study involved an intervention phase of nine months that spanned the primary/secondary transition point. The study incorporated a variety of self and peer evaluation and assessment strategies and a pre- and post-intervention 'benchmarking' assessment task, a version of the *Assessing Design Innovation* task. Included in the intervention was a focus on the learner's own judgements and responses, their perception of what they had learned, and explicit identification of further learning they felt they required in order to progress.

Accordingly, a key focus of the study was on the tools and strategies that were devised to encourage the learner to recognise, and place value on, their achievements and have the potential to contribute towards processes that support 'personal learning planning' (LTS, 2004). The motivation behind this was, in part, the belief that the purposes of assessment should be extended to include the preparation of students for sustainable assessment, where:

> Sustainable assessment encompasses the abilities required to undertake those activities that necessarily accompany learning throughout life in formal and informal settings. Assessment acts need both to meet specific and immediate goals, as well as establishing a basis for students to undertake their own assessment activities in the future. (Boud, 2000, p. 158)

The pilot study involved learners from seven schools, in two different Scottish Local Authorities. Pre- and post-intervention data were gathered from all participants through three instruments, which together provided an integrated summative and formative data set:

- a 'learner attitudes towards creativity' questionnaire (see Stables, 2006);
- an 'unpickled portfolio' type activity (Stables & Kimbell, 2000) based on the *Assessing Design Innovation* task, but cut down to three hours in length; and
- a 'learner evaluation' questionnaire (Kimbell et al., 2004) undertaken after the baseline and follow up activities.

Three categories of research cohorts were created to take a quasi-experimental approach: cohort 1 received full intervention, supported by one of the research team; cohort 2 received intervention via their teachers who devised their own strategies based on an understanding of the project; and cohort 3 where there was no intervention and who acted as a 'control'.

The intervention strategies and learner evaluation questionnaire were designed with sustainable assessment practice in mind and in line with key factors outlined in *Beyond the Black Box* (Assessment Reform Group, 1999):

- the provision of effective feedback to learners;
- the active involvement of learners in their own learning;
- a recognition of the profound influence assessment has on the motivation and self-esteem of learners, both of which are crucial influences on learning; and
- the need for pupils to be able to assess themselves and understand how to improve.

The learner evaluation questionnaire asked the learners to record their agreement/ disagreement with given statements that explored their response to the activity in terms of why they liked/disliked it and what they felt the activity had allowed them to demonstrate. They were also asked to record what they had liked the best, and what they had found easiest and most difficult. To probe further, learners were asked to note three things they felt they had learned and three things they would like to get better at. In addition, each participant was asked to complete open-ended text requests prompted by the following stem sentences: I was best at ..., The easiest thing was ..., The most difficult thing was ..., Today I learned ..., I want to get better at ...

An examination of profiles of consistently high performers and those whose performance improved most (across the pre- and post-intervention assessment tasks) indicated that these learners evaluated the experience using a greater variety of factors than the poor performers. The higher performers seemed to have a higher awareness of their own strengths (Bain & McLaren, 2006). The words 'thinking', 'choosing', 'designing', and 'planning' also feature more frequently in this group, with thinking skills, creativity, processes, issues, and key learning involved more likely to be mentioned. The use of such words suggests that these learners were relating their experience to the thinking skills, creativity, processes, issues, and key learning involved in the tasks.

A key feature of this case study was that performance data indicated that learners who were part of the intervention groups progressed more than those who were not (see McLaren et al., 2006). Their performance (assessed using a holistic rubric as described above) was better, and they (through self assessment of their learning) showed increased awareness of what contributed to improving their performance. In short, through the intervention process learners appeared to have become more active partners in learning. This was done through involving learners in monitoring their own learning, sharing what is to be learned and what is expected of them, engaging them in thinking about what needs to be done to improve their learning, and how to go about it (Hayward, Kane, & Cogan, 2000).

In addition, higher performing learners demonstrated a better understanding of technological capability. Interestingly, this is in line with research into teachers which indicates that when teacher understanding of technological capability is limited, there is a tendency for the teacher to focus on the actual production of a product rather than the thinking skills, creativity, processes, issues, and key learning involved (e.g., Harlen, 1997; Harlen, Holroyd, & Byrne, 1995; Jones & Compton, 1998). In this case study, it was the lower performers who focus on production of the product rather than thinking skills, etc.

The performance and learner evaluation responses indicated that the learning and teaching intervention strategies, by incorporating aspects of teacher, self and peer assessment with a view to aiding progression, contributed to establishing a basis for students to undertake their own assessment activities in the future (Boud, 2000). The findings were thus consistent with the expectations around the development of learning and teaching interventions that might help develop a sustainable assessment skill-set. They also indicated that such a 'learner-centred' approach to assessment results in high consequential validity in terms of learner progress.

THE *e-scape* PROJECT

The *e-scape* project (Kimbell, Wheeler, Shepard, Brown-Martin, Perry, Hall, et al., 2005; Kimbell, Wheeler, Miller, & Pollitt, 2007) also grew directly from *Assessing Design Innovation*, but in this case built on the initial research (notably linked to the photo storyline) to explore further the potential for using digital tools in assessment. After experimenting with a range of digital tools (digital pens, voice recognition software, handheld computers), an activity derived directly from *Assessing Design Innovation* was trialled using handheld computers (PDAs). These offered the greatest combined range of response modes to the learners (drawing, writing, voice recording, photography, video, mind mapping) enabling us to support a broader range of learning styles. Importantly, they also facilitated a shift in the locus of control for data collection from the assessor to the learner with, for example, the learner taking and selecting their own photographs of modelling developments.

Particularly significant for a 'fit for purpose' assessment system was the opportunity for learners to make explicit their thinking in respect of their designing in a range of modes, for example, by adding voice memos to the photographs of their developing work. Through this, we (the assessors) literally heard the voice of the learner, which was liberating for them and extremely informative for us. By providing a range of response 'tools', not only were we enabling learners with different learning and designing styles (Lawler, 1999), we were also creating assessment data in a range of modes. For example, to convey the design idea and the thinking behind it, a learner could potentially:

- draw on the PDA screen using 'pocket painter' and then annotate by writing freehand on the screen;
- take a photograph of a sketch model and annotate by adding a voice memo;
- use the video facility to 'fly through' a sketch model, or show it in action, using voice-over;
- draw an idea on paper, photograph it, and add a voice memo;
- draw and annotate an idea on paper, then photograph it;
- photograph a sketch model, and then sketch or annotate freehand on top of the photo image on screen; or
- use any other combination of the various techniques above.

In addition, the PDA activity enabled the learners to review the work once completed and add comments or further information, thereby taking advantage of hindsight within the assessment period. Through the software that is used to support the *e-scape* system, the learners' work is progressively (virtually in real time) uploaded into their individual e-portfolio on a web space. These data offer massive potential in terms of summative assessment (see Chapter 42 by Richard Kimbell) and also present exciting opportunities for post-activity review and discussion – amongst learners and with the teacher.

The *e-scape* project is currently in its third phase, which includes the adaptation and development of the activity 'authoring' software to be placed in the hands of teachers as a tool for them to develop their own learning, teaching, and assessment activities. We are working with a small group of six schools to explore this potential. Developments are at an early stage, but indications are that teachers are seeing the assessment potential of the digital unpickled portfolio far more in terms of formative than summative assessment, particularly in the context of collaboration and self and peer reflection. The system also provides for virtually 'live' ongoing electronic monitoring of learners' work. This, in turn, provides opportunities for formative feedback and dialogue that, due to the development of a distinctly different approach to summative assessment can genuinely focus on learning and progression.

TOWARDS DEVELOPING SUSTAINABLE ASSESSMENT PRACTICES

Threading through these case studies, both explicitly and implicitly, are three themes that are central to a move towards sustainable assessment practices: peer and self assessment; real-time (immediate) feedback; and placing the learner in control of their assessment, thus democratising the process.

Peer and Self Assessment

Learning with and from peers is the dominant mode of informal and everyday learning (Falchikov, 2007). The role of peers in assessment is explicitly valued and understood in many theories and philosophies of learning. Amongst others, Dewey (1897), Bruner, (1960), and Lave (1999) believe that knowledge needs an authentic content and structure requiring social collaboration and action. Peer-assessment can help provide opportunities for this kind of authentic collaboration (Falchikov, 2001). Consistent with QCA's 'big picture', this type of assessment benefits students' critical thinking and social skills, personal and intellectual development, and understanding of the assessment process itself (Tan, 2007).

Peer-assessment provides an important complement to, and may even be a pre-requisite for, effective self assessment (Black, Harrison, Lee, Marshall, & Wiliam, 2003). The case studies examined above highlight the self-awareness that peer and self assessment have afforded the learners in a range of ways. We have seen the direct impact on the development of the learners' ideas. We have also seen their growing insight into their own approach to designing and to interpersonal skills, for example, through the young learners in *CAPITTAL-DT* who declared that what they had learnt was "not to rush", that "you should always try something out", and that "you can have many good ideas and not know it". This was further reflected in what they wanted to get better at, such as: "putting ideas into words", "commenting on others' ideas", and "thinking up my ideas and planning them" (McLaren et al., 2006, pp. 20–21). Giving the learners the facility to reflect using voice memo in the *e-scape* project was liberating for some learners and produced levels of honesty and self critique one would not expect in an assessment situation, as the following transcription from one 15 year old demonstrates:

> Well, I can't really further develop it 'cos we're finished, but if I did have extra time I would sort out the cogs that are still in, 'cos it just looks like something that's just fallen out the sky and hit the ground at a very high speed and has managed to fall and glue itself together. And it's just wrong. I don't know how I did it. It's just a disgrace. But I would sort that out, yeh. (*e-scape* Phase 2 trial)

In all of the above examples the self-awareness of the learners is clearly evidenced. In the context of summative assessment, the insights provided are useful for the assessor but represent a dead end for the learner. If used in a formative assessment context, they can become part of the learning journey, empowering learners in knowing when to seek further help, and from whom, thus moving toward truly sustainable assesment practice.

Feedback

Perhaps central to this is the role of feedback. For many working in the field of assessment, a central role for students in the targeting, generating, interpreting, and communicating of feedback has been vigorously advocated (Black & Wiliam, 1998b; Boud, 1995; Falchikov, 2001; Hounsell, 2007). The examples above illustrate how valuable feedback can be provided through self and peer-assessment. However, they also hint at the danger of this feedback not being mediated in some way by the teacher. The learner above, who declared his work to be a 'disgrace', could have been demotivated – something that might have been countered by an experienced teacher providing additional constructive feedback to speed him on to further development.

Formative assessment in general, and self and peer-assessment in particular, place importance on feedback as central to learning and high attainment (Black & Wiliam, 1998b; Hounsell, 2007), yet Askew (2000) asserts that feedback needs to engage learners *and* teachers in collaborative and reflexive dialogue. This is quite different to feedback as part of the dominant discourse of assessment, which constructs learners as the passive recipients of information where "the teacher identifies what is needed and provides this information to the student" (Boud, 2007, p. 18). Opportunities to initiate this type of dialogue have often lain latent in the 'unpickled portfolio' because of its predominantly summative use. The strategies employed have been effective in helping the learner articulate their ideas and thinking, but in this the teacher has been mute.

Current trials of the *e-scape* system with teachers using the resource (including the authoring tool) to facilitate formative self and peer-assessment *and* teacher feedback will hopefully yield insights into this, but for now this must be left as 'watch this space'.

Democratising Assessment

The final, and possibly implicit, theme running through this text is that of the central place the learner plays in the development of sustainable assessment, and the belief that we need to move towards a democratising of assessment to achieve this. The historic discourse of assessment, as outlined by Waters (2007) in the introduction to this chapter, appears to construct learners as passive subjects where they conform to the rules and procedures of others. Here, students "have no role other than to subject themselves to the assessment acts of others, to be measured and classified" (Boud, 2007, p. 17).

However, through an increasing emphasis on self and peer-assessment, and of learners identifying for themselves their own learning goals (as they did most effectively in the *CAPITTAL-DT* project) the locus of assessment control can shift from the assessor to the learner. With assessment approaches such as those embedded in the 'unpickled portfolio' approach, and with pedagogies that support the formative and metacognitive dimensions of learning, we believe this to be possible in day-to-day assessment for learning experiences.

But what of 'high stakes' (Knight, 1995) summative assessment – might there be a role for learners in this? The *e-scape* system is, perhaps, presenting a scenario where this too is manageable. Ultimately, the presence of anonymised work on the web, and a new approach to summative assessment means it would be possible to involve not only teachers in assessing their own learners and those of other teachers' elsewhere in the country (see Chapter 42), but this process might be extended to involve the learners themselves – something that is being trialled in the current phase of *e-scape*. While this might seem too much of a 'brave new world', it would again present a win-win situation. Involving both teachers and learners in this way has huge potential for maximising learning, as – in order to engage in this process – explicit understanding of the holistic qualities being assessed would need to be developed and shared amongst all those involved.

In conclusion, the unpickled portfolio assessment approach has shown itself to be valuable in a summative assessment context (assessment *of* learning). Through the case studies presented throughout this chapter, we have explored its flexibility and adaptability for assessment *for* learning. While the unpickled portfolio certainly does not constitute the ultimate, multipurpose assessment tool, we suggest it is a model that might be used in a range of contexts, supporting the development of a diverse range of knowledge, skills, and understandings by providing an approach to assessment that we believe can measure up as 'fit for purpose'.

REFERENCES

Askew, S. (Ed.). (2000). *Feedback for learning*. London: RoutledgeFalmer.

Assessment Reform Group. (1999). *Assessment for learning: Beyond the black box*. Cambridge, UK: University of Cambridge, School of Education.

Atkinson, S. (1994). Key factors which affect pupils performance in technology project work. In J. Smith (Ed.), *IDATER 94: The International Conference on Design and Technology Educational Research and Curriculum Development* (pp. 30–37). Loughborough, UK: Loughborough University.

Atkinson, S. (2000). Does the need for high levels of performance curtail the development of creativity in design and technology project work? *International Journal of Technology and Design Education, 10*(3), 255–281.

Bain, J. (2005). Photo-stories from Durham: A case study on assessing design innovation. In E. W. L. Norman, D. Spendlove, & P. Grover (Eds.), *Inspire and educate: DATA International research conference 2005* (pp. 19–34). Wellesbourne, UK: DATA.

Bain, J., & McLaren, S. (2006). Sustainable assessment: Exploring a learner centred approach in practice. In H. Middleton, M. Pavlova, & D. Roebuck (Eds.), *Values in technology education 4th biennial international conference on technology education research* (pp. 1–10). Brisbane, Queensland: Griffith University.

Ball, S., & Vincent, C. (1998). 'I heard it on the grapevine': 'Hot' knowledge and school choice. *British Journal of Sociology of Education, 19*(3), 377–400.

Black, P., & Wiliam, D. (1998a). Assessment and classroom learning. *Assessment in Education, 5*(1), 7–74.

Black, P., & Wiliam, D. (1998b). Inside the black box: Raising standards through classroom assessment. *Phi Delta Kappan, 80*, 139–149.

Black, P., Harrison, C., Lee, C., Marshall, B., & Wiliam, D. (2003). *Assessment for learning. Putting it into practice*. Maidenhead, UK: Open University Press.

Boud, D. (1995). Assessment and learning: Contradictory or complimentary? In P. Knight (Ed.), *Assessment for learning in higher education* (pp. 35–48). London: Kogan Page in association with the Staff and Educational Development Association.

Boud, D. (2000). Sustainable assessment: Rethinking assessment for the learning society. *Studies in Continuing Education, 22*(2), 151–176.

Boud, D. (2007). Reframing assessment as if learning were important. In D. Boud & N. Falchikov (Eds.), *Rethinking assessment in higher education: Learning for the longer term* (pp. 14–25). Abigndon: Routledge.

Bruner, J. (1960). *The process of education*. Cambridge, MA: Harvard University Press.

Costa, A., & Kallick, B. (1993). Through the lens of a critical friend. *Educational Leadership, 51*(2), 49–51.

Davies, T. (2000). Confidence! Its role in the creative teaching and learning of design and technology. *Journal of Technology Education, 12*(1), 18–31.

Dewey, J. (1897). My pedagogic creed. *School Journal, 54*(3), 77–80.

Department for Children, Schools and Families [CSF]. (2008). *Major reforms to school accountability including an end to compulsory national tests for fourteen year olds*. Retrieved January 10, 2009, from http://www.dcsf.gov.uk/pns/DisplayPN.cgi?pn_id=2008_0229

DfES/QCA. (1999). *Design and technology: National Curriculum for England*. London: DfES.

Falchikov, N. (2001). *Learning together. Peer tutoring in higher education*. London: RoutledgeFalmer.

Falchikov, N. (2007). The place of peers in learning and assessment. In D. Boud & N. Falchikov (Eds.), *Rethinking assessment in higher education: Learning for the longer term* (pp. 128–143). Abingdon, UK: Routledge.

Harlen, W. (1997). *Confidence rising slowly: Primary teachers' feelings about science and technology*. Edinburgh, UK: Scottish Council for Research in Education

Harlen, W., & Deakin-Crick, R. (2003). Testing and motivation for learning. *Assessment in Education, 10*(2), 169–207,

Harlen, W., Holroyd, C., & Byrne, M. (1995). *Confidence and understanding in teaching science and technology in primary schools*. Edinburgh, UK: Scottish Council for Research in Education.

Hayward., L, Kane, J., & Cogan, N. (2000). *Improving assessment in Scotland*. Report of the consultation on the review of assessment. Pre-school and 5-14. Retrieved October 1, 2006, from http://www.scotland.gov.uk/library3/education/5-14-00.asp

Hounsell, D. (2007). Towards more sustainable feedback to students. In D. Boud & N. Falchikov (Eds.), *Rethinking assessment in higher education: Learning for the longer term* (pp. 101–113). Abingdon, UK: Routledge.

Jones, A., & Compton, V. (1998). Towards a model for teacher development in technology education: Research to practice. *International Journal of Technology and Design Education, 8*(1), 51–65.

Jones, A., & Moreland, J. (2005). The importance of pedagogical content knowledge in assessment for learning practices: A case study of a whole school approach. *The Curriculum Journal, 16*(2), 193–206.

Kimbell, R. (2000). Creativity in crisis. *The Journal of Design and Technology Education, 5*(3), 206–211.

Kimbell, R., Miller, S., Bain, J., Wright, R., Wheeler, T., & Stables, K. (2004). *Assessing design innovation: A research and development project for the Department for Education & Skills (DfES) and the Qualifications and Curriculum Authority (QCA).* London: Goldsmiths, University of London.

Kimbell, R., & Stables, K. (2007). *Researching design learning: Issues and findings from two decades of research and development.* Dordrecht, The Netherlands: Springer.

Kimbell, R., Stables, K., Wheeler, T., Wozniak, A., & Kelly, A. V. (1991). *The assessment of performance in design and technology.* London: SEAC/HMSO.

Kimbell, R., Wheeler, T., Miller, S., & Pollitt, A. (2007). *e-scape portfolio assessment: A research and development project for the Department for Education & Skills (DfES) and the Qualifications and Curriculum Authority (QCA). Phase 2 report.* London: Technology Education Research Unit, Goldsmiths, University of London.

Kimbell, R., Wheeler, T., Shepard, T., Brown-Martin, G., Perry, D., Hall, P., et al. (2005). *e-scape portfolio assessment: A research and development project for the Department for Education and Skills (DfES) and the Qualifications and Curriculum Authority (QCA). Phase 1 report.* London: Technology Education Research Unit, Goldsmiths, University of London.

Knight, P. (Ed.). (1995). *Assessment for learning in higher education.* London: Kogan Page.

Lave, J. (Ed.). (1999). *Everyday cognition: Its development in social context.* Lincoln, NE: iUniverse Inc.

Lawler, T. (1999). Exposing the gender effects of design and technology project work by comparing strategies for presenting and managing pupils' work. In P. H. Roberts & E. W. L. Norman (Eds.), *IDATER 99: International conference on design and technology educational research and curriculum development* (pp. 130–137). Loughborough, UK: Loughborough University of Technology.

Lawton, D. (1992). Whatever happened to the TGAT report? In C. Gipps (Ed.), *Developing assessment for the National Curriculum* (pp. 95–103). London: Kogan Page.

Linn, R. L., Baker, E. L., & Dunbar, S. B. (1991). Complex, performance-based assessment: Expectations and validation criteria. *Educational Researcher, 20*(8), 15–21.

LTS. (2004). *Personal learning planning: A framework for development.* Retrieved January 10, 2009, from http://www.ltscotland.org.uk/assess/images/plpframework_tcm4-168238.pdf

McLaren, S. V., Stables, K., & Bain, J. (2006). *Creativity and progression in transition through assessment for learning in design and technology education: Report to Determined to Succeed Division.* Glasgow, Lanarkshire: University of Strathclyde.

McLaren, S. V., Stables, K., & Bain, J. (2007). Exploring creativity and progression in transition through 'assessment is for learning'. In J. R. Dakers, W. J. Dow, & M. J. de Vries (Eds.), *PATT 18: Teaching and learning technological literacy in the classroom* (pp. 143–151). Glasgow, Lanarkshire: University of Glasgow.

Mentkowski, M. (2000). *Learning that lasts: Integrated learning, development and performance in college and beyond.* San Fransisco: Jossey-Bass.

Messick, S. (1989). Validity. In R. L. Linn (Ed.), *Educational measurement* (pp. 13–104). New York: Macmillan.

National Foundation for Educational Research [NFER]. (2002 to present). *Review of curriculum and assessment frameworks (INCA) internet archive project.* Retrieved from www/inca.org.uk/

Newton, P. (2003). The defensibility of national curriculum assessment in England. *Research Papers in Education, 18*(2), 101–127.

Nuttall, D. (1990). *Differences in examination performance.* London: Inner London Education Authority Research & Statistics Branch.

Prest, D. (2002). *An analysis of the attainment target level descriptors and associated programme of study in relation to the design and technology mission statement.* London: Department for Education and Skills, Design and Technology Strategy Group.

Quality and Curriculum Authority [QCA]. (2008). *Curriculum 'big picture'.* Retrieved 10 January, 2009, from http://www.qca.org.uk/qca_5855.aspx

SOED. (1996). *Achievement for all.* Edinburgh, UK: HMSO.

SOED. (1998). *Achieving success: A report on the review of provision in S1/S2 by HM Inspectors of Education*. Retrieved January 10, 2009, from www.scotland.gov.uk/library3/education/as12-05.asp

Stables, K. (2006). Valuing creativity: The perceptions of learners and teachers and how these relate to performance. In H. Middleton, M. Pavlova, & D. Roebuck (Eds.), *Values in Technology Education 4th Biennial International Conference on Technology Education Research* (pp. 432–439). Brisbane, Queensland: Griffith University.

Stables, K., & Kimbell, R. (2000). The unpickled portfolio: Pioneering performance assessment in design and technology. In R. Kimbell (Ed.), *D&T International Millennium Conference: Learning from Experience: Modelling new Futures* (pp. 195–202). Wellesbourne, UK: DATA.

Stables, K., & Kimbell, R. (2007). Evidence through the looking glass: Developing performance and assessing capability. In L. Taxén, (Ed.), *The 13th International Conference on Thinking. Linköping Electronic Conference Proceedings, No. 21* (pp. 175–182). Norrköping, Sweden: Linköping University Electronic Press. Retrieved January 10, 2009, from http://www.ep.liu.se/ecp/021/vol1/ 175 – 182

Tan, K. (2007). Conceptions of self-assessment: What is needed for long-term learning? In D. Boud & N. Falchikov (Eds.), *Rethinking assessment in higher education: Learning for the longer term* (pp. 114–127). Abingdon, UK: Routledge.

UNESCO, (2005). *Education for all global monitoring report 2005*. Retreived January 10, 2009, from http://portal.unesco.org/education/en/ev.php-URL_ID=35939&URL_DO=DO_TOPIC&URL_SECTION=201.html

Waters/QCA. (2007). Retrieved September 18, 2007 http://curriculum.qca.org.uk/organising-your-curriculum/principles_of_curriculum_design/

Wiliam, D. (2003). National curriculum assessment: How to make it better. *Research Papers in Education, 18*(2), 129–136.

Wilson, D., Croxson, B., & Atkinson, A. (2006). What gets measured gets done: Headteachers' responses to the English Secondary School Performance Management System. *Policy Studies, 27*(2), 153–171.

Kay Stables
Jenny Bain
Technology Education Research Unit (TERU)
Department of Design
Goldsmiths, University of London, UK

RICHARD KIMBELL

42. PERFORMANCE PORTFOLIOS . . . PROBLEMS, POTENTIALS, AND POLICY

INTRODUCTION

Assessment is about placing value on things. Thus, at the outset, assessment demands a discussion about the nature of education and its practices so as to be confident that we are valuing the right things. In this chapter, I explore the nature of performance assessment in design and technology (d&t) and specifically the nature of portfolios that typically act as the basis on which such assessments are made. I question current practices and describe how their evident failings launched a series of research projects in the Technology Education Research Unit (TERU) at Goldsmiths University of London. I describe the approach adopted through this research, which might be characterised as transforming the portfolio from being what it currently typically is (a pretty end-product) to being what I believe it ought to be (the 'trace-left-behind' by purposeful activity).

Our attempts to do this took a step forward when we embraced a new generation of technology to create e-portfolios. There was no ethical or conceptual change in research direction – but the digital transition empowered a fundamental development that has the potential to transform completely the assessment world. I outline this development and discuss some of its implications. It is important to recognise that it is not the digital technology itself that is important – but the ways in which it enables us (i) to capture and (ii) to value qualities of capability that have formerly proved so elusive.

PRIORITISING *PERFORMANCE*

So – as a starting point – what are the things that I value in students? At the head of a long-ish list would be things like:

- are they inquisitive and imaginative?
- are they prepared to take on challenges?
- do they get stuck in and do their best – or hang back and follow others?
- do they have sticking power when things start to go pear-shaped?

I realise that a lot of this is not about *knowing* stuff – or even about being skilled in *doing* stuff. It's about having the right *emotional attitude* towards it. And I recognise that this only gets you so far ... for somewhere along the line it is necessary to know things and be able to do things. The fact is, however, that the vast majority of assessments that our learners endure in schools (and certainly the high-stakes assessments) seldom even attempt to address those critical qualities of the individual that are so central to effective performance – not least in design and technology education.

So there we have it. I am interested in *performance*; in the ability of learners to tackle tasks and bring them to effective resolution. I am not interested in abstracted tests of knowledge or skill, not even if those tests purport to cover areas of knowledge and skill that are specifically technological.

A.T. Jones and M.J. de Vries (eds.), International Handbook of Research and Development in Technology Education, 509–522.
© 2009 Sense Publishers. All rights reserved.

Rather I am interested in the qualities (as well as the knowledge and skills) that learners bring to bear on a task … on the rigour and quality of their response to it … and on their preparedness to be open and self-critical about their performance (for how else will they improve?).

This is why *performance assessment* appeals to me – not just as a supplement to other forms of assessment but as the only thing that is worth doing. As has been argued elsewhere:

> Of the many priorities that we might hold concerning the assessment challenge, ours has always been to find ways of assessing the **performance** of learners in design & technology. Given a description of what design & technology capability entails … our priority has been to find ways of assessing how well learners are able to demonstrate this capability. … This is performance assessment. It takes as its starting point the description of capability, and embodies it in a 'real' task; real in the sense that the activity represents a believable, authentic design & technology challenge. We then assess how well learners manage themselves through the task and how successfully they manage to resolve it. This approach is often described in the literature as 'authentic' assessment. (Kimbell & Stables, 2007, p. 53)

PERFORMANCE AND *CAPABILITY*

If learner *capability* in technology education is to be our starting point, where do we find descriptions of this capability? Fortunately the England and Wales National Curriculum statement was explicit about this, and not just for design & technology. Each subject in that curriculum has an explicit statement about what it is that makes it important for learners, and why it should be included in the curriculum. Fortunately they are now all available on-line, and they are illustrated in the following claims that are made for science.

> Science stimulates and excites pupils' curiosity about phenomena and events in the world around them. It also satisfies this curiosity with knowledge. Because science links direct practical experience with ideas, it can engage learners at many levels. Scientific method is about developing and evaluating explanations through experimental evidence and modelling. This is a spur to critical and creative thought. Through science, pupils understand how major scientific ideas contribute to technological change – impacting on industry, business and medicine and improving quality of life. Pupils recognise the cultural significance of science and trace its worldwide development. They learn to question and discuss science-based issues that may affect their own lives, the direction of society and the future of the world. (http://www.nc.uk. net/nc/contents/Sc-home.htm)

This is a terrific capability statement – but sadly not one that bears much relationship to typical science assessments. "Exciting curiosity … evaluating explanations through modelling … creative thought … improving the quality of life … questioning science-based issues and the direction of society …" ... It doesn't look much like the science exams I have seen lately. But then – to be fair – the science community has seldom claimed that performance assessment is their core concern. They are up to their armpits in content, which inevitably somewhat closes down their options for assessment.

We should hope that technology does a bit better in this regard, and I believe that it does. For those interested in the d&t statement, just follow the web reference above and make the necessary adjustments to get to the d&t domain. But the key to understanding technology was summarised decades ago by Bronowski (1973):

> Among the multitude of animals that scamper, fly, burrow and swim around us, man is the only one who is not locked into his environment. His imagination, his reason, his emotional subtlety and toughness make it possible for him not to accept the environment but to change it. (p. 19)

In the animal kingdom, we are alone in being able to say, "I don't like getting cold and wet, so I'll make some clothes, build a shelter and make up a fire." We don't just accept the world as it is, we are driven to change it … to improve it.

> The subject matter of design & technology is our *made* world; our clothes, our food, our means of travel, our shelters, our communication systems. It is a study of our material culture. But, more than that, design & technology is about *creating change* in the made world; about understanding the processes of change and becoming capable in the exercise of change-making. (Kimbell & Perry, 2001, p. 3)

I have always taken the view that it is this capability that should be at the heart of any assessment.

REAL PERFORMANCE AND SHAM PERFORMANCE

I was for many years a supporter of coursework assessment as a means for judging the performance and technological capability of learners. It seemed to me that if we are concerned with how well learners can tackle tasks and bring them to resolution, then the obvious approach is to let them get on and do it … and watch them through the process. We might then be able to make some reasonable assessment of their performance against any number of criteria – including those tricky ones like 'stickability'.

Over the last decade, however, I have become increasingly disillusioned about this coursework assessment process for the following reason: In that time we have seen the rise and rise of accountability mechanisms in schools in the UK, by far the most pernicious of which has been the operation of the Ofsted inspection body in combination with published league-tables of schools based primarily on GCSE pass rates.

This evil combination has conspired to create a situation in which learners' examination performance is even more important to the school than it is to the learner, since being higher in the tables than other local schools means that more parents will seek to get their children into the school. It was inevitable that this bald ranking process would have a trickle-down effect in schools, and indeed it is now commonplace for schools to disaggregate their league table data to departments – and even to individual teachers. The 'successful' ones are then well placed in terms of salary and promotional advantage. We have in effect re-created the Victorian notion of 'payment by results', which was formally abolished in 1897. Richards (1999) recounts the experience of Edmond Holmes, an HM Inspector of Schools, reflecting in 1911 that the "deadly system of 'payment by results' seems to have been devised for the express purpose of arresting growth and strangling life, and bound us all, myself included, with links of iron" (p. vii). He went on to speak of teachers "who had drilled themselves into passivity and helplessness" (p. 55).

Having abolished this practice because of the 'strangling', 'drilling', 'passivity', and 'helplessness' that it engenders, we have taken about another 100 years to re-create it, and all sorts of extremely questionable practices have arisen. Consider the following practice, bearing in mind that league tables are based on the percentage of learners getting 5 or more GCSE passes at grades A-C. When 'mock' exams are taken six or 12 months ahead of the real thing, learners are treated very differently based on their 'mock' score. Those scoring A-C are left to carry on as normal; those scoring E and F are left to get on as normal (i.e. struggling and failing); but those scoring D are given special coaching to get them over the hurdle into the C bracket. The easiest way to influence the A-C statistic is to fiddle with that D category – so this becomes formal policy in some schools. Which seems to me to be ethically bankrupt and educationally unforgivable. Never mind about stretching the capabilities of the very best performers, or about supporting the least capable … lets spend our time with the ones that will have most impact on the schools' league-table position.

I fear that design & technology teachers are not immune from these pressures, and whilst 'drilling' might be too strict a word to describe the process, there is no doubt that coursework portfolios have changed from being manifestations of learners' capability into measures of how well teachers can manage and massage learners' portfolios to hit all the buttons in the mark scheme. When they were going through this process, my own two sons brought home instructions from their d&t teachers about their portfolio. Page 1 will have …, page 2 will have …, etc. Careful teacher control of the process ensured good grades, and in the process disguised and distorted my boys' real levels of capability. The result of all this was that the candidates getting A grades were not the best designers, but rather were the learners who had two other qualities: (i) teachers that best understood how to play the game, and (ii) learners with the patience and preparedness to play it.

By 2002 this problem was well understood. Ive, the senior HMI, regularly condemned the 'neat nonsense' that he observed (see Barlex & Trebell, 2008) and Prest (2002) undertook a specific analysis of it for the Design and Technology Strategy Group that had been set up by the Department for Education and Skills (DfES). The upshot of the resulting report, *Building on Success* (Barlex, 2003) was that TERU at Goldsmiths was commissioned to find a way to overcome this problem. Given that we were not empowered to abolish Ofsted and league-tables, we were forced to look at the process of constructing portfolios to see if we could devise an assessment methodology that could improve on the current practice.

REAL-TIME PERFORMANCE PORTFOLIOS

The approach originated in our Assessment Performance Unit (APU) research of the 1980s and was developed by my colleague Kay Stables and myself through a series of subsequent projects. We christened it the 'un-pickled portfolio' approach (see Stables & Kimbell, 2000) in the sense that performance was quick-fire, with the portfolio emerging as the real-time 'trace-left-behind' by the activities of the learner. The transition from 'timeless' portfolios (in d&t examinations in England and Wales, frequently prepared over many months) to 'real-time' portfolios is significant in several ways, but centrally the issue concerns direct evidence of process-centred capability.

Real-time performance data have been used for some years in the worlds of management and computer science, where system evaluation (and indeed assessment) has entailed the forensic analysis (step-by-step/minute-by-minute) of sub-elements of the performance of systems (e.g., Calvez, Wyche, & Edmundson, 1993; Mohd Adnan, 1993). It seemed to us that – since we claim that *process* performance is central to what we do and value – we might usefully explore this forensic approach to the creation of design portfolios.

The first step in this process is to create an activity through which learners progress from a task statement towards a design prototype. The choreography of the activity is informed partly by pedagogic priorities (what will work well as an activity) and partly by assessment priorities (what trace will be left behind by it – and how can we enrich it). Sometimes these priorities work well together and sometimes they conflict … leading to a good deal of trialling and optimisation in the classroom/studio/workshop.

In November 2003 we launched the first Assessing Design Innovation prototype of an assessment activity using a format evolved into a 6-hour task: two consecutive mornings of three hours. In that time, students start with a task and work through from an initial concept to the development of a prototype solution – a 'proving' model to demonstrate that their ideas will work. The six hours are run by the teacher following a script that choreographs the activity through a series of sub-tasks, each of which is designed to promote evidence of students' thinking in relation to their ideas. These 'steps' in the process all operate in designated spaces in a pre-prepared booklet. The following structure is characteristic of the activities so far developed:

- establish the task
- explore a series of 'idea-objects' in a handling collection designed to promote ideas about the task
- put down first ideas in a designated box in the booklet
- working in groups of three, students swop their booklets and each team-mate adds ideas to the original
- team-mates swop again so that each team member has the ideas of the other two members
- booklets return to their 'owner' and team members discuss the ideas generated
- the teacher introduces the modelling/resource kit that can be used throughout the two mornings
- students develop their ideas in the booklet – and/or through modelling with the resources
- students stop to reflect on the user of the end product and on the context of use
- photographs are taken at approximately 1-hour intervals to develop a visual story line to illustrate the evolution of models & prototypes
- at the end of the first morning, students – and their team members – reflect on the strengths and weaknesses of their evolving ideas
- the second morning starts with a celebration of the work emerging from day 1
- further prototype development
- regular hourly photos and pauses for reflective thought on strengths and weaknesses
- final team reflections when (in turn) team members review each others' ideas and progress
- individually, students then 'fast-forward' their idea illustrating what the product will look like when completely finished and set-up in context
- they finally review their work from start to finish.

The finished booklet is a folding A2 sheet that ends up packed with ideas, drawings, notes, and photos. It is an immensely rich data source and a vivid record of the students' experience of six hours of (often) frenetic activity. Using various versions of this structured worksheet approach, we undertook trials of 10 different tasks ranging from product/packaging design and fashion/textile design, to electro-mechanical systems design and food product design. In total we acquired un-pickled portfolio responses from approx 300 students in 11 schools spread across the UK. Our approach to this project has been described in detail in a number of places (e.g., Kimbell 2007, Kimbell & Stables, 2007; Kimbell, Wheeler, Miller, Bain, Wright, & Stables, 2004).

Perhaps the greatest contribution of this work was the way in which it liberated learners from the clutches of their teachers. This might sound odd given the fact that the whole process was choreographed by the teacher script. Throughout the development process, learners kept reporting that they loved the collaborative element of design development – sharing ideas in their group at the outset and reflecting on them later. They also appreciated the story-line of digital photos that had the effect of showing them their own progress and enjoyed working at a fast pace through a concentrated six hour activity and having the encouragement to develop really interesting (and somewhat risky) ideas with the confidence that they would not be judged by the craft quality of the final outcome. But above all – time and time again – they referred to the 'freedom' that they felt in the process. Indicative responses include:

AB119 - Being able to use my own ideas how I wanted to
AB1111 - These have been the best two days in school

AB216 - You can make things you didn't expect
AB2210 - Starting from scratch to make you think
AB314 - Using your own ideas; making models instead of drawing - they work better
AB319 - Seeing your idea develop
AB323 - Being able to make something by myself and what I wanted to make
AB324 - I realised that I could do more than I thought I could
AB411 - I liked getting my imagination go wild
AB423 - It made you think about how you could do things and with different materials

There is no sense here of teachers controlling learners' ideas or work, and indeed the teachers were very uncomfortable about the lack of control that they could exercise. They could require the group to "spend the next few minutes developing your ideas", but they could not say how to do it, or what the ideas should be. The script was utterly procedural, leaving the learners to the make the critical conceptual and technical decisions that would amount to their way of tackling the task.

At the end of this project (2004) we were not entirely clear whether we had created a new kind of *examination* or a new form of structured *coursework*. It sat somewhere between the two. But what we did know was that we had evolved a form of working that was liberating for learners and very indicative of the qualities of capability that they possessed. In six hours, we could put learners through a design and development process that left behind a highly valid trace of their real-time performance. These were not the doctored, 'prettied-up' portfolios of conventional GCSE type. Rather, they provided terrific evidence of the raw 'un-pickled' performance of learners grappling with a design task in real-time.

The success of the project can be measured in the fact that one of the national GCSE Awarding Bodies adopted the system as one of the assessment devices for their new Product Design specification (Oxford Cambridge and RSA examinations [OCR], n.d.).

The next step in the development to our current work arose from a teacher in a trial school commenting on the work of two learners who were having an argument about their work in a corner of the workshop. He observed that "if we could hear what they were saying ... we would know so much more about what they understand about their work". This, along with some other influences, prompted us to start exploring other elements of digital performance-capture (in addition to the photo story-line). Thus was born project *e-scape* (**e-s**olutions for **c**reative **a**ssessment in **p**ortfolio **e**nvironments).

PROJECT E-SCAPE AND E-PORTFOLIOS

The central transformation of *e-scape* (moving on from *assessment design innovation*) is that the paper-based work has been replaced by work on hand-held computers. The strength of the PDA-like tool lies in its multi-functionality. It can be used for drawing, note-writing (based on handwriting or virtual keyboarding), taking photos and videos, and recording voice memos. Moreover, the devices have wi-fi capability, enabling us to link a class set into a local area network run from a teachers' laptop.

The devices all have the *e-scape* software loaded into them in the form of 23 linked screens through which the activity progresses. The administrator controls the activity through the laptop. As s/he activates "box 1" on the administrator interface, the signal is sent (via wireless router) to all the devices and they 'come live' with box 1. Learners can draw/make notes on the devices, and at the end of the allotted time for box 1 a warning screen pops up prompting learners to 'tap-here' to save their work. This is then sent back through the router and stored in the administrator's laptop, as well as in the memory on each individual device. In fact, this was merely a back-up 'save' routine since -every few seconds throughout the activity learners' work (including drawings, notes, photos, and sound files) is automatically transmitted (by wi-fi) to the administrator's lap-top and from there into a secure web-space.

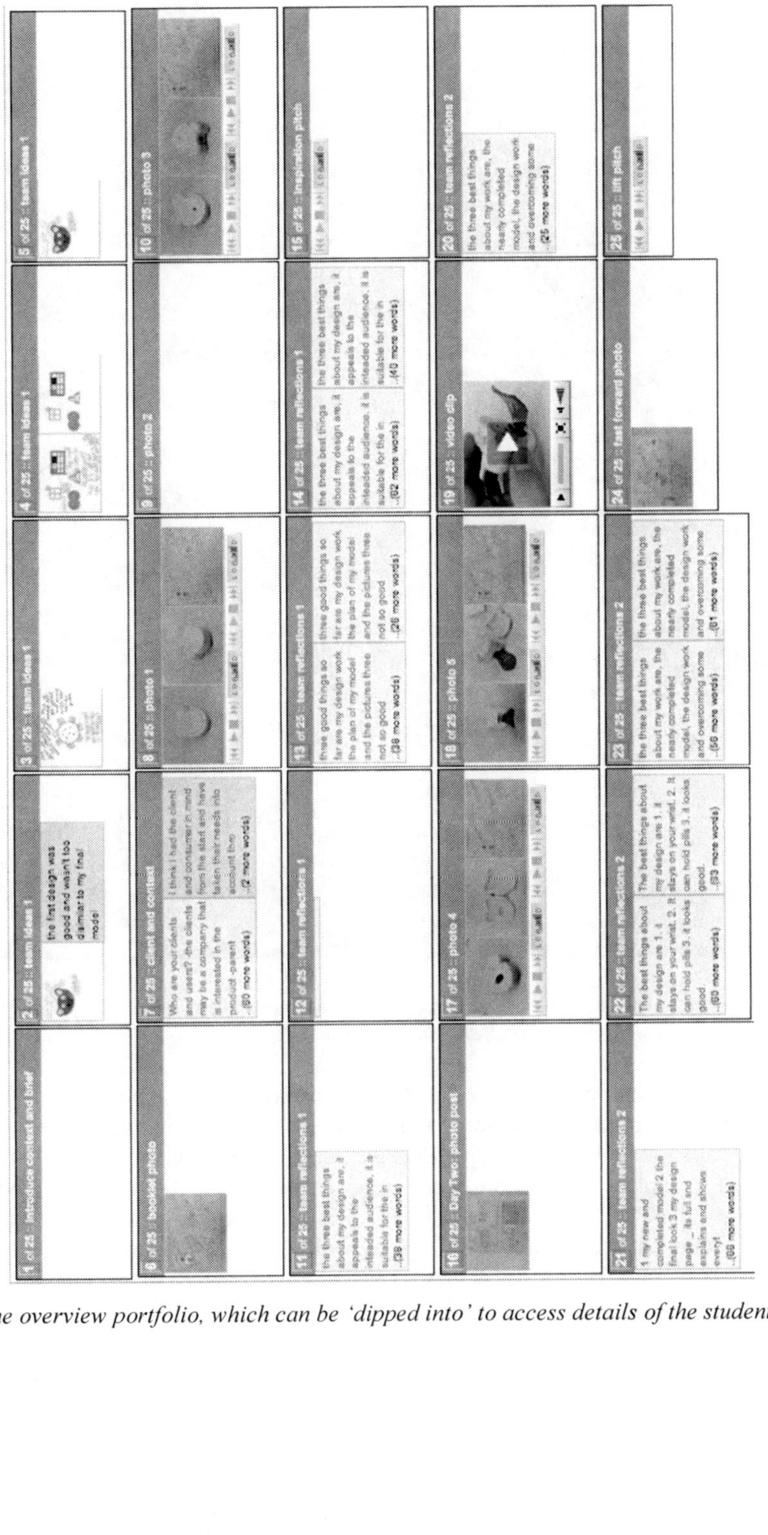

Figure 1. The overview portfolio, which can be 'dipped into' to access details of the student's learning.

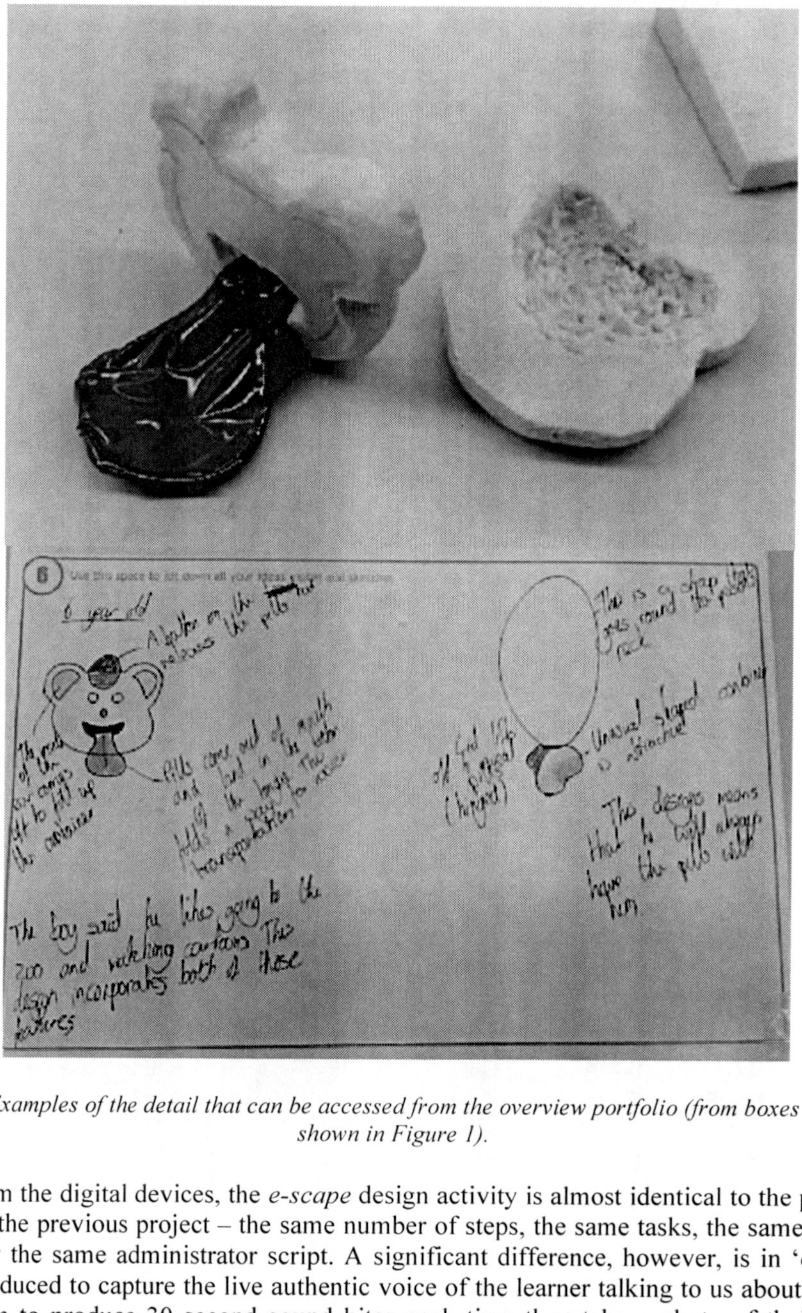

Figure 2. Examples of the detail that can be accessed from the overview portfolio (from boxes 18 and 24 shown in Figure 1).

Apart from the digital devices, the *e-scape* design activity is almost identical to the paper-based activities of the previous project – the same number of steps, the same tasks, the same time-scale, and virtually the same administrator script. A significant difference, however, is in 'design-talk' that we introduced to capture the live authentic voice of the learner talking to us about their work. We ask them to produce 30 second sound bites each time they take a photo of their emerging modelling, based on the two questions: What's going well? What needs more work?

Undoubtedly the biggest difference with the move to digital working was in terms of the output from the activity, since the paper-based learner booklet was replaced by an e-portfolio in a website. The portfolio is structured through the 25 sub-tasks distributed through the six-hour activity, with response modes of various kinds and with both individual and team-based work. Like the paper-portfolios that were the precursor to *e-scape*, these web-screens provide a very real picture of the learners' evolving prototype and their thoughts about it along the way.

The scrutiny of the portfolios starts with an overview screen (see Figure 1), but the whole thing is 'live' in the website, which enables the viewer to dip into it for the details. Clicking on a photo or drawing renders it full screen (Figure 2). Mousing over the text boxes brings them to full size. Clicking on the sound files makes them play, and we hear the authentic voice of the learner responding to the two questions. Taken together, these multi-media data provide a real insight into the learners' understanding of their work. It is important to note that this 'photo and sound file' routine recurs throughout the activity – essentially once an hour throughout the six hours. At least three significant things result from this. First, the learners get better – more articulate – in describing their work and the circumstances surrounding it. Second, the routine – taken together – leaves a real-time visual/audio evidence trail that is quite unique in the assessment of performance in design & technology. Third, learners' approaches to the task are enriched as they are more motivated, braver (take more risks), and think more deeply about what they are doing.

Finally, we offer learners an opportunity to reflect on their work – but it is a reflection made at the *very end* of the six hours of activity. Looking back over their whole work, we invite them to think about what they might have done differently if they had known then what they know now. Sometimes these responses are descriptive, and sometimes learners are deeply analytic of their own performance. The two examples shown in figure 3 are illustrative of these reflective moments, and they appear in the portfolio as yellow boxes.

THE CHALLENGE OF ASSESSMENT

Conventionally, portfolio assessment is undertaken by identifying performance criteria, allocating a block of marks to each criterion, and judging individual pieces of work to decide how many marks to award for each criterion. Our previous work at APU (Kimbell et al., 1991, onwards) has illustrated the value of a prior (holistic) judgement of the *whole* piece of work, but this too is a criterion-based approach that requires the marker to make a judgement about the whole integrated performance.

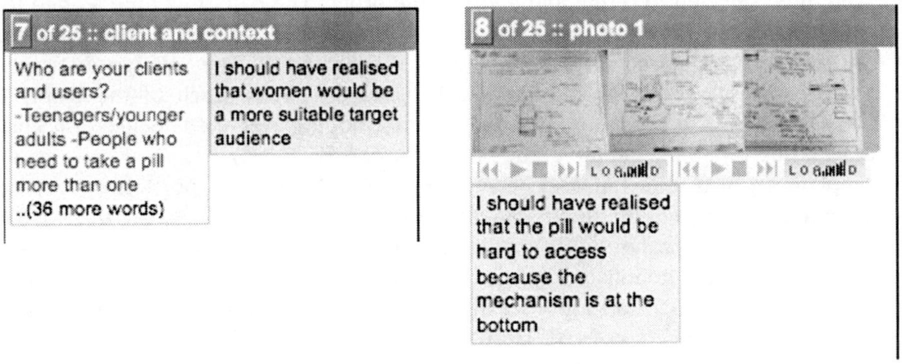

Figure 3. If I had known then what I know now. Students reflect on their work.

I was intrigued when I was referred recently to a book on the psychology of judgement (Laming, 2004), for he suggests that "... there is no such thing as absolute judgment. All judgments are comparisons of one thing with another." What he means is that when I decide that for criterion A (out of 15 points) this piece of work scores 8 points, I am *either* saying

- it's the same quality as another piece that I have already scored 8, *or*
- I am comparing it with a 'notional' piece of work that I hold in my head as representing 8-ness.

In other words, judgements are *relative*. And in that case why not compare the work *directly*?

Based on this idea, an alternative approach emerged from the psychophysical research of Louis Thurstone, and specifically on his Law of Comparative Judgement (Thurstone, 1927). The essential principle in this law is that whenever a judge compares two portfolios (using their own personal 'standard' or internalised criteria) the judge's personal standard cancels out. The greater the true difference between the quality of the two portfolios the more likely it is that the better one will 'win' each time they are compared. Thus a large set of comparisons does more than just generate a rank order; the relative frequency of success of one performance against another also indicates how far apart they are in quality.

> Statistical analysis of a matrix of comparative judgements of 'scripts' can construct a measurement scale expressing the relative value of the performances. The result of comparisons of this kind is objective relative measurement, on a scale with a constant unit. Furthermore, if a few scripts that have already been agreed to represent grade boundaries – perhaps from a previous sitting of the examination – are included in the comparisons, the whole process of marking, grading and comparability of standards can be replaced by the collection and analysis of paired comparative judgements. (Pollitt, 2006, p. 11)

Based on this idea, the judging process for the main body of *e-scape* portfolios was designed in three rounds using seven judges. Round 1 involved each of the judges reviewing 140 pairs and deciding on the 'winner' in each case. The general response of judges was that the early pairs (say the first 20 or 30) took as long as 10 minutes per pair to decide, but gradually we got quicker. This speeding-up resulted in part from being more skilled in working our way around the web-based portfolio, and in part from the fact that the pairings inevitably threw up repeats. Having got properly inside a piece of work at the first time of asking, it took only a much briefer scan second time around to remind us of its qualities. By the end of the 140 pairs we were typically doing each pair in 2 minutes.

Having completed Round 1 judging, the results were analysed into a rank order and Round 2 pairings were then selected to refine and confirm the order. The pairs this time tended to be closer together, comparing (for example) something like a B with a C, whereas Round 1 (being random) had just as frequently required us to compare an A with an E. The relative difficulty of these Round 2 decisions was offset by the familiarity (by now) with much of the work. Generally, Round 2 was quicker than Round 1. Round 3 of the judging was then focused on enriching the data (so at to assure reliability) at the grade boundaries (A/B/C/D/E).

The complete set of results can best be seen in Figure 4. In it, the portfolios have been sorted into order and are shown with their standard errors. In formal statistical terms, 68% of the portfolios' 'true' values will lie within one standard error of the reported value. Lines are drawn through the notional grade boundaries (A-E) to show how many pupils would fall into each grade. The analysis of the judgements also gives a traditional indication of the quality of the measurement process and the key figure here is the reliability coefficient of 0.93.

This figure allows for unreliability between markers *as well as* for lack of internal consistency within the examination. Most traditional reliability coefficients only allow for one of these. Very few current GCSEs can achieve this level of reliability – and certainly not those with qualitatively rich portfolios.

ISSUES ARISING

A number of issues arise from this work, which is currently being continued with the support of the DCSF, Becta, GCSE Awarding Bodies, software developers, and numerous teachers and learners in schools throughout England and Wales. Many of the issues arising concern teaching and learning, but in this chapter – with its focus on assessment – it seems sensible to concentrate on these.

Holistic Judgement and Criteria

Since the assessment process appears to involve a simple comparison of one portfolio with another, it has been suggested that this amounts to 'normative' assessment – with the implication that it is not therefore criterion-based. Such a criticism is misguided, for several reasons. First, the activity through which we diagnose learner capability has been constructed with a clear view of the criteria of performance with which we are concerned. The activity has to be able to provoke the evidence of capability in learners' responses. Second, the holistic judgement is made in the light of a set of specified criteria that *together* make up capability. These criteria are discussed and exemplified with the judging team. The difference is that we do not then ask the team to *score* each learner portfolio in respect of each of these criteria. Rather, we ask them to make a balancing judgement about which of the two portfolios – overall – represents the more effective combination of these criteria. Finally – once all the judgements have been completed – we arrive at a rank-order and subsequently we conduct what amounts to a conventional 'awarding' meeting at which the rank is divided into grades (say, A-E). This process requires the identification of grade boundaries and the establishment that those portfolios in the A band have the performance characteristics that justify the award on an A according to the rules of the Awarding Body, and similarly for all the other grade boundary bands.

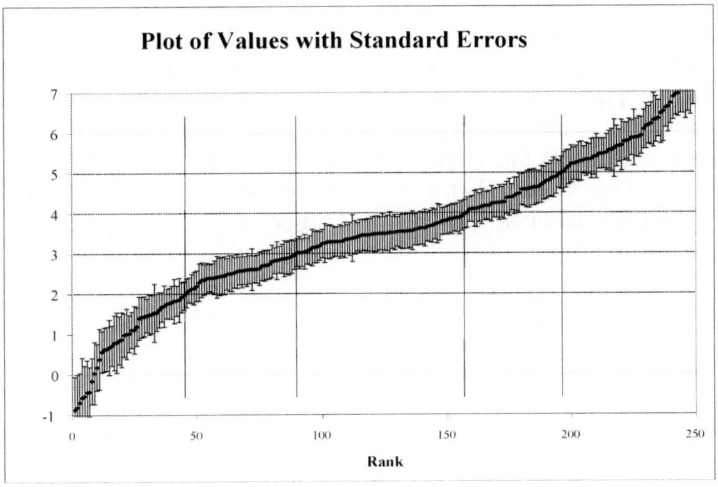

Figure 4. Performance distribution of the 250 portfolios.

Reliability and Error

One of the truly amazing features of this process (as it unfolded before us as we continued to make our judgements) was its diagnostic power linked to its reliability. If the reader refers back to the chart of 'values with standard error' (Figure 3), you can see that each 'score' has a + and – 'tail' attached to it. Moreover, some of these tails are longer than others. The four or five portfolios with particularly long tails tell us that they are portfolios about which the judging team is in disagreement. Some of us were saying it beats its comparators, while others were saying it didn't. Rather than being a problem, what this does it focus attention on the tricky pieces – enabling us to pick them out and subject them to special moderation processes. The judging process thus automatically puts the finger on where we should concentrate our moderation processes. Moreover, the system generates a 'misfit' statistic both about each of the portfolios (is their Standard Error acceptably small?) and about the judging team (are they sufficiently consensual?). Wayward judges (i.e. those making judgements that conflict with the 'centre of gravity' of the judging team) can either be taken out and shot at dawn, or – more constructively – they can be engaged in a professional discussion about what it is they have seen in this work that other judges have missed. They might be right.

The upshot of this whole process is that assessments result from a *collective* professional judgment about the work. Each portfolio is compared to many other portfolios (typically 20) and the judgements are made by many judges. The resulting reliability of the assessment data far exceeds anything that currently exists in GCSE coursework assessment – which is typically the result of a single teachers' scoring of their own students' work.

Political and Professional Interest

This is one of the reasons for the level of political interest in the system. Having effectively abolished the former coursework procedures because of their unreliability, Ministers can see that the *e-scape* form of 'controlled assessment' deals with both of the central problems. The real-time creation of the portfolio eliminates any doubt about whose work it is – and the reliability statistic attached to the assessment outcome (because of the multiple comparisons and the collective professional judgements) is astonishingly high.

It is easy to see the potential benefit to the teaching community as a whole. When this becomes a normal GCSE assessment process, the people doing the judging will be the teachers who have put learners in for the exam. At the moment these teachers are expected to assess their own learners' coursework, but in the new *e-scape* arrangement they will join the collective judging team. One of the real benefits of this is that they will be seeing work from across the spectrum of national performance. Thus, as a teacher, I don't just judge work in my school. Rather, I judge a selection of work from the national pool and in the process I get to see what performance looks like nationally.

Transferability to Other School Subjects

Portfolio performance is not restricted to design & technology. Part of the brief that we received from the Ministry for the current round of e-scape is that we demonstrate how the system operates in other disciplines – in our case science and geography – and we have on our own behalf also undertaken experiments in mathematics and English. The driving requirement in each case is that we are involved in *performance* assessment – so there must be some performance to assess.

In science this centres on the learners making up a research team to better understand traffic accidents. Since car brakes, steering, tyres, headlights, etc., are all getting better, why are there still so many accidents? The task focussed learners on 'driver-distraction' (specifically their reaction time) and all kinds of fascinating investigations were designed and conducted, concerning, for example, radio/music players (does heavy rock reduce driver reaction-time more than soporific concertos?), mobile phones, traffic signs, etc. Reaction times in various settings were measured, conclusions drawn, and the limitations of the data carefully considered. These were real science performances that took place over a three-hour morning, and we currently have 60 portfolios being judged by a team of science teachers.

The additional interest in the geography activity/performance was that it took place over a single day that started in the classroom, then had the learners out of school – off site – collecting data live in the environment, and then moved back into the classroom where learners worked on data analysis, interpretation, and recommendations. The e-scape system – operating on hand held computers – allowed this off-site work to be supported digitally (including with GPS).

Validity and Potential

In research contract terms, the third phase of *e-scape* has been focussed on two overriding requirements: *scalability* (is the system capable of scaling up from interesting research to nationally administered examinations?) and *transferability* (is the system capable of dealing with all subjects?).

We were always confident that these challenges would prove manageable, and our real motivation through this phase has been tangential to these two – but central to the challenge of assessment: *validity*. There is no point in investing in more portfolios (even in the e-form) if they suffer from the same limitations that have caused current coursework assessment in England and Wales to become discredited. Accordingly, whilst tackling the scalability and transferability challenges, our real ambition has been to create a system that produces portfolios that (when viewed by subject experts) have the stamp of authenticity. Is this portfolio good science? Does this portfolio demonstrate capable geographers? Are the most capable design & technology learners producing the most authoritative portfolios?

We have created a vehicle for performance assessment that facilitates the real-time, emergent, raw, authentic voice (literally) of the learner to speak through the portfolio. The evidence suggests that the subsequent collective professional judgements that are made of the portfolios produce not only *reliable* outcomes, but outcomes that teachers are happy to authenticate. The best performers are no longer the ones who are prepared to spend countless hours doctoring their portfolios to hit all the buttons. The best performers are now – quite simply – the best performers.

From September 2009, the *e-scape* system will be used by Awarding Bodies and real awards (GCSEs and Diplomas) will be given as a result.

REFERENCES

Barlex, D. (2003). *Building on success.* Retrieved February 1, 2009, from http://www.data.org.uk/generaldocs/free/dt_strategy_group_report.pdf

Barlex, D., & Trebell, D. (2008). Design without make: Challenging the conventional approach to teaching and learning in a design & technology classroom. *International Journal of Technology and Design Education, 18*(2), 119–138.

Bronowski, J. (1973). *The ascent of man.* London: British Broadcasting Corporation.

Calvez, J. P., Wyche, A., & Edmundson, C. (1993). *Embedded real-time systems.* New York: John Wiley & Sons.

Kimbell, R. (2007). Assessment of design and technology in the U.K.: International approaches to assessment. In M. Hoepfl & M. Lindstrom (Eds.), *Assessment of technology education. The 56th yearbook of the council for technology teacher education* (pp. 181–203). New York, Columbus, OH; Chicago, and Woodlands Hill, CA: McGraw Hill Glencoe.

Kimbell, R., & Perry, D. (2001). *Design and technology in a knowledge economy*. London: Engineering Council.

Kimbell, R., & Stables, K. (2007). *Researching design learning*. Dordrecht, The Netherlands: Springer.

Kimbell, R., Wheeler, T., Miller, S., Bain, J., Wright, R., & Stables, K. (2004). *Assessing design innovation*. Goldsmiths University of London. Retrieved from http://www.goldsmiths.ac.uk/teru/projects.php

Kimbell, R., Stables, K., Wheeler, T., Wozniak, A., & Kelly, A. V. (1991). *The assessment of performance in design and technology*. London: SEAC/HMSO.

Laming, D. (2004). *Human judgement: The eye of the beholder*. London: Thomson.

Mohd Adnan, M. (1993). The role of performance assessment in real-time system management. *Journal of Irrigation and Drainage Systems*, *7*(4), 224–236.

Oxford Cambridge and RSA examinations [OCR]. (n.d.). *Design and technology: Product design*. Retrieved from http://www.ocr.org.uk/qualifications/gcse/design_and_technology_product_design/

Richards, C. (1999). *Primary education at a hinge of history*. London: Falmer.

Pollitt, A. (2006). *Paired-comparison scale creation for e-scape: A report from Alistair Pollitt to TERU*. Published in the e-scape phase 2 report available at http://www.goldsmiths.ac.uk/teru/projects.php

Stables, K., & Kimbell, R. (2000). The unpickled portfolio: Pioneering performance assessment in design and technology. In R. Kimbell (Ed.), *Design and technology international millennium conference* (pp. 195–202). Wellesbourne: DATA.

Thurstone, L. L. (1927). A law of comparative judgement. *Psychological Review*, *34*, 273–286.

Richard Kimbell
Technology Education Research Unit (TERU)
Department of Design
Goldsmiths, University of London
UK

SECTION 7

TECHNOLOGY TEACHER EDUCATION

P. JOHN WILLIAMS

43. TEACHER EDUCATION AND PROFESSIONAL DEVELOPMENT

INTRODUCTION

The role of formal education in any country is fundamental to its continuity and development. Education serves to transmit the prevailing cultural norms to the next generation, and equip its participants with the skills needed to participate in economic and service roles to maintain the society. At the same time it addresses a range of personal needs to enable each individual to achieve to their fullest potential. Given such a significant domain of activity, it is perplexing that the role of the teachers who conduct this domain is not of a higher status and treated as more significant.

The prevailing culture that is transmitted through education is increasingly technological, the participatory skills needed by individuals are often technologically dependent, and people need a basic level of technological literacy in order to realise their potential. Hence technology education is a significant component of the core educational curriculum, but it is not uncontested. Few educational systems, and even fewer universities, place technology education and technology teacher education in a secure place. As curriculum revisions take place, it is often the case that the rationale for technology education needs re-arguing in order to ensure its retention. There is some evidence of cycles in technology education coinciding with economic cycles: in difficult economic times technology education receives more attention (Williams, 1996).

A consistent theme throughout this section of the *Handbook* is the complexity of the activity of technology education. For example, I discuss in Chapter 44 the diversity of the varied foci of technology teacher education programmes, which consequently become the philosophical underpinning of the graduating teacher's practice and Steve Keirl utilises the essential dynamism of the profession, both in its content and its pedagogy, as one of the themes of his discussion of professional development in Chapter 45.

The application of the varied constituents of good teaching is to at least some extent the product of individual perception, as Frank Banks indicates anecdotally at the beginning of Chapter 46. This sense of the individual frame of practice reinforces the complexity of the technology education profession, both in its preparation and sustenance. Philip Reed and John Ritz elaborate on this theme of the individual with their thesis, presented in Chapter 48, that professionalism is an attitude, a state of mind, and as such takes the professional teacher well beyond competencies in the skills of teaching to a level of dedication and caring for students and colleagues.

The adequacy of the attempt in the UK to reduce the art of teaching to a list of specific competencies to be mastered by those entering the profession, and then the addition of further competencies to be mastered on the path to becoming an excellent teacher, is questioned by Banks in Chapter 46 as a method to capture the complexity of professional competence. He makes the point that while it is blatantly possible to reduce elements of professional performance to a list of specific competencies, the whole is greater that the sum of the parts and it is questionable whether

A.T. Jones and M.J. de Vries (eds.), International Handbook of Research and Development in Technology Education, 525–529.
© 2009 Sense Publishers. All rights reserved.

newly qualified teachers can re-integrate all the competencies into an effective approach to teaching after having achieved mastery of each separate part.

In his discussion of the progression of the Bologna Process in Europe, Jacques Ginestié speculates in Chapter 47 on the difficulties in developing a unified structure across countries whose educational systems essentially represent the historical diversity of different approaches to technology education. Long traditions of technology and entrenched structures of education combine to form a complex and diverse pattern across the European Union that conflicts with the proposed unification of the Bologna Process.

Both Williams (Chapter 44) and Keirl (Chapter 45) refer to a range of literature in emphasising the complexity of the role expectations of teachers. Teachers are expected to be entrepreneurial, a leader-learner, a technician, to reflect on their practice, engage in collegial teamwork, critically enquire into their profession and society, adhere to professional standards, be effective change agents, and take risks (Fullan & Stiegelbauer, 1991; Hextall & Mahony, 1998; Lingard, Hayes, Mills, & Christie, 2003; McWilliam, 2008; Sachs, 2003; Schon, 1987; Smyth, Dow, Hattam, Reid, & Shacklock, 2000; Yost, Sentner, & Forlenza-Baily, 2000). Education and professional development encompassing such scope and diversity is a theme that threads through this section.

Glimmers of the development of a new positivism about technology teaching also appear in this section, a positivism derived from recognition of the deficiencies of standards and competencies, the loss of creativity and innovation in classrooms, the futility of international competition based on fraught measures of international excellence, and the importance of technology education in the curriculum.

Keirl cites Hargreaves and Goodson (2003) in injecting optimism into his discussion of professional development:

> Amid all this despair and danger though, there remains great hope and some reasons for optimism about a future of learning that is tied in its vision to an empowering, imaginative and inclusive vision for teaching as well ... Governments almost everywhere are beginning to speak more positively about teachers and teaching – bestowing honour and respect where blame and contempt had prevailed in the recent past. (p. viii)

Working against this new positivism remains a management culture of documentation, reporting, and measurement in the name of transparent accountability that implies a distrust of teachers as competent and trustworthy designers of educational experiences for their students.

An alternative positive outcome is cited by Banks through the notion that successful performance by teachers, as judged by a competency framework, will reward them with the freedom to have more control over the curriculum with which they are involved. The jury is probably still out over the possibility of this outcome.

Banks cites Storey and Hutchinson (2001), predicting that engagement with teacher education competencies will provide opportunities in "wresting a greater share of the contested terrain of education through participation [and] may, ironically, be one tangible and possible outcome beyond the current era's accent on quality control" (p. 51). They propose that competent performance against the standards framework will enable those teachers so judged as successful to be rewarded with more autonomy in the school context.

The context within which the contestation of technology education plays out is varied, and a number of such contexts arise in the chapters in this section. One such context arises from a systemic preoccupation with testing and measurement procedures and outcomes in the so-called basics of education. In the face of such high-stakes activities with significant implications, those areas of the curriculum considered non-core – such as technology education – become contested within the institutional timetable. One outcome of this type of contestation is to seek an alignment with those areas of the curriculum that have support and advocacy, and so consequently technology is taught for the literacy skills it can promote or for the numeracy competencies it can reinforce, rather than goals that are more a reflection of the nature of technology.

Keirl refers to Layton (1994) in elaborating on the haphazard arbitration of competing economic, industrial, professional, political, and educational pressures that bear on design and technology education. For example, "... the politics of technological literacy – who creates and controls the meanings of the phrase, how the imposition of meaning is attempted – is a central concern of technology education today" (p. 13).

An additional area of contestation not commonly mentioned in this section but nevertheless prevalent is enacted through the distribution of budgets, both at school and university levels. Given the generally lower numbers of design and technology students and staff than other faculties within an institution, and higher costs associated with equipment, materials and workshops, a pure economics/business model does not sit well with technology education programmes. This competition for budgets has resulted in the closures of some programmes and at least the need for continual justification of those that remain.

In the introductory chapter to teacher education, Chapter 43, I begin with an indicator of the international – and often national – diversity in technology teacher education by summarising some of the qualifications that are available as a means to become a teacher. One of the themes developed in this chapter is the contrast between a complex process of teacher education, a complex content area of technology, and the instrumentalisation of teaching through its reduction to competencies that can be measured and reported on, which consequently and by default defines the role of the teacher. This is elaborated through a discussion of three aspects of technology teacher education. The first is the focus of teacher education programmes, which is a significant determinant of the philosophy of the beginning teacher. The difficulty is that teacher education programmes must have a multi-focussed approach that matches the needs of the market for which the teachers are being prepared, and makes sense to the individual who is becoming a teacher. The second aspect is the extent to which the elements of a teacher education programme are integrated. Successful programmes integrate a significant number of variables in order to deliver a consistent message about good teaching. Finally, the balance of a course between content and pedagogy, represented by an accommodating holistic schema, will encourage a sound vision of technology education.

Keirl frames his discussion of professional development in the context of the re-building of a boat which must be undertaken while on the voyage rather than bringing it into dry dock and doing a thorough job (Chapter 45). Thus, the building process continues while all the external pressures remain, as does the expectation that it continue a successful voyage. An analysis of the nature of a profession and 'being professional' follows with the accompanying recognition that any construct of professionalism is determined by the contextual climate, the meanings applied to it, and the nature of its members. From this emanates some principles of professional development, namely its basis in adult learning theory, active participant engagement, research-based principles, and a sound educational rationale.

Through a critique of the competency based teacher training system in the UK, and an analysis of the assumptions implicit in such a system, Banks uses an historical commentary and a country summary as the basis for asking if competencies progress or diminish the profession (Chapter 46). It may be indicative that the introduction of competencies accompanied a move in teacher education away from universities to include more school-based activities. Banks concludes by referring to a number of authors who, on the one hand, see in the introduction of competencies a process of de-skilling and de-professionalisation of teachers (Harris, 1997), and others who see future opportunities of greater autonomy developing from effective engagement with the framework of standards (Storey & Hutchinson, 2001).

In Chapter 47, Ginestié selects the countries of Germany, Spain, France, and Greece to use as exemplars in describing the effects of the Bologna process within the European Union on the training of technology teachers. The Bologna Process was initiated in 1998 as a way to modernise, reform, and bring unity into the higher education sector of the member countries of the European Union by 2010. The implication of Ginestié's discussion is that consistency in teacher education in the EU is unlikely to be a reality by 2010. The reasons for the historically developed diversity across countries are fundamental and philosophical, and relate to questions of the balance of content and pedagogical knowledge and the context and sequence in which knowledge and skills are best developed.

In Chapter 48, on graduate study as professional development, Reed and Ritz begin with an examination of the nature of professionalism, advocating an approach that goes beyond competencies to a broader attitude, and dedication toward the range of activities that are necessary to support and advance a profession. They utilise a discussion of Myers' (1973) characteristics of a profession to conclude that one essential characteristic of any professional is life long learning. This provides the introduction to research conducted into graduate study in design and technology education (Ritz & Reed, 2007) as a form of professional life long learning. Aspects of the research, such as the nature of graduate programmes and the areas of research undertaken, were used as a practical basis for the proposal of a model of master's and doctoral studies in technology education.

The chapters in this section provide a particular rather than representative view of the global diversity in technology teacher education and professional development. However, the themes of a complex and contested area of the curriculum that can look forward with renewed positivism to manifold opportunities is widely applicable.

REFERENCES

Fullan, M., & Stiegelbauer, S. (1991). *The new meaning of educational change.* NewYork: Teachers College Press.

Hargreaves, A., & Goodson, I. (2003). Series editors' preface. In B. Lingard, D. Hayes, M. Mills, & P. Christie (Eds.), *Leading learning. Making hope practical in schools* (p. viii). Maidenhead, UK: Open University Press.

Harris, A. (1997). The deprofessionalisation and deskilling of teachers. In K. Watson, C. Modgil, & S. Modgil (Eds.), *Educational dilemmas: Debate and diversity 1—Teachers, teacher education and training* (pp. 57–65). London: Cassell.

Hextall, I., & Mahony, P. (1998). Effective teachers for effective schools. In R. Slee, G. Weiner, & S. Tomlinson (Eds.), *School effectiveness for whom? Challenges to the effectiveness and school improvement movements* (pp. 128–143). London: Falmer.

Layton, D. (Ed.). (1994). *Innovations in science and technology education* (Vol. V). Paris: UNESCO.

Lingard, B., Hayes, D., Mills, M., & Christie, P. (2003). *Leading learning: Making hope practical in schools.* Maidenhead, UK: Open University Press.

McWilliam, E. (2008) Making excellent teachers. In A. Phelan & J. Sumison (Eds.), *Critical readings in teacher education* (pp. 33–44). Rotterdam, The Netherlands: Sense.

Myers, D. (1973). *Teacher power – professionalism and collective bargaining.* Toronto, ON: Lexington Books.

Ritz, J. M., & Reed, P. A. (2007). Graduate study in technology education: State-of-the-art. In J. R. Dakers, W. J. Dow, & M. J. de Vries (Eds.), *Teaching and learning technological literacy in the classroom* (pp. 406–414). Glasgow, Scotland: University of Glasgow.

Sachs, J. (2003). *The activist teaching profession.* Buckingham, UK: Open University Press.

Schon, D. (1987). *Educating the reflective practitioner.* San Francisco: Jossy-Bass.

Smyth, J., Dow, A., Hattam, R., Reid, A., & Shacklock, G. (2000). *Teachers' work in a globalizing economy.* London: Falmer.

Storey, A., & Hutchinson, S. (2001). The meaning of teacher professionalism in a quality control era. In F. Banks & A. Shelton Mayes (Eds.), *Early professional development for teachers* (pp. 41–53). London: David Fulton.

Williams, A. (1996). An introduction to technology education. In P. J. Williams & A. J. Williams (Eds.), *Technology education for teachers*. Melbourne, Victoria: Macmillan.

Yost, D. S., Sentner, S. M., & Forlenza-Baily, A. (2000). An examination of the construct of critical reflection: Implications for teacher education programming in the 21st century. *Journal of Teacher Education, 51*(1), 39–49.

P. John Williams
Edith Cowan University
Perth, Western Australia
and University of Waikato
Hamilton, New Zealand

P. JOHN WILLIAMS

44. TEACHER EDUCATION

INTRODUCTION

A significant amount of diversity exists in technology teacher education programmes around the world. It is not the intent of this chapter to try and provide a summary of such diversity, but rather to attempt to generalise from both examples of, and research about, technology teacher education, to conclude with some principles of good practice.

There are many variables that come together to provide an effective technology teacher education programme. The thesis of this chapter is that the extent to which all these variables are effectively integrated to provide a clear, explicit focus determines the durability of the desired philosophy of the teacher graduates. This philosophy will essentially vary from country to country as the unique role of the educational system is addressed. No value judgements can be made about a decontextualised teacher education programme; evaluation is derived from the extent to which it meets the needs of the related educational system, and to do this effectively there are structural elements which are necessary.

In order to establish the extent of the diversity that does exist in technology teacher education programmes, some examples will be provided. Entry into a technology teacher education programme can vary from the completion of the fifth year of secondary education with no specific prerequisites to a competitive examination after three years of tertiary education. In Botswana, students can gain entry into a design and technology teacher education programme at a Teachers College upon successful completion of 'O' Level examinations after five years of secondary schooling. In France, after completing three years of tertiary education, students must sit an examination in order to gain entry into the two year technology teacher education programme.

Similar variability exists in the duration of technology teacher education courses. These vary from one year to five years, but this can be done at undergraduate diploma or postgraduate master's level. The titles of courses consequently also vary, and have different meanings in different countries. For example, both diploma and bachelors courses can be undergraduate or postgraduate. Some examples follow.

Two Year Diploma of Education

As a result of increasing standards, this course is becoming less common as a pathway into design and technology teaching. The course consists of an integrated mix of curriculum and content subjects with blocks of practice teaching experience in schools.

Four Year Bachelor of Education

In many countries this is the traditional model of teacher training and many institutions still offer this course, although it is becoming less common. Education, curriculum, pedagogy, and content

A.T. Jones and M.J. de Vries (eds.), International Handbook of Research and Development in Technology Education, 531–540.
© 2009 Sense Publishers. All rights reserved.

units are studied concurrently throughout the four years. School practice experiences begin in the first or second year, and generally conclude with an extensive period (for example, one term of 10 weeks) working in a school assuming the full teaching load of a teacher.

In this model the majority of the units in the course are taught by the faculty of education, although some specialised areas may be contracted out to other areas within the university or to external institutions such as secondary schools or technical colleges. In some universities, all practical work is done at either a secondary school or technical college.

Applicants may receive advanced standing into the Bachelor of Education degree if they have relevant experience or qualifications in a relevant area, such as a technical college diploma or a trade qualification. Students may also gain entry to this degree programme through completion of an Associate Degree, which has less demanding entry qualifications.

Four Year Double Degree

This award is either two degrees undertaken concurrently, or a 2+2 structure where one degree is completed before the other begins, but either pattern is completed within four years. One degree is typically offered by the faculty of education, and the other by the area in which the student specialises. The education degree is generally a Bachelor of Education or a Bachelor of Teaching, and examples of the other degree include a Bachelor of Technology, Bachelor of Design, and Bachelor of Science.

Two Year Bachelor of Teaching or Bachelor of Education

This is an integrated post-graduate course offered to students from varying backgrounds. Applicants must have an initial three year degree in a technology-related area such as engineering, industrial design, architecture, or computing. Students focus on education, curriculum, and pedagogical studies with the assumption that their content expertise is derived from their initial degree, although there may be the option to undertake study in technology subject areas other than those covered in their first degree. An alternative entry may be a one-year bridging course for technical college graduates who would typically have a trade background.

One Year Graduate Certificate or Diploma of Education

The prerequisites for entry into the Graduate Certificate or Diploma are generally a three year initial degree in a technology-related area. Students then focus on education, curriculum, and pedagogical studies, and spend a significant proportion of course time in schools developing pedagogical skills and school knowledge. The difficulty training design and technology teachers in this context is the absence of undergraduate degrees that encompass the breadth of the learning area; consequently, time is often required to develop proficiency in content.

Two Year Graduate Diploma of Education

Entry into this award can be through a number of means, for example, substantial relevant work experience, including an apprenticeship, or a two year technical college qualification plus work experience. The prerequisite qualifications and experience again provide the area of content expertise, so the course units focus on education, pedagogy, and curriculum studies.

One and Two Year Master of Education Degree

It is uncommon that this is an entry qualification into teaching, but where this is the case the course is a combination of curriculum, pedagogy, and educational studies, with the content studies, again, being derived from the relevant prerequisite undergraduate course.

Commonalities in Diversity

In the face of such diversity the question arises, what are the commonalities? Certainly not structure, duration, or content. What is common is the need to prepare students for increasingly complex roles as teachers. While retaining a focus on teaching that ensures effective learning and student enquiry, teachers are also expected to reflect on their practice, engage in collegial teamwork, critically enquire into their profession, adhere to professional standards, be effective change agents, and take risks (Fulland & Stiegelbauer, 1991; McWilliam, 2008; Schon, 1987; Yost, Sentner & Forlenza-Baily, 2000).

All this is expected to take place in a context in which, for many educational systems, education is equated with training students in skills deemed relevant for economic purposes (Kane, 2002). Politicians in many countries lament the literacy and numeracy crisis, which they directly relate to economic performance. The role of the teacher has consequently been instrumentalised to a person who delivers a set curriculum and manages assessment and reporting in order to comply with a market driven regime (Banks, 2006). Such instrumentalisation of teaching, together with the vocationalisation of the curriculum, is in some respects dichotomous to the reflective and critically-engaged practitioner that is expected of teacher education programmes, and certainly diminishes any social justice agenda teacher education may have.

The framework for terminology used in this chapter has been derived from research by Banks and Barlex (1999) and Hansen (1993). Banks and Barlex' research related to the development of a conceptual model of the knowledge that is important for successful technology teaching and is based on the work of Shulman (1986) and Gardner (1983, 1991). Hansen's conceptual framework is an attempt to identify the central tasks in the preparation of technology teachers. From this research and other models of technology teacher training programmes, the following terminology will be utilised in forming a framework:

- Education Studies: theories of education, educational psychology, general teaching and learning.
- Pedagogical Studies: instructional strategies specifically related to technology (design, problem solving, technological method).
- Content Studies: the development of expertise in areas of technology.
- Purpose Studies: rationales for teaching technology, personal philosophies of technology and technology education.
- School and Curriculum Studies: understanding of curriculum, curriculum development, practical experiences in schools, study of syllabuses and support material, prescribed assessment requirements.

Research indicates that there are a number of factors that enable teacher education programmes to have a significant impact on the initial conceptions of prospective teachers, and are effective in supporting student teacher learning and graduate teachers who remain with the profession beyond the first few years. From these factors will be derived some issues for consideration in the design of technology teacher education programmes.

FOCUS

The first factor I would like to consider is the focus of teacher education programmes – which may be self determined or imposed – but either way, is a significant determinant of the philosophy of the beginning teacher. The difficulty of course is that teacher education programmes must have a multi-focussed approach that matches the needs of the market for which the teachers are being prepared.

A high proportion of beginning teachers leave the profession after a few years. The transition from teacher education, even that which includes a significant school-based component, to a sustainable teaching career is insurmountable for many teachers. This would tend to indicate that there is a deficiency in teacher preparation programme focus on reflective, metacognitive analysis of themselves as teachers. Despite well developed skills in behavioural management and curriculum development and delivery, and successful 'practice' teaching experiences, teachers are not well prepared for the reality of their first full time experiences in the classroom (Russell & McPherson, 2001). It would seem that teachers should begin their career having developed a deep personal understanding of themselves as teachers, and their training programme must have this as a focus.

An additional focus of teacher preparation programmes that may help ease the transition into the teaching workforce is the development of school knowledge. This is proposed as an element of the teacher's personal subject construct together with subject knowledge and pedagogical knowledge, which has come to be known as the DEPTH framework (Developing Professional Thinking for Technology Teachers) (Banks, Barlex, Jarvinen, O'Sullivan, Owen-Jackson, & Rutland, 2004). The DEPTH research developed Verret's (1975) thesis that school knowledge grows out of a general body of knowledge by recognising that school knowledge is "informed by the local school ethos, common practices and the authenticity of the activities that pupils are required to undertake during their work in technology lessons" (Banks, 2008, p. 226). There is a limit to the extent that a teacher can gain school knowledge prior to becoming a teacher, but to that extent, the development of such knowledge must be a focus of teacher education. This knowledge is gained through practical experiences in schools and consolidated through integration with other aspects of teacher education studies.

The pressure of the school curriculum is an undeniably valid constraint as it forms the basis of the teachers' future professional activity. The school curriculum, and consequently its basis and approach, is then either vicariously adopted by the teacher education programme, or is accommodated within its broader philosophy. Countries in which a national school curriculum is implemented include England and Wales, New Zealand, France, and Russia. The implication for these countries is that the content of teacher education programmes is relatively similar within each country, but there is still significant diversity in the approaches taken to train teachers.

However, most qualifications granted at the conclusion of a teacher education programme are internationally transferable, and a programme that incorporates too specific a focus on the local curriculum may detract from the transferability of the teacher. A curriculum focus that is developed from a sound philosophical base, together with a framework that positions a critical approach to technology education will help ensure principled transferability.

Given current world events, it is increasingly less difficult to argue that a focus of technology education should be sustainability and the environment (see also Chapter 21). Such a focus provides a relevant rationale for technology education, a point of integration for the activities in a teacher education programme, and the criteria for the selection of relevant content. The relationship between technology and the social and physical environment is daily confrontational, through the media, politics, and increasingly individual lifestyle choices. A technology teacher education programme that focuses on developing an awareness of such relationships, an informed knowledge base from which to critique technological and social developments, and the practical mechanisms through which students can develop a relevant form of technological literacy would be appropriate.

More than many other areas of the curriculum, traditional technology teaching is grounded in an epistemology of technical rationality that construes teachers as technicians delivering a prescribed curriculum that gives little credence to learning through experience (Kane, 2002), while at the same time the basis of technology education is experiential. This seeming dichotomy is resolved through the acknowledgement that the experiences are teacher-centred rather than student-centred, so while students are learning through experience, it is an experience that is predetermined by the teacher.

Many teachers who have a primary focus on skill development and believe strongly in the importance of pre-vocational experiences. The majority of technology teacher education programmes continue to battle against this legacy of technology education as a non-academic area of study dealing with a narrow range of skills and materials in a vocational or prevocational context. This is contrasted with strong evidence that a focus of technology education that is distinctly general, broad, and considers the social and environmental implications of technology can exist without losing the important legacy of thinking in a practical context.

The emphasis on the so-called basics of education – the 'No Child Left Behind Act' in the USA, the national testing programme in literacy and numeracy in Australia, and similar neo-liberal agendas in many other countries – has the effect of threatening elective curriculum subjects such as technology education. While this has little direct effect on teacher education, it does tend to marginalise technology education as a less important area of education, particularly in elementary schools where the focus on literacy and numeracy is most pronounced and where teachers are generally less well equipped to deliver technology education. The focus on literacy and numeracy, the evaluation of education for economic ends, and the vocationalisation of technology education result in adverse forces on a culturally-sensitive type of technology teacher education who may desire to deal with maximising individuality, social justice, and diversity, thus constricting the education of socially and environmentally conscious teachers.

In many countries the focus of teacher preparation is moving toward a school-based apprenticeship model, particularly in countries where teacher accreditation is by way of the achievement of a set of competencies, which can be developed in any context: school, university, or other accredited provider. (Frank Banks explores this in the context of the UK in Chapter 46.) This allows real opportunities for students to practice their teaching and test their theories, but maybe fewer opportunities for reflection on themselves as teachers or for developing a conceptual base for their profession. Both aspects are important in preparing teachers who are effective at their craft and secure in their philosophy, and so teacher preparation programmes must include a balance of real practice and critical professional reflection.

The focus of support for beginning teachers by teacher education programmes should continue beyond graduation and into the first years of teaching. Research indicates that the first years of teaching are the most difficult (Watson, Steele, Vozzo, & Aubusson, 2007) of both the pre-service and in-service components. To provide teachers with additional forms of support during this period in order to extend the pre-service approaches to collegiality, reflection, and application into the early teaching years will reinforce and render more effective the pre-service programme. An additional outcome will be the retention of teachers in their profession and the consequent cost savings this will have for the employer (Riley, 2008).

INTEGRATION

The second factor for consideration when designing technology teacher education is the extent to which the elements of a programme are integrated. Successful programmes integrate a significant number of variables in order to deliver a consistent message about good teaching. There include variables that are significant in any teacher education programme: education studies, pedagogical studies, content, curriculum, and practical experiences in schools. In addition, there are those variables unique to technology education that must be presented in an integrated manner: theory and practice, design, creativity, and innovation.

In instances where trainee teachers are faced with a range of conflicting or inconsistent messages about the fundamentals of their teaching, there is a tendency for each message to interfere with or even nullify the effects of another (Wideen, Mayer-Smith, & Moon, 1998). This is problematic in technology teacher education for a number of reasons. One is the trend in some countries (e.g., Hong Kong, USA, Australia, Israel) for the teacher education programme to span a number of faculties within an institution. This is most often a change from the situation where the entire course was within one faculty, often a Faculty of Education. Other faculties involved may include engineering, industrial design, or art and design, where the students receive some of the technology content knowledge. The philosophical approach and rationale for learning about technology is different in these other faculties and so presents the trainee teachers with mixed messages. For example, the rationale for learning about technology in faculties other than education is often vocational, an approach which may not be transferable to education. An additional outcome of this trend is reduced technology education faculty involvement, which of course impacts on research and general professional development in technology education.

Ensuring a consistent and reinforcing message from both the university and the schools in which trainee teachers practice is difficult, particularly with regard to content and philosophy. Teacher education programmes that seek to effect change by questioning the delivery of traditional content are often contradicted by practical experiences in schools. While this may provide the opportunity for trainees to examine a range of approaches to this particular issue, evidence is that a type of structured fragmentation of approach is the result (Gore & Zeichner, 1991).

The contextual influence of schools is naturally more determining after graduation in the early years of teaching. While a teacher may graduate full of enthusiasm for design-focussed, innovative approaches to learning, the initial pressures to survive and successfully manage classes lapse into a comfort zone of conformity, resulting in the perpetuation of the school tradition. The maintenance of focus on an innovative and well integrated technology teacher education programme into the early years of teaching can be enhanced by continued professional development support after graduation.

In some educational systems, generally where technology is not an element of the core school curriculum, links have to be made with other subject areas in order to ensure the continued viability of technology education. This has happened in some parts of the USA and is becoming evident in the UK, for example, where links have been made with mathematics and science, and more recently engineering. This has resulted in changes in the teacher training programme, and again has the potential to present to teacher trainees an inconsistent approach to technology education, maybe seeing it as applied science or pre-engineering.

The dichotomous philosophical positions with which teacher trainees are confronted can generally be characterised as the vocational and the general. Technology teacher education courses are most often focussed on inculcating a general philosophy of technology education – that it is important for all students to be technologically literate and have the opportunity to be innovative and creative in a practical environment. This may not, however, be the focus in schools this may not be the focus, where a vocational or pre-vocational focus might prevail. A vocational *and* general approach to technology education, while existing in many schools, are concurrently incompatible in terms of pedagogies, content, and assessment. The presentation of this dual focus to teacher trainees is confusing and does not support the grounding of teaching within a consistent philosophy.

CONTENT

The balance in a technology teacher education courses between teaching content and content-related pedagogies varies significantly. However, there are some indicators which guide the nature and presentation of content in training effective teachers.

Whilst much research in the past has focussed on *how* teachers learn, it seems that *what* they learn is just as important in shaping their teaching practice. For example, a relationship has been found between teacher effectiveness and the amount of training teachers have received in content-specific teaching methods (Monk, 1994). It also seems that the content and pedagogies used to teach that content has a greater influence on teacher practice than the length of the professional learning experiences (Kennedy, 1998). This may give cause to question the increasing proportion of time spent in school-based experiences as a part of many teacher education programmes.

It also seems clear that if teacher learners are provided with a conceptual framework or cognitive map – that is, they begin their studies with an holistic overview of technology education – it deepens their understanding and allows them to integrate what they learn into their schema as they progress through their course to become teachers. A course that is constructed to enable progressive development of such an holistic schema will facilitate individual integration and place all the 'bits and pieces' into a pattern.

A succession of subjects in a course that focuses on the development of skills in a range of technologies will never become more that the 'sum of the parts' unless the framework is developed to accommodate all the parts into a greater 'wetlanschung'. Teachers will be left with a succession of discrete activities that do not relate to each other and that will, by default, become their inadequate vision of technology education. All aspects of technology education hang off a particular philosophical position, and unless they do relate to a bigger picture, each aspect remains as a disjointed segment. So the danger becomes, for example, that CAD/CAM has little to do with design, assessment is simply a series of boxes to be checked, and woodwork is a sequence of skills to be mastered.

The derivation of content for the purposes of technology education is basically philosophical. Answering epistemological, axiological, and metaphysical questions provides parameters for technological content knowledge:

- What do we know?
- What do we value?
- What is real?

But even after having answered these questions within the domain of technology, the resultant content is too broad to be covered in a course; consequently additional organisers of content must be introduced. These organisers may be quite pragmatic, such as education legislation, which can have a significant determining influence on the nature of technology teacher education programmes. For example, the Bologna Process agreement for standardisation of university programmes across the European Union is having real impacts on teacher education programmes in countries like Germany and Russia. In England, legislation detailing the competencies for entry into teaching has a very deterministic impact on both the structure and content of technology teacher education, and in the US, standards have become increasingly rigorous.

An additional content organiser is of course the school curriculum for which the teacher trainees are being prepared. Even in instances where the teacher education course is designed to promote graduates who have a transformative approach to their teaching (Kalantzis & Cope, 2008), they still have to function in a range of school environments. Consequently certain content must be covered in order to ensure new teachers can 'hit the ground running' during their first school appointment.

Mitchell, Weber, and Yoshida (2008) ponder the implications for the content of teacher education courses if pre-service teacher educators thought of themselves as youth workers and the students as youth activists. They suggest three possible implications. The first relates to community service. Based on the notion that "service-learning reflects the belief that education can contribute to developing social responsibility and can help to prepare young people to be involved citizens in democratic life" (p. 148), an encounter with social critique and community action in a pre-service course would increase the potential for schools to become hubs of community development. The second implication is that young teachers can be assisted to resolve intergenerational tensions when they are often closer in age to their students than to the other teachers in the school. Young teachers often want to champion the causes of their students, and the development of mechanisms to this end in pre-service courses may result in a heightened level of meaningful engagement with youth. In addition, engagement with participatory, visual, and other creative approaches provides opportunities to address a range of issues including bullying, gender violence, street youth, and community development. In addition to tools of expression such as hip-hop, graffiti, puppetry, and street theatre there are the range of generational familiar technologies aligned with Web 2.0 (Williams, 2007) that provide heretofore unequalled participatory access to media technologies such as YouTube, Facebook, MySpace, etc. These forms have meaning and relevance to young people and provide the mechanism for personal action and group empowerment.

The manner in which content is presented, the place it occupies within a broader philosophical framework, and the links that are established with context as the medium in a teacher education programme have a significant influence on the development of technology teachers.

ISSUES

A number of issues are currently so significant to technology teacher education in many different countries that they are worthy of mention.

Funding – The recent change in status of universities in Japan to independent corporations is indicative of a new commercialisation in universities in many countries. There is typically greater competition between universities for students and research funding, and emphasis on the development of income-generating programmes. These pressures – in Japan, Hong Kong and Australia, for example – have resulted in institutional integration. The resulting impact on technology teacher education programmes has been a focus on economics. Given generally lower numbers of students and staff than other university faculties and higher costs associated with equipment, materials and workshops, a pure economics/business model does not sit well with technology teacher education programmes. This has resulted in the closures of some programmes and at least the need for continual justification of remaining programmes.

Teacher shortage – Many countries are experiencing a shortage of technology teachers. Despite the logic that guaranteeing employment should be a useful marketing tool to attract more students, this does not seem to be the case and recruitment into technology teacher education remains difficult. Smaller classes, which result from declining enrolments, can affect the viability of courses from a university management perspective. Economies of scale generated through large lecture groups are not consistent with a practical workshop approach to technology teacher education. Other impacts include the downsizing of technology teacher education departments and the closing of some departments, for example in Germany and the USA. The overall result is a general weakening of the technology education profession.

CONCLUSION

The provision of an effective technology teacher education programme is a complex process dealing with multiple factors. The measure of effectiveness itself is fraught with many variables – for example, initial teacher support and subject and school knowledge – and as a measure is also variable and may relate to length of tenure, student performance, or student evaluations of teachers.

While there is no 'best' way of organising experiences for an effective technology teacher education programme, there seem to be some common considerations. Studies of teacher learning effectiveness indicate that exposure to mutually-reinforcing ideas and skills, experienced across a range of learning contexts and grounded in a common focus, result in the development of enduring expertise.

REFERENCES

Banks, F., Barlex, D., Jarvinen, E-M., O'Sullivan, G., Owen-Jackson, G., & Rutland, M. (2008). DEPTH: Developing professional thinking for technology teachers: An international study. *International Journal of Technological and Design Education, 14*(3), 141–157.

Banks, F. (2006). Technology teacher education in the United Kingdom. In P. J. Williams (Ed.), *International technology teacher education, 55th Council on Technology Teacher Education yearbook* (pp. 215–239). Peoria, IL: Glencoe/McGraw Hill.

Banks, F. (2008). Learning in DEPTH: Developing a graphical tool for professional thinking for technology teachers. *International Journal of Technological and Design Education, 18*(3), 221–229.

Banks, F., & Barlex, D. (1999). No one forgets a good teacher! What do 'good' technology teachers know? *The Journal of Design and Technology Education, 4*(3), 223–229.

Fullan, M., & Stiegelbauer, S. (1991). *The new meaning of educational change*. New York: Teachers College.

Gardner, H. (1983). *Frames of mind: The theory of multiple intelligences*. New York: Basic Books.

Gardner, H. (1991). *The unschooled mind*. New York: Basic Books.

Gore, J., & Zeichner, K. (1991). Action research and reflective teaching in preservice teacher education: A case study from the United States. *Teaching and Teacher Education, 7*(2), 119–136.

Hansen, R. (1993). A technological teacher education program planning model. *Journal of Technology Education, 5*(1), 21–28.

Kalantzis, M., & Cope, B. (2008). *New learning: Elements of a science of education*. New York: Cambridge University Press.

Kane, R. (2002). How we teach the teachers: New ways to theorize practice and practice theory. *Prospects, 32*(3), 347–364.

Kennedy, M. (1998). *Form and substance in inservice teacher education. Research Monograph, 13*, National Institute for Science Education.

McWilliam, E. (2008). Making excellent teachers. In A. Phelan & J. Sumison (Eds.), *Critical readings in teacher education* (pp. 33–44). Rotterdam, The Netherlands: Sense.

Mitchell, C., Weber, S., & Yoshida, R. (2008). Where are the youth?: Reframing teacher education within the context of youth participation. In A. Phelan & J. Sumison (Eds.), *Critical readings in teacher education* (pp. 139–154). Rotterdam, The Netherlands: Sense.

Monk, D. (1994). Subject matter preparation of secondary mathematics and science teachers and student achievement. *Economics of Education Review, 13*(2), 125–145.

Riley, P. (2008, January 28). Keeping our teachers. *The Age*, p. 11.

Russell, T., & McPherson, S. (2001, May). *Indicators of success in teacher education: A review and analysis of recent research.* Paper presented at the Pan-Canadian Education Research Agenda (PCERA) Symposium on Teacher Education/Educator Training, Quebec.

Schon, D. (1987). *Educating the reflective practitioner*. San Francisco: Jossy-Bass.

Shulman, L. S. (1986). Those who understand: Knowledge growth in teaching. *Educational Researcher, 15*(2), 4–14.

Verret, M. (1975). *Le temps des etudes.* Paris: Librarie Honore Champion.

Wideen, M. F., Mayer-Smith, J., & Moon B. (1998). A critical analysis of the research on learning-to-teach. *Review of Education Research, 68*(2), 130–178.

Watson, K., Steele, F., Vozzo, L., & Aubusson, P. (2007). Changing the subject: Retraining teachers to teach science. *Research in Science Education, 37*(2), 141–154.

Williams, P. J. (2007, June). *Digital democracy—A new role for technology education.* Paper presented at PATT-18: Pupils' Attitudes Towards Technology International Design & Technology Education Conference, Glasgow, Scotland.

Yost, D. S., Sentner, S. M., & Forlenza-Baily, A. (2000). An examination of the construct of critical reflection: Implications for teacher education programming in the 21st century. *Journal of Teacher Education, 51*(1), 39–49.

P. John Williams
Edith Cowan University
Perth, Western Australia
and University of Waikato
Hamilton, New Zealand

STEVE KEIRL

45. DOING THE ODYSSEY IN NEURATH'S BOAT: DESIGN AND TECHNOLOGY PROFESSIONAL DEVELOPMENT JOURNEYS AND TEACHER PROFESSIONAL JUDGEMENT

INTRODUCTION

In this chapter I draw on the rich discourse of general teacher professional development literature in order to inform that of Design and Technology (D&T) education. Because of the literature's breadth and depth, practitioners wanting to extrapolate and rationalise from it must apply, I argue, considered professional judgement. I also argue that professional development must be thought of as needing both individual and collective professional judgement.

It is not uncommon to use the metaphor of 'odyssey' when writing about curriculum and teachers' professional journeys in particular. The odyssey is often cast as a series of difficult challenges – which D&T teaching certainly entails – but it is in its sense of optimistic travelling into the future that I see the professional teaching journey. That is, I write in full recognition of continuous change and that things cannot stay the same. Despite dangerous winds and currents, monsters and seduction to distraction, there is a greater purpose and intent to teachers' work. It seems to me that our professional development is both *of ourselves* and *of our profession* – as, like technology, it can be framed as both tool and as existential practice.

However, despite the many diversions represented in the Odyssian metaphor (with both troubled and happy landings) I also value the analogy offered by Austrian philosopher Neurath, who suggested: "We are like sailors who must rebuild their ship on the open sea, never able to dismantle it in dry-dock and to reconstruct it there out of the best materials" (cited in Honderich, 1995, p. 97). Such is the reality of teaching – but of D&T teaching especially when rapid technological change, economic trends, and political fads all impact on our curriculum practice.

The chapter opens with an overview of the socio-political climate in which much education finds itself operating. There follows some exploration of constructions of 'profession', 'professional', and 'professional development' where, as will be seen, there are direct relationships between political circumstances and the shaping of the profession and its practice. There is a rich generic professional development literature that I draw upon to inform considerations about D&T itself. It is my contention that the contradictions and tensions extant in the D&T field demand deep practitioner reflection if effective professional development is to be generated.

THE GLOBAL POLITICAL AND PROFESSIONAL SCENE

At any time, in any society, education is a tool of, and for, that society. It promotes the ideals of the society and it prepares its citizenry. Usually, the kind of political system is reflected in the kind of education. Thus, totalitarianism does not value free- and critical-thinking students whereas democracies claim to espouse such thinking. I say 'claim' because democracies vary and, as has been argued, the camouflaging of ideology and the merging of economic imperatives with

A.T. Jones and M.J. de Vries (eds.), International Handbook of Research and Development in Technology Education, 541–554.
© 2009 Sense Publishers. All rights reserved.

education are strong determinants of educational practice in many societies today (Apple, 2001; Galbraith, 2004; Layton, 1994; Postman; 2000).

It is common to talk of the (rather tautologous) 'globalised world', the most common view of which is economic globalisation. However, globalisation is not merely economic and/or political but takes many forms of socio-cultural and technological networks and alliances. Thus, individuals, communities, or societies may be willing or unwilling participants, to varying degrees, in any of a number of 'globalised' entities. There is not *one way* but *many possibilities*. In such climates, the education profession and curricula are under pressure and teachers are often positioned in ways that call for (re)action. Fortunately, many researchers recognise that education is a matter of the public good and that teachers cannot be left the whole responsibility for its success – or its failure (Darling-Hammond, 1997).

Hargreaves (2003), a world leader in the research of teachers' professional lives, documents the developments, internationally, that have occurred in education since the 1980s. He utilises an OECD analysis that offers "six likely scenarios for the future of state education in the knowledge society" (Hargreaves, 2003, p. 126):

> Two of these presume an unravelling of existing arrangements that will lead either to more entrenched bureaucracy in school systems, or to increasing emphasis on market and choice-based solutions as people's dissatisfaction with state education spreads. A second pair of options presumes a shrinkage of state schooling either by atrophy as teacher shortages and a desperate proliferation of innovations create panic and 'meltdown' in educational policy, or by investment in alternatives outside the school in e-learning and non-formal learning.
>
> Just two scenarios, which the OECD labels Re-Schooling, assume that state schooling can be saved and improved for the better. One . . . sees the school being reinvented as a focused learning organization that emphasises learning for the knowledge society . . . The other envisions schools as focal points for broader community relationships and networks, developing pupils' social capital and enabling them to live well and work productively in the knowledge society. This last pair amounts to teaching for the knowledge society and beyond it. (pp. 127–128)

Hargreaves' closing phrase "and beyond it" signals, of course, that the term 'knowledge society' is a transient one and that the education we provide is the foundation for a future which may have a different socio-economic profile. In common with aspects of the OECD Re-schooling scenarios, Smyth, Dow, Hattam, Reid, and Shacklock (2000) advocate socially critical schools and describe five features of the socially critical educator: courageous educational vision, teacher scholarship-debate, dialogue with the community, student voice on both micro- and macro-level issues, and curriculum debate for critical literacies.

At this point, one must put alongside any political profiling something of the grounded nature of teaching in contexts of, and for, change. Hoban (2002) concludes that a hallmark of teaching as a profession is that it cannot be 'mastered'. He utilises Shulman's observation of any profession's 'inherent and inescapable uncertainty', quoting Shulman:

> Teachers can teach in the same manner to three classes in a row and experience different consequences each time. Professions (like teaching) deal with that part of the universe where design and chance collide. One cannot resolve that uncertainty by writing new rules. The way forward is to make that collision, that unpredictability in our fields, itself an object of individual and collective investigation. (p. 174)

Here, we witness teaching's inherent uncertainty, the fact that structures or rules will not resolve such uncertainty, and recognition that teachers themselves must navigate such uncertainty. As change remains a fact of life and certainty is elusive, so we are ever in Neurath's boat.

To summarise, teachers' professional journeys are spent working in powerful and changeable international climates, with what 'can't be mastered', with the clash of 'design and chance', and with the professional question of how to 'develop' in such circumstances. This is not to invoke a pessimistic analysis – these factors are the reality of teaching. Despite the huge 'performance' pressures of recent decades, international researchers are suggesting that new change is offering optimism:

> Amid all this despair and danger though, there remains great hope and some reason for optimism about a future of learning that is tied in its vision to an empowering, imaginative and inclusive vision for teaching as well. The educational standards movement is showing visible signs of overreaching itself as people start to complain about teacher shortages in schools, and the loss of creativity and inspiration in classrooms. . . . Governments almost everywhere are beginning to speak more positively about teachers and teaching – bestowing honour and respect where blame and contempt had prevailed in the recent past. (Hargreaves & Goodson, 2003, p. viii)

JUST WHAT IS THE 'PROFESSION'?

It cannot be assumed that the notion of profession has a universal meaning or that it can be usefully applied to teachers as a group. For example, the literature offers, inter alia, analyses of teacher as: worker (Reid, 1998); leader-learner (Lingard, Hayes, Mills, & Christie, 2003); ethical (Campbell, 2003); entrepreneur (Sachs, 2003); technician (Hextall & Mahony, 1998); activist professional (Sachs, 2003); reflective practitioner (Killen, 2006; Schon, 1987); critical practitioner (Blackmore, 2002); socially critical (Smyth et al., 2000); and continuing learner-worker (Groundwater-Smith, Brennan, McFadden, & Mitchell, 2001). Such a collage of constructions provides insight to the kinds of tensions and contradictions extant in the field and illustrates the challenge for the 'profession' in determining either its identity or its professional development.

One thread that emerges in the literature on the profession addresses the significance of creativity. Florida (2003) categorises occupational classes, and places teachers in the 'super-creative core' – a rare accolade given the media-generated negativity towards education in many jurisdictions. (Surely, amongst teachers, D&T teachers are necessarily of the most creative.) But it is teachers' creativity that has suffered in recent policy climates:

> Knowledge-driven organizations depend on effective brainpower – on understanding, reflection, ingenuity and creativity. But standardised reforms have taken away teachers' time to think; and their imposed, prescriptive requirements have replaced creativity with compliance. An over-examined professional life is producing an unexamined, unreflective one. (Hargreaves 2003, pp. 82–83)

It is interesting to temper Florida's analysis with that of Goodson (2003), who documents the effects of policy shifts on teachers and their creative potential:

> … it is the most creative and resourceful of our teachers who are the most disenchanted with new prescriptions and guidelines … It is instructive to view any profession or workforce not as a monolithic entity but as made up of a number of segments. (p. 85)

Goodson distinguishes three such segments:

- An elite or vanguard (10-20%) who are creative, motivated, and

  > ... help define, articulate and extend the 'mission of teaching' generally, and of a school in particular. Their commitment to change and reform is a basic prerequisite for successful implementation: their disenchantment and disengagement leave change and reform as a hollow rhetoric. This is not least because of their mentorship and leadership of the mainstream group of teachers (p. 85)

- a mainstream 'backbone' (60-70%) – 'honest, hardworking professionals' whose interplay of leadership and mentorship with the elite is vital in motivating and defining the workforce; and
- a borderline group (10-20%) who are minimally involved in what is 'just a job' and who border on the competence level.

On a key political note for anyone considering how their profession might 'develop', Goodson points out that 'reforms' (media campaigns come to mind, too) are targeted at the borderline group but that, for them, little changes: "However, and paradoxically, the world is transformed for the elite and the backbone. By attacking small substandard groups, which all professions contain, many of the reforms have encountered a colossal downside by demotivating the vanguard and the backbone" (p. 85).

From what has been said it can be seen that, in order to validate the idea of a teaching profession, it is necessary to know the climates in which it must exist, the collage of constructions that may be applied to it, and the profile of the people who will constitute its membership.

ON BEING PROFESSIONAL ...

Sachs (2003) identifies "the complexity and the lack of agreement regarding definitions of professionalism" and argues that we move "beyond what Friedson (1994) calls a commonsense idea of professionalism, that is one developed passively and which is not elaborated, systematised or refined self-consciously but rather grows out of everyday social use" (p. 28). She compares 'old' and 'new' professionalism and demonstrates that the comparison is not straightforward. Studies of the former tend to debate the case of teaching as a profession, whereas the latter build on such debates as well as assuming "a changed analytical perspective" (p. 7). As she recognises, and as illustrated above, matters of power distribution, ideology, flexibility, change, and choice all interplay:

> ... the concept of teacher professionalism is not static, nor is it a concept whose meaning is understood or shared by the whole profession. If anything, the term 'teacher professionalism' is a site of struggle over meaning among its various constituencies. Its meaning changes in response to external pressures, public debates and developments in the scholarly field. (p. 17)

Sachs also describes two types of professionalism – managerial and democratic – and argues that each positions teachers quite differently. As Codd (2005) puts it:

> There is now a dominant managerialist culture within our educational institutions that competes with the traditional democratic culture. This dominant culture is more concerned with what can be recorded, documented and reported about teaching and learning than it is with the educative process itself. Knowledge, experience, understanding, and especially imagination, are recognised only if they can be reduced to something observable, or to some performance outcome that can be specified in advance of the educational moment. Managerialism ... treats teachers as functionaries ... and ... diminishes their autonomy and commitment to the values and principles of education. (p. xv)

Campbell (2003) argues the case for the ethical teacher and ethical professionalism, seeing the teacher as moral agent. She advances ethical culture-building – to be developed by teachers and not externally imposed as ethical codes and 'standards'. She sees the teacher and the profession as agents and communities practising an ethics of self-determination. Goodson (2003) argues that a new professionalism is required for the 21st Century – a 'principled professionalism' comprising seven components, including moral and social purposes, discretionary judgement, collaborative cultures, occupational heteronomy, and continuous learning. By extension, such components inform the shaping of professional development and, in the complexities of D&T's curriculum contestation, it is our own professional group who might offer new futures through 'communities practising an ethics of self-determination'.

Sachs (2003) elaborates a 'transformative professionalism' of benefit to *all* education stakeholders (not just teachers) and typified by certain characteristics: inclusive membership, public ethical code of practice, collaborative and collegial, activist orientation, flexible and progressive, responsive to change, self-regulating, policy-active, enquiry-oriented, and knowledge-building.

To reflect on even this briefest of portrayals of where research is taking our understandings of what it is to be 'professional' is to also recognise that our sense of 'professional development' for ourselves and for D&T education must be seen as problematic.

PROFESSIONAL DEVELOPMENT FOR ALL

Commonly, professional development is thought of as something that is focused on the individual, whether at their personal instigation or at that of the system. However, it can justifiably refer to the development of the profession – very much in the spirit of the authors in the section above. If having an enhanced teaching profession would enhance the quality and standing of education in society, then a logical pathway is to *develop* that profession accordingly. Thus, in developing ourselves as D&T professionals, we can also conceive of the development working for and with our immediate colleagues, our schools, our professional associations, and for education as a whole.

Another sense might be that the development is, itself, *professional.* This is not mere wordplay. Here, the question is begged of the nature and style of the development. Is it determined by the participants for the best educational reasons? Is it externally imposed? Is it merely tokenistic and passively accepted?

Analysis and caution are advisable, as the idea that D&T professional development is 'obvious' or a 'given' for the profession should not go unquestioned. It can also be seen that formalised approaches to professional development may not bring about the intended outcomes. The literature acknowledges that teacher involvement in the *determination* of the nature and quality of professional development is essential to its success: "[T]eachers tend to participate strongly in activities in which they see themselves as having significant control over the direction and the processes of the activity" (Groundwater-Smith et al., 2001, p. 233).

Professional development can also be distinguished in some ways from professional learning, which for Hargreaves (2003) is about gaining knowledge and skills relevant to one's work – for example, learning a new CAD package. Professional development, he says, is about

> … more than a slick, self-managed portfolio of certificates and achievements accumulated as individual credits … It is through professional and personal development that teachers build character, maturity and other virtues in themselves and others, making their schools into moral communities … Professional development … is a personal path towards professional integrity and human growth. (p. 48)

Thus, professional development addresses the needs of the whole educational community through the growth of its teachers. Good professional D&T development will bring with it whole-school benefits. Organisationally, the initiative may come from the individual, a department, a professional association, or a whole school. As Groundwater-Smith et al. (2001) suggest, professional networks are key enablers and D&T associations play a huge role here.

Hargreaves (2003) offers a useful critique of the developments and the options available to the profession. In broad terms, he suggests that we have (been) moved from 'cultures of knowledge and experience' to 'contracts of performance'. In the former, local teacher knowledge, expertise, and judgement were trusted and respected. In the latter, the system seeks contracts in return for performativity: "[l]earning standards for pupils, performance standards for teachers, an increasingly casualised teaching force, more and more contracting out of professional development and other support services ..." (p. 127).

However, Hargreaves (2003) cautions against nostalgia and oversimplification with regard to the past. Much was unsatisfactory with a system that enabled cronyism, biased leadership, inadequate collaboration, incompetence, and patchy professional development. Rather, he offers a useful framework for understanding the positioning of teachers and how their respective professional development models might be shaped. As such, he maps six possible future scenarios for teaching in the knowledge society, and these offer serious food for thought for D&T professional practice. Drawing on Hargreaves, 2003, we find:

- Permissive individualism
  - the old model – teachers teach alone, without scrutiny
  - qualifications licensed teacher autonomy
  - innovations (there were many) arose from voluntary commitments – and did not last
  - further and in-service education a solo affair with little impact on colleagues
- Collaborative cultures
  - collaboration encouraged by effective leadership
  - working with others, sharing resources and planning
  - in-service education directed at teams, often on-site
  - potentially 'comfortable', social, but not effective
  - often without any external reference points to benchmark quality
- Contrived collegiality
  - Hierarchically-generated strategy
  - forced and artificial
  - the what, where, with whom, and when all determined by others – it is constrained by a 'prison of micromanagement'
  - teacher-initiated learning, inquiry, research, and planning inhibited or unwelcome
  - can "produce temporary surges of teamwork under pressure, but it rarely produces sustainable improvement" (p. 130)
- Corrosive individualism
  - a global phenomenon, since the 1990s, of contracts and competitive individualism
  - paradoxical combination of "choice that is supposed to promote diversity, with standards that impose uniformity" (p. 131)
  - teachers and schools compete with each other, don't share resources
  - social division and inequity greatly expanded
  - teachers' time consumed in meeting short-term performance targets
  - a positive legacy – achievement data and other evidence taken seriously by schools and teachers
- Professional learning communities

  - contracts and cultures combine
  - involve collaborative work that improves teaching and learning, uses evidence and data for improvement and for tackling whole-school problems
  - data used to inform curriculum action rather than embarrass/blame teachers/schools
  - do not flourish in standardised systems
  - include strong professional learning communities, teacher networks, and action research groups – they utilise cultural and social capital
  - are not about 'quick fix' but develop sustainably over time
  - utilise professional development that is continuous, shared, job-embedded, and closely connected to teaching and learning
- Performance training sects
  - pedagogy and learning are the key focus and these are largely applied to high profile areas such as literacy and mathematics
  - ambitious performance targets across a whole system and seeking large gains with rapid success
  - priority targeting of low achievement to narrow the performance gap
  - prescribed pedagogical programmes for teachers to follow
  - intensive training of teachers to levels of high competence
  - strong peer coaching, trainer and, consultant support
  - alignment of improvements in teaching and learning with evaluation and testing.

I'm sure most readers can recall a range of experiences falling under the banner of 'professional development' and can apply this framework reflectively – and critically. For the purposes of this chapter, it is the last two of these regimes that particularly help inform our perspectives of D&T teacher professional development.

PROFESSIONAL DEVELOPMENT FORMS AND CULTURES

When it comes to the provision of professional development, there are many possibilities and combinations available. For example: oneself, a D&T department, a professional association or union, a school, an employer, an independent consultant, or a university. (See Chapter 48 for a model of graduate study as professional development.) However, decisions about the kind of professional development and the provider(s) warrant careful consideration. Well-known are the guru approaches of a 'specialist' addressing a large number of people, perhaps conducting some workshops and then leaving the group inspired for a few days but impotent in meaningfully applying their learning. Research also cautions against the one-stop or 'quick-fix' approach and just-in-time training. These have their places, but not in developing rich professional development.

As is the case with pedagogy in general, appropriate participant engagement matters. If a transmissive approach imbues passivity and has no follow-through and a transactional approach is tokenistic, then it is the genuinely transformative that will bring richer and lasting outcomes. For this to happen, teachers are better served when they are participants in not just a professional development 'activity' but also in the activity's very design, planning, and follow-up. This way, professional development is culture- and curriculum-building rather than something that is 'done' by one party to another. Groundwater-Smith et al. (2001) review a range of professional development practices and initiatives and note that all "involve some form of investigation and documentation of teachers' workplaces. [And] … where teachers are enabled to investigate their workplaces their participation in workplace renewal, professional understanding and collegiality are generally increased" (p. 233).

It is, therefore, important that professional learning activities take account of adult learning theory (Knowles, Holton, & Swanson, 2005). Pedagogical (or andragogical) styles should be understood, as should learning style theory when sharing professional practice. Some research has shown the importance of *applying* learning theory meaningfully in practical settings in the professional development of further education teachers (Smith & Dalton, 2005).

No profession today can operate without continuous reference to its growing body of knowledge, and the education profession is no exception. Willinsky (2001) draws from the medical profession in presenting the case for evidence-based education, that is, using research well to enhance both practice and outcomes whether at leadership or classroom level. Indeed, Lingard et al. (2003) discuss the interplay of the concepts of leadership and learning at *all* levels of education and articulate both learned teacher leadership (teachers as leaders, whether in or beyond the classroom) and leadership itself as pedagogy. Certainly, in light of D&T's advances in the fields of design and creativity, where new pedagogies are emerging, D&T critical-reflective practitioners have much to offer as curriculum leaders, and the professional development challenges here need to be well grounded in the research.

This leads to the second sense of viable research and that is action research. As Le Cornu, Peters, and Johnson (1999) caution, there are many perspectives on what constitutes 'critically reflective practice', and that collegiality is important in countering the potential emotional cost of deep critical reflection. They also nominate action research as a key support in achieving quality critical reflection. They note, as with professional development in general, that it is unhelpful if critical reflection is 'pushed' onto teachers. Parsons and Brown (2002) promote the development from reflective practitioner (often in non-systematic ways) to action researcher (applying data-gathering and systematic observation) and then, importantly, utilising both in one's professional development.

A challenge for developing both a culture and the organisation of professional development is "focussing the efforts of a loosely-coupled infrastructure for professional development" (Darling-Hammond, 1997, p. 259). Here, Darling-Hammond refers to the sometimes competing intentions and agendas of a variety of players wishing to influence professional development, and D&T is vulnerable. Thus, the 'professional learning communities' referred to by Hargreaves (2003) are more than teachers alone, but bring together educational, and other, professionals to enquire into practice. For him, "Teachers … need to be committed to and continually engaged in pursuing, upgrading, self-monitoring and reviewing their own professional learning" (p. 16). He nominates:

- face-to-face and virtual professional learning networking;
- adopting continuous professional development portfolios;
- consulting and critically applying the evidence of educational research to their own practice;
- undertaking action research and inquiry of their own; and
- connecting professional learning with levels of reward in teacher pay.

For professional growth to occur, Hargreaves articulates the respective attributes of professional learning communities (PLC) and performance training sects (PTS) (see Table 1).

While these two are not mutually exclusive, the respective power distribution in them is readily apparent, with one creating teacher dependence and the other autonomous professionals. We can reflect on D&T professional developmentoptions and models and critique how they position us individually and/or collectively. Hargreaves (2003) points out, however, that while the PLC is the goal to work for, the evidence shows that it may be necessary to move from a PTS position to achieve the PLC. But he also highlights the phenomenon of 'professional development apartheid' wherein the PTS approach is used for the majority of schools (those deemed to be 'failing') while PLC becomes the norm for the few 'high-flyer' schools. Instead, he advances the options through what he terms 'complementary growth' wherein development might, in early stages of a particular locality, be predominantly PTS and less so PLC. In time, the proportionate representation reverses and becomes largely a PLC (the desired goal) with some intermixed PTS. Through the transition, a key goal is that both teacher and school are developing professional autonomy – the 'sect' ethos dissipates, or is at least recognised for what it is and used appropriately.

Table 1. Professional learning communities and performance training sects (Hargreaves, 2003, p. 147).

Professional learning communities	*Performance training sects*
Transform knowledge	Transfer knowledge
Shared inquiry	Imposed requirements
Evidence informed	Results driven
Situated certainty	False certainty
Local solutions	Standardized scripts
Joint responsibility	Deference to authority
Continuous learning	Intensive training
Communities of practice	Sects of performance

When Hargreaves (2003) talks of advanced achievement (for both teachers *and* schools) resulting from a higher capacity PLC, he articulates the importance to the PLC of incentive, investment, excitement, and support – all of which are equally valid for student learning. Thus comes the recognition of a rich learning community as being one of shared enterprise and purpose. I appreciate that such talk may seem only marginally relevant to a subject group such as ours but, apart from our own in-house curriculum struggles, D&T increasingly has its role in general education policy initiatives such as literacy, numeracy, civics, sustainability, and values education.

DESIGN AND TECHNOLOGY EDUCATION IN THE GLOBAL PICTURE

In turning to Design and Technology as a component of a defensible curriculum for the 21st Century it is unwise to consider professional development for D&T practitioners without also reminding ourselves of the context in which we operate. As Layton (1994) pointed out, Design and Technology is prone to the pressures of many competing economic, industrial, professional, political, and educational interests. The arbitration of these competing claims remains as haphazard as ever. His words of 1994 seem apposite today: "… the politics of technological literacy – who creates and controls the meanings of the phrase, how the imposition of meaning is attempted – is a central concern of technology education today" (p. 13).

As the knowledge economy and the market economy move on, one might turn to the Organisation for Economic Co-operation and Development [OECD], as agent for these constructs, for some direction (albeit one that is economy-driven), or even anticipation of its place for Design and Technology in the curriculum. But the focus remains on the triumvirate of English, maths, and science and a particular version of problem-solving (OECD, 2003, 2004). Nowhere are to be found design, creativity, innovation, or enterprise – all catchcries of enlightened views of a knowledge society and of many governments seeking to 'stay ahead' in economic terms.

While much of the political pressure on schools has targeted literacy and numeracy (with associated calls for more testing and analysis) it could be that D&T's potential is to take advantage of not being under the microscope and slip under the net of political attention by continuing to establish its place and value in any quality curriculum. However, the fact is that within our field, there remains much curriculum and pedagogical contestation within our field.

It would seem that the determination of what Design and Technology itself is, and what it is to become, is intertwined with two things. First, its determination might best be led by the very professionals engaged with its delivery. Second, that same group may need to develop itself in new ways in order to enact a determination that usefully centres technological and design literacy in the curriculum. Thus, two genera of professional development emerge for the field: first, professional development that develops the D&T profession (their knowledge, vision, and collegiality) and, second, professional development that contributes towards the articulation and public recognition of the educational value of technological literacy for all.

With the research literature in mind, one can see that the old view of professional development for self (alone) remains untenable. Hopefully all D&T professional development will enhance, in some way, the field, its purveyors, student learning outcomes, and education for the common good. Such an approach builds on the research literature presented. In turn, this begs the question of how D&T professional development can operate and what its inhibitors might be. If the research is well applied then the profession can anticipate ways of 'developing' that serve and sustain them individually, collectively, and (in the strongest ethical sense [Keirl, 2006]) serve and sustain democratic education in the 21st century.

However, the realities of the issues peculiar to D&T cannot be ignored. I would suggest that these often remain unexamined and unchallenged, yet they can serve to inhibit professional development (in the sense of professional progress). It would thus be imprudent to ignore some of the contradictions and tensions that exist within the field.

DESIGN AND TECHNOLOGY'S CONTRADICTIONS AND TENSIONS

Whether individually or collectively, D&T teachers periodically find themselves offering some sort of defence of the subject/field/area. It even faces demise at times under curriculum pressure. At such times personal values are exposed and the search for collective professional expression of shared values emerges. The nuanced values differences within the field need not be seen as a negative but, rather, celebrated as the field's strength. In taking this position, I suggest that the range of contradictions and tensions that exist signal the dynamic and the holistic rather than the static and the narrow. They are, at once, the challenges of the odyssey and the materials for rebuilding our boat. To maintain rigour and vibrancy we can, first, acknowledge the contradictions and tensions and, second, engage with them rather than ignore them.

Thus, professional learning, development, *and* discourse opportunities present themselves and contradictions might become dialectics offering new directions while tensions may dissolve. From such opportunities, the D&T profession – individually and collectively – develops, and it is possible to better defend the field in public and global arenas – something advocated by many of the authors cited. What then are the kinds of contradictions and tensions that inform and shape (or distort) our field? In offering the following candidates, I suggest they can be seen as topics for focussed professional development in their own right, or it may be that they are in some way informed vicariously by professional development focus in a quite different area.

The technicist-creativity balance. The term 'skilling' is common to D&T and often appears as part of political rhetoric (e.g., skills shortage in industry) when what is meant by skill and just what skills are really needed is rarely clarified. Further, the position is held that D&T is fundamentally about skills (often trade) – in direct contrast to one advocating creative and innovative design practices.

Knowledge forms, currency and durability. Which knowledge(s), of what type (propositional, procedural, etc.), and the extent to which they maintain currency is a continuous concern for D&T. Balancing traditional and emergent technologies is an issue here.

Pedagogical repertoires. The field of D&T education can call upon the spectrum of teaching and learning theory and practice. Depending on the blend of design, skilling, standards, creativity, enterprise, etc. being taught, D&T teachers have high demands to give quality performance. Developing their repertoire is very much a matter of professional development.

Curriculum identity – specialist (subject) and general (cross-curricular) relationships. Claims for subject status are one means of survival but they can position D&T in comparison to other 'subjects'. Subjects increasingly have cross-curricular obligations to such foci as values, sustainability, literacy, numeracy, and civics. There are also questions about subject identity loss in times of emergent curriculum frameworks articulating 'learning areas', 'rich tasks', and 'essential learnings'. The OECD initiatives are already influencing national and local curricula in ways that have ramifications for the D&T practitioner.

Teacher-system expectations. Over recent decades there have been increased tensions between teachers and employing bodies over matters of assessment, accountability, and resourcing. While, as noted, this general climate may be changing, teacher efficacy it terms of curriculum voice is an issue of some contention for the D&T practitioner.

Interpretations of technological literacy. Perhaps the single all-embracing area of contradiction is that of technological literacy and its multiple potentials (see, for example, Dakers, 2006). As the philosophy and sociology of the field develop, so the debates refine but it cannot be said that the intellectual tensions are dissipating yet. There is much for D&T practitioners to engage with that, for many, was not part of their original training (e.g., de Vries, 2005).

Such candidates not only offer rich potential for professional learning, development, and discourse – they amount to the internal politics of D&T curriculum. Thus, there is the case for professional development about professional development itself – about the resolution or, at least life with – such contradictions and tensions. Professional development about the politics of curriculum, managing change, curriculum leadership, and ethical self-determination for the individual and the profession (Campbell, 2003) may be valid as a starting point for clarifying aims and directions. Ultimately, of course, the call is on teachers' professional judgement – again, professional both in nature and in purpose. And good judgement depends on continuous professional learning.

SPHERES OF PRACTICE

For some D&T teachers, professional development amounts to a 'kit and can-do' approach. It's about learning how to use the latest equipment – which is valid if this ultimately fits with the school's (not just the teacher's) vision and aims for the students. Thus, educational rationale underpins any form of professional development. But if such learning is copying demonstrations from a sales representative and repeating the demonstrations with students, the educational aim or the pedagogy might be questioned.

The spheres in which professional development can be practiced are numerous but generally depend on the participants and the focus. As has been shown, the participants may be a sole practitioner, a team, a whole school, or more. Focus may be on an individual's practice, on a D&T department's curriculum, perhaps a whole-school technological literacy policy. Ultimately, the literature is well worth consulting and something along the lines of the professional learning community outlined above would seem the most ethically defensible approach.

D&T teachers are not islands. Nor should they be members of just one team or group – this is how professional life can be rich. Membership of several professional development communities (some as sub-sets of others) is viable. Individual or collective interaction with agencies in the broader community supports the two-way flow of learning as well as enhancing the profession beyond the school boundaries. If good professional practice is to be sustained and sustainable, then a passive or 'recipe' approach to development will be unrewarding. Perhaps a good benchmark is to ask: If teaching is the learning profession are we, ourselves, learning?

It is naïve to advance D&T professional development without an eye on the broader global and societal ideologies. They *do* impact on education (recently, in powerful ways) and have arguably de-professionalised the profession. But optimism, ethical purpose, and vision all have their place in shaping professional development too. The political dimensions of teachers' professional lives are the same political dimensions that shape other lives in the community, and by engaging the community in the very tensions and contradictions outlined above, the community learns about the complexity and challenges facing the profession. Many authors (viz. Goodson, 2003; Hargreaves, 2003; Hoban, 2002; Sachs, 2003; Smyth et al., 2000) articulate the case for taking our professional development into the public arena – warts and all – for the very reason of strengthening and giving rigour to our professional practice. Here lies a serious challenge for the development of the D&T profession.

CONCLUSION

> No one teacher knows enough to cope or improve by themself. It is vital that teachers engage in action, inquiry and problem-solving together in collegial teams or professional learning communities. Through such teams, teachers can undertake joint curriculum development, respond effectively and creatively to external reform imperatives, engage in collaborative action research, and analyse pupil achievement data together in ways that benefit their pupils' learning. (Hargreaves, 2003, p. 17)

D&T teachers have their own voyages to navigate but there is a collective journey too. The D&T field, perhaps more than any other in the curriculum, has much work to do: in continuing to overcome the ignorance that surrounds it (people in and beyond schools resorting to tired stereotypes about skills, ICT, vocationalism, etc.); in resolving as many of its internal contradictions and tensions as possible; and in determining the professionally defensible course through all the stakeholder interests.

It is a fact of life that D&T practitioners are in Neurath's boat and have to work on their deconstruction and reconstruction as they go. Through iterative research and investigation, theory is used to build new theory. But, to return to Florida (2003), creativity, design, and innovation are Design and Technology's bread and butter and some sense of Sach's (2003) activism is needed in framing professional development while its practitioners voyage. The tools of scholarship, enquiry, critical reflection, and action research can all be put to use to achieve this but not without, as the literature suggests, ethical professional courage and vision. In times of heavy work pressure on teachers the first reaction for many is to say how nice professional development would be if only there was time. I have used the term 'professional breathing' (Keirl, 2007) to suggest that time must be made – to discuss, to establish networks and communities, to argue our cases. A good breath is the minimum we need to get started. In time, an ongoing culture of professional development would be the D&T profession's oxygen, the wind in its sails.

REFERENCES

Apple, M. W. (2001). *Educating the "right" way: Markets, standards, God and inequality*. New York: Routledge Falmer.

Blackmore, J. (2002). Speaking out for critical professionalism and education. *Curriculum Perspectives: The Journal of the Australian Curriculum Studies Association, 22*(3), 60–62.

Campbell, E. (2003). *The ethical teacher*. Maidenhead, UK: Open University Press.

Codd, J. (2005). Introduction: Is there a 'third way' for education policy? In J. Codd & K. Sullivan (Eds.), *Education policy directions in Aotearoa New Zealand* (pp. vii–x). Southbank, Victoria: Thomson Dunmore.

Dakers, J. R. (Ed.). (2006). *Defining technological literacy: Towards an epistemological framework*. Basingstoke, UK: Palgrave Macmillan.

Darling-Hammond, L. (1997). *The right to learn: A blueprint for creating schools that work*. San Francisco: Jossey-Bass.

de Vries, M. J. (2005). *Teaching about technology: An introduction to the technology of philosophy for non-philosophers*. Dordrecht, The Netherlands: Kluwer.

Florida, R. (2003). *The rise of the creative class. And how it's transforming work, leisure, community and everyday life*. North Melbourne, Victoria: Pluto.

Galbraith, J. K. (2004). *The economics of innocent fraud*. London: Penguin.

Goodson, I. (2003). *Professional knowledge, professional lives: Studies in educational change*. Maidenhead, UK: Open University Press.

Groundwater-Smith, S., Brennan, M., McFadden, M., & Mitchell, J. (2001). *Secondary schooling in a changing world*. Melbourne, Victoria: Thomson.

Hargreaves, A. (2003). *Teaching in the knowledge society: Education in the age of insecurity*. Maidenhead, UK: Open University Press.

Hargreaves, A., & Goodson, I. (2003). Series editors' preface. In B. Lingard, D. Hayes, M. Mills, & P. Christie (Eds.), *Leading learning. Making hope practical in schools* (p. viii). Maidenhead, UK: Open University Press.

Hextall, I., & Mahony, P. (1998). Effective teachers for effective schools. In R. Slee, G. Weiner, & S. Tomlinson (Eds.), *School effectiveness for whom? Challenges to the effectiveness and school improvement movements* (pp. 128–143). London: Falmer.

Hoban, G. F. (2002). *Teacher learning for educational change: A systems thinking approach*. Maidenhead, UK: Open University Press.

Honderich, T. (1995). *The Oxford companion to philosophy*. Oxford, UK: Oxford University Press.

Keirl, S. (2006). Ethical technological literacy as democratic curriculum keystone. In J. R. Dakers (Ed.), *Defining technological literacy: Towards an epistemological framework* (pp. 81–103). Basingstoke, UK: Palgrave Macmillan.

Keirl, S. (2007). The politics of technology curriculum. In D. Barlex (Ed.), *Design and technology for the next generation: A collection of provocative pieces, written by experts in their field, to stimulate reflection and curriculum innovation* (pp. 60–73). Shropshire, UK: CliffeCo.

Killen, R. (2006). *Effective teaching strategies: Lessons from research and practice*. South Melbourne, Victoria: Thomson Social Science Press.

Knowles, M. S., Holton, E. F., & Swanson, R. A. (2005). *The adult learner*. Burlington, MA: Butterworth-Heinemann/Elsevier.

Layton, D. (Ed.). (1994). *Innovations in science and technology education* (Vol. V). Paris: UNESCO.

Le Cornu, R., Peters, J., & Johnson, B. (1999, July). *Shaping learning communities: The challenge of critical reflection*. Paper presented at the 9th International Conference of the International Study Association on Teachers and Teaching, Dublin.

Lingard, B., Hayes, D., Mills, M., & Christie, P. (2003). *Leading learning: Making hope practical in schools*. Maidenhead, UK: Open University Press.

Organisation for Economic Co-operation and Development [OECD]. (2003). *The PISA 2003 assessment framework: Mathematics, reading, science and problem solving knowledge and skills*. Paris: Author.

OECD. (2004). *Problem solving for tomorrow's world – First measures of cross-curricular competencies from PISA 2003*. Paris: Author.

Parsons, R., & Brown, K. (2002). *Teacher as reflective practitioner and action researcher*. Belmont, CA: Wadsworth.

Postman, N. (2000). *Building a bridge to the eighteenth century: How the past can improve our future*. Carlton North, Victoria: Scribe.

Reid, A. (1998). Regulating the education market: The effects on public education workers. In A. Reid (Ed.), *Going public: Education policy and public education in Australia* (pp. 57–66). Canberra, ACT: ACSA.

Sachs, J. (2003). *The activist teaching profession*. Buckingham, UK: Open University Press.

Schon, D. A. (1987). *Educating the reflective practitioner: Toward a design for teaching and learning in the profession.* San Francisco: Jossey-Bass.
Smith, P. J., & Dalton, J. (2005). *Learning styles: Relevance and good practice in VET.* Adelaide, SA: NCVER.
Smyth, J., Dow, A., Hattam, R., Reid, A., & Shacklock, G. (2000). *Teachers' work in a globalizing economy.* London: Falmer.
Willinsky, J. (2001). Extending the prospects of evidence-based education. *Insight, 1*(1), 23–41.

Steve Keirl
School of Education
University of South Australia
Adelaide, Australia

FRANK BANKS

46. COMPETENCY BASED TRAINING IN THE UK

INTRODUCTION

Some years ago, as part of an Open University course team, we set an activity where student-teachers were asked to think back and note the characteristics of a teacher that they remembered from school. To see if the activity might work, some of us tried the activity ourselves. When comparing notes, most of us said that we became teachers because of the influence of the one particular teacher that we had recalled, either because they were particularly good – 'I would like to be like him' – or because they were not so good – 'I went in to teaching because I was convinced I could do a much better job than her!'

In my case, I remembered a good teacher. Mr Field taught me physics for six years and he also encouraged me and my classmates to develop an interest in electronics and a wide range of technical issues. He was the reason that I was motivated to study technical subjects at university and to go on to become a technology teacher. I thought he was a good teacher and I wanted to become a good teacher like him. Mr Field explained difficult concepts using metaphors and anecdotes, and was a good performer in the classroom – lively, well organised, and fun. However, he had his favourites – favourite classes and favourite pupils – and always seemed to ensure he only taught the high-attaining pupils. He was passionate about physics but very dismissive of other school subjects and put considerable pressure on pupils to choose his subject when option choices were made.

Was Mr Field really such a good teacher? Certainly not all my fellow pupils thought so. What criteria should we use to come to such a view? How should we judge teaching competence in general, especially the teaching of those who are about to enter the profession? In observing lessons in India recently it became very clear to me that what constitutes 'good' teaching there is very different to what would be considered 'good' teaching in England. But even in one cultural context, is it possible to agree on aspects of knowledge, skills, and understandings about pedagogy that should be exhibited by *all* competent teachers? Moreover, how can more experienced teachers demonstrate enhanced expertise?

The United Kingdom (UK) includes the countries of England, Northern Ireland, Scotland, and Wales, along with a number of other minor states within the British Isles such as the Channel Islands and the Isle of Man. Since 1999, responsibility for education has been increasingly devolved to these separate countries as national assemblies and – in the case of Scotland, a new parliament independent of London – were created. In practice, Northern Ireland and Scotland have been autonomous in the design and structure of their education systems for many years, and Wales is increasingly moving away from the curriculum model prevalent in England to allow the distinctive culture of Wales to be emphasised in the *Curriculum Cymreig.*

Competence, or standards-based teacher education and training has been a feature of initial teacher assessment in all the nation states of the UK for at least 15 years, and now underpins career-long professional development. This chapter analyses the assumptions implicit in such

A.T. Jones and M.J. de Vries (eds.), International Handbook of Research and Development in Technology Education, 555–567.
© 2009 Sense Publishers. All rights reserved.

assessment regimes, explores the criticisms, and considers the advantages of the different approaches to teacher standards in the different nations of the UK. The Design and Technology Association (DATA) uniquely among subject associations constructed subject-specific competences at the initial teacher education level to which all teacher education institutions in England subscribe. This gives a framework for student-teachers to analyse how they are progressing to meet what is considered fundamental professional knowledge for all technology teachers and what is specific to certain specialist teachers within the rather broad curriculum area that constitutes technology.

BACKGROUND TO TEACHING STANDARDS

In 1992, teaching standards (then called teacher competences) were first introduced in England and Wales as assessment requirements for entry to teaching (Department for Education [DfE], 1992). A year later Scotland had produced teacher competences for initial teacher training and induction (Scottish Office, 1993) and Northern Ireland had gone one stage further and designated teacher competences for initial induction and continuing professional development (Department for Education Northern Ireland [DENI], 1993), and indeed have stayed with the term 'competences'. This was not a move initiated by teacher educators. In fact there was resistance from many to the notion that teaching could be distilled into such a set of 'behaviours'. There were social and political reasons for the move, but also powerful educational motivations, as Moon and Shelton Mayes (1995) identified:

> From an *educational* perspective the broadening of the concept of assessment in schools has inevitably created a momentum for change in higher education. Advocates for assessment as a formative and diagnostic process, as well as a summation of students' achievement, have been influential in this process (Brown, 1989). From a *social*, as well as educational perspective there has been increasing disillusion, if not distrust, with normative assessment processes. While class lists and 'concurs' may be appropriate when assessment serves rigorous selection purposes, they sit less well in the world of vocational and professional qualifications. The emphasis is not how one student compares with another, but rather on whether they can do the job, or meet the requirement required within an occupational grouping (Jessup, 1990). Finally from a *political perspective* there has, in a number of countries, been an increased concern about the quality of entrants to the teaching profession, and therefore, the means by which students are deemed to achieve qualified status. (p. 232)

Many teacher educators were interested in constructing courses more in tune with addressing the identified needs of the student-teachers and developing criterion-referenced rather than normative assessment systems (Whitty & Willmott, 1991). Thus, in contrast to much of the rhetoric at the time, the pressure for change was not wholly from politicians concerned with the quality of teachers and wishing to impose a system to control entry to the profession, although this undoubtedly provided the impetus towards common nation-wide assessment criteria.

In the 1970s, competence-based teacher assessment was pioneered in the USA where it was developed from an analysis of the functional tasks. This type of approach to teacher assessment met the political requirements of government, concerned with raising the quality of schools by, in its terms, raising the quality of teachers. British educators, however, were circumspect in their analysis of the possible advantages and disadvantages that might arise from introducing a competence-based approach to assessment. Critics of competences suggested they promoted a narrow model of teacher as instructor, drawing on behavioural and technicist models of teaching, and warned that competences would lead to a tick-box approach to teacher assessment. They cited the USA example of 121 separate behaviours checked off by an independent classroom observer and fed into a computer to produce a competence level (Gitlin & Smyth, 1989, reported in Whitty & Willmott, 1991). Much later, Ted Wragg (2001) would make the same criticism of the qualifying standards for primary teachers:

> The initial teacher training of teachers, especially for the primary phase, has now reached the stage where it is totally driven by the fulfilment of hundreds of objectives. It is the ultimate horror, preparing people brilliantly for a life of ticking boxes. ... In order to become certificated as a teacher you have to complete 851 objectives. ... Of course teaching can be broken down into weeny bits. But can it be stitched back together again? More pertinently, can teachers be trained and certificated objective by objective? ... This view of teaching as a purely mechanical activity is criminal. It distils out the intelligence and the imagination, drains human warmth, reducing it to the robotic – fine for Daleks, useless for teachers. (p. 40)

Similarly, Hyland (1993) warned that competence-based programmes of teacher training would "lead to teachers ill-equipped for further professional development, uncritical of educational change and largely ignorant of the wider cultural, social and political context in which the role of teachers need to be located" (p. 130). Hyland takes the view that competences, rather than being a minimum requirement, come to define the total teacher-education curriculum, squeezing out wider issues not covered by the nation-wide framework.

Others were more optimistic, lending cautious support to the development of broadly constructed competences that would incorporate knowledge, understanding and attitudes, as well as skills. For example, Hextall, Lawn, Menter, Sigwick, and Walker (1991, quoted in Whitby 1995) argued that even the quality of reflection could be formulated in a series of competences that could be monitored:

> – A reflective practitioner can articulate and defend his/her own purpose as teacher and relate this to other professional opinion.
> – A reflective teacher treats teaching as an experimental process, recognising the necessity of turning reflection into action, choosing between alternatives, and critically evaluating the process. (p. 203)

Other benefits were suggested, such as:

- clearer goals for professional development;
- opportunities for systematic training as one useful strategy for teacher education;
- greater confidence in employers and society as to what teachers can do;
- helping to structure professional judgements on teacher performance; and
- providing a common language for teachers about pedagogy (see Furlong, 1995; Moon & Shelton Mayes, 1995; Whitty & Willmott, 1991).

The benefits of 'providing a common language for teachers about pedagogy' through the introduction of a set of competences was very important. This was because in all UK countries except Scotland, hand in hand with the introduction of teacher competences at pre-service level was a move to reduce the time spent in Higher Education and increase the time spent in schools. In England, the use of the word 'trainee' rather than 'student' emphasises this work-based training view of teacher preparation, which has increased in its stridency in recent years. With the appointment of school-based mentors (experienced teachers in the school) the need for a clear and common understanding of the facets of good teaching that should be considered when assessing a novice teacher – made explicit and shared between student (trainee), school mentor, and the University tutor – was welcomed.

In Scotland, Higher Education resisted this move to the use of 'mentors' in school-based training, although a set of agreed competency areas was introduced. A major stumbling block to whole-scale policy reform was the reluctance of school teachers to become mentors with the extra work load required. In Northern Ireland, the competency model was introduced with refinements. Instead of a list of aspects of teaching all of which needs to be demonstrated *before* entry to the profession, the competences are also used during the initial induction years; some competences are stressed in pre-service, others during the first year of employment.

CURRENT USE OF COMPETENCES

The teaching standards or competences used in different parts of the UK have some things in common. Most of the standards serve a gate keeping function, setting a quality threshold that teachers must meet in order to progress. An example is qualified teacher status, without which teachers cannot work in maintained schools. However, a closer look at the standards in different parts of the UK reveal that although they serve the same purpose there are fundamental differences that imply a difference in the *model of teaching* that underpins the standards. For example, in contrast to Wragg's (2001) criticisms mentioned above, about reducing teaching merely to the completion of 'tick-sheets' in England, the working group set up to produce the first teacher competences in Northern Ireland reports:

> However, as our thinking progressed we began to see that a simple list of competences, no matter how comprehensive and well structured, could not convey the totality of what we wanted to say about the professional competence of a teacher. ... one of the criticisms often made of competence-based approaches to education and training, and to professional education and training in particular, is that the atomisation of professional, knowledge, judgement and skill into discrete competences inevitably fails to capture the essence of professional competence. ... the definition of competences which we wish to offer is therefore rather more complex [and we have set out] underlying qualities of the teacher which enable him or her to pull the individual competences together and apply them in a professional context. (DENI, 1993, p. 7)

Yet even this more enlightened view of competences resulted in 92 separate competence statements.

In 2008, over 15 years later, the use of standards or competence-based approaches to teacher assessment is firmly embedded in the UK system of teacher assessment. Moreover, it has been extended in all UK nation states from an initial gate keeping function at entry into the profession to a series of professional milestones that are intended to guide a teacher's personal development, and includes standards for school leaders such as middle leaders and headteachers. However, due to the different models of teaching underlying the milestones in each country, the competences attached to them are different as is illustrated by the following brief overview of the standards or competences for career-long classroom teachers only, across the different nation states.

England

A range of new teacher professional standards was introduced in September 2007 (Training and Development Agency for Schools [TDA], 2007). Earlier in the decade, research was commissioned by the Department for Education and Employment into the characteristics of 'teacher effectiveness', with Hay McBer (2000) looking at 35 micro-behaviours of teachers in England. Work such as this contributed directly to the standards for teachers on the upper pay scale, and significantly to a framework where the model of teaching is to promote what research has concluded as 'effective' teaching strategies. Consequently, the framework of professional standards for classroom teachers defines the characteristics of teachers at each career stage as shown in Table 1. The professional standards are intended to provide clarity of the expectations at each career stage and arranged as a framework in three interrelated sections covering:

Table 1. Teacher professional levels – England.

1. Qualified Teacher Status (QTS) (**Q**)
2. Teachers on the main pay scale (Core) (**C**)
3. Teachers on the upper pay scale (Post Threshold Teachers) (**P**)
4. Excellent Teachers (**E**)
5. Advanced Skills Teachers (ASTs) (**A**)

- professional attributes;
- professional knowledge and understanding; and
- professional skills.

Some standards from 'professional knowledge and understanding' are exemplified in Table 2.

Interestingly, the number of standards is not the same across the different professional milestones. There are 33 standards to be achieved for Qualified Teacher Status, but to pass probation and remain on the main scale of teachers' pay, a teacher must exhibit 41 standards in the same broad areas but at an increased level of sophistication. To cross the professional 'threshold' and be awarded higher pay, a teacher must gather evidence for a further 10 standards, and an excellent teacher will be rewarded for demonstrating they have achieved a further 15 standards. Finally, an Advanced Skills Teacher (AST) demonstrates their competence through achieving three further standards and it is intended that such expertise is shared with other schools. This 'cradle to grave' approach to professional development through a series of specific milestones is unique to England, but other nations of the United Kingdom also have some differentiation in standards.

Table 2. Examples from the standards for professional knowledge and understanding – England (TDA, 2007, p. 9).

Those recommended for the award of QTS (**Q**) should: Q10 Have a knowledge and understanding of a range of teaching, learning and behaviour.
All teachers (**C**) should: C10 Have a good, up-to-date working knowledge and understanding of a range of teaching, learning and behaviour management strategies and know how to use and adapt them, including how to personalise learning to provide opportunities for all learners to achieve their potential.
Post Threshold Teachers (**P**) should: P2 Have an extensive knowledge and understanding of how to use and adapt a range of teaching, learning and behaviour management strategies, including how to personalise learning to provide opportunities for all learners to achieve their potential.
Excellent Teachers (**E**) should: E3 Have a critical understanding of the most effective teaching, learning and behaviour management strategies, and including how to select and use approaches that personalise learning to provide opportunities for all learners to achieve their potential.

Northern Ireland

The General Teaching Council for Northen Ireland (GTCNI) has published new professional competences for teachers in the province. The title of the document 'Teaching: The reflective profession' (GTCNI, 2008) indicates the different emphasis to England. One of the principles which underpins Northern Ireland's concept of competence is thus the centrality of reflective practice:

> It is the Council's view that competence is developed through reflection on practice and through dialogue with colleagues. This process of reflective practice has recently been highlighted by the Education and Training Inspectorate (ETI), in its publication, *The Reflective Teacher,* in which it is noted that the teacher's role is to:
> - reflect on the provisions made for pupils;
> - assess the quality of provision within their own classroom; and
> - decide how the quality of teaching is contributing to improvement throughout the school. (p. 12)

There are four professional milestones in Northen Ireland (see Table 3). Northen Ireland teachers prefer to use the term 'competence' rather than 'standard', and 27 teacher competences have been organised within three broad areas:

- professional values and practice;
- professional knowledge and understanding; and
- professional skills and application -
  - planning and leading;
  - teaching and learning; and
  - assessment.

Some competences on 'teaching and learning' (competence 20) are shown in Table 4. The use of competences across both Initial Teacher Education and Induction has been a feature of the competence framework in Northern Ireland since the early 1990s, but the extension of the framework to embrace Continuing Professional Development is very new and not yet fully embedded in the provision offered by CPD providers such as the Education and Library Boards.

Table 3. Teacher professional levels – Northern Ireland.

1. Initial Teacher Education
2. Induction (into the profession)
3. Early Professional Development
4. Continuing Professional Development, Collaborative and School Improvement.

Table 4. Examples from the competences for teaching and learning – Northern Ireland (GTCNI, 2008, p. 36).

Initial Teacher Education Use a range of teaching strategies appropriate to the age, ability, interests and experiences of pupils.
Induction Use a range of teaching approaches effectively to promote high levels of learning and achievement.
Early Professional Development Teaching reflects increasingly the judicious and effective use of a range of teaching strategies and is informed by self-evaluation.
Continuing Professional Development, Collaborative and School Improvement. Support others in the use of a range of teaching strategies that enable learning to take place.

Scotland

The General Teaching Council for Scotland (GTCS) was the first teacher professional registration in the UK and one of the first Teaching Councils in the world, having been established in 1963. There are a number of milestones available, and *Standards for Initial Teacher Education* (SITE) specifies what is expected of a student teacher at the end of Initial Teacher Education in order to receive provisional registration with the GTCS. The other standards for classroom teachers are the *Standard for Full Registration* (SFR) and, for those with at least five years teaching experience, the *Standard for Chartered Teacher*. Like in England, there is no obligation to progress from 'core' standards that all teachers should exhibit throughout their career, such as the *Standard for Full Registration* (see below). The Chartered Teacher programme requires work at Masters level and currently such programmes are offered by Scottish Higher Education Institutions. It is voluntary, but gaining Chartered Teacher Status leads to a pay increase just as passing the 'threshold' standards does in England.

In 2006, the GTCS specified similar areas of teaching in the standards for both Initial Teacher Education and Full Recognition, namely:

- professional knowledge and understanding;
- professional skills and abilities; and
- professional values and personal commitment (GTCS, 2006).

This is mirrored in the standards for the Charter Teacher, which contain the key components:

- professional values and personal commitments;
- professional knowledge and understanding;
- professional and personal attributes; and
- professional action.

The *Standard for Chartered Teacher* is related to, and builds on, the SFR. Like the SFR, it consists of *professional knowledge and understanding*; it draws on a wide range of *professional attributes and abilities*; and issues in appropriate forms of *professional action*. Table 5 shows an example of this progression for 'professional skills and abilities'. The SFR, achievable after one year's experience as a teacher, may be said to represent a consolidation of the Standard for Initial Teacher Education. However, the *Standard for Chartered Teacher*, marking, as it does, substantial experience as a teacher, represents a significant enhancement of professional skill and understanding beyond the SFR (Scottish Executive, 2002, p. 4).

The establishment of the GTCS was due to concern that, following the Second World War, unqualified teachers were teaching in schools in Scotland. Consequently, from its inception, the GTCS been keen to ensure that teachers are properly educated for the phase in which they teach. Recently, some flexibility has been introduced through two schemes: *Professional Recognition* and *Professional Registration*. The former is recognition of particular expertise demonstrated by a teacher and confirmed through the annual professional review and development process. Through the recommendation of a headteacher to the GTCS, a teacher can be awarded a certificate for five years that recognises particular expertise in an area of the curriculum. *Professional Recognition* is more substantive. Through this system, a teacher can demonstrate that they have enough credit in a subject area outside their current teaching discipline to be registered as a teacher of that extra subject. For example, a secondary teacher of English might have university credit in Geography and so register to teach that subject too. Similarly, a primary teacher could show they have enough credit to be able to teach lower secondary level and so by teaching across phases, improve the transition arrangements. The model of teaching prevalent in Scotland therefore is one that recognises the importance of pedagogy, and certainly expects more experienced teachers to become reflective and researchers in their classrooms through the Chartered Teacher Programme, but is firmly based in the prime importance of subject knowledge.

Table 5. Examples from the standards for teaching and learning – Scotland (GTCS, 2006, p. 10).

Standard for Initial Teacher Education 2.1.1 Plan coherent, progressive teaching programmes which match their pupils' needs and abilities, and justify what they teach.
Standard for Full Registration 2.1.1 Registered teachers are able to plan coherent, progressive and stimulating teaching programmes which match their pupils' needs and abilities, and they can justify what they teach.
The Chartered Teacher *should demonstrate through his or her work a critical understanding of:* current approaches to teaching and learning; current research on teaching and learning; and the school curriculum, its knowledge base, and how the curriculum relates to life in the community, including citizenship and the world of work.

Wales

The National Assembly for Wales (NAW) was created in 1999 and since then education has become increasingly distinct to that of England. Even before then, standards for entering the profession were not the same as those in England. There is another key difference. In March 2001, the Department for Education and Employment (DfEE) in England published 'Learning and teaching: A strategy for professional development', which stated:

> We will pilot Early Professional Development for teachers in their second and third years of teaching in a sample of schools, building on the experience of the induction year. … This will encourage a firm base for career-long professional development. (p. 9)

The pilot work was highly successful and demonstrated that nurturing teachers in the early years of their career increased the likelihood of teachers remaining in the profession. Early Professional Development has not become a significant stage in England. However, since 2004 teachers in their second and third years do receive earmarked funding in Wales:

> Years two and three of teaching in Wales are recognised as a period of Early Professional Development (EPD). EPD enables new teachers to extend their professional knowledge and understanding, and develop their teaching skills and professional characteristics through evaluating and reflecting on their own practice and that of others. The support provided during ITT and Induction equips new teachers to develop professionally and to establish the practices of reflection and self-evaluation. (NAW, 2004, p. 5)

Unlike in Northern Ireland, however, there are currently no standards published for EPD. Rather, it is a period of guaranteed broad professional development due to National Assembly funding.

The professional milestones for classroom teachers in Wales are shown in Table 6. The current standards for Qualified Teacher Status (QTS) were revised in 2006 (NAW, 2006) and The Chartered Teacher Programme is newer than that of Scotland as the standards were only produced in 2007 (General Teaching Council for Wales [GTCW], 2007). The Chartered Teacher Programme in Wales is thus in its pilot phase at the time of writing, with different models of delivery being trialled. However, there is a significant policy difference between the Chartered Teacher Scheme in Scotland to that of Wales, where award of Chartered Teacher status does not lead to a statutory increase in pay (although it may make the teacher more attractive to a prospective employer).

Table 6. Teacher professional levels – Wales.

Qualified Teacher Status (QTS)
Induction
Early Professional Development (EPD)
Chartered Teacher

Compared with other UK nations, the links between the standards for different professional milestones in Wales are not so explicit. For example, there are 42 standards for QTS (and some of these are broken down further by teaching phases such as primary and secondary) in the areas of:
- professional values and practice;
- knowledge and understanding; and
- teaching
  - planning, expectations, and targets;
  - monitoring and assessment; and
  - teaching and class management

There are 15 End of Induction standards in the different areas of
- professional characteristics;
- knowledge and understanding;
- planning, teaching and learning, and class management; and
- monitoring, assessment, recording, and reporting.

For Chartered Teacher there are14 standards split across:
- professional knowledge and understanding;
- professional skills in practice; and
- professional commitment.

However, despite categorising teaching under different headings, some commonalities can be seen as shown in Table 7.

The philosophical underpinning to the standards seems less clear in Wales than in the other nation states and that may be due to its initial closeness to England, combined with an increasing desire to be different. The Welsh QTS standards seem to be similar to those in England, promoting specific teacher behaviours, and yet the rhetoric for further development in the profession is rather one of increasing reflection on practice.

Table 7. Examples from the standards for teaching and learning – Wales (GTCW, 2007, p. 5).

QTS S3.1.3 They select and prepare resources, and plan for their safe and effective organisation, taking account of pupils' interests and their language and cultural backgrounds, with the help of support staff where appropriate.
End of Induction 3.2 demonstrate increasing proficiency in selecting and using a broad range of teaching and learning strategies and available resources, which they evaluate critically in terms of pupils' learning.
Chartered Teacher 2.1.1 Advise and assist colleagues to plan for effective teaching by setting goals, and designing and implementing up-to-date learning programmes, and by selecting appropriate learning resources.

COMPETENCES FOR TECHNOLOGY TEACHER EDUCATION

All teachers of technology are judged against the various standards or competences pertinent to the different professional milestones they have reached for their particular nation in the UK. That is the over-arching professional competence assessment tool. But unlike most other teachers, teachers of technology are working in a relatively new curriculum area that has also developed rather differently in each of the nation states. Consequently, the professional development strategies for technology teacher education reflect those curriculum variations. (See Banks, 2006, for an extensive description of technology teacher education across the UK.)

The breadth of school design and technology (D&T) is such that many technology teachers who were traditionally craft teachers, domestic science teachers or science teachers do not have the subject knowledge background, even in their specialist area, to teach the entire current school technology curriculum with confidence. In 1994, DATA surveyed the training needs of D&T teachers in England and Wales (DATA, 1994). That survey assumed that the most important issue to address was the particular subject elements in the national curriculum for D&T in which teachers were either untrained, or considered themselves to be inadequately trained to teach. There was a given assumption that it was only the 'rectification' of their subject knowledge 'deficit' that was required. New pedagogical strategies still have a relatively low priority for D&T teachers. This is very different from teachers of history or science, where the common expectation is that in-service courses will lead to novel pedagogy such as techniques for promoting discussion or group solutions to problems.

This emphasis on skill or subject knowledge rectification has had a profound impact on pre-service technology teacher education. Many teacher preparation courses, even those of just one-year duration which follow a degree in a technological subject, might spend half or more of the time introducing the student teachers to the subject content of school technology. This 'school knowledge' would include, for example, expertise in general instruction in the particular CAD/CAM packages to be found in schools as well as more focused subject development to 'plug the gaps' identified on a personal subject audit.

In 1995, the subject association for Design and Technology in England and Wales set out for England some subject knowledge guidelines for common areas of designing and graphical communication, and for specialist areas of resistant materials, systems & control, food technology, and textiles technology (DATA, 1995). This was revised in 2003 (DATA, 2003). A secondary school teacher entering the profession is expected to teach *two* of these material areas adequately to pupils of 14 years of age (Key Stage 3) and *one* specialist area to students of 18 years of age (Key Stage 4 and beyond). Figure 1 sets out the current subject knowledge expectations.

All newly qualified teachers should be able to exhibit the 'Core Competences' such as:

C.1 understand and use a range of strategies and approaches to identify and clarify design problems. (e.g. analyse an existing product with the view of improving it; observe a product being used with the view of improving its performance for the user; use a modern or smart material as a starting point for a new product – e.g. an EL panel);

C.3 research a wide range of information sources appropriate to the problem. Analyse and select this information to inform their designing (e.g. research an art and design movement, conduct 'taste testing', seek the views of an end user/expert);

C.7 use techniques, processes and procedures appropriate for each of the specialist fields to manufacture products and systems;

C.8 when planning and conducting design and technological activities give due regard to Health and Safety of their pupils, themselves and other adults and be aware of current, relevant Health and Safety responsibilities, legislation and liability;

C.10 an awareness of industrial methods and approaches to design, manufacture and quality control on production (e.g. batch, mass and continuous flow production);
C.11 have an understanding of citizenship and values and how these relate to Design and Technology (e.g. sustainable design such as recycling materials, designing products with dual functions, GM foods, alternative/renewable energy sources, fair trading).
C.12 critically use ICT to enhance teaching and learning in Design and Technology using modern technologies (e.g. Interactive whiteboards, scanners, digital cameras, and presentation applications);
C.15 nurture a creative teaching and learning environment where pupils feel confident and safe to experiment, explore and take risks. (DATA, 2003, pp. 6–7)

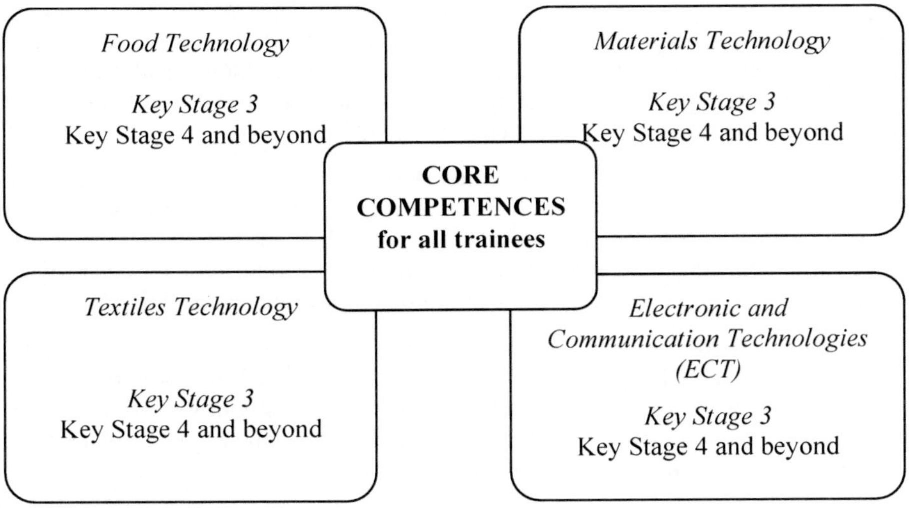

Figure 1. Minimum subject competence areas for secondary school technology teachers in England (DATA, 2003).

Although following these DATA guidelines is not a formal requirement, virtually all of the teacher education institutions have endorsed this subject knowledge framework from the subject association, although they rarely take on student teachers in all four material areas. Secure subject knowledge is a standard in all professional development frameworks in the UK. That rightly continues to be important, but historically manual subjects were largely based on teaching practical skills and adopted pedagogy not so different to the 'master-apprentice' model of a medieval guild. The pupil would be given a job – to produce a pipe-rack (now occasionally seen as a CD rack!), coat-hook, pincushion, or sponge cake – and shown all the skills and techniques necessary to produce a (more or less) satisfactory outcome.

The safe manufacture of quality products is still important but it is only one – restricted – aspect of technological capability. It is clear that this traditional model of teaching is now inadequate as it teaches making skills without any underlying understanding or development of other skills. In the UK, the development of professional standards has encouraged the use of a wider range of teaching strategies which better match the broader aims that the subject has developed.

COMPETENCES – DEPROFESSIONALISM OR REPROFESSIONALISM?

Storey and Hutchinson (2001) argue that the idea of 'the teaching profession' has served as shorthand for a number of values that mark out a liberal profession. These include

> access to and application of a systematic and codified body of specialist knowledge; an exclusive jurisdiction; regulation and certification by a professional body; and adherence to a set of professional norms which transcend pecuniary gain and which stem from a code of ethics supported by public trust. (p. 41)

They further argue that this concept implies "the exercise of degrees of self-regulation both in terms of individual autonomy and judgement and also in terms of professional issues" (ibid.).

This old version of professionalism, they contend, is in conflict with the Westminster government's intent on driving through changes relating to consistency, higher standards, and accountability in the education system. This has been interpreted as an attempt to de-skill and de-professionalise teachers (Harris, 1997; Hyland, 1993). However, Storey and Hutchinson (2001) reach a different conclusion about the re-shaping of what it means to be a professional in teaching in the 21st Century. They see opportunities for all teachers to:

> take the content, processes and structures of the modernisation scenario and seek to shape them in specific educational contexts in order to reprofessionalise and to write their own version of an educational vision through dialogue and action. Wresting a greater share of the contested terrain of education through participation may, ironically, be one tangible and possible outcome beyond the current era's accent on quality control. (p. 51)

They present an optimistic future where there is greater autonomy for technology teachers and schools who have performed competently against a framework of standards that are now more subtle and wide-ranging descriptions of what effective teaching might be. I like to think that Mr Field, who influenced me to become a teacher, would himself have greatly benefited from the greater participation in structured professional development.

ACKNOWLEDGEMENT

I would like to acknowledge the contribution of Professor Ann Shelton Mayes, University of Northampton, to the original Open University course on the use of standards for personal professional development that has informed this chapter.

REFERENCES

Banks, F. (2006). Technology teacher education in the UK. In P. J. Williams (Ed.), *International technology teacher education. 55th council on technology teacher education yearbook* (pp. 215–239). Peoria, USA: CTTE, McGraw Hill-Glencoe.

Brown, S. (1989). *Assessment: A changing practice*. Edinburgh, UK: Scottish Academic Press.

Department for Education [DfE]. (1992). *Initial teacher training (secondary phase)*. Circular 9/92 (35/92 in Wales). London: Author.

Department for Education and Employment [DfEE]. (2001). *Learning and teaching: A strategy for professional development.* Circular 71/2001. London: Author.

Department for Education Northern Ireland [DENI]. (1993). *Review of initial teacher training (ITT) in Northern Ireland: Report of the Northern Ireland Working Group on Competences (Working Group I)*. Bangor, UK: Author.

Design and Technology Association [DATA]. (1994). *A survey of the qualifications and training needs of D&T teachers in secondary schools. Research paper No. 2.* Wellesbourne, UK: Author.

DATA. (1995). *Minimum competences for trainees to teach design and technology in secondary schools. Research paper No. 4*. Wellesbourne, UK: Author.

DATA. (2003). *Minimum competences for trainees to teach design and technology in secondary schools. Research Paper No. 4*. Wellesbourne, UK: Author.

Furlong, J. (1995). The limits of competence: A cautionary note on Circular 9/92. In T. Kerry & A. Shelton Mayes (Eds.), *Issues in mentoring* (pp. 225–231). London: Routledge.

General Teaching Council for Northern Ireland [GTCNI]. (2008). *Teaching: The reflective profession*. Belfast, County Antrim: Author.

General Teaching Council for Scotland [GTCS]. (2006). *Standard for initial teacher education and standard for full registration cross-referencing*. Edinburgh, UK: Author.

General Teaching Council for Wales [GTCW]. (2007). *The chartered teacher standards*. Cardiff, South Glamorgan: Author.

Gitlin, A., & Smyth, J. (1998). *Teaching and learning through critical reflective practice*. Lewis, UK: Falmer.

Harris, A. (1997). The deprofessionalisation and deskilling of teachers. In K. Watson, C. Modgil, & S. Modgil (Eds.), *Educational dilemmas: Debate and diversity 1 – Teachers, teacher education and training* (pp. 57–65). London: Cassell.

Hay McBer. (2000). *Research into teacher effectiveness*. Norwich, UK: Her Majesty's Stationery Office.

Hextall, I., Lawn, M., Menter, I., Sigwick, S., & Walker, S. (1991). *Imaginative projects: Arguments for a new teacher education*. London: Goldsmith's College.

Hyland, T. (1993). Professional development and competence-based education. *Educational Studies, 19*(1), 123–132.

Moon, B., & Shelton Mayes, A. (1995). Integrating values into the assessment of teachers in initial education and training. In T. Kerry & A. Shelton Mayes (Eds.), *Issues in mentoring* (pp. 232–242). London: Routledge.

National Assembly for Wales [NAW]. (2004). *Induction for newly qualified teachers in Wales. Circular 19/04*. Cardiff, South Glamorgan: Author.

NAW. (2006). *Qualified teachers standards. Legislation 49/06*. Cardiff, South Glamorgan: Author.

Scottish Executive. (2002). *Standard for chartered teacher*. Edinburgh, UK: Stationery Office.

Scottish Office. (1993). *Guidelines for teacher training courses*. Edinburgh, UK: Author.

Storey, A., & Hutchinson, S. (2001). The meaning of teacher professionalism in a quality control era. In F. Banks & A. Shelton Mayes (Eds.), *Early professional development for teachers* (pp. 41–53). London: David Fulton.

Training and Development Agency for Schools (TDA). (2007). *Professional standards for teachers in England*. London: Author.

Whitty, G. (1995). Quality control in teacher education. In T. Kerry & A. Shelton Mayes (Eds.), *Issues in mentoring* (pp. 200–207). London: Routledge.

Whitty, G., & Willmott, E. (1991). Competence-based teacher education: Approaches and issues. *Cambridge Journal of Education, 21*(3), 309–318.

Wragg, T. (2001, 23 February). Welcome to the Dalek factory. *Times Educational Supplement*, p. 40.

Frank Banks
Centre for Research in Education and Educational Technology
The Open University
UK

JACQUES GINESTIÉ

47. TRAINING TECHNOLOGY TEACHERS IN EUROPE

Putting the Bologna Process into Action

INTRODUCTION

The establishment of the European Union has had profound impacts, including the harmonisation of national legislation and citizens' rights. This construction, from the European Economic Community to the union of the citizens, is widely based on the harmonisation of the educational policies of the Member States; it privileges the mobility of the students, teachers, and executives of education. However, school and university systems differ extensively from country to country. For example, the passage from elementary school to secondary education does not occur at the same age in different countries. The same diversity can be seen at the university level, including, for example, the duration of studies necessary to complete a doctorate degree (from six to nine years). The Bologna process has been established to bring about European harmonisation in university courses in each of the Member States. The history of this process, and its current status two years before its official completion, are presented in the first part of this chapter.

The training of teachers in Europe is particularly diverse, with some trained in faculties of education, others in vocational schools, and others learning their role through on-the-job training. To this diversity in terms of place and duration of courses is added differences in curricula as well as the different kinds of institutions which organise this formation. The difference between courses for teachers training in a faculty of education and those offered to new teachers not having received any formal teacher education before finding themselves in front of a class characterises the value attributed to the job of teacher and to the professional competencies required to practice. Compared to specific educational training in the first case, the role of 'teacher' is too often one of 'transmitter of knowledge' in the second case, where only a solid academic knowledge of the field is required: it is necessary (and it is considered sufficient) to be good in mathematics to be a good teacher of mathematics. Such general considerations significantly impact views of the role of teachers, for which there is a real disparity between the various European countries.

This chapter presents examples of teacher education from four European countries – Germany, Spain, France, and Greece – that have needed to undergo significant reforms as a result of the Bologna process. The examples demonstrate the role and place occupied by schools in each country and, consequently, the role and place granted to teachers and their training. From Germany, where each Lander (or region) has its own Minister of Education, and Spain, where each region is more autonomous; to the delocalised structure of Greece and the French academic institutes constrained within an old Jacobin tradition, we will examine how technology teachers' education and training are organised, and through this which values and social images of the job are promoted. How the Bologna process is implemented will then be considered.

A.T. Jones and M.J. de Vries (eds.), International Handbook of Research and Development in Technology Education, 569–580.
© 2009 Sense Publishers. All rights reserved.

THE BOLOGNA PROCESS

The Bologna process, begun in May 1998 at a conference at the Sorbonne in Paris, targets the construction of a European teaching system in higher education by 2010 (Eurydice, 2007). It is the result of a series of political decisions taken during a number of conferences (Paris, Bologna, Prague, Berlin, Bergen, and London). Agreement upon the structure of higher education in Europe features in the Sorbonne declaration, signed by four countries (Germany, France, Italy, and the United Kingdom). It constituted the framework of the Bologna declaration and aimed to:

- heighten international recognition of various areas of study and of diplomas by aiming for a common framework regarding qualifications and studies;
- facilitate the integration of students and teachers into the European labour market; and
- create a common system for diplomas in degree, masters, and doctorate study programmes.

Heavily influenced by the Sorbonne declaration, the Bologna declaration (June 1999) regarding higher education in Europe was ratified by 29 European countries (the fifteen member states; the eleven candidates for entry into the EU; and the three associated countries, Iceland, Norway, and Switzerland). The intention was to establish a common structure for modernising and reforming higher education teaching in Europe. This plan is known as the Bologna process. The aim is to unify the different teaching methods in further education in different countries by establishing similar requirements for diplomas and giving them value by means of organising them in two main cycles or levels (degree and masters). Obtaining these diplomas is based on a credits system, known as ECTS credits. Through their validity and the organisation into identical cycles, ECTS credits aim to make life easier for students, teachers, and researchers. The key is to promote a European dimension within higher education, notably through cooperation between establishments; numerous EU programmes support this development and the Bologna process. Clearly, the ultimate aim of the Bologna process is to heighten the international competitiveness of European higher education and to ensure that such a system occupies a key position across the world (Eurydice, 2007).

At the Prague Conference (May 2001), four countries (Cyprus, Croatia, Liechtenstein, and Turkey) joined the initial group. The goal of this conference was to lay down a marker and define the key elements of the process. Hence, a group was created to oversee the process development. Presided over by the member state occupying the presidency of the European Union, it is made up of representatives from each of the countries involved and of the European Commission, and numerous institutions and organisations in an observatory capacity. The statement published following the conference emphasises three points: The importance of learning and training throughout the course of a lifetime, the implications of this for higher education establishments and students, and the need to make European further education more attractive.

Seven new states (Albania, Andorra, Bosnia-Herzegovina, Macedonia, Russia, Saint-Siege, and Serbia-Montenegro) joined the group during the Berlin Conference (September 2003). Apart from this increase in the number of members, this conference marked an important development in the process objectives over the two years that followed:

- promotion of quality at institutional, national, and European levels;
- implementation of the dual cycle study structure (degree and masters);
- mutual recognition of diplomas and study periods from 2005 onwards;
- creation of a global framework for qualifications in further education;
- inclusion of the third cycle (doctorate); and
- development of stronger links between further education teaching and research.

The development group presented the requested report at the Bergen conference (May 2005). The number of countries involved had now reached 40, with the arrival of Armenia, Azerbaijan,

Georgia, Moldova, and Ukraine, and common reference points and directives for further teaching were adopted in order to ensure the quality of the organisation. Priorities were widened to strengthen the social dimension and to eradicate any functional obstacles. It was also at this conference that national certification structures, the awarding and recognition of diplomas, and the putting in place of flexible courses of study in higher education, with the notable inclusion of skills acquisition, were decided.

The London conference (May 2007) considered all the policies integrated into the legislation of all the countries involved. The European Commission insisted upon the necessity to concentrate on implicating the reforms required by the Bologna process. Assessment of the progress of the process revealed that more than half the outlined objectives had been met. The aim was to construct an everyday reality from these reforms for students and teachers.

TEACHER TRAINING WITH REGARD TO THE BOLOGNA PROCESS

At the end of the 1990s, the situation regarding teacher training varied significantly from one European country to another. On the one hand were two kinds of training institutions – education faculties and professional training schools – and on the other hand, two main processes for training – simultaneous training comprising teacher training and acquisition of academic knowledge, and a successive training scheme that concentrated firstly upon acquiring the academic know-how before learning the job. The crossover of these two criteria – type of institution and training structure – resulted in a wide variety of possible training programmes. In addition, technology education itself takes very different forms from one country to the next and so it is logical that training structures vary from one country to another, even from one region to another; indeed, teacher training institutions aim to train teachers to teach their discipline such as it is organised in the school system in which they operate. This multi-faceted diversity complicates the putting in place of the Bologna process for teacher training and goes some way to explain the heterogeneous nature of situations in Europe at the present time (Chatoney & Ginestié, 2006). In order to examine this, the methods for training technology teachers in several countries will be examined.

Teacher Training in Germany

In Germany, teacher training is based upon skills recognised by the state and is sanctioned by two national certificates; a teacher is recognised as such upon successful completion of the second certificate. The first certificate is for university skills, studies are organised in accordance with rules designated by each region, which controls teaching content and the required levels of demand. The second certificate is instituted by the federal educational authorities. Teachers are trained according to the type of school in which they are going to teach. Studies vary in duration – seven semesters for teachers in the 'Grundschule', 'Hauptschule', and 'Realschule'; and ten semesters for teachers in high schools and professional schools. Generally, prospective teachers are required to study two scholarly disciplines. The study in each discipline is always divided into three components: an academic part, a didactic part, and a practical part. Studies also include specialised teaching modules centred upon pedagogy, psychology, sociology, philosophy, political sciences, as well as work experience in schools (Theuerkauf & Graube, 2006).

The training of technology teachers for general teaching is based principally upon the understanding of technical aids and procedures. This study is correlated to the economical, social, and ecological environment. It must render prospective teachers capable of tackling, based on scientific principles, technical subjects and problems within the framework of the aforementioned environments.

The scientific component. Specialised scientific teaching addresses the concepts of materials, energy, information, and socio-technical systems through the principles of transformation, production, distribution, and recycling. The scientific approach of the methods and the technical processes of conception, production, construction, and diffusion constitute an essential element (Theuerkauf & Mager, 2004). Thus, we can also find teachings based upon the mathematical and physical elements of technical systems, as well as others relying on human sciences (philosophy and history of technique) or on economical sciences (management procedures, etc.). Other, more specialised forms deal with conception, construction, fabrication, processes for the transformation and transmission of energy, processes for the treatment and transfer of information, or the impact of technological developments upon societies (the planning of urban transport links and ecological impact, for example).

The first university stage is largely dedicated to studying the theory of systems in view of use of matter, energy and information. The second stage deals with specialisation and heightened understanding through projects or case studies that examine interactions between humans and natural and man-made environments.

The didactic component. Didactic study deals with the choice of teaching materials and learning processes through planning, putting into practice, and evaluating. This study relates to the basis of teaching processes, the foundations of technology education, and the link between teaching and learning. Modules on the development of curricula, their usage and evaluation are also taught (Graube, Dyrenfurth, & Theuerkauf, 2003). These modules are supplemented by work experience in schools.

The interdisciplinary component. A significant part of the teaching is linked to interdisciplinary dimensions. Such modules serve to aid the putting into practice of interdisciplinary scientific projects, notably through curricular application. In technology education, references to the economy or the organisation of work are given priority. Links with the natural sciences for aspects related to the impact of technologies on the environment, are also given credence. Additionally, information technology plays a key role in all teachers' training and thus constitutes a strong interdisciplinary focal point (Theuerkauf & Mager, 2004).

In general terms, teacher training in Germany is a simultaneous training system that alternates between a mastery of knowledge of the subject area, and competence in the teaching of this knowledge. However, this training is widely centred on the role of the teacher in the classroom rather than in the school, interacting with the other teachers; the latter is scaled down to a few interdisciplinary matters and the integration of the pupils into the working world of the future. This general organisation is evolving to integrate the guidelines for degree/master study programmes heralded by the Bologna process. Apart from the final certification, which is the responsibility of the federal authorities, the whole of the teacher training programme is governed entirely by each university, which organises training curricula in accordance with directives and orientations halted at the level of each region.

Teacher Training in Spain

Teacher training in Spain has undergone numerous important changes over the course of the last few years since the passing of the fundamental law of 23 December 2002 on the quality of education. Highlighting the role of teachers in the evolution of the education system, this law (and the consequent applicatory decrees) insists upon the duty of regional education authorities to promote teacher training in accordance with a dynamic of constant improvement. This evolution of the original training scheme is an adaptation of the university system for diplomas for higher education teaching in Europe.

That said, initial teacher training remains successive: the degree phase is dedicated to disciplinary (subject area) knowledge acquisition; professional training for teaching is developed afterwards. This succession is organised in the conventional framework between the university establishments conducting the training, and education authorities (Ginestié, 2006c).

The responsibility of the university within the first phase is that of a traditional university in which modules geared towards training teachers can be put in place. Generally speaking, these units are optional and have little impact upon the students' course of study. Passing them rarely figures in the awarding of qualifications and they carry little weight. In the vast majority of cases, content is based upon a general knowledge of the education system and the role of the teacher within this system; mostly, this content leans towards a social approach to education and elements of psycho-pedagogy. Rather than helping students who wish to become teachers begin to construct a professional identity, these teaching units place the emphasis more on the stakes of contemporary education and a grouping of stereotypical procedures that are supposed to equip students to tackle teaching situations. The courses are largely cut off from problems linked both to the teaching of disciplinary (subject) knowledge and to the use of teaching materials in a class.

After obtaining the degree, students wishing to become teachers join a public training centre affiliated to a university in order to undergo a post-degree, two year training course in didactics. This post-graduate training is obligatory for teachers in general secondary education or in professional teaching (of the trades, for example). It hinges, on the one hand, upon the scientific foundations of the subject area chosen by the candidate (a continuation of what was studied at degree level) and, on the other hand, upon the didactic basis of the said discipline or subject area. It comprises university teaching and practical work experience in scholarly institutions.

Hence, training leading to becoming a teacher in Spain involves two phases: a period of theoretical training directly linked to the candidate's area of further study, and a period of hands-on professional teacher training. It is this second period that allows for the evaluation of all teachers who are training, whether in the public or private sector. At the end of this phase of the training students obtain the qualification of specialised teacher of a given subject area. This post-graduate system also leads to a Masters degree in professional teaching, in accordance with the Bologna process.

Training within the structure of this post-graduate cycle is a scheme which, as we have seen, organises university teaching and work experience in schools. This general organisation is developed within the guidelines of conventions between the university and the education authority concerned. These conventions form a partnership between the training centre and schools, which welcome the students. Each student is henceforth under the tutelage of an experienced teacher who will act as supervisor throughout the two year training period (Ginestié, 2007b).

In terms of the training content, the grouping together of, on the one hand, knowledge and methods linked to sciences and, on the other hand, specific didactic approaches, constitute one of the key elements. This approach favours the development of a specifically educative focus, notably through aspects of coordination, orientation, and pupil tutoring. Managing pupil diversity, specifically from the point of equality between girls and boys, is an essential part of this training (Ginestié, 2006b). At the same time, mastering information technologies, communication, and a foreign language, and developing an aptitude in educative research and innovation form part of the key skills for all teachers and must be acquired through this training.

More globally, teacher training in Spain is marked by the principles of regional autonomy, on the one hand, and by a longstanding university tradition of academic training on the other. This continuous training forms part of the development of professional schools, which relied on the National Education Ministry for university training. The five year structure should not pose any particular problem in integrating this training according to the Bologna process guidelines.

Teacher Training in France

Teachers in France are hired by means of a national concourse granting access to public services; ministerial rulings define the objectives and the levels demanded for these concourses. For practically all teachers working in France, such training demands the same minimal requirements fixed by the Ministry for Education. The Instituts Universitaires de Formation des Maîtres (IUFM) are responsible for this training. Formed in 1991, they are becoming integrated into universities (bearing some resemblance to the faculty model) (Ginestié, 2006d).

Like Spain, teacher training in France is successive. First, students wishing to become teachers choose a subject to study at university up to degree level. Once they have attained this, students transfer into the IUFM for a two year training period. The first year is dedicated to preparing to sit the teaching concourse; the second year is one of professional (practical, hands on) training. The IUFMs are charged with ensuring the successful passage of students from specific subject study at university through to becoming a teacher (Ginestié, Brandt-Pomares, Chatoney, Martinez, & Oliveri, 2006).

First year at the IUFM. The first year of training at the IUFM is geared towards, on the one hand, helping students to prepare for the concourse to teach in the public domain, and on the other hand, starting to consider the professional aspects needed to be a teacher. As a consequence of the highly selective nature of the concourse and the nature of the exams which form it, academic preparation for the concourse forms the major focus, with comparatively less emphasis given to the knowledge and competence necessary to become a teacher – pedagogy, epistemology, teaching-learning processes, etc. The first year of training at the IUFM is geared towards, on the one hand, helping students to prepare for the concourse to teach in the public domain, and on the other hand, starting to consider the professional aspects needed to be a teacher. As a consequence of the highly selective nature of the concourse and the nature of the exams which form it, academic preparation for the concourse forms the major focus, with comparatively less emphasis given to the knowledge and competence necessary to become a teacher – pedagogy, epistemology, teaching-learning processes, etc. In addition, students come from a diverse range of subject expertise and this requires a harmonisation of knowledge based upon an epistemology of technology education – learning about the existence of technical objects and the study of social contexts in which such objects exist. The absence of predefined university courses, or of degree programmes geared towards this specific professional domain, clearly highlights the limitations of the sequential teacher training system.

Organisation of teaching adopts the main traits of the concourse exams. The first, written, part of the concourse – admissibility – takes the form of two written exams; the first is to check knowledge of the discipline, the second is an analysis of a complex technical system (a drinks machine or photocopier, for example) with regard to its technical, economical, and society aspects. The second, oral part – admission – is made up of three exams; the first is another test to examine the student's knowledge of the discipline, the second is a presentation of an industrial project undertaken by the student during the training year, and the third is a practical exam that aims to gauge the student's skills in using tools, machines, and instruments generally used in technology teaching. Preparing for these last two exams allows the professional element to be clearly introduced into the first year curriculum.

Teaching of specific disciplines brings together the basic elements of electronics, mechanics, business management, and industrial projects. These are all applied to technology teaching. These three disciplinary areas are integrated into a global technological approach by teaching how to conduct a project for producing technical objects or systems. This teaching hinges upon the links between conception, production, and utilisation of an object or system, and also upon operation, function, structure, form, and materials. It is based on students undertaking a project to produce a technically, economically, and commercially evaluated product complimented by a study file detailing the procedures followed, successfully completed phases, methods and tools used, problems to be solved, solutions found, and the method(s) of achieving them. This file must also serve as a study for a project which could be put in place in middle school technology teaching; its pedagogical use in this context is an essential element to be taken into account. The file and the prototype of the object produced back up the concourse oral exam.

Second year at the IUFM. The aim of second year of training is to bring together the three areas of expertise required for teaching: mastery of disciplinary knowledge (i.e. subject areas), competent teaching of this knowledge, and getting to grips with one's role as a participant in the education system (Ginestié, 2007a). Of course, the first level does not concentrate on specific modules, with students having (supposedly) worked hard on the subject area content earlier in their courses.

Second year training is of an alternating nature, switching between lessons at the IUFM and work experience in schools. At the end of the year, the IUFM validates the training and it is on this basis that the Minister for National Education decides whether or not to employ the prospective teacher in the public sector. Passing this training period hinges on three elements: professional skills acquired and used during work experience, teaching modules, and a professional dissertation.

Work experience in academic establishments. Future technology teachers undertake three distinct types of work experience during their second year of teacher training: hands on teaching, supervised teaching and company relations. Hands on (responsibility) training relates to the alternating part of the course. For the duration of the new academic year, students teach for six to eight hours in two or three classes, with varying tasks such as any full time teacher in the establishment would have. Thus, the student is required to plan, prepare, and evaluate his/her teaching in collaboration with the school's technology teachers, and to find a viable place for it in the teaching team's and the establishment's curriculum. The student also has to organise pupils' assessment, participate in staff meetings, and contribute to the individualised follow-up of the pupils.

During this period, the student is aided by a teacher/advisor, chosen from the experienced teachers in the school, who helps him/her with aspects of pedagogy, didactics, and relationships relevant to teaching. This supervised phase involves constructing sequences, observation, and analysis of practice; advice on communication and of a technical nature; development of resources; etc. The teacher/supervisor is in contact with the IUFM through the intermediary of a trainer, who takes on the role of a tutor. In the case of there being no unforeseen difficulties, the relationship between the tutor and the academic advisor are based on an exchange of information regarding the development of practices used by the trainee. Such exchanges are intensified the moment one of the partners identifies a problem. In such cases, the tutor can choose to visit the trainee in his or her school classroom.

The evaluation of this work experience phase, which is one of the three elements evaluated, takes into account the crossover in points of view (of the pedagogical advisor, tutor, head of the school, and a valuator jury member) regarding the practical/professional development of the prospective teacher. If needed, and depending on the student's personal experiences, observational or practical supervision work experience in other classes or schools may be required.

Knowledge of the industrial milieu and associated professional organisations is one of the key components in the building of community links for technology education. This knowledge is distinctive when considered from a teacher's point of view, and even if the intern/trainee has previous professional experience it is often necessary to ask him/her to modify, broaden, or diversify his/her point of view about the role and placing of the company selected as one of the community links. This is the objective of the company relations period.

Teaching modules. Numerous teaching modules are offered to trainees in either a compulsory or optional manner. These different modules form part of a single common evaluation. Teaching of didactics and subject areas deals with didactic aspects and pedagogical setups in the context of specialised subject area knowledge. This allows for the placing of, on the one hand, professional practice in relation to interactions between teacher, pupil, and subject knowledge; and on the other hand, curricular organisation put into practice.

Professional training groups organise the coherence of the different components of teacher training (work experience, modules, and the professional dissertation) and bring a sense of meaning to the teaching experience. The analysis of such practices, undertaken in these groups, leads to the discovery of the different tasks and functions which make up the job of 'teacher', helping students to form their own professional identity. Analysis of these practices is organised under such thematic considerations as evaluation, citizenship, safety, and teamwork.

Interdisciplinary and crossover training groups, together with training components, deal with the role of teacher beyond the transmission of subject area knowledge. It is organised into conferences and workshops aiming to develop the identity of students as actors within the education system in which they are going to work.

The integration of information and communication technologies (ICT) in teaching practice is an important part of training which deals with the everyday use of access to distance learning resources; contemporary means of communication and treatment of information; and current designing, inventing, and creating systems either with the help of computers or with processes for operating complex systems such as measurement chains or automated systems.

The evaluation of teaching modules is based upon content studied during the year. It is done before a jury and involves the presentation of the teaching plan or sequence that was designed, developed, and put into practice by the student in a classroom context. The assessment is based upon the student's ability to make links between teaching theory and practice.

The professional dissertation aims to answer a question asked with regard to practical activity, and to form a response in accordance with a rigorous methodology. The idea is to take a step back and take into account the possibility of forming a rational analysis of a point with a contentious and thus problematic nature. In this sense, the student is able to use his/her own practices, through, on the one hand, looking at research already dealing with the point in question, and on the other hand, a constructed analytical strategy allowing for the formation of a response without the need for emotional spontaneity. The dissertation is presented in public and individually before a jury composed, at the very least, of the dissertation tutor and a member of the validation committee.

The way teachers are recruited to work in the French public service is the main obstacle to students' enrolment on the training course based on the Bologna process. Currently, the continuous training system highlights numerous limitations, apart from the fact that such training cannot be sanctioned by a Masters degree. First, recruitment by concourse, and the placing of this concourse in the fourth year of tertiary study, totally disrupts the organisation of the practical training; it is the concourse exams that determine access to further training, and any attempt to avoid thorough preparation for these exams is pointless. In the same way, validation of the diploma is not compatible with the recruitment concourse exams; the one is based on the level of knowledge and competence acquisition of the students, the other is based on employment opportunities within the state. There is thus confusion between an evaluation of knowledge and skill levels, and a selection process based on the number of available jobs. As a result, students place more importance on securing a job than on being awarded the diploma and, of course, this raises the question of the status of teachers and the recognition of their professionalism.

Second, the consecutive training system is not devised from a professionalisation perspective except for the consideration that only a mastery of subject area knowledge will guarantee teacher competence. It is clear that the job of teacher does not hinge only on mastering academic knowledge, which is organised and used in an entirely different perspective; professionalisation of university courses in the context of the Bologna process implies a redefinition of content taught at university. Thus, what subject knowledge does a technology teacher need? Not necessarily the same knowledge as that which is needed by an engineer or a researcher. The specification of university courses through the identification of the area of study is something which still needs a lot of work in France.

Teacher Training in Greece

In Greece, training for technology teachers is provided in advanced schools for pedagogy and education technology ("Anotati Sholi Pedagogikis ke Tehnologikis Ekpedefsis", or ASPAITE), created in June 2002. In addition to the centre in Athens, five sites (Heraklion, Ioannina, Patras, Thessaloniki, and Volos) serve the whole of the country (Papoutsakis, 2006).

An integral part of the education system for technology teaching involves l'ASPAITE technological universities, which train engineers, electricians, mechanics, electronics experts, etc., and an area of general study which includes pedagogy. The pedagogy department offers pedagogical training programmes geared towards technology teaching, either in a general context, or in its more specialised forms for professional training.

Students obtain a place at l'ASPAITE technology universities through a national entrance exam at the age of eighteen, following secondary education. Study at the l'ASPAITE technology schools lasts for four years and corresponds to cycle A. They offer pedagogical, scientific, and technological components for future engineers, electricians, mechanics, electronics experts, and civil, infrastructural, and buildings engineering. At the end of the four years and on condition that all exams have been passed, students wishing to pursue a career in technology teaching move on to a one year cycle B specialisation in order to qualify as either a technology teacher in general teaching, or as a specialised teacher in a specific technical domain for professional teaching.

Completing cycle A allows students to obtain a diploma equivalent to those awarded by a Technology Education Institute (TEI). Teaching is split into four groups: general teaching (20% to 30% of the overall programme), specific teaching (25% to 40% of the overall programme), specialised teaching (25% to 35% of the overall programme), and a final component grouping together management, economics, law, and human sciences (10% to 20% of the overall programme). The Ministry for National Education controls the training offered in different establishments.

ASPAITE training is successive because up until the end of their studies, students have the possibility of opting for a career as an engineer; teacher training only comes into play in the fifth year. Available study courses have a considerable bearing on economics, and numerous relations between schools and businesses are established. The setting up of a careers office in each centre has played a key role in business development and hence in the national economy. It is a place students can go to in order to get advice on choosing specialisations with greatest career opportunities. This office informs companies of the areas of specialisation developed by the ASPAITE and the qualifications of its students, and also informs students and graduates about job opportunities, graduate skills required, and factors concerning the labour market.

It is clear that in Greece, high knowledge and skill levels relative to a technological domain relate closely to being a teacher. For this reason, the placing of this course of study within the guidelines of the Bologna process poses no particular problems other than that posed by an attempt to bring together teacher training and engineering courses.

DISCUSSION AND CONCLUSION

The four examples above demonstrate that teacher training is aligning itself with the Bologna process, albeit much more slowly than other university programmes because of the unique requirements of teacher training and the corresponding structure of teacher training courses. The question of awarding a Masters degree should probably be quite easy to address. As we have seen, such opportunities are already in place in Germany and the transformation of post-graduate courses in Spain and Greece ought not to pose any real problems; this will probably take a little longer in France because it would be necessary to distinguish between the public sector recruitment process through a concourse and professional (practical) teacher training. However, it is clear that European unity is still far from being achieved. Besides the particularities of training in one country or another, the definitions of professional training allowing one to become a technology teacher poses several problems.

The teaching question highlights the extremely heterogeneous concepts through which professionalism can be characterised. The notion that a teacher's ability depends merely upon mastery of subject area knowledge remains present in each of the four countries described. It is the main focus in successive training schemes like in Spain, France, and Greece, but also present in Germany. Whilst no-one can deny that a teacher has to be knowledgeable about the subject he/she is going to teach, but this condition, whilst being necessary, is not sufficient.

There are also differences regarding the definition of fundamental knowledge from one country to another. Indeed, the traditions of technology education come from distinct backgrounds, and the curricula have been developed in different sociocultural contexts (Ginestié, 2006a). Germany places great emphasis on building and has consequently developed an understanding of industrial production organisations. Social and human aspects are of less importance than processes for organising industrial systems and associated technical procedures. Important differences also exist between one region and another. Spain bases its curricula on a tradition incorporating both manual labour and strong reference to the notion of artistic creation. The technological aspects linked to industrial production systems are considered in the context of social and cultural approaches, balancing creativity in the designing of industrial products, and the impact of technological advances on society and the environment. French curricula swing between an approach centred on the processes of design, production, and distribution of manufactured material objects, and the discovery of how different jobs are organised. Greece gives priority to aspects linked to the use of objects and systems from the point of view of both the impact on society, and the putting to use of complex systems. Such diversity spreads beyond these four countries to the whole of the European Union. From this perspective, it is difficult to imagine training programmes based on an ensemble of recognisably identified content in each of the Union's twenty-seven countries, and in the forty-five adhering nations.

Mastering the teaching of subject area knowledge constitutes the second phase of practical teacher training. This aspect is hugely important in Germany and Spain, these two countries tending to attach greater importance to the link between knowledge and the ability to teach this knowledge than Greece, which gives priority to mastering subject area knowledge, or France, which attaches great importance to the role of the teacher in the educative system. A more detailed study of the tools and methods for the teaching of technology education in different countries would highlight considerable heterogeneity of resources and, beyond that, of teaching approaches. Thus, ways in which teacher training is organised differs, each approach springing from the conditions and contexts of teaching technology education in the country in question.

To consider that a teacher's job is not merely to pass on knowledge of his/her subject area is a view shared by many, even if it is not firmly placed into the way training is organised or, more precisely, if this placing varies greatly from one country to another. The diversity is related to the working conditions of the teaching profession and what is expected of a teacher. Thus, a teacher's professional/practical training must allow him/her to integrate into the relevant education system as quickly as possible. An immediate problem with such integration relates to language barriers between the native language and regional language (like in Spain), or the student's mother language; this difficulty applies not only to the academic components, but also to the practical components in schools.

Given such heterogeneity across regions and countries, how can one envisage a coherent grouping together of teaching units allowing a Greek student, for example, to train for one semester in Germany and another in France? How does one help students to succeed by validating these teaching units in the form of credits that fit within the Bologna process? As things stand currently, student mobility – for example, the Erasmus programmes provide financial support to students who go to study in other European university – is largely limited to observational work experience in schools accompanied by modules on knowledge of the respective education systems. Few experiences go beyond this to target a crossover of knowledge of teaching units that work towards a harmonising of training content. This is the work which needs to be done to enable the Bologna process can be applied both to general teacher training and to the training of technology teachers in particular.

REFERENCES

Chatoney, M., & Ginestié, J. (2006). La formation universitaire et professionnelle des enseignants dans le LMD: Exemples d'intégration et de structuration des cycles universitaires dans les domaines scientifiques et technologiques. In N. Benjelloun (Ed.), *Symposium international: Formation, apprentissage et évaluation en sciences et techniques à l'université*. Fez (Maroc): Université Sidi Mohammed Ben Abdallah – Faculté des sciences Dhar El Mehraz.

Eurydice. (2007). *Focus sur les structures de l'enseignement supérieur en Europe. Évolutions nationales dans le cadre du processus de Bologne – édition 2006/2007*. Bruxelles: Eurydice.

Ginestié, J. (2006a). Analysing technology education through the curricular evolution and the investigation themes. In M. de Vries & I. Mottier (Eds.), *International handbook of technology education: Reviewing the past twenty years* (pp. 133–145). Rotterdam, The Netherlands and Taipei, Taiwan: Sense.

Ginestié, J. (2006b). The training of technology teachers considering the organizations of their own training courses: Aspects of its problematics and research follow ups (C. Benson, Trans.). In J. Ginestié (Ed.), *Teacher training:*

Preparing young people for their future lives. An international study in technology education (pp. 7–16). Santiago, Chile: Editions Los Salesianos.

Ginestié, J. (2006c). *Teacher training: Preparing young people for their future lives. An international study in technology education* (C. Benson, Trans.). Santiago, Chile: Éditions Los Salesianos.

Ginestié, J. (2006d). Technology education teacher training in France. In P. J. Williams (Ed.), *International technology teacher education. 55th Yearbook* (pp. 45–67). New York: Glencoe/McGraw Hill.

Ginestié, J., Brandt-Pomares, P., Chatoney, M., Martinez, Y., & Oliveri, G. (2006). Educational system and teacher training in France (C. Benson, Trans.). In J. Ginestié (Ed.), *Teacher training: Preparing young people for their future lives. An international study in technology education* (pp. 63–76). Santiago, Chile: Editions Los Salesianos.

Ginestié, J. (2007a). *The cultural transmission of artefacts, skills and knowledge: Eleven studies in technology education.* Rotterdam, The Netherlands: Sense.

Ginestié, J. (2007b). Les relacions entre el prescriptor i l'operador: La formació professional del professorat a françа. *Temps d'educacio, 31*, 25–54.

Graube, G., Dyrenfurth, M., & Theuerkauf, W. (2003). *Technology education international concepts and perspectives.* Frankfurt am Main: Peter Lang.

Papoutsakis, H. (2006). Educational system and teacher training in Greece. In J. Ginestié (Ed.), *Teacher training: Preparing young people for their future lives. An international study in technology education* (pp. 89–111). Santiago, Chile: Editions Los Salesianos.

Theuerkauf, W., & Graube, G. (2006). Educational system and teacher training in Germany. In J. Ginestié (Ed.), *Teacher training: Preparing young people for their future lives. An international study in technology education* (pp. 77–87). Santiago, Chile: Editions Los Salesianos.

Theuerkauf, W., & Mager, W. (2004). Internetgestütztes steuern und Regien. *Unterrichtsfach Arbeit und Technik, 22*, 52–55.

Jacques Ginestié
Gestepro – UMR ADEF
Aix-Marseille Université
IUFM Aix-Marseille – Université de Provence
Marseille, France

PHILIP A. REED AND JOHN M. RITZ

48. GRADUATE STUDY AS PROFESSIONAL DEVELOPMENT

INTRODUCTION

Graduate study is the primary means of developing and maintaining a profession. In education, advanced coursework and research extend the practice of classroom teachers, administrators, and teacher educators through a continuous cycle of teaching and learning. The key to this cycle is the creation of knowledge. Graduate study is primarily designed to make practitioners creators of knowledge, whereas pre-service programmes, for the most part, consider practitioners consumers of knowledge. This chapter discusses the keys to sound graduate study in design and technology education by looking at professionalism and the characteristics of a profession. Additionally, research is presented to show the current state of graduate study and to provide a model graduate programme for professional development.

PROFESSIONALISM

Professionalism is a mental state directed toward exemplary practice of a career. In education, it is a term that has been brought to our attention through literature or conference attendance (see Gilberti & Rouch, 1999). It is a state of being that focuses on attitudes toward the advancement of a person's career (theoretical – what one believes), and yet it is based upon one's specific knowledge related to his/her particular career (methodological – how one acts).

Through reflection, we might suggest that the trait of professionalism was instilled into most through family, mentors, or professional organisations with which we have become involved. According to Rouch (1999), "Professionalism is not a set of competencies but predominantly an attitude; a real professional is a technician (one who knows their work) who cares" (in Gilberti & Rouch, 1999, p. 18).

As this definition suggests, professionalism is a state of mind that is associated with intrinsic rewards and energies exerted beyond the job, toward the good of the career field. For those in education, it is a dedication to the teaching field (knowledge-based content and practice) as well as the people who are learning about the field and/or serve the field as colleagues (teachers, professors, association representatives, and administrators). Members of the profession are considered competent in their occupation and they continue to work earnestly toward becoming increasingly competent – they undertake professional development. These individuals may take university or workshop courses, attend conferences, read relevant journals and books, and participate in electronic information exchanges that focus on their subject area. Additionally, a long-standing view of professionals is that they volunteer to work on committees that attempt to better the profession, and do not talk against their profession or its members (Rouch, 1999; Vollmer & Mills, 1966). Keirl discusses further the concepts of professionalism and professional development in Chapter 45. These concepts, however, are at the core of graduate study and will be used to frame our discussion of professional development in this area.

A.T. Jones and M.J. de Vries (eds.), International Handbook of Research and Development in Technology Education, 581–592.
© 2009 Sense Publishers. All rights reserved.

GRADUATE STUDY AND THE CHARACTERISTICS OF A PROFESSION

On a continuum, most careers can be evaluated to determine if they represent professional work, or rather just work that leads to payment of wages. For example, one would expect medical doctors to be on one end of the spectrum and day labourers at the other. What knowledge do we use to make these judgments toward some careers to be professions, while others do not hold this stature?

Myers (1973) associates professions with a number of characteristics. These include:

- Knowledge based upon scientific theory;
- service orientation;
- function;
- standards of education and training;
- long adult socialisation period;
- licensure and admission boards;
- legislation;
- income, power, and prestige ranking;
- norms, member evaluation, and control;
- identification and affiliation with the profession; and
- lifelong occupation.

It is interesting to review these characteristics, since they serve as filters through which we reflect our thoughts based upon personal experiences. Of course, some views will vary based upon the circumstances for the region where you work as a professional. For example, the licensure and evaluation components will vary from region to region and with the alignment that one's country has established in the development of its educational and qualification processes. Each of these characteristics will be reviewed in hopes of establishing a grounding for graduate study as professional development.

Knowledge Based upon Scientific Theory

People in a profession have different knowledge and skills than those not practicing the profession. Educators may figure out how to operate a needle and hypodermic syringe, but probably could not administer a shot as well as a physician or nurse. Can an engineer properly prepare, deliver, and assess a unit of instruction? This is an important consideration as design and technology education wrestles with the direction of its knowledge base. Will the profession focus on technological literacy, engineering, or the integration of science, technology, engineering, and mathematics (STEM)?

Graduate study in design and technology education must include coursework and research that covers all of these areas, and more. Professional practice "rests upon some branch of knowledge to which the professionals are privy by virtue of long study and by initiation and apprenticeship under masters already members of the profession" (Everett, 1963, p. 656). Professionals use knowledge that practitioners have codified through research, testing, and application. Accumulating this knowledge through guided study, professionals learn how to act when they encounter different situations in the practice of theory in their professions.

Different countries impose different laws and professional regulations established by government or the professions themselves (e.g., the granting of licenses). For example, in the U.S.A. the federal government (Office of the Press Secretary, 2002), has called for K-12 education to be based on practice that is scientifically determined. In other words, it requires professional educators to conduct scientific studies that determine what should be taught based on empirical data (Shavelson & Towne, 2002). Graduate study lends itself to this model by its very nature.

Graduate faculty and students create new knowledge by becoming familiar with current practice and the profession's body of knowledge. This body of knowledge is questioned and investigations are undertaken to fill in identified gaps. These studies are often published in scholarly journals or as theses and dissertations so that they may inform other practitioners. However, in design and technology education whilst graduate research makes up the bulk of our new knowledge, it often does not get published (Reed, 2002).

The teaching of reading possibly best represents a school subject that has sufficiently established research-based data for best practice. In design and technology education, some have data to support what we teach and how best to teach the content. For examples, presenters at conferences such as PATT have shared some of their research findings. Dugger (International Technology Education Association [ITEA], 2000), through *Standards for technological literacy*, used scientific practice to codify and verify the content that should be experienced by students as they study design and technology. This content was validated the by National Research Council, making it more credible, and *Standards for technological literacy* has been translated into several languages other than English. However, our profession continues to base many decisions on what and how to teach from a philosophical standpoint.

Service Orientation

One major characteristic of a profession is its service orientation. People join together to promote and shape their profession. The membership shares joint responsibility for guiding the profession (i.e. establishing goals), determining activities, and electing members to represent the profession.

If an organisation or organisations is/are formed by those who practice the profession, the role of the organisation is to best represent its clients and members. The profession then usually establishes a code of ethics that its members follow. An example of this is exhibited in ITEA's (n.d.) document, *This we believe*, in which design and technology education:

- Should be a part of the learning experiences of all students at all levels of grade and ability, in order that they may understand, function, and control their industrial/technological environment.
- Requires the highest level of competence from its instructional staff. Teachers must possess creativity and ingenuity, enjoy working with people, and maintain a high degree of personal and professional integrity.
- Uniquely contributes to students of all learning ability levels of both sexes and regardless of career choices. It provides equal opportunities for those of high or low economic status, and those who may choose a professional life or a future as an industrial worker. It's equally important to everyone as all members of society must learn to be aware of and learn effectively in tomorrow's technological society.
- Fosters an awareness of industry and enterprise and their place in the world culture. It also provides opportunities for learners to discover their talents and abilities in the areas of technology and applied science in the world of industry.
- Activities form a continuum with other visual and applied arts, ranging from the free expressive forms to the more exacting demands of machine tools and applied sciences.
- Is an organization of subject matter that provides opportunities for experiences concerned with developing insights into technology, its evolution, utilization, and significance; and industry, its organization, personnel, systems, techniques, and products; and their social/cultural impact.
- Provides technical skills and knowledge basic to most occupations and professions. Technology education enables the future scientist and engineer to solve technical problems, and the future craftsperson or technician to develop knowledge, skills, and the ability to obtain technical information.

- Provides wholesome changes in learners. These may take the form of a developed interest in the human-made world – its materials, products, and processes. These changes may also involve self-evaluation of attitudes towards constructive work and how this work can be utilized for health and recreation, as well as economic value; they may involve the development of a favorable attitude toward creative thinking, and toward character improvement – knowing and making the most of one's environment.
- Employs actual involvement of tools, machines, and materials that reinforce the written and spoken word. It enables all students to derive meaning from concrete experiences that aid in the understanding of abstract ideas and the development of concepts. (p. 1)

A service orientation such as *This we believe* is adopted by the members and becomes a part of professional indoctrination.

Function

Each professional organisation fulfils a function for its members. What is the function of the organisation – to serve its members? Does it provide for standards under which to operate? Does it provide for recognition of its members? Does it provide a forum for its members to band together to try to influence legislation? Professional organisations describe their functions through their bylaws or other documented forms.

There are many organisations in design and technology education that compliment graduate study professional development. Conferences held by the ITEA and Pupils' Attitudes Towards Technology (PATT) provide a forum for graduate faculty and students to present professional research and concepts. The Council on Technology Teacher Education (CTTE) has also created the *21st Century Leaders* programme to mentor graduate students on how to become professionally engaged. Additionally, student organisations such as the Technology Education Collegiate Association (TECA) allow graduate students the opportunity to participate in design and technology education competitions and the chance to judge undergraduate work.

Standards of Education and Training

Professional organisations at times make recommendations to legislative bodies that might improve the way that the profession is conducted. Some legislative bodies establish boards that assist with these decisions. Usually members of the professional bodies are invited to serve on boards. Examples of this occur in the U.S.A. when professionals serve on boards for the National Academy of Engineering (NAE), the National Research Council (NRC), or the American Association for the Advancement of Science (AAAS). In design and technology education, content standards have been developed to define what to teach (ITEA, 2000) and for professional development (ITEA, 2003a). If government accepts these standards, then the individuals entering these professions should be trained to understand and teach the standards via training. Our profession does this through university programmes for initial qualification/certification. Practicing professionals read about the changes and attend graduate courses or workshops to learn about new content standards and improved means of delivery. Additionally, the CTTE has worked with the largest accrediting body in the U.S.A., the National Council for Accreditation of Teacher Education (NCATE), to develop accreditation standards for design and technology education programmes at the initial and graduate levels (ITEA, 2003b).

Long Adult Socialisation Period

Since some professions are based upon scientifically-established knowledge, it takes time for a neophyte to learn all of the knowledge and the theory that guides best practice. During initial preparation only the minimal amounts of knowledge can be learned. The norms and practices of the profession must also be learned. This is why it takes many years for most to become masters of their professions.

Graduate design and technology education programmes are designed to indoctrinate students to the profession. For example, a master's degree typically takes one to one-and-a-half years to complete beyond the bachelor's degree. Coursework at this level is typically more theoretical, involving extensive study of the philosophy and history of the profession. The doctoral degree is typically three years beyond the master's degree and involves a thorough review of the literature and intensive research activities to help bolster the knowledge-base of the profession. During this socialisation period, the practitioner typically serves a residency that immerses them within the professional environment. Professional service activities increase at this level, professional networks are established, and a research agenda is initiated.

Licensure, Admissions Boards, and Legislation

Initial preparation for members of a profession involves formal education. During learning of the standards of the profession through education and practice, neophytes develop the competence needed to earn a license. Most professions, including design and technology education, have internships where their skills are honed. Government bodies often establish the criteria for licensing a professional to practice, but active professional organisations generally have members that sit on these boards to draft the criteria that are assessed to determine if the neophyte earns the license. If the member does not follow the standards set forth to retain their license to practice, the license might be revoked.

Graduate students preparing for administration or the professoriate may need a license depending on provincial or national requirements. Graduate design and technology education programmes have admissions standards that must be met for the institution as well as the programme. These are often regulated by government bodies and/or accrediting organisations.

Income, Power, and Prestige Rankings

Professionals, *for the most part*, earn higher salaries than those who are not in occupations considered professions. (*For the most part* is italicised because supply and demand influences what people will pay.) Teaching is a lower paying profession since it does not directly generate income (an early deficit for a community, but it pays off in the long term by providing higher quality citizens). This is similar for fire fighters and police. Professions such as physicians and lawyers earn higher incomes, since clients are willing to pay higher wages for their services.

There is an aura of power that surrounds individuals who seek careers in professions (i.e. positional authority). Professionals generally have higher educations and can gain power by the numbers of members they serve. Professionals can band together when talking to legislative bodies and be heard as one voice. At times, professional groups pledge the support of their members to the election of political figures.

As a result of income and power, professionals gain prestige when compared to other working classes. In some countries, there is much respect given to one who becomes a teacher and higher respect given to a professor. Professions work to gain prestige for their members.

Norms, Member Evaluation, and Control

Professionals meet and establish the norms of practice of their professions. They develop documents that are used to guide the training of individuals who seek to enter their profession. For example, the ITEA and CTTE each have bylaws that guide their professional activities. Members elect the officers and their duties are carried out according to the mission of the organisation.

Graduate design and technology education programmes are accountable to laws, regulations, and accreditation guidelines. Programmes must show that their graduates are prepared to practice the norms of the profession according to these policies. Therefore, graduate programmes work with professional organisations and regulatory agencies to ensure norms are established, evaluation is conducted by educated members, and control of the profession is not relegated to those that do not know the field.

Identification and Affiliation with the Profession

Professional associations are the mechanisms used for members of professions to work toward their common good. These organisations provide members the opportunity for advancement in their careers (learning and recognition) and a social setting for affiliation (networking and learning). Associations enable members to affiliate with others who believe in similar concepts about their specialties. They also provide for protection of values, norms, and prestige for those who affiliate. In education, many members identify with associations for all educational subjects, but they might be more closely aligned with professions that focus on the specific content they teach. Some organisations are international in focus and professionals affiliate with these groups so they can learn how individuals with similar interests practice their profession in other regions of the world.

Lifelong Occupation

Since it takes an extended amount of time to learn the knowledge to practice a profession, and in so doing higher salaries and status can be gained, professions are usually held as lifetime endeavours by its members. With an array of occupations available to those with mastery of their knowledge-base, individuals have the ability to change careers within professions, but many choose to continue practicing the same profession throughout life even though they may change jobs. Some refer to this as a 'calling' for their career. In graduate education, the higher one goes the more specialised one becomes. Someone who receives a doctoral degree, for example, is considered an expert in the area in which they conducted their dissertation. This may seem limiting, but in actuality it establishes the individual's research agenda and provides them a career pathway to hone and through which they can share their expertise.

Summary of a Profession

Knowledge, education, and training are key characteristics tied to professionals. Life-long learning is a characteristic found in professions (i.e. a long adult socialisation period). Life-long learning requires continued formal education for the professional. Such continued professional development allows one to grow in his or her profession (Reeve, 1999). Graduate education is a natural pathway for professional development, and embodies each aspect of a profession.

STATE-OF-THE-PROFESSION: GRADUATE DESIGN AND TECHNOLOGY EDUCATION

Graduate programmes must embody the characteristics of a profession noted above if they wish to remain productive and relevant. Professional development occurs when practitioners learn what others within the field are doing and reflect upon scientifically-based research and contemporary practice. Many graduate programmes exist within technology education, and sound professional development must be based on these existing programmes.

Ritz, Deal, Ndahi, and Reed (2006) surveyed institutional members of the ITEA to determine the outlook of graduate study. There were 46 institutional members of this professional association. Additionally, a second group of 14 international institutions was surveyed using mail and electronic means (Ritz & Reed, 2007). Of the two groups, the following return rates were achieved: Thirty-seven of the ITEA institutional members were returned for an 80.4% return rate; ten of the international institutions were returned for a 71.4% return rate. Of the 47 returned instruments, 36 institutions offered graduate education in design and technology education. These degrees and their number were reported as follows:

- 35 Masters (masters of science, masters of art in teaching, masters of education).
- 6 Educational Specialist (a degree that is 30 or more credits beyond the masters).
- 11 Doctor of Philosophy (Ph.D.).
- 8 Doctor of Education (Ed.D.). (Ritz & Reed, 2007, p. 407)

A primary question asked of participants was to identify non-design and technology education required coursework in the areas of foundations, curriculum, instruction, research, and other areas. It was felt that a review of the required coursework would reveal programme autonomy (e.g., whether programme control was at the program/department level or at the college/university level). Foundations courses, for example, were identified to be either general (e.g., Advanced Educational Psychology) or specific to design and technology education (e.g., History/Philosophy of Design and Technology Education). Most curriculum courses were focused toward design and technology education (e.g., Implementing Design and Technology Education, Curriculum Planning in Occupational Studies). Courses in instruction, on the other hand, varied widely. Many institutions had broad courses (e.g., Instructional Strategies & Innovations), while others focused on areas such as leadership development.

Requirements for research at the master's level varied widely and were often not clearly defined. A thesis was not required by many programmes; a research paper, project, or portfolio was often required instead. Research coursework at the master's level was typically 3-6 credit hours (one or two courses) per week over 1–1½ years, commonly with a course sequence consisting of statistics and research design/methodology. At the doctoral level, 12 hours of research was common per week for an Ed.D., and 15–18 hours per week for a Ph.D. over a three-year period Most of the doctoral institutions had specific qualitative course offerings and some institutions required doctoral students to take qualitative courses.

Several respondents indicated that they had themes running through their programmes for required non-design and technology education coursework. Common themes include science, technology, engineering, and mathematics (STEM) integration, workforce development, and instructional technology. Other programmes require coursework in supervision and administration, or working with special populations, so candidates are prepared (i.e. licensed/endorsed) for positions in these areas as well as design and technology education positions.

Participants were asked to list the specific design and technology education graduate courses that they offered. Professional and technical courses were frequently cited as required design and technology education courses. Seminars, readings, leadership, philosophy, technological systems, and programme development were cited as common professional courses. Technical coursework included technical systems and problem solving. Almost all institutions require courses in instructional technology. The minimum hours of required design and technology education coursework per programme was 6 hours per week at the master's level and 18 hours at the doctoral level. Specific course titles included: Introduction to Technology; History and Philosophy of Technology Education; Technological Systems; Program Development in Technology Education; Readings in Technology Education; Cognition in Engineering and Technology Education; Design Thinking; Engineering Design and Analysis; Methods of Integrating Science, Technology, Engineering and Mathematics; Elementary Technology Education; and Appropriate Technological Development (Ritz & Reed, 2007).

Delivery of Graduate Design and Technology Education

The proliferation of distance learning has resulted in several recent studies on the delivery of graduate design and technology education (Baltzer, Lazaros, & Flowers, 2007; Ndahi, 2006; Ndahi & Ritz, 2002; Ritz & Reed, 2007). At many institutions, students are still required to be present on campus at some point for their graduate education. This prohibits some from taking advanced course work in technology education if they are not located near a suitable campus. Some universities offer classes at off-campus locations, where the professor travels to the areas where a group of teachers are working. An additional method is to have adjunct faculty employed at remote sites to facilitate coursework. The use of adjunct faculty, however, can create problems for university accreditation if the adjunct faculty member does not possess a terminal degree. Because of these limitations, universities are continuing to develop more courses via distance learning. Some programmes conduct traditional correspondence courses using mailed out assignments by the faculty (some on CD/DVD) and mailed in assignments by the student. However, most programmes have moved to advanced technologies such as two-way, audio-video, televised courses, and web-based video streaming technologies (synchronous and asynchronous).

Worldwide research on graduate design and technology education programme delivery found that 32 out of 36 institutions offered campus-based programmes. Twenty offered courses via distance learning technologies, nine of which had distance learning courses at the doctoral level. Additionally, nine universities reported offering classes from extended campus locations with faculty or adjunct faculty teaching off-campus, mostly at K-12 sites or university regional centres (Ritz & Reed, 2007). Five of 36 institutions reported using televised, two-way video, and audio instruction; seven reported using video-streaming over the internet; four used CD/DVD technologies; and 13 reported using non-video, web-based delivery (Ritz & Reed, 2007).

An earlier study in the United States investigated the use of distance learning techniques by design and technology education faculty (Ndahi & Ritz, 2002). The results of this study indicated that 60.4% of U.S. universities that teach design and technology education and industrial technology were using distance learning technology to deliver their programmes. More recently, Ritz and Reed (2007) showed a 55.5% offering of distance learning technologies. Several universities indicated that whilst they had developed courses, they could not sustain a student population large enough to continue offering the courses using distance learning capabilities.

These findings suggest that distance learning is regularly utilised, but also indicate that further research is needed to effectively target broader audiences. However, Ritz and Reed (2007) did ask those universities involved in graduate design and technology education who were not using distance learning technologies if they intended to use these course delivery techniques in the near future. Four universities (out of 36) planned to implement distance learning in the near future, while eight universities had no short range plans to begin using distance learning techniques.

Research in Design and Technology Education Graduate Programmes

The first characteristic of a profession, as noted above, is knowledge based upon scientific theory. Professional development at the graduate level actively immerses the learner in the process of creating knowledge using scientific methods. The resulting new knowledge *should* enter the teaching and learning cycle to further the aims and goals of the profession. The term *should* is used because, unfortunately, much of the knowledge created in design and technology education during graduate study is not shared with the profession. This relegates the graduate research to a mere academic exercise as opposed to scholarship that is presented to the profession and published (Reed & Sontos, 2006).

A second consideration regarding graduate research is the fact that students are becoming increasingly selective. Many students select a university graduate programme based upon the line of research that the faculty are pursuing. International research has been conducted to determine if universities have themes or topics that graduate students would research as part of the degree requirements. Twenty-eight out of 36 programmes (77.8%) reported that theme-based research studies were not required of their graduate students. Eight (22.2%) of the programmes required theme-based research, including such themes as:

- Engineering education
- Effects or value of technology education program
- Action research – technology education
- Class activities to support technology education
- Technological literacy
- Cognition and student learning through technology education. (Ritz & Reed, 2007, p. 410)

An alternative method of determining the new knowledge that is being created by a profession is to look at the graduate studies that have been conducted. Table 1 classifies recent design and technology education doctoral dissertation topics that have been completed in the United States. The majority of studies have focused on instruction. Interestingly, this is a change from the last major review by Zuga (1994) when curriculum was the leading research topic from 1987–1993. A second compelling finding is that almost 20% of the dissertations were classified as 'foreign', meaning that the researcher was looking at technology education in a country/countries other than the United States.

Table 1. Dissertation research in the United States, 2000-2005 (Sontos, 2005).

Categories	Number of Studies	Percentage
Attitudes	7	12%
Instruction (how)	17	29%
Curriculum (what)	5	8%
Continuing Education	2	3%
Professional Development	8	14%
Foreign	11	19%
Work-based Education	9	15%

A third important consideration for graduate research involves the dissemination of student work through a clearinghouse (i.e. University Microfilms International [UMI/ProQuest]). Globally, Ritz and Reed (2007) had 15 respondents (45.9%) that indicated that they do require students to submit their research to UMI. Due to the limited number of research studies in technology education, such a requirement could greatly increase access to research if this requirement is made mandatory (Reed, 2002).

A MODEL FOR GRADUATE TECHNOLOGY EDUCATION

Graduate study is one of the primary sources for professional development in design and technology education. The preceding sections have illustrated how the field is using graduate study to further professionalism and embody the characteristics of a profession. To carry on, the profession must continually analyse practice and ensure that there is consistency. The following models are designed to provide guidance toward that end. Research conducted in the United States (Ritz et al., 2006) and furthered at the international level (Ritz & Reed, 2007) form the basis for recommended study at the master's and doctoral levels.

Graduate Study at the Master's Level

A master's degree usually consists of 30–36 semester hours of study beyond the baccalaureate. Coursework should include a minimum of 3–6 hours of study per week over a semester in each of the four following areas: foundations, curriculum, instruction, and research. Foundations courses may be general (e.g., Educational Psychology) or focused on design and technology (e.g., Philosophy of Technology). Curriculum courses need to engage the learner in the planning and evaluation of instructional units, not just individual lessons. Instruction courses may be general (e.g., Instructional Technology) or focused on design and technology (e.g., Instructional Strategies and Innovations in Design and Technology). The research courses should include at least one course in social science methods (e.g., Statistics) and one course in research design. It is highly recommended that master's students be engaged in the process of designing, conducting, and disseminating scholarly research as opposed to developmental projects such as portfolios or curriculum materials.

Additional coursework at the master's level may involve electives that focus on broad trends and issues or lines of inquiry that are specific to the institution. For example, many universities are now focusing on specific lines of teaching and research such as: science, technology, engineering, and mathematics (STEM) integration; action research; technological literacy; or cognition. Faculty are encouraged to advise students to take classes and conduct focused research that adds to the professional body of knowledge. Ultimately, coursework in specific areas will help further the goals of the institution and should lead to increased scholarship for the profession at large.

A comprehensive examination is recommended for students at this level. Questions should seek critical reflection on key issues with the intention of eliciting extended responses that are supported by research. Responses should be read by multiple reviewers.

Graduate Study at the Doctoral Level

Doctoral programmes usually consist of 60 semester hours of study. Like the master's level, coursework should include study in each of the four following areas: foundations, curriculum, instruction, and research. It is important for students at this level to work out a plan of study with their advisor that is based on the student's professional goals for teaching, scholarship, and service. Research indicates the distinction between a Doctor of Education (Ed.D.) and a Doctor of Philosophy (Ph.D.) is often determined in the research sequence (Ritz & Reed, 2007). Ed.D. students typically take 12 semester hours of research courses and Ph.D. students take 15–18 semester hours. This is based on tradition, where the Ed.D. candidate has typically been groomed to re-enter primary, elementary, or secondary education. The Ph.D., on the other hand, has traditionally been trained to enter post-secondary education. Regardless of the type of degree, research courses at this level should include at least one course in each of the following areas: advanced social science methods (e.g., Statistics), advanced research design, and qualitative research design.

The second component in doctoral study is the comprehensive examination. Like the master's level, questions should seek critical reflection on key issues with the intention of eliciting extended responses that are supported by research. At least one question should require the use of social science methods to address a research issue. Ideally, the comprehensive examination at this level should be focused on the student's research and scholarship interests. Responses should be read by multiple reviewers.

The dissertation is the third phase of doctoral study. Programmes should allocate a minimum of 12 semester hours for the completion of a dissertation. Ideally, the advisor should work with the student at programme entry so that the dissertation is based on the student's professional goals for teaching, scholarship, and service. The advisor and committee should also work with the student on professional presentations and publications related to their dissertation.

The preceding models have been offered as general recommendations that are based on current practice. Graduate programmes must also consider content standards (i.e. *Standards for technological literacy* [ITEA, 2000]) as well as professional development and programme standards (i.e. *Advancing excellence in technological literacy* [ITEA, 2003a]). Additionally, national and state/ provincial requirements will have an impact on the shape of graduate professional development.

REFERENCES

Baltzer, H., Lazaros, E., & Flowers, J. C. (2002). Review of doctoral programs in technical education. *Journal of Industrial Teacher Education, 44*(2), 37–59.

Everett, C. H. (1963, Fall). Professions. *Daedalus, Journal of the American Academy of Arts and Sciences*, 656.

Gilberti, A. F., & Rouch, D. L. (1999). *Advancing professionalism in technology teacher education.* Peoria, IL: Glencoe/McGraw-Hill.

International Technology Education Association [ITEA]. (2000). *Standards for technological literacy: Content for the study of technology.* Reston, VA: Author.

ITEA. (2003a). *Advancing excellence in technological literacy: Student assessment, professional development, and program standards.* Reston, VA: Author.

ITEA. (2003b). *ITEA/CTTE/NCATE curriculum standards: Initial program in technology teacher education.* Reston, VA: Author.

ITEA. (n.d.). *This we believe.* Retrieved October 8, 2008, from http://www.iteaconnect.org/AboutITEA/ThisWeBelieve.pdf

Myers, D. (1973). *Teacher power – Professionalism and collective bargaining.* Toronto, ON: Lexington Books.

Ndahi, H. B. (2006). The use of innovative methods to deliver technology education laboratory courses via distance learning: A strategy to increase enrolment. *Journal of Technology Education, 17*(2), 34–43.

Ndahi, H., & Ritz, J. (2002). Distance learning in industrial teacher education programs. *The Journal of Technological Studies, 28*(1), 64–69.

Office of the Press Secretary. (2002). *Fact sheet: No child left behind act.* Retrieved March 30, 2008, from www.whitehouse.gov/news/releases/2002/01/20020108.html

Reed, P. A. (2002). Research in technology education: Back to the future. *Journal of Technology Education, 13*(2), 66–70.

Reed, P. A., & Sontos, D. A. (2006). *A United States perspective: 112 years of graduate research in technology education.* Paper presented at the PATT-16 Conference, Baltimore, MD. Retrieved April 1, 2008, from http://www. iteaconnect.org/

Reeve, E. M. (1999). Professional associations, organizations, and other growth opportunities. In A. F. Gilberti & D. L. Rouch (Eds.), *Advancing professionalism in technology education* (pp. 69–96). Peoria, IL: Glencoe/McGraw-Hill.

Ritz, J., Deal, W., Ndahi, H., & Reed, P. A. (2006, November). *Graduate technology education.* Paper presented at the Southeastern Technology Education Conference, Nashville, TN.

Ritz, J. M., & Reed, P. A. (2007). Graduate study in technology education: State-of-the-art. In J. R. Dakers, W. J. Dow, & M. J. de Vries (Eds.), *Teaching and learning technological literacy in the classroom* (pp. 406–414). Glasgow, Scotland: University of Glasgow.

Rouch, D. L. (1999). Professionalism and leadership in technology education. In A. F. Gilberti & D. L. Rouch (Eds.), *Advancing professionalism in technology education* (pp. 3–14). Peoria, IL: Glencoe/McGraw-Hill.

Shavelson, R. J., & Towne, L., (Eds.). (2002). *Scientific research in education.* Committee on Scientific Principles for Education Research. Washington, DC: Center for Education. Division of Behavioral and Social Sciences and Education, National Research Council, National Academy Press.

Sontos, D. A. (2005). *Directions of dissertation research at universities preparing future technology education teacher educators.* Unpublished graduate research paper, Norfolk, Virginia: Old Dominion University. Retrieved March 25, 2008, from http://www.lib.odu.edu:8000/dspace/handle/123456789/2

Vollmer, H. M. & Mills, D. L. (1966). *Professionalism.* Englewood Cliffs, NJ: Prentice-Hall, Inc.

Zuga, K. (1994). *Implementing technology education: A review and synthesis of the research literature.* Information series no. 356. ERIC Clearinghouse on Adult, Career, and Vocational Education, Columbus, Ohio. ED 372 305.

Philip A. Reed
John M. Ritz
Darden College of Education
Old Dominion University
Norfolk, Virginia
USA

SECTION 8

THEORETICAL AND PRACTICAL APPROACHES TO RESEARCH

ALISTER JONES

49. DEVELOPING RESEARCH APPROACHES TO TECHNOLOGY EDUCATION

As highlighted in section 1 of the *Handbook*, there has been an emerging and growing research culture in technology education in the last 20 years. This section explores specific research approaches. In introducing the section, it is important consider past research agendas in relation to where we are now. However, in exploring the past in order to consider what needs to be done in the future, there is the danger that the field becomes understood by the research undertaken – not by what needs to be done.

In the earlier days of technology education as it began to develop its own distinct research area, Pupils' Attitudes Towards Technology (PATT) played a key role and the research instruments developed to explore student and teacher understandings and attitudes have been widely adapted and used around the world. The conferences have also brought researchers in technology education together for over 20 years. PATT has thus been an important initiative, both in setting the research agenda for technology education as well as building an international forum for the research.

In the 1990s as the field of technology education research began to define itself more clearly, a number of people proposed research agendas for the developing field. For example, Olson and Henning Hansen (1994) identified two strands in technological education research – vocational preparation and the social significance of technology. Gilbert's (1994) agenda is based on past work in science education, although this can also limit aspects that are examined. This partiality of what work has been done is then reflected in material compiled for postgraduate study courses, which defines the field for those who come to study it. For example, McCormick, Sparkes, and Newey (1992) and McCormick, Murphy, and Harrison (1992) explored technological capability, learning in technology, and the implementation of teaching and learning technology. Again there are strong links with science in many of these themes. In other reviews of research in technology, in particularly North American research from 1987–1993, Zuga (1997) found that the studies were skewed in favour of curriculum issues with little in relation to learning. In a meta-analysis of articles in the *Journal of Technology Education*, Petrina (1998) noted that there was little published in the area of critical theory, arguing that the field had thus far been conservatively inclined. De Vries (2000) also notes that the majority of articles in the *International Journal of Technology and Design Education* in the previous 10 years had been dominated by articles related to defining the subject, particularly with an emphasis on design and problem solving, although this has more recently been changing.

In all of the research reviews it would appear that there is a need to define technological literacy and set out a research plan related to the key characteristics related to student technological literacy. Lewis (1999) argues that there needs to be a multi-dimensional, sustained programme of work in the area. Similarly, in a major review of 600 articles on curriculum, pedagogy, and assessment (Carr, McGee, Jones, McKinley, Bell, Barr, & Simpson 2000) it was found that there needed to be much more research undertaken in classrooms, and in particular more long-term studies. Technology education is no different; classroom-based research must become the focus to enhance

A.T. Jones and M.J. de Vries (eds.), International Handbook of Research and Development in Technology Education, 595–598.
© 2009 Sense Publishers. All rights reserved.

our understanding of technology in classrooms, pre-service and inservice education, and curriculum development.

If student learning in technology is to be enhanced then this requires research into the role and place teacher knowledges play in enhancing student learning (Black, 1998). A greater exploration on researching teachers teaching technology needs to be undertaken. Teacher conceptualisation of technology is complex and requires an understanding of the factors that influence it. Some factors have included how willing teachers are to change their own concepts of technology and technology education, background experience, subject sub-culture, and the level of support given to teachers during any change process. The wealth of research data on students' existing concepts in science indicates that conceptual knowledge significantly affects students' predictions, explanations, and perceptions of phenomena and problems (Hennessy, 1993). This means that students' and teachers' concepts of technology, including concepts of technological knowledge and processes, will impact on student learning in technology. Greater emphasis on researching students' learning in technology education, including ways in which this learning can be enhanced, is required.

Technological knowledge is socially constructed and context dependent, and human mental processes are situated within their historical, cultural and institutional setting (Wertsch, 1991). Technology therefore is essentially an activity that involves not just the social context, but also the physical context, with thinking being associated with and structured by the objects and tools of action. This then leads to research in technology classrooms being strongly influenced from a sociocultural perspective. This is highlighted in the following chapters, which represent a shift from past research approaches to explore new ways of developing more in-depth understandings of student and teacher understandings of technology and of teaching and learning in the classroom context. Although research from different theoretical perspectives is presented, a sociocultural perspective dominates.

In Chapter 50, Richard Kimbell and Kay Stables explore the interconnection between design and research in technology education. Richard and Kay build on their work for the Assessment of Performance Unit (APU) (Kimbell, Stables, Wheeler, Woznaik, & Kelly, 1991), arguing that research and design can be seen as parallel worlds and that moving between the two allows new, productive ways of approaching research in technology education to be identified. Operational research such as research for a client, as is the case for many research units and centres, is similar to design where a brief is articulated and appropriate research designs and approaches are developed. In research, just as in design, there are constraints and trade-offs. For example, in research for a client whilst there may be the potential to examine something in-depth, cost and time are often inhibiting factors, as is the type of data required by the client. Richard and Kay provide a number of interesting examples of the process of developing research approaches from their research projects at Technology Education Research Unit (TERU) at Goldsmith's College, London. By highlighting a number of research projects in this way, they provide the reader with possible strategies to approach their own research in technology education. A clear argument is made for mixed methods approaches to problems, both qualitative and quantitative, and formal and informal. This breaks down past binaries to research and allows us to adapt the most appropriate methods to enhance our understanding of technology education.

In taking this theme forward, Carolyn Jones argues in Chapter 51 that if we are to understand young children's experiences and learning in technology then we need to be innovative in our research and data collection approaches. She illustrates ways of recording both children's action and thoughts, as well as finding out about the home influences, or funds of knowledge, that children bring to a new learning situation.

In Chapter 52, Bronwen Cowie and Judy Moreland highlight methodological considerations for classroom research, specifically when using a sociocultural lens. They point out that classroom research has changed its orientation over time; that different views of learning exist, from individual to social; and that there is a changing political context that influences education research, from what counts as evidence through to research-informed policy. They also explore how, when thinking about research in the classroom, it is not only issues related to classroom research approaches that need to be considered, but also methodological considerations related to technology as a discipline and the classroom activities related to technology. They conclude that technology education needs to be seen as a multimodal activity that involves multiple perspectives and is accomplished over time. Units of analysis need to be linked to these.

Wolff-Michael Roth in his chapter on studying communication in technology classrooms (Chapter 53) also highlights that students communicate over and about artefacts, including drawings. To this end, he outlines a method for tracking this communication that takes into account the multileveled and distributed nature of technology activities where communication is more than just words, but also includes gestures, body position, utterances, and tone. Michael also highlights the strong temporal aspect, providing rich examples to illustrate the concepts that he is developing. In the subsequent chapter (Chapter 54), he provides an example of students undertaking a technological activity that he then analyses through a cultural-historical activity theory (CHAT) lens. Michael develops the background to cultural-historical activity theory, arguing that this theoretical frame provides a sound basis from which to explore teaching and learning in technology, moving on to consider third generation activity theory in action. Throughout his chapter, classroom data is provided to demonstrate the points made.

Marilyn Fleer builds on Michael's chapter by providing in Chapter 55 an extensive literature review related the growing use of CHAT in technology education research, providing examples of its operationalisation in classroom research contexts.

The developing research approaches illustrate the growing maturity of the field of technology education research, which now has the potential to inform other fields of educational research rather than – as we have tended to do historically – appropriating methods from other fields.

REFERENCES

Black, P. (1998) *Testing: friend or foe? Theory and practice of assessment and testing.* London: Falmer.

Carr, M., McGee, C., Jones, A., McKinley, E., Bell, B., Barr, H., & Simpson, T. (2000). *The effects of curricula and assessment on pedagogical approaches and on education outcomes.* Wellington, NZ: Ministry of Education.

de Vries, M. (2000, December). *Can we train researchers and teachers to make a team? Win-win strategies in technology education.* Paper presented to the 1st Biennial International Conference on Technology Education Research, Gold Coast, Australia.

Gilbert, J. K. (1994). The interface between science and technology in schools. In A. Jones, A. Begg, B. Bell, F. Biddulph, M. Carr, M. Carr, et al. (Eds.), *SAMEpapers* (pp. 144–164). Hamilton, NZ: Centre for Science, Mathematics and Technology Education Research, University of Waikato.

Hennessy, S. (1993). Situated cognition and cognitive apprenticeship: Implications for classroom learning. *Studies in Science Education, 22*, 1–41.

Kimbell, R., Stables, K., Wheeler, T., Woznaik, A., & Kelly, A. V. (1991). *The assessment of performance in design and technology.* London: SEAC/HMSO

Lewis, T. (1999). Research in technology education – Some areas of need. *Journal of Technology Education, 11*(1), 41–56

McCormick, R., Murphy, P., & Harrison, M. (1992). *Teaching and learning technology.* London: Addison-Wesley.

McCormick, R., Sparkes, J., & Newey, C. (Eds.). (1992). *Technology for technology education.* London: Addison-Wesley.

Olson, J., & Henning Hansen, K. (1994). Research on technology education. In D. Layton (Ed.), *Innovations in science and technology education* (Vol. V, pp. 225–239). Paris: UNESCO.

Petrina, S. (1998). The politics of research in technology education: A critical content and discourse analysis of the *Journal of Technology Education*, Volumes 1–8. *Journal of Technology Education, 10*(1), 27–57.

Wertsch, J. V. (1991). *Voices of the mind: A sociocultural approach to mediated action.* Cambridge, MA: Harvard University Press

Zuga, K. (1997). An analysis of technology education research in the United States based upon an historical overview and review of contemporary curriculum research. *International Journal of Technology and Design Education,* *7*(3), 203–217.

Alister Jones
School of Education
University of Waikato
Hamilton, New Zealand

RICHARD KIMBELL AND KAY STABLES

50. RESEARCH AS A DESIGN TASK

INTRODUCTION

In this chapter we explore the interconnections between designing and researching processes. Both are intensely value-laden activities, and typically in relation to 'wicked' or indeterminate tasks. Designing is a client-centred process, as is research. In both, needs, interests, and priorities are required to be understood and addressed. Additionally, both are iterative processes that evolve through interacting ideas for addressing the task (design or research) with issues that need to be resolved.

One of the unique strengths of designing is that it forces abstract thought processes to become explicit. Through design practice, possible futures are explored and developed into tangible realities. In this tangible form they allow us to model and test our perceptions against the priorities that initiate the task. So research processes – seen through the eyes of designers – can become tangible manifestations of research purposes, for example, designing the research design and creating new research instruments, new ways of visualising and communicating data.

Using examples drawn from a range of research projects we have conducted, we will illustrate the parallels between processes of designing and researching and demonstrate the potential of taking a designerly approach to constructing research methodology. In the process we will discuss the formal research field of 'Design Experiments Research' (Kelly, 2003) and the potential for creating a stronger design methods research culture in taking forward research in technology education.

Research could be described as understanding 'what is …' – creating new knowledge to understand current realities. Design, however, is more 'what if …' – creating new ideas and futures. In order to explore the relationship between the two, we begin with design.

DESIGN IS . . .

Designers are people who think with their hearts (James age 10).

We won't make things cheaper than the Far East nations – but we can make things better (Terence Conran).

Only one company can be cheapest … all the others must use design (Rodney Fitch).

I would like to be a designer because you could make things that help people (Jade age 10).

Design is not just about survival and profitability, it is about making a difference – and it ought to be a difference for the better (Charles Handy).

Design is a way of changing life and influencing the future (Ernest Hall – Dean Clough).

A.T. Jones and M.J. de Vries (eds.), International Handbook of Research and Development in Technology Education, 599–613.
© 2009 Sense Publishers. All rights reserved.

> If there was no design, there would be nothing to do and nothing would progress or get better (Anna age 11). (Design Council, 1997)

The Design Council published this collection of insights about design and its role in the world in 1997, following their 1995 Little Black Book of *Definitions of Design* (Design Council, 1995). It is hardly a cross section of public opinion, but it goes some way towards capturing a set of insights that have been formed through experience of the designing activity.

Two recurrent themes shine through from this brief selection. First, that design is about the *new* and, second, that design is about the *better*. Designing results in new things that influence how we live in the world and (we would hope – or at least James, Jade, and Anna appear to hope), that these things improve our lot in the world.

We recognise of course that this notion of 'improvement' is a very tricky one, because someone designed traffic speed-cameras, and even aerial torpedoes and cluster bombs. None of these has substantially improved our lot over recent years. Indeed it could be argued that for every bit of 'the new' there are winners and losers. A new bypass aids the motorist but destroys a few more acres of former green-space; a new Nike trainer enables the purchaser to send all the appropriate good messages to his/her peers, but the polymer processing – and the conditions of work – renders this advantage at a high environmental and personal cost (quite apart from the retail price). So we are not dewy-eyed about the 'improvement' that is inevitably rendered by design. Our point is only that the motive that underlies any designing activity is one of improvement – for someone – measured against some scale. So the cluster-bomb is an improvement only in the sense that it dispenses death more effectively than its predecessor. And this is presumably seen as an improvement by the arms manufacturer or by the client that commissioned it.

Designing is essentially about purposeful change in the made world – driven by the motive of improvement (of something for someone) – and the modus operandi of designers is worth a brief digression:

> … designers … are concerned with what might be, the conception and realisation of 'the form of things unknown'. In describing their work, they talk of 'seeing with the mind's eye'. This is literally a visionary act, a mode of thought which is non-verbal and which has been a characteristic of design and technology throughout its history … imaging finds its representation in drawings, diagrams, plans, models, prototypes and computer displays and simulations, before its eventual realisation in a product, which may be an artifact, system or environment … (DES/WO, 1988, pp. 4–5)

We have spent a considerable amount of time over the last 20 years of research in the Technology Education Research Unit (TERU) at Goldsmiths reflecting on and attempting to describe this task-based activity, for example, in our first major collaboration for the Assessment of Performance Unit (APU) (see Table 1). This project precipitated the need to describe the process we were aiming to assess young people engaging in, as we had rejected the more formulaic, linear models existing in the technology education literature of the time. In the final report of the project, we described the process as quoted below and presented in Figure 1.

> When engaged in a task, ideas are inevitably hazy if they remain forever in the mind, and this inhibits their further development. By dragging them out into the light of day as sketches, notes, models or the spoken word, we not only encourage the originator to sharpen up areas of uncertainty, but we also lay them open to public scrutiny … the act of expression pushes ideas forward...and the additional clarity that this throws on the idea enables the originator to think more deeply about it, which further extends the possibilities of the idea … concrete expression (by whatever means) is therefore not merely something that allows us to see the

designers ideas, it is something without which the designer is unable to be clear what the ideas are. (Kimbell, Stables, Wheeler, Wozniak, & Kelly, 1991, p. 142)

Table 1. Outline of the Assessment of Performance in Design & Technology (APU D&T) Project, *1985–1991.*

Context: The Assessment of Performance Unit (APU) was the major research arm of the Department for Education and Science (DES) during the 1970s and 1980s. Following exploratory work undertaken in 1983 by the National Centre for School Technology, a contract was awarded to Goldsmiths to undertake a full survey.

Approach: The brief was to assess the D&T capability of a 2% sample of UK-based 15 year olds (10,000 learners). A pilot was conducted in 1987 and the full survey in 1988. Three forms of assessment were developed:

- 90 minute 'paper and pencil' tests;
- 150 minute 'modelling' tests; and
- case studies of full, extended D&T projects.

Outcomes: A major outcome was the development and validation of a model of designing based on imaging, modelling, and the iteration of thought and action. In addition, an approach to authentic performance assessment (subsequently labelled the 'unpickled portfolio') was created. This was based around short, focused holistic design activities choreographed by an unfolding assessment booklet and standardised by an administrator's script. 24 activities were devised using this structure. A further outcome was an approach to the holistic assessment of capability.

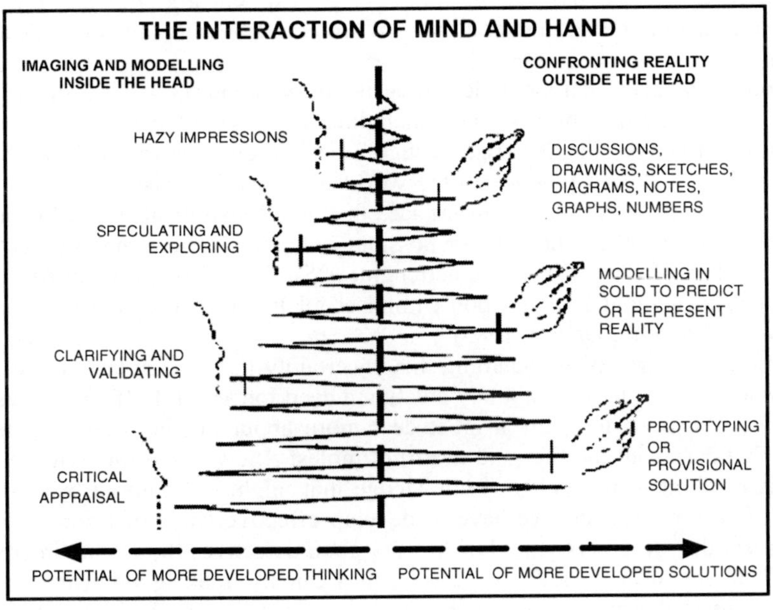

Figure 1. APU model of the design and technology process (Kimbell et al., 1991).

It was in deliberating on this process that we saw the possibility that the process might apply to a wider range of activities – well beyond the scope of 'made-world' outcomes. What if we designed a new reading book for an emergent reader? Could we argue that this is design & technology? The outcome is a product – albeit a printed-paper one – maybe with lots of interesting illustrations, and perhaps some paper engineering elements. And if this argument were to be accepted, what if we stretch it a bit further? What if there is no illustration … and no paper engineering … just a jolly good story. Have we designed it? Could we not argue that we have investigated the market, and researched themes that excite that market sector, and then created some characters, and given them a context to operate in, and created a draft (prototype), and tested it all out with a user group, and refined the draft and finalised the book? But have we designed it? Was Dickens a designer?

What is being highlighted by all this is the (potentially) ubiquitous nature of 'the designerly' as a part of everyday life – both personal and professional. But the 'potentially' part of this relates to the attitude and approach to everyday life that is taken. It is just as evident that people behave in a 'consumerly' rather than designerly way – and do this in varying degrees. For example, most of us engage in reading rather than writing novels. More are inclined to create a new interior when moving house rather than live with someone else's colour scheme and choice of wallpaper, while a small minority will completely remodel the interior, or even design and build their own house.

And, so it seems to us, this can also be applied to research – some accept the research outcomes of others; some undertake their own research, applying the tried and tested approaches and methodologies of others; and some take a more fundamental (and designerly) approach, working from first principles with 'need to know' as the driver for creating new research models, new research instruments, new ways of handling data, and new ways of presenting findings.

In what started tacitly but has become increasingly explicit and strategic, this has been the approach we have taken to research over a range of projects, spanning the last 20 years (Kimbell & Stables, 2007).

Our assumption has been that since designing is always about creative decision-making, then wherever one needs to be making decisions, including in a research context, they might be more creative and more effective if undertaken through a design lens. Central to this idea of the more general applicability of design practice to the world of research decision-making is our take on the notion of uncertainty, stemming from the idea that any design challenge has countless possible approaches and (as a result) innumerable possible outcomes. There are no right answers, or perhaps there are lots of them. The designer/engineer Sir Ted Happold claimed: "I really have, perhaps, one real talent; that is that I don't mind at all living in the area of total uncertainty" (Cross, 1990, p. 130, quoting Davies, 1985).

But one thing is for sure. When handling this uncertainty in a design task, there will be things that it is necessary to find out more about – creating a need for research. If we are designing a new tennis racket we will need to know a great deal more about playing tennis; a new mole trap requires us to find out about the life (and death) of moles; a new tap to find out about the flow of water. There is no limit to the fields of knowledge that might be required by the designer and, since we don't know everything, we have to develop effective ways of finding out about those things. We suspect that designers would make very good detectives (there's an interesting idea for a new project!), and by the same token we believe they can also make good researchers.

So this brings us to the punch-line for this introduction. It is possible for us to conceive of all task-based activities as being essentially designerly. Being designerly requires us (quite apart from the creative visualising and modelling that is characteristic of design behaviour) to have a confident and purposeful view of delving into the unknown – finding out what we need to know.

When we undertake a new research task in TERU we do so as designers since it is our view that research is essentially a designing activity.

Research and design are in many ways parallel worlds and we move now to explore the nature of these parallels and illustrate how, in our research, we have worked between the two to exploit the productive nature of doing so, achieving new ways of approaching research in design and technology education.

RESEARCHING IS LIKE DESIGNING – IDENTIFYING THE PARALLELS

We have already identified above the way in which the 'need to know' is as important a driver of researching as it is of designing – and indeed the very substance of researching is responding to a 'need to know' something. This takes us to another feature of both designing and researching – the fact that designers and researchers alike swim about in a pool of indeterminacy – where problems are 'wicked problems' (Buchanan, 1995) that contain no prescribed 'right' or 'wrong' answers. We have quoted Ted Happold above who speaks of the skill of feeling comfortable with uncertainty. In the world of designing, operating effectively in an indeterminate situation is in part based on drawing on experience, repertoire, or 'precedent' (Lawson, 2004), and in part having the confidence to explore and experiment based on an 'intuitive hunch'. So too are experience, repertoire, intuition, and hunch important to researchers. Many researchers, including Einstein, are prepared to admit to a mix of methodology and hunch in taking forward research enterprises, although for most the hunch is seen as a trigger for the trajectory of the 'real' research. Seago (1994) writes of discovery as "a combination of rigorous methodology and the following up of intuitive 'hunches'" (p. 5). But as designers, engaged by definition in a creative practice, the intuitive hunch is more than something to 'follow up' – a trigger, or hypothesis – for the research outcome. It is more an attitude that is embedded in all aspects of research processes, as creative, imaginative, and intuitive ideas are seeded continually throughout the length of a research project. Rosenberg (2000) argues strongly for this process to not be seen merely as a staging post within conventional research approaches:

> The 'hunch' is subsumed by 'methodology'. ... The input of the imagination has been so reduced that it is considered only in terms of a point of outcome ... foreclos[ing] on the notion of research ... as imaginative process in as much as the hypothesis is only valid at the moment at which it is proved through 'experimentation'. (p. 2)
>
> The action of the imagination needs to be supported and promoted as research in itself and not just read as just a 'hunch'. We need to take a position, which argues against the reification of conventional research methodologies, and most especially the requirement to conform to the formalizing protocols of their epistemologies, which in their absolute application are unsympathetic to creative practice. (p. 13)

Rosenberg is arguing his case particularly in the case of design practitioner research. But what we are tagging as an embedded attitude of designerly approaches needs defending in the same spirit – taking a creative, intuitive approach to all aspects of research allows us to ask and respond to research questions creatively – what we describe in what follows as 'what if' time, in ways that seed new research methodologies, models, and instruments in much the same way as a designer may create new products.

As with designing, being prepared to take creative approaches as a researcher requires the confidence to handle uncertainty and risk. This needs to be done in a highly professional way, as good research practice is synonymous with ethical research practice and more often than not involves other people – from the research 'subjects' to the clients who are paying the bill. This in turn raises further issues, particularly those of values and conflict, as the client – whether in a design or a research context – may neither recognise the complexities involved, nor the value of taking a creative approach, expecting rather that existing, rational approaches will be used. This dilemma is characterised by Schön (1987):

> These indeterminate zones of practice – uncertainty, uniqueness, and value conflict – escape the canons of technical rationality. When a problematic situation is uncertain, technical problem solving depends on the prior construction of a well-formed problem – which is not in itself a technical task. When a practitioner recognises a situation as unique, she cannot handle it solely by applying theories or techniques derived from her store of professional knowledge. . . . It is just in these indeterminate zones of practice, however, that practitioners and critical observers of the professions have come to see with increasing clarity ... as central to professional practice. (pp. 6–7)

One might add of the designer, or researcher.

ILLUSTRATING OUR DESIGNERLY APPROACH TO RESEARCH

Having drawn attention, in theory, to the parallels, we will now turn to indicate the potential of exploiting these parallels in practice – in taking forward innovative and appropriate approaches to researching our relatively fledgling field of technology education. The research projects that we will discuss below illustrate how we have undertaken these projects *as designers*. It's not just that we happen to be designers – it's more that we deliberately set out to be designerly in our approach to research. At the outset – 20 years ago – this might well have been simply the result of our intuitive response (or hunch) to the need to find out what we, at that time, needed to know. But as time has gone on we have come to see the deliberate value in being designerly as we undertake our research. What follows is a brief analysis of some of the elements of that value.

The Challenge of Values

Most of the research projects we have run in TERU have a sponsor; an organisation that needs to know something, that has a budget capable of launching the research, and that sees us (the researcher team) as capable of producing the required outcome. We have undertaken such projects for government departments, charitable bodies, quangos, industry, museums, professional organisations, as well as for conventional research funding bodies.

At the outset of any project we have found it necessary to behave exactly as one does at the outset of a design project – to specify the brief and to get as much clarity as possible on the matter of what might count as an acceptable outcome. This is not so that we just tell people what they want to hear, but rather so that we know in advance what *kind* of outcome is sought.

> We offer up tentative solution-types to gauge reaction and thereby get a better grip on what is really wanted. Are they looking for a statistic that will convince a policy body; or a collection of case study examples to illuminate practice; or do they seek to shape that practice in particular ways. Not infrequently the client will say 'yes' 'yes' and 'yes'... we'll have all of that. At which point it is our turn to point out that everything is not an option unless there is lots of time and money. So we help them to prioritise what they really want, and what might be a nice added extra. These underlying value debates then directly shape what we might do in the research. (Kimbell & Stables, 2007, p. 302)

Research Design

This is the area of research that we find most fun and is probably most designerly. It's 'what-if' time that equates directly to the kinds of open-ended concept work that typically initiates any design project. Essentially our early concept discussions are typically conducted round a big table with lots of sheets of scribbling paper, or around a white-board with lots of coloured pens. Ideas are knocked around and possibilities considered. Critically, these 'what-if' discussion merge into 'what-then' discussions about the consequences for data collection, time, resource, and all the practicalities of research design. We typically arrive at a few strong starting points, and – bearing in mind the values issue discussed above – share them with the client. These discussions help us to refine and detail a research plan.

Whilst we have no direct data on this matter, it seems reasonable to suggest that our experience of this process as designers has been immensely helpful to us in replicating the process as researchers.

Instrument Design

Having arrived at a research design (and therefore having arrived at some agreement about what will count as data for the purposes of this or that project) it becomes necessary to deal with the realities of data collection. We describe below – albeit briefly – the *Understanding Technological Approaches* (UTA) project (see Table 2) that we undertook for the Economic and Social Sciences Research Council, to illustrate how data collection is, quite genuinely, a product design challenge.

Our research design discussions led us to the (very expensive) conclusion that we needed to observe projects in action – minute by minute – as the work unfolded. This committed us to hours of workshop/studio observation for every one of the 20 case studies. But then …

> … the question that dominated our thinking was 'what is it we are going to observe'? A class full of learners and a teacher over a period of 48 hours (which one KS4 project ran to) can generate a prodigious amount of "stuff" to be observed. Are they smiling or frowning? Talking or silent? Working in groups or alone? With numbers or drawings or words? Engaged or off-task? Undertaking interesting or banal work? Mechanical or visual or digital? And so on ad infinitum. (Kimbell & Stables, 2007, p. 292)

Table 2. Outline of the Understanding Technological Approaches (UTA) *project, 1992–1994.*

Context: The APU project (see Table 1) only provided findings on 15 year olds and no detailed data on how learners work through full projects. We applied for and received funding from the Economic and Social Sciences Research Council (ESRC) to research to fill this gap.

Approach: We had two main research aims:

- to understand the detail of learner's approaches, across the age range from 5-15 years, to teacher-set design and technology classroom projects; and
- to explore the relationship between the data derived and the findings of the APU project.

Outcomes: The project resulted in 80 case studies of learners built on highly detailed observation of how they engaged in D&T projects. It also resulted in an innovative observation tool and approach to analysing and presenting data.

Designing effective observation typically involves a set of trade-offs, just as product development does in design. The key trade off is flexibility against data efficiency. A blank sheet of paper might be thought to be a good data collection device, so long as you are prepared to write down everything that goes on. In reality this is time-consuming and clumsy and by working out, ahead of time, what the key things are to be observed these can be pre-designed into the sheet.

Having worked up a rough observation form (a prototype) we took it into classes and attempted to use it – and having done so we gave the completed sheet to another team member to see if they could re-describe the activity that had been observed:

> Having observed only a few lessons it became obvious to us that we needed some measure of the learners' engagement with the task. We wanted to distinguish between learners who were disenchanted or disengaged or just off task, from those that were fully engaged, crashing ahead purposefully and at pace. And in trying to record these data we identified a middle category who were in what might be termed 'tick-over' mode; doing enough to be seen (by the teacher) to be working, but more going through the motions than making real progress. We evolved a categorisation of these behaviours into a three-point tick box:
>
> – stationary - going no-where/off task;
> – poddling - in tick-over mode;
> – motoring - fully engaged, making real dynamic progress.
>
> With intervening categories (e.g. between poddling & motoring) we had a 5-point scale to capture this level of engagement. It proved very easy and reliable to note and the resulting data rendered really valuable insights into learning and teaching practices. (Kimbell & Stables, 2007, pp. 294–295)

This process of prototyping went through many iterations before we arrived at our final form (Figure 2). This form represented a very careful balance of pre-coded data (completed just with a tick) with sections of free descriptive text. In this way we tracked very effectively the work of four learners – every five minutes – through their multi-hour projects in each of 20 schools. The resulting database remains probably the biggest single database of observed performance in design & technology anywhere in the world.

Bringing Data Alive

Having a big database gets you nowhere if its depth and complexity make it inaccessible (and therefore meaningless) for users. Our approach to reporting data is – at every level – informed by our design experience. We have continually striven to make sure that any messages in the data are made crystal clear and presented in a form that is immediately understandable to clients. But first they have to become clear to us, and our starting point is typically to find ways to represent the data graphically – enabling us to look for patterns that might suggest significant trends. A good example can be seen in the *Attitudes of Potential Teachers of D&T* project that we conducted for the DFEE in the late 1990s (Kimbell & Miller, 2000) (see Table 3).

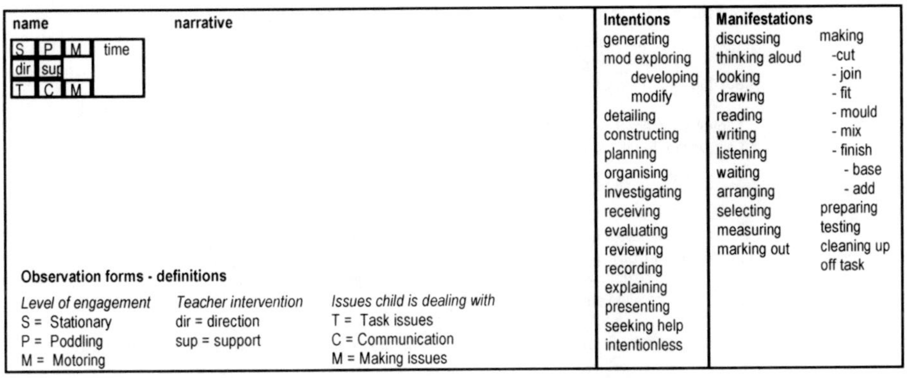

Figure 2. UTA observation tool – a five-minute segment.

Table 3. Outline of the Attitudes of Potential D&T Teachers *project, 1999-2000.*

Context: By the mid 1990s extreme concern was being expressed about poor levels of teacher recruitment, particularly from graduates from D&T related subjects. The Department for Education and Employment (DfEE) commissioned TERU to research the perceptions of D&T teaching held by such graduates and their attitudes towards becoming D&T teachers.
Approach: The team identified three groups – 'potential recruits' in the final year of D&T related degree programmes, their tutors, and students undertaking post graduate pre-service teacher programmes. Data were collected through interviews and questionnaires.
Outcomes: The project demonstrated a distinct mismatch between the aspirations of the graduates and their perceptions of teachers. In gathering and presenting these data, the team created new research instruments that enabled the collection of highly value-laden data in both qualitative and quantitative forms.

From a series of trials we identified a list of 15 qualities that were central to graduates' choices of employment and we invited the graduates to rank the 15 items (most-to-least important) in terms of their 'ideal' job. Later, in interviews, we invited them to repeat the process in terms of their view of a job teaching Design and Technology. The fits and misfits can then readily be identified by joining the dots, and the glaring misfit qualities jump off the page (see Figure 3).

This and many other presentational techniques have enabled us to make the messages from our research very transparent for clients and users. And again it seems reasonable to us to suggest that our visual design-based background has helped us to originate and develop these techniques.

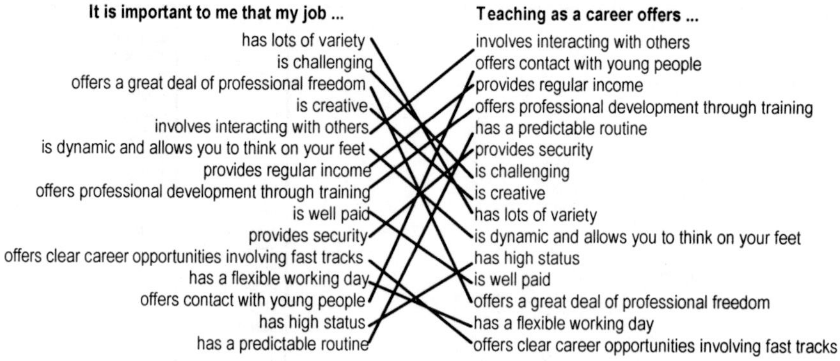

Figure 3. Contrasts in job aspirations of graduates between design and teaching D&T.

The following example is drawn from the APU D&T project (see above) and illustrates not just how we communicated research data to others, but also how we used graphical means to interrogate and understand the data within the research team itself. Figure 4 shows the raw data for average (mean) holistic assessment scores for girls and boys across the whole matrix of D&T test activities that were undertaken in the APU survey. As a set of figures it is hard to 'read'. Figure 5 presents this differently. Taking the same raw data, it shows graphically the differences in performance between girls and boys across the design contexts we used – girls performing well in the 'people' context, neither gender performing better in the 'environment' context, and a mixed result showing up in the 'industry' context. Figure 6 then takes a sub-section of the raw data to explore the mixed results – illustrating how the gender effect was further affected by the 'test structure' – in which the 'tighter' structure of the 'early ideas' test benefited girls, whilst the 'looser' structure of the 'developing solutions' test benefited boys. Providing all three images allows the raw data to be available for close scrutiny, and the graphic representations enable the overall differences to be seen with the 'naked' eye.

APU holistic performance by context (A, C & E) and gender

* indicates 5% sig.
** indicates 1% sig.

	People	Environment	Industry
T.1 Boys	2.19	2.03	1.71
T.1 Girls	2.42 *	2	1.95*
T.2 Boys	1.89	1.77	1.99
T.2 Girls	2.02	1.85	2.21**
T.3i Boys	1.97	2.23	1.87**
T.3i Girls	2.06	2.19	1.58
T.3ii Boys	1.96	1.8	2.41
T.3ii Girls	2.08	1.77	2.29
T.4 Boys	2.15	2.09	1.96
T.4 Girls	2.5**	2.40**	2.29**
Mod Boys	2.29	2.55	2.65*
Mod Girls	2.38	2.37	2.38

Figure 4. APU D&T Project raw data for holistic performance scores.

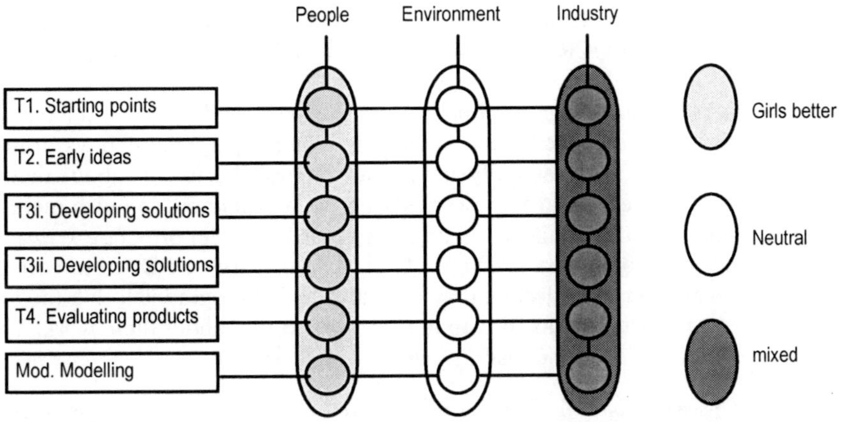

Figure 5. Presenting the data from Figure 4 to illustrate the overall gender effect across contexts.

Figure 6. Presenting a section of the data (cf. Figures 4 and 5) to illustrate the overlapping effects of context and test structure.

Going Digital

Our current project – *e-scape* (see Chapter 42) – provides our final two examples.

In a previous project, *Assessing Design Innovation* (Kimbell et al., 2004), we had been working on the challenge of enabling assessment to overcome the well-known problem of what Mike Ive (former chief HMI for D&T in English schools) called 'neat nonsense' portfolios in which learners dress up their work (often with the explicit support of their teachers) to maximise the hits against assessment criteria. This problem frequently resulted in the most innovative students *not* getting the best marks – since these high marks went instead to those who were most attentive to (or who had teachers who were most attentive to) ticking all the boxes. A linked aim of the research project therefore was to find ways of explicitly rewarding innovative designing.

We developed an approach to 'real-time' portfolios that originated in our APU experience but that we developed into six-hourly structured tasks. The project resulted in a new form of assessment, best described as half way between an exam and a piece of coursework. It was adopted by the Awarding Body OCR and exists as the 'innovation challenge' in their new Product Design GCSE.

The *e-scape* project (see Table 4) grew directly from this work and arose from a passing remark from a teacher who was running one of the activities. Watching a pair of boys arguing in one corner of the workshop (they were actually having an intense discussion about their work) he pointed out that if we could hear what they were saying we would know so much more about their work and their understanding of it. This set us on a trail that led us (in project *e-scape*) to digitise the activity using hand-held computers (Personal Digital Assistants, or PDAs) to collect all kinds of data. Specifically – in addition to collecting drawings, text, and photographs to log the evolution of learners' design ideas – we used audio-files to get learners to talk about their work. This 'design talk' proved a revelation and a fantastic discriminator of performance.

Table 4. Outline of the e-scape *project (e-solutions for creative assessment in portfolio environments), 2005 and continuing.*

Context: Assessing Design Innovation introduced approaches to digital data capture, indicating the possibilities in embedding further digital tools into the activity and creating portfolios on handheld computers that could be uploaded onto the internet. This afforded opportunities for learners and for new approaches to assessment, including controlled coursework assessment.
Approach: The project (still underway) is being conducted in four phases:
Phase 1 – creating a digital approach to the 'unpickled portfolio'; proof of concept
Phase 2 – develop and trial the methodology in D&T classrooms
Phase 3 – create and pilot a system that will work for national assessment in D&T and trial the approach in Geography and Science
Phase 4 – pilot a full national assessment system
Outcomes: The project is resulting in digital portfolio-led assessment in a range of subjects for GCSE examinations. It has also resulted in a unique approach to assessment based on 'pairs judging'. In addition it has spawned a new project exploring the potential of the approach for formative assessment.

We have reflected on the process that led us to the research solution that is encapsulated in *e-scape*. We were behaving as researchers – trying to maximise the research data that would inform us about the capability of the learners. But equally we were behaving as designers – tweaking the activity through endless iterations to encourage better design performance. We have concluded that the two aspects of our role were inseparable.

The same applies to our final example, which also arose within *e-scape*. It concerns the assessment process itself once all the web-portfolios have been completed by learners and logged into the secure *e-scape* website. The example reflects a pure piece of 'what-if' thinking. We were in discussion with the assessment experts at an examination board who were using a Thurstone Pairs technique (Thurstone, 1927) to check the reliability of some of their previously conducted assessments. Essentially it's necessary to know that an A grade means more or less the same standard across all the examination boards. To do this they take a small sample of scripts from each board and do a series of comparisons (e.g., 1 with 23, 2 with 16, 3 with 8), just recording which is better in each case. The result of many such comparisons can be shown as a rank-order with the best work at the top (they win every comparison) and the least good at the bottom (they always lose). It's a standard piece of examination board cross-moderation.

But what if the work had never been marked in the first place? What if we just looked at portfolios for the first time as sets of pairs? Would a Thurstone analysis provide a rank-order of performance? This is not normally a question that can be asked, because of the logistic difficulties that would be involved in comparing hundreds of different scripts/portfolios. But *e-scape* portfolios are web-based and can be called up at will … by anyone with the passwords … sitting at a computer anywhere in the world. Wow.

To cut a long story short, it does work, and with astonishing reliability, because the comparisons can be undertaken by many judges and because each portfolio is compared to many (20+) others (see Chapter 42). This combination stands in stark contrast to normal GCSE marking, which is undertaken only by the teacher and only in the light of that school's work. The full details of this can be read in the *e-scape* research report (see TERU website: www.goldsmiths.ac.uk/teru).

Based on this idea (that was first tested in the *e-scape* sample of 2006) – and being designers at heart – we have now commissioned a new piece of software. The Pairs Engine will select pairs of portfolios and present them to judges. In the new *e-scape* sample of 2008 we will have 350 portfolios – each compared to at least 20 others and seen by 28 judges. The Pairs Engine will work dynamically, responding to the input of each paired judgement and updating itself to select the next pair to be judged. The process will run continuously and stop only when the standard error on judgments has been reduced to an acceptable figure. We anticipate that it won't take long for the judging process to improve on the standard error of GCSE and most conventional assessments. And all the judges have to do is to read the portfolios and make a holistic balancing judgement about which is the better piece of work; which candidate displays the greater capability.

Of course there is a whole debate to be had here about holistic judgement, and its advantages over the more normal atomistic variety. As we write, this debate, and this project, continue.

DESIGNERLY RESEARCH AND DESIGN-BASED RESEARCH METHODOLOGY

As is clear from the case we make in this chapter, we have a strong belief in the value that design holds for research. Whilst it is an approach that we have both tacitly and explicitly employed for some 20 years, more recently we have become conscious of groups of researchers beyond the world of design who hold the same view – those who actively engage in 'Design-Based Research Methodology' (e.g., Kelly, 2003; Design Based Research Collective, 2003). This group of researchers, predominantly working in educational research, recognise the way design methodologies deal with complex, real situations in an empirical, robust, and 'pragmatic' (Gorard & Taylor, 2004) way that translates well into educational research. They speak in a language that we easily understand:

> Good design is iterative. The process of creating something to address a goal is repeated many times as the designed artifact or process is tested, observed, and refined. The iterative nature of design is often missing in research, but is vital in testing our interventions. By repeatedly creating, implementing, enacting, and improving our interventions, one begins to understand intuitively and empirically what works and what doesn't, and also which features of the design are essential and which are irrelevant to the goals. In typical design … this type of refinement is an informal way of doing research. … The sage researcher uses mixed methodologies combining informal and formal methods according to costs and benefits. (Hoadley, 2002, pp. 453–454)

There is a clear sense in the literature surrounding this approach to research that it is developing into an important research paradigm in educational research because it is happy to used mixed methods, informal and formal, qualitative and quantitative, driven by the 'need to know' of the research questions, in much the same way as we have described above. In the words of Gorard and Taylor (2004):

> Once on the road to conduct research, everything is potentially informative and the researcher becomes a 'Hoover' of data, as far as possible. The research starts with draft research questions, and continues with an attempt to answer them by whatever means it takes. It is difficult to imagine why anyone would do anything very different. (p. 5)

So, at a time when, in TERU, we are beginning to take our 'designerly' approach to research and apply it to other educational settings, other researchers are coming to meet us from the other direction. Maybe this indicates an opportunity for collaboration – an opportunity for researchers in technology education more generally to be sharing new ideas and approaches to research not just with each other, but also with a wider research community. Hake (2004), in writing about the parallels between design-based research and physics education research, writes of the "deplorable insularity of educational research" (p. 3) – an accusation that we in technology education are not immune to and should not ignore.

Our work in TERU continues to shimmy between designing and researching. With the latest developments in the *e-scape* project, researching the activity involves redesigning it, and redesigning it presents new research opportunities, such as pairs judging, that in turn lead us to new design developments, such as the new software that enables us to redesign the assessment process. And so it rolls on. The new assessment 'Pairs Engine' has taken us into the more generic world of assessment research and has prompted huge interest in fields outside design and technology education. We are now involved in re-designing activities in English, geography, and science. But it's research with English, geography, and science through the eyes of designers – an approach to research we believe has value within and beyond the world of technology education.

REFERENCES

Buchanan, R. (1995). Wicked problems in design thinking. In V. Margolin & R. Buchanan (Eds.), *The idea of design* (pp. 3–20). Cambridge, MA and London: MIT Press.

Cross, N. (1990). The nature and nurture of design ability. *Design Studies, 11*(3), 127–140.

Design Based Research Collective. (2003). Design-based research: An emerging paradigm for educational inquiry. *Educational Researcher, 32*(1), 5–8.

Design Council. (1995). *Definitions of design*. London: Author.

Design Council. (1997). *Insights* (Vol. 1). London: Author.

DES/WO. (1988). *National Curriculum Design and Technology Working Group interim report*. London: Department of Education and Science and the Welsh Office.

Gorard, S., & Taylor, C. (2004). *Combining methods in educational and social research*. Maidenhead, UK: Open University Press.

Hake, R. R. (2004). *'Design-based research': Old PER wine in a new bottle*. Submitted on 16 July 2004 to the PERC 2004 Physics Education Research Conference, 4–5 August 2004, Sacramento, California. Retrieved February 1, 2009, from http://www.physics.indiana.edu/~hake/PERC2004-Hake6.pdf

Hoadley, C. (2002). Creating context: Design-based research in creating and understanding CSCL. In G. Stahl (Ed.), *Computer support for collaborative learning: Foundations for a CSCL Community. Proceedings of CSCL, 2002, Boulder, CO*. (pp. 453–462). New Jersey: Lawrence Erlbaum Associates.

Kelly, A. (2003). Research as design. *Educational Researcher, 31*(1), 3–4.

Kimbell, R. A. (1991). Tackling technological tasks. In B. Woolnough (Ed.), *Practical science* (pp. 138–151). Buckingham, UK: Open University Press.

Kimbell, R., & Miller, S. (2000). *Attitudes of potential teachers of design & technology* (Vol. 194). London: Department for Education and Employment (DfEE).

Kimbell, R., Miller, S., Bain, J., Wright, R., Wheeler, T., & Stables, K. (2004). *Assessing design innovation: A research and development project for the Department for Education & Skills (DfES) and the Qualifications and Curriculum Authority (QCA)*. London: Goldsmiths University of London.

Kimbell, R., & Stables, K. (2007). *Researching design learning: Issues and findings from two decades of research and development*. Dordrecht, NL: Springer

Kimbell, R., Stables, K., Wheeler, T., Wozniak, A., & Kelly, A. V. (1991). *The assessment of performance in design and technology*. London: SEAC/HMSO.
Lawson, B. (2004). *What designers know*. Oxford, UK: Elsevier.
Rosenberg, T. (2000). 'The reservoir': Towards a poetic model of research in design. *Working Papers in Art and Design, 1*, 1–12.
Schön, D. (1987). *Educating the reflective practitioner: Towards a new design for teaching and learning in the professions.* San Francisco: Jossey-Bass.
Seago, A. (1994). Research in art and design; Conflicts and contradictions. *Royal College of Art Research Papers, 1*(3).
Thurstone, L. L. (1927). A law of comparative judgement. *Psychological Review*, *34*, 273–286.

Richard Kimbell
Kay Stables
Technology Education Research Unit (TERU)
Department of Design
Goldsmiths, University of London

CAROLYN JONES

51. BEING REFLECTIVE AND RESPONSIVE IN EARLY CHILDHOOD RESEARCH

An Example of a Technology Education Episode

INTRODUCTION

In considering the provision of supportive learning environments for students it is important to acknowledge and build on the complexity of each student's world. The work described in this chapter is strongly embedded in a sociocultural framework. This approach, discussed in more detail in Chapter 52, expands the notion of learning beyond one of cognition only, to one that includes environment, people, context, etc. The implication of this for research is that new tools are then required for data collection. The purpose of this chapter is to explore what those tools might be in an early childhood research context, both in terms of methodology and methods. A technological episode is highlighted to illustrate this research approach.

The context for this chapter is the *Dispositions in the Social Context* (DISC) project, a three-year project that tracked 27 children from the age of around three years nine months, in their early childhood settings (Phase 1), through Phase 2 just before the children started school, to when the children were about five and a half and had been at school for a few months (Phase 3).[1] The project examined the relationship between the learning dispositions of the children and the learning architecture of the contexts they moved within, for example, their families, the early childhood settings, and later their schools (Carr, Smith, Duncan, Jones, Lee, & Marshall, in press). Exploring very young children's experiences and learning in technology is not easy (Mawson, 2007) and this chapter highlights fruitful research approaches.

METHODOLOGICAL APPROACH

The methodology was designed to frame our research with young children in such a way to allow an investigation into the development and mobility of the children's learning dispositions, or habits of mind, across place and time. The methodology was based within an interpretive framework. That is, we were aware that there could be more than one interpretation of what we were observing and that our methodology and methods would have to reflect this by taking into account the multiple voices of the teachers, the children, the families, and our own. In order to do this we based our work on the Mosaic approach developed by Clark and Moss (2001), taking cognisance of the way "The Mosaic approach acknowledges the importance of a framework for listening, which does not exclude the perspectives of practitioners and parents; a culture of listening should extend to all involved with an early childhood institution" (Clark, 2005, p. 39).

This fitted well with a second methodological aspect, which built on the concept of funds of knowledge as discussed by González, Moll, and Amanti (2005) and recognised the value of a

A.T. Jones and M.J. de Vries (eds.), International Handbook of Research and Development in Technology Education, 615–623.
© 2009 Sense Publishers. All rights reserved.

family's knowledge of their child as well as their child's experience in settings outside of the early childhood centre or school classroom.

Finally, underpinning all of the above it was also important to be respectful and responsive to the needs and responses of the children, families, early childhood centres, and classrooms during the course of data collection and to constantly be reflective about the effectiveness of our research methods in gathering the type of data we were seeking. A large part of this involved being responsive to what the children were telling, and not telling, us.

Cohen, Manion, and Morrison (2000) suggest that "researchers must provide a credible and meaningful explanation of their research intentions . . . and that children must be given a real and legitimate opportunity to say that they do not want to take part" (p. 52). Brooker (2001) adds: "While other professionals may attempt to cajole children into participating in activities, researchers must be scrupulous in offering children the right to opt out if they choose, before *or* during an interview" (p. 166).

Aspects of the Mosaic approach, funds of knowledge, and being responsive will be reflected on in the sections that follow, alongside a discussion of specific methods of data collection.

METHODS OF DATA COLLECTION

A range of methods were chosen to gather as much information as possible about not only what the child was doing and experiencing, but also how this connected with and reflected or built on their learning experiences in other contexts across both time and place. As discussed above, the Mosaic approach pointed to the importance of identifying methods that would involve input from all of those central to the child, that is, the child, the teachers, other children perhaps, and of course the families.

Recording Children's Voices Using Wireless Microphones

Margaret Carr, involved in the DISC project, and a colleague, Jenny Young-Loveridge, had previously worked on projects that entailed recording children's voices via remote microphones placed inside jackets worn by the children. This became the starting point for considering how the children's voices might be captured during the early childhood phase of this project and this proved to be a major source of data. Small, sleeveless, light calico jackets were made with adjustable side straps to enable them to fit a fairly wide range of sizes of young children. A pocket was sewn onto the back into which a small transmitter for the microphone was able to easily be inserted and removed. A second sleeveless jacket was attached over the plain base jacket and it was this jacket that the children were really interested in, as a range of materials were used in the jackets' construction. Some over-jackets were made of bright orange fabric with silver tape placed to make them look like builders' reflective jackets, others were made of fabric with pink sequins, Pasifika designs, black fabric with stars that a magician might wear, and a number of other designs.

About 15 base jackets were made with a larger number of over jackets. There were two reasons for this. Although we were only planning to record two or three children at any one time we felt it was important to have sufficient jackets so any child in the early childhood centre who wanted to wear one could do so. This would mean that the case study children would not necessarily feel different or stand out in any way. It also was a way of acknowledging the specialness of the other children in the setting. Secondly, this approach was seen as a way of empowering the children in terms of giving them a choice of over jacket that they wanted to wear, and that this would hopefully engage them in wanting to wear the jackets. The fabric designs were chosen to reflect as wide a range of tastes as possible and included designs that we felt would be familiar to the children. We wondered if the wearing of the jacket would influence the nature of the games the children played, for example, if wearing a magician's or builder's jacket would perhaps lead them in that direction of play, but there was very little evidence of this.

The jackets were not left at the centre in between visits in order to maintain the novelty of wearing them. However, this meant that we had to arrive at the setting early enough to set everything up and necessitated a number of bags of equipment. In one setting a cry would go up from some of the children as we were spotted arriving, "The jacket-ladies are here!" and there would sometimes be a scramble for the favourite jackets.

Most of the children, most of the time, were happy to co-operate and wear the jackets with the microphones in them when we asked them to. However, there were times when they were less inclined to do so. This raised the issue outlined above about balancing the need to have the appropriate data in order to be able to draw robust conclusions and the need to empower and respect the children in terms of ethical research. It also highlighted for us the need to be responsive to the children's ideas. We decided to find an alternative way of having the children carry the microphones and transmitters around and found some bum bags (also known as fanny packs, belt packs, or hip packs) with Tigger, Piglet, or The Wiggles[2] on them. The transmitters fitted perfectly and the children were very excited by them. The children's demands for the bags out-did the supply available and turn-taking had to be negotiated. This was done on a daily basis, rather than taking turns during the day, to enable a target child to be followed for the whole day. However, when at times a child did not want to wear a jacket or bum-bag at all this was respected and observation notes (see below) were taken without the audio recordings.

Using the wireless microphones allowed us to capture the children's voices as they played, giving us greater insight into the connections they were making with people, places, and things. It also allowed us to capture the voices of those children playing with those wearing the microphones, and all their interactions. Some of the children became very interested in the way that we were recording their voices via the microphones onto tape recorders in the staff office. We talked them through the process and showed them all the equipment and how it worked and they tested it out for themselves. Sometimes, during the course of their play, they would stop and hold the microphone close to their mouths and talk into it, usually something like "testing 1 2 3", before running off again. However, these incidences became quite rare after the initial period once they had become very familiar with the jackets and microphones.

During the school phase of the project it was decided that this method of data collection would be too intrusive and disempowering for the children as at times our target child would be the only child we were observing in a classroom. Instead we carried small hand-held tape recorders and where possible placed these on a table near to the child in order to capture the conversations.

Capturing Children's Discussions

Another method used to try to access the children's ideas about what they were doing was to take digital photos of the children during the day and then ask them to tell us about the photos. This was seen as a way of triangulating the data from our observation notes to strengthen the internal validity of the project. We knew the children were familiar with having photos taken while in the educational settings, as the teachers would use this as part of their teaching and assessment practices. In Phase 1, when the children were around the age of four, we began the process by showing pictures to the children of what they had done the previous day and asking them to talk to us about it while we recorded these conversations, either on a small tape recorder on the table or through the remote microphones if they were wearing a jacket. These conversations were done at the early childhood setting. At this stage the children did not always say very much. After reflecting on this, we asked a teacher to sit with the children who had been in the photos, the target child and perhaps one other child, and talk to them about it, thinking the child would be more relaxed with an adult with whom they were familiar. This was more successful, especially in centres where teachers and children revisited their learning portfolios (a collection of Learning Stories[3] written about each child as an assessment tool and sometimes discussed by the teacher with the child).

Our next approach was to print out copies of the photos with space around them and to give these to the families and we asked the families to talk to the children about what they had been doing in the photos. This was the most generally successful, with the families writing the children's responses around the photos. We also gave the families copies of the photos to keep.

In considering the initial lack of success of this method it is useful to note the reflections of Graue and Hawkins (2005) who document their experiences of interviewing school children and note importance of the context in which the interview is carried out. In their reflections, they try to imagine the interview experience from the perspective of the child and by doing so identify how we are asking children to participate in something where they are not sure what it is we actually want and therefore are unsure of how to respond. This is supported by Brooker (2001), who writes:

> On the whole, researchers agree that the limitations to young children's competence as respondents are generally the limitations of those who interview them. Well-known studies of young children's language (Tizard and Hughes, 1984; Wells, 1985) confirm the commonsense observation of anyone working with the very young, that children's utterances are better in every way (longer, clearer, more complex, more thoughtful) when the children are in a familiar environment, with familiar adults. (p. 167)

When we began our conversations with the children, we thought that conducting these conversations in a familiar environment would be enough to enable the children to feel comfortable talking to us, but this was not very successful. We then thought that perhaps talking with a familiar person, such as their teacher, may help, especially as in some cases the children had English as a second language and for some children the teacher was able to converse with them in their home language. This was also not as successful as we had hoped as these were not the usual conversations they had with their teacher and they were perhaps unsure of what was required of them. From the child's perspective the teacher had been there and had seen what had happened so what was it they were asking? By giving the families the photographs and asking them to discuss the pictures at home, the conversations were potentially a lot more natural, familiar, and meaningful to the child. This also empowers and involves the family in the research process. In this instance, the family were the best-placed people to elicit the type of information we were seeking, namely, the children's stories about what was happening in the photos.

It must be noted, however, that because of the type of information being sought and the need at times for an experienced interviewer, it is not always desirable or practical to ask others to conduct interviews, or to involve the families in the way that we did. However, our experience suggests three aspects that enhanced the possibility of eliciting high quality responses from the young children in our study: The environment should be familiar and one they are comfortable in; the person conducting the interview or conversation should also be someone they are at least familiar with; and the conversation itself should be in a format that to them can appear authentic and familiar in relation to the person they are talking to Perhaps an alternative strategy to encourage children to reflect on the activities they were involved in would have been to sit with them and write a Child's Voice story (an aspect of the Learning Stories referred to earlier) about the episode of interest, as this would be a much more authentic and familiar process or conversation, in this setting, with this person, for the child.

Observations

Another method of data collection that we used was the observation of children in their early childhood setting. Each child was observed for about three days per phase. Observation notebooks containing running records were kept for each child with a timescale in the margins to allow us to match the observation to the remote microphone tape recording. As a team we decided not to each focus on particular case study children but instead we took turns observing each of the children, although at times one person would do more of a certain child's observations and less of another's. Informal debriefing sessions between the researchers, which were tape-recorded, were held after the observation sessions and we would discuss the children who had been observed that day, what we had observed, and how the observations fitted with previous observations made by other team members. This debriefing ensured that continuities and discontinuities in each child's learning experiences could be identified across different observation periods. It also allowed us to mitigate, to some degree, against observer subjectivity by discussing what we had observed and what our interpretation of it was. In her work on direct observation Rolfe (2001) describes this as follows:

> Different observers, watching the same child at the same time, can generate different observational records … The reason for the differences resides very much in the subjectivity of observation, particularly if the observer is attempting to write down all that is seen. As human observers, it is inevitable that our own feelings and interpretations influence what we see or don't see. The behavioural record reflects, to a greater or lesser extent, our 'biases, preconceptions and emotional responses' (Nicolson and Shipstead, 1998, p. 13). In other words, we see the world around us through a lens; what I see and how I interpret it may not be the same as what you see, or how you interpret it. We may notice different behaviours, focus on certain events or interactions more than others or simply 'miss' certain behaviours because we are distracted or interrupted, even for a brief period of time. (p. 231)

Notes were made about a number of aspects: where the child was in the setting; who they were with; what we were interpreting the play to be about (which was considered later with reference to the tape recordings to try to ascertain the child's conversation); how long they stayed in any one place; their mode of participation (were they watching, joining in on the periphery, actively engaged in an episode of joint attention with either another child or children or an object, for example, a jigsaw); what were some of the possible affordances or constraints that had led to what they were doing (e.g., they were following a friend around to various activities, art materials were easily accessible at all times, the parents tended to congregate around the area where the jigsaws were kept which meant there was an adult easily accessible there).

In addition, anecdotal records, in the form of Learning Stories, were also written to describe and analyse each episode and the corresponding learning in more depth. Copies of these were also placed in the children's assessment portfolios or given to the families. Digital photographs also proved a very useful method of collecting data in connection with the observation notes. These allowed us to capture the context of what was happening, who was involved, where it happened, etc.

The children became very familiar with the researchers, especially after the home visits (see below), and in general did not seem to find their presence in any way intrusive while the observations were being carried out. When, on the odd occasion where it seemed to us that a child had become increasingly sensitive to us being nearby, we would stop the observations and move to another child. Again, this was an important part of being ethical in our research with young children and allowing them the opportunity to stop the observation.

Photography

The taking of photographs was embedded throughout the other methods described above: they were used to elicit responses from the children about what they had been doing, and they were used throughout the observations as an additional way to record what the children were doing as well as the context of each episode.

In addition, in Phase 2 the families were given disposable cameras prior to a scheduled visit by the researchers, and were asked to take photos of what the children did outside of their time at the early childhood setting over a two – three week period. It was suggested that this might include the places they went to, the people they visited or who visited them, the things they like to do, or the organised activities they attended, such as swimming lessons. The cameras were then collected and the photographs developed. A copy of all pictures was given to each family during the next time. They were used as triggers with the family and child to discuss the sorts of things they liked to do and the community in which they lived. A more detailed description of these visits is given below.

As part of our ethics application, we had felt it was important to maintain the anonymity of the children, the early childhood settings, and the families. We had sought permission from all the families – not just those of the target children – at the early childhood settings and primary school classrooms to take photos on the understanding that these would only be used for data collection and analysis purposes and that no one would be identifiable in pictures that may be used for publication or dissemination. We were very careful throughout to ensure that whilst we might take pictures of the children from the front to assist in our data analysis, we also took photos from behind the children where feasible, to enable us to have images we could use for a wider audience.

Family Interviews

The families were interviewed in their homes as part of all three phases of the project. The interviews were carried out at a time suitable to the families, which meant some were done during the day, over a parent's lunch break, and some in the evenings. In one early childhood setting the head teacher also came to the interviews at home, something the families seemed to very much appreciate as the teacher was careful to show that she saw it simply as a way of getting to know the family better. The culture of the early childhood settings were very inclusive of parents and families, and in one centre the families came from a wide range of nationalities. As a way of building an initial relationship, the researchers attempted where possible to be at the early childhood setting when the children were dropped off so that informal conversations could occur with the families.

Two or three researchers would attend each visit and because one researcher often talked more to the child, in the same room, the families did not seem to be overwhelmed by this, especially once they were more familiar with us. On several occasions members of the extended family, who lived with the child's family, were also present and a number of conversations would happen simultaneously.

In Phase 1, the interviews had slightly more of a getting-to-know-each-other focus and because the families did not really know us at this stage we did not feel it was appropriate to record the interviews. In the later phases the number of people present often made the recording and transcribing of the interviews very difficult. However, the transcripts were used as a backup to the debriefing that occurred after each visit, which again were often tape recorded.

By Phase 2 the children were more familiar with the researchers, having previously had them visit at home and seen them regularly at their early childhood education setting. In addition the conversations were carried out in the child's home with their parents present. The mode of the conversation was also familiar, in this context; it was a genuine discussion about places and events that the researchers genuinely didn't know about and about which the child was expert (these interviews took place after the family had taken photographs at home with the digital camera). Finally, the children were also nearly a year older than the first time we visited. Because of these factors this method of data collection was very effective. At times the parents or other family members would chip in and add their voice to the conversation, often when the child couldn't remember something or to clarify something that the child said. Because we were mainly seeking information about the things the child did at home or in his or her community, and were not focussing in particular on the child's view of these things, this sort of interruption seemed natural and enhanced the quality of the data we obtained. We ensured we made space in the conversation for the child to continue their commentary on the photographs and the interruptions did not seem to be off-putting for them.

AN EXAMPLE IN A TECHNOLOGY SETTING: DAVID'S CONCRETE

As discussed above, the Mosaic approach to data collection involved all the participants in the gathering of the data. The example that follows clearly demonstrates the value of the data gathered by the family, and the information that only the family held about the particular child, David (a pseudonym). The methods we used helped us to build our understanding of the complexity underpinning the particular learning identity we were exploring in this instance – that of being an expert. David demonstrated this learning identity in the episode described below, at his early childhood centre, and later in another episode in a completely different context at school. The analysis of the rich data for the episode described below, which relates to David's experience making concrete, and a later episode at school, can be found in Duncan, Jones, and Carr (2008).

One morning, during Phase 2 of the project when the children were aged about 4 years and 9 months, word arrived at the early childhood setting that a concrete truck had arrived at a house that was being built a couple of houses away. The children were quickly organised and we all set off to watch the concrete floor being poured. David was especially interested. At this point his jacket and microphone were removed as they would not work at a distance, but a researcher was nearby, notebook in hand, recording as much as possible about his engagement and comments. He was talking to the teacher using words such as 'cement' and noting the use of water. He seemed to know quite a lot about the process.

His jacket was put back on when he got back to the centre and he immediately went out to the sandpit with his friend and the value of microphone became evident as we were able to capture his whole conversation, his absolute focus on wanting to make concrete, and his extensive knowledge on the subject. The observation notes describe the intense look of concentration and the digital photos taken of the scene document those children who stayed with David over a period of time, and those who came and went, as well as the ever increasing complexity of the environment as more resources are added.

This was a very sustained, involved episode for David and lasted close to two hours, ending only when it was time for everything to be packed up. During this time, he interacted with other children and adults, showing them exactly how to mix the concrete; and he brought in all kinds of other resources to the sandpit, including a bag of flour that was substituted for concrete, water for mixing his concrete, and pipes to add to the construction. He had negotiated space for his construction and had not been distracted from his task when others around him had wandered off or changed their game to something else. It had been a very meaningful activity for him.

However, the full complexity and richness of the episode did not become apparent until additional data were considered. Conversations that David had with teachers around the morning-tea table on other days, again picked up on the jacket microphone, showed that David and his father often did activities together at home. For example, he described how they baked biscuits together, and knew the ingredients that went into them. When the photographs from the disposable camera were developed there were several photos of David with his father and another adult mixing concrete using a concrete mixer and building a wall together at home. The grin on David's face in the photos said it all! These photos, and telling the family the story of David making concrete in the sandpit at the early childhood setting, provided a focus for further discussion with the family and from this we were able to explore the connections that David was making between activities he did at home and what was happening in the early childhood setting – or, as Wenger (1998) describes it in his dimensions of educational design, "*the local and the global* – educational experiences [connecting] to other experiences" (p. 264).

CONCLUSION

Exploring student learning in technology is concerned with not just the cognitive, but also includes manipulation of artefacts in relation to environments, people, and contexts over time. This chapter has explored ways in which data may be collected about not only what a child might be doing and experiencing at a particular time, but how this is connected with and builds on their learning experiences in other contexts across time and place. In exploring and understanding young, including very young, children's broad learning experiences, recognition of funds of knowledge (González et al., 2005) is fundamental. As well as viewing a family's skills and knowledge as resources the teacher can tap into, as advocated by González et al., it is also possible to see the family as a mediating factor in terms of the child's interactions with the various learning communities the child is involved in (Jones, 2006). This highlights the necessity of developing reciprocal and responsive relationships between the family and the educational centre or classroom.

Research with young children needs also to be respectful and responsive to the needs of the children, their families, and the learning environment, including the right to opt out. This chapter also highlights ways in which children and families take an active part in the research. Photography is not only a way of recording children's experiences but can also act as a trigger for discussion with children and their families. In addition, establishing long term relationships children, their families, and educational centres enhances the validity of the data and builds an understanding of the complexity of learning.

NOTES

1 The DISC project was jointly headed by Professor Anne Smith from the Children's Issues Centre, University of Otago, and Professor Margaret Carr from the Wilf Malcolm Institute of Educational Research, the University of Waikato. Whilst very similar data collection methods, based on the same methodology, were used by the team members based at Waikato and Otago, the techniques differed slightly in how they were managed because of the need to be reflective and responsive to the children and the environment, in this case early childhood education settings and Year 1 primary school classrooms. The Otago-based team have reflected on their methods in Smith, Duncan, and Marshall (2005). The experiences of the Waikato team form the focus of this chapter.

[2] The Wiggles are an Australian group who sing and perform for young children and who were popular at the time of the project.

[3] The Learning Story approach to assessment in early childhood was developed by Margaret Carr (2001), who describes learning stories as

> ... written vignettes of individual children displaying one or more of the five target domains of learning disposition. The five domains of disposition are translated into actions: taking an interest, being involved, persisting with difficulty or uncertainty, expressing an idea or a feeling, and taking responsibility or taking another point of view. (p. 96).

REFERENCES

Brooker, L. (2001). Interviewing children. In G. Mac Naughton, S. Rolfe, & I. Siraj-Blatchford (Eds.), *Doing early childhood research* (pp. 162–177). Crows Nest, Australia: Allen & Unwin.

Carr, M. (2001). *Assessment in early childhood settings.* London: Paul Chapman.

Carr, M., Smith, A., Duncan, J., Jones, C., Lee, W., & Marshall, K. (in press). *Learning in the making: Disposition and design in the early years.* Rotterdam: Sense.

Clark, A. (2005). Ways of seeing: Using the Mosaic approach to listen to young children's perspectives. In A. Clark, A. Trine Kjϕrholt, & P. Moss (Eds.), *Beyond listening: Children's perspectives on early childhood services* (pp. 29–50). Bristol, UK: Policy Press.

Clark, A., & Moss, P. (2001). *Listening to young children: The Mosaic approach.* London: National Children's Bureau for the Joseph Rowntree Foundation.

Cohen, L., Manion, L., & Morrison, K. (2000). *Research methods in education* (5th ed.). London. Routledge/Falmer.

Duncan, J., Jones, C., & Carr, M. (2008). Learning dispositions and the role of mutual engagement: Factors for consideration in educational settings. *Contemporary Issues in Early Childhood, 9*(2), 107–117.

González, N., Moll, L., & Amanti, C. (2005). *Funds of knowledge: Theorizing practices households, communities and classrooms.* London: Lawrence Erlbaum Associates.

Graue, E., & Hawkins, M. (2005). Relations, refractions, and reflections in research with children. In C. Lewis, P. Enciso, & E. Birr Moje (Eds.), *Power & voice in research with children* (pp. 45–54). New York: Peter Lang.

Jones, C. (2006). Continuity of learning: Adding funds of knowledge from the home environment. *Set. Research information for teachers, 2*, 28–31.

Mawson, B. (2007). Factors affecting learning in technology in the early years at school. *International Journal of Technology and Design Education, 17*(3), 253–269.

Rolfe, S. (2001). Direct observation. In G. Mac Naughton, S. Rolfe, & I. Siraj-Blatchford (Eds.), *Doing early childhood research* (pp. 224–239). Crows nest, Australia: Allen & Unwin.

Smith, A., Duncan, J., & Marshall, K. (2005). Children's perspectives on their learning: Exploring methods. *Early Child Development and Care, 175*(6), 473–487.

Wenger, E. (1998). *Communities of practice: Learning, meaning and identity.* Cambridge: Cambridge University Press.

Carolyn Jones
Wilf Malcolm Institute of Educational Research
University of Waikato
Hamilton, New Zealand

BRONWEN COWIE AND JUDY MORELAND

52. METHODOLOGICAL CONSIDERATIONS IN STUDYING CLASSROOM INTERACTIONS IN TECHNOLOGY EDUCATION

INTRODUCTION

The research orientation for classroom studies has changed over time as researchers have contributed to new and different views of knowledge and learning, accessed new technologies for data collection and analysis, and sought to accommodate a changing political landscape. Early classroom studies identified patterns of interaction and contributed to understandings of the active nature of the learner in sense-making processes. Later studies highlighted the social and situated nature of learning. Running alongside this work, studies in the sociology and philosophy of technology have led to a deeper understanding of the nature of the work of technologists – their goals, motivations, and practices, as well as the situated cultural, economic, political, and value-laden nature of their work. Debates about what counts as 'evidence' and challenges to the value of educational research coupled with the push for research-informed policy and evidence-based practice have further impacted on approaches to classroom-based research. Combined, these shifts have led to new and different research questions, practices, involvements, and dissemination processes.

In this chapter we explore the methodological implications of sociocultural approaches for the study of classroom interactions in technology education. The considerations that flow from this focus include the value of student and teacher perspectives, the need for a multi-modal data set, the requirement to generate data on the process as well the product of the learning, and the need for a multi-layered analysis. There is also a need to reconsider issues around interactions between teacher and researcher roles and responsibilities. We illustrate these points with examples from our own classroom-based research (Bell & Cowie, 2001; Cowie, 2000; Cowie & Moreland, 2008; Cowie, Moreland, Jones, & Otrel-Cass, 2008; Jones & Moreland, 2004; Moreland, 2003).

FRAMING OUR RESEARCH APPROACH FOR TECHNOLOGY EDUCATION

A researcher's theoretical orientation towards knowledge and learning shapes the research questions they pose, the data they collect, the unit of analysis they use, and how they substantiate and report on findings (Erickson, 1986). It also informs their conceptions and concerns around teacher and researcher research roles, relationships, and responsibilities; and their view of the relation between research and change in classrooms. Because of this, classroom-based research approaches need to provide for coherence over conceptions of knowledge (in this case the discipline of technology), learning, and methodology.

A.T. Jones and M.J. de Vries (eds.), International Handbook of Research and Development in Technology Education, 625–635.
© 2009 Sense Publishers. All rights reserved.

The Nature of Technology

Technologists seek to intervene in the made world to extend human capabilities. Technology comprises the network of people, organisations, knowledge, processes, and cultural meanings that are involved in creating and operating technological artefacts as well as the artefacts themselves (Pearson & Young, 2002). Technological practice therefore is an activity in which capability lies in the interaction of knowledge, skills, and values, and not in the aggregation of these (Kimbell, Stables, Wheeler, Wosniack, & Kelly, 1991). This meld is encompassed in the current focus on the development of student technological literacy as the key outcome of technology education. In New Zealand, from where the illustrative examples in this chapter are drawn, the curricular emphasis is on the interplay of student understanding of the nature of technology, student technological knowledge, and student practical capability (see Chapter 9).

A Sociocultural Orientation Towards Learning

The term 'sociocultural' encompasses a rich and diverse cluster of concepts and practices. What sets these apart is that they all promote the vision of human thinking as essentially social in its origins and as inextricably dependent on historical, cultural, and situational factors. The basic assumption of the sociocultural perspective is that "human mental functioning is inherently situated in social interactional, cultural, institutional and historical contexts" (Wertsch, 1991, p. 86). Seen this way, student and teacher goals, motives, and interests are viewed as part of the learning process.

Learning involves more than knowledge acquisition; it also involves changes in a person's identity and in their participation in the valued activities of their community. Changes in the ways individuals participate shape and are shaped by changes in the collective and the setting; they influence and mediate each other and combine to a whole. Put another way, learning is viewed as mediated and the mediational means used by teachers and students – which include talk, tools, and artefacts and routines – are of interest. A sociocultural perspective provokes the need for a multi-layered analysis that attends to the way change and development occurs for individuals, social groups, communities, and institutions, both moment-by-moment and over the longer term. In relation to technology education, this means that learning and knowing are situated interactional accomplishments.

Technology, and what it means to be technologically literate, may be constructed differently in different classrooms and in the same classroom in different ways at different times depending on the resources (conceptual and material) that the teachers and students bring to the overarching technological task.

METHODOLOGICAL ISSUES FOR A SOCIOCULTURAL ORIENTATION TO TECHNOLOGY EDUCATION

Adopting a sociocultural orientation towards the learning and research in technology education poses a number of methodological challenges. The most compelling of these, in both a theoretical and practical sense, is that of what constitutes an appropriate unit of analysis. The unit of analysis relates to what it is you want to be able to say something about (Patton, 1990). The considerations associated with the unit of analysis when working within a sociocultural frame in technology classrooms include the value of multiple perspectives (student, teacher); the need for a multimodal data set; the requirement to generate data on the process as well the product of the learning; the need to consider what constitutes the appropriate grain of focus for analysis; and the need to review researcher and teacher roles, relationships, and responsibilities.

Technology Learning is an Intentional Activity: The Need for Multiple Perspectives

Once people are considered to be intentional participants in the activities of their communities it becomes important to understand teacher and student purposes and motivations for action and interaction, as well as their perspectives of consequences. Interviews are the usual means for doing this. From a sociocultural perspective interviews have the advantage that they are a familiar activity. Teachers and students will almost certainly have seen and heard interviews in the media and may have participated in parent-teacher-student interviews. The audio-tape technology used for recording interviews is in common use and so this aspect is also familiar. Nevertheless, interviews are occasions for the social construction and display of identity and information for all participants. The setting, the power differentials (perceived and real) between participants, participants' proficiency with the language of the interview topic and the means used to mediate talk along with established, or anticipated, interpersonal elements such as trust shape the interview conversation. One-to-one interviews can be easier for the interviewer to manage and have the benefit that what is discussed can be kept relatively confidential. In contrast, group interviews are more difficult to manage and the participants may be influenced by each other's answers and the nature of their relationships. However, group interviews provide more opportunities for interviewees to set the level and context of the interview. Our classroom studies make active use of both individual and group interviews with teachers and students.

Teacher beliefs about teaching, learning, students, and technology, as well as their more immediate intentions for student learning and their reflections on lessons, are important when the goal is to develop an understanding of a teacher's contribution to the evolution of teaching and learning within a particular classroom context. These aspects can be elicited via interviews before lessons and through the collection and discussion of teacher planning documents. If teachers use their plans as working documents it is important to collect a record of any changes and additions. Finding time to interview teachers immediately after a lesson can be challenging but the effort is usually worth it in terms of the insights gained into teacher motivations, intentions, and analysis of consequences. We have talked to teachers as they prepare for the next lesson, as undertake road-crossing duty, and while they supervise students over the lunchtime. We have also talked to them on the telephone later in the day. For some teachers, email proved to be a useful tool to exchange thoughts on initial descriptions and analysis of observed lessons. Primary teachers have also tended to talk with us during lessons. This usually happens 'on-the-fly' and provides insights into the challenges they are facing and the dynamics of their thoughts-in-action. We consider this dynamic two-way interaction to enhance the trustworthiness of the processes and outcomes of our research. In a recent study, collaborative analysis and writing up of key classroom incidents also proved to be an enjoyable and enriching experience for us and the teacher participants.

In our work we acknowledge that students are active participants in the teaching and learning process and that their motivations and interests shape the way a subject comes to be construed and experienced within a particular classroom, albeit that students generally have less influence on this than their teacher. We have found students are eager to be involved in research that has the goal of enhancing teaching and learning, both for themselves and for other students. In establishing this as a shared focus we have found it is important to explain carefully, and reiterate at regular intervals, the goals of the research and what will happen to their data – these discussions occurring in addition to formal ethical consent processes. We take care to ensure that the formal informed consent material is in age appropriate language and we ask students, not just their parents or teacher, for informed consent to take part in a study. We usually use props and probes to mediate talk in a manner consistent with the call for new tools to extend our understanding of children's perspectives (Clark & Moss, 2001). Student workbooks also provide a useful resource for students and the interviewer where the goal is to reflect on experiences and what has been learned. Photographs, objects, and student-produced artefacts including prototypes, models, drawings, and action plans can serve a similar role. Stimulated video recall interviews allow students to re-engage with earlier activities and, while time consuming, can elicit detailed and in-depth flow of consciousness descriptions and analysis of past events. For example, sitting beside students asking them what and why they are doing a particular thing while they are working is one strategy we have used successfully to elicit student perspectives. Students quickly became accustomed to us asking why they are undertaking particular actions. Once our presence is accepted, students are more than willing to talk with us after lessons and to take part in more formal impromptu interviews about their technology ideas and learning. On our arrival to class, students often note and relay to us events and discussions that have taken place outside of the scheduled observations.

Group interviews are useful for interviewing children who are used to working together. Group interviews can also be useful in encouraging shy students to participate in a research project. In addition, they go some way towards addressing the contextual issues associated with an adult interviewing a child/student in a school setting. We have found they allow interviews to become more like conversations as students negotiate the meaning of the researcher's questions and build on each other's answers. We have observed, for example, how students talking together about their technology projects were able to qualify and modify views and voice agreement. They brought to the fore important ideas and issues and they sometimes challenged each other's views. Group interviews also allow more students to engage actively in the research process. We have found that students beyond the group we have planned to work with ask to be interviewed, and a group interview is an efficient way to accommodate their desire to contribute.

A persistent issue for all of the above strategies is that they necessarily take place after the event and so they rely on the recall of events, attitudes, and understandings. They also place the researcher in control of the data collection. Working with students as co-researchers offers the possibility for students to influence all aspects of the research design – from the formulation of research questions, to data collection, analysis, and reporting. This is an area worthy of further exploration as new technologies become available that allow students to document and represent their learning (see Chapter 42).

Technology Learning as Mediated: The Need for Multimodal Data

Whilst talk has been the dominant focus for classroom interaction research, a sociocultural perspective focuses attention on the full range of mediational means people use to express and negotiate meanings. Attention to the multiple modes children employ during classroom interactions enables consideration of the distributed and situated nature of technology learning and knowing. Technologists, for instance, use a wide range of media and modes in the generation, development, and testing of ideas and solutions. These include talk, drawing, the examination of artefacts, and the building of models and prototypes. Students developing discernment and skills in the use of multiple communicative resources – both singly and in combination – as they generate, represent, develop, and test out their ideas and thinking is an important aspect of technological literacy (Roth, 1998).

In concord with others who have conducted classroom studies based on the belief that design technology is a social process (Anning, 1997; Hennessy & Murphy, 1999; Roth, 1997) we have examined the ways teachers and students make and express meaning through a combination of media and modes – drawing, talking, writing, action as modeling, gesture and dramatisation, and the use and production of artefacts. For instance, the teachers and students in one of our studies (Cowie et al., 2008) made substantial use of drawing, to the extent that it can be usefully conceptualised as a visual mode of thinking for both communication and the co-construction of ideas. Like other researchers who work with young children (e.g., Fleer, 2000; Gustafson, MacDonald, & Gentilini, 2007) our work indicates children need support to see the connections between different modes and media. For example, they may prefer to make a 3-dimensional model rather than create 2-dimenstonal drawings to generate, explore, and test out ideas. They do not always see the connection between their drawn plan and the final physical artefact, making the children's reasoning around this a useful vehicle for tracing learning.

We use digital video, digital photography, and digital audio-taping to generate a multimodal data set that represents and reflects the multimodal nature of teacher and student engagement with technological practises. Video is useful for capturing the flow of activity although it is important to remember that the lens frames and therefore only captures part of the whole. We have found digital photographs a useful complement to the video record; a person using a small digital camera usually has greater ease of movement and can be less obtrusive. Our experience of giving students disposable cameras to depict what they conceive of as technology (Moreland & Cowie, 2005) indicates that children can effectively use this technology and that they enjoy being in control of the data generation process. The choice of objects and activities in their photographs surprised us. Their photographs proved to be a fruitful means for mediating discussion of their ideas about technology in their world. Documentary data are also important. Where possible we collect the products of teacher and student action and interaction such as completed class brainstorms, records of whole class discussions, student books and worksheets, and students' physical products/ prototypes as well as copies of any materials posted on the classroom walls.

A focus on the visual through the use of video and photographs raises a number of ethical, research design, and interpretation issues. Ethical issues include matters of potential harm arising from making images public. While we are careful in the selection of images for use in reports and presentations, and we gain express permission for the use of a specific image in which participants can be identified, the possibility of uncontrolled use of images from our research is of concern all the more so given the ubiquity and reach of the Internet.

The positioning of the video camera and the extent to which its presence might distort the usual pattern of classroom interaction and relationships is a research design issue although many teachers and students in New Zealand are familiar with these technologies. Video and digital photographs are often used as part of day-to-day work as a means for recording teaching and learning for later joint reflection and analysis and for representing learning to those outside the classroom, including parents and teacher colleagues.

Issues both practical and conceptual can arise in the coordination and collation of multiple data sets if the researchers wish to access, transcribe, and analyse large amounts of multimodal data as a coherent whole. While a number of theorists have proposed frameworks for the analysis of talk (e.g., Mercer, 1995) researchers are only beginning to develop the tools for analysing images, gesture, image-text relations, and so on (Kress & Van Leeuwen, 1996). (See also Chapter 53.)

Technology Learning is Accomplished over Time: The Need to Generate Data on the Process as well as the Outcome

A sociocultural orientation draws attention to the temporality of learning and knowing (Lemke, 2000): "As one moves to a distributed perspective it is the complex and dynamic intersection of individual, environment, and activity over time that constitutes the unit of analysis" (Barab, Hay, & Yamagata-Lynch, 2001, p. 64). This aspect is particularly pertinent in the case of technology education where the nature of technology tasks means that knowledge and learning are dynamic, contingent, and emergent. Children's successful negotiation of a technology task requires that they identify and work on problems 'on-the-spot'. How students access and develop the knowledge and expertise they need at the time they need it in order pursue the task is of interest (Kimbell, 2002). Adding to the complexity, students face a challenge in communicating their thinking as 'thought in action' in a way that can be witnessed, evidenced, and interpreted by others (McLaren, 2007). The research implication of this is that researchers need to accumulate multimodal evidence of student learning over time so that they are able to trace the *process*, not just the product of student learning. Furthermore, when learning is viewed as a situated, distributed process it is important for children to have opportunities to make sense of an idea across a range of contexts (Greeno, 1997). Researchers therefore need to somehow track the emergence, evolution, and diffusion of practices, conceptual understanding, resources, and student-constructed artefacts across extended time frames and contexts (Roth, 1996). Tracing student learning to understand it in the moment and over time requires ongoing and prolonged engagement with teachers and students.

Our work with teachers has highlighted the value of ongoing and prolonged data collection; our preferred practice is to work with teachers over two to three years. In our classroom work we always try to observe the full teaching sequence: the way the teacher sets the scene and motivates the students, how she progresses the task and the learning, and how she concludes the teaching sequence. We have found that when students are observed in action as they grapple with the processes associated with learning technology we are not just considering what students can and are willing to disclose or make public, but also how students come to understand – the same apparent product can have very different trajectories of development conceptually and materially in terms of ideas and artefacts developed and tested. For us as researchers, actions at one point in time can only be understood with respect to what has come before and the longer-term expectations and goals of participants. However, primary teachers conducting teaching sequences over the course of eight or more weeks pose a logistical challenge to our participation across the full time span of a teaching sequence. Over the course of our work with primary teachers we have encouraged them to consider that for a technology task to foster student creative and critical learning it does not always need to be loosely structured and of long duration (five weeks or more). Nested within the wider context of learning for the term or year and/or carefully linked to student interests, it can be as short as four sessions on four consecutive days. In this case we usually work more intensively with the teacher during the planning and preparation (Cowie et al., 2008). This said, in New Zealand at least, teachers are keen to respond to student questions and interests and sometimes we find that when we arrive to a scheduled session, teachers and students have already discussed a key technological practice. This might have happened, for example, as part of the practice of children sharing news from home or in response to a book or activity used within another curriculum area. As teachers come to understand the goal of the research they also take the time to share these events with us, pointing to the importance of shared understanding and respect that can only be developed over time.

As noted above, generating data on student learning over the longer term is important from a sociocultural perspective. It is also congruent with the current political context, which emphasises the development of student dispositions and skills as lifelong learners. Studying the development of individual *teacher* practice over time can be problematic. In our experience, teachers who engage in research are often proactive in seeking other professional opportunities and may move schools for career, and personal, reasons. Similarly, we have found it difficult to work with students over more than a year; students, particularly those in lower socioeconomic areas, tend to move schools quite frequently. Tracing teacher and student trajectories over time is, and is likely to remain, a persistent issue for those who wish to engage in sustained classroom work.

Technology Learning is Multifaceted: The Need for a Multilayered Analysis

Deciding on the temporal and situational dimensions of the unit of analysis is an important consideration when we are seeking to understand 'What's happening here?' for the purposes of contributing new theoretical understandings, and at the same time generating knowledge that will be useful for teachers. A focus at any particular level of activity foregrounds some aspects of activity and backgrounds others. Engeström (1999) therefore suggests it is fruitful to move from the analysis of individual actions to the analysis of the broader context and back again. In the case of technology, technological literacy involves students being able to bring together technological knowledge and capabilities in dynamic relations; working sometimes as individuals, sometimes in small groups, and sometimes as a class; and over time frames that range from minutes to days or months. A focus at any of these levels is potentially interesting.

At the micro level, it is possible to explicate the ways teachers and students use talk, writing, drawing, action, and artefacts to express ideas and co-construct meaning in and through their moment-by-moment interactions. In particular, video makes it is possible to study how technology is talked, meant, and portrayed to students moment-by-moment. Moreover, the dynamics of action and interaction are available for repeated examination and reconsideration by individuals and teams. We have found a commercially available video analysis tool, which allows for the multi-coding of chunks of video, to be useful for this.

However, moment-by-moment interaction is grounded in longer-term understandings of taken-for-granted and valued ways of acting and interacting. This necessitates analysis of sequences of episodes, learning tasks, and lessons, along with linkages and transitions. A mid-time frame analysis provides insight into how knowledge, expertise, and authority are distributed within a classroom. In particular, what sources of knowledge and expertise are welcomed, invoked, and warranted (stories from home; whānau, hapu or iwi stories[1] and lived experiences; practical experiments; teacher talk; information accessed from the Internet; and so on)? At this mid-time scale the focus is not so much on the nuances of interaction but rather on patterns of activity within the same and across different tasks and sub-tasks and the transitions between them. In our work, for example, consideration of these mid-term aspects has led us to focus more on how ideas, tasks, and lessons are linked and made coherent over time (Moreland & Cowie, 2007; Roth, 1998). This seems especially important given the long-term nature of many technological activities and the need for teachers to assist students to maintain a focus on the overall goals of the technology task while working through a series of inter-related sub-tasks. The understandings to be gained thus differ from those obtained from analysis at the micro level, and yet both are needed to understand a complex phenomenon such as classroom activity.

In our experience, the description and analysis of student learning in the long term has been somewhat serendipitous, in part because of difficulties in generating data over the school day, over the full course of a school year, and across the years. On occasion, students have discussed with us what they have learned prior to our research. For example, a six year old boy introduced the idea of using a drawbridge as a door for a model house in a unit during the third school term. He explained he had learned about drawbridges in the first term as part of a social studies unit on castles. Teachers have also discussed with us instances of students using earlier work in a current unit, such as knowing how to draw a diamond, which had ostensibly been learned during mathematics lessons.

Changing Research Roles, Relationships, and Responsibilities

When learning is construed as increasingly knowledgeable participation in the valued activities of a community it is not possible to study learning as a dispassionate observer outside the process: understanding teaching and learning is inextricably intertwined with participating in the process (Langemeyer & Nissen, 2005). Rather than attempting to decontextualise it, knowledge about learning needs to be judged by the ecological validity of the research that generates it – that is, by its conduct in a real-life situation that "maintains its integrity", by its alignment with the larger sociocultural context of the participants, and by its consistency with the participants' understandings (Cole, 1996, p. 226). Hence, sociocultural views of learning provoke new and different questions about what constitutes an appropriate relationship between researchers and teachers, and also about what constitutes quality in educational research. Acknowledging this, our research has been based on the notion that we participate in the research process as researchers-in-support-of-teaching and teachers participate as teachers-in-support-of-research, both communities learning about and gaining from the other.

For us, the process of mutual learning is operationalised through negotiated intervention (Jones & Simon, 1991). This process acknowledges that researchers come to a research project with particular knowledges, interests, and skills; similarly, teachers come to a research project with their own knowledges and expertise. Within a negotiated intervention approach the aim is that together the two professional groups will negotiate, in an ongoing and responsive manner, the course of the research, albeit within the framework agreed on at the inception of the study. We have found iterative cycles of classroom work and joint meetings to reflect on and plan for further classroom work and the ways the teachers and researchers work together in the classroom to be effective in promoting a melded community of practice. For example, a technology-specific planning framework was developed and then refined in an iterative way through an alternating series of planning meetings and classroom work (Cowie et al., 2008) served as a forum and focus for the negotiation of a shared purpose, the articulation of knowledge, and the setting of common goals. The planning framework and its associated practices created a means for researcher and teacher roles to evolve to encompass active and mutual (albeit different) support for teaching and research. We have also found that the combination of classroom work and joint meetings allows us to work with teachers in a manner consistent with our understanding of learning (of the student, teacher, and researcher) as a sociocultural process. These practices resonate with those proposed by Wenger (1998) as key components of a community of practice: mutual engagement, joint enterprise, and a shared repertoire of practice.

Considering the implications of a sociocultural view on teacher forms of research participation, it is important that teachers come to understand and participate in the culture of research. In our case, joint out-of-school meeting provide a forum for the discussion of the theoretical underpinnings of our work, including the nature of pedagogical content knowledge and assessment for learning. Theory was linked with practice through a process that fostered movement between the general and the particular, and description and analysis/explanation. This was important in the development of a shared understanding of and commitment to the goals of the research, and for the collaborative construction of knowledge.

Such collaborative work has the potential to contribute a warrant for relevance and at the same time be a rich source of diverse ideas and experiences (Edwards, Sebba, & Rickinson, 2007). Our teacher participants raised the question of how to represent our findings to non-participant teachers – both their immediate colleagues and the wider professional group. A sociocultural perspective highlights the need to provide sufficient detail so that teacher-readers can participate in what Eisner (2002) calls "educational criticism", that is, making judgments about the worth of research findings through the exercise of "educational connoisseurship" (p. 189). Working with teacher participants as teachers-in-support-of-the-research-process helps us to maintain a focus on the 'how' as well as the 'what' and 'why' of classroom interactions, meaning our work is more likely to be of interest and value to other teachers (Boaler, 2003). We do not therefore aim to provide generalised advice but rather to provide a rich description of the dimensions of the local context and to develop local instructional theories in a manner that supports others in adapting lessons learned in one context to their local contingencies, as recommended by Barab and Krisher (2001).

CONCLUSIONS

The focus of classroom research has shifted over time in response to changing conceptions of how learning is accomplished and how learning and knowing are made visible. At this time, a sociocultual view of learning offers the greatest generative potential for researchers seeking to build understanding and develop theory, and at the same time generate findings that have the potential to inform and enhance teacher practice and student learning across the broad range of outcomes now deemed necessary for students to be active participants in 21st century society. This social turn has contributed new and different research questions, practices, involvements, and dissemination processes. It has posed the need for changed researcher and teacher roles and relationships, the need for researchers to actively take account of and engage with student and teacher perspectives, the need for a multi-modal data set, the requirement to generate data on the process as well the product of the learning, and the need for a multi-layered analysis.

This shift has benefited from and at the same time gained momentum from developments in the technologies available to document and analyse classrooms processes. In the case of technology education, the dynamic, multidimensional, multimodal nature of technological activity appears to lend itself to a sociocultural framing. Nevertheless, persistent questions remain. The most compelling of these concern the selection of the events and instances for in-depth analysis, the nature of the most pertinent unit of analysis, and how to attend to its most likely dynamic nature. The selection of a particular, or even multiple, theoretical framework(s) to guide the overall research process is key in addressing this.

Many of the challenges posed by micro-level analysis are being addressed by developments in digital technologies. However, the longer-term logistical, organisational, and conceptual challenges to researchers – and to teachers and policy makers – associated with extending and mapping student learning pathways over time persist. We hope we have provided some insights into how classroom- based work underpinned by a sociocultural view might begin to address some of these concerns.

ACKNOWLEDGEMENTS

This chapter was informed by the Classroom InSiTE project (Cowie et al., 2008), undertaken as part of a Teaching and Learning Research Initiative, a competitive fund managed by the New Zealand Council for Education Research.

NOTES

[1] Whānau, hapu, and iwi refer to concepts of extended family, subtribes, and tribes within the Maori population in New Zealand.

REFERENCES

Anning, A. (1997). Drawing out ideas: Graphicacy and young children. *International Journal of Technology and Design Education, 7*(3), 219–239.

Barab, S. A., & Kirshner, D. (2001). Methodologies for capturing learner practices occurring as part of dynamic learning environments. *Journal of the Learning Sciences, 10*(1&2), 5–15.

Barab, S. A., & Landa, A. (1997, March). Designing effective interdisciplinary anchors. *Educational Leadership*, 52–55.

Barab, S., Hay, K., & Yamagata-Lynch, L. (2001). Constructing networks of action-relevant episodes: An in-situ research method. *The Journal of the Learning Sciences, 10*, 63–112.

Bell, B., & Cowie, B. (2001). *Formative assessment in science education*. Dordrecht, The Netherlands: Kluwer.

Boaler, J. (2003). Studying and capturing the complexity of practice – The case of the 'dance of agency'. *Proceedings of the 2003 Joint Meeting of PME and PMENA* (Vol. 1, pp. 3–16). Great Britain.

Clark, A., & Moss, P. (2001). *Listening to young children*. London: National Children's Bureau and Rowntree Foundation.

Cole, M. (1996). *Cultural psychology: A once and future discipline*. Cambridge, MA: Harvard University Press.

Cowie, B. (2000). *Formative assessment in science classrooms*. Unpublished doctoral dissertation, the University of Waikato, New Zealand.

Cowie, M., Moreland, J., Jones, A., & Otrel-Cass, K. (2008). *The classroom InSiTE project: Understanding classroom interactions to enhance teaching and learning in science and technology*. Teaching and Learning Research Initiative, Final Report. Wellington: New Zealand Council for Education Research.

Edwards, A., Sebba, J., & Rickinson, M. (2007). Working with users: Some implications for educational research. *British Educational Research Journal, 33*(5), 647–661.

Eisner, E. W. (2002). *The educational imagination: On the design and evaluation of school programs*. Upper Saddle River, NJ: Merrill.

Engeström, Y. (1999). Activity theory and individual and social transformation. In Y. Engeström, R. Miettinen, & R. Punamäki (Eds.), *Perspectives on activity theory* (pp. 19–38). Cambridge, UK: Cambridge University Press.

Erickson, F. (1986). Qualitative methods in research on teaching. In M. C. Wittrock (Ed.), *Handbook of research on teaching* (pp. 119–161). New York: Macmillan.

Fleer, M. (2000). Interactive technology: Can children construct their own technological design briefs? *Research in Science Education, 30*(2), 241–253.

Gustafson, B., MacDonald, D., & Gentilini, S. (2007). Using talking and drawing to design: Elementary children collaborating with university industrial design students. *Journal of Technology Education, 19*(1), 19–34.

Greeno, J. G. (1997). On claims that answer the wrong questions. *Educational Researcher, 26*(1), 5–17.

Hennessy, S., & Murphy, P. (1999). The potential for collaborative problem solving in design and technology. *International Journal of Technology and Design Education, 9*(1), 1–36.

Jones, A., & Moreland, J. (2004). Enhancing practicing primary school teachers' pedagogical content knowledge in technology. *International Journal of Technology and Design Education, 14*(2), 121–140.

Jones, A., & Simon, S. (1991). Strategies for educational change: The work of the OPENS project in the context of a new national curriculum for science. In A. Begg, B. Bell, F. Biddulph, M. Carr, M. Carr, J. McChesney, & J. Young Loveridge (Eds.), *SAMEpapers 1991* (pp. 15–35). Auckland, NZ: Longman Paul.

Kimbell, R. (2002). Assessing design and innovation: The famous five and the terrible two. *Journal of Design and Technology Education, 7*(3), 172–180.

Kimbell, R., Stables, K., Wheeler, T., Wozniak, A., & Kelly, V. (1991). *The assessment of performance in design and technology—The final report of the APU design and technology project*. London: SEAC/COI.

Kress, G., & Van Leeuwen, T. (1996). *Reading images: Grammar of visual design*. Routledge: London.

Langemeyer, I., & Nissen, M. (2005). Activity theory. In B. Somekh & C. Lewin (Eds.), *Research methods in the social sciences* (pp. 188–196). London: Sage.

Lemke, J. (2000). Across the scales of time: Artefacts, activities, and meanings in ecosocial systems. *Mind, Culture and Activity, 7*(4), 273–290.

McLaren, S. (2007). An international overview of assessment issues in technology education: Disentangling the influences, confusion and complexities. *Design and Technology Education: An International Journal, 12*(2), 10–24.

Mercer, N. (1995). *The guided construction of knowledge*. Clevedon, UK: Multilingual Matters Ltd.

Moreland, J. (2003). *Becoming effective technology teachers: Enhancing assessment practices in primary classrooms*. Unpublished doctoral dissertation, University of Waikato, New Zealand.

Moreland, J., & Cowie, B. (2007). Making masks for Maui: Keeping the macro task in mind. *Australian Early Years Research Journal, 14*(1), 81–93.

Moreland, J., & Cowie, B. (2005). Exploring the methods of auto-photograph and photo-interviews: Children taking pictures in science and technology. *Waikato Journal of Education, 11*(1), 73–88.

Patton, M. Q. (1990). *Qualitative evaluation and research methods* (2nd ed.). Newbury Park, CA: Sage.

Pearson, G., & Young, A. T. (Eds.). (2002). *Technically speaking: Why all Americans need to know more about technology*. Washington, DC: National Academies Press.

Roth, W.-M. (1996). Knowledge diffusion in a grade 4-5 classroom during a unit on civil engineering: An analysis of a classroom community in terms of its changing resources and practices. *Cognition and Instruction, 14*(2), 179–220.

Roth, W.-M. (1997). Interactional structures during a grade 4-5 open-design engineering unit. *Journal of Research in Science Teaching, 34*(3), 273–302.

Roth, W.-M. (1998). *Designing communities*. Dordrecht: Kluwer.

Wenger, E. (1998). *Communities of practice: Learning, meaning, and identity*. Cambridge, MA: Cambridge University Press.

Wertsch, J. V. (1991). *Voices of the mind: A sociocultural approach to mediated action*. Cambridge, MA: Harvard University Press.

Bronwen Cowie
Judy Moreland
Wilf Malcolm Institute of Educational Research
The University of Waikato
Hamilton, New Zealand

WOLFF-MICHAEL ROTH

53. STUDYING COMMUNICATION DURING TECHNOLOGICAL DESIGN ACTIVITIES

> Language isn't a superior kind of conduct. It is the element in which conduct confirms itself as conduct of sense. (Nancy, 2003, p. 187)

Technological design activities have a central advantage over lecture-based courses: they allow students to communicate *over* and *about* the material artefacts and design drawings they produce. This immediately leads to a distributed nature of communication because what one person says cannot be understood independent of what other people have said and possibly will say and from how they will respond to what has been said. But communication does not consist in words alone; it is not even, if we follow the philosopher of the introductory quote, a superior kind of conduct. Rather, it is but one among many ways in which human beings make and mark sense. There are other resources in making and marking sense during communication, including gestures, body orientations, body positions, and prosody (pitch, speech intensity, and speech rate). Drawing on data from a design experiment in a split sixth-/seventh-grade (elementary school) technological design curriculum, I articulate in this chapter a method for tracking communication over and about technological design drawings and artefacts. On a micro-level (seconds, minutes), communication is distributed over different semiotic (meaning-making) resources including participants, materials (tools) in the setting, orientation and positioning of participants, gestures, and prosody. Temporal changes in these semiotic resources and in their saliencies – for example, from artefacts to gestures, from gestures to talk (Roth & Lawless, 2002a) – provide indices of learning and development. What is salient to communication participants at any moment in time emerges from the interaction and mediation of all semiotic resources; what is salient also changes with the nature of the setting so that minor changes in the material configuration of the setting. In this chapter a model for studying and theorising communication in design activity is provided, which takes into account the multi-levelled and distributed nature of human activities generally and that of communication in particular.

A THEORY FOR STUDYING LEARNING IN TECHNOLOGY EDUCATION

Theoretical frames are important to research on learning and instruction because they mediate the nature of the data we collect to make sense of what teachers and learners do. To study human activities in contexts as different as formal schooling (elementary, middle, high school and university), work (electricians, hatchery workers), environmentalism, and scientific research I have come to use cultural-historical activity theory because of its holistic nature for understanding phenomena related to cognition. This begins with the understanding of the *activity* concept, which is conceived of as denoting any societally motivated and organised structures that lead to the satisfaction of collective needs and, thereby, the survival of the society. Examples of activities include producing food (e.g., farming, fishing, hunting, gathering), producing tools (e.g., manufacturing tractors, scythes, hoes, baskets), and producing clothing (e.g., tailoring). The term

A.T. Jones and M.J. de Vries (eds.), International Handbook of Research and Development in Technology Education, 637–650.
© 2009 Sense Publishers. All rights reserved.

activity does not pertain to such phenomena as doing a mathematics problem, building a simple machine as part of an engineering curriculum, or writing an essay. Rather, all these are *tasks* by means of which the *activity* of schooling is realised.[1]

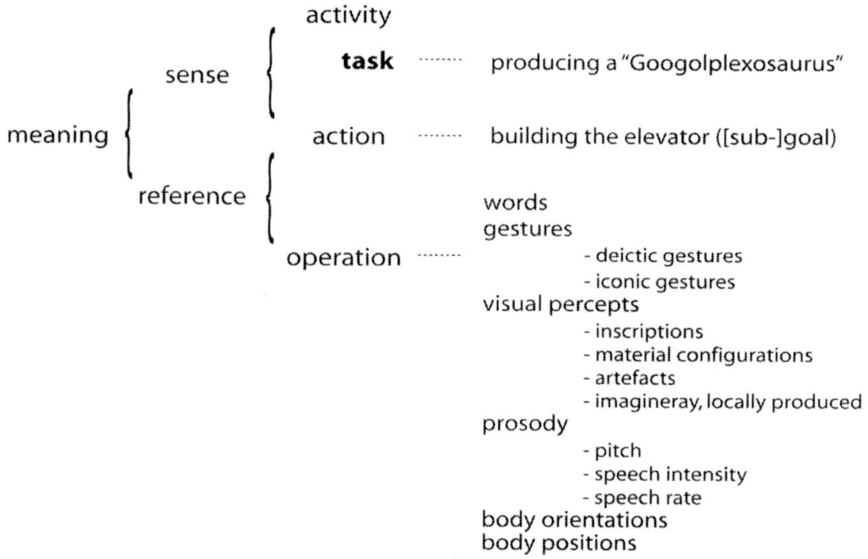

Figure 1. In cultural-historical activity theory, any event is analysed in terms of three major levels of events that are associated with sense, reference, and meaning concretised here with materials from an engineering design curriculum (right).

Activities and tasks are realised concretely by means of goal-directed actions, which, in turn, presuppose the activities that they contribute to bringing forth: The two levels therefore are mutually constitutive (dialectical) and make for a relation that I denote by the term *sense* (see Figure 1). A different sense is produced by the same action when it realises a different activity. The same utterance, "Leave me alone!" is associated with a different sense and does different interactional work when uttered on the playground by one child pestered by another one than when uttered to a teacher in the classroom. Possible and real actions therefore have a collective nature because, in being produced, they always presuppose their own intelligibility.[2]

Actions can be analysed in terms of chains of unconscious operations. In contrast to actions that realise conscious goals, operations are determined by current conditions. Thus, while hammering a nail, we do not think about how our fingers clasp the handle and thereby hold the tool; similarly, while participating in everyday (classroom) conversations, we do not consciously select the words and gestures by means of which we produce the utterances intended to express something (our opinion concerning the design of an artefact). The current state of the action with respect to the goal, the material and social setting, the current bodily states (including emotions), and so on mediate the type and quality of operations we produce from which our actions spring forth. This means that whereas actions are produced through the enchainment of operations, the latter are produced only because of the existing goal realised by the unfolding action. The relation between the two levels therefore also is a dialectical one: an action presupposes the operations that realise it, but the operations presuppose the actions as a context that brings them forth. This relationship is termed *reference*.

Sense grounds our goal-directed actions in societal activity and therefore arises from our social nature; reference grounds actions in our bodies and therefore arises from our material and psychological nature. Actions, which are produced at the interface of culture and body, therefore are both social and psychological, ideal and material. The relation between these two dimensions sense and reference is again dialectical: I denote it by the term *meaning*. Therefore, in each action, meaning is both produced actively in and through the body of the person, and reproduced (presupposed) as something recognisable by and intelligible to the recipient of a communicative production.

The picture may appear slightly more complicated when we attempt to answer the question, "Why should an action make sense given that it is composed of singular operations?" The answer is that operations are crystallised actions, which have disappeared or never entered consciousness. Actions become operations and thereby disappear. For example, learning to drive using a stick-shifted car first requires us to consciously attend to the precise timing of clutch and stick shift until a point when we shift gears upward and downward without ever being conscious of shifting or the current gear engaged. Other operations may be learned by means of mimesis, as we reproduce actions during participation in activity without ever thinking about how we do what we do. But because we have acquired the competency by means of tacit observation (i.e. mimesis), we also acquire the operations in culturally meaningful, embedded and embodied form.

In the past, researchers of learning and cognition have focused on the words produced during person-person transactions independent of everything else available in the setting for making sense of and producing and reproducing meaning. Grounded in sociocultural theory, one psycholinguistic approach recognises that in addition to words, the (unconscious) gesticulations a person produces are important in making sense of what is being communicated (McNeill, 2002). Importantly, what is being expressed by gestural means cannot be reduced to language and, reciprocally, language cannot be reduced to the gesticulations that are co-produced with utterances. Rather, the two forms are expressions of the same higher-order communicative unit, which they can represent only in one-sided form. The phenomenon of light presents us with a good analogy for this relation: Neither its wave nor its particle character expresses the full extent of what light is, and neither characteristic can be reduced to (i.e. explained by) the other.

More recent work on the nature of real-time communication in everyday settings including school classrooms shows that there are other aspects in communicative situations that speakers and listeners use as resources for producing utterances and for producing and reproducing (making) sense (Roth & Pozzer-Ardenghi, 2006). A complete theoretical model for analysing communication therefore includes the three levels of events, with a particular focus on the operations that realise actions (Figure 1). To understand the technological design decisions and trajectories of design work and design products, any method for studying communication in unfolding and emergent activity needs to keep track of – i.e. collect data about – the different types of events (movements, changes) at the three levels. Thus, even though the students under investigation may not talk about the task at all, researchers must not lose sight of and take into account the motive/goal for doing what learners in technology classrooms do. (Realistically, the research needs to account for the activity level, which is schooling; we only understand why some students – e.g., working class, African American, indigenous people – are less successful when we take into account society while studying classroom learning.) Thus, foremost students [and teachers] produce and reproduce schooling as societal object/motive.) At the next level, students realise the task by setting goals, for example, building an elevator into their Rube Goldberg machine (called 'Googolplexosaurus').[3] These actions are realised and accomplish through actions, constituted by unconscious operations.

01 B: ↑ * ‘or (0.33)

02 * [my ‘brother
[((RH points to “elevator”))
03 (0.22)
04 A: uh hm?

05 * (1.50)

06 L: ((nods))
07 B: ↓he has a parking lot *
08 (0.85)
09 B: um:
10 (0.45)
11 [you can ‘take this part out];
12 [((repeatedly moves up and down
“elevator”))]

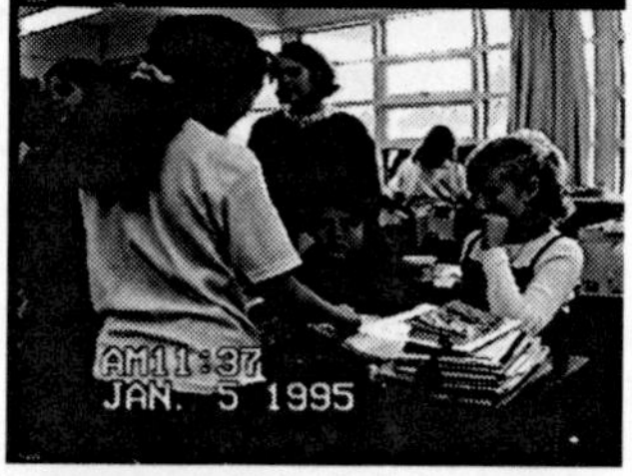

13 (0.32)
14 then you pull like [this. *
15 [((pulls motion
along the tower part))
16 (0.45)

17 * then put some [batteries in it]; and it’ll
works.
18 L: [((nods repeatedly))]
19 (1.12)
20 L: although we can’t do it

Communication is a complex phenomenon and consists of more than words. In fact, when there are only words, as the experience of many individuals with e-mail has shown, many misunderstandings emerge. For the study of communication, therefore, especially in environments rich with artefacts and technologies (e.g., in scientific laboratories and fieldwork, workplaces, and technology-rich curricula), i have developed an approach that tracks very different resources for making sense and meaning. Thus, i track words, gestures, visually available material, (ideal) structures (e.g., inscriptions, artefacts), prosody, body orientations, and body positions (e.g., roth & pozzer-ardenghi, 2006). These different resources are constituted at the (unconscious) operational level, which makes salient the fact that we do not stop and reflect on (a) every (or even any) word in everyday conversation, (b) the gesticulations that we produce while talking, (c) the perceptually available structures we know to be present, (d) the pitch, speech intensity, and speech rate that characterises our utterances at a particular moment, (e) the way we orient our bodies to facilitate communication, or (f) the way in which our bodies are positioned with respect to others in the situation.

TRACKING COMMUNICATION IN AND OVER REAL TIME

Societal phenomena integrate over themselves, that is, as they emerge in cultural and historical settings they are like snowballs that increase in their size and importance. A conversation at one point in time may draw on all the resources available to interaction participants both *at this moment* and at *any moment preceding* it. To understand what is happening in a concrete situation, we need to know the knowledge that interaction participants presuppose and take to be common but leave unarticulated. Thus, in my research I always begin by spending time in the relevant community (classrooms, staff rooms, parent councils), sometimes up to a year to understand the situation *prior to studying an intervention.* The extensive understanding that I develop in the process allows me to interpret the intervention in its cultural and historical context. This also means that – as any human being – I do not have available events and conversations that occur some time in the future. Thus, what will be happening in the future is available neither to participants nor to researcher-participants, so that the latter may not use that information to *teleologically* explain the course of conversations and events. Rather, I interpret each moment in the classroom only with respect to what has happened but never in terms of where the lessons ultimately end up. I am keen to understand, as any other participant, why the people in a situation do and say what they do and say: I can draw only on past and present information but not on what will have been.

Throughout the lessons, each of the two or three cameras I use follows a particular student group, usually for the duration of a curriculum segment such as the design (conceptualisation) to prototype construction and presentation to classroom community. The person holding the camera – who may be a graduate student doing his or her MA or PhD in the context of the same study, or who may be a non-academic research assistant – also makes notes of observations not likely to be captured on the videotape. I prepare maps of the classroom to note where students are at particular moments in time (see Figure 2). This allows me to construct maps of where the students have been and whom they have interacted with during a particular lesson. For example, on January 5, 1995, the three girls Amanda, Bella, and Leanne began working on their prototype at their desk (position 1, video episodes #1–7) before moving to the table reserved for working with power tools, saws, hammers, and so on (Figure 2). Here at position 2 have been recorded video clips #8 and #9 before the girls moved to the other end of the table (position 3, video clips #12–18). They continued their work while moving between their regular seats and the workstation before ending up doing some gluing work, which has had its own workstation. In addition to recording their trajectory, I also note with whom they interact in the course of the day. For example, in position 5 (Figure 2) they took the opportunity to speak to Jessica, who was spending some time with them while they were working on the tubular "shoot".

To exemplify how I study communication, I draw on the following exemplary episode, which I subsequently interpret in terms of the different aspects that can be identified and heuristically isolated from the communicative whole.

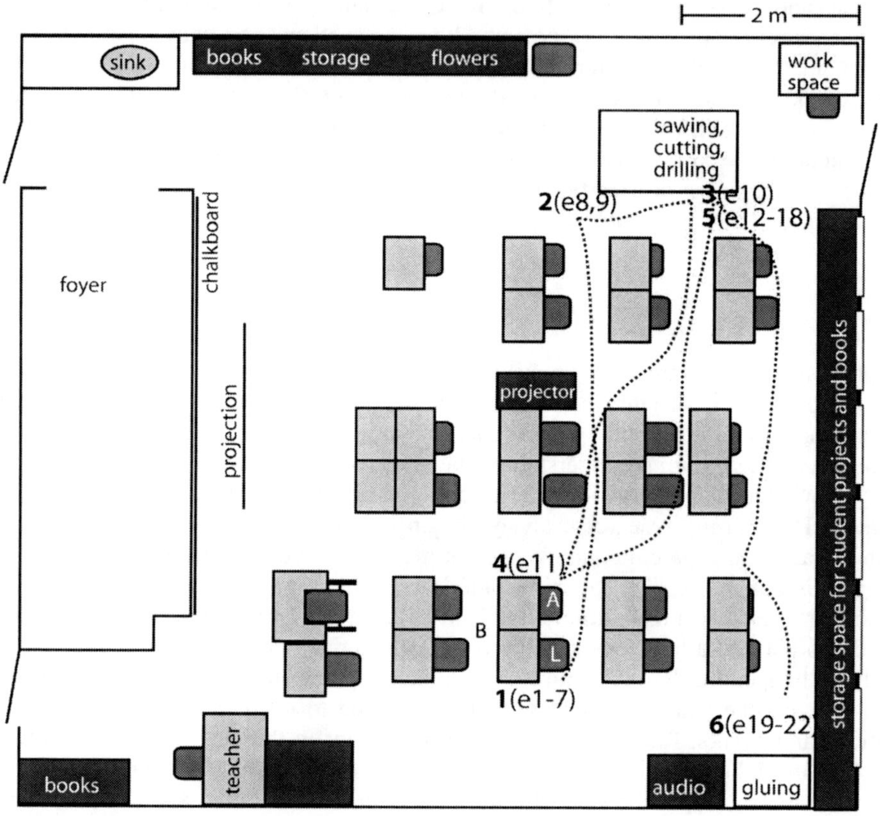

Figure 2. Maps allow tracking student–student interactions in time and space.

A Moment in Technological Design

The following episode had been recorded at the moment when the three girls were deliberating how to go about building the tower that would hold an elevator, the first element in their design of a Rube Goldberg machine (Figure 3). Leanne (far left in offprints) previously had explained that they needed to build a two-sided tower; Amanda (far right) and Bella (centre) went along so that it appeared that there was agreement on this design. Following Leanne's suggestion, the three had decided to cut the wood needed to build that part of their prototype. At one moment, Leanne talked about cutting two pieces of wood for the two sides of the structure. This is where the following episode begins, just after Leanne has articulated what they were going to do and was locating the different tools, raw materials, and diagrams that they needed for taking the first steps in the construction of the prototype. (I first provide the raw transcript[4] and then a verbal description of the events.)

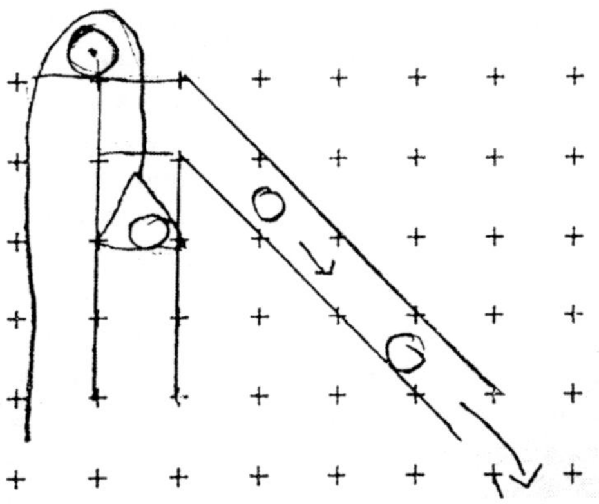

Figure 3. The students had decided to have a tower and a chute as the first two elements in their Googolplexosaurus, a complicated device that would accomplish the simple task of feeding a cat. The episode concerns the construction of the tower part.

Leanne has proposed to use a piece of wood that Amanda had apparently shown her as the "shoot", but Amanda proposes some other piece (inaudible). At this point, Bella begins to talk. She proposes that they could take a piece out of her brother's "parking lot". Throughout her talk, Bella used gestures. Initially, she had rested her upper body on her right hand and arm (line 01). By turn 02, her hand had moved from the table to the diagram pointing to it while she was talking about her brother. Her hand has moved away, arranging the chair beneath her, and then moves back to the diagram between them as she comes to the end of turn 07. Her hand moved repeatedly up and down a tower-like part of the drawing (Figure 3) while she utters that the part could be taken out (turns 10–11). She makes a gesture that shows what they would do with the part. Her final gesture hovers above the drawing and her arm rocks twice back and forth while her thumb and index produce a gesture similar to the rubbing that is used to indicate "money" in everyday conversations (turn 15).

The Different Moments of Communication

Dialectical units, including *communication* as I present it here, cannot be broken down into parts because they already constitute the smallest possible subdivision that still has the properties of the whole. This has the consequence that whereas it is possible to identify words and gestures, they cannot be understood independently from another; they also do not represent the whole communicative unit unless in reduced and one-sided form. In dialectics, these smaller parts into which a unit seemingly can be subdivided are called *moments*. Such moments cannot be understood independently but require each other for a complete understanding of communication. It is therefore necessary to provide visual information in a research report rather than translating visual information – as provided in gestures, for example – into words, which do not do justice to the *real* role gestural imagery plays in communication. In the following, I exhibit some of the possible ways in which communication in technology education may be analysed.

Words. With the conjunctive "Or" (turn 01), Bella grammatically produces an alternative. It is a contrast to what has been proposed before, when they had asked for the materials. Bella was responsible for bringing a pulley, and this responsibility has been inscribed into the diagram, at the bottom, where they noted the materials needed. Subsequently, Bella admitted that she did not bring one or have one. The two other girls talked about the "shoot" (see Figure 3). The "or" sets up a difference, a contradiction to what they had done or were presently doing. In this episode, Bella then develops the different idea. It takes shape in her talk and action, but at the same time retains its ephemeral nature, for talk and gestures 'vanish' as soon as they have been produced, they recede into the past, increasingly so, unless they are reproduced in subsequent actions and talk. Continuous reproduction of words can lead to a certain stability of an idea, a vision, even though it may not have taken a material form.

The three girls in fact are aligned twice, both in the sense that they could draw on their personal traces of the conversation from the previous day but also from their (culturally) based understanding of "this part" of a parking garage. This part they could take out of her brother's garage and put into their own design. There is a conflation in the sense that Bella already pointed to the diagram and said that they could take this part (in diagram) out of her brother's garage. That is, this part in their design and the parking garage had a common element, and all three were aligned to the fact that in this case the diagram does not refer to their design but to the parking garage.

The limitations that arise when analysts focus solely on words immediately are evident from this episode and my account thereof. At no point in her story does Bella articulate anything about an elevator, about the fact that the piece of material she suggests using is to come from the toy parking garage, or what the particular part is supposed to do. All this other information is made available in and through other modes much better suited than language to articulate the idea – especially at this stage when ideas are in development (Roth & Lawless, 2002a). I now turn to some of the other resources speakers and listeners make available for each other and for the analyst to produce and reproduce sense.

Gestures and perceptions. Gestures are an important part of communication because they allow communication participants to make salient visual-perceptual moments that words cannot render, or do so only in a difficult and cumbersome way. Visual-perceptual expressions can be made salient as a constitutive part of communication by means of two types of (unconsciously produced) gestures: pointing (i.e. *deictic*) gestures and *iconic* gestures.[5] By means of pointing gestures, speakers make salient, highlight, and can bring into communication perceptually available structures in the setting. (When people are very familiar with one another and the events that they are attuned to, gestures and words are needed only when there is a sense that participants are out of alignment [Roth, 2004].) For example, in turn 03, Bella points to the design drawing in front of Leanne while uttering "my brother". In this way, she links her brother to the particular part of the design drawing, the "elevator". That is, her story is about the elevator. She does not have to articulate this in words but a simple pointing gesture makes the aspect salient. A little later, after bringing into the conversation her brother's (toy) parking lot, she points again to the elevator in their drawing and suggests, "You can take this part out" (turn 11). Here, Bella's pointing gesture 'layers' the elevator in the drawing onto the narrative about her brother's garage so that we can understand what is being communicated as taking the elevator out of her brother's parking garage. This creates a new possibility namely that the story can run also the other way around: the garage part could be used to realise the elevator in their Googolplexosaurus (a googolplex is a certain, very large number in mathematics; this is how we named giant things that are so giant that their size exceeds all imagination). This is precisely how trajectory of the narrative unfolds, the garage elevator is battery operated and makes their design ("it") "work".

As her story unfolds, Bella also produces iconic gestures to show how the elevator in their design and in the garage works: she moves her hand up and down along the elevator part (Figure 2), thereby providing perceptual dimensions of her thinking design. The moment s envisioned, visualised, and articulated for others in and through the hand movement, which stands in a perceptual (i.e. *iconic*) relation to the material thing.

It is immediately evident that it is dangerous to describe the content of gestures in words. Thus, we do not really know *what* Bella points to *precisely* other than that the pointing gesture is toward the general area of the elevator. It may turn out that what is perceptually salient to her at the moment is not the same thing as what is salient to her audience (or the observing researcher). The future work within the small group will always show whether the assumption that they have looked at and understood the same thing can be upheld. If not, as is sometimes the case, interaction participants articulate for one another their different understandings and then revise what they really meant to indicate and say. Similarly, the aspect of an iconic gesture that the audience has to attend to is not made clear with its production. Thus, it may turn out that the listeners saw a different aspect of the moving hand and arm, in which case they will realign their understandings once it has become evident that they have been misaligned for some time.

These considerations demonstrate that technology education, where students are exposed to using artefacts and design drawings to a greater degree than in other school subjects, gestures can play an important role in communication. In fact, elsewhere I show how students develop complex understandings when their early communicative effort in a new theoretical and experiential domain is supported by the presence of artefacts, because these allow students to communicate in the absence of a developed verbal repertoire (Roth & Lawless, 2002a). As students develop their communicative competence, the reliance on the actual materials and artefacts decreases and the iconic gestural and verbal modes of communication increase to the point that students are competent to provide descriptive and theoretical accounts of what they have done in the absence of the material things about which they are communicating.

Prosody. How humans produce their utterances, that is, the quality of their voice in uttering words, is an important means to communicate emotions in addition to highlighting what is to be taken as more salient than other aspects of the situation. Prosody generally, and the voice parameters pitch (F_0), speech intensity, speech rate, and the first formant (F_1), are important ways of (unconsciously) expressing emotions such as anger, impatience, and joy (Scherer, 1989). Analyses of prosody have shown great promise to understand what is going on in science and technology classrooms generally and student–student and student–teacher interactions more specifically (Roth, 2006). Prosody also is an important resource for the interaction participants in the exemplary feature episode of this chapter.

In addition to using the adverbial “or”, which grammatically presents what follows as an alternative to what has been said before, Bella speaks in a considerably higher pitch range than normal until the completion of the word “brother” (turn 02). Following Amanda’s much lower pitched “uh hm” (turn 04) and a lengthy pause, Amanda continues to speak but with a pitch in her normal range. Jump in pitch to a higher level, increases high pitch contours (as in “brother”, which mounts to a maximum of 527 Hz), is associated with conflict and difference, not only in this database but in research on children’s interactions generally (e.g., Goodwin, Goodwin, & Yaeger-Dror, 2002). However, conflict often emerges when the next speaker “ups the ante” and increases pitch levels again. In the present situation, this is not the case and Bella develops her alternative without bringing about conflict. The non-conflictual nature of the present interaction also is available from the prosodic information (Figure 4). In contrast to other situations where difference in opinion is associated with conflict and increases in pitch (F_0) levels at a turn boundary and within a turn, Leanne rejects Bella’s proposal without accompanying her negative response by an increase in pitch. In fact, when she starts speaking, her pitch is the same as that of Bella when the latter ends

(Figure 4). Semantic differences uttered at the same or lower pitch levels generally lead to constructive discussions; similar changes in F_1 levels and speech intensity also are expressions of specific emotions and orientations toward the communicative partner (Roth & Tobin, in press). The prosodic resources therefore allow the three girls to discuss several alternatives in sequence without experiencing a rejection of a proposal as a put-down, as would have been in the case of sharply rising pitch (F_0), increasing speech intensity, and increasing F_1 levels. Figure 4 also shows how waveforms provide resources to the researcher for measuring the precise length of pauses and the relative rates of utterances, here indicated in terms of an expansion and compression of kerning (letter distances).

Body position and orientation. Body position and body orientation of both speakers and audiences play a crucial role in the successful production of communication. Although most people are not aware of it, listeners continually produce resources for speakers to understand that the former are following or not following the presentation – for example, head nods or turning head and body toward speaker. Similarly, speakers (most often unconsciously) position and orient their bodies to draw the attention of the audience to a particular feature of the setting and there are a number of studies that show how changes in the genre of teacher talk are correlated with significant changes in the orientation of the entire body (e.g., Roth & Lawless, 2002b).

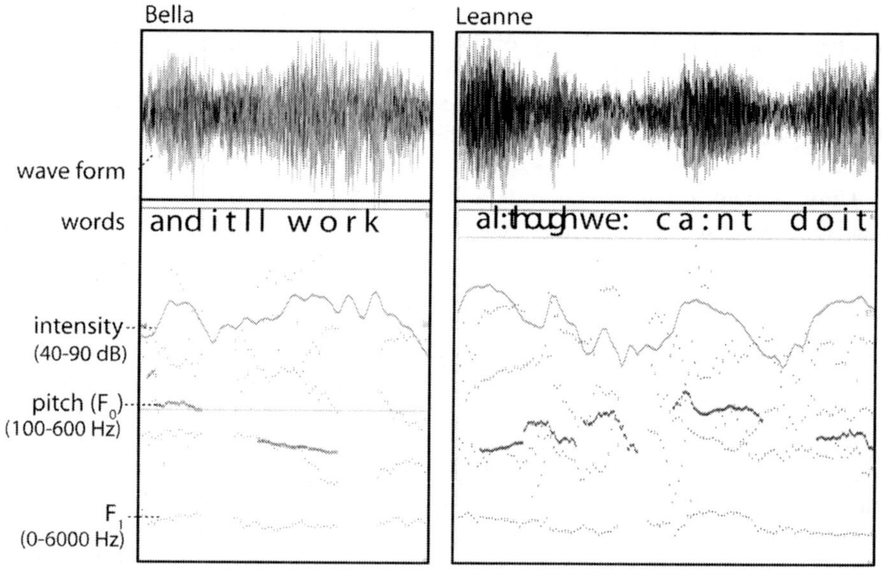

Figure 4. Key prosodic parameters such as speech intensity (scaled from 40–90 dB), pitch (scaled from 100–600 Hz), speech rate (changes available in proximity of letters), and F_1 levels (scaled from 0–6,000 Hz) can be obtained from the PRAAT software. Represented are the end of turn 17 (left panel) and turn 20 (right panel).

In the present feature episode, while Bella is developing the alternative design, or rather, the particular implementation of the "elevator" part, Amanda and Leanne provide her with evidence that they are attuned with the design suggestions that the former unfolds for them. At the very beginning of the episode (see offprint associated with turn 01), Amanda still is turned toward Leanne, who just has finished uttering her proposal. While Bella utters "my brother" (turn 02),

Amanda slightly brings up her shoulder and turns her head until she squarely faces Bella. Similarly, during the pause (turn 05), Leanne reorients her gaze from the design drawing to Bella's face (see change in offprints from turn 02 to turn 05). Both thereby provide Bella with resources for understanding that her peers are attending to her designing-aloud of an alternative for the elevator part. Subsequently, Leanne repeatedly nods her head, which also signals to Bella that she is attending to the unfolding design proposal. In fact, we may understand that the pause in turn 05 has a specific conversational function: it both seeks Leanne's attention, and provides her with a resource to signal attention. Thus, when there appears to be evidence that Bella does not continue while attention appears to be focused elsewhere, Leanne signals alignment by turning her gaze toward Bella. In other words, we can see this episode in this way: after uttering "brother" (turn 02), Leanne still is looking downward at the drawing. Her gaze moves up to meet that of Bella only 0.97 seconds after Amanda has completed her "uh hm". The pause therefore may have had the function to provide time until alignment has been signalled. By the time Bella has utters "lot", Leanne is gazing at the diagram as if following the pointing finger, but Amanda was still gazing at Bella. The latter's continuation precisely falls together with the point in time when Amanda, too, has directed her gaze at the diagram. At "this part" both listeners were looking at the diagram until Bella has finished the explanation of what to do with the elevator part from her brother's garage. Both simultaneously move their gaze to look Bella squarely into the face. Amanda continued to gaze at Bella, whereas Leanne nodded repeatedly. As soon as Bella has completed her utterance, Leanne, still facing Bella, begins to talk and Amanda shifts her gaze to the next speaker after having briefly dropped it downward in the direction of the design.

In the present episode, we also observe one significant change in body position. When Bella utters "or", she is about one step away from the desk. As she utters, her right leg moves forward toward the desk, which allows her finger to land on the elevator part of the diagram. During the pause, Bella moves her leg and left part of the body backward, only to bring it forward again. It is because she has brought her whole body forward that her hand and index finger are enabled to reach the diagram and to bring about a more specific deictic gesture that is precisely directed to the elevator rather than more generally to the diagram as a whole. That is, the change in body position here is associated with and enables the production of another communicative resource. Without taking into account this change in body position, we could not answer the question, "What are the resources that enable the successful production of communication at this moment?"

In the same way that we observe with gestures, technology education is in a privileged situation with respect to enabling additional modes of communication: the presence of materials and technological artefacts configures and structures the physical space, which then becomes a resource in the production and reproduction of sense. Thus, in an extensive study of cognition in this classroom, I have been able to exhibit the interactional effects of group size (whole class, small group), (presence/absence) of focal artefacts, and spatial configuration on the nature and quality of classroom conversations (Roth, McGinn, Woszczyna, & Boutonné, 1999).

UNDERSTANDING COMMUNICATION AS UNFOLDING IN TIME

In the foregoing section, I present the different communicative resources that speakers and listeners make available during an episode in a classroom where students learn scientific (physical) concepts by designing technological artefacts. The sequential presentation, however, may make it appear as if we (researchers) could consider the different modes on their own. This, however, is not the case. Like the physicists' light – which expresses itself sometimes as a wave and sometimes as a particle without being understandable as wave *or* as light, or as a combination of wave *and* light – the technology educator's communication cannot be reduced to the words students utter or the gestures they produce, the gazes that they change, the prosody that they deploy, or any combination of these modes. More so, in the same way that the physicist cannot express the particle character of light in terms of a wave, the technology educator *ought not* attempt to reduce prosody, gesture, perceptual information, and so on to words. There is a different *kind* of information made available in and through the different modes; and this *kind* has to be preserved to render justice to what is happening in communicative exchanges.

It is also important that the technology educator or researcher analysing communicative events attend correlated changes in the different communicative modes (Figure 1). Signification is achieved precisely when something becomes apparent; and making apparent usually occurs by means of a sign that points to some (relevant) thing, which may be material or ideal (i.e. a concept made available in verbal, gestural, perceptual etc. mode). Bella achieves signification by changing her body position, correlated with a verbal production ("or") and a deictic (pointing) gesture *toward* a diagram. These changes together, produced against an unchanging background, make salient the structural aspects of communication. In other words, signification arises from and is marked by the production and reproduction of variant and invariant communicative structures.

Figure/ground relations are dialectical, as change (variation) always is change against an invariant background. It is therefore easy to see why the relation between the different communicative modes also is dialectical. For example, a deictic (pointing) gesture is directed toward an elevator part, and thereby makes the latter salient. But the pointing gesture not only motivates the salience of the diagram part but is motivated itself by the latter. This is even more evident when we consider iconic gestures, for example, Bella's up-down movement with her right index along the elevator. The up-down movement makes salient (motivates) something in the diagram/elevator, but it is the elevator that motivates the gesture. That is, the correlated ephemeral (gesture) and ink traces produce salience. Thus, if there is a particular iconic gesture without an evident counterpart, we cannot make sense; if there is a pointing gesture without something in the area that can be perceptually isolated as being related to the current topic, we cannot make sense. Further, it is precisely when a rise in pitch (increase in speech intensity, slowing of speech rate) is co-produced with a particular word that the word becomes salient (we hear it as something emphasised). Again, it is the co-presence of two communicative resources that makes something stand out; and sense precisely is the process (conduct) of marking and making stand out (Nancy, 2003). But the co-presence does not inherently privilege the thing or the sign pointing to it, because both require a material basis and both require this material basis to be structured – words really are structured and recurrent sounds, prosody denotes qualities of sound production, gestures require hands and other body parts. The communication of even the most abstract and ideal concept – ethics, the mathematics of string theory – requires a structured material substrate. In this chapter, I present a method of understanding communication in technology education by extending the analyses to material substrates other than the structural aspects of sound that we hear as words.

Another important aspect to be considered in the analysis of communication is the fact that grammar, as much as sense, does not *predetermine* communication but itself is an outcome (product) of the ongoing communicative effort. Thus, for example, it is more productive to view Bella as thinking/designing aloud rather than having a fully specified idea of what her design looks like. This does not prevent us from thinking that she may have a more or less developed but nevertheless vague idea, some kernel idea that begins to emerge somewhere around the moment she utters "or". This kernel – from a Vygotskian sociocultural perspective, a 'growth point' (McNeill, 2002) – comes to be developed and articulated in the course of Bella's designing aloud. The process is punctuated and therefore grammaticalised by prosodic means, for example, by the contours of the pitch lines. Thus, in turn 11, the pitch slightly falls, indicated by the semi-colon in the transcript following the end of the utterance "you can take this turn out". Such a slight fall normally is heard as the end of a clause or sentence in an unfinished account. Thus, even though there is a pause unfolding (turn 13), the other two do not speak thereby providing Bella with an opportunity to continue, which she does, until the idea is fully and completely developed – at least for the moment. If Bella has not fully thought out what she is going to say, at what point will an utterance specifically and the story more generally end? Here, the unfolding utterance (story) itself provides the context for stopping whenever the participants have the sense that what has been said is complete. That is, the emergent product of communication provides a resource for grammar to act as a sort of *stop order* (McNeill, 2002), a measure for when something is complete. Thus, the grammar of the utterance is an emergent product and a tacit (usually nonformal) understanding of formal grammar provides a measure of when the emergent utterance (story) is complete.

CODA

In the course of the past decade, I have conducted and published numerous studies on aspects of communication other than words, including gestures, body positions and body orientations, and prosody. These studies most frequently involved technological design and the design of experiments and investigations. All of these studies point to the particular affordances (opportunities) that arise when there are artefacts and inscriptions present, which not only allow additional communicative modes but also communicative modes that express certain aspects of the overarching communicative unit (sense, meaning) much better than the verbal mode – even though they still do so one-sidedly. The artefacts present in a communicative situation actively support and constitute part of the communication, making it easier for learners to communicate without having to express everything in words. That is, technological artefacts and representations thereof provide students with special opportunities for communication and, therefore, learning. In this chapter, I present a method of analysing communicative events that goes far beyond the simple analysis of verbal utterances and therefore constitutes a method that parallels in its potential for researchers the com municative possibilities technology provides to students.

NOTES

1. The English language does not have the means to distinguish between two types of phenomena that the German and Russian languages – in which activity theory was developed – do. Thus, in German, activity theory is *Tätigkeitstheorie*, and *Tätigkeit* is distinguished from *Aktivität* (activity), which simply means being busy, doing something, without implying the societal motivation that orients *Tätigkeit*. In Russian, the same distinction is made by means of the concepts *deyatel'nost'* and *aktivnost'*, respectively.
2. This is consistent with both philosophical analyses, which state that in each observed action we recognise something we could have done ourselves (Hegel, 1806/1977) and with recent neuroscientific research on the role of mirror neurons in both bringing about and recognising actions (e.g., Gallese, 2003).
3. Rube Goldberg was a cartoonist who drew complicated machines that accomplished the most mundane everyday actions. Constructing a Rube Goldberg machine is a popular design activity of middle school students around the world. More information is available at http://www.rube-goldberg.com/.
4. To get a better sense for the changes in body positions and hand gestures involved, readers should visually scan the images before reading the transcript. The following transcription conventions have been used in line with traditional conversation analysis enhanced with features for transcribing prosody: (0.41) – time in seconds; [two sides] – square brackets enclose speech overlapped by gesture or other body movement; ((rH, lH moves up)) – description of gesture (rH, lH = right and left hand, respectively), body movement; * – the asterisk aligns speech and video offprints; .,?! – punctuation is used to indicate speech features, such as rising intonation heard as a question, or falling intonation to indicate the end of an idea unit (sentence); ‘ – the diacritical sign indicates rising pitch within an utterance; ↑↓ – the arrows indicate upward and downward jumps of the pitch range.
5. There are also beat and metaphoric gestures, which, because of space limitations, I cannot address here.

REFERENCES

Gallese, V. (2003). The roots of empathy: The shared manifold hypothesis and the neural basis of intersubjectivity. *Psychopathology, 36*, 171–180.

Goodwin, C., Goodwin, M. H., & Yaeger-Dror, M. (2002). Multi-modality in girls' game disputes. *Journal of Pragmatics, 34*, 1621–1649.

Hegel, G. W. F. (1977). *Phenomenology of spirit* (A. V. Miller, Trans.). Oxford, UK: Oxford University. (First published in 1806).

McNeill, D. (2002). Gesture and language dialectic. *Acta Linguistica Hafniensia, 34*, 7–37.

Nancy, J.-L. (2003). *A finite thinking*. Stanford, CA: Stanford University.

Roth, W.-M. (2004). Perceptual gestalts in workplace communication. *Journal of Pragmatics, 36*, 1037–1069.

Roth, W.-M. (2006). Conversation analysis: Deconstructing social relations in the making. In K. Tobin & J. Kincheloe (Eds.), *Doing educational research: A handbook* (pp. 235–267). Rotterdam, The Netherlands: Sense.

Roth, W.-M., & Lawless, D. (2002a). Signs, deixis, and the emergence of scientific explanations. *Semiotica, 138*, 95–130.

Roth, W.-M., & Lawless, D. (2002b). When up is down and down is up: Body orientation, proximity and gestures as resources for listeners. *Language in Society, 31*, 1–28.

Roth, W.-M., McGinn, M. K., Woszczyna, C., & Boutonné, S. (1999). Differential participation during science conversations: The interaction of focal artifacts, social configuration, and physical arrangements. *Journal of the Learning Sciences, 8*, 293–347.

Roth, W.-M., & Pozzer-Ardenghi, L. (2006). Tracking situated, distributed, and embodied communication in real time. In M. A. Vanchevsky (Ed.), *Focus on cognitive psychology research* (pp. 237–261). Hauppauge, NY: Nova Science.

Roth, W.-M., & Tobin, K. (in press). Solidarity and conflict: Prosody as a transactional resource in intra-and intercultural communication involving power differences. *American Journal of Sociology*.

Scherer, K. R. (1989). Vocal correlates of emotion. In H. L. Wagner & A. S. R. Manstead (Eds.), *Handbook of psychophysiology: Emotion and social behavior* (pp. 165–197). London: Wiley.

Wolff-Michael Roth
Lansdowne Professor of Applied Cognitive Science
University of Victoria
Canada

WOLFF-MICHAEL ROTH

54. LEARNING AND TEACHING THROUGH TECHNOLOGY AND DESIGN

A Cultural-Historical Activity Theoretic Perspective

INTRODUCTION

> Engineering requires a large knowledge base to make choices and those choices are not made by chance, but logical reason based on previous experience. … A building cannot be designed or constructed in real life without logical reasoning using basic engineering principles. (Aerospace engineer, personal communication, June 12, 1994)

Engineers, such as the aerospace engineer who commented to me about my description of engineering design, often assume that students first have to memorise technical engineering knowledge before they can design technological artefacts; in fact, students *should not be allowed to* engage in design before they have not acquired all the prerequisite engineering and physics knowledge said to underlie the design. Although some design educators hold up this claim because of apparent correlations between knowledge and creativity (Christiaans & Venselaar, 2005), they are immediately confronted with the transfer problem that distinguishes between knowledge and its application, a separation that makes very little sense for a situated or enactive conceptualisation of knowledge. On the other hand, as the excerpt from a fifth-grader's (Jordan) presentation of a bridge design (Figure 1) to his peer and teacher critics shows, even young children can evolve technological artefacts that embody engineering knowledge before learning formal physics and engineering. In the process of working toward some goal, for example, to build a bridge that can carry as much weight as possible, my study shows that students will seek out any knowledge and skills that *expand their room to manoeuvre* and, provided some assistance from the teacher, attempt to find suitable knowledge, skills, and tools just in time and as needed.

Seeing that some engagement will expand the current possibilities for acting is inherently motivating so that the associated learning is one form of what cultural-historical activity theorists call 'expansive learning'. As part of his emergent design and learning, Jordan has also produced two entries for his glossary in which he illustrates, using arrows, the forces that act in two different bridge designs (Figure 2). This form of expansive learning occurs not just with respect to science, which has always been my interest, but also with respect to technology, which constitutes a form of knowledge itself because of its focus on design, problem-solving, problem-posing, project planning, and testing-optimisation (e.g., Ankiewicz, de Swardt, & de Vries, 2006).

A.T. Jones and M.J. de Vries (eds.), International Handbook of Research and Development in Technology Education, 651–668.
© 2009 Sense Publishers. All rights reserved.

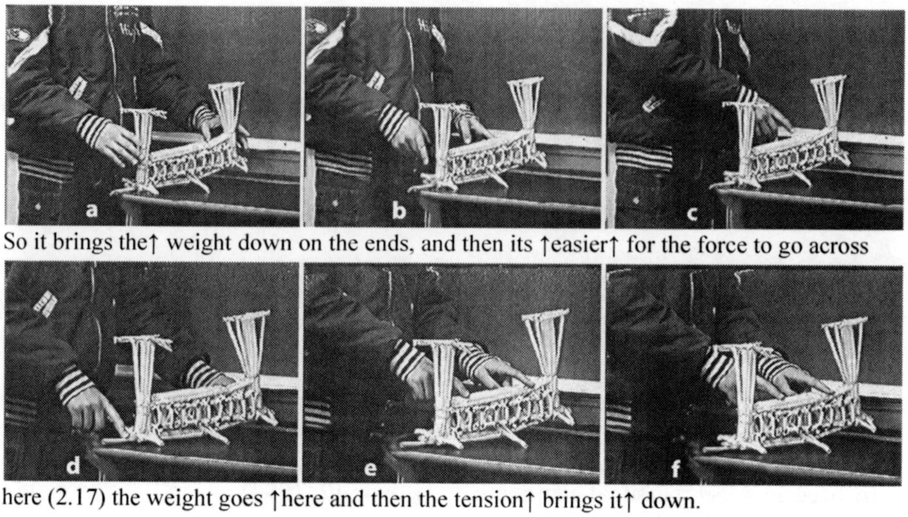

Figure 1. Jordan, fifth-grade student, shows how forces act in his bridge design. (Arrows match text with corresponding images above.)

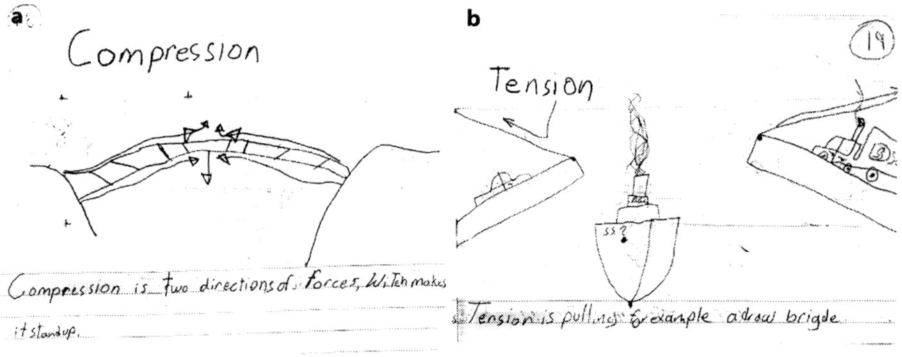

Figure 2. Jordan's two glossary entries that show how compression and tension forces act in two bridge designs.

In this chapter, I describe learning and teaching technology and design through the lens of cultural-historical activity theory (CHAT). This theory has shown great promise, especially for learning in technology and in complex systems (e.g., Miettinen, Lehenkari, & Tuunainen, 2008) because of its consideration of and sensitivity to the cultural and historical contexts of knowing, doing, and talking technology and artefact design. It is particularly appropriate for technology education because of its conceptualisation of activity on *mutually presupposing* material and ideal (mental) levels. Thus, CHAT theory recently has been proposed as a useful basis for the design of instruction that aims at and integrates the various purposes of design and technology education (Stevenson, 2004); the theory also addresses an important lacuna in the theorising of learning in technology education.

In contrast to the drawings Jordan produced during the course, the prototype model Jordan constructed and then talks about is physical, which affords new communicative possibilities, especially to deictic (pointing) and iconic (depicting) gestures (see Chapter 53). In his first description, the hands rest at the upper part of the pillar until the utterance of "the weight" (Figure 1a). Then, while uttering "down on the end", the hand moves down alongside the pillars, stopping at the end until completion of the verbal part of the meaning unit. The first gesture along the horizontal deck falls together with the utterance of "it's easier for the force" (Figure 1b, c). At this point, the verbal utterance is not yet completed. Thus, in the process of completing the verbal meaning unit ("to go across here"), the hand moves twice more back and forth along the truss.

Such gestures over and about engineering design drawings and artefacts are central to optimising communication during professional technological design (Luff & Heath, 1993). In the present case, the gestures represent the forces, an important physics concept. From a physics perspective, forces are vector rather than scalar quantities. Technically, therefore, they cannot be pointed to, but have to be indicated in terms of magnitude and direction. Here, in the first part of the episode, Jordan gestures the direction in which the weight of the pillars acts. The gestures are iconic conceptualisations and their extensions in space are paralleled by the time it takes for the utterance of the corresponding meaning unit. Perhaps because the gestural meaning unit is completed before the verbal meaning unit (at Figure 1c), the right hand repeats the action of the horizontal force. These gestures arise directly out of the practical work and become, consistent with cultural-historical activity theory, the foundation on which more abstract forms of knowing (e.g., talking, discoursing) are subsequently established.

Jordan then reiterates the meaning unit of the relationship between pillar weight and tension in the upper part of the truss. In this, the effect of the weight down and the force across becomes even clearer when Jordan actually pushes down on the ends of the bridge deck. Again, the downward motion of the hands begins with the utterance of "weight". Now, however, Jordan actually pushes down onto the ends of the bridge (Figure 1d). This straightens the bridge deck. At the same time, although small, the bridge pushes back so that Jordan has feedback from the reaction force to the weight. He then releases the bridge deck (thereby allowing the bridge to bounce back) and, while uttering "and then", brings his right hand into the position for explicating what the consequence of the weight is. (The left hand simply holds the bridge, which begins to rotate as the right hand moved along the truss.) The beginning of the horizontal gesture then coincides with the utterance of "tension". In this situation, the end of the verbal meaning unit falls together with the end of the horizontal gesture and there is no repetition as in the earlier case.

This episode shows how, at the end of a three-week investigation and construction of bridges, Jordan has developed a sophisticated understanding of forces in bridges. Although he is not entirely correct in suggesting that the tension brings the deck down, his gestures enact the forces and their relationship. His gestures therefore enact an explanation of the form weight → tension: the downward weight implies or causes the horizontal tension. Here, Jordan's verbal scientific argumentation coincides with the corresponding iconic gestures, consistent with the literature on the developmental trajectories of gesture and talk (e.g., Hayward & Tarr, 1995). However, these gestures should not be considered ancillary to the verbal meaning unit. Because of the vectorial nature of the physical concept of a force, from the perspective of the theoretical position developed here, each – words and gestures – are only one-sided expressions of the entire meaning unit (Vygotsky, 1986). It is actually the downward gesture (and the force Jordan can feel from the bridge deck tension in the truss resisting to bending) that expresses the appropriate topology of this phenomenon.

From an audience perspective, Jordan's gestures in the first part of his explanation are similar to the gestures that one might observe when students talk in the presence of technological design drawings such as those in Figure 2. The three-dimensional object, however, makes the identification of the physically significant elements more difficult: whereas in a diagram only the physics-relevant elements are drawn, the artefact does not embody such discourse-relevant selections. On the other hand, Jordan's second explanation (the gesture in which the hand actually pushes on the bridge) shows the effect that the downward weight has on the bridge deck. Thus, Jordan's utterances can be compared directly with the changes in the technological system he designed. He thereby exhibits tremendous physics and engineering knowledge.

Although there had been only facilitation, with no direct instruction (via lectures or demonstrations) on this topic on the part of the two teachers of this class (Brigit and Christina), Jordan and his classmates developed sophisticated forms of discourse (Roth, 1998). My various tests of knowing conducted during and following the unit (e.g., Roth, 1996a) show that the students competently used scientific representations (drawings, language) to explain tension and compression (e.g., Figure 2). Jordan's verbal explanation for compression suggests that it arises from forces in two directions. His drawing elaborates the nature of these forces. First, they act in the bridge from left to right and right to left along the bridge structure. In the centre, then, the upward ends of the forces that act in the structure itself counterbalance the downward force of the bridge weight. Similarly, the text of the "tension" glossary entry identifies it as something that arises from pulling. Furthermore, both tension forces are indicated by arrows parallel to the ropes (cables) in which they act. I emphasise that, in both instances, gestures such as those exhibited in Figure 1 have been precursors to the signifying arrows drawn later, consistent with the literature on gesture–speech development in other domains and age ranges (Roth & Lawless, 2002). The arrows are abstractions from the earlier actions, where Jordan's hands and index finger moved through horizontal and vertical trajectories now represented in the arrows. The multiple arrows appear to reconstruct the repeated movements of his index finger across the truss.

LEARNING THROUGH TECHNOLOGY AND ENGINEERING DESIGN

During the 1990s, much of my research focused on learning science in and through technology-focused design of artefacts. During that decade I developed, predominantly grounded in a theoretical framework based on situated cognition, a detailed understanding of why and how students learn science by designing technology. A second major focus had been my interest in *inscriptions*, that is, forms of (external) representations that scientists and engineers use to articulate their understanding of the system they are studying or want to build. I found science and technology to share two fundamental similarities: (a) the production and transformation of representations and (b) the performative idiom describing the two domains. Many of the cues for my interest actually came from a discipline other than education: social studies of science and technology (Küchler, 2008), which presents and explains engineering, technology, and design in ways that are very different to what one might find in textbooks on the topic. The *production* and *transformation* of objects and the world generally, and of representations specifically, has been a core interest to cultural-historical activity theorists since its inception, that is, of all three generations that I discuss in this chapter. The performative idiom, that is, the languages we use to get the productive and transformative work done, have been of interest especially to first-generation theorists including Lev S. Vygotsky.

Interests and goals are central to people's meaningful everyday activities; this is why, in activity theory, the object orientation of collectively motivated endeavours takes a central role. Understanding *how* people pursue goals is critical to understanding knowing and learning.

Problem-posing/solution-finding is especially promising as a learning environment because it allows abstract knowledge to emerge from the concrete engagement with the world in a way that capitalises on students' prior knowledge (Barak & Shachar, 2008). Design is a form of problem solving where interests and goals bear on processes and outcomes in essential ways; and, by their very nature, design processes are about the new, uncertain, and ill-defined (Bucciarelli, 1994). Because design is the central activity in technology, it fits well with an emerging interest in science-technology-society (STS) issues, and in technology as context for learning science (e.g., Layton, 1994). Technology education constitutes a constraint for all those who might be tempted to pry science from its context and encourages us to take a situated perspective where knowledge always exhibits itself in action (McCormick, 2004).

Designing – the creation of artefacts that specify how a system should be organised – is one of four types of learning environments that make learning-in-practice a central element (Schank, Fano, Bell, & Jona, 1994). Design tasks offer opportunities for learning in school precisely because they are the epitome of open-ended and ill-structured problem solving and allow students to construct and test knowledge by incorporating in the design their ideas and feelings (e.g., Kafai, 1994). Learning through design can be distinguished from professional designing. Learning through design is not exclusively represented in final products (artefacts), but in the process as well (art, performative aspects). In contrast, most professional design concentrates on the product as the essential outcome; learning to and through design makes *process* the central issue of education. Yet, as I show in the next section, the product has an important orienting function in the design process because it co-constitutes the object/motive; and when it is associated with subsequent use rather than a superfluous entity cast away after receiving a grade, then learning through technological design becomes particularly (inherently) motivating. However, we need to keep in mind that learning is always only partially reflected in the artefacts that result from the design process. In a contradictory manner, these artefacts, though representing activity only one-sidedly and incompletely, traditionally are taken as the sole basis for the evaluation of learning. In fact, the resulting artefacts are one-sided and therefore partial representations of the entire system within which design unfolds as shown in the next section. But they are marked by all the moments of the activity system.

Based on the results of previous studies on children's designing and my own work on student-centred design of science experiments, I identified the *Engineering for Children: Structures* curriculum that engages students in the design of artefacts. *Engineering for Children* activities are not designed to *transmit* legitimated and canonical engineering knowledge. Rather, they provide students with opportunities to (a) explore and experience some critical (but not necessarily pre-specified) issues about designing structures, (b) learn to manage the complexity of open-ended and ill-structured design situations, and (c) learn to collectively design, make sense, negotiate, plan, and execute group projects. In engaging with the designing tasks, the children can develop a practical consciousness on the basis of which they may evolve a theoretical consciousness of designing, and of design artefacts as well.

Although I had originally grounded my work in science and technology studies, situated cognition, and phenomenology, I have found in cultural-historical activity theory a much more powerful framework for understanding human acting and thinking and for designing learning environments that allow students to identify with motives that already exist in their community. The theory turned out to be an important tool in my research because it focuses on productive and transformative work, mediated by such entities as the tools/instruments, relations to others (division of labour), and so on. This focus on material production is one that technology educators, as those social scientists concerned with material productions more generally, can affiliate with and see concretised in their own work. CHAT has been shown to be particularly useful for understanding work processes in technological design because it does not abstract knowledge and practices from the context (e.g., Engeström & Escalante, 1996), without which we would not even be able to understand the slightest phenomenon of consciousness.

CULTURAL-HISTORICAL ACTIVITY THEORY

In human history, practical understanding always preceded theoretical understanding. In construction, a separation between masons and master-masons-become-architects occurred around the time when the great Gothic cathedrals were built (Turnbull, 1993). A need was felt for going from ad hoc practices characteristic of masons to more planful practices of individuals who could envision the ultimate outcome. Nowadays, as our own research among workers and technicians has shown, both practitioners and theoreticians show some contempt for the practices of the other. Cultural-historical activity theory has been developed to overcome the gap that historically has developed between everyday, practical, and sensuous activity (e.g., stone masons) on the one hand, and theoretical considerations said to occur in the head (e.g., architects) on the other hand. Karl Marx always held that practical activity and social interactions preceded the development of a theoretical, explicit understanding of activity and the concepts it involved. This undergirded both my decision to teach physics concepts through technological design and to begin this chapter with concrete, 'real-world' examples of technological design in elementary schools. In this section, I briefly review the evolution of CHAT theory through its three generations and then show how it concretises itself in my study of technological design in elementary schools.

Historical Perspective on Cultural-Historical Activity Theory

Grounded in dialectical materialist philosophy – Karl Marx and Friedrich Engels developed a strongly historically oriented activity theory on the basis of Georg W.F. Hegel's work – the social psychologist Vygotsky (e.g., 1978) explicitly drew on his philosophical predecessors' work to argue for the sign- and tool-mediated nature of human activity designed to fulfil personal goals and human needs. Today, Vygotsky is referred to as having brought about the first generation of activity theory in pointing out the *mediated nature* of material work and of the practical consciousness that goes with it. Thus, in the same way that the means of production (tools, instruments) mediate manual labour, which cannot be understood independently from the mediation, signs (words, inscriptions, physical models) mediate knowing and therefore learning. Although crucial to human consciousness, science and technology educators frequently do not sufficiently attend to language in their concerns of thought and thought structures that are said to be independent of language. For Vygotsky (1986), however, such an approach would not have made sense because language is central to consciousness, which always is consciousness for another as much as it is consciousness for the self. More so, thought and language (speech) are not independent, but thoroughly intertwined *processes* that mutually presuppose each other; and they are related by means of a third, irreducible process that he called "word meaning". This word meaning cannot be understood independent of speaker, audience, and activity; it is the way reality appears in the collective consciousness that participants have in the societal context that they presently share. Thus, what Jordan and Jonathan design is a function of the language and drawings that they have at their hand, but speaking and drawing further elaborates their thinking. Concretely designing technological artefacts, therefore, allows Jordan and Jonathan to develop on microgenetic scales as they engage with the task at hand. The bridge artefact mediates Jordan's communicating design in the opening episode, providing special opportunities for mobilising and understanding gestures, and these expressions allow further development of thinking to occur. This microgenetic development mediates what they learn and therefore their ontogenetic development. That is, the activity leads their development, which is a thought central to the work of Vygotsky and his co-workers as well as students of the second-generation theory.

Vygotsky (1986) makes another crucial point subsequently taken up in second-generation activity theory but which has since then gotten lost: the central role of emotion in human activity generally, and in human thinking more specifically. That is, science and technology educators theorise learning and practise teaching without considering affect as a core moment of human activities. Vygotsky held that without affect "the thought process appear[s] as an autonomous flow of 'thoughts thinking themselves', segregated from the fullness of life from the personal needs and interests, the inclinations and impulses of the thinker" (p. 10). We cannot understand why Jordan built his bridge in *this* way, supporting an amazing 356 wooden blocks; and we cannot understand why, at the same time, Shannon was building a beautiful bridge "for people and dogs", which she represented with respective toy figures. It is not as much that emotions affect thinking and learning from the outside, as additional factors that diminish what thought could have done on its own. Rather, for Vygotsky, there would not be thought, consciousness, or culture without emotion. This would be a central implication that one line of third-generation activity theorists would take up from Vygotsky and his students and co-workers, who created second-generation activity theory.

Vygotsky's students and co-workers not only took over his agenda, but also further developed the approach in the sense of theorising affect at a collective (social, societal) level. This then constituted the second generation of cultural-historical activity theory. Following the original idea that all higher psychological processes are found in concrete interactions within a society as a whole, Aleksei Nikolaevich Leont'ev (1978) suggested that (a) for humans, society has become the unit of analysis and (b) correspondingly, collective motives drive human activities. Because in developed societies there is a division of labour to meet collective needs (e.g., shelter, food, safety), individual human beings select the activity in which they want to participate (e.g., farming, manufacturing, schooling) and through their participation realise the activity in a concrete way. They choose concrete goals to orient their own actions and thereby realise collective motives such as control over the environment, acquisition/production of food, or construction of shelters. These actions therefore not only are constitutive moments of an activity but also presuppose the activity that the actions prospectively are oriented to realise. In exchange for contributing to the collective activity, individuals take what they produce or receive (income, salary) to barter or buy what they need.

This theory immediately requires us to consider student learning not in itself but as part of a system, 'schooling', which is designed for the production and reproduction of society. This has the immediate consequences that there are problems when students do not buy into and take up the collective motive: the resistance to schooling observed among some working class (lads) or some American African students can immediately be explained. From my perspective, any technology design activity at school has to be understood as nested within schooling, for otherwise we easily fail to understand why working class kids might find themselves more frequently in technology education classes than students from middle class and affluent home environments (Jones, 2003; McKenzie, 1992).

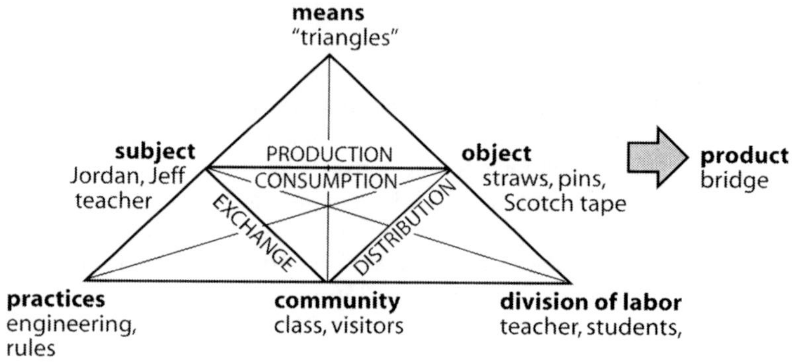

Figure 3. The structural moments of an activity system.

Second generation theory was expanded into what now is known as third-generation activity theory (Roth & Lee, 2007), known largely through an iconic triangle that depicts the structural relations within an activity system (Figure 3). This activity system has to be thought of as connected to other activity systems, from which it receives or to which it gives its moments (Engeström, 1987). Thus, the tools and materials (objects) that the children use to produce their design come from other activity systems in which they had been the product/outcome. The model highlights the fact that knowing and learning are mediated and *cannot be understood separate from material relations of actually producing things.* The model emphasises that it makes little sense to think about knowledge *independent of* the practical activity producing such things as designs and prototypes of bridges; consciousness, according to Leont'ev, as for his teacher Vygotsky, is integrally tied to the material practice.

Our own work has shown that as a depiction of structural relations, the model fails to highlight the essentially changing nature of human activity not only on cultural-historical timescales (phylogeny) but also, and especially relevant in schools, at the microgenetic (minute-to-minute) and ontogenetic levels (individual development). The depiction of the structural relations also fails to make salient the fundamentally agential aspects of human activities and actions, which can be understood only as continuously emerging from an agency | structure dialectic (Sewell, 1992). More so, many newcomers to the theory and its triangular depiction think in terms of its individual *moments* – 'subject', 'object', 'means', and so on – as *elements* that together compose the activity system. But elements can be understood on their own, which is not the case with the internal structure of an activity: the latter is the smallest "further unanalysable" unit that "retains the properties of the whole" (Vygotsky, 1986, p. 5). Materialist dialectics therefore uses the term 'moment' to refer to those structural features that can be identified but not understood independent of other structural features (e.g., subject, object, tools).

Another form of third-generation activity theory that allows us to understand the *reasons why* people generally and students specifically do what they do, that is, the grounds of their actions, developed as 'Critical Psychology' in Germany at the Free University of Berlin. Directly taking up the programme Leont'ev (1981) had outlined concerning the ideas of collective object/motives and individual goals, and the role of affect in individual and collective human endeavours, these social psychologists worked on establishing a "science of the subject" (e.g., Holzkamp, 1983; Holzkamp-Osterkamp, 1990). Most saliently, Klaus Holzkamp and his colleagues focused on the gap that is created when individual subjects do not take up the object/motive (see Figure 3) characteristic for the activity system. Accordingly, students may engage in tasks and learn what it takes to complete them to avoid penalisation and punishment (low grades, suspensions) and thereby engage in 'defensive learning' (Holzkamp, 1993). What they learn may not at all fall together with the curriculum objectives. On the other hand, 'expansive learning' describes the situation where students recognise that doing something expands their control over conditions and action possibilities; and such expansion has positive emotional valence. In our work, both versions of post-Leont'ev activity theory have proven useful.

Third-Generation Activity Theory in Action

When we use this framework for investigating what Jordan (in the introductory example) has done, we first notice that he and his partner Jonathan have subscribed to the overarching task, designing (including the prototyping of) a bridge that carries a maximum load. Not only were the two students to build a prototype, they also would have to present and defend their designs in a forum where the teachers, the remainder of the class, and any visitors to the class would constitute a critical audience. They did not just work towards satisfying themselves or their teachers by building a strong bridge and getting a good grade for it, but rather they designed for a larger public that would evaluate what they had done. To understand any instant during their design praxis (productive activity) we have to take into account the entire system, of which Jordan, Jonathan, their teacher, and the entire class are constitutive parts (Figure 3). The object/motive of the activity includes the materials they have at their hand, the vision of the final product they intend to design (which is a continually unfolding and changing one), and the knowledge of the material and social conditions of the context (Saari & Miettinen, 2001). But the relationship between the subject of action (Jordan, Jonathan) and their object/motive is not direct: it is mediated by other moments of the activity as a whole. First, what actions we will see, and therefore what the *outcomes* of the actions are, both are functions of the available *tools*. Thus in the present instance, the two teachers decided to 'constrain' what students were doing, allowing them only needle pins to fix two or more straws to one another; they wanted students to draw on the idea of braces (in this class they were called "triangles"). But some students began using glue guns, which they then after the objective fact negotiated with the teacher, a co-participant who takes his or her place in the (hierarchical) division of labour on the way leading to the final artefact. The use of glue guns not only changed the individual actions but also (and radically so) the nature and structural properties of the resulting designs.

There were other mediating moments as well. Thus, we cannot understand what Jordan and Jonathan designed independent of the student–teacher transactions in which they were taking part. In highlighting any (hierarchical or horizontal) *division of labour* that might occur, cultural-historical activity theory emphasises the role of others not only in the productive process but also in the artefacts that result (Figure 3). Thus, if a teacher were to ask, "What is the selling feature of this design?" then students might think about how to make the "selling feature" more salient. Or if a teacher suddenly were to allow the use of glue guns (*means* of production), then we might observe – which I actually have – the proliferation of glue guns and related cultural *tools* and associated *practices* (lower left in triangle, Figure 3). This leads us directly to the (ontogenetic, socio-genetic) development of observable practices, which also mediate what the products will look like. This expansion is another form of *expansive learning*, one as seen by the activity system as a whole (Engeström, 1987). Thus, as the children in Jordan's classroom began using glue guns, they not only employed them in joining two pieces with a drop of liquefied glue, but also evolved entirely new practices such as burning holes into one straw and inserting another.

The "*rules* of the game" further mediate what can be done, and further negotiations with the teacher may change the rules. Here, too, I made interesting observations consistent with the theory.

One student, Adam, had left the classroom to build an artefact in the entry hall to the classroom. He was very pleased with his work until the instant when he re-entered the classroom, where, unbeknownst to him, first one student and then everyone else had (re)negotiated the rules for the current task. As a result, the students in the class had come up with much more advanced designs than was possible under the original conditions. When Adam returned and saw what "everyone else" had done, he put his design away and even refused presenting what he had done because, as he explained, everyone else had done something so much better than what he had accomplished. Here we can also see the central role of emotion in the evaluation of work and product and its mediation by the activity system.

Finally, the *community* (users) for which the products are ultimately designed – the one that "consumes" the outcomes – also mediates the productive actions that one can observe. Thus, Jordan and Jonathan continuously developed their bridge because they knew they would have to present it and defend what they had done, for example, specific design features. They suspended weights and when it collapsed with only 56 building blocks hanging from its centre, they sought recourse to their design board – containing examples of different techniques they and their peers had developed before. Jordan and Jonathan subsequently talked about adding braces between the layers of cross beams to strengthen the deck in its vertical direction (see bridge in Figure 1). It is in and for this community that the students produced their bridges, which, in addition, came to be exhibited at the local science museum alongside the results of a design competition for a new bridge to replace an existing landmark in the city. This orientation clearly mediated what students had done because the *bridges* were important in and of themselves rather than as a tool to get what normally is the product of student and teacher work: *grades*.

The structural representation in Figure 3 allows us to understand one of the advantages that comes with technological design tasks: the relationship between learning and the object/motive of the activity, which is all to often misunderstood. Thus, for example, the students like Jordan and Jonathan learned about *structural strength* and *stability* not because these were the objects (goals) of the task. Rather, the students came to learn and understand structural strength and the stability of designs as they attempted to build *strong* bridges. That is, these students learned physics concepts even though the task did not orient them toward physics but rather toward some other object/motive in the pursuit of which the conceptual knowledge turned out to expand their room to maneuver, giving them more options and better designs. This is consistent with experimental evidence on learning through technological designs, which A.N. Leont'ev produced during the 1930s.[1]

In the Pioneer Palace (1933–1934) at Charkow, the organisers had set up the possibility for youth to build model airplanes. Although the participating children were very enthusiastic and eager, they did not show any interest in the theoretical side of airplane construction, all the while engaging in very difficult aspects of constructing "nice models" – as Shannon had done in my technological design unit – such as bending wood to make ribs (used in an airplane's body and wings). Characteristically, after completing their models they could not answer questions about why and how a plane flies, about wind resistance, or the angle of attack – which would be the equivalents to articulating the forces as Jordan had done in and with his bridge design. The children did not attend to the advice that there was a need to know more about the flight of a plane from a theoretical perspective. In a similar way, a number of the children in my classroom focused on making "nice" bridges with little people and animals on them but were unconcerned about the mechanical strength of their bridges. These students had not taken up the motive and instead pursued a different one with different subordinate goals, so that it did not come as a surprise that their understanding of the underlying physics was less well developed than Jordan's or Jonathan's.[2]

When the organisers in the Pioneer Palace talked to Leont'ev he advised to reorganise the activity by changing the motive. Rather than asking students to build "good models", the new task asked them to make model planes fly a pre-determined distance. It turned out that after the first few crashes, the children recognised the need for further theoretical knowledge about flying and airplane design. They now followed the instructor, who used vectors to show how forces operate on the plane and wings (in the way Jordan had done in his glossary diagrams and enacted by means of his gestures), which make the plane fly, rise, or crash. Now this knowledge would constitute an expansion of their building and design capabilities, leading them to a rapid development of interest in theory. More so, this change in the object/motive changed the activity so radically that it became of interest generally, enrolling an increasing number of youth. Thus, within only 12 days, the average daily attendance in the model-building workshop rose from 6.6 to 40.7 children.

Leontjew (1982) reported similar success from a laboratory task involving electromechanical phenomena, where the restructuring of the motive led to a drastic increase in children's theoretical interests: From reading posters with theoretical explanations for a short period of time (60% of children spent less than 3 minutes, another 30% read for between 5 and 10 minutes) to substantially longer periods of time (50% read theory for between 10 and 20 minutes, 30% read these materials for 20+ minutes). That is, the motive mediated the orientation that the children took with respect to their actions. Leont'ev therefore suggested:

> To raise interest, one must not show the goal [i.e., theoretical knowledge] and then attempt to justify the action directed toward this goal. Rather one has to do the reverse, that is, create the motive and then reveal the possibility to determine the goal in its objective content. (p. 276, my translation)

AFFORDANCES OF TECHNOLOGICAL DESIGN

In this section, I provide further examples from the technology curriculum in which students learned important scientific concepts through designing technological artefacts. As the teacher, I prepared tasks that offered related science concepts (e.g., language, diagrams) as resources. My research generally has shown that technology activities provide an ideal context for learning science. Here I have to rely on a few illustrative examples; an entire book on the various ways in which science discourse can emerge – or is constrained from emerging – from technology-based activities in all elementary school grades is provided elsewhere (Roth, Tobin, & Ritchie, 2001). The affordances of technological design come from the fact that it explicitly brings together its material and ideal aspects in the way that cultural-historical activity theory understands it: Figure 3 in fact constitutes two planes, a material and an ideal one, so that each moment exists twice (Roth & Lee, 2007).

Technology Affords Making Ideas Inspectable and Arguable

There are two important aspects for learning through technological design. First, when students begin designing technological artefacts, they have to communicate their ideas to peers. Because of the material basis of technology, design drawings and design prototypes are part of the discourse. These drawings make students accountable for their ideas that have become materially embodied, publicly availably representations (e.g., Roth, 1996c). As such, they are *inspectable, challengeable,* and *arguable*. Second, the shared material basis of these representations (compared to the more ephemeral ones of spoken utterances) brings in another dimension in addition to that of topic. Inscriptions, such as the drawings in the glossary (Figure 2), can become the backdrop to conversations; deictic utterances ("it", "they", "there") and deictic (pointing) and iconic gestures provide the link between the two. As such, in the example of Jordan elaborated above, inscriptions became part of the conversation as indexical ground. In this case, it makes more sense to speak of a distributed communicative act that consists of words, inscriptions, and gestures as the introductory episode and associated discussion have shown (see also Chapter 53).

Thinking in the World

The introductory example shows Jordan in the process of describing the features of his bridge and in the process of explaining why it had developed into what he now held in his hands. In the process of being constructed, the bridge actually constituted a way of thinking in the world. Rather than just visualising something privately for himself – only a specific case of visualisation that is ongoing as we interact with the world (Leont'ev, 1978) – Jordan and Jonathan were thinking together, for each other, and with one another. In this way, consciousness always was as much *consciousness-for-others* as it was *consciousness-for-the-self* (Vygotsky, 1986), always something as much external as internal. Without processes internal to the brain, human activity would not occur; but without the concrete material and social world from which all the tools for thinking have come and to which the thoughts return, there would not be internal processes: Mind and society interpenetrate each other, mind being as much in society as society is in the mind (Vygotsky, 1978). It therefore is not useful to separate thinking into internal and external aspects, because the two are forever intertwined into a unit of consciousness, which always is knowing (Lat. *sciere*) together with (Lat. *con-*) others. Technological design, because of its explicit material nature, therefore supports processes that simultaneously and irreducibly occur on the material and ideal planes.

Thinking with scribbling, making sketches, or doing detailed drawings are important activities during design. However, these activities are limited in that the dynamic nature of the constitutive elements and their relations is not specified. That is, an individual may look at a series of pulleys or cogwheels and suggest that the last one in a series should move clockwise. Neither the image the person envisions nor the drawing can provide feedback whether this movement would occur in the corresponding physical device. I propose that 'designing' should actually refer to the entire set of engagements that lead from initial, vague ideas (which may exist verbally ["we are making an Eiffel Tower"] or visually), through the construction of initial prototypes that are tested, to the final working model. In this case, the later stages of the process, which involve building and testing prototype models, constitute an epistemic shift to actively situate parts of knowing into the environment. (This shift, of course, involves both the performative and representational dimensions of human actions that are central to my theoretical and methodological framework.) Prototyping constitutes an important stage in, and process of, designing for it affords 'thinking in the world' by means of entities and their relationships that are represented in real objects and gestures. Design artefacts allow the materialisation and concretisation, that is, literally, the *object*ification of ideas. As such, ideas become accessible. Design artefacts therefore constitute starting points for critical reflection and a rich ground for even more ideas. This is precisely the fundamental underpinning of dialectical materialist psychology from Vygotsky through Leont'ev to the present day: the ideal and material are irremediably intertwined.

In sum, therefore, designing technological artefacts allows students to move part of their cognitive effort into the world in two ways. First, designing itself occurs partially in the world as ideas are concretised into material elements and their relation and thinking is done in concord with these materials. Second, cognition related to communication partially rests in the environment because entities that are visually available 'go without saying' and serve as background to, and constrain, meaning of gestures. Choosing, as I do, *activity* as the unit of analysis prevents researchers employing the unfortunate and problematic distinction between concrete and abstract as they relate to cognition.

Toward Design Communities

Technology educators have suggested 'community' to be an important part where students can become acculturated to and participate in cultures of talking, thinking, and doing technology (McCormick, 2004). From the perspective of cultural-historical activity theory, there are actually two ways in which we can think about community. As Figure 3 shows, the *community* (of practice) mediates the relationship between the subject and the object, the forms of engagement, the rules employed, and so forth. But there is also another community, the one for which the outcomes of the activity are produced and which "consumes" the products (Engeström, 1987). In the present instance, Jordan, Jonathan, and their peers produce their bridges ultimately for a bridge exhibition in the local science museum so that we can think of its visitors as the "consumers" of the children's products. But thinking more immediately of the classroom events, Jordan and Jonathan were part of a community of practice of designers. In fact, my own contribution to theorising 'communities of practice' and the manner in which they produce and reproduce forms of knowledge has come out of the study of the learning processes in these students' classroom (e.g., Roth, 1996b). Why might the community mediate learning?

Technological design activities inherently make available participation structures recommended by a variety of educators ever since John Dewey. For example, when students are called on to develop their design ideas, the lessons automatically start at developmentally appropriate points for each student. In traditional science lessons, where students are at the moment in their thinking is never quite clear; thus, teachers never quite know whether students are ready today for a concept like strength or force. However, asking students to produce artefacts of their own design inherently allows them to start where they are at – their present developmental stage, likely to be different for each student in a single class. Furthermore, because this form of design asks for students' own ideas, a motive is inherently intrinsic to what Jordan and Jonathan are doing; and (collective, therefore intelligible) motives drive emotions and motivation. Jordan and Jonathan are motivated because they want their bridge to carry a lot of weight, and to achieve the goals they set in the realisation of the motive ("bridge carrying lot of weight"), they engage in learning because it expands their designing possibilities in the direction of making a strong bridge. When they achieve what they set out to achieve, they feel rewarded, proud, and elated – the resulting artefacts are tied to the individual assessment of success, an important regulator of engagement and *expansive learning* (Holzkamp, 1993). They want to feature their ideas and thereby contribute in the first instance to the cultural production and only in a second instance to cultural reproduction of science. Finally, because there likely are many artefacts with different underlying scientific principles, students are exposed to and are challenged to engage with a variety of ideas and scientific principles. (An important aspect of the teacher's work is to assist students in making salient those structures that are relevant to technology, design, and science. This is the crucial point for any science-technology link to succeed.) This also provides students with grounds to develop a critical stance to others and their own work.

In Jordan and Jonathan's class, students' contributions to conversations became increasingly independent of teacher scaffolding, an observation which was especially apparent in whole-class discussions. This is reflective of the design culture that developed together with the associated community of practice. Here, an additional element was introduced to the conversations: student questions. These lengthened the exchanges, which were not simply responses to teacher questions but self-sustaining conversations to which Brigit (the teacher) also contributed on an equal footing with students. (From my cultural-historical activity theoretic perspective, Brigit's interactions with her students could not be studied independently because of Christina's presence. However, Christina (the second teacher) had shifted to similar patterns of interactions due to their collaborative effort to improve questioning.) An example of sustained student-centred discussion is featured in the following episode, extracted from the same lesson segment in which Jordan and Jonathan presented their bridge.[3]

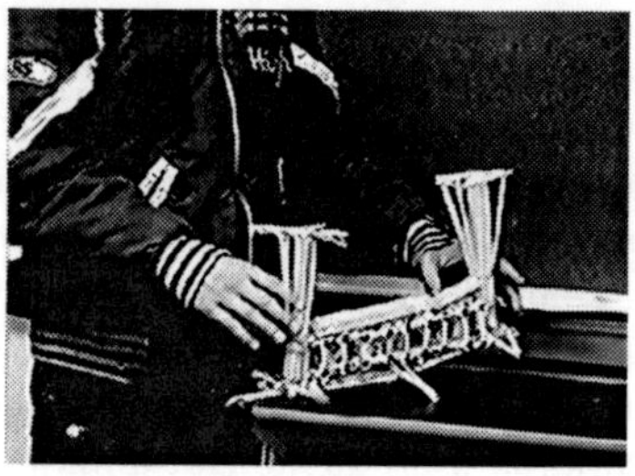

01 Rafferty: Did you write anything in the description thing about (.) about that it held that many, but upside down?
02 Jordan: No, we didnt write upside down.
03 Christina: It would be an interesting thing to add. (.) Dan.
04 Dan: Is there people or cars on your bridge?
05 Jordan: We dont know, it doesnt really matter if it is for people or cars
06 Jonathan: Before=
07 Christina: =Its the idea of a bridge, right?
08 Jordan: Yes.
09 Jonathan: Before we put the braces on, we were gonna make people through there and cars go up on top, but it wouldnt be strong enough, so we put the braces=
10 Jordan: =Cause, cause like that it would be like an elevator to bring [people up]
11 Christina: [Oh, cool]
12 Jordan: But now we say, well at least it's a bridge and we dont care what bridge it is. (3 s). Stan?
13 Stephen: When you will display it, are you going to put it upside down or the way you got the three hundred and forty=seven blocks or you=re gonna put it?
14 Jordan: In the case, we actually are going to put it like this, I think, that would make it look better. (.) Cause if we have it like this, and you=re just passing by, you probably think it is like a flight control or something.

This episode is but a small part of the entire sequence surrounding the question of the orientation in which Jordan and Jonathan's bridge should be displayed at the local science museum. Several peers had concerns with the procedure Jordan and Jonathan had chosen to test their bridge by putting it upside down. Throughout the 13-minute presentation and discussion of the project, students brought this topic up repeatedly. In this excerpt, both Rafferty and Stephen were concerned with the discrepancy between making the claim of the bridge's strength (347 blocks, line 13) when it was tested upside down in the way Jordan had held it during the presentation (line 01, Insert, line 13), and displaying it right side up (at the museum). The teachers' contributions were requests for further elaborations or comments (line 03), or interventions to guarantee the flow of the conversation (line 07) – particularly in Christina's case, who wanted neither presenters to be embarrassed nor conflicts of any sort. However, the students themselves often posed the important questions (e.g., lines 01, 04, 13). In terms of content, questions and answers addressed issues relating to testing procedures (line 01), the design's purposes, and user specifications. Brigit's questions often were requests for elaborations or justifications of topics raised in students' accounts.

SYNTHESIS

In the preceding parts of this chapter, I provide illustrations from one technology-centred classroom in which students learned about material strength, stability, and structural engineering in increasingly scientific ways. The curriculum is consistent with cultural-historical activity theory because it focused on the object/motive oriented nature of human activities and the means employed in the production of artefacts; the unit focuses on the performative dimensions of learning. The episodes are representative of technology-centred classrooms that lead not only to technological and design learning but also to science learning, embodying some of the central features of all my studies. First, as material objects, artefacts and their representations provide students with concrete topics of talk and with specific, action guiding motives. They also provide students with a background against which gestures can be meaningfully used and as an additional form of communicative representation; in this, they mediate communication in the way Vygotsky had theorised. Depending on the type of representations (discourse, inscriptions) and how accessible these are to students, teachers can constrain the evolution of discursive patterns to become increasingly scientific (e.g., Roth, Tobin, & Ritchie, 2001).

Second, because students externalise their ideas in concrete form, they open themselves to critique from and negotiations with others. The concrete form of design artefacts – which literally constitutes an *object*ification of thinking that is required for a consciousness to be critical of itself (Hegel, 1806/1977) – also allows students to distance themselves from their own ideas, which literally lie in front of them in concrete form (sketch, building plan, or material artefact). This distancing facilitates students in critically reflecting on their own ideas, now partially and one-sidedly represented in the artefact that they are looking at. Designing artefacts actually moves thinking partially into the world and thereby frees up mental resources in developing increasingly complex ideas.

Third, when students present, discuss, argue, and critique their designs in a public forum (e.g., whole-class discussion), further opportunities are provided for developing a discourse. Because this task structure makes students' own products the focus of the lessons, there is likely to be sustained interest in the topic and the conversation is more likely to be maintained. Participation in practice, however, is known to be constitutive of learning the practice and of constructing the identity of a practitioner. In making these statements, I am resting my case not just on these illustrations and on my understanding from this classroom; rather, my understanding is consistent with evidence from an extended database established during a decade of research and on studies published in a variety of domains. In this chapter, I present a – to technology educators novel – way for theorising those research findings.

Fourth, the role of the teacher no longer is as central as it would be in the direct-teaching model. Rather, the teacher is but another subject in the activity who takes a particular role in the division of labour toward the produced artefacts (outcomes), which themselves have a different role than the outcomes of normal school tasks because they come to be re-circulated in the classroom, school, and society more broadly.

CODA: TECHNOLOGICAL DESIGN AND EXPANSIVE LEARNING

Over the years, my studies have shown what kind of technological knowledge students evolve in and through design activities and how they develop this knowledge, which, as many colleagues have told me, exceeds the knowledge direct teaching of two- and three-year older students have evolved (Ken Tobin, personal communication, April 1995). This knowledge is fundamentally of the technological type, valuable in and for itself (Ankiewicz et al., 2006). In fact, my personal predilection would be to teach technology rather than traditional science, because repairing a bicycle, a home appliance, or a lawnmower has more direct usefulness in and to life.[4] But there is more to technological design activities, and this 'more' arises from their close relation to scientific knowledge.

Here at the end, I therefore return to the question that constituted the central concern for me in doing the research that I have done: "What do technology-related activities bring to the learning situations that do not exist in teaching science through textbook activities?" To me, the most important aspect of a more complex answer is this. In technology activities focused on design, students are deeply involved in creating and transforming artefacts and representations in both of the directions traditionally conferred to science and technology. The students learn physics not because they focus on learning physics but because the design challenges are such that when the students set specific goals – building sturdier bridges, more stable towers, machines that decrease effort – they come to realise that their room to manoeuvre expands when they seek recourse to already-existing forms of cultural knowledge and skills. They then engage in expansive learning – in contrast to defensive learning in which students often engage in to avoid punishment or low grades. That is, students learn physics not because it is the articulated goal to orient students' endeavour but rather because they cognise a personally relevant need for expanding their possibilities to reach their goals. Expansive learning is inherently motivating because students know that they will be able to do more than before in the pursuit of a personally meaningful goal that they have set and want to achieve.

Students are also deeply involved in activity, and therefore in the performative aspects of knowledge. In artefacts, individual elements and their relations can be described and explained in terms of a performative idiom. This, however, is not the only possible idiom. Other idioms may be concerned solely with the functioning of the artefact, its artistic value, etc. This appears to be the crucial point for teaching science through technology-related activities. Because technological school tasks provide grounds for the development of discourses that bear family resemblance with scientific discourse, there is legitimate reason to employ such activities in science education. Technological design tasks are therefore prime contenders for an integrated approach to teaching, for they could be the hub where students engage in a variety of discourses that bear resemblance with those traditionally associated with the arts, technological design, mathematics, science, language arts, and so forth. Above all, as cultural-historical activity theory suggests, teachers have to frame the collective motive and students have to buy into this motive in their concrete goal setting such that precisely what *should be learned* constitutes a resource that expands students' action possibilities, who therefore *want to learn* because it allows them to reach what they have set as their goals.

NOTES

1 This example is actually provided in the German translation (Leontjew, 1982), as an additional chapter exemplifying his theoretical work in action.

2 The two teachers Christina and Brigit were explicitly opposed to have design challenges because they believed it would turn off the girls, who predominantly made "nice" designs.

3 The following transcription conventions have been used. The transcription is phonetic and the punctuation indicates flow of intonation toward the end of a unit, slightly rising, strongly rising, slightly falling, strongly falling (,?;.). (.) – indicates brief pause. = – marks "latching", that is, there is no pause between two utterances. [] – brackets indicate overlapping talk. (3 s) – Time in seconds.

4 Similarly, with respect to the sciences I prefer doing environmentalism or gardening with students because these already constitute ongoing activities in the lives of the children's families and communities so that what they do integrates them into existing activity systems rather than preparing them for those systems (e.g., Roth & Barton, 2004).

REFERENCES

Ankiewicz, P., de Swardt, E., & de Vries, M. (2006). Some implications of the philosophy of technology for science, technology and society (STS) studies. *International Journal of Technology and Design Education, 16*(2), 117–141.

Barak, M., & Shachar, A. (2008). Projects in technology education and fostering learning: The potential and its realization. *Journal of Science Education and Technology, 17*(3), 285–296.

Bucciarelli, L. L. (1994). *Designing engineers.* Cambridge, MA: MIT.

Christiaans, H., & Venselaar, K. (2005). Creativity in design engineering and the role of knowledge: Modelling the expert. *International Journal of Technology and Design Education, 15*(3), 217–236.

Engeström, Y. (1987). *Learning by expanding: An activity-theoretical approach to developmental research.* Helsinki, FL: Orienta-Konsultit.

Engeström, Y., & Escalante, V. (1996). Mundane tool or object of affection? The rise and fall of the postal buddy. In B. Nardi (Ed.), *Context and consciousness: Activity theory and human-computer interaction* (pp. 325–374). Cambridge, MA: MIT.

Hayward, W. G., & Tarr, M. J. (1995). Spatial language and spatial representation. *Cognition, 55*, 39–84.

Hegel, G. W. F. (1977). *Phenomenology of spirit* (A. V. Miller, Trans.). Oxford, UK: Oxford University Press. (First published in 1806).

Holzkamp, K. (1983). *Grundlegung der Psychologie.* Frankfurt a/M.: Campus.

Holzkamp, K. (1993). *Lernen: Subjektwissenschaftliche Grundlegung.* Frankfurt a/M.: Campus-Verlag.

Holzkamp-Osterkamp, U. (1990). *Grundlagen der psychologischen Motivationsforschung 2. Die Besonderheit menschlicher Bedürfnisse–Problematik und Erkenntnisgehalt der Psychoanalyse* (4th ed.). Frankfurt a/M.: Campus.

Jones, A. (2003). The development of a national curriculum in technology for New Zealand. *International Journal of Technology and Design Education, 13*(1), 83–99.

Kafai, Y. B. (1994). *Minds in play: Computer game design as a context for children's learning.* Hillsdale, NJ: Lawrence Erlbaum Associates.

Küchler, S. (2008). Technological materiality: Beyond the dualist paradigm. *Theory, Culture & Society, 25*, 101–120.

Layton, D. (1994). *Innovations in science and technology education.* Paris: UNESCO.

Leont'ev, A. N. (1978). *Activity, consciousness and personality.* Englewood Cliffs, NJ: Prentice Hall.

Leont'ev, A. N. (1981). *Problems of the development of the mind.* Moscow: Progress.

Leontjew, A. N. (1982). *Tätigkeit, Bewusstsein, Persönlichkeit.* Köln: Studien zur Kritischen Psychologie.

Luff, P., & Heath, C. (1993). System use and social organization: Observations on human-computer interaction in an architectural practice. In G. Button (Ed.), *Technology in working order: Studies of work, interaction, and technology* (pp. 184–210). London: Routledge.

McCormick, R. (2004). Issues of learning and knowledge in technology education. *International Journal of Technology and Design Education, 14*(1), 21–44.

McKenzie, D. (1992). The technical curriculum: Second class knowledge? In G. McCulloch (Ed.), *The school curriculum in New Zealand: History, theory, policy & practice* (pp. 29–39). Palmerston North, NZ: Dunmore.

Miettinen, R., Lehenkari, J., & Tuunainen, J. (2008). Learning and network collaboration in product development: How things work for human use. *Management Learning, 39*, 203–219.

Roth, W.-M. (1996a). Engineering talk in a grade 4/5 classroom: A pre- and post-unit assessment. *International Journal of Technology and Design Education, 6*(2), 107–135.

Roth, W.-M. (1996b). Knowledge diffusion in a grade 4–5 classroom during a unit on civil engineering: An analysis of a classroom community in terms of its changing resources and practices. *Cognition and Instruction, 14*(2), 179–220.

Roth, W.-M. (1996c). Thinking with hands, eyes, and signs: Multimodal science talk in a grade 6/7 unit on simple machines. *Interactive Learning Environments, 4*(2), 170–187.

Roth, W.-M. (1998). *Designing communities.* Dordrecht, The Netherlands: Kluwer.

Roth, W.-M., & Barton, A. C. (2004). *Rethinking scientific literacy.* New York: Routledge.

Roth, W.-M., & Lawless, D. (2002). Signs, deixis, and the emergence of scientific explanations. *Semiotica, 138*, 95–130.

Roth, W.-M., & Lee, Y. J. (2007). "Vygotsky's neglected legacy": Cultural-historical activity theory. *Review of Educational Research, 77*(2), 186–232.

Roth, W.-M., Tobin, K. G., & Ritchie, S. M. (2001). *Re/constructing elementary science.* New York: Peter Lang.

Saari, E., & Miettinen, R. (2001). Dynamics of change in research work: Constructing a new research area in a research group. *Science, Technology, & Human Values, 26*(3), 300–321.

Schank, R. C., Fano, A., Bell, B., & Jona, M. (1994). The design of goal-based scenarios. *The Journal of the Learning Sciences, 3*, 305–345.

Sewell, W. H. (1992). A theory of structure: Duality, agency and transformation. *American Journal of Sociology, 981*, 1–29.
Stevenson, J. (2004). Developing technological knowledge. *International Journal of Technology and Design Education, 14*(1), 5–19.
Turnbull, D. (1993). The ad hoc collective work of building gothic cathedrals with templates, string, and geometry. *Science, Technology, & Human Values, 18*(3), 315–340.
Vygotsky, L. S. (1978). *Mind in society: The development of higher psychological processes.* Cambridge, MA: Harvard University Press.
Vygotsky, L. S. (1986). *Thought and language*. Cambridge, MA: MIT Press.

Wolff-Michael Roth
Lansdowne Professor of Applied Cognitive Science
University of Victoria
Canada

MARILYN FLEER

55. CULTURAL-HISTORICAL ACTIVITY RESEARCH "FRAMING INTELLECT AND ACTION" IN DESIGN AND TECHNOLOGY EDUCATION

Methodological Directions

INTRODUCTION

Scholarship in technology and design education has progressively evolved as research within this field has matured (see Middleton, 2008; Middleton, Pavlova, & Roebuck, 2004). Through continued research and scholarship, we now know a great deal more about the knowledge base of technology than we did just twenty years ago (Jones, 2003). In this chapter, dialectical materialism is explored in relation to research in design and technology education through an examination of the works of both Vygotsky and Leontiev. A wholeness approach to research (see Hedegaard, 2008) for re-imagining how design and technology education research can be conceptualised will be presented, and the technique of digital video data generation and analysis will be discussed to illustrate dialectical materialism.

Technology and design education has been perceived and enacted differently across the global community. However, research in this field has tended to follow a narrower path (Latour, 2003). For instance, Bakhurst (2007) in drawing attention to the philosophical division between mind and body, points to a longstanding dualism in technology and design education – 'working with your hands' and 'working with your mind'. As argued by Glick (1995), this "... relates to a privileging of the 'intellectual forms' over the other forms often associated with 'doing'" (p. 361). In citing Alfred Shon-Rethel (1978), Glick argues that Western forms of "social/economic organization of capitalism" have established "a value hierarchy between those activities that are 'intellectual' (cognitive) and those that are 'merely activity' (manual)" (p. 361). Essentially, this view positions particular forms of knowledge more favourably than others. Although this is not a new argument (see Layton, 1993; MacKenzie & Wajcman, 1985), tensions continue to surface (Petrina, 1998) and new research needs arise (see Stein, McRobbie, & Ginns, 2000). Cartesian logic within research in technology education still prevails as the norm, unwittingly guiding the directions of the research design. Active resistance to these reinforcing dualisms and the need to generate new forms of knowledge production are becoming increasingly important in technology education. In discussing the contribution of other worldviews in psychology, Bakhurst (2007) draws attention to Vygosky's critique of modern psychology, where Vygotsky (1933) argues that research "cannot find a way to understand the real sensible tie between our thoughts and feelings on the one hand, and the activity of the body on the other hand" (pp. 196-197, as cited in Bakhurst, 2007, p. 59). Bakhurst continues: "He [Vygotsky] was also well aware of the shadow Descartes' dualism cast over the history of psychology. Many attempts to develop a monistic picture had failed, Vygotsky argued, because they had allowed Descartes to define the terms of the debate and tried to reduce mind to matter conceived mechanistically."

A.T. Jones and M.J. de Vries (eds.), International Handbook of Research and Development in Technology Education, 669–682.
© 2009 Sense Publishers. All rights reserved.

It will be shown in this chapter that the argument put forward by Vygotsky equally applies to the area of technology and design education. In the first section, the technology education literature is examined in relation to the foundational philosophical perspective (unity of mind and body) underpinning the field. This is followed by a theoretical discussion of dialectical logic. In the final part of the chapter a detailed account of how digital video can be used as a tool for research in technology education is presented alongside a theoretical discussion of a wholeness approach.

THE UNITY OF THINKING AND DOING IN TECHNOLOGY AND DESIGN EDUCATION

The Legacy of Cartesian Logic

Research in technology and design education has attempted to move beyond Cartesian logic through invented constructs such as "hands-on, minds-on technology" (see Swindell, Richmond, & Waller, 2002), "holistic technology education" (Seemann, 2002; 2003), "working and thinking technologically" (Paige, 2002), "design in action" (Macgregor & Zeegers, 2004), "technacy" (Seemann & Talbot, 1995), "collective designing" (Roth, 2001), and "innovation in practice" (Stevenson, 2004). These terms deliberately integrate thinking and doing and position technology education as not just simply 'practical', but also as a 'conceptual' area of learning (see Walmsley, 2002).

Important conceptual work has been undertaken in the UK (e.g., Hill & Anning, 2001; Kimbell, 2004), Canada (e.g., Roth, 2001), New Zealand (e.g., Jones, 2003), Taiwan (e.g., Wei & Huang, 2004), the US (e.g., Martin & Middleton, 2003), and Australia (e.g., Campbell, 2004; Herbert, 2004a,b; Seemann, 2002) seeking to make explicit the importance of content knowledge in design and technology education. For instance, a series of national studies led by Jones (see Compton & Jones, 1998; Jones & Compton, 1998; Moreland, Jones & Northover, 2001) demonstrate a direct relationship between the knowledge base of teachers and the conceptual outcomes for children in technology and design education. This work makes visible for the first time how primary teachers think about concepts in technology education – both in context (formative) and after working with children for extended periods of time (summative).

Others have focussed their research attention on the nature of integrated knowledge and the significance of undertaking technology education through practical activity and, through this, generating better learning outcomes for students in all curriculum areas being studied (Venville, Rennie, & Wallace, 2004). Fine grained studies, such as those of Norton, McRobbie, and Ginns (2007), show mixed results due to teacher beliefs about student autonomy and responsibility in open-ended learning tasks compared with tightly-framed, procedurally-oriented classrooms.

Research in technology and design education has also focussed on identifying the relations between classroom activity and work-based activity (e.g., Hill & Anning, 2001; Milne, 2004), with some looking to frame their research around 'values' (Bondy, 2002) or 'creativity' (Barlow, 2004; Webster, 2004; Wei & Huang, 2004) and others seeking to better understand how real world technology takes place and how this may inform pedagogy and the content of the technology curriculum (Fox-Turnbull, 2002; Mawson, 2002). Others have investigated the everyday technological lives of young children and how their experiences at home influence their technological learning in school (e.g., Bers, New, & Boudreau, 2004; Jane & Robbins, 2006; Robbins & Jane, 2006).

Collectively, these studies point to a continued need to better understanding the relations between intellect and activity in technology and design education.

Early Childhood Technology Education

Evidence for the unity of action and thinking can be seen nowhere as strongly as in the early childhood period. As argued by Fichtner (1999): "The child is not able to separate itself, its mental experiences, from the perception of material objects" (p. 60). There is an abundance of early infant research which shows that "mental life is an undifferentiated, unstructured experience, an amalgam of affects, drives, and sensations" (ibid.). Research into pre-speech infants points to "the important role of a special kind of infant activity in the development of psychic life" (Lektorsky, 1999, p. 110).

Interactions with human-made things, and a culture of infant-adult interaction with things, enables the generation of meaning within social contexts. Lektorsky (1999) states that:

> Culture has created modes of actions with these [human made] things and with the help of them (a baby is shown concrete examples in order to teach him/her these actions)... So the means of these actions can be understood by analogy with the meanings of language expression. … I think that it is possible to speak not only about language games, but also about practical games (with correspondent rules) at the pre-speech level of psychic development. (pp. 110–111)

Thus, a strong relational link is established for the infant when action and language are brought together in social contexts. These contexts are imbued with social meaning, and there is no differentiation between action and thinking for the infant: "An object is seen, touched, felt, tasted, and smelled. By means of this communication, the child experiences heaviness and lightness, hardness and softness, wetness and dryness, and so on" (Fichtner, 1999, p. 62). In essence, the infant uses its whole body to investigate, gaining meaning through the social framing generated by the significant adults in the child's world. Fichtner suggests that the infant's interactions have a more or less *theoretical quality* [italics added]" (p. 62).

The theoretical quality of infant interactions can also be likened to technology and design education. For example, significant theoretical intellectual and doing knowledge are built when children examine materials for their useability in their projects (such as testing for strength or flexibility) or when they explore new materials to see how they behave. As they explore and test out materials, they build theoretical knowledge. In addition, through their active explorations they extend and build theoretical-dialectical knowledge, where "one type of knowledge is complementary to the other so that if a change takes place it will be reflected in all the central relations of the system" (Hedegaard, 2007, p. 252). For instance, when a child uses a particular joining technique to apply two different surfaces together, the child uses a theoretical model about how the particular materials will behave in relation to the adhesive or technique applied. As argued by Hedegaard:

> This kind of knowledge can be found in theories and models that can be used to understand events and situations and to organize and experiment with actions (*concrete life activities*) [italics added]. This type of knowledge can also be found in all professional work (e.g., engineering, city planning, professional cooking or nursing, steering a ship, dress designing), *where persons have a theory and models of their work* [italics added]. (p. 252)

It is through the interactions of the infant and the child with their world that they build theoretical models of how the world works. In design and technology education, teachers actively support children to build theoretical-dialectical knowledge as they design, make, and appraise. These models build over time and the concrete actions that underpin their models become implicit and may become taken-for-granted. The taken-for-granted tacit doing knowledge is infused into models and relational understandings.

A further early-life example that is directly relevant to technology and design education is given by Fichtner (1999):

> The eye and the hand cooperate with each other. At the intersection of these two heterogeneous senses a peculiar sort of intimacy establishes itself – an intimacy which constitutes the objectivity of the world that *we adults take so much for granted* [italics added]. The production of objectivity is apparently contingent upon it being reflected in self-awareness. (p. 63)

As Fichtner shows, the relations between acting on the world and thinking about that action (consciousness) are clearly mutually constituted. Vygotsky (1933) put forward a theory to explain this dialectical relationship between the mind and the social world. The writings of Vygotsky, and the philosophical standpoint from which they are derived, differ to Cartesian logic and provide great possibilities for research in design and technology education. Most notably, they provide a way of exemplifying how, in technology education, children move to higher planes of consciousness about how they think and act upon the world. As discussed by Leontiev and Luria (2005):

> Having developed out of initially undifferentiated forms of relations to the reality, only in man [sic], with his historically established forms of consciousness, *do skill and intellect enter into the relations of a higher unity* [italics added] that comprises the specifically human core of psychological development and allows man [*sic*], in mastering nature, for the first time to become master of his [*sic*] own conscious activity as well. (p. 47)

There is a growing interest in research which uses different theoretical approaches or techniques, such as those discussed by Leontiev and Luria. This can be seen through some of the titles of papers that have emerged over the past five years, such as: "Technology teacher education: It is time to do a Maslow?" (Barlow, 2002); "Basic principles of holistic technology education" (Seemann, 2003); and "Sustainable technology learning by theatre-playing" (Haupt, Wehling, & Langkau, (2002). These examples illustrate how the field is grappling to re-frame the dualistic legacy of Descartes, and seeking to develop a new philosophical and theoretical basis for research in technology and design education. It is thus timely that a new view of 'activity' (Glick, 1995) within technology and design education be generated, and new and different approaches to research be created. In the next section, an alternative theoretical view is discussed.

DIALECTICAL MATERIALISM AND ACTIVITY THEORY

Latour (2003) argued that early in philosophical history, Western thinking followed the path of Cartesian logic and disbanded the potential of dialectical materialism. In order to understand the potential of this philosophy and the theoretical approaches that have spawned from this particular worldview, it is important to examine the concept of 'motives' within an 'activity' and 'the dialectical relations between people, objects and society' (see Hedegaard, 2008). These constructs will be examined below in order to establish a foundational basis for a discussion on the worth of the philosophical standpoint of dialectical materialism for informing research within technology and design education. It will be argued that activity theory holds much potential for technology and design education, and that dialectical materialism provides an alternative philosophy for framing research in this area.

According to Leontiev (1978), Vygotsky introduced to the research community the concept of analysing activity as a method of scientific human enquiry. Vygotsky's research methodology sought to examine within a given task the concept of tooled (instrumental) operations, the concept of purposes, and the concept of motives ("motivational sphere of consciousness"). However, Leontiev argued that Vygotsky proposed a concept of activity that held a general collective meaning which was too abstract and theoretical, and therefore was never developed in his research. According to Tulviste (1999), Leontiev concretised Vygotsky's cultural-historical theory. For example, Leontiev suggested that in Vygotsky's general collective meaning of activity, researchers must consider specific activities, "each of which answers a definite need of the subject, is directed toward an object of this need, is extinguished as a result of its satisfaction, and is produced again perhaps in other, altogether changed conditions" (p. 62). Leontiev (1978, 2005a, b, c, d) developed Vygotsky's original concepts into a remarkable theory of activity, where:

> The main thing that distinguishes one activity from another, however, is the difference of their objects. It is exactly the object of an activity that gives it a determined direction. According to the terminology I have proposed, the object of an activity is its true motive. It is understood that the motive may be either material or ideal, either present in perception or existing only in the imagination or in thought. The main thing is that behind activity there should always be a need, that it should always answer one need or another. (Leontiev, 1978, p. 62)

The Significance of 'Motive'

In Leontiev's (1978) theory of activity, he proposed the view that an activity cannot exist without a motive. However, the motive may be subjectively or objectively hidden. Thus, in technology education, a child who is tinkering with a series of clocks is engaged in a task where the motive may be hidden. For instance, the teacher may have introduced the clocks and a series of tools to the child in order to generate a motive for learning about tools. The motive for learning about tools, in this example, does not necessarily relate to an explicit need of the child, but rather is a societal goal that children will know about tools and will have some competence in their use for later home or industry work. The societal goal for learning about tools at school is not necessarily immediate or obvious. Indeed, specialised tool use will only be one set of expertise within a collective, where others may learn other skills and where the overall effect is that the collective has the required skills to work effectively to produce artefacts and processes for general human survival (e.g., food production). These particular concepts are inherent in Leontiev's (1978) activity theory and are based upon Karl Marx's writings on labour and production (see, Marx, 1976, 1978, 1981).

Tulviste (1999) argued that often 'activity' is considered as 'context' rather than as an explanatory principle. For example, researchers might use activity to explain the various ways of thinking that are used in a range of contexts, rather than to ask "why and how the ways of thinking come into being, or what is their origin" (p. 75). Tulviste has argued that activity theory can be viewed as a weak or strong research tool:

> Either (a) various ways of thinking come into being independently; culture just decides which ways of thinking are used at all and in which particular situations, or (b) culture, including activity, is responsible for the coming into being of various ways of human thinking functionally related to activities. (p. 75)

When this broader view of activity is considered in technology and design education classrooms and the everyday lives of children, it becomes apparent that research methods need to expose the relations between motives, needs, goals, and actions of people and objects. As suggested by Jensen (1999):

> … we live in a world populated by objects produced in techno-scientific practice (e.g., Latour, 1987), or constructed in accordance with procedures of abstraction and idealization of experimental practice. These 'fossils' of human practice, which together with our ongoing practices, mediate our relation with the world and our intersubjective relations. (p. 86)

An analysis of the "fossils of human practice" give directions for research in technology and design education. We need research tools which allow researchers to map the relations between the objects, their objectification, and the subjectivities that have been realised by particular societal practices. These tools allow for the possibility of providing a gateway through which children's technological thinking as enacted in schools, can be made visible. It is argued by Leontiev and Luria (2005) that schooling positions children to act volitionally and consciously in relation to their environment. They argue that "education creates completely new relationships between" a child's actions and motives and "forms the conditions for the emergence of new structures and functional features of psychological processes" (p. 44).

Complexity within Technology and Design Education Research

As children grow, they "increasingly enter into connections with historically established human experience, and come to know objective reality with increasing breadth and depth" (Leontiev & Luria, 2005, p. 47). Hedegaard (2008) has developed an approach which explains the relations between the individual child, their family, the community, and the society at large. Her work is useful to technology education because she argues for a dialectical relation between the societal goals for technology education, the institutional contexts (i.e. classrooms and schools), and the motives that are generated for the child through their engagement in the teachers' practices within classrooms. The foregrounding of the relational dimensions of society, school, family, and the child in Hedegaard's model means that it is possible to research and analyse holistically what happens in technology education classrooms. As shown by Fichtner (1999): "Activity is not [only] what an organism does [action]. Rather, the organism consists in its activity. Activity is the mode of existence by which organisms establish themselves as subjects of their life processes" (Fichtner, 1999: 55). As such, it is important that the often hidden relations between society, schools, and individuals are researched in order to better understand technology and design education in classrooms.

Dialectical materialism as a philosophical standpoint locates the research lens on the relations between people and their environment – like two sides of the same coin. They cannot be separated – they each constitute the other. In line with this standpoint, activity theory concentrates research attention on activities (the child's perspective) and practices (the teacher's perspective) which realise motives for design and technology education (see Hedegaard, 2008).

Norton, McRobbie, and Ginns (2007) have used activity theory for analysing how teacher goals in the area of mathematics, science, and technology influenced the actions of children during a series of 20 robotics lesson. Activity theory allowed the researchers to map the complexity of children's problem solving, creativity, and technological drafting actions in relation to the challenges set by the teacher. These data were analysed alongside the teachers' beliefs about how learning should be tightly or loosely organised in maths, science, and technology lessons. In the study, Norton et al. demonstrate how activity theory can keep intact the dynamic context of robotic education and generate relational understandings of creativity, problem solving, scientific cognition, and drafting/planning in technology education.

Complexity in technology education has been captured by researchers drawing upon other dimensions of cultural-historical activity theory, such as Lave's original work on situated cognition (e.g., McGrath, 2002; Roth, 2001); Wertsch's translations of Vygotsky's work, which he called sociocultural theory (e.g., Jane & Robbins, 2004; Rowell, 2002), Leontiev's writing on activity theory (Norton, et al., 2007; Roth, 2001), and – unusually – Hegel's writings on dialectical materialism and dialectical realism (e.g., Seemann, 2002). These theories can be collectively classified under the banner of cultural-historical activity theory (CHAT), and originate from the seminal ideas of Vygotsky. (See also Chapter 54 by Wolff-Michael Roth). In the next section, CHAT will be concretised through a discussion of how digital video technology can generate data and frame analyses in a non-dualistic way, keeping whole the unity of intellect and action.

A WHOLENESS APPROACH TO RESEARCH IN TECHNOLOGY AND DESIGN EDUCATION

Hedegaard's (2008) wholeness perspective builds on the works of Vygotsky, Leontiev, and other contemporary cultural-historical and activity theory discussed in the previous section. The philosophical and theoretical basis for CHAT is significant for generating new research approaches in technology and design education. In this section, I will illustrate some of the key constructs of CHAT and discuss their potential for realising the research method of using digital visual technology for capturing the complexity of the higher unity of action and intellect discussed previously. CHAT constructs relevant for technology education research that will be discussed here are:

- capturing the wholeness of the technological activity;
- foregrounding the dialectical relations between participants' perspectives; and
- dynamic research methodology – the researcher is part of the research context.

Capturing the Wholeness of the Technological Activity

In technology education classrooms and centres, children are involved in not just one activity, but rather may be simultaneously involved in numerous events and tasks. For example, a child designing a bird feeder may be exploring different design techniques (3-D drawing, plan view, cross-sections) whilst also being engaged in building social relationships (e.g., reinforcing friendships through engaging peers in a conversation about their favourite birds). Tulviste (1999) suggests that "people involved in different kinds of activities, and therefore evolving different kinds of tasks, will have different semiotic means or tools provided by the society, and by using different tools, they would think in different ways" (p. 69). A CHAT approach to studying children participating in multiple tasks and engaging in different ways of thinking seeks to capture the wholeness of the situation (Hedegaard, 2008). That is, when one dimension is examined, it is analysed in relation to all of the other events that are occurring at that time. As Rogoff, Radzizewska, and Masiello (1995) state:

> The concept of activity focuses attention on sociocultural events as the unit of analysis, seeking a unit of analysis that preserves the inner workings of events of interest, rather than separating an event into elements that no longer function as does the living unit. (p. 127)

A CHAT analysis also seeks to explore things that are not immediately obvious or that may be taken-for-granted. For example, what materials and tools are presently available to the child as she/he is designing the bird feeder? Similarly, what past events has the child engaged in to generate the motive for participating in the design and construction of a bird feeder (e.g., did they make observations of the birds in the school ground, noting that they were hungry over winter?)?

Digital video technology supports a wholeness approach because it is possible to position the camera to capture as many of the interactions as possible between the child and the child's classroom. Electronic visual material, when broadly captured in this way (assuming good sound quality), has the advantage of allowing multiple approaches to the analyses as required. Through the repeated viewing of the videotapes, microanalysis is possible. As noted by Graue and Walsh (1998):

> Being able to view the interactions repeatedly and at times to review them frame by frame was invaluable for doing microanalysis of children's interactions. Unlike adults, whose language abilities are well developed, children's interactions are marked by nonverbal signs and body language. For example, spatial distance often is an important clue for understanding children's peer relationships. Having a video record also allowed me to ask the children to view their interactions and to comment on them. (p. 149)

Building upon these arguments, a CHAT perspective would suggest that these repeated viewings of video tapes allows researchers to notice both the teaching practice and the child's activity whilst at the same time mapping learning and the development of motives for engagement in technology education.

Video observations are particularly useful to CHAT researchers, who seek to examine the dialectical relations between participants, the social setting, and the institutional practices, and who need to revisit their material many times in order to make interpretations from a range of perspectives in order to understand the child's technological development. For example, an analysis frame can be applied to the data set, making possible multiple ways of examining the material. An example is provided in Table 1, with the final row explicitly foregrounding the dialectical relations between categories.

It is possible to see that researchers following a cultural-historical tradition dialectically build video images of 'reality' and from those video images generate interpretations about the *motives, goals* and *values* of the participants within the practice traditions of different *institutions* – that is, the institutions of schooling and family life (see Hedegaard, 2008). These interpretations must speak back to reality, but give new insights about both practice and activity. An understanding of the children's technological activity in relation to the teacher's teaching can be interpreted and new insights about technology education generated. Figure 1 shows the relations between technological practices, activities, video recordings, and interpretations.

Table 1. A wholeness frame for analysis.

What were the technological tasks the children were involved in?	What activity generated a *motive* for designing?	What *institutional* practice was being promoted?	What are the societal *goals* at play here?	What activity is the child participating in to satisfy a *need*?	What *family values* and practices are relevant here?
Designing a bird feeder	Previously Mary found a dead bird in the school grounds and the teacher developed and implemented a bird watching programme	The teacher was interested in developing the children's design skills	Being able to design or know how to read a design is an important life and professiona l skill for Australian citizens	Mary spends a great deal of time talking about the dead bird. Mary draws a series of coffins and discusses with her peers the burial ritual previously enacted.	Mary recently lost her grand-mother.
What are the relations between these?					

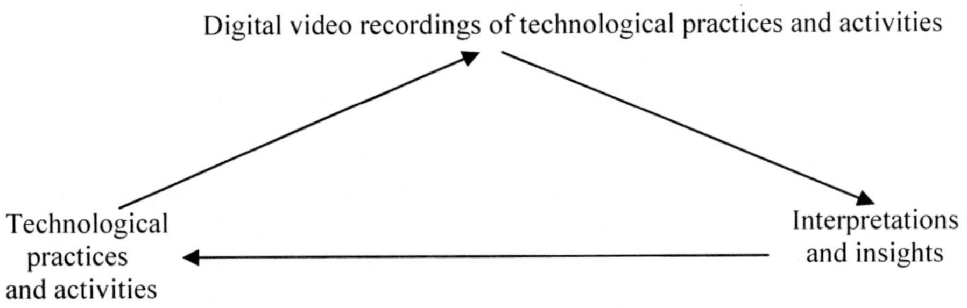

Figure 1. The relations between practices, activities, video recordings, and interpretations (adapted from Fleer, 2008).

Foregrounding the Dialectical Relations between Participants' Perspectives

A child's motives may not always be in tune with that of the teacher's motives. How is it possible to determine not just the motives that are expected to be generated within a particular technology and design task, but the disparity that may arise between different motives and expectations between children and their teacher? Research methods which seek to analyse the perspectives of all the participants provide the possibility of determining the multiple motives that may be being generated through the technological task.

Hedegaard (2008) has theorised how multiple perspectives can be dialectically framed within a study design. For example, when the teacher's perspective is being considered, the data are analysed in relation to the practices that are evident. These practices include practice traditions which are privileged and supported by the institution, such as how the classroom is configured to allow for technological activity (i.e. is there enough space to work safely?), or if a budget is available for the purchase of materials or specialist technological tools for children to utilise. On the other hand, a child's perspective can be uncovered through examining the activities that a child generates for herself/himself within the practice tradition organised by the school. This is possible to achieve when digital video technology is used, as the video can capture both the activities of the children and the practices of the teacher, whilst at the same time recording the institutional environment.

Hedegaard (2008) has identified the following questions as being useful for guiding researchers following a CHAT perspective:

- What kind of institutional practices do children participate in?
- What activities dominate the institutional practices?
- What demands do these put on children?
- What are the possibilities for acting and how do children act in these activities?
  - conflicting demands and motives
  - transitions.

She argues that it is often during transition from one activity to another or when there is disparity or conflict between the teacher and the child that important insights can be gained when these data are examined. For example, in Table 2 it is possible to note that the teacher wishes Mary to develop her design skills through the designing and making of a bird feeder. However, what she observes Mary doing is creating drawings and undertaking conversations that do not relate to the task she has set. There is a disparity between what the teacher wants and what the child does. As researchers following a CHAT tradition, these points of disparity provide opportunities for learning more about the motives that underpin technological activity. In this particular example, the researchers needed to learn more about what Mary was bringing to the technological activity – something that could be gained through acquiring the family perspective (another institutional perspective) – that Mary had experienced the death of her grandmother, and the death of the bird generated a need around learning about and dealing with death (rather than designing a bird feeder).

In a dialectical research approach, it is impossible to consider the child's perspective without also considering the practice traditions organised by the teacher. That is, you cannot have schools if there are no students, and you cannot have students if there are no schools. They each constitute the other. As such, it is important to pay attention to the broader practice traditions when researching the technological learning of the pupils and, conversely, it is important to consider the children's actions, values, needs, and motives when studying the pedagogical practices of teachers in technology education classes.

Table 2. The dialectical relations between different perspectives.

Teacher's perspective	Child's perspective	Research perspective	Family perspective
What *institutional* practice was being promoted?	What activity is the child participating in to satisfy a *need*?	What is the research question and the theoretical perspective of the researcher?	What *family values* and practices are relevant here?
The teacher was interested in developing the children's design skills	Mary spends a great deal of time talking about the dead bird. Mary draws a series of coffins and discusses with her peers the burial ritual previously enacted.	The researchers were interested in finding out how young children engage in the design process and how they develop their motives? CHAT guided their study design.	Mary recently lost her grandmother.

Dynamic Research Methodology: The Researcher Is Part of the Research Context

In many study designs the role of the researcher is quite problematic (Rogoff, Radziszewska, & Masiello, 1995). For example, Flick (2002) expressed concerns in relation to the researcher contaminating the data by being 'in the research site'. To address this, some research – mostly found within traditional psychology – ensures that the research assistant is blind to the research question, often being given an interview script and a static research site to videotape (such as a table and three chairs where participants are directed to sit), and with no knowledge of the problem area being investigated. This is in contrast with CHAT, where the researchers always have pre-concepts that they use to guide them with framing and capturing digital video observations. The whole research team must know the research question and understand the theoretical approach, even if they are only employed to simply operate the second video camera for the principal researcher.

Knowing about the details of the study is important because the team must point the camera at what they perceive to be relevant interactions that may be occurring within the technology classroom or centre. For example, when Mary is talking about death and dying, it would be important to continue to video tape these conversations rather than to move the camera to another area within the classroom. As discussed earlier, transitions and conflicts are likely to introduce important data and, therefore, continuing to video tape disparities between activities and the teacher's practices are required when undertaking CHAT research in technology education.

CHAT research is highly interactive and requires that the researcher's role be clearly defined. As shown in Table 2, the researcher has a perspective and a specific role when entering the research site. When in the technology classroom, the researcher is not 'the teacher' and is also 'not a participant in the lesson'. But this does not mean that the researcher does not interact with the children. The researcher may be asked to assist a child, for example, in holding a cylinder to a box as the child fastens it. The researcher may talk to the children about their work, but would not take on the role of actively 'teaching the children'. In this way, the researcher is a visible member of the classroom community but with their own 'research role' of documenting what is taking place and asking questions in order to find out more about the technological learning and teaching being documented through digital video recordings. Although the researcher is not there to teach, their interactions can have a positive effect because he/she is asking questions or highlighting particular interactions as important (for example, by simply pointing the camera on what Mary is engaged in, attention is drawn to the activity), and this level of interest facilitates the technological pertinent activity and – importantly for the researcher – captures the nature of technological development for the children being researched (i.e. in the formation of the higher unity of intellect and action).

As noted by Lektorsky (1999) "a researcher (in particular, an experimenter) and a subject under investigation do not discover something that existed entirely before their communication, but in the process of a dialogue jointly create or construct a new social reality..." (p. 111). CHAT research in technology education by its very nature captures over time the dynamic activity that is being generated – with and without researcher interaction. Learning is not a static process, to be captured at a particular 'end point'. Rather, the dynamic relations between 'objects' and 'people in action' becomes the focus of attention. Rogoff et al. (1995), who use the term 'sociocultural' in positioning themselves within CHAT, state that "in a sociocultural perspective, learning and development are regarded as processes of changing participation in cultural activity" (p. 131).

Capturing the changing nature of children's participation in design and technology education is greatly enhanced through digital video recording. Many of the previously discussed studies in design and technology education that have actively sought to capture complexity, have used video techniques for data generation and analysis. These studies lend support for the use of CHAT as a theoretical approach for design and technology education.

CONCLUSION

In this chapter, I have outlined how the philosophical and theoretical assumptions within CHAT can usefully inform research within technology and design education. I have drawn upon the foundational theorists (Vygtosky, Leontiev) who have provided us with a theoretically-robust discussion of the key concepts within cultural-historical and activity theory. With a new philosophical foundation from which to generate theoretically different study designs in technology education, I have argued that the unity between intellect and action can be preserved and deeper understandings realised. Digital video technology as a method for data generation and computer analyses (e.g., *Imovie*) following relational categories (depicted in Table 1) provide an approach that sits well with dialectical materialism and CHAT. Taken together, it is possible to see how complexity in technology and design education can be kept intact, and how better understandings can be gained, through research following a cultural-historical and activity theory tradition. It is timely that we consider such alternative approaches. As Rogoff et al. (1995) point out: "It will help progress in research to pay greater attention to the assumption systems of our research methods and questions" (pp. 145–146).

REFERENCES

Bakhurst, D. (2007). Vygotsky's demons. In H. Daniels, M. Cole, & J. V. Wertsch (Eds.), *The Cambridge companion to Vygotsky* (pp. 50–76). New York: Cambridge University Press.

Barlow, J. (2002). Technology teacher education: Is it time to do a Maslow? In H. Middleton, M. Pavlova, & D. Roebuck (Eds.), *Learning in technology education: Challenges for the 21st Century* (Vol. 1, pp. 9–16). Queensland: Griffith University.

Bers, M. U., New, R. S., & Boudreau, L. (2004). Teaching and learning when no one is expert: Children and parents explore technology. *Early Childhood Research Practice*, *6*(2). Retrieved from http://ecrp.uiuc.edu/

Bondy, A. (2002). Building 'flax bridges' in technology education to uphold the Treaty of Waitangi partnership principle. In H. Middleton, M. Pavlova, & D. Roebuck (Eds.), *Learning in technology education: Challenges for the 21st Century* (Vol. 1, pp. 35–45). Queensland: Griffith University.

Campbell, C. (2004). The interweaving of technology education into a thematic environmental program. In H. Middleton, M. Pavlova, & D. Roebuck (Eds.), *Learning in technology education: Challenges for the 21st Century* (Vol. 2, pp. 104–112). Queensland: Griffith University.

Compton, V., & Jones, A. (1998). Reflecting on teacher development in technology education: Implications for future programmes. *International Journal of Technology and Design Education*, *8*(2), 151–166.

Fichtner, B. (1999). Activity revisited as an explanatory principle and as an object of study – old limits and new perspectives. In S. Chaiklin, M. Hedegaard, & U. J. Jensen (Eds.), *Activity theory and social practice* (pp. 51–65). Denmark: Aarhus University Press.

Fleer, M. (2008). The relations between cultural-historical theory, methodology and digital video methods, In M. Hedegaard & M. Fleer (Eds.), *Studying children: A cultural-historical approach* (pp. 104–117). London: Open University Press.

Flick, U. (2002). *An introduction to qualitative research* (2nd ed.). London: Sage.

Fox-Turnbull, W. (2002). The relationship between out of context assessment tasks in technology education and that of the same task embedded in authentic classroom technological practice. In H. Middleton, M. Pavlova, & D. Roebuck (Eds.), *Learning in technology education: Challenges for the 21st Century* (Vol. 1, pp. 128–141). Queensland: Griffith University.

Glick, J. (1995). Intellectual and manual labor: Implications for developmental theory. In L. M. W. Martin, K. Nelson, & E. Tobach (Eds.), *Sociocultural psychology. Theory and practice of doing and knowing* (pp. 357–382). Cambridge, UK: Cambridge University Press.

Graue, M. E., & Walsh, D. J. (1998). *Studying children in context. Theories, methods, and ethics*. London: Sage.

Haupt, W., Wheling, J., & Langkau, T. (2002). Sustainable technology learning by theatre-playing. In H. Middleton, M. Pavlova, & D. Roebuck (Eds.), *Learning in technology education: Challenges for the 21st Century* (Vol. 1, pp. 170–177). Queensland: Griffith University.

Hedegaard, M. (2007). The development of children's conceptual relations to the world, with a focus on concept formation in preschool children's activity. In H. Daniels, M. Cole, & J. V. Wertsch (Eds.), *The Cambridge companion to Vygotsky* (pp. 246–275). New York: Cambridge University Press.

Hedegaard, M. (2008). Basic conceptions in a cultural-historical theory of children's development. In M. Hedegaard & M. Fleer (Eds.), *Studying children: A cultural-historical approach*. London UK: Open University Press.

Herbert, L. (2004a). Can technology education, as a separately taught subject, assist student learning in other subject areas? In H. Middleton, M. Pavlova, & D. Roebuck (Eds.), *Learning for innovation in technology education* (Vol. 2, pp. 40–49). Queensland: Griffith University.

Herbert, L. (2004b). Knowledge transfer: Transfer of the design process from technology education to other contexts, In H. Middleton, M. Pavlova, & D. Roebuck (Eds.), *Learning for innovation in technology education* (Vol. 2, pp. 50–59). Queensland: Griffith University.

Hill, A. M., & Anning, A. (2001). Comparisons and contrasts between elementary/primary 'school situated design' and 'workplace design' in Canada and England. *International Journal of Technology and Design Education*, *11*(2), 111–136.

Jane, B., & Robbins, J. (2004). Block play as a potential for learning for innovation in technology education in early childhood classrooms. In H. Middleton, M. Pavlova, & D. Roebuck (Eds.), *Learning for innovation in technology education* (Vol. 2, pp. 60–67). Queensland: Griffith University.

Jane, B., & Robbins, J. R. (2006). Grandparents teaching science and technology in informal contexts. In S. Yoong, M. Ismail, A. Nurulazam, M. D. Zain, F. Salleh, F. S. Fook, L. C. Sam, & M. N. L. Yan (Eds.), *Proceedings of IOSTE Symposium* (pp. 539–544). Penang, Malaysia: Malaysia School of Educational Studies, University Sains Malaysia.

Jensen, U. J. (1999). Categories in activity theory: Marx's philosophy just-in-time. In S. Chaiklin, M. Hedegaard, & U. J. Jensen (Eds.), *Activity theory and social practice* (pp. 79–99). Denmark: Aarhus University Press.

Jones, A. (2003). The development of a national curriculum in technology for New Zealand. *International Journal of Technology and Design Education, 13*(1), 83–99.

Jones, A., & Compton, V. (1998). Towards a model for teacher development in technology education: From research to practice. *International Journal of Technology and Design Education, 8*(1), 51–65.

Kimbell, R. (2004). Assessment in design and technology education for the department of education and skills, UK. In H. Middleton, M. Pavlova, & D. Roebuck (Eds.), *Learning for innovation in technology education* (Vol. 2, pp. 99–112). Queensland: Griffith University.

Latour, B. (2003). Do you believe in reality? News from the trenches in the science wars. In R. C. Scharff & V. Dusek (Eds.), *Philosophy of technology. The technological condition. An anthology* (pp. 126–137). Oxford, UK: Blackwell.

Latour, B. (1987). *Science in action.* Cambridge, MA: Harvard University Press.

Layton, D. (1993). *Technology's challenge to science education.* Buckingham, UK: Open University Press.

Leontiev, A. N. (1978). *Activity, consciousness, and personality* (M. J. Hall, Trans.). New Jersey, US: Prentice-Hall.

Leontiev, A. N. (2005a). Study of the environment in the pedological works of L.S. Vygotsky. A critical study. *Journal of Russian and Eastern European Psychology, 43*(4), 8–28.

Leontiev, A. N. (2005b). Paper presented at the all-union institute of experimental medicine. *Journal of Russian and Eastern European Psychology, 43*(4), 48–57.

Leontiev, A. N. (2005c). Lecture 13. Language and consciousness. *Journal of Russian and Eastern European Psychology, 43*(5), 5–13.

Leontiev, A. N. (2005d). Lecture 14. The structure of consciousness. Sensory fabric, meaning, personal sense. *Journal of Russian and Eastern European Psychology, 43*(5), 14–24.

Leontiev, A. N., & Luria, A. R. (2005). The problem of the development of the intellect and learning in human psychology. *Journal of Russian and East European Psychology, 43*(4), 34–47.

Lektorsky, V. A. (1999). Historical change of the notion of activity: Philosophical presuppositions. In S. Chaiklin, M. Hedegaard, & U. J. Jensen (Eds.), *Activity theory and social practice* (pp. 100–113). Denmark: Aarhus University Press.

MacKenzie, D., & Wajcmann, J. (Eds.). (1985). *The social shaping of technology.* Milton Keynes, UK: Open University Press.

Macgregor, D., & Zeeger, Y. (2004). Design in action: An investigation into the teaching of design skills in primary classrooms. In H. Middleton, M. Pavlova, & D. Roebuck (Eds.), *Learning for innovation in technology education* (Vol. 2, pp. 142–150). Queensland: Griffith University.

Martin, G., & Middleton, H. (Eds.). (2003). *Initiatives in technology education: Comparative perspectives. Proceedings of the American-Australian technology education forum held at the Sheraton Marina Mirage, Gold Coast, Australia.* Technical Foundation of America and the Centre for Technology Education Research Griffith University.

Marx, K. (1976). *Capital* (Vol. 1). London: Penguin.

Marx, K. (1978). *Capital* (Vol. 2). London: Penguin.

Marx, K. (1981). *Capital* (Vol. 3). London: Penguin.

Mawson, B. (2002). A very different experience: Teacher practice and children's learning. In H. Middleton, M. Pavlova, & D. Roebuck (Eds.), *Learning in technology education: Challenges for the 21st Century* (Vol. 2, pp. 8–17). Queensland: Griffith University.

McGrath, A. (2002). The model of situated cognition: Lessons for technology education. In H. Middleton, M. Pavlova, & D. Roebuck (Eds.), *Learning in technology education: Challenges for the 21st Century* (Vol. 2, pp. 43–51). Queensland: Griffith University.

Middleton, H. (Ed.). (2008). *Researching technology education: Methods and techniques.* Rotterdam, The Netherlands: Sense.

Middleton, H., Pavlova, M., & Roebuck, D. (Eds.). (2004). *Learning for innovation in technology education* (Vols. 1–3). Queensland: Griffith University.

Milne, L. (2004). Where have the fish gone? A case study exploring an example of biotechnology related work in a New Zealand primary school. In H. Middleton, M. Pavlova, & D. Roebuck (Eds.), *Learning for innovation in technology education* (Vol. 2, pp. 196–205). Queensland: Griffith University.

Moreland, J., Jones, A., & Northover, A. (2001). Enhancing teachers' technological knowledge and assessment practices to enhance student learning in technology: A two-year study. *Research in Science Education, 31*(1), 155–176.

Norton, S. J., McRobbie, C. J., & Ginns, I. S. (2007). Problem solving in a middle school robotics design classroom. *Research in Science Education, 37*(3), 261–277.

Paige, K. (2002). Primary teacher beliefs and practices in working and thinking scientifically and technologically. In H. Middleton, M. Pavlova, & D. Roebuck (Eds.), *Learning in technology education: Challenges for the 21st Century* (Vol. 2, pp. 88–95). Queensland: Griffith University.

Petrina, S. (1998). Multidisciplinary technology education. *International Journal of Technology and Design Education, 8*, 103–138.

Robbins, J., & Jane, B. (2006, July). *Intergenerational learning: Grandparents supporting young children's learning in science and technology*. Paper presented at the 37th Annual Conference of the Australasian Science Education Research Association, Canberra, ACT.

Rogoff, B., Radziszewska, B., & Masiello, T. (1995). Analysis of developmental processes in sociocultural activity. In L. M. W. Martin, K. Nelson, & E. Tobach (Eds.), *Sociocultural psychology. Theory and practice of doing and knowing* (pp. 125–149). Cambridge, UK: Cambridge University Press.

Roth, W.-M. (2001). Modeling design as situated and distributed process. *Learning and Instruction, 11*, 211–239.

Rowell, P. M. (2002). Talking a design into existence: Making model parachutes. In H. Middleton, M. Pavlova, & D. Roebuck (Eds.), *Learning in technology education: Challenges for the 21st century* (Vol. 2, pp. 145–154). Queensland: Griffith University.

Seemann, K. (2002). Holistic technology education. In H. Middleton, M. Pavlova, & D. Roebuck (Eds.), *Learning in technology education: Challenges for the 21st century* (Vol. 2, pp. 164–173). Queensland: Griffith University.

Seemann, K. (2003). Basic principles in holistic technology education. *Journal of Technology Education, 14*(2), 28–39.

Seemann, K., & Talbot, R. (1995). Technacy: Towards a holistic understanding of technology teaching and learning among aboriginal Australians. *Prospects, XXV*(4), 761–775.

Stein, S. J., McRobbie, C. J., & Ginns, I. (2000). Recognising uniqueness in the technology key learning area: The search for meaning. *International Journal of Technology and Design Education, 10*(3), 105–123.

Stevenson, J. (2002). Transforming learners through technology education. In H. Middleton, M. Pavlova, & D. Roebuck (Eds.), *Learning in technology education: Challenges for the 21st century* (Vol. 2, pp. 203–210). Queensland: Griffith University.

Swindell, R., Richmond, J., & Waller, J. (2002). Hands-on, minds-on technology to reinforce understanding of fundamental science and technology principles. In H. Middleton, M. Pavlova, & D. Roebuck (Eds.), *Learning in technology education: Challenges for the 21st century* (Vol. 2, pp. 222–229). Queensland: Griffith University.

Tulviste, P. (1999). Activity as explanatory principle in cultural psychology. In S. Chaiklin, M. Hedegaard, & U. J. Jensen (Eds.), *Activity theory and social practice* (pp. 66–78). Denmark: Aarhus University Press.

Venville, G., Rennie, L., & Wallace, J. (2004). Decision making and sources of knowledge: How students tackle integrated tasks in science, technology and mathematics. *Research in Science Education, 34*(2), 115–135.

Vygotsky, L. S. (1933). *The teaching about emotions: Historical-psychological studies*. In *The collected works of L.S. Vygotsky* (Vol. 6, pp. 69–235). New York: Plenum Press.

Walmsley, B. (2002). Technology education and higher-order thinking. In H. Middleton, M. Pavlova, & D. Roebuck (Eds.), *Learning in technology education: Challenges for the 21st century* (Vol. 2, pp. 245–253). Queensland: Griffith University.

Webster, A. (2004). Teacher and peer assessment of creativity in technology education. In H. Middleton, M. Pavlova, & D. Roebuck (Eds.), *Learning for innovation in technology education* (Vol. 3, pp. 178–186). Queensland: Griffith University.

Wei, Y.-S. J., & Huang, C.-S. J. (2004). Linkages between science, technology and creativity in Taiwan's national curriculum. In H. Middleton, M. Pavlova, & D. Roebuck (Eds.), *Learning for innovation in technology education* (Vol. 3, pp. 187–197). Queensland: Griffith University.

Marilyn Fleer
Centre for Childhood Studies
Monash University
Melbourne, Australia

SECTION 9

CONCLUSION

ALISTER JONES

56. LOOKING BACK TO LOOK FORWARD: THE FUTURE OF TECHNOLOGY EDUCATION

INTRODUCTION

This last chapter of the Handbook reflects on the development of the field of technology education, particularly in relation to research and development, and considers how to nurture and grow the field – how to *act* like a field. From reviewing how the field has developed and its current strengths, consideration is given to where the field might go and how it can be supported in that process.

This Handbook has provided a unique coverage of the developing field of technology education, from consideration of different approaches in different countries, through thinking about the nature of technology, relationships between science and technology, teaching and learning, assessment, teacher development, and developing research approaches. This not only provides us with an overview the field as it currently is, but also provides us scope to consider how we might create a sustainable field.

Although the crafts and technical aspects have been in curricula internationally since the 1800s and probably before, it is only recently that technology has been considered potentially to be part of a 'curriculum for all'. Probably the most compelling reason for studying technology is that it is a major, and some would argue a determining feature, of the world we inhabit. Young people need to be introduced to it so they can understand its nature and be able to participate in it at some level or other. Technology therefore stands alongside other ways we represent culture (science, mathematics, music, etc.) This makes it more difficult to represent (and to agree on its nature), but it is no less an aspect of culture than the major areas of knowledge and understanding that are represented in other school subjects. If technology is indeed a determining feature of the world we inhabit, it follows that young people, as future citizens, need to understand how it shapes the world. This is because technology is a value-laden activity, and citizens need to understand and control many of the decisions that are made about it. It follows that education must prepare them to do this by dealing with the technical, social, political, and economic issues that underlie technological processes. This will allow them to take part as active citizens of their society (Jones, 2008).

The different sections within this Handbook have indeed argued that technology is an essential part of education for all. They included illustrations of ways in which this has occurred, reflections on the discipline itself, ways in which we might teach and assess it to students as well as pre-service and in-service teachers, and how we might carry out research to enhance our understanding of technology education. The development of professional communities for supporting research, and development are key to defining the field of technology education and moving it forward.

A.T. Jones and M.J. de Vries (eds.), International Handbook of Research and Development in Technology Education, 685–692.
© 2009 Sense Publishers. All rights reserved.

MOVING FORWARD

Developing National and International Communities and Networks

Wenger's (1998) notion of communities of practice as a framework for considering individuals and organisations is useful in understanding and articulating the field of technology education and where we might go. For individuals in the field, it is in the engaging and contributing; for communities it is refining and enhancing practice and ensuring new generations; and for organisations it is in relation to sustaining interconnected communities of practice. Wenger argues that no practice is actually global, and this is true of technology education. Although common themes emerge in terms of defining technology – the desire for technological literacy, etc. – the way the practice and research is actually defined is local to the country, or indeed the region, a particular school, or even a particular research group. Technology education is therefore defined at the local level.

Key to the development of technology education has been the development of technology teachers, professional development programmes, and research organisations. The development of professional organisations is also a key aspect of the development of a field or discipline as a discipline is recognised by the professional bodies that surround and support it. Such organisations play a key role in moving the field forward, and it is evident that they have done so up until now. Some of these organisations have been as straight forward as an irregular international conference, such as the second Jerusalem International Science & Technology Education Conference (JISTEC), held in 1996. This conference was hugely instrumental in bringing together key players in technology education and has led to significant international collaborations. In fact, the origins of this Handbook can be traced back to that conference. PATT (Pupils' Attitudes to Technology Teaching) has played a long-term and expanding role as being a research-focused conference, as has TERC (Technology Education in Research Consortium) more recently, particularly for Australasia.

Teacher and professional conferences have long played a part in sharing new ideas, curriculum enhancement, resource development, and teacher professional development. These organisations have adapted as technology has broadened from a technical focus, through industrial arts, craft, and design and technology, to a broadening to technological literacy. As we have seen in Section 1, each educational jurisdiction has developed its own technology curriculum, teaching approaches, and resources consistent within its local education context. More localised teacher and professional organisations that have since expanded to have strong international collaborations, such as International Technology Education Association (ITEA) in the US, Design and Technology Association (DATA) in the UK, and Technology Education New Zealand (TENZ), all of which provide support, advice, and subject leadership. The Design and Technology Associations in states in Australia support local state initiatives. Other professional organisations support the development of technology education in their respective countries, not only by defining the field, but also by developing research agendas and influencing teacher education.

The on-going development of technology education is going to not only require sustainable professional organisations in each country, but also sustainable collaborations between the different organisations.

The Emerging Place of Technology Education in the Wider Curriculum

In Section 1, describing the history and context of technology education in different countries, demonstrates that how the development of this field is set within the historical, cultural, and political environment of countries and regions. As highlighted, curriculum development, teacher education, and even educational research does not sit in isolation from the country's historical and cultural context. Further, it is not only the framework for technology education at the national and regional level that needs to be considered, but the way in which the curriculum is interpreted at the local level that has the major impact on teachers and students.

This section highlighted the journey that a curriculum takes when moving from one that is marginalised or narrowly defined for a limited number of students, to one that is mandated as an essential part of education for all. Curriculum history indicates that one of the crucial determiners as to whether a curriculum has high-stakes value is whether it has recognition for university entrance requirements. It would appear that although there are exceptions, as a subject in school, technology is yet to reach an appropriate level internationally to reflect its importance as an essential part of general education. In many places its existence is still somewhat fragile, particularly as many educational jurisdictions struggle to balance literacy and numeracy initiatives with other competing subject areas. Elsewhere, technology – having claimed a place for itself as a discipline in its own right – is now being seen as part of integrated studies or as a subset of other disciplines. In such cases, it is crucial that teachers and students develop an understanding of the nature of technology so that it does not become lost in any integration – or indeed disintegration – process.

Where technology has been more successfully introduced into general education there are some key messages that can be highlighted. These are in terms of a clear articulation of the discipline; a strong teacher education and professional development environment; clear pathways from early childhood and elementary education through to secondary to tertiary education, including recognition for university entrance; strong professional networks; and a developing research culture, including graduate and postgraduate programmes. It is crucial, therefore, that in developing technology education for the 21st Century we are cognisant of the aspects that create robustness rather than fragility.

The Philosophy of Technology Education

Technology education as a field is moving forward, but at the same time as defining the boundaries there is some blurring as governments examine its place in the curriculum. In the development of technology education, it has been crucial that it has been defined as an area of study in its own right with its own knowledge and skill base. Much of the early work in technology education thus focused on exploring the nature of technology and attempting to define the field for technology education.

In moving towards a coherent theory of practice, it is important to define the discipline both from historical and philosophical perspectives as well as from an analysis of current technological practice. The second section in the Handbook defines technology for technology education and provides a strong basis and rationale from which to consider the characteristics of technology and technological developments in order to develop a possible framework for technology education in schools. This defining of the discipline is crucial if technology education is going to develop its own space in the curriculum, as well as a robust research culture. The section on the nature of technology is thus pivotal in defining what can fruitfully be explored in technology education, and how we move forward. In the past technology education was defined very generally, for example, problem solving, design, and creativity. If these general terms are used to define a discipline then it is not surprising that technology education can lose its place in the curriculum and as a developing field. For example, the place of values and beliefs is an important aspect that can be overlooked in curriculum development and implementation, as can environmental issues. Outlining the characteristics and boundaries of a discipline as well as its intersection with other disciplines is also crucial. The translation of the historical and philosophical aspects of technology for technology education continues to be an important part of ongoing development of the field.

The way in which a subject is defined in practice depends on how it is defined by a range of stakeholders, including politicians, the community, teachers, and students. Section 3 of the *Handbook* explores the perceptions that teachers and students have of technology education. This is critical, since perceptions are one of the determining factors that influence curriculum adoption and student learning. For example, if technology is only perceived as applied science then this perception will influence the type of learning activities that are planned and implemented, as well as their place in the curriculum. It has been found that students' perceptions of technology also influence the way they undertake technological activities and their own learning in technology education (Jones, 1997). Therefore, in moving forward it is important that we understand how others view the field so that we may start from where their understanding is and develop strategies that may enhance perceptions of technology enhance at the different levels in the community.

Although it is crucial to define technology and technology education as areas or disciplines in their own right, where there have been close alliances or possible future curriculum alliances then these should be explore. Science, Technology, and Society (STS) and Science, Technology, Society, and Environment (STSE) are examples of this. As was highlighted in Section 1, STEM (Science, Technology, Environment and Mathematics) is also an area of collaboration between disciplines. However, in thinking about these collaborations it is important that a robust understanding of technology is understood and articulated; otherwise, technology can be perceived as a subset of other disciplines, such as science. As technologies become more sophisticated – for example, biotechnology and nanotechnology – then there are close relationships with the sciences. STS approaches can be used to engage students in science and technology, and have been demonstrated to enhance general scientific and technological literacy. If technology is becoming increasingly squeezed in the curriculum, then STS can provide a space. In Section 4, approaches to STS as ways of linking science and technology are highlighted. In particular, this section illustrates ways in which collaborative approaches to science and technology can contribute to general education, and that technology and science do not need to be always separately defined in the curriculum or educational research.

Teaching, Learning, and Assessment

Teaching, learning, and assessment are at the heart of the implementation of technology education, whether it be defining the subject, framing the curriculum, or setting the research agenda. As we move forward in defining technological literacy for 21st Century learners, then the enactment of this is in the classroom. This includes thinking not only about compulsory schooling, but also about technology education as part of early childhood and the contribution it might make to life-long learning.

As technological literacy has become more broadly defined, we have needed to think about what this means for teaching, learning, and assessment. Associated with this has been increased recognition of the relationship between the nature of the discipline and appropriate pedagogies, notions of authentic assessment, and how students make progress in technology education as well as how teachers can enhance this progress.

This raises questions about how we might think about not only how students learn in technology but what is important for students to learn if they are to develop a broad notion of technological literacy. Indeed, there is a need for a framework for technological knowledge and processes to be developed, and for both concepts of the nature of technology and technological concepts to be explicitly identified and understood. From this framework, we need to then develop a range of authentic pedagogical and assessment processes that enhance student learning.

Teacher knowledges are an emerging theme that needs to be addressed from the view of teaching and assessment in technology education, and the notion of pedagogical content knowledge (PCK) needs to be more fully developed in this context. Although PCK has been highlighted in the section on assessment (Section 5), effective pedagogical knowledge in technology education needs to be explored more fruitfully and fully if the field is to make a contribution to enhancing the practice of technology education in the classroom. It has been argued that teacher knowledge of their subject, how students learn, and the interaction of these are essential to support learning. Greater articulation of this interaction is required for technology education and we cannot just rely on what we know from other areas of education.

A clear message in this *Handbook* has been that if students are to progress in technology then they must be involved in self and peer assessment as well as experiencing authentic technological activities. The importance of assessment – both formative and summative, and thinking about formative assessment as being much more in terms of student-teacher and student-student interactions – has been highlighted. A key feature of moving forward is thus to enhance our understanding of effective classroom practices, and greater cognisance is required of teaching, learning, and assessment in context.

The Role of Teacher Education and Professional Learning

If, as has been argued, teachers are central to the learning process and teacher knowledges are a key part of this, then the nature of initial or pre-service education and on-going professional learning for teachers needs to be considered. Teacher education can be seen in a number of ways. Firstly, there are generalist teachers for whom technology education is only one curriculum area amongst others. Secondly, there are those for whom technology education is of special interest; these may act as technology curriculum specialists but teach other subjects. Thirdly, there are specialist technology education teachers. For generalist teachers, knowledge in technology and technology education has to be developed generally in a very short time frame. Consideration then has to be given to what knowledge is most important for these teachers to know, given that they are expected to have content knowledge, curriculum knowledge, knowledge about learners, as well as other knowledges that effective teachers require.

Two influences have converged to highlight the crucial role of the teacher in educational reform. These are changes in how learners and learning are viewed; and an appreciation of the relationship between teaching, learning, assessment, and curriculum. Research into the knowledge required for teaching has demonstrated the complexity of the knowledges teachers bring into the classroom teaching environment (Hiebert, Gallimore, & Stigler, 2002; Shulman, 1987). Teaching is thus a complex process that that cannot be dichotomised into knowledge and action (Boaler, 2003). Rather, teachers require rich pedagogical content knowledge (PCK), a specialised form of knowledge about a subject. The development of appropriate PCK for technology teachers at all levels of the curriculum is crucial if student learning in technology is to progress.

PCK is concerned with how teachers shift from knowing subject matter/content to being able to translate and communicate that content so their students can learn it. This translation act is an active process where content knowledge and knowledge of practice have a symbiotic relationship (Rodrigues, Marks, & Steel, 2003). A teacher needs to understand subject matter and be able it "in new ways, reorganise and partition it, clothe it in activities and emotions, in metaphors and exercises, and in examples and demonstrations, so that it can be grasped by students" (Shulman, 1987, p. 13). Black and Wiliam (1998) identified distortions to classroom interactions when there is minimal subject knowledge and gaps in how that subject knowledge is represented and developed for students. For technology, Jones and Moreland (2005) detail the distortions that can occur when there is lack of teacher knowledge around technology as well as a lack of understanding of the nature of technology. There is emerging evidence that introducing pre-service teachers to concepts of the nature of technology can fast forward their thinking about technology.

When broader notions of technology were introduced 10–15 years ago in schools, most of the teachers teaching this 'new' subject had already been teaching for a number of years and would continue to do so. Therefore, teacher professional development and strategies to enhance in-service professional learning are key to enhancing the development of technology education in schools. For teachers to teach technology in a manner that supports and extends student learning, they may need to develop new understandings and make changes to their current practices. Teachers' beliefs about the subject area, teaching, their students, and the curriculum influence how they interact with curriculum materials. In technology education, the influence of existing subject subcultures on the interpretation of a new curriculum has been noted. These subject subcultures also influence classroom activities and student learning (Jones, 2006).

As already pointed out, a significant amount of research has examined teachers' perceptions and the way these potentially influence the interpretation and implementation of any new curriculum. Teacher professional learning needs to take these perceptions into account if it is to move teachers forward. Effective professional learning programmes for teachers include opportunities for the developing the following: robust concepts of technology and technology education, including the nature of technology; technological knowledge and skills in a number of areas of technology; understandings of student learning in technology and appropriate pedagogical and assessment processes; and ways to develop effective technology education programmes for students. Educative curriculum materials also have potential as vehicles to support such learning in that they highlight concepts, tasks, procedures, and teaching approaches and can be used over time in classroom contexts (Ball & Cohen, 1996; Schneider & Krajcik, 2002).

The work on developing and enhancing in-service teachers' PCK in technology education provides a fruitful base from which further work can be undertaken. Early indications are that when teachers' PCK in technology is enhanced, teachers are more confident and student learning is promoted (Jones and Moreland, 2005). Research and development in PCK in technology is therefore key in developing technology education further.

Building Research Capability

If technology education is to advance as a curriculum area of worth and as a focus of research, then much more of our research effort must focus on student and teacher learning and building the notion of a robust academic field in technology education. A number of research approaches have been highlighted in Section 8 of this Handbook. The fact that we are developing research approaches, which reflect technology education in its various forms, illustrates the developing nature and maturity of the field. As the field moves forward, however, a number of issues need to be considered.

First, we need to consider the sustainability of the field and ways in which we might more closely align research, development, and practice. Moving to a research-informed field is a key part of this process. As already pointed out, the early work in technology education spent much time defining the field and the curriculum. More recently, there has been an increase in the amount of research that has the potential to inform practice, evidenced by the increased number of classroom-based studies. However, although research may be seen to inform practice, how it gets translated into practice is another matter. Involving teachers as research partners rather 'the researched' is a way of breaking down some of these barriers. Other ways are to create a research praxis nexus within the technology education community or through publications that are based on research but written for teachers. A number of these are appearing in technology education around the world. This Handbook has also indicated the power of developing a classroom research culture in technology education as a means to move the field forward. A number of theoretical and practical approaches have been provided which can provide a sound basis for moving forward.

As has been noted through a number of the chapters, a graduate and post-graduate research culture is beginning to emerge and the number and range of doctoral programmes is increasing in a number of countries. This emerging culture is crucial to building a sustainable research culture in technology. This Handbook can also provide a basis for the consideration of graduate course material, doctoral research, and new research questions and approaches. These graduates then move into a range of research and academic positions, adding momentum to the field.

As an emerging field of research, a number of obstacles exist. These include the notion that funding for technology education research is limited. Funding for educational research generally is limited to that compared with the sciences, and for an emerging area such as technology education there are further limitations. Another obstacle is the increasing number of research assessment or research quanta exercises that are being implemented in the number of countries. These include the Research Assessment Exercise in the United Kingdom and the Performance Based Research Fund in New Zealand. These reward researchers publishing in perceived high-level journals. As an emerging research area within education, newer journals may not be as recognised as those in more traditional areas. This can also limit the publication of research that is focused in technology education. However, as more high quality research publications are developed and critiqued, this will become less of an issue.

As the research culture in technology education expands, it also becomes possible to build up an international research agenda, even if funded from individual countries. As the 'language' of technology education research becomes richer and more shared, there are increased opportunities not only for collaboration but an increasingly shared vision of what is possible and what might be done. To develop a shared vision we need to know where we have come from, where we are now, and illumination of future possibilities. This Handbook contributes to such an understanding and provides the foundation for developing such a vision.

REFERENCES

Ball, D. L., & Cohen, D. K. (1996). Reform by the book: What is – or might be – the role of curriculum materials in teacher learning and instructional reform? *Educational Researcher, 25*(9), 6–14.

Black, P., & Wiliam, D. (1998). Assessment and classroom learning. *Assessment in Education, 5*, 7–74.

Boaler, J. (2003). Studying and capturing the complexity of practice – the case of the 'dance of agency'. In N. A. Pateman, B. J. Dougherty, & J. T. Zilliox (Eds.), *Proceedings of the 2003 Joint Meeting of PME and PMENA* (pp. 3–16). University of Hawai'i: PME.

Hiebert, J., Gallimore, R., & Stigler, J. (2002). A knowledge base for the teaching profession: What would it look like and how can we get one? *Educational Researcher, 31*(5), 3–20.

Jones, A. (1997). Recent research in student learning of technological concepts and processes. *International Journal of Technology and Design Education, 7*(1–2), 83–96.

Jones, A. (2006). The role and place of technological literacy in science teacher education. In K. Appleton (Ed.), *Elementary science teacher education* (pp. 197–218). New Jersey: Lawrence Erlbaum.

Jones, A. (2008). Reflecting on the development of a new school subject: The development of technology education in New Zealand. *Waikato Journal of Education, 13*, 273–294.

Jones, A., & Moreland, J. (2005). The importance of pedagogical content knowledge in assessment for learning practices: A case study of a whole school approach. *The Curriculum Journal, 16*(2), 193–206.

Rodrigues, S., Marks, A., & Steel, P. (2003). Developing science and ICT pedagogical content knowledge: A model of continuing professional development. *Innovation in Education and Teaching International, 40*(4), 386–394.
Schneider, R. M., & Krajcik, J. (2002). Supporting science teacher learning: The role of educative curriculum materials. *Journal of Science Teacher Education, 13*(3), 221–224.
Shulman, L. S. (1987). Knowledge and teaching: foundations of the new reform. *Harvard Educational Review, 57*, 1–22.
Wenger, E. (1998). *Communities of practice: Learning, meaning and identity.* Cambridge: Cambridge University Press.

Alister Jones
School of Education
University of Waikato
Hamilton, New Zealand

LIST OF AUTHORS

Steve Alsop is a Professor of Education at York University, Canada. He is a former primary and secondary school teacher and coordinator of the Centre for Learning and Research in Science Education at Roehampton Inst (University Surrey, UK). Steve's Ph.D. is in philosophy of education with a specialisation in science, technology, and risk. His current research interests include the social and political organisation of scientific and technological knowledge; environmental and sustainability education; place-based, community, and popular education; and education within new social movements.

Jenny Bain worked in industry before moving into secondary Design and Technology teaching in 1994 and then to Goldsmiths, University of London in 2001. Her primary research interest is assessment practice at both secondary and higher education levels, and she has been involved in a number of research projects as part of the Technology Education Research Unit (TERU) team.

Frank Banks is Director of International Development in Teacher Education at The Open University, UK. After 11 years as a school teacher, he joined the staff of the University of Wales, Swansea, before moving to the Open University in 1992. He has authored or edited twelve books and handbooks for teachers and over eighty academic papers, mainly in the field of teacher professional knowledge.

Clare Benson is director of the Centre for Research In Primary Technology (CRIPT) in the Faculty of Education, Birmingham City University, England. She taught in a wide range of primary schools both in England and overseas before moving into Higher Education. She has written extensively about primary design and technology, and her current areas of research include Early Years and children's perceptions' of design and technology.

Cathy Buntting has a background in biochemistry and biotechnology and a PhD in science and technology education. She holds a research position with the Wilf Malcolm Institute of Educational Research, University of Waikato, New Zealand, and managed the initial development of the New Zealand Biotechnology Learning Hub. Her research interests focus on using web-based resources to support science and technology teaching and learning.

Angela Calabrese Barton is an associate professor in the College of Education at Michigan State University. Her research focuses on issues of equity and social justice in urban centres, with primary attention on understanding the learning experiences of low-income youth. Drawing primarily from feminist and sociocultural theories, she is deeply committed to researching with teachers and youth to build equitable and place-based learning environments and opportunities to gain access to and learn science in ways that support who youth are and want to be.

Dennis W. Cheek is Vice President for Education at the Ewing Marion Kauffman Foundation in Kansas City, MO, USA. He is former Editor-in-Chief of the *Journal of Technology Studies* and has published or edited over 700 publications and multimedia products. He holds a Ph.D. in science education from Pennsylvania State University and Ph.D. in theology from the University of Durham.

Sugra Chunawala is Associate Professor at Homi Bhabha Centre for Science Education (HBCSE), Tata Institute of Fundamental Research, Mumbai. Her research interests include gender issues in education; students' and teachers' perceptions of science, scientists, and technology; and students' alternative conceptions of science concepts. She is currently involved in developing design and technology education units for middle schools in India.

Vicki Compton is a research director with UniServices, University of Auckland. She has been involved in research, policy, curriculum, and national qualifications development for technology education in New Zealand since its initial development as a learning area. Recent research has focused on defining constructs for technological knowledge, technological practice, and the nature of technology, as well as formative assessment tools. She is involved in a range of initiatives to support technology education nationally and internationally.

Deborah Corrigan is an associate professor in the Faculty of Education, Monash University. She heads the Science Education Research Group in the Centre for Science, Mathematics and Technology Education at Monash and has previously been Associate Dean (Teaching) within the Faculty. Her current research focuses on science curriculum, policy and practice, the contextual basis necessary for learning science of relevance, as well as the evidence base for professional science teachers in facilitating learning of science. She was a Chemistry and Biology Teacher in Victorian schools for 10 years before commencing at Monash.

Bronwen Cowie is the director of the Wilf Malcolm Institute of Educational Research and past director of the Centre for Science and Technology Education Research at The University of Waikato, Hamilton, New Zealand. Her research interests are in assessment for learning, classroom interactions in science and technology education, and student perspectives, all within a sociocultural frame. She has led a number of national and international projects focused on enhancing student learning.

John R. Dakers lectures at the University of Glasgow. His research interests are the philosophy of technology and technology education. John is involved in several international collaborations on issues relating to technology and science education. His particular focus is on the development of technological literacy in technology education curricula. His recent book, *Best Practices in Technology Education* (2008), with Marc de Vries, Rod Custer and Gene Martin, won the prestigious Silvius-Wolansky Award for the Outstanding Scholarly Publication in Technology Education.

Marc J. de Vries is Professor of science and technology education at Delft University of Technology, the Netherlands, and an affiliate professor of reformational philosophy of technology at the same institute. He is currently the editor-in-chief of the *International Journal of Technology and Design Education* (Springer), the series editor of the *International Technology education Series* (Sense Publishers), and co-organiser of the international PATT (Pupils' Attitudes Towards Technology) conferences.

Justin Dillon is a Senior Lecturer in Science and Environmental Education and Head of the Science and Technology Group at King's College London. After teaching in London schools for 10 years, he joined King's in 1989. Justin was elected President of the European Science Education Research Association (ESERA) in 2007 and is one of the editors of the *International Journal of Science Education.*

Bangping Ding is a professor of comparative education at the School of Education, Capital Normal University, Beijing. His interest in international science and technology education has developed over the past decade. As a comparative educator, he has introduced to Chinese science educators theories and reforms in science education that have emerged recently in western countries.

William E. Dugger, Jr., is currently the Senior Fellow at the International Technology Education Association (ITEA). From 1994 to 2005, he was the director of ITEA's Technology for All Americans Project. He is an Emeritus Professor at Virginia Tech, and served as a Professor of Education and the programme area leader for technology education in the College of Education at Virginia Tech from 1972 to 1994. He was a president of ITEA (1984–85), has served on the Phi Delta Kappa International (PDK) Board of Directors.

Leo Elshof has worked in the fields of engineering, science, technological, and environmental education for over 25 years. His formal educational background includes training in Electronics Engineering Technology, an Honours degree in science from the University of Waterloo, a Master of Science degree from McMaster University, a Bachelor of Education degree from the University of Western Ontario and a Doctor of Philosophy degree from the University of Toronto. Leo's research involves primarily work in environmental sustainability and technological education.

Marilyn Fleer holds the Foundation Chair of Early Childhood Education at Monash University, and is President of the International Society for Cultural Activity Research (ISCAR). Her research interests include early childhood science education, childhood development, cultural studies, and early childhood pedagogy.

Over the years she has studied extensively seminal texts in cultural-historical theory, making important links with the Vygotsky Institute (headed by Vygotsky's granddaughter, Professor Elena Kravtsova) at the Russian State University for the Humanities in Moscow.

Andrew Feenberg is Canada Research Chair in Philosophy of Technology at Simon Fraser University in Vancouver, Canada. Among his research interests are philosophy of technology, Critical Theory, the Internet, and Japanese intellectual history, and he is recognised as an early innovator in the field of online communication.

Jacques Ginestié is director of the IUFM (Academic Institute of Teachers' Training) in the University of Provence, Aix-Marseilles Université, France. Coordinator of the laboratory Gestepro, his research interests relate to the study of teaching learning in technology education through analysing pupil activities, the role of technical languages in the learning process, and pupil-teacher interactions.

Chris Harrison is a Senior Lecturer in Science Education and Head of the Assessment for Learning Research Group at King's College London. After teaching science in London schools for 13 years, she worked on various curriculum projects before joining King's in 1993. She is well known for her work on research/developmental projects with teachers on assessment for learners across a wide range of subject areas and age groups.

Ann Marie Hill holds Design Certifications from Ecole Alyne Larin and Ecole Cotnoir Capponi in Montreal, a B.Ed. degree from McGill University in Montreal, teaching qualifications from the Canadian provinces of Québec and Ontario, and a Ph.D. degree from The Ohio State University in the United States. She has taught technological education in Québec at both high school and college levels and is currently Professor in Technological Education at the Faculty of Education, Queen's University, Kingston, Ontario, Canada. Her work in technological education has extended to development in Burkina Faso and Kenya through the Canadian International Development Agency (CIDA), to Bangladesh, and to participation in NATO workshops.

Derek Hodson has 40 years experience in science education: 10 years in secondary schools in the UK, 8 in pre-service teacher education at the University of Manchester, 5 at the University of Auckland, and 17 at the Ontario Institute for Studies in Education as Professor of Science Education, retiring in June 2008. He was Director of the Centre for Studies in Science, Mathematics and Technology Education at OISE (1998–2006) and Editor of the *Canadian Journal of Science, Mathematics and Technology Education* (2001–2007).

Gill Hope is senior lecturer in the Primary Education Department at Canterbury Christ Church University. Her Ph.D. thesis (2003) investigated the way in which young children (ages 5-9 years) used drawing as a means of generating and developing design ideas. She is a member of the UK Design & Technology Association's editorial board and also referees papers for the British Educational Research Association's journal and conferences.

Don Ihde is distinguished professor in the Department of Philosophy and Director of the Technoscience Research Group at The State University of New York at Stony Brook. His research focuses on the philosophy of both science and technology, with a growing interest in imaging technologies. He is also interested in work on intercultutral perception and plural cultural patterns.

Alister Jones is the Dean of the School of Education at The University of Waikato, Hamilton, New Zealand. He is a past director of the Wilf Malcolm Institute of Educational Research and the Centre for Science and Technology Education Research at the University. His research interests are in curriculum and teaching and learning in technology education.

Carolyn Jones is the Research Manager at the Wilf Malcolm Institute of Educational Research at the School of Education, University of Waikato, New Zealand. Her particular research interests are in early childhood education and transition to school and she was involved in the development of the New Zealand Early Childhood Exemplar Development Project. She also has wider interests in general education research approaches and the design and implementation of research projects.

Tapani Kananoja began his career as a primary school teacher before teaching woodwork at junior secondary school level and then being appointed as Chief Inspector for technical work subjects for Finland's National Board of General Education. Before retiring he lectured at Turku and Oulu Universities in Finland and was also actively involved in development projects in Africa. His publications include textbooks, curriculum documents, in-service training materials, as well as more than 200 research reports.

Steve Keirl lectures at the University of South Australia in curriculum and Design and Technology where he works with undergraduate, postgraduate, and in-service primary and secondary school teachers. He is a researcher with the University's Centre for Research in Education, Equity and Work and with the Hawke Research Institute for Sustainable Societies. His research interests include the relationships amongst ethics, curriculum, pedagogy, democracy and Design and Technology.

Richard Kimbell founded the Technology Education Research Unit (TERU) at Goldsmiths University of London in 1990. Since that time, research sponsors include research councils, industry, government departments, as well as professional and charitable organisations. Richard has published widely in the field of technology education research and written and presented television programmes. He is a consultant to the National Academy of Engineering and the National Science Foundation in the USA, and is a Visiting Professor at the University of Stockholm.

Julie Lunt is an independent consultant for primary design and technology based in Wales, UK. She has a background in primary teaching and is an experienced provider of in-service training for primary teachers. She works closely with The Centre for Research and Curriculum Development in Primary Design and Technology (CRIPT) based at Birmingham City University. Her research interests include children's perspectives of learning and teaching, writing in design and technology, and children's designing.

Robert McCormick is an Emeritus Professor in the Faculty of Education and Language Studies at the Open University, UK. He is internationally respected in technology education, making major keynotes to conferences and publishing widely with over 100 books, articles, and papers. He has directed a number of research projects in technology education, including on problem solving, the use of mathematics in design and technology, and electronics in the curriculum. Other research includes ICT in education and learning how to learn.

Ros McLellan is a lecturer in Teacher Education and Development/Pedagogical Innovation at the Faculty of Education, University of Cambridge. She has ten years of research experience working on a number of projects within the Faculty. Prior to this she taught secondary mathematics and psychology for eight years. Her research interests are in creativity, student motivation (the subject of her doctoral research), attitudes and self-concept, and issues relating to gender and achievement.

Howard Middleton is Associate Professor in the Faculty of Education at Griffith University. He created the Technology Education Research Unit at Griffith and established the biennial Technology Education Research Conference. Howard's research interests include the nature of complex problem-solving in design and the kinds of research methods appropriate for researching technological activity. He is currently on study leave after two years as Foundation Head of the School of Education and Professional Studies.

David Mioduser is Associate Professor in the Tel Aviv University's School of Education, where he heads the Department of Math, Science and Technology Education and the Knowledge Technologies Lab. His research interests focus on cognitive and learning issues in technology education, particularly with young children. His work examines the development of conceptions, and knowledge construction and skills acquisition processes concerning the human-mind-made world.

Judy Moreland is a Senior Research Fellow at the Wilf Malcolm Institute of Educational Research, the School of Education at The University of Waikato, Hamilton, New Zealand and was previously at the Centre for Science and Technology Education Research at the University. Her research interests include primary technology education, children's views of learning and teaching, multimodal approaches in assessment for learning, and sociocultural research approaches.

Chitra Natarajan is Professor at Homi Bhabha Centre for Science Education, Tata Institute of Fundamental Research, Mumbai, although she began her career in the R&D sector, working on surface analytical instruments and the science of surfaces. Her research interests include surface science and plasma technology, S&T policy, environmental and developmental issues, and research in design and technology and environmental education. She is also a member of the "multilingualism network" that engages with issues of sustaining multiple modes of expression: languages and technologies.

Bill Nicholl is a lecturer at The Faculty of Education, University of Cambridge, where he set up the course component responsible for training teachers of Design & Technology. His research interests are in creativity, in particular pupil's generation of early design ideas, and pupil voice. As well as publishing in academic journals, he has authored a number of school textbooks. He also taught for nine years in two inner London state comprehensive schools.

Stephen Norton teaches and conducts research in mathematics education (and teacher education) at Griffith University, Australia. He examines ways to help students understand mathematics from early elementary to senior secondary years. Some of his recent research has focused upon the potential of studying mathematics in applied and contextual settings including across curricula. Stephen has also investigated the use of materials to teach algebra and educational games which focus on developing student understand of the common structures across number and algebra.

Jonathan Osborne holds the California Chair of Science Education in the School of Education, Stanford University. He started his career as a teacher of physics in London in 1973, joined King's College in 1985, and held the Chair in Science Education there between 2004–8. He has conducted research on policy in science education, attitudes to science, and argumentation.

Eric Parkinson has taught in the primary, secondary, and higher education sectors in the UK and a number of developing countries. He is a Principal Lecturer in Education, and Learning and Teaching Co-ordinator for the Faculty of Education at Canterbury Christ Church University, Kent, England. He has a special research interest in the use of construction kits for creative building tasks undertaken by children in primary education.

Gloria Quiñones is a PhD student in the Faculty of Education at Monash Universtity, Melbourne, Australia. Her research interests focus on the relationship between school life and everyday life in early childhood settings in Mexico and Australia.

Philip A. Reed is an associate professor and technology education programme leader at Old Dominion University in Norfolk, Virginia, USA. His expertise is in curriculum development and evaluation. His research covers graduate technology education, distance learning, and the research base of technology education.

Michael Reiss is Assistant Director and Professor of Science Education at the Institute of Education, University of London, Chief Executive of Science Learning Centre London, Honorary Visiting Professor at the University of York, Director of the Salters-Nuffield Advanced Biology Project, a member of the Farm Animal Welfare Council and editor of the journal *Sex Education*. His research interests are in science education, bioethics, and sex education.

Stephen Ritchie is an Associate Professor in the School of Mathematics, Science and Technology Education at Queensland University of Technology, Australia. Steve conducts research in engaging diverse learners and the professional transitions of beginning science teachers. Recent studies have interpreted (classroom) interactions dialectically from a sociocultural perspective. Specifically, this research has focussed on emotion, identity, leadership, and collaboration. He is also interested in engaging middle school students in socio-scientific topics through writing hybridised text that includes technical information and narrative. He is the Chief Editor of *Research in Science Education*.

John M. Ritz is a professor and department chairperson at Old Dominion University in Norfolk, Virginia, USA. His expertise is in technology education programme development and technological literacy. He has undertaken research to determine state-of-the-art of graduate technology education, goals for technological literacy, and the effect of instruction on technological literacy of university populations.

Wolff-Michael Roth is Lansdowne Professor of Applied Cognitive Science at the University of Victoria in Canada. Much of his research is concerned with embodied, cultural-historical, and linguistic aspects of knowing and learning in mathematics, science, and technology. He has taught quantitative and qualitative research methods since working at the university level in 1992.

Jim Ryder is a Senior Lecturer in Science Education, at the Centre for Studies in Science and Mathematics Education, School of Education, University of Leeds, UK. His research addresses the role of teaching and learning about the nature of science within formal science education settings.

Kay Stables started her career in London as a textiles teacher, moving to Goldsmiths, University of London in 1986 as a researcher. As a founding member of the Technology Education Research Unit (TERU) she has directed and contributed to a broad range of research projects in the area of design and technology education, including primary and secondary age learners both in the UK and overseas.

Andrew Stevens is a lecturer in technology education at Rhodes University, South Africa. He has taught mathematics at high school level and was the principal of one of the first non-racial schools in South Africa before joining the ORT-STEP Insititute, an NGO which pioneered the introduction of technology education in South Africa. He is interested in the development of technology education both in South Africa and internationally, particularly through the expansion of teacher education and research.

Edna Tan is a postdoctoral fellow with the Department of Teacher Education in the College of Education at Michigan State University. Her research interests focus on the teaching and learning of science in urban schools and after-school programs, especially those of urban minority girls from high poverty neighbourhoods.

Pierre Vérillon is a research engineer at the French National Institute for Pedagogical Research (INRP), currently working with the UMR ADEF joint research unit in Marseille and with the Centre for Research on Work and Development (CRTD) in Paris. As a psychologist, his work and publications focus mainly on cognition in technological settings. Through the concept of instrument-mediated activity, he has developed a cultural and epistemological approach to learning and development in contexts of material and semiotic tool-use, notably in technological and vocational education.

John Williams lectures in the School of Education at Edith Cowan University in Perth, Western Australia. He is the Director of Secondary Programs, and coordinates undergraduate, postgraduate and offshore programs in technology education. Apart from Australia, he holds an adjunct position at the University of Waikato in New Zealand and has worked and studied in a number of African countries and in the United States. His current research interests include design collaboration at a distance and electronic assessment of performance.

INDEX

D

Lightning Source UK Ltd.
Milton Keynes UK
25 January 2010

149055UK00001B/2/P

9 789087 908775